A History of the Theories of AETHER & ELECTRICITY

A History of the Theories of
AETHER
&
ELECTRICITY

Volume I
The Classical Theories

Volume II
The Modern Theories 1900–1926

Two Volumes Bound as One

Sir Edmund Whittaker

Dover Publications, Inc.
Mineola, New York

Bibliographical Note

This Dover edition, first published in 1989 and reissued in 2017, is an unabridged and unaltered republication in one volume of the works originally published by Thomas Nelson & Sons, London, in 1951 and 1953.

Library of Congress Cataloging-in-Publication Data

Whittaker, E. T. (Edmund Taylor), 1873–1956.
A history of the theories of aether & electricity / Sir Edmund Whittaker.
p. cm.
Reprint. Originally published: London ; New York : T. Nelson, 1951–1953.
Contents: v. 1. The classical theories—v. 2. The modern theories, 1900–1926.
ISBN-13: 978-0-486-26126-3
ISBN-10: 0-486-26126-3
1. Ether (of space)—History. 2. Electricity—History. 3. Electromagnetic theory—History. 4. Physics—History. I. Title.

QC177.W63 1989
530.1'09—dc20

89-23588
CIP

Manufactured in the United States by LSC Communications
4500057112
www.doverpublications.com

Preface

In 1910 I published a work under the title *A History of the Theories of Aether and Electricity, from the age of Descartes to the close of the nineteenth century.* When the original edition was exhausted, I felt that any new issue should describe the origins of relativity and quantum-theory, and their development since 1900. My opportunities were however not sufficient to enable me to prepare an accurate and fully-documented account of this very creative period, and I was compelled to lay the plan aside. Retirement from my professorial chair has made it possible for me to take up this project again ; it will occupy three volumes, of which this, the first, deals with the classical theories. The volume of 1910 has been to a considerable extent rewritten, with the incorporation of much additional material ; and in the second volume, the story will be continued to 1926.

A word might be said about the title *Aether and Electricity.* As everyone knows, the aether played a great part in the physics of the nineteenth century ; but in the first decade of the twentieth, chiefly as a result of the failure of attempts to observe the earth's motion relative to the aether, and the acceptance of the principle that such attempts must always fail, the word 'aether' fell out of favour, and it became customary to refer to the interplanetary spaces as 'vacuous' ; the vacuum being conceived as mere emptiness, having no properties except that of propagating electromagnetic waves. But with the development of quantum electrodynamics, the vacuum has come to be regarded as the seat of the 'zero-point' oscillations of the electromagnetic field, of the 'zero-point' fluctuations of electric charge and current, and of a 'polarisation' corresponding to a dielectric constant different from unity. It seems absurd to retain the name 'vacuum' for an entity so rich in physical properties, and the historical word 'aether' may fitly be retained.

My grateful thanks are due to Professors E. T. Copson of the University of St Andrews and J. M. Whittaker of the University of Liverpool for help in reading the proofs.

E. T. Whittaker

48 George Square
Edinburgh, April 1951

The intellectual love of God

SPINOZA

Memorandum on Notation

VECTORS are denoted by letters in black type, as **E**.

The three components of a vector **E** are denoted by E_x, E_y, E_z; and the magnitude of the vector is denoted by E, so that

$$E^2 = E_x^2 + E_y^2 + E_z^2.$$

The *vector product* of two vectors **E** and **H**, which is denoted by [**E** . **H**], is the vector whose components are

$$(E_yH_z - E_zH_y, \quad E_zH_x - E_xH_z, \quad E_xH_y - E_yH_x).$$

Its direction is at right angles to the direction of **E** and **H**, and its magnitude is represented by twice the area of the triangle formed by them.

The *scalar product* of **E** and **H** is $E_xH_x + E_yH_y + E_zH_z$. It is denoted by (**E** . **H**).

The quantity $\frac{\partial E_x}{\partial x} + \frac{\partial E_y}{\partial y} + \frac{\partial E_z}{\partial z}$ is denoted by div **E**.

The vector whose components are

$$\left(\frac{\partial E_z}{\partial y} - \frac{\partial E_y}{\partial z}, \quad \frac{\partial E_x}{\partial z} - \frac{\partial E_z}{\partial x}, \quad \frac{\partial E_y}{\partial x} - \frac{\partial E_x}{\partial y}\right)$$

is denoted by curl **E**.

If V denote a scalar quantity, the vector whose components are $\left(-\frac{\partial V}{\partial x}, \; -\frac{\partial V}{\partial y}, \; -\frac{\partial V}{\partial z}\right)$ is denoted by grad V.

The symbol ∇ is used to denote the vector operator whose components are $\frac{\partial}{\partial x}, \frac{\partial}{\partial y}, \frac{\partial}{\partial z}$.

VOLUME I:

The Classical Theories

Contents

Chapter I

THE THEORY OF THE AETHER TO THE DEATH OF NEWTON

THE earliest science, that of the ancient Greeks, had its origin in the speculations of philosophers regarding the nature of Being and Becoming. Modern science likewise was at first closely related to philosophy, and a history of its beginnings must be prefaced by some account of the philosophical situation at the close of the Middle Ages.

As was pointed out by Aristotle,[1] there are fundamentally two types of philosophy, according as the approach to philosophical problems is made dialectically (λογικῶς) or 'naturally' (φυσικῶς). To the first type belong all those philosophers who have speculated in the grand manner, from Plato to Hegel ; to the second belong Democritus, Aristotle himself, and most of the modern men of science who have written on philosophical subjects.

The philosophers of the first type have produced schemes, more or less independent of experience, which have claimed to comprehend all Being in their framework ; those of the second type proceed more cautiously—their fundamental concepts are obtained by systematising their experience of the external world. In other words, they proceed to their general philosophy by extending their conceptual analysis of natural science (that this is true of Aristotle is obvious when his *Metaphysics* is compared with his *Physics*). Of course such a method implies the principle that philosophical truth is attained only progressively, and that none of the all-embracing systems of the first type is likely to be true.

Aristotle's plan for constructing a philosophy was, then, one which would be fully approved by a modern man of science. But unfortunately the cosmological and physical notions from which he started were entirely false. He believed that there was a radical difference in kind between the heavenly bodies on the one hand and terrestrial objects on the other : the former were changeless, uncompounded and eternal, moving in circular paths, while the latter were corruptible, synthetic, transient and moved naturally in straight lines. The earth was at rest, while the stars were carried round on celestial spheres, whose motions were directed by immaterial

[1] *De generatione et corruptione*, i, 2, 316*a*

intelligences. His word for 'matter' (ὕλη) did not connote matter in our sense, but meant something out of which something else could be made by imposing on it a further condition or structural principle, called a *form*. Ordinary bodies he supposed to be constituted of four elements, fire, air, water, earth; but he believed that these could be transmuted into each other, and were therefore merely different 'forms' associated with a 'prime matter' which was itself inaccessible to experience. His explanation of the motion of bodies was, that the 'substantial form' or 'nature' of each body impels it to some particular place: thus, fire moves upwards, while other bodies tend to move towards the centre of the earth. All things, in fact, seek an 'end,' or, in Aristotelian terminology, a *final cause*; and the 'final cause' of a movement in space is some definite location. It may be added that Aristotle studied nothing quantitatively; it is only fair to him to say that he insisted on the necessity for observation and experience, but the injunction was lost sight of by many of his followers in later ages.

A typical example of the Aristotelian natural philosophy is its interpretation of light in terms of the metaphysical notions of *potency* (δύναμις) and *act* (ἐνέργεια or ἐντελέχεια). A transparent body has a 'potency' for transmitting light, but it does not become actually transparent until light is passed through it and thereby brings the transparency into action. Therefore light was defined by Aristotle as 'the act of a transparent body, inasmuch as it is transparent.'

Under the influence chiefly of St Thomas Aquinas, Aristotelianism was adopted generally throughout western Europe in the thirteenth century. Like Aristotle, St Thomas declared his philosophy to be of what we have called the second type. 'The origin of our knowledge,' he said,[1] 'is in the senses, even of those things that transcend sense;' and 'metaphysics has received its name, i.e. *beyond physics*, because to us, who naturally arrive at the knowledge of things immaterial by means of things sensible, it offers itself by rights as an object of study after physics.'[2] Metaphysics is 'the last of the parts of philosophy to be studied,' because 'it is necessary to have previous knowledge of many things.'[3]

All this is very sound, but it was impossible to extract a true metaphysics from the false Aristotelian physics. The evil was aggravated by the medieval schoolmen, who regarded Aristotelianism as an encyclopaedic body of knowledge, in no need of being supple-

[1] *Summa contra gentiles*, i, ch. 12

[2] *Opusc. lx* (ed. 1571), *in Boet. de Trinit.*, qu. 5, art. 1

[3] *Summa contra gentiles*, i, ch. 4

mented by further experiments. They abandoned St Thomas' view of the relation between physics and metaphysics. 'It is not,' they said, 'the province of physics to theorise on its own facts and laws, or to undertake a reconstruction of cosmology or metaphysics,'[1] and 'if a physical theory is inconsistent with received metaphysical teaching, it cannot be admitted, because metaphysics is the supreme natural science, not physics.'[2] They interpreted the external world only by applying formal logic in order to make deductions from obscure and sterile principles, which really represented a petrifaction of Aristotle's erroneous physics : a procedure which yielded nothing but prolix sophistry.[3] As Erasmus said, 'There are innumerable niceties concerning *notions*, *relations*, *instants*, *formalities*, *quiddities* and *haecceities*, which no-one can pry into, unless he has eyes that can penetrate the thickest darkness, and there can see things that have no existence whatever.'[4] Until this tradition could be overthrown, there was no possibility of any right understanding of nature.

For the birth of modern science, then, a necessary condition was an emancipation from Thomist philosophy. This was brought about chiefly as a result of the work of William of Ockham, an English Franciscan friar, who lived in the first half of the fourteenth century. In metaphysics, Ockham introduced principles which need not concern us here, but which undermined Aristotelianism and led ultimately to that general rejection of medieval philosophy which was associated with the Renaissance ; and it was he who made a beginning of rational dynamics, which after his death was developed by his followers, and on which all subsequent progress was founded. The most decisive advances were the new conception of the relation between the earth and the sun, proposed by Copernicus in 1543, and the success of Kepler, in the early years of the next century, in disproving the Aristotelian principle that celestial bodies are different in kind from terrestrial. But the age of decadent scholasticism was long drawn out, and nearly three centuries after Ockham the Aristotelians still had enough influence to procure the condemnation of Galileo.[5]

It was a younger contemporary of Galileo, René Descartes, who

[1] T. Harper, *The Metaphysic of the School* (London, 1879–84), i, p. 41

[2] ibid. ii, p. 249

[3] This is not true of Robert Greathead and of Roger Bacon, but they were senior to St Thomas in age and may be said to belong to the pre-Aristotelian era ; and it is not true of the followers of William of Ockham, but they were not Thomists.

[4] Erasmus, *Moriae Encomium* (1509)

[5] The decree of condemnation contains the following : *Che la terra non sia centro del mondo, nè immobile, ma che si moue etiandio di moto diurno, è propositione assurde, e falsa in filosofia.*

first attempted a general reconstruction of ideas regarding the physical universe, and it is with an account of his work that the history properly begins.

Descartes was born in 1596, the son of Joachim Descartes, Counsellor to the Parliament of Brittany. As a young man he followed the profession of arms, and served in the campaigns of Maurice of Nassau and the Emperor ; but his twenty-fourth year brought a profound mental crisis, apparently not unlike those which have been recorded of many religious leaders, and he resolved to devote himself thenceforward to the study of philosophy.

The age which preceded the birth of Descartes, and that in which he lived, was marked by events which greatly altered the prevalent conceptions of the world. The discovery of America, the circumnavigation of the globe by the companions of Magellan, the invention of the telescope, the overthrow of the Ptolemaic system of astronomy and the general dissatisfaction with scholasticism, all helped to loosen the old foundations and to make plain the need for a new structure. It was this that Descartes set himself to erect. His aim was nothing less than to create from the beginning a theory of the universe, worked out as far as possible in every detail.

Of such a system the basis must necessarily be metaphysical ; and it is by this part of Descartes' work that he is most widely known. Here we shall refer only to those of his philosophical principles which determined his representation of the external world.[1]

The first step was to discard the futile methods of the Middle Ages, the attempts to interpret Nature in terms of act and potency, matter and form, substance and accident, the ten categories, and the like. This had already been proposed by many philosophers, and new ways had been indicated, particularly by Francis Bacon (1561–1626) and Galileo Galilei [2] (1564–1642). They rejected altogether the Aristotelian practice of referring everything in physics to Final Causes,[3] and asked, *How?* rather than *Why?* They dwelt on the necessity for observing the external world, and moreover insisted that besides the study of such occurrrences as are naturally presented to us,

[1] Of the works which bear on our subject, the *Dioptrique* was published (anonymously) in 1637, the *Météores* in 1638, and the *Principia Philosophiae* at Amsterdam in 1644, six years before the death of its author. The *Discours de la méthode*, published in 1637, was an essay prefatory to the *Dioptrique* and the *Météores*.

[2] The predominant part played by Galileo in destroying the Aristotelian-Thomistic conception of the world is very well brought out by Dr A. C. Crombie in a monograph, 'Galileo's "Dialogues concerning the two principal systems of the world,"' *Dominican Studies*, 1950, 34 pp.

[3] *Causarum finalium inquisitio sterilis est*, Bacon, *De Augment* (1623), iii, 5. Bacon's principal philosophical works were written about eighteen years before those of Descartes.

there should be an interrogation of Nature by means of carefully devised experiments. (This recommendation was not superfluous, for experimentation had hitherto been associated chiefly with alchemy, and had no good reputation.) Bacon's teaching was in one important respect defective, for he failed to appreciate the importance of the quantitative in contradistinction to the qualitative aspects of phenomena ; the same charge cannot be made against Galileo, but he was an experimental rather than a mathematical physicist, and it was a third contemporary, Johannes Kepler (1571–1630) who laid stress on the supreme importance of mathematics in the study of Nature.

Kepler's teaching provided the chief inspiration of Descartes, whose researches were dominated by a conviction that the theorems of mathematics had a precision, an indubitability and a universal acceptance, which were not to be found in other fields of study. So to these features he attached the highest importance, laying it down as an axiom that clarity and certainty were marks of all genuine knowledge. Now one of the problems of natural philosophy was to account for actions transmitted between bodies not in contact with each other, such as those indicated by the behaviour of magnets, or by the connection between the moon's position on the one hand and the rise and fall of the tides on the other. To accept these as 'occult' influences would have been contrary to his principles ; and he concluded that they must be effected by the agency of the only types of action between bodies which were perfectly intelligible, namely, pressure and impact. This implied that bodies can act on each other only when they are contiguous ; in other words, he denied action at a distance ; [1] and this had the further consequence, that the space between the moon and the earth, and indeed the whole of space, could not be void. It is occupied partly by ordinary material things—air and tangible bodies ; but the interstices between the particles of these, and the whole of the rest of space, must be filled with particles of a much more subtle kind, which everywhere press upon, or collide with, each other : they are the contrivance introduced in order to account for all physical happenings. Space is thus, in Descartes' view, a *plenum*, being occupied by a medium which, though imperceptible to the senses, is capable of transmitting force, and exerting effects on material bodies immersed in it—the *aether*, as it is called. The word [2] had meant originally the blue sky or upper

[1] The doctrine that force cannot be communicated except by pressure or impact had been maintained by the ancient Greek atomists and also by Aristotle and St Thomas, but had been denied by Duns Scotus and his followers. [2] αἰθήρ

air (as distinguished from the lower air at the level of the earth), and had been borrowed from the Greek by Latin writers,[1] from whom it had passed into French and English in the Middle Ages. In ancient cosmology it was sometimes used in the sense of that which occupied celestial regions ; and when the notion of a medium filling the interplanetary void was introduced, *aether* was the obvious word to use for it. Before Descartes, it had connoted merely the occupancy of some part of space : he was the first to bring the aether into science, by postulating that it had mechanical properties. In his view, it was to be regarded as the solitary tenant of the universe, save for that infinitesimal fraction of space which is occupied by ordinary matter.

Descartes assumed that the aether particles are continually in motion. As however there was no empty space for moving particles to move into, he inferred that they move by taking the places vacated by other aether particles which are themselves in motion. Thus the movement of a single particle of the aether involved the motion of an entire closed chain of particles ; and the motions of these closed chains constituted *vortices*, which performed important functions in his picture of the cosmos.

Holding, then, that the effects produced by means of contacts and collisions were the simplest and most intelligible phenomena in the external world, Descartes admitted no other agencies. He did not claim for his scheme an accurate quantitative agreement with experiment, in the spirit of a modern scientific textbook. He trusted more to the clarity and distinctness of a speculation than to its consistence with observed fact [2] : rather is his work to be regarded as a stupendous effort of the imaginative intellect—a myth, in the same sense as the *Timaeus* of Plato is a myth—intended to show that all the contents and happenings of the universe might be exhibited as parts of a logically co-ordinated mechanical scheme, depending solely on very elementary types of physical action, and having (once the premisses were accepted) the complete certitude of deduction that had hitherto been associated only with mathematics. Descartes was the originator of the *Mechanical Philosophy*, i.e. of the doctrine that the external inanimate world may for scientific purposes be regarded as an automatic mechanism, and that it is possible and desirable to imagine a mechanical model of every physical phenomenon.

Such a conception could not well have originated earlier than the Renaissance, because the ancients and the men of the Middle Ages

[1] e.g. *innubilis aether* (cloudless sky), Lucr. 3, 21

[2] That explains why he preferred his own account of the circulation of the blood to Harvey's.

had little or no experience of mechanisms that were self-contained and could work independently of human direction; they were acquainted only with tools, which in order to achieve definite purposes required intelligent guidance. Any manifestation of regularity in performance seemed to our forefathers to indicate, as its directing cause, an activity of mind; indeed it was precisely the order and harmony observed in the movements of the heavenly bodies that led the Greek thinkers to regard them as possessing souls.[1] Movements such as the fall of a material body to the ground were accounted for, as we have seen, by supposing that heavy matter tended to seek its natural place, the centre of the universe. This explanation became unsatisfying when the Copernican theory of the solar system was accepted, since the earth was now in motion in infinite space, and no point could be identified as the centre of the universe. It was at this juncture that Descartes put forward his revolutionary suggestion, that the cosmos might be thought of as an immense machine. This entailed the general principle that all happenings in the material world can be predicted by mathematical calculation, and this principle has proved the element of greatest value in the Cartesian philosophy of nature.

Descartes himself seems to have gone further, and to have held the doctrine of *epistemological rationalism*, that is, the assertion that physics can, like Euclidean geometry, be derived entirely from *a priori* principles, without any dependence on observation and experiment. At any rate, his general practice was to represent phenomena as the effects of preconceived dispositions and causes. In this respect he departed from the sound doctrines that had been preached a generation earlier by Francis Bacon and Galileo, and he drew down upon himself the criticism of Huygens[2]: 'Descartes,' wrote Huygens, 'who seemed to me to be jealous of the fame of Galileo, had the ambition to be regarded as the author of a new philosophy, to be taught in academies in place of Aristotelianism. He put forward his conjectures as verities, almost as if they could be proved by his affirming them on oath. He ought to have presented his system of physics as an attempt to show what might be anticipated as probable

[1] I. may be remarked that modern machines can simulate intelligence in a wonderful way. For instance, long division, as taught to schoolchildren, requires an act of intelligent choice, in order to obtain each new digit in the quotient; but machines now exist which perform long division without any human guidance during the process: when the machine is *set*, that is to say, furnished with the problem to be solved, then it is necessary only to press a button which switches on an electric current, and presently the required quotient appears on a dial.

[2] It was given in his annotations on Baillet's *Life of Descartes* (a book which appeared in 1691), which are printed in V. Cousin's *Fragments philosophiques*, tome ii, p. 155.

in this science, when no principles but those of mechanics were admitted : this would indeed have been praiseworthy ; but he went further, and claimed to have revealed the precise truth, thereby greatly impeding the discovery of genuine knowledge.'

In putting forward an all-embracing theory of the universe before he had studied any of its processes in detail, Descartes was continuing the tradition of the ancient Greeks, rather than treading in the new paths struck out by Tycho, Kepler and Galileo : he never really grasped the principle that true knowledge can only be acquired piecemeal, by the patient interrogation of nature. A further weakness in his system was involved in the assumption that force cannot be communicated except by actual pressure or impact, a principle which compelled him to provide an explicit mechanism in order to account for each of the known forces of nature. This task is evidently much more difficult than that which lies before those who are willing to admit action at a distance as an ultimate property of matter.

The many defects of Descartes' method led to the rejection of almost all his theories in less than a century. It must be said, however, that the grandeur of his plan, and the boldness of its execution, stimulated scientific thought to a degree before unparalleled. ' Give me matter and motion,' he cried, ' and I will construct the universe.'

Matter, in the Cartesian philosophy, is characterised not by impenetrability, or by any quality recognisable by the senses, but simply by extension ; extension constitutes matter, and matter constitutes space. The basis of all things is a primitive, elementary, unique type of matter, boundless in extent and infinitely divisible. In the process of evolution of the universe three distinct forms of this matter have originated, corresponding respectively to the luminous matter of the sun, the transparent matter of interplanetary space, and the dense opaque matter of the earth. ' The first is constituted by what has been scraped off the other particles of matter when they were rounded ; it moves with so much velocity that when it meets other bodies the force of its agitation causes it to be broken and divided by them into a heap of small particles that are of such a figure as to fill exactly all the holes and small interstices which they find around these bodies. The next type includes most of the rest of matter : its particles are spherical, and are very small compared with the bodies we see on the earth ; but nevertheless they have a finite magnitude, so that they can be divided into others yet smaller. There exists in addition a third type exemplified by some kinds of matter—namely, those which, on account of their size and figure, cannot be so easily moved as the preceding. I will endeavour to show that all the bodies of the visible world are composed of these three forms of matte., as of three distinct elements ; in fact, that the sun and the fixed stars are formed of the first of these elements, the interplanetary spaces of the second, and the earth, with the planets and comets, of the third. For, seeing that the sun and the fixed stars emit light, the heavens transmit it, and the earth, the planets and the comets reflect it, it appears to me that there is ground for using these three qualities of luminosity, transparence and opacity, in order to distinguish the three elements of the visible world '.[1]

According to Descartes' theory, the sun is the centre of an immense vortex formed of the first or subtlest kind of matter.[2] The vehicle of light in interplanetary space is matter of the second kind or element, composed of a closely packed assemblage of globules whose size is intermediate between that of the vortex matter and that of ponderable matter. The globules of the second element, and all the matter of the first element, are constantly straining away from the centres around which they turn, owing to the centrifugal force of the vortices [3] ; so that the globules are pressed in contact with each other and tend to move outwards, although they do not actually so move.[4] It is the transmission of this pressure which constitutes light ; the action of light therefore extends on all sides round the sun and fixed stars, and travels instantaneously to any distance.[5] In the *Dioptrique*,[6] vision is compared to the perception of the presence of objects which a blind man obtains by the use of his stick ; the transmission of pressure along the stick from the object to the hand being analogous to the transmission of pressure from a luminous object to the eye by the second kind of matter.

Descartes supposed the ' diversities of colour and light ' to be due to the different ways in which the matter moves.[7] In the *Météores*,[8] the various colours are connected with different rotatory velocities of the globules, the particles which rotate most rapidly giving the sensation of red, the slower ones of yellow and the slowest of green and blue—the order of colours being taken from the rainbow. The assertion of the dependence of colour on periodic time is a curious foreshadowing of a great discovery which was not fully established until much later.

[1] *Principia*, pt. iii, § 52

[2] It is curious to speculate on the impression which would have been produced had the spirality of nebulae been discovered before the overthrow of the Cartesian theory of vortices.

[3] ibid. §§ 55–9 [4] ibid. § 63 [5] ibid. § 64 [6] *Discours premier*

[7] *Principia*, pt. iv, § 195 [8] *Discours huitième*

The general explanation of light on these principles was amplified by a more particular discussion of reflection and refraction. The law of reflection—that the angles of incidence and reflection are equal—had been known to the Greeks. Kepler in his *Dioptrics* (1611) discussed refraction through lenses ; he had found experimentally that for glass, when the incidence is almost perpendicular, the angles of incidence and refraction are nearly in the ratio of 3 to 2 ; and assuming this value, he had found the correct result that the focus of a double convex lens, whose two faces have the same curvature, is the centre of curvature of the side nearest the object. He did not, however, succeed in discovering the general law of refraction—that the sines of the angles of incidence and refraction are to each other in a ratio depending on the media—which was now published for the first time.[1] Descartes gave it as his own : but he seems to have been under considerable obligations to Willebrord Snell (1591–1676), Professor of Mathematics at Leyden, who had discovered it experimentally (though not in the form in which Descartes gave it) about 1621. Snell did not publish his result, but communicated it in manuscript to several persons, and Huygens affirms[2] that this manuscript had been seen by Descartes.

Descartes presents the law as a deduction from theory. This, however, he was able to do only by the aid of analogy ; when rays meet ponderable bodies, ' they are liable to be deflected or stopped in the same way as the motion of a ball or a stone impinging on a body ' : for ' it is easy to believe that the action or inclination to move, which I have said must be taken for light, ought to follow in this the same laws as motion.'[3] Thus he replaces light, whose velocity of propagation he believes to be always infinite, by a projectile whose velocity varies from one medium to another. The law of refraction is then proved substantially as follows[4] :

Suppose that a ray of light is refracted across a plane interface from one medium into another. Let a light corpuscle, whose velocity in the first medium is v_i, be incident on the interface, making an angle i with the normal to the interface, and let it be refracted at an angle r into the second medium, in which its velocity is v_r. Descartes assumed that the ratio of the velocities v_i and v_r depends only on the nature of the media : $(v_r/v_i) = \mu$, say. He assumed

[1] *Dioptrique, Discours second*
[2] cf. Cousin's *Fragments philosophiques*, tome ii, p. 162
[3] *Dioptrique, Discours premier*
[4] ibid. *Discours second*

also that the component of velocity parallel to the interface is unaffected by the refraction, so we have

$$v_i \sin i = v_r \sin r$$

Combining these equations, we have

$$\sin i = \mu \sin r$$

which is the law of refraction.

These equations imply that if $i > r$ (e.g. if the refraction is from air into glass) the velocity is greater in the second or denser medium. As we shall see, this consequence of the corpuscular theory in its primitive form is in contradiction with experimental fact.[1]

Crude though the Cartesian system was in many features, there is no doubt that by presenting definite mechanical conceptions of physical activity, and applying them to so wide a range of phenomena, it stimulated the spirit of inquiry and in some degree prepared the way for the more accurate theories that come after. In its own day it met with great acceptance;[2] the confusion that had resulted from the destruction of the old order was now, as it seemed, ended by a reconstruction of knowledge in a system at once credible and complete. Nor, as we shall see later, did its influence quickly wane.

So far as the theory of light was concerned, Descartes' conceptions rapidly displaced those which had been current in the Middle Ages. The validity of his explanation of refraction was, however, called in question by his fellow-countryman, Pierre de Fermat (1601–65),[3] and a controversy followed, which was kept

[1] Descartes' idea can however be modified so as to avoid erroneous consequences in the following way. Instead of considering the *velocities* of the light-corpuscle in the two media, consider its *momenta*, say p_i and p_r (for a light-corpuscle, as will be seen later, the momentum is not measured by the product of the mass and the velocity). Let us assume that the ratio of p_r to p_i depends only on the nature of the media, and that the component of momentum parallel to the interface is unaffected by the transition from one medium to the other, then as in Descartes' proof we have

$$(p_r/p_i) = \mu$$
$$p_i \sin i = p_r \sin r$$

and therefore

$$\sin i = \mu \sin r$$

These equations are correct; the momentum is actually greater in the optically denser medium.

[2] Though it appears from the *Optica promota* of James Gregory, which was published in 1663, and which contains an account of the reflecting telescope known by his name, that Gregory had not heard of the discovery of the law of refraction, which had been given in Descartes' *Dioptrique* twenty-five years earlier.

[3] *Renati Descartes Epistolae, Pars tertia* (Amsterdam, 1683). The Fermat correspondence is comprised in letters xxix to xlvi.

up by the Cartesians long after the death of their master. Fermat eventually introduced a new fundamental law, from which he proposed to deduce the paths of rays of light. This was the celebrated *Principle of Least Time*, enunciated [1] in the form, 'Nature always acts by the shortest course.' From it the law of reflection can readily be derived, since the path described by light between a point on the incident ray and a point on the reflected ray is the shortest possible consistent with the condition of meeting the reflecting surfaces.[2] In order to obtain the law of refraction, Fermat assumed that 'the resistance of the media is different,' and applied his 'method of maxima and minima' to find the path which would be described in the least time from a point of one medium to a point of the other. In 1661 he arrived at the solution.[3] 'The result of my work,' he writes, 'has been the most extraordinary, the most unforeseen, and the happiest, that ever was; for, after having performed all the equations, multiplications, antitheses, and other operations of my method, and having finally finished the problem, I have found that my principle gives exactly and precisely the same proportion for the refractions which Monsieur Descartes has established.' His surprise was all the greater, as he had supposed light to move more slowly in dense than in rare media, whereas Descartes had (as we have seen) been obliged to make the contrary supposition.

Although Fermat's result was correct, and, indeed, of high permanent interest, the principles from which it was derived were metaphysical rather than physical in character, and consequently were of little use for the purpose of framing a mechanical explanation of light. The influence of Descartes' theory was therefore scarcely at all diminished as a result of Fermat's work. But a new attack soon opened from another quarter: nothing less, indeed, than the rise of a new school of philosophy, hostile to Aristotelianism and to Cartesianism alike. Its author was Pierre Gassendi (1592–1655), Professor at the Collège de France in Paris, a follower

[1] *Epist.* xlii, written at Toulouse in August 1657 to Monsieur de la Chambre; reprinted in *Œuvres de Fermat* (ed. 1891), ii, p. 354.

[2] That reflected light follows the shortest path was no new result, for it had been affirmed in a Commentary on Aristotle's *Meteor*, iii, by Olympiodorus, who lived at Alexandria in the middle of the sixth century A.D.: and also in the *κεφάλαια τῶν ὀπτικῶν* of Damianus, a pupil of Heliodorus of Larissa, a work of which several editions were published in the seventeenth century. Damianus gives in his first book the optical work of Hero of Alexandria, to whom he attributes the theorem regarding the path of reflected light.

[3] *Epist.* xliii, written at Toulouse on 1 Jan. 1662; reprinted in *Œuvres de Fermat*, ii, p. 457; i, pp. 170, 173

of Copernicus and Galileo, who re-introduced the doctrine of the ancient atomists, namely that the universe is formed of material atoms, eternal and unchangeable, moving about in a space which except for them is empty. The most formidable objection to the new teaching was that in the Graeco-Roman world it had been associated with the moral and theological views of Epicurus and Lucretius. Gassendi, who was himself a priest, worked hard to show that it had no necessary connection with religious error, and that it could be accepted as a basis for physics by Christian men. After a sharp controversy with Descartes in 1641–6, he had such success that, as we shall see, his doctrine was accepted not long afterwards by Newton, and in fact became the departure point for all subsequent natural philosophy.

One great advantage of it was, that since space was not now a plenum, there was no need for moving particles to form closed chains, and the Cartesian vortices could be done away with. As a matter of fact, most of the later developments of the theory postulated an aether (or sometimes more than one), and so in a sense reverted to the idea of space as a plenum; but these aethers were conceived as something which offered no resistance to the motion of ordinary matter through them, so that material particles could be treated as if they were situated in a vacuum.

The next event of importance in the theory of light was the publication in 1667 [1] of the *Micrographia* of Robert Hooke (1635–1703).[2] Hooke, who had begun his scientific career as the assistant of Robert Boyle, and who was ultimately secretary of the Royal Society, made two experimental discoveries which concern our present subject; but in both of these, as it appeared, he had been anticipated. The first [3] was the observation of the iridescent colours which are seen when light falls on a thin layer of air between two glass plates or lenses, or on a thin film of any transparent substance. These are generally known as the 'colours of thin plates,' or 'Newton's rings'; they had been previously observed by Boyle.[4] Hooke's second experimental discovery,[5] made after the date of the *Micrographia*, was that light in air is not propagated exactly in straight lines, but that there is some illumination within the geometrical shadow of an opaque body. This observation had been published in 1665 in a posthumous work [1] of Fr. Francesco Maria Grimaldi

[1] The *imprimatur* of Viscount Brouncker, *P.R.S.*, is dated 23 Nov. 1664.
[2] On Hooke, cf. the excellent Wilkins Lecture by E. N. da C. Andrade, printed in *Proc. R.S.*, cci (1950), p. 439
[3] *Micrographia*, p. 47
[4] Boyle's *Works* (ed. 1772), i, p. 742
[5] Hooke's *Posthumous Works*, p. 186

(1613–63), a Jesuit, who had given to the phenomenon the name *diffraction*.

Hooke's theoretical investigations on light were of great importance, representing as they do the transition from the Cartesian system to the fully developed theory of waves. He begins by attacking Descartes' proposition, that light is a tendency to motion rather than an actual motion. 'There is,' he observes,[2] 'no luminous Body but has the parts of it in motion more or less'; and this motion is 'exceeding quick.' Moreover, since some bodies (e.g. the diamond when rubbed or heated in the dark) shine for a considerable time without being wasted away, it follows that whatever is in motion is not permanently lost to the body, and therefore that the motion must be of a to-and-fro or vibratory character. The amplitude of the vibrations must be exceedingly small, since some luminous bodies (e.g. the diamond again) are very hard, and so cannot yield or bend to any sensible extent.

Concluding, then, that the condition associated with the emission of light by a luminous body is a rapid vibratory motion of very small amplitude, Hooke next inquires how light travels through space. 'The next thing we are to consider,' he says 'is the way or manner of the *trajection* of this motion through the interpos'd pellucid body to the eye: And here it will be easily granted:

'First, that it must be a body *susceptible* and *impartible* of this motion that will deserve the name of a Transparent; and next, that the parts of such a body must be *homogeneous*, or of the same kind.

'Thirdly, that the constitution and motion of the parts must be such that the appulse of the luminous body may be communicated or propagated through it to the greatest imaginable distance in the least imaginable time, though I see no reason to affirm that it must be in an instant.

'Fourthly, that the motion is propagated every way through an *Homogeneous medium* by *direct* or *straight* lines extended every way like Rays from the centre of a Sphere.

'Fifthly, in an *Homogeneous medium* this motion is propagated every way with *equal velocity*, whence necessarily every *pulse* or *vibration* of the luminous body will generate a Sphere, which will continually increase, and grow bigger, just after the same manner (though indefinitely swifter) as the waves or rings on the surface of the water do swell into bigger and bigger circles about a point of it, where by the sinking of a Stone the motion was begun, whence

[1] *Physico-Mathesis de lumine, coloribus, et iride* (Bologna, 1665), i, prop. i

[2] *Micrographia*, p. 55

it necessarily follows, that all the parts of these Spheres undulated through an *Homogeneous medium* cut the Rays at right angles.'

Here we have a fairly definite mechanical conception. It resembles that of Descartes in postulating a medium as the vehicle of light; but according to the Cartesian hypothesis the disturbance is a statical pressure in this medium, while in Hooke's theory it is a rapid vibratory motion of small amplitude. In the above extract Hooke introduces, moreover, the idea of the *wave-front*, or locus at any instant of a disturbance generated originally at a point, and affirms that it is a sphere, whose centre is the point in question, and whose radii are the rays of light issuing from the point.

Hooke's next effort was to produce a mechanical theory of refraction, to replace that given by Descartes. 'Because,' he says, 'all transparent *mediums* are not *Homogeneous* to one another, therefore we will next examine how this pulse or motion will be propagated through differingly transparent *mediums*. And here, according to the most acute and excellent Philosopher *Des Cartes*, I suppose the sine of the angle of inclination in the first *medium* to be to the sine of refraction in the second, as the density of the first to the density of the second. By density, I mean not the density in respect of gravity (with which the refractions or transparency of *mediums* hold no proportion), but in respect only to the *trajection* of the Rays of light, in which respect they only differ in this, that the one propagates the pulse more easily and weakly, the other more slowly, but more strongly. But as for the pulses themselves, they will by the refraction acquire another property, which we shall now endeavour to explicate.

'We will suppose, therefore, in the first Figure, ACFD to be a physical Ray, or ABC and DEF to be two mathematical Rays,

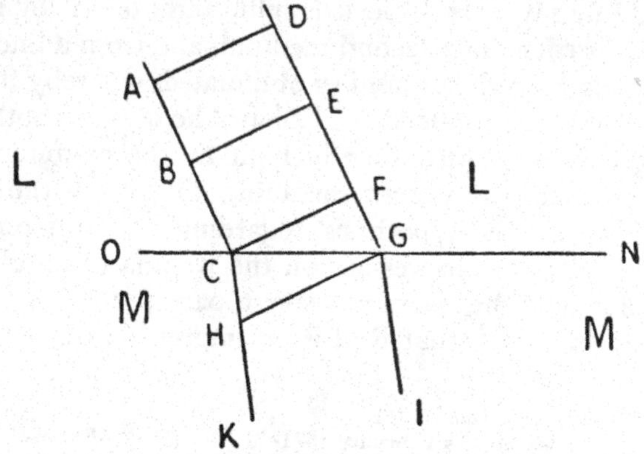

trajected from a very remote point of a luminous body through an *Homogeneous* transparent medium LL, and DA, EB, FC, to be small portions of the orbicular impulses which must therefore cut the Rays at right angles, these Rays meeting with the plain surface NO of a *medium* that yields an easier *transitus* to the propagation of light, and falling *obliquely* on it, they will in the medium MM be refracted towards the perpendicular of the surface. And because this medium is more easily *trajected* than the former by a third, therefore the point C of the orbicular pulse FC will be moved to H four spaces in the same time that F, the other end of it, is moved to three spaces, therefore the whole refracted pulse to H shall be *oblique* to the refracted Rays CHK and GI.'

Although this is not in all respects successful, it represents a decided advance on the treatment of the same problem by Descartes, which rested on a mere analogy. Hooke tries to determine what happens to the wave-front when it meets the interface between two media; and for this end he introduces the correct principle that the side of the wave-front which first meets the interface will go forward in the second medium with the velocity proper to that medium, while the other side of the wave-front, which is still in the first medium, is still moving with the old velocity; so that the wave-front will be deflected in the transition from one medium to the other.

This deflection of the wave-front was supposed by Hooke to be the origin of the prismatic colours. He regarded natural or white light as the simplest type of disturbance, being constituted by a simple and uniform pulse at right angles to the direction of propagation, and inferred that colour is generated by the distortion to which this disturbance is subjected in the process of refraction. 'The Ray,' [1] he says, 'is dispersed, split, and opened by its Refraction at the Superficies of a second medium, and from a line is opened into a diverging Superficies, and so obliquated, whereby the appearances of Colours are produced.' 'Colour,' he says in another place,[2] 'is nothing but the disturbance of light by the communication of the pulse to other transparent mediums, that is by the refraction thereof.' His precise hypothesis regarding the different colours was [3] 'that Blue is an impression on the Retinia of an oblique and confus'd pulse of light, whose weakest part precedes, and whose strongest follows. And, that Red is an impression on the Retina of

[1] Hooke's *Posthumous Works*, p. 82
[2] To the Royal Society, 15 February 1671–72
[3] *Micrographia*, p. 64

an oblique and confus'd pulse of light, whose strongest part precedes, and whose weakest follows.'

Hooke's theory of colour was completely overthrown, within a few years of its publication, by one of the earliest discoveries of Isaac Newton (1642–1727). Newton, at that time a scholar of Trinity College, Cambridge, had in the beginning of 1666 obtained a triangular prism, 'to try therewith the celebrated Phaenomena of Colours.' For this purpose, 'having darkened my chamber, and made a small hole in my window-shuts, to let in a convenient quantity of the Sun's light, I placed my Prisme at his entrance, that it might be thereby refracted to the opposite wall. It was at first a very pleasing divertisement, to view the vivid and intense colours produced thereby; but after a while applying myself to consider them more circumspectly, I became surprised to see them in an *oblong* form, which, according to the received laws of Refraction, I expected should have been *circular*.' The length of the coloured spectrum was in fact about five times as great as its breadth.

This puzzling fact he set himself to study; and after more experiments the true explanation was discovered—namely, that ordinary white light is really a mixture [1] of rays of every variety of colour, and that the elongation of the spectrum is due to the differences in the refractive power of the glass for these different rays.

'Amidst these thoughts,' he tells us,[2] 'I was forced from Cambridge by the intervening Plague'; this was in 1666, and his memoir on the subject was not presented to the Royal Society until five years later. In it he propounds a theory of colour directly opposed to that of Hooke. 'Colours,' he says, 'are not *Qualifications of light* derived from Refractions, or Reflections of natural Bodies (as 'tis generally believed), but *Original and connate properties*, which in divers Rays are divers. Some Rays are disposed to exhibit a red colour and no other: some a yellow and no other, some a green and no other, and so of the rest. Nor are there only Rays proper and particular to the more eminent colours, but even to all their intermediate gradations.

'To the same degree of Refrangibility ever belongs the same

[1] The word *mixture* must not be taken to imply that the rays of different colours, when compounded together, preserve their separate existence and identity unaltered within the compound, like the constituents of a mechanical mixture. On the contrary, as was shown by Gouy in 1886 (*Journal de Physique*, (2) v, p. 354), natural white light is to be pictured, in the undulatory representation, as a succession of short *pulses*, out of which any spectroscopic apparatus such as a prism *manufactures* the different monochromatic rays, by a process which is physically equivalent to the mathematical resolution of an arbitrary function into periodic terms by Fourier's integral theorem.

[2] *Phil. Trans.* vi (19 February 1671–2), p. 3075

colour, and to the same colour ever belongs the same degree of Refrangibility.

'The species of colour, and degree of Refrangibility proper to any particular sort of Rays, is not mutable by Refraction, nor by Reflection from natural bodies, nor by any other cause, that I could yet observe. When any one sort of Rays hath been well parted from those of other kinds, it hath afterwards obstinately retained its colour, notwithstanding my utmost endeavours to change it.'

The publication of the new theory gave rise to an acute controversy.[1] As might be expected, Hooke was foremost among the opponents, and led the attack with some degree of asperity. When it is remembered that at this time Newton was at the outset of his career, while Hooke was an older man, with an established reputation, such harshness appears particularly ungenerous ; and it is likely that the unpleasant consequences which followed the announcement of his first great discovery had much to do with the reluctance which Newton ever afterwards showed to publish his results to the world.

In the course of the discussion Newton found occasion to explain more fully the views which he entertained regarding the nature of light. Hooke charged him with holding the doctrine that light is a material substance. Now Newton had, as a matter of fact, a great dislike of the more imaginative kind of hypotheses ; he altogether renounced the attempt to construct the universe from its foundations after the fashion of Descartes, and aspired to nothing more than a formulation of the laws which directly govern the actual phenomena. His desire in regard to optics was to present a theory free from speculation as to the hidden mechanism of light. Accordingly, in reply to Hooke's criticism, he protested[2] that his views on colour were in no way bound up with any particular conception of the ultimate nature of optical processes.

Newton was, however, unable to carry out his plan of connecting together the phenomena of light into a coherent and reasoned whole without having recourse to hypotheses. The hypothesis of Hooke, that light consists in vibrations of an aether, he rejected for reasons which at that time were perfectly cogent, and which indeed were not successfully refuted for over a century. One of these was the incompetence of the wave-theory to account for the rectilinear propagation of light, and another was its inability to embrace the facts—

[1] The astronomer John Flamsteed explained his objections to it in a letter to Collins of date 17 April 1672, printed in Rigaud's *Correspondence of scientific men of the XVIIth century* ii, p. 134.

[2] *Phil. Trans.* vii (1672), p. 5086

discovered, as we shall presently see, by Huygens, and first interpreted correctly by Newton himself—of polarisation. On the whole, he seems to have favoured a scheme of which the following may be taken as a summary [1]:

All space is permeated by an elastic medium or *aether*, which is capable of propagating vibrations in the same way as the air propagates the vibrations of sound, but with far greater velocity.

This aether pervades the pores of all material bodies, and is the cause of their cohesion; its density varies from one body to another, being greatest in the free interplanetary spaces. It is not necessarily a single uniform substance: but just as air contains aqueous vapour, so the aether may contain various 'aethereal spirits,' adapted to produce the phenomena of electricity, magnetism and gravitation.

The vibrations of the aether cannot, for the reasons already mentioned, be supposed in themselves to constitute light. Light is therefore taken to be 'something of a different kind, propagated from lucid bodies. They, that will, may suppose it an aggregate of various peripatetic qualities. Others may suppose it multitudes of unimaginable small and swift corpuscles of various sizes, springing from shining bodies at great distances one after another; but yet without any sensible interval of time, and continually urged forward by a principle of motion, which in the beginning accelerates them, till the resistance of the aethereal medium equals the force of that principle, much after the manner that bodies let fall in water are accelerated till the resistance of the water equals the force of gravity. But they, that like not this, may suppose light any other corporeal emanation, or any impulse or motion of any other medium or aethereal spirit diffused through the main body of aether, or what else they can imagine proper for this purpose. To avoid dispute, and make this hypothesis general, let every man here take his fancy; only whatever light be, I suppose it consists of rays differing from one another in contingent circumstances, as bigness, form or vigour.' [2]

In any case, light and aether are capable of mutual interaction; aether is in fact the intermediary between light and ponderable matter. When a ray of light meets a stratum of aether denser or

[1] cf. Newton's memoir in *Phil. Trans.* vii (1672); his memoir presented to the Royal Society in December 1675, which is printed in T. Birch's *History of the Royal Society of London* (4 vols., 1756–7), iii, p. 247; his *Opticks*, especially queries 18, 19, 20, 21, 23, 29; the Scholium at the end of the *Principia*; and a letter to Boyle, written in February 1678–9, which is printed in Horsley's *Newtoni Opera*, p. 385, and in Rigaud's *Correspondence*, ii, p. 407.

In the *Principia*, bk. i, § xiv, the analogy between rays of light and streams of corpuscles is indicated; but Newton does not commit himself to any theory of light based on this.

[2] Royal Society, 9 Dec. 1675 (Birch iii p. 255)

rarer than that through which it has lately been passing, it is, in general, deflected from its rectilinear course; and differences of density of the aether between one material medium and another account on these principles for the reflection and refraction of light.[1] The condensation or rarefaction of the aether due to a material body extends to some little distance from the surface of the body, so that the inflection due to it is really continuous, and not abrupt; and this further explains diffraction, which Newton took to be 'only a new kind of refraction, caused, perhaps, by the external aether's beginning to grow rarer a little before it came at the opake body, than it was in free spaces.'

Although the regular vibrations of Newton's aether were not supposed to constitute light, its irregular turbulence seems to have represented fairly closely his conception of heat. He supposed that when light is absorbed by a material body, vibrations are set up in the aether, and are recognisable as the heat which is always generated in such cases. The conduction of heat from hot bodies to contiguous cold ones he conceived to be effected by vibrations of the aether propagated between them; and he supposed that it is the violent agitation of aethereal motions which excites incandescent substances to emit light.

Assuming with Newton that light is not actually constituted by the vibrations of an aether, even though such vibrations may exist in close connection with it, the most definite and easily conceived supposition is that rays of light are streams of corpuscles emitted by luminous bodies. Although this was not the hypothesis of Descartes himself, it was so thoroughly akin to his general scheme that the scientific men of Newton's generation, who were for the most part deeply imbued with the Cartesian philosophy, instinctively selected it from the wide choice of hypotheses which Newton had offered them; and by later writers it was generally associated with Newton's name. A curious argument in its favour was drawn from a phenomenon which had then been known for nearly half a century: Vincenzo Cascariolo, a shoemaker of Bologna, had discovered, about 1630, that a substance, which afterwards received the name of *Bologna stone* or *Bologna phosphorus*, has the property of shining in the dark after it has been exposed for some time to sunlight; and the storage of light which seemed to be here involved was more easily explicable on the corpuscular theory than on any other. The evidence in this quarter, however, pointed the other way when it was found that

[1] Newton's proof of the law of refraction, in *Opticks*, i, prop. 6, does not differ greatly in principle from Descartes' proof.

phosphorescent substances do not necessarily emit the same kind of light as that which was used to stimulate them.

In accordance with his earliest discovery, Newton considered colour to be an inherent characteristic of light, and inferred that it must be associated with some definite quality of the corpuscles or aether vibrations. The corpuscles corresponding to different colours would, he remarked, like sonorous bodies of different pitch, excite vibrations of different types in the aether ; and ' if by any means those [aether vibrations] of unequal bignesses be separated from one another, the largest beget a Sensation of a *Red* colour, the least or shortest of a deep *Violet*, and the intermediate ones, of intermediate colours ; much after the manner that bodies, according to their several sizes, shapes and motions, excite vibrations in the Air of various bignesses, which, according to those bignesses, make several Tones in Sound.' [1]

This sentence is the first enunciation of the great principle that homogeneous light is essentially *periodic* in its nature, and that differences of period correspond to differences of colour. The obvious analogy with sound was expressly mentioned by Huygens in his *Traité de la lumière*, of which more hereafter [2] : and it may be remarked in passing that Newton's theory of periodic vibrations in an elastic medium, which he developed [3] in connection with the explanation of sound, would alone entitle him to a place among those who have exercised the greatest influence on the theory of light, even if he had made no direct contribution to the latter subject.

Newton devoted considerable attention to the colours of thin plates, and determined the empirical laws of the phenomena with great accuracy. In order to explain them, he supposed that ' every ray of light, in its passage through any refracting surface, is put into a certain transient constitution or state, which, in the progress of the ray, returns at equal intervals, and disposes the ray, at every return, to be easily transmitted through the next refracting surface, and, between the returns, to be easily reflected by it.' [4] The interval between two consecutive dispositions to easy transmission or ' length of fit ' varies, as he found, with the colour, being greatest for red light and least for violet. If then a ray of homogeneous light falls on a thin plate, its fortunes as regards transmission and reflection at the two surfaces will depend on the relation which the length of

[1] *Phil. Trans.* vii (1672), p. 5088

[2] The hypotheses that the colour of monochromatic light depends on the period of vibration, and that the brightness increases with the amplitude of the vibration, seem to have been first clearly enunciated in 1699 by N. de Malebranche (1638–1715).

[3] Newton's *Principia*, ii, props. xliii–l

[4] *Opticks*, ii, prop. 12

fit bears to the thickness of the plate; and on this basis he built up a theory of the colours of thin plates. It is evident that Newton's 'length of fit' corresponds in some measure to the quantity which in the undulatory theory is called the wave-length of the light; but the theory of *fits of easy transmission and easy reflection*, though it served to explain successfully phenomena which were not known until after Newton's time, was abandoned after the triumph of the wave-theory in the nineteenth century.[1]

Newton's theory of light led to a correspondence with Fr. Ignace-Gaston Pardies, S.J. (1636–73),[2] Professor of Mathematics in the Collège de Clermont (later known as the Collège Louis-le-Grand) at Paris. As a result, Pardies, who was a Cartesian, seems to have become convinced of the truth of Newton's doctrine. He was, however, strongly in favour of the wave-hypothesis (which was not irreconcilable with it), and before his early death wrote a dissertation on wave-motion which seems to have influenced Huygens, who saw it in Paris. Pardies adumbrated the notion of aberration, which was not discovered until more than half a century later, and he rejected Descartes' supposition of an infinite velocity for light.

At the time of the publication of Hooke's *Micrographia*, and Newton's theory of colours, it was not known whether light is propagated instantaneously or not. An attempt to settle the question experimentally had been made many years previously by Galileo,[3] who had stationed two men with lanterns at a considerable distance from each other; one of them was directed to observe when the other uncovered his light, and exhibit his own the moment he perceived it. But the interval of time required by the light for its journey was too small to be perceived in this way; and the discovery was ultimately made by an astronomer. In 1675 Olaf Roemer[4] (1644–1710), a Dane who was at the time resident in Paris as professor of mathematics to the Dauphin of France, observed that the eclipses of the first satellite of Jupiter were apparently affected by an unknown disturbing cause; the time of the occurrence of the phenomenon was retarded when the earth and Jupiter, in the course of their orbital motions, happened to be most remote from

[1] It was however a remarkable anticipation of the twentieth-century quantum-theory explanation: the "fits of easy transmission and easy reflection" correspond to the *transition probabilities* of the quantum theory.

[2] Two letters from Pardies are printed in *Phil. Trans.* vii (1672), pp. 4087, 5012, followed by Newton's comments.

[3] *Discorsi e dimostrazioni matematiche*, p. 43 of the Elzevir edition of 1638

[4] *Mém. de l'Acad.* x (1666–99), p. 575. Roemer, who was afterwards appointed Director of the Copenhagen Observatory, was the inventor of many astronomical instruments, notably the equatorial and the transit instrument.

each other, and was accelerated in the contrary case. Roemer explained this by supposing that light requires a finite time for its propagation from the satellite to the earth ; and by observations of eclipses, he calculated the interval required for its passage from the sun to the earth (the *light equation* as it is called) to be 11 minutes.[1]

Shortly after Roemer's discovery the wave-theory of light was greatly improved and extended by Christiaan Huygens (1629–95). Huygens, who at the time was living in Paris, communicated his results in 1678 to Cassini, Roemer, De la Hire and the other physicists of the French Academy, and prepared a manuscript of considerable length on the subject. This he proposed to translate into Latin, and to publish in that language together with a treatise on the optics of telescopes ; but the work of translation making little progress, after a delay of twelve years he decided to print the work on wave-theory in its original form. In 1690 it appeared at Leyden,[2] under the title *Traité de la lumière où sont expliquées les causes de ce qui luy arrive dans la réflexion et dans la réfraction. Et particulièrement dans l'étrange réfraction du cristal d'Islande. Par C.H.D.Z.*[3]

The truth of Hooke's hypothesis, that light is essentially a form of motion, seemed to Huygens to be proved by the effects observed with burning-glasses ; for in the combustion induced at the focus of the glass, the molecules of bodies are dissociated ; which, as he remarked, must be taken as a certain sign of motion, if, in conformity to the Cartesian philosophy, we seek the cause of all natural phenomena in purely mechanical actions.[4]

The question then arises as to whether the motion is that of a medium, as is supposed in Hooke's theory, or whether it may be compared to that of a flight of arrows, as in the corpuscular theory. Huygens decided that the former alternative is the only tenable

[1] It was soon recognised that Roemer's value was too large ; and the astronomers of the succeeding half-century reduced it to 7 minutes. Delambre, by an investigation whose details appear to have been completely destroyed, published in 1817 the value $493{\cdot}2^s$, from a discussion of eclipses of Jupiter's satellites during the previous 150 years. Glasenapp, in an inaugural dissertation published in 1875, discussed the eclipses of the first satellite between 1848 and 1870, and derived, by different assumptions, values between 496^s and 501^s, the most probable value being $500{\cdot}8^s$. Sampson, in 1909, derived $498{\cdot}64^s$ from his own readings of the Harvard Observations, and $498{\cdot}79^s$ from the Harvard readings with probable errors of about $\pm\ 0{\cdot}02^s$. The inequalities of Jupiter's surface give rise to some difficulty in exact determinations.

[2] Huygens had by this time returned to Holland.

[3] i.e. Christiaan Huygens de Zuylichem. The custom of indicating names by initials was not unusual in that age. English translation by S. P. Thompson, London, 1912.

[4] It would seem from Berkeley's First Dialogue between Hylas and Philonous that this opinion was favoured in England in 1713. 'Hylas : I tell you, Philonous, external light is nothing but a thin fluid substance, whose minute particles, being agitated with a brisk motion, communicate different motions to the optic nerves.'

one, since beams of light proceeding in directions inclined to each other do not interfere with each other in any way.

Moreover, it had previously been shown by Torricelli that light is transmitted as readily through a vacuum as through air ; and from this Huygens inferred that the medium or aether in which the propagation takes place must penetrate all matter, and be present even in all so-called vacua. Light, therefore, consists of disturbances, propagated with great velocity, in a highly elastic medium composed of very subtle matter.

The process of wave-propagation he discussed by aid of a principle which was now [1] introduced for the first time, and has since been generally known by his name. It may be stated thus : consider a wave-front,[2] or locus of disturbance, as it exists at a definite instant t_0 ; then each surface-element of the wave-front may be regarded as the source of a secondary wave, which in a homogeneous isotropic medium will be propagated outwards from the surface-element in the form of a sphere whose radius at any subsequent instant t is proportional to $(t-t_0)$; and the wave-front which represents the whole disturbance at the instant t is simply the envelope of the secondary waves which arise from the various surface-elements of the original wave-front. The introduction of this principle enabled Huygens to succeed where Hooke and Pardies had failed, in achieving the explanation of refraction and reflection. His method was to combine his own principle with Hooke's device of following separately the fortunes of the right-hand and left-hand sides of a wave-front when it reaches the interface between two media. The actual explanation for the case of reflection is as follows :

Let AB represent the interface at which reflection takes place, AHC the incident wave-front at an instant t_0, GMB the position which the wave-front would occupy at a later instant t if the propagation were not interrupted by reflection. Then by Huygens' principle the secondary wave from A is at the instant t a sphere RNS of radius equal to AG ; the disturbance from H, after meeting the interface at K, will generate a secondary wave TV of radius equal to KM, and similarly the secondary wave corresponding to any other element of the original wave-front can be found. It is obvious that the envelope of these secondary waves, which constitutes the final wave-front, will be a plane BN, which will be

[1] *Traité de la lum.*, p. 17

[2] It may be remarked that Huygens' ' waves ' are really what modern writers, following Hooke, call ' pulses ' ; Huygens never considered true wave-trains having the property of periodicity.

inclined to AB at the same angle as AC. This gives the law of reflection.

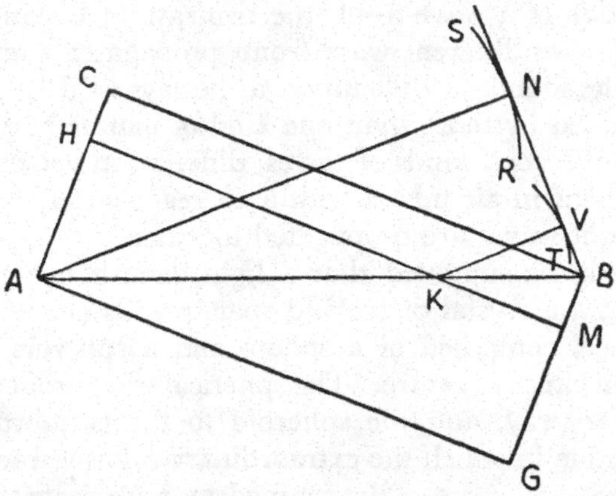

The law of refraction is established by similar reasoning, on the supposition that the velocity of light depends on the medium in which it is propagated. Since a ray which passes from air to glass is bent inwards towards the normal, it may be inferred that light travels more slowly in glass than in air.

Huygens offered a physical explanation of the variation in velocity of light from one medium to another, by supposing that transparent bodies consist of hard particles which interact with the aethereal matter, modifying its elasticity. The opacity of metals he explained by an extension of the same idea, supposing that some of the particles of metals are hard (these account for reflection) and the rest soft; the latter destroy the luminous motion by damping it.

The second half of the *Traité de la lumière* is concerned with a phenomenon which had been discovered a few years previously by a Danish philosopher, Erasmus Bartholin (1625–98). A sailor had brought from Iceland to Copenhagen a number of beautiful crystals which he had collected in the Bay of Röerford. Bartholin, into whose hands they passed, noticed [1] that any small object viewed through one of these crystals appeared double, and found the immediate cause of this in the fact that a ray of light entering the crystal gave rise in general to *two* refracted rays. One of these rays was subject to the ordinary law of refraction, while the other, which was called the *extraordinary* ray, obeyed a different law, which Bartholin did not succeed in determining.

[1] *Experimenta cristalli Islandici disdiaclastici*, 1669

The matter had arrived at this stage when it was taken up by Huygens. Since in his conception each ray of light corresponds to the propagation of a wave-front, the two rays in Iceland spar must correspond to two different wave-fronts propagated simultaneously. In this idea he found no difficulty ; as he says : ' It is certain that a space occupied by more than one kind of matter may permit the propagation of several kinds of waves, different in velocity ; for this actually happens in air mixed with aethereal matter, where sound-waves and light-waves are propagated together.'

Accordingly he supposed that a light disturbance generated at any spot within a crystal of Iceland spar spreads out in the form of a wave-surface, composed of a sphere and a spheroid having the origin of disturbance as centre. The spherical wave-front corresponds to the ordinary ray, and the spheroid to the extraordinary ray ; and the direction in which the extraordinary ray is refracted may be determined by a geometrical construction, in which the spheroid takes the place which in the ordinary construction is taken by the sphere.

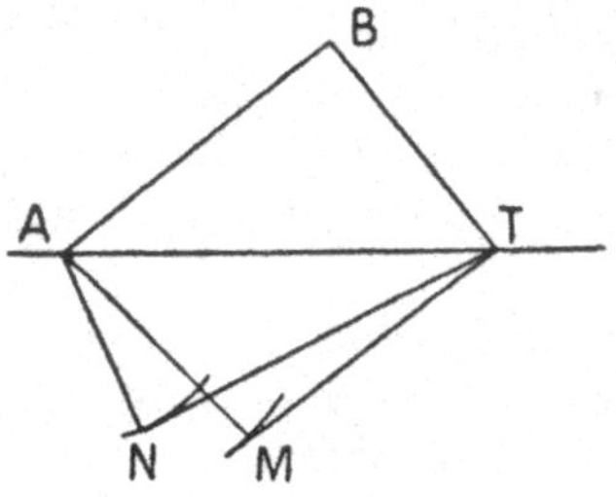

Thus, let the plane of the figure be at right angles to the intersection of the wave-front with the surface of the crystal ; let AB represent the trace of the incident wave-front ; and suppose that in unit time the disturbance from B reaches the interface at T. In this unit interval of time the disturbance from A will have spread out within the crystal into a sphere and spheroid ; so the wave-front corresponding to the ordinary ray will be the tangent plane to the sphere through the line whose trace is T, while the wave-front corresponding to the extraordinary ray will be the tangent plane to the spheroid through the same line. The points of contact N and M will determine the directions AN and AM of the two refracted rays[1] within the crystal.

Huygens did not in the *Théorie de la lumière* attempt a detailed physical explanation of the spheroidal wave, but communicated one later in a letter to Papin,[2] written in December 1690. ' As to the kinds of matter contained in Iceland crystal,' he says, ' I suppose one composed of small spheroids, and another which occupies the

[1] The word *ray* in the wave-theory is always applied to the line which goes from the centre of a wave (i.e. the origin of the disturbance) to a point on its surface, whatever may be the inclination of this line to the surface-element on which it abuts ; for this line has the optical properties of the ' rays ' of the emission theory.

[2] Huygens' *Œuvres* (ed. 1905), x, p. 177

interspaces around these spheroids, and which serves to bind them together. Besides these, there is the matter of aether permeating all the crystal, both between and within the parcels of the two kinds of matter, just mentioned ; for I suppose both the little spheroids, and the matter which occupies the intervals around them, to be composed of small fixed particles, amongst which are diffused in perpetual motion the still finer particles of the aether. There is now no reason why the ordinary ray in the crystal should not be due to waves propagated in this aethereal matter. To account for the extraordinary refraction, I conceive another kind of waves, which have for vehicle both the aethereal matter and the two other kinds of matter constituting the crystal. Of these latter, I suppose that the matter of the small spheroids transmits the waves a little more quickly than the aethereal matter, while that around the spheroids transmits these waves a little more slowly than the same aethereal matter. . . . These same waves, when they travel in the direction of the breadth of the spheroids, meet with more of the matter of the spheroids, or at least pass with less obstruction, and so are propagated a little more quickly in this sense than in the other ; thus the light disturbance is propagated as a spheroidal sheet.'

Huygens made another discovery [1] of capital importance when experimenting with the Iceland crystal. He observed that the two rays which are obtained by the double refraction of a single ray afterwards behave in a way different from ordinary light which has not experienced double refraction ; and in particular, if one of these rays is incident on a second crystal of Iceland spar, it gives rise in some circumstances to two, and in others to only one, refracted ray. The behaviour of the ray at this second refraction can be altered by simply rotating the second crystal about the direction of the ray as axis ; the ray undergoing the ordinary or extraordinary refraction according as the principal section of the crystal is in a certain direction or in the direction at right angles to this.

The first stage in the explanation of Huygens' observation was reached by Newton, who in 1717 showed [2] that a ray obtained by double refraction differs from a ray of ordinary light in the same way that a long rod whose cross-section is a rectangle differs from a long rod whose cross-section is a circle ; in other words, the properties of a ray of ordinary light are the same with respect to all directions at right angles to its direction of propagation, whereas a ray obtained by double refraction must be supposed to have *sides*, or properties

[1] *Théorie de la lumière*, p. 89

[2] The second edition of Newton's *Opticks*, query 26

related to special directions at right angles to its own direction. The refraction of such a ray at the surface of a crystal depends on the relation of its sides to the principal plane of the crystal.

That a ray of light should possess such properties seemed to Newton [1] an insuperable objection to the hypothesis which regarded waves of light as analogous to waves of sound. On this point he was in the right; his objections are perfectly valid against the wave-theory as it was understood by his contemporaries,[2] although not against the theory [3] which was put forward a century later by Young and Fresnel.

Although Huygens' chief researches in optics came later than Newton's discoveries of 1666–72, Huygens was actually senior to Newton by thirteen years, and he was, as regards his physical conceptions, in the main a Cartesian. His aether was, like Descartes', constituted of subtle particles; and he attempted to interpret gravity in terms of it, supposing it to be in rapid rotation in the region of space surrounding the earth—to form, in fact, a Cartesian vortex—and, by the effect of its centrifugal force, to displace terrestrial objects towards the centre of the earth. As we have seen, however (p. 7), he was well aware of the errors of principle involved in Descartes' method of approach to Nature.

The correct law of gravitation, which is valid for the whole universe, and which made it possible to calculate the motions of all the heavenly bodies, was published in 1687 in Newton's *Principia*: namely, that any two particles are attracted to each other with a force whose magnitude is proportional to the product of their masses and the inverse square of their distance apart. Newton claimed nothing more for his discovery than that it provided the necessary instrument for mathematical prediction, and he pointed out that it did not touch on the question of the mechanism of gravity. As to this, he conjectured that the density of the aether might vary from place to place, and that bodies might tend to move from the denser parts of the medium towards the rarer; but whether this were the true explanation or not, at any rate, he said, to suppose 'that one body may act upon another at a distance through a vacuum, without the mediation of anything else, . . . is to me so great an absurdity, that I believe no man, who has in philosophical matters a competent faculty for thinking, can ever fall into.'

[1] *Opticks*, query 28

[2] In which the oscillations are performed in the direction in which the wave advances.

[3] In which the oscillations are performed in a direction at right angles to that in which the wave advances.

The work of Newton was of such unparalleled brilliance that one is surprised to learn how slowly it displaced Descartes' theory of vortices. Long after the publication of the *Principia*, the text-book of natural philosophy used at Cambridge continued to be a translation (into Latin from French) of the *Physics* of Rohault, a work entirely Cartesian. Whiston has recorded that, having returned to Cambridge after his ordination in 1693, he resumed his studies there, 'particularly the Mathematicks and the Cartesian Philosophy, which was alone in Vogue with us at that Time. But it was not long before I, with immense Pains, but no Assistance, set myself with the utmost Zeal to the study of Sir Isaac Newton's wonderful Discoveries.'[1] The change in the character of the official teaching was brought about in a very curious manner. Dr Samuel Clarke, a zealous Newtonian, published about the year 1718 a new translation of Rohault, with a running commentary of notes which, while avoiding the language and appearance of controversy, actually constituted a complete refutation of the text. This edition superseded the older one in current use, and the younger generation peacefully adopted the new knowledge.

The first British university to incorporate Newtonianism into its formal instruction seems to have been Edinburgh, where David Gregory, a friend of Newton, was professor from 1683–90. To quote Whiston again, 'He caused several of his Scholars to keep Acts, as we call them, upon several Branches of the Newtonian Philosophy; while we at Cambridge, poor Wretches, were ignominiously studying the fictitious Hypotheses of the Cartesian.'[2]

On the continent, the change took place still more slowly. 'A Frenchman who arrives in London,' wrote Voltaire in 1730 'will find Philosophy, like everything else, very much changed there. He had left the world a *plenum*, and now he finds it a *vacuum*.'[3] For this he gave a most surprising explanation. 'It is,' he said,[4] 'the language used, and not the thing in itself, that irritates the human mind. If Newton had not used the word *attraction* in his admirable philosophy, everyone in our Academy would have opened his eyes to the light; but unfortunately he used in London a word to which an idea of ridicule was attached in Paris; and on that alone he was judged adversely, with a rashness which will some day be regarded as doing very little honour to his opponents.'[5]

[1] W. Whiston's *Memoirs* (1749), i, p. 36 [2] ibid. p. 36

[3] Voltaire, *Lettres philosophiques*, quatorzième lettre

[4] In a letter to M. de la Condamine, of date 22 June 1734, printed in *Œuvres complètes de M. de Voltaire* (Deux-Ponts, 1791–2), tome 79, p. 219

[5] Presumably the Cartesian word *impulsion* would have satisfied them.

In Germany, Leibnitz described the Newtonian formula as a return to the discredited scholastic concept of *occult qualities* ; [1] and as late as the middle of the eighteenth century Euler and two of the Bernoullis based the explanation of magnetism on the hypothesis of vortices.[2]

The rejection of the inverse-square law of gravitation by the French Cartesians antagonised the younger disciples of Newton to such an extent that the latter hardened into an opposition not only to the vortices but to the whole body of Cartesian notions, including the aether. In the second edition (1713) of the *Principia*, there is a preface written by Roger Cotes (1682–1716), in which the Newtonian law of action at a distance is championed as being the only formulation of the facts of experience which does not introduce unverifiable and useless suppositions. The principle which Cotes now affirmed was that the aim of theoretical physics is simply the prediction of future events, and that everything which is not strictly needed for this purpose, and which is not directly deducible from observed facts, should be pruned away. In this assertion there is clearly a question involved which is fundamental in the philosophy of nature, and which it will be worth while to look at more closely.

The investigations of the theoretical physicist are concerned partly with events which can actually be observed—these will be called phenomena—and partly also with events which cannot be detected in themselves, but which are assumed to exist in order to set up continuity between separated events that are in fact observed ; continuity is felt to be necessary if the picture of the world is to be logically coherent. These hypothetical events are called by modern writers [3] *inter-phenomena*. An illustration may be drawn from optics : light cannot be perceived except when it impinges on matter : these impacts are phenomena ; but since the velocity of light is finite, there are intervals of time between successive encounters with matter ; and if we adopt the belief that the luminous disturbance continues to exist in some form between its leaving one point and appearing at another, then during this intermediate state it is classified as an inter-phenomenon. Another illustration is provided by gravity : assuming (what is in fact now established) that its speed of propagation is finite, the question as to what is happening while it is in course of propagation belongs to the realm of inter-phenomena.

[1] S. Clarke, *Papers which passed between Mr Leibnitz and Dr Clarke* (1717), p. 265

[2] Their memoirs shared a prize of the French Academy in 1743, and were printed in 1752 in the *Recueil des pièces qui ont remporté les prix de l'Acad.*, tome v.

[3] e.g. H. Reichenbach

Cotes's principle recommends that inter-phenomena should be ignored, on the ground that they are unobservable and that formulae of prediction can be established without making use of them; that they belong to metaphysics rather than to physics. On the other hand, the school of which Descartes and Huygens were the first illustrious members, and to which almost all the physicists of the next two centuries belonged, regarded the understanding of inter-phenomena as one of the primary aims of natural philosophy; as the carrying of ideas beyond phenomena into regions which were more profound, and which still had ontological validity.

In Cotes's day the propagation of gravitation was assumed to be instantaneous, and consequently there was no span of time to be bridged by inter-phenomena. When the eighteenth-century natural philosophers found by experience that the Newtonian law was marvellously powerful, yielding formulae by which practically every observable motion in the solar system could be predicted, while on the other hand the search for an explanation of inter-phenomena led to no practical result, opinion set in favour of Cotes's attitude, which came to prevail widely; and in the middle of the century R. G. Boscovich (1711–87), a Croatian Jesuit, who was the first exponent of Newtonian ideas in Italy, attempted to account for all known physical effects in terms of action at a distance between point particles.[1] The luminiferous aether was abandoned, in spite of Huygens' brilliant development of the wave-hypothesis, and a corpuscular theory of light was adopted almost universally.

Even during this age, however, the hope of understanding inter-phenomena was never entirely abandoned. A curious attempt to Cartesianise Newtonianism was made by a French-Swiss, George Louis Le Sage,[2] who proposed to account for gravitation by means of an aether of Descartes' original type, that is to say, a cloud of excessively minute particles, *ultra-mundane corpuscles* as he called them. These, which resembled the neutrinos of the modern atomic physicist, he supposed to exist in great numbers in all parts of space, and to be moving with great speed in all directions; their diameters were further assumed to be so small in comparison with their distances apart that collisions between them were exceedingly rare. Then two particles of ordinary matter would to some extent screen each other from bombardment by the corpuscles, each particle receiving fewer

[1] Boscovich, *Theoria Philosophiae Naturalis*, Venice, 1763

[2] *Mém. de Berlin* for 1782 (Berlin, 1784), p. 404

impacts on the side facing the other than on the reverse side; and Le Sage showed that this effect would be equivalent to a force of attraction between them, varying inversely as the square of their mutual distance; which force he identified with that discovered by Newton.

Chapter II

ELECTRIC AND MAGNETIC SCIENCE PRIOR TO THE INTRODUCTION OF THE POTENTIALS

The ancients were acquainted with the curious properties possessed by two minerals, amber (ἤλεκτρον) and magnetic iron ore (ἡ λίθος Μαγνῆτις). The former, when rubbed, attracts light bodies; the latter has the power of attracting iron.

The use of the magnet for the purpose of indicating direction at sea was not known in classical antiquity; and the questions as to where, when and by whom the compass was invented, cannot be answered with complete certainty. Until recent years the general opinion was that it originated in China, and was brought by the Arabs to the Mediterranean and so came to the knowledge of the Crusaders. This, however, was not so; the Chinese were acquainted with the directive property of a magnet by the end of the eleventh century, but did not apply it to purposes of navigation until at least the end of the thirteenth. There is no evidence that the Arabs had any share in the invention or first transmission of knowledge regarding the compass. The earliest reference to it is in a work written, most probably in 1186, by Alexander Neckam (1157–1217), a monk of St Albans, and he does not refer to it as something new. There seems little doubt that it was known in north-western Europe, probably in England, earlier than elsewhere.[1]

Magnetism was one of the few sciences that made any progress during the Middle Ages; for in the thirteenth century Pierre de Maricourt, a native of Picardy, made a discovery of fundamental importance.[2]

Taking a natural magnet or lodestone, which had been rounded into a globular form, he laid it on a needle and marked the line along which the needle set itself. Then laying the needle on other parts of the stone, he obtained more lines in the same way. When the entire surface of the stone had been covered with such lines their general

[1] The early history of magnetism is treated very fully and with great learning by A. Crichton Mitchell, *Terrestrial Magnetism and Atmospheric Electricity*, xxxvii (1932), 105; xlii (1937), 241; xliv (1939), 77; li (1946), 323.

[2] His *Epistola Petri Peregrini de Maricourt de magnete* was written in 1269. At that time both St Albert the Great and St Thomas were living, but they and their immediate successors do not refer to it.

disposition became evident ; they formed circles which girdled the stone in exactly the same way as meridians of longitude girdle the earth ; and there were two points at opposite ends of the stone through which all the circles passed, just as all the meridians pass through the north and south poles of the earth.[1] Struck by the analogy, Peregrinus proposed to call these two points the *poles* of the magnet ; and he observed that the way in which magnets set themselves and attract each other depends solely on the position of their poles, as if these were the seat of the magnetic power. Such was the origin of those theories of poles and polarisation which in later ages have played so great a part in natural philosophy.

The observations of Peregrinus were greatly extended not long before the time of Descartes by William Gilberd or Gilbert [2] (1540–1603). Gilbert was born at Colchester ; after studying at Cambridge he took up medical practice in London, and had the honour of being appointed physician to Queen Elizabeth. In 1600 he published a work [3] on magnetism and electricity, with which the modern history of both subjects begins.

Of Gilbert's electrical researches we shall speak later ; in magnetism he made the capital discovery of the reason why magnets set in definite orientations with respect to the earth ; which is, that the earth is itself a great magnet, having one of its poles in high northern and the other in high southern latitudes. Thus the property of the compass was seen to be included in the general principle, that the north-seeking pole of every magnet attracts the south-seeking pole of every other magnet, and repels its north-seeking pole. Gilbert went further, and conjectured that magnetic forces were capable of accounting for the earth's gravity and the motions of the planets.

Descartes attempted [4] to account for magnetic phenomena by his theory of vortices. Adopting a suggestion of Gilbert's, he postulated a vortex of fluid matter round each magnet, the matter of the vortex entering by one pole and leaving by the other : this matter was supposed to act on iron and steel by virtue of a special resistance to its motion afforded by the molecules of those substances.

In the seventeenth century several Jesuits—N. Cabeo, A. Kircher, V. Léotaud—published works on magnetism which, while acknow-

[1] 'Procul dubio omnes lineae hujusmodi in duo puncta concurrent sicut omnes orbes meridiani in duo concurrunt polos mundi oppositos.'

[2] The form in the Colchester records is Gilberd.

[3] Gulielmi Gilberti *de Magnete, Magneticisque corporibus, et de magno magnete tellure*, London, 1600. An English translation by P. F. Mottelay was published in 1893.

[4] *Principia*, pt. iv, § 133 sqq.

ledging the discoveries of Gilbert, attempted to bring the subject within the framework of the Aristotelian-scholastic philosophy.[1]

The magnetic discoveries of Pierre de Maricourt and Gilbert, and the vortex hypothesis by which Descartes had attempted to explain them, had raised magnetism to the rank of a separate science by the middle of the seventeenth century. The kindred science of electricity was at that time in a less developed state; but it had been considerably advanced by Gilbert, whose researches in this direction will now be noticed.

For two thousand years the attractive power of amber had been regarded as a virtue peculiar to that substance, or possessed by at most one or two others.[2] Gilbert proved [3] this view to be mistaken, showing that the same effects are induced by friction in quite a large class of bodies, among which he mentioned glass, sulphur, sealing-wax and various precious stones.

A force which was manifested by so many different kinds of matter seemed to need a name of its own; and accordingly Gilbert gave to it the name *electric*, which it has ever since retained.[4]

Between the magnetic and electric forces Gilbert remarked many distinctions. The lodestone requires no stimulus of friction such as is needed to stir glass and sulphur into activity. The lodestone attracts only magnetisable substances, whereas electrified bodies attract everything. The magnetic attraction between two bodies is not affected by interposing a sheet of paper or a linen cloth, or by immersing the bodies in water; whereas the electric attraction is readily destroyed by screens. Lastly, the magnetic force tends to arrange bodies in definite orientations; while the electric force merely tends to heap them together in shapeless clusters.

These facts appeared to Gilbert to indicate that electric phenomena are due to something of a material nature, which under the influence of friction is liberated from the glass or amber in which under ordinary circumstances it is imprisoned. In support of this view he adduced evidence from other quarters. Being a physician, he was well acquainted with the doctrine [5] that the human body

[1] It should be said for Cabeo, however, that in his commentary on Aristotle's *Meteorology* he frequently insists on the desirability of following experience rather than Aristotle.

[2] Bede in his *Ecclesiastical History* mentions that jet, when warmed by rubbing, 'holds fast whatever is applied to it, like amber.'

[3] *De Magnete*, lib. ii, cap. 2

[4] The substantive *electricity* was introduced by Sir Thomas Browne in his *Pseudodoxia epidemica* (1646), p. 79.

[5] The doctrine of humours had been borrowed by the physicians of western Europe from the Arabs, who in turn had derived it from the Greeks.

contains various humours or kinds of moisture—phlegm, blood, choler and melancholy—which, as they predominated, were supposed to determine the temper of mind,[1] and when he observed that electrifiable bodies were almost all hard and transparent, and therefore (according to the ideas of that time) formed by the consolidation of watery liquids, he concluded that the common menstruum of these liquids must be a particular kind of humour, to the possession of which the electrical properties of bodies were to be referred. Friction might be supposed to warm or otherwise excite or liberate the humour, which would then issue from the body as an effluvium and form an atmosphere around it. The effluvium must, he remarked, be very attenuated, for its emission cannot be detected by the senses.

The existence of an atmosphere of effluvia round every electrified body might indeed have been inferred, according to Gilbert's ideas, from the single fact of electric attraction. For he believed that matter cannot act where it is not; and hence if a body acts on all surrounding objects without appearing to touch them, something must have proceeded out of it unseen.

The whole phenomenon appeared to him to be analogous to the attraction which is exercised by the earth on falling bodies; for in the latter case he conceived of the atmospheric air as the effluvium by which the earth draws all things downwards to itself.

Gilbert's theory of electrical emanations commended itself generally to such of the natural philosophers of the seventeenth century as were interested in the subject; among whom were numbered Niccolo Cabeo (1585–1650), who was perhaps the first to observe that electrified bodies repel as well as attract;[2] the English royalist exile, Sir Kenelm Digby (1603–65); and the celebrated Robert Boyle (1627–91). There were, however, some differences of opinion as to the manner in which the effluvia acted on the small bodies and set them in motion towards the excited electric; Gilbert himself had supposed the emanations to have an inherent tendency to reunion with the parent body; Digby likened their return to the condensation of a vapour by cooling; and other writers pictured the effluvia as forming vortices round the attracted bodies in the Cartesian fashion.

There is a well-known allusion to Gilbert's hypothesis in Newton's *Opticks*.[3]

[1] *Phlegmatic*, *sanguine*, *choleric* and *melancholic* were terms describing excess of the different humours.

[2] The first publication of this phenomenon seems to have been by O. von Guericke in 1663.

[3] query 22

'Let him also tell me, how an electrick body can by friction emit an exhalation so rare and subtle,[1] and yet so potent, as by its emission to cause no sensible diminution of the weight of the electrick body, and to be expanded through a sphere, whose diameter is above two feet, and yet to be able to agitate and carry up leaf copper, or leaf gold, at a distance of above a foot from the electrick body?'

It is, perhaps, somewhat surprising that the Newtonian doctrine of gravitation should not have proved a severe blow to the emanation theory of electricity; but Gilbert's doctrine was now so firmly established as to be unshaken by the overthrow of the analogy by which it had been originally justified. It was, however, modified in one particular about the beginning of the eighteenth century. In order to account for the fact that electrics are not perceptibly wasted away by excitement, the earlier writers had supposed all the emanations to return ultimately to the body which had emitted them; but the corpuscular theory of light accustomed philosophers to the idea of emissions so subtle as to cause no perceptible loss; and after the time of Newton the doctrine of the return of the electric effluvia gradually lost credit.

Newton died in 1727. Of the expositions of his philosophy which were published in his lifetime by his followers, one at least deserves to be noticed for the sake of the insight which it affords into the state of opinion regarding light, heat and electricity in the first half of the eighteenth century. This was the *Physices elementa mathematica experimentis confirmata* of Wilhelm Jacob 'sGravesande (1688–1742), published at Leyden in 1720. The Latin edition was afterwards reprinted several times, and was, moreover, translated into French and English; it seems to have exercised a considerable and, on the whole, well-deserved influence on contemporary thought.

'sGravesande supposed light to consist in the projection of corpuscles from luminous bodies to the eye; the motion being very swift, as is shown by astronomical observations. Since many bodies, e.g. the metals, become luminous when they are heated, he inferred that every substance possesses a natural store of corpuscles, which are expelled when it is heated to incandescence; conversely, corpuscles may become united to a material body; as happens, for instance, when the body is exposed to the rays of a fire. Moreover, since the heat thus acquired is readily conducted throughout the

[1] 'Subtlety,' says Johnson, 'which in its original import means exility of particles, is taken in its metaphorical meaning for nicety of distinction.'

substance of the body, he concluded that corpuscles can penetrate all substances, however hard and dense they be.

Let us here recall the ideas then current regarding the nature of material bodies. After the publication in 1661 of *The Sceptical Chemist*, by Robert Boyle (1626–91) it had been recognised generally that substances perceptible to the senses may be either *elements* or *compounds* or *mixtures*; the compounds being chemical individuals, distinct from mere mixtures of elements. But the substances at that time accepted as elements were very different from those which are now known by the name. Air and the calces [1] of the metals figured in the list, while almost all the chemical elements now recognised were omitted from it; some of them, such as oxygen and hydrogen, because they were as yet undiscovered, and others, such as the metals, because they were believed to be compounds.

Among the chemical elements, it became customary after the time of Newton to include light corpuscles.[2] That something which is confessedly imponderable should ever have been admitted into this class may at first sight seem surprising. But it must be remembered that questions of ponderability counted for very little with the philosophers of the period. Three-quarters of the eighteenth century had passed before Lavoisier enunciated the fundamental doctrine that the total weight of the substances concerned in a chemical reaction is the same after the reaction as before it. As soon as this principle came to be universally applied, light parted company from the true elements in the scheme of chemistry.

We must now consider the views which were held at this time regarding the nature of heat. These are of interest for our present purpose, on account of the analogies which were set up between heat and electricity.

The various conceptions which have been entertained concerning heat fall into one or other of two classes, according as heat is represented as a mere condition producible in bodies, or as a distinct species of matter. The former view, which is that universally held at the present day, was advocated by the great philosophers of the seventeenth century. Bacon maintained it in the *Novum Organum*: 'Calor,' he wrote, 'est motus expansivus, cohibitus, et nitens per partes minores.' [3] Boyle [4] affirmed that the 'Nature of

[1] i.e. oxides

[2] Newton himself (*Opticks*, p. 349) suspected that light corpuscles and ponderable matter might be transmuted into each other; much later, Boscovich (*Theoria*, pp. 215, 217) regarded the matter of light as a principle or element in the constitution of natural bodies.

[3] *Nov. Org.*, Lib. II, Aphor. XX

[4] *Mechanical Production of Heat and Cold*

Heat' consists in 'a various, vehement, and intestine commotion of the Parts among themselves.' Hooke [1] declared that 'Heat is a property of a body arising from the motion or agitation of its parts.' And Newton [2] asked: 'Do not all fixed Bodies, when heated beyond a certain Degree, emit light and shine; and is not this Emission performed by the vibrating Motion of their Parts?' and, moreover, suggested the converse of this namely, that when light is absorbed by a material body, vibrations are set up which are perceived by the senses as heat.

The doctrine that heat is a material substance was maintained in Newton's lifetime by a certain school of chemists. The most conspicuous member of the school was Wilhelm Homberg (1652–1715) of Paris, who [3] identified heat and light with the *sulphureous principle*, which he supposed to be one of the primary ingredients of all bodies, and to be present even in the interplanetary spaces. Between this view and that of Newton it might at first seem as if nothing but sharp opposition was to be expected.[4] But a few years later the professed exponents of the *Principia* and the *Opticks* began to develop their system under the evident influence of Homberg's writings. This evolution may easily be traced in 'sGravesande, whose starting-point is the admittedly Newtonian idea that heat bears to light a relation similar to that which a state of turmoil bears to regular rectilinear motion; whence, conceiving light as a projection of corpuscles, he infers that in a hot body the material particles and the light corpuscles [5] are in a state of agitation, which becomes more violent as the body is more intensely heated.

'sGravesande thus holds a position between the two opposite camps. On the one hand he interprets heat as a mode of motion; but on the other he associates it with the presence of a particular kind of matter, which he further identifies with the matter of light. After this the materialistic hypothesis made rapid progress. It was frankly advocated by another member of the Dutch school, Hermann Boerhaave [6] (1668–1738), Professor in the University of Leyden, whose treatise on chemistry was translated into English in 1727.

[1] *Micrographia*, p. 37

[2] *Opticks*, Query VIII

[3] *Mém. de l'Acad.* (1705), p. 88

[4] Though it reminds us of a curious conjecture of Newton's: 'Is not the strength and vigour of the action between light and sulphureous bodies one reason why sulphureous bodies take fire more readily and burn more vehemently than other bodies do?' (*Opticks*, Query VIII)

[5] I have thought it best to translate 'sGravesande's *ignis* by 'light corpuscles.' This is, I think, fully justified by such of his statements as *Quando ignis per lineas rectas oculos nostros intrat, ex motu quem fibris in fundo oculi communicat ideam luminis excitat.*

[6] Boerhaave followed Homberg in supposing the matter of heat to be present in all so-called vacuous spaces.

Somewhat later it was found that the heating effects of the rays from incandescent bodies may be separated from their luminous effects by passing the rays through a plate of glass, which transmits the light, but absorbs the heat. After this discovery it was no longer possible to identify the matter of heat with the corpuscles of light; and the former was consequently accepted as a distinct element, under the name of *caloric*.[1] In the latter part of the eighteenth and early part of the nineteenth centuries [2] caloric was generally conceived as occupying the interstices between the particles of ponderable matter —an idea which fitted in well with the observation that bodies commonly expand when they are absorbing heat, but which was less competent to explain the fact [3] that water expands when freezing. The latter difficulty was overcome by supposing the union between a body and the caloric absorbed in the process of melting [4] to be of a chemical nature; so that the consequent changes in volume would be beyond the possibility of prediction.

As we have already remarked, the imponderability of heat did not appear to the philosophers of the eighteenth century to be a sufficient reason for excluding it from the list of chemical elements; and in any case there was considerable doubt as to whether caloric was ponderable or not. Some experimenters believed that bodies were heavier when cold than when hot; others that they were heavier when hot than when cold. The century was far advanced before Lavoisier and Rumford finally proved that the temperature of a body is without sensible influence on its weight.

Perhaps nothing in the history of natural philosophy is more amazing than the vicissitudes of the theory of heat. The true hypothesis, after having met with general acceptance throughout a century, and having been approved by a succession of illustrious men, was deliberately abandoned by their successors in favour of a conception utterly false, and, in some of its developments, grotesque and absurd.

[1] Scheele in 1777 supposed caloric to be a compound of oxygen and phlogiston, and light to be oxygen combined with a greater proportion of phlogiston. Euler alone among the philosophers of the mid-eighteenth century refused to regard heat as a material substance.

[2] In spite of the experiments of Benjamin Thompson, Count Rumford (1753–1814), in the closing years of the eighteenth century. These should have sufficed to re-establish the older conception of heat.

[3] This had been known since the time of Boyle.

[4] Latent heat was discovered by Joseph Black (1728–99) in 1762. Black was the first to make a clear distinction between *quantity of heat* and *temperature*, although G. D. Fahrenheit (1686–1736) had devised a thermometer which was practically useful as a measuring instrument in 1708–17.

We must now return to 'sGravesande's book. The phenomena of combustion he explained by assuming that when a body is sufficiently heated the light corpuscles interact with the material particles, some constituents being in consequence separated and carried away with the corpuscles as flame and smoke. This view harmonises with the theory of calcination which had been developed by Becher and his pupil Stahl at the end of the seventeenth century, according to which the metals were supposed to be composed of their calces and an element *phlogiston*. The process of combustion, by which a metal is changed into its calx, was interpreted as a decomposition, in which the phlogiston separated from the metal and escaped into the atmosphere; while the conversion of the calx into the metal was regarded as a union with phlogiston.[1]

'sGravesande attributed electric effects to vibrations induced in effluvia, which he supposed to be permanently attached to such bodies as amber. 'Glass,' he asserted, 'contains in it, and has about its surface, a certain atmosphere, which is excited by Friction and put into a vibratory motion; for it attracts and repels light Bodies. The smallest parts of the glass are agitated by the Attrition, and by reason of their elasticity, their motion is vibratory, which is communicated to the Atmosphere above-mentioned; and therefore that Atmosphere exerts its action the further, the greater agitation the Parts of the Glass receive when a greater attrition is given to the glass.'

The English translator of 'sGravesande's work was himself destined to play a considerable part in the history of electrical science. Jean Théophile Desaguliers (1683–1744) was an Englishman only by adoption. His father had been a Huguenot pastor, who, escaping from France after the revocation of the Edict of Nantes, brought away the boy from La Rochelle, concealed, it is said, in a tub. The young Desaguliers was afterwards ordained, and became chaplain to that Duke of Chandos who was so ungratefully ridiculed by Pope. In this situation he formed friendships with some of the natural philosophers of the capital, and amongst others with Stephen Gray,

[1] The correct idea of combustion had been advanced by Hooke. 'The dissolution of inflammable bodies,' he asserts in the *Micrographia*, 'is performed by a substance inherent in and mixed with the air, that is like, if not the very same with, that which is fixed in saltpetre.' But this statement met with little favour at the time, and the doctrine of the compound nature of metals survived in full vigour until the discovery of oxygen by Priestley and Scheele in 1771–5. In 1775 Lavoisier reaffirmed Hooke's principle that a metallic calx is not the metal minus phlogiston, but the metal plus oxygen; and this idea, which carried with it the recognition of the elementary nature of metals, was generally accepted by the end of the eighteenth century.

an experimenter of whom little is known [1] beyond the fact that he was a pensioner of the Charterhouse.

In 1729 Gray communicated, as he says,[2] 'to Dr Desaguliers and some other Gentlemen' a discovery he had lately made, 'showing that the Electrick Vertue of a Glass Tube may be conveyed to any other Bodies so as to give them the same Property of attracting and repelling light Bodies as the Tube does, when excited by rubbing; and that this attractive Vertue might be carried to Bodies that were many feet distant from the Tube.'

This was a result of the greatest importance, for previous workers had known of no other way of producing the attractive emanations than by rubbing the body concerned.[3] It was found that only a limited class of substances, among which the metals were conspicuous, had the capacity of acting as channels for the transport of the electric power; to these Desaguliers, who continued the experiments after Gray's death in 1736, gave [4] the name *non-electrics* or *conductors*.

After Gray's discovery it was no longer possible to believe that the electric effluvia are inseparably connected with the bodies from which they are evoked by rubbing; and it became necessary to admit that these emanations have an independent existence, and can be transferred from one body to another. Accordingly we find them recognised, under the name of the *electric fluid*,[5] as one of the substances of which the world is constituted. The imponderability of this fluid did not, for the reasons already mentioned, prevent its admission by the side of light and caloric into the list of chemical elements.

The question was actively debated as to whether the electric fluid was an element *sui generis*, or, as some suspected, was another

[1] Those who are interested in the literary history of the eighteenth century will recall the controversy as to whether the verses on the death of Stephen Gray were written by Anna Williams, whose name they bore, or by her patron Johnson.

Now, hoary Sage, pursue thy happy flight
With swifter motion haste to purer light
Where Bacon waits with Newton and with Boyle
To hail thy genius, and applaud thy toil;
Where intuition breaks through time and space
And mocks experiment's successive race;
Sees tardy science toil at Nature's laws
And wonders how th' effect obscures the cause.

[2] *Phil. Trans.* xxxvii (1731), pp. 18, 227, 285, 397

[3] Otto von Guericke (1602–86) had, as a matter of fact, observed the conduction of electricity along a linen thread; but this experiment does not seem to have been followed up. cf. *Experimenta nova magdeburgica*, 1672.

[4] *Phil. Trans.* xli (1739), pp. 186, 193, 200, 209; *Dissertation concerning Electricity*, 1742

[5] The Cartesians defined a fluid to be a body whose minute parts are in a continual agitation. The word *fluid* was first used as a noun in English by Boyle.

manifestation of that principle whose operation is seen in the phenomena of heat. Those who held the latter view urged that the electric fluid and heat can both be induced by friction, can both induce combustion and can both be transferred from one body to another by mere contact; and moreover, that the best conductors of heat are also in general the best conductors of electricity. On the other hand it was contended that the electrification of a body does not cause any appreciable rise in its temperature; and an experiment of Stephen Gray's brought to light a yet more striking difference. Gray,[1] in 1729, made two oaken cubes, one solid and the other hollow, and showed that when electrified in the same way they produced exactly similar effects; whence he concluded that it was only the surfaces which had taken part in the phenomena. Thus while heat is disseminated throughout the substance of a body, the electric fluid resides at or near its surface. In the middle of the eighteenth century it was generally compared to an enveloping atmosphere. 'The electricity which a non-electric of great length (for example, a hempen string 800 or 900 feet long) receives, runs from one end to the other in a sphere of electrical *Effluvia*,' says Desaguliers in 1740[2]; and a report of the French Academy in 1733 says[3]: 'Around an electrified body there is formed a vortex of exceedingly fine matter in a state of agitation, which urges towards the body such light substances as lie within its sphere of activity. The existence of this vortex is more than a mere conjecture; for when an electrified body is brought close to the face it causes a sensation like that of encountering a cobweb.'[4]

The report from which this is quoted was prepared in connection with the discoveries of Charles-François du Fay (1698–1739), superintendent of gardens to the King of France. Du Fay[5] accounted for the behaviour of gold leaf when brought near to an electrified glass tube by supposing that at first the vortex of the tube envelopes the gold leaf, and so attracts it towards the tube. But when contact occurs, the gold leaf acquires the electric virtue, and so becomes surrounded by a vortex of its own. The two vortices, striving to extend in contrary senses, repel each other, and the vortex of the tube, being the stronger, drives away that of the gold leaf. 'It is then certain,' says du Fay,[6] 'that bodies which have become electric

[1] *Phil. Trans.* xxxvii, p. 35
[2] *Phil. Trans.* xli, p. 636
[3] *Hist. de l'Acad.* (1733), p. 6
[4] This observation had been made first by Hauksbee at the beginning of the century.
[5] *Mém. de l'Acad.* (1733), pp. 23, 73, 233, 457; (1734), pp. 341, 503; 1737, p. 86; *Phil. Trans.* xxxviii (1734), p. 258
[6] *Mém. de l'Acad.* (1733), p. 464

by contact are repelled by those which have rendered them electric; but are they repelled likewise by other electrified bodies of all kinds? And do electrified bodies differ from each other in no respect save their intensity of electrification? An examination of this matter has led me to a discovery which I should never have foreseen, and of which I believe no-one hitherto has had the least idea.'

He found, in fact, that when gold leaf which had been electrified by contact with excited glass was brought near to an excited piece of copal,[1] an *attraction* was manifested between them. 'I had expected,' he writes, 'quite the opposite effect, since, according to my reasoning, the copal and gold leaf, which were both electrified should have repelled each other.' Proceeding with his experiments he found that the gold leaf, when electrified and repelled by glass, was attracted by all electrified resinous substances, and that when repelled by the latter it was attracted by the glass. 'We see, then,' he continues, 'that there are two electricities of a totally different nature—namely, that of transparent solids, such as glass, crystal, &c. and that of bituminous or resinous bodies, such as amber, copal, sealing-wax, &c. Each of them repels bodies which have contracted an electricity of the same nature as its own, and attracts those whose electricity is of the contrary nature. We see even that bodies which are not themselves electrics can acquire either of these electricities, and that then their effects are similar to those of the bodies which have communicated it to them.'

To the two kinds of electricity whose existence was thus demonstrated, du Fay gave the names *vitreous* and *resinous*, by which they have ever since been known.

An interest in electrical experiments seems to have spread from du Fay to other members of the Court circle of Louis XV; and from 1745 onwards the *Memoirs of the Academy* contain a series of papers on the subject by the Abbé Jean-Antoine Nollet (1700–70), afterwards preceptor in Natural Philosophy to the Royal Family. Nollet attributed electric phenomena to the movement in opposite directions of two currents of a fluid, 'very subtle and inflammable,' which he supposed to be present in all bodies under all circumstances.[2] When an electric is excited by friction, part of this fluid escapes from its pores, forming an *effluent stream*; and this loss is repaired by an *affluent stream* of the same fluid entering the body from outside.

[1] A hard transparent resin, used in the preparation of varnish.

[2] cf. Nollet's *Recherches* (1749), p. 245. Nollet's lectures in *Experimental Philosophy* were translated by John Colson in 1748.

Light bodies in the vicinity, being caught in one or other of these streams, are attracted or repelled from the excited electric.

Nollet's theory was in great vogue for some time; but six or seven years after its first publication, its author came across a work purporting to be a French translation of a book printed originally in England, describing experiments said to have been made at Philadelphia, in America, by one Benjamin Franklin. 'He could not at first believe,' as Franklin tells us in his *Autobiography*, 'that such a work came from America, and said it must have been fabricated by his enemies at Paris to decry his system. Afterwards, having been assured that there really existed such a person as Franklin at Philadelphia, which he had doubted, he wrote and published a volume of letters, chiefly addressed to me, defending his theory, and denying the verity of my experiments, and of the positions deduced from them.

We must now trace the events which led up to the discovery which so perturbed Nollet.

In 1745 Pieter van Musschenbroek (1692–1761), Professor at Leyden, attempted to find a method of preserving electric charges from the decay which was observed when the charged bodies were surrounded by air. With this purpose he tried the effect of surrounding a charged mass of water by an envelope of some non-conductor, e.g. glass. In one of his experiments, a phial of water was suspended from a gun barrel by a wire let down a few inches into the water through the cork; and the gun barrel, suspended on silk lines, was applied so near an excited glass globe that some metallic fringes inserted into the gun barrel touched the globe in motion. Under these circumstances a friend named Cunaeus, who happened to grasp the phial with one hand, and touch the gun barrel with the other, received a violent shock; and it became evident that a method of accumulating or intensifying the electric power had been discovered.[1]

Shortly after the discovery of the *Leyden phial*, as it was named by Nollet, had become known in England, a London apothecary named William Watson (1715–87) [2] noticed that when the experiment is performed in this fashion the observer feels the shock 'in no other parts of his body but his arms and breast'; whence he inferred that in the act of discharge there is a transference of something

[1] The discovery was made independently (and probably somewhat earlier) by Ewald Georg von Kleist, Dean of Kammin in Pomerania. Kleist did not publish it, and it was first described in a work *Geschichte d. Erde in d. allerältesten Zeit* (1746) by J. G. Krüger, Professor of Medicine at Halle.

[2] Watson afterwards rose to eminence in the medical profession, and was knighted.

which takes the shortest or best conducting path between the gun barrel and the phial. This idea of transference seemed to him to bear some similarity to Nollet's doctrine of afflux and efflux ; and there can indeed be little doubt that the Abbé's hypothesis, though totally false in itself, furnished some of the ideas from which Watson with the guidance of experiment, constructed a correct theory. In a memoir [1] read to the Royal Society in October 1746 he propounded the doctrine that electrical actions are due to the presence of an ' electrical aether,' which in the charging or discharging of a Leyden jar is *transferred*, but is not created or destroyed. The excitation of an electric, according to this view, consists not in the evoking of anything from within the electric itself without compensation, but in the accumulation of a surplus of electrical aether by the electric *at the expense of some other body*, whose stock is accordingly depleted. All bodies were supposed to possess a certain natural store, which could be drawn upon for this purpose.

' I have shewn,' wrote Watson, ' that electricity is the effect of a very subtil and elastic fluid, occupying all bodies in contact with the terraqueous globe ; and that every-where, in its natural state, it is of the same degree of density ; and that glass and other bodies, which we denominate electrics *per se*, have the power, by certain known operations, of taking this fluid from one body, and conveying it to another, in a quantity sufficient to be obvious to all our senses ; and that, under certain circumstances, it was possible to render the electricity in some bodies more rare than it naturally is, and, by communicating this to other bodies, to give them an additional quantity, and make their electricity more dense.'

In the same year in which Watson's theory was proposed, a certain Dr Spence, who had lately arrived in America from Scotland, was showing in Boston some electrical experiments. Among his audience was a man who already at forty years of age was recognised as one of the leading citizens of the English colonies in America, Benjamin Franklin of Philadelphia (1706–90). Spence's experiments ' were,' writes Franklin,[2] ' imperfectly performed, as he was not very expert ; but, being on a subject quite new to me, they equally surprised and pleased me.' Soon after this, the ' Library Company ' of Philadelphia (an institution founded by Franklin himself) received from Mr Peter Collinson of London a present of a glass tube, with some account of its use. In a letter written to Collinson on 11 July

[1] *Phil. Trans.* xliv, p. 718. It may here be noted that it was Watson who improved the phial by coating it nearly to the top, both inside and outside, with tinfoil.

[2] Franklin's *Autobiography*

1747,[1] Franklin described experiments made with this tube, and certain deductions which he had drawn from them.

If one person A, standing on wax so that electricity cannot pass from him to the ground, rubs the tube, and if another person B, likewise standing on wax, passes his knuckle along near the glass so as to receive its electricity, then both A and B will be capable of giving a spark to a third person C standing on the floor ; that is, they will be electrified. If, however, A and B touch each other, either during or after the rubbing, they will not be electrified.

This observation suggested to Franklin the same hypothesis that (unknown to him) had been propounded a few months previously by Watson : namely, that electricity is an element present in a certain proportion in all matter in its normal condition ; so that, before the rubbing, each of the persons A, B and C has an equal share. The effect of the rubbing is to transfer some of A's electricity to the glass, whence it is transferred to B. Thus A has a deficiency and B a superfluity of electricity ; and if either of them approaches C, who has the normal amount, the distribution will be equalised by a spark. If, however, A and B are in contact, electricity flows between them so as to re-establish the original equality, and neither is then electrified with reference to C.

Thus electricity is not created by rubbing the glass, but only transferred to the glass from the rubber, so that the rubber loses exactly as much as the glass gains ; *the total quantity of electricity in any insulated system is invariable*. This assertion is usually known as the *principle of conservation of electric charge*.

The condition of A and B in the experiment can evidently be expressed by plus and minus signs : A having a deficiency $-e$ and B a superfluity $+e$ of electricity. Franklin, at the commencement of his own experiments, was not acquainted with du Fay's discoveries; but it is evident that the electric fluid of Franklin is identical with the vitreous electricity of du Fay, and that du Fay's resinous electricity is, in Franklin's theory, merely the deficiency of a stock of vitreous electricity supposed to be possessed naturally by all ponderable bodies. In Franklin's theory we are spared the necessity for admitting that two quasi-material bodies can by their union annihilate each other, as vitreous and resinous electricity were supposed to do.

Some curiosity will naturally be felt as to the considerations which induced Franklin to attribute the positive character to vitreous rather than to resinous electricity. They seem to have been founded on a comparison of the brush discharges from

[1] Franklin's *New Experiments and Observations on Electricity*, letter ii

conductors charged with the two electricities ; when the electricity was resinous, the discharge was observed to spread over the surface of the opposite conductor 'as if it flowed from it.' Again, if a Leyden jar whose inner coating is electrified vitreously is discharged silently by a conductor, of whose pointed ends one is near the knob and the other near the outer coating, the point which is near the knob is seen in the dark to be illuminated with a star or globule, while the point which is near the outer coating is illuminated with a pencil of rays ; which suggested to Franklin that the electric fluid, going from the inside to the outside of the jar, enters at the former point and issues from the latter. And yet again, in some cases the flame of a wax taper is blown away from a brass ball which is discharging vitreous electricity, and towards one which is discharging resinous electricity. But Franklin remarks that the interpretation of these observations is somewhat conjectural, and that whether vitreous or resinous electricity is the actual electric fluid is not certainly known.

Regarding the physical nature of electricity, Franklin held much the same ideas as his contemporaries ; he pictured it as an elastic [1] fluid, consisting of 'particles extremely subtile, since it can permeate common matter, even the densest metals, with such ease and freedom as not to receive any perceptible resistance.' He departed, however, to some extent from the conceptions of his predecessors, who were accustomed to ascribe all electrical repulsions to the diffusion of effluvia from the excited electric to the body acted on ; so that the tickling sensation which is experienced when a charged body is brought near to the human face was attributed to a direct action of the effluvia on the skin. This doctrine, which, as we shall see, practically ended with Franklin, bears a suggestive resemblance to that which nearly a century later was introduced by Faraday ; both explained electrical phenomena without introducing action at a distance, by supposing that something which forms an essential part of the electrified system is present at the spot where any electric action takes place ; but in the older theory this something was identified with the electric fluid itself, while in the modern view it is identified with a state of stress in the aether. In the interval between the fall of one school and the rise of the other, the theory of action at a distance was dominant.

The germs of the last-mentioned theory may be found in Franklin's own writings. It originated in connection with the explanation of the Leyden jar, a matter which is discussed in his third letter to

[1] i.e. repulsive of its own particles

Collinson, of date 1 September 1747. In charging the jar, he says, a quantity of electricity is taken away from one side of the glass, by means of the coating in contact with it, and an equal quantity is communicated to the other side, by means of the other coating. The glass itself he supposes to be impermeable to the electric fluid, so that the deficiency on the one side can permanently coexist with the redundancy on the other, so long as the two sides are not connected with each other ; but, when a connection is set up, the distribution of fluid is equalised through the body of the experimenter, who receives a shock.

Compelled by this theory of the jar to regard glass as impenetrable to electric effluvia, Franklin was nevertheless well aware [1] that the interposition of a glass plate between an electrified body and the objects of its attraction does not shield the latter from the attractive influence. He was thus driven to suppose [2] that the surface of the glass which is nearest the excited body is directly affected, and is able to exert an influence through the glass on the opposite surface ; the latter surface, which thus receives a kind of secondary or derived excitement, is responsible for the electric effects beyond it.

This idea harmonised admirably with the phenomena of the jar ; for it was now possible to hold that the excess of electricity on the inner face exercises a repellent action through the substance of the glass, and so causes a deficiency on the outer faces by driving away the electricity from it.[3]

Franklin had thus arrived at what was really a theory of action at a distance between the particles of the electric fluid ; and this he was able to support by other experiments. 'Thus,' he writes,[4] 'the stream of a fountain, naturally dense and continual, when electrified, will separate and spread in the form of a brush, every drop endeavouring to recede from every other drop.' In order to account for the attraction between oppositely charged bodies, in one of which there is an excess of electricity as compared with ordinary matter, and in the other an excess of ordinary matter as compared with electricity, he assumed that 'though the particles of electrical matter do repel each other, they are strongly attracted by all other matter' ; so that 'common matter is as a kind of spunge to the electrical fluid.' It absorbs electricity until saturation is reached, after which any further excess of electricity must lie on or near the surface of the body.

These repellent and attractive powers he assigned only to the

[1] *New Experiments* (1750), § 28
[2] ibid. (1750), § 34
[3] ibid. (1750), § 32
[4] Letter v

actual (vitreous) electric fluid; and when later on the mutual repulsion of resinously electrified bodies became known to him,[1] it caused him considerable perplexity.[2] As we shall see, the difficulty was eventually removed by Aepinus.

In spite of his belief in the power of electricity to act at a distance, Franklin did not abandon the doctrine of effluvia. 'The form of the electrical atmosphere,' he says,[3] 'is that of the body it surrounds. This shape may be rendered visible in a still air, by raising a smoke from dry rosin dropt into a hot teaspoon under the electrified body, which will be attracted, and spread itself equally on all sides, covering and concealing the body. And this form it takes, because it is attracted by all parts of the surface of the body, though it cannot enter the substance already replete. Without this attraction, it would not remain round the body, but dissipate in the air.' He observed, however, that electrical effluvia do not seem to affect, or be affected by, the air, since it is possible to breathe freely in the neighbourhood of electrified bodies; and moreover a current of dry air does not destroy electric attractions and repulsions.[4]

Regarding the suspected identity of electricity with the matter of heat, as to which Nollet had taken the affirmative position, Franklin expressed no opinion. 'Common fire,' he writes,[5] 'is in all bodies, more or less, as well as electrical fire. Perhaps they may be different modifications of the same element; or they may be different elements The latter is by some suspected. If they are different things, yet they may and do subsist together in the same body.'

Franklin's work did not at first receive from European philosophers the attention which it deserved; although Watson generously endeavoured to make the colonial writer's merits known,[6] and inserted some of Franklin's letters in one of his own papers communicated to the Royal Society. But an account of Franklin's discoveries, which had been printed in England, happened to fall into the hands of the naturalist Buffon, who was so much impressed that he secured the issue of a French translation of the work; and it was this publication which, as we have seen, gave such offence to Nollet. The success of a plan proposed by Franklin for drawing lightning from the clouds soon engaged public attention everywhere; and in a short time the triumph of the *one-fluid theory* of electricity, as the hypothesis of Watson and Franklin is generally called, was

[1] He refers to it in his Paper read to the Royal Society, 18 Dec. 1755.
[2] cf. letters xxxvii and xxxviii, dated 1761 and 1762
[3] *New Experiments* (1750), § 15 [4] Letter, vii, 1751 [5] Letter v
[6] *Phil. Trans.* xlvii, p. 202. Watson agreed with Nollet in rejecting Franklin's theory of the impermeability of glass.

complete. Nollet, who was obdurate, 'lived to see himself the last of his sect, except Monsieur B—— of Paris, his élève and immediate disciple.'[1]

The theory of effluvia was finally overthrown, and replaced by that of action at a distance, by the labours of one of Franklin's continental followers, Franz Ulrich Theodor Aepinus[2] (1724–1802). Aepinus, after taking his doctor's degree in 1747 at the university of his native city of Rostock, was appointed in 1755 a member of the Berlin Academy of Sciences. In 1751 a young Swede named Johan Carl Wilcke[3] (1732–96) came to Rostock and attended Aepinus' lectures there, following him to Berlin in 1755. Here they set up house together, and came very much under the influence of Euler, who was in Berlin from 1741 to 1766. In 1757 Aepinus was translated to St Petersburg, where the rest of his active life was spent, while Wilcke in the same year returned to Sweden.

The doctrine that glass is impermeable to electricity, which had formed the basis of Franklin's theory of the Leyden phial, was generalised by Aepinus[4] and Wilcke into the law that all non-conductors are impermeable to the electric fluid. That this applies even to air they proved by constructing a machine analogous to the Leyden jar, in which, however, air took the place of glass as the medium between two oppositely charged surfaces. The success of this experiment led Aepinus to deny altogether the existence of electric effluvia surrounding charged bodies[5]: a position which he regarded as strengthened by Franklin's observation, that the electric field in the neighbourhood of an excited body is not destroyed when the adjacent air is blown away. The electric fluid must therefore be supposed not to extend beyond the excited bodies themselves. The experiment of Gray, to which we have already referred, showed that it does not penetrate far into their substance; and thus it became necessary to suppose that the electric fluid, in its state of rest, is confined to thin layers on the surfaces of the excited bodies. This being granted, the attractions and repulsions observed between the bodies compel us to believe that electricity acts at a distance across the intervening air.

Since two vitreously charged bodies repel each other, the force

[1] Franklin's *Autobiography*

[2] This philosopher's surname had been hellenised from its original form *Hoeck* or *Hoch* to *αἰπεινός* by one of his ancestors, a distinguished theologian.

[3] cf. C. W. Oseen, *Johan Carl Wilcke, Experimental-Fysiker*, Uppsala, 1939

[4] F. V. T. Aepinus, *Tentamen Theoriae Electricitatis et Magnetismi*, St Petersburg, 1759

[5] This was also maintained about the same time by Giacomo Battista Beccaria (1716–81) of Turin.

between two particles of the electric fluid must (on Franklin's one-fluid theory, which Aepinus adopted) be repulsive ; and since there is an attraction between oppositely charged bodies, the force between electricity and ordinary matter must be attractive. These assumptions had been made, as we have seen, by Franklin ; but in order to account for the repulsion between two resinously charged bodies, Aepinus introduced a new supposition—namely, that the particles of ordinary matter repel each other.[1] This, at first, startled his contemporaries ; but, as he pointed out, the 'unelectrified' matter with which we are acquainted is really matter saturated with its natural quantity of the electric fluid, and the forces due to the matter and fluid balance each other ; or perhaps, as he suggested, a slight want of equality between these forces might give, as a residual, the force of gravitation.

Assuming that the attractive and repellent forces increase as the distance between the acting charges decreases, Aepinus applied his theory to explain a phenomenon which had been more or less indefinitely observed by many previous writers, and specially studied a short time previously by John Canton[2] (1718–72) and by Wilcke[3]—namely, that if a conductor is brought into the neighbourhood of an excited body without actually touching it, the remoter portion of the conductor acquires an electric charge of the same kind as that of the excited body, while the nearer portion acquires a charge of the opposite kind. This effect, which is known as the *induction* of electric charges, had been explained by Canton himself and by Franklin[4] in terms of the theory of electric effluvia. Aepinus showed that it followed naturally from the theory of action at a distance, by taking into account the mobility of the electric fluid in conductors ; and by discussing different cases, so far as was possible with the means at his command, he laid the foundations of the mathematical theory of electrostatics. Wilcke followed up his work of 1757 by presenting to the Swedish Academy in 1762 a memoir[5] in which he described many experiments relating to electric induction (including a method of generating any desired amount of statical electricity by this means), and also discussed the physics of non-conductors, adumbrating the theory proposed long afterwards by Faraday, that a dielectric

[1] It will be seen that the relations of 'electricity' and 'ordinary matter' in Aepinus' theory correspond closely to those of 'electrons' and 'atomic nuclei' in modern atomic physics.

[2] *Phil. Trans.* xlviii (1753), p. 350

[3] *Disputatio physica experimentalis de electricitatibus contrariis*, Rostock, 1757

[4] In his paper read to the Royal Society on 18 Dec. 1755

[5] *K. Vetensk. Acad. Handlingar*, vol. xxiii, 1762

exposed to an electric field is thereby thrown into a state of electric polarisation.

Aepinus did not succeed in determining the law according to which the force between two electric charges varies with the distance between them; and the honour of having first accomplished this belongs to Joseph Priestley (1733–1804), the discoverer of oxygen. Priestley, who was a friend of Franklin's, had been informed by the latter that he had found cork balls to be wholly unaffected by the electricity of a metal cup within which they were held; and Franklin desired Priestley to repeat and ascertain the fact. Accordingly, on 21 December 1766, Priestley instituted experiments, which showed that, when a hollow metallic vessel is electrified, there is no charge on the inner surface (except near the opening), and no electric force in the air inside. From this he at once drew the correct conclusion, which was published in 1767.[1] 'May we not infer,' he says, 'from this experiment that the attraction of electricity is subject to the same laws with that of gravitation, and is therefore according to the squares of the distances; since it is easily demonstrated that were the earth in the form of a shell, a body in the inside of it would not be attracted to one side more than another?'

This brilliant inference seems to have been insufficiently studied by the scientific men of the day; and, indeed, its author appears to have hesitated to claim for it the authority of a complete and rigorous proof. Accordingly we find that the question of the law of force was not regarded as finally settled for eighteen years afterwards. In 1769 Dr John Robison (1739–1805) of Edinburgh determined the law of force by direct experiment, and found it to be that of the inverse 2·06th power of the distance for the repulsion of two like charges, while for the attraction of two unlike charges he found the inverse of a power less than the second: he conjectured that the correct power was the inverse square.[2]

By Franklin's law of the conservation of electric charge, and Priestley's law of attraction between charged bodies, electricity was raised to the position of an exact science. It is impossible to mention the names of these two friends in such a connection without reflecting on the curious parallelism of their lives. In both men there was the same combination of intellectual boldness and power with moral earnestness and public spirit. Both of them carried on a

[1] J. Priestley, *The History and Present State of Electricity, with Original Experiments* (London, 1767), p. 732. That electrical attraction follows the law of the inverse square had been suspected by Daniel Bernoulli in 1760; cf. Socin's Experiments, *Acta Helvetica*, iv, p. 214.

[2] Robison's *Mechanical Philosophy*, vol. iv, pp. 73–4

long and tenacious struggle with the reactionary influences which dominated the English Government in the reign of George III; and both at last, when overpowered in the conflict, reluctantly exchanged their native flag for that of the United States of America. The names of both have been held in honour by later generations, not more for their scientific discoveries than for their services to the cause of religious, intellectual, and political freedom.

The most celebrated electrician of Priestley's contemporaries in London was the Hon. Henry Cavendish (1731–1810), whose interest in the subject was indeed hereditary, for his father, Lord Charles Cavendish, had assisted in Watson's experiments of 1747.[1] In 1771 Cavendish[2] presented to the Royal Society an 'Attempt to explain some of the principal phenomena of Electricity, by means of an elastic fluid.' The hypothesis adopted is that of the one-fluid theory, in much the same form as that of Aepinus. It was, as he tells us, discovered independently, although he became acquainted with Aepinus' work before the publication of his own paper.

In this memoir Cavendish makes no assumption regarding the law of force between electric charges, except that it is 'inversely as some less power of the distance than the cube'; but he evidently inclines to believe in the law of the inverse square. Indeed, he shows it to be 'likely, that if the electric attraction or repulsion is inversely as the square of the distance, almost all the redundant fluid in the body will be lodged close to the surface, and there pressed close together, and the rest of the body will be saturated'; which approximates closely to the discovery made four years previously by Priestley. Cavendish did, as a matter of fact, rediscover the inverse-square law shortly afterwards; but, indifferent to fame, he neglected to communicate to others this and much other work of importance. The value of his researches was not realised until the middle of the nineteenth century, when William Thomson (Lord Kelvin) found in Cavendish's manuscripts the correct value for the ratio of the electric charges carried by a circular disk and a sphere of the same radius which had been placed in metallic connection. Thomson urged that the papers should be published; which came to pass[3] in 1879, a hundred years from the date of the great discoveries which they enshrined. It was then seen that Cavendish had anticipated his successors in several of the ideas which will presently be discussed—amongst others, those of electrostatic capacity and specific inductive capacity.

[1] *Phil. Trans.* xlv (1750), p. 67
[2] *Phil. Trans.* lxi (1771), p. 584
[3] *The Electrical Researches of the Hon. Henry Cavendish*, edited by J. Clerk Maxwell, 1879

In the published memoir of 1771 Cavendish worked out the consequences of his fundamental hypothesis more completely than Aepinus ; and, in fact, virtually introduced the notion of electric potential, though, in the absence of any definite assumption as to the law of force, it was impossible to develop this idea to any great extent.

One of the investigations with which Cavendish occupied himself was a comparison between the conducting powers of different materials for electrostatic discharges. The question had been first raised by Beccaria, who had shown [1] in 1753 that when the circuit through which a discharge is passed contains tubes of water, the shock is more powerful when the cross-section of the tubes is increased. Cavendish went into the matter much more thoroughly, and was able, in a memoir presented to the Royal Society in 1775,[2] to say : 'It appears from some experiments, of which I propose shortly to lay an account before this Society, that iron wire conducts about 400 million times better than rain or distilled water—that is, the electricity meets with no more resistance in passing through a piece of iron wire 400,000,000 inches long than through a column of water of the same diameter only one inch long. Sea-water, or a solution of one part of salt in 30 of water, conducts 100 times, or a saturated solution of sea-salt about 720 times, better than rain-water.'

The promised account of the experiments was published in the volume edited in 1879. It appears from it that the method of testing by which Cavendish obtained these results was simply that of physiological sensation ; but the figures given in the comparison of iron and sea water are remarkably exact.

While the theory of electricity was being established on a sure foundation by the great investigators of the eighteenth century, a no less remarkable development was taking place in the kindred science of magnetism, to which our attention must now be directed.

The law of attraction between magnets was investigated at an earlier date than the corresponding law for electrically charged bodies. Newton, in the *Principia*,[3] says : 'The power of gravity is of a different nature from the power of magnetism. For the magnetic attraction is not as the matter attracted. Some bodies are attracted more by the magnet, others less ; most bodies not at all. The power of magnetism, in one and the same body, may be increased and diminished ; and is sometimes far stronger, for the quantity of matter, than the power of gravity ; and in receding from the magnet, decreases not in the duplicate, but almost in the

[1] G. B. Beccaria, *Dell' elettricismo artificiale e naturale* (Turin, 1753), p. 113

[2] *Phil. Trans.* lxvi (1776), p. 196

[3] Bk. iii, prop. vi, cor. 5

triplicate proportion of the distance, as nearly as I could judge from some rude observations.'

The edition of the *Principia* which was published in 1742 by Thomas Le Seur and Francis Jacquier contains a note on this corollary, in which the correct result is obtained that the directive couple exercised on one magnet by another is proportional to the inverse cube of the distance.

The first discoverer [1] of the law of force between magnetic poles was John Michell (1724–93), at that time a young Fellow of Queens' College, Cambridge,[2] who in 1750 published *A Treatise of Artificial Magnets ; in which is shown an easy and expeditious method of making them superior to the best natural ones.* In this he states the principles of magnetic theory as follows [3] :

'Wherever any Magnetism is found, whether in the Magnet itself, or any piece of Iron, etc., excited by the Magnet, there are always found two Poles, which are generally called North and South ; and the North Pole of one Magnet always attracts the South Pole, and repels the North Pole of another : and *vice versa*.' This is of course adopted from Gilbert.

'Each Pole attracts or repels exactly equally, at equal distances, in every direction.' This, it may be observed, overthrows the theory of vortices, with which it is irreconcilable. 'The Magnetical Attraction and Repulsion are exactly equal to each other.' This, obvious though it may seem to us, was really a most important advance, for, as he remarks, 'Most people, who have mention'd any thing relating to this property of the Magnet, have agreed, not only that the Attraction and Repulsion of Magnets are not equal to each other, but that also, they do not observe the same rule of increase and decrease.

'The Attraction and Repulsion of Magnets decreases, as the Squares of the distances from the respective poles increase.' This great discovery, which is the basis of the mathematical theory of

[1] A suggestion of the inverse-square law of attraction between magnetic poles had however been made by the Cardinal Nicholas of Cusa (1401–64) in 1450 ; cf. *Nicolai de Cusa Opera Omnia*, ed. L. Bauer (Lipsiae, 1937), v, p. 127. Of course it was the *statical* notion of force (measured by balancing weights) that was involved, as the dynamical notion was not yet discovered.

[2] Michell had taken his degree only two years previously. Later in life he was on terms of friendship with Priestley, Cavendish and William Herschel. The plan of determining the density of the earth, which was carried out by Cavendish in 1798 and is generally known as the 'Cavendish Experiment,' was due to Michell. Michell was the first inventor of the torsion balance ; he also made many valuable contributions to astronomy. In 1767 he became Rector of Thornhill, Yorks, and lived there until his death. cf. *Memoirs of John Michell* by Sir A. Geikie, Cambridge, 1918.

[3] loc. cit. p. 17

Magnetism, was deduced partly from his own observations, and partly from those of previous investigators (e.g. Dr Brook Taylor and P. Musschenbroek), who, as he observes, had made accurate experiments, but had failed to take into account all the considerations necessary for a sound theoretical discussion of them.

After Michell the law of the inverse square was maintained by Tobias Mayer [1] (1723–62) of Gottingen, better known as the author of Lunar Tables which were long in use; and by the celebrated mathematician, Johann Heinrich Lambert [2] (1728–77).

The promulgation of the one-fluid theory of electricity, in the middle of the eighteenth century, naturally led to attempts to construct a similar theory of magnetism; this was effected in 1759 by Aepinus,[3] who supposed the 'poles' to be places at which a *magnetic fluid* was present in amount exceeding or falling short of the normal quantity. The permanence of magnets was accounted for by supposing the fluid to be entangled in their pores, so as to be with difficulty displaced. The particles of the fluid were assumed to repel each other, and to attract the particles of iron and steel; but, as Aepinus saw, in order to satisfactorily explain magnetic phenomena it was necessary to assume also a mutual repulsion among the material particles of the magnet.

Subsequently *two* imponderable magnetic fluids, to which the names *boreal* and *austral* were assigned, were postulated by the Hollander, Anton Brugmans (1732–89) and by Wilcke. These fluids were supposed to have properties of mutual attraction and repulsion similar to those possessed by vitreous and resinous electricity.

The writer who next claims our attention for his services both to magnetism and to electricity is the French physicist, Charles Augustin Coulomb [4] (1736–1806). By aid of the torsion balance, which was independently invented by Michell and himself, he verified in 1785 Priestley's fundamental law that the repulsive force between two small globes charged with the same kind of electricity is in the inverse ratio of the square of the distance of their centres. In the second memoir he extended this law to the attraction of opposite electricities.

Coulomb did not accept the one-fluid theory of Franklin, Aepinus and Cavendish, but preferred a rival hypothesis which had

[1] Noticed in *Göttinger Gelehrter Anzeiger* (1760); cf. Aepinus, *Nov. Comm. Acad. Petrop.* (1768) and Mayer's *Opera Inedita*, herausg. von G. C. Lichtenberg

[2] *Histoire de l'Acad. de Berlin* (1766), pp. 22, 49

[3] In the *Tentamen*, to which reference has already been made

[4] Coulomb's First, Second and Third Memoirs appear in *Mém. de l'Acad.*, 1785; the Fourth in 1786, the Fifth in 1787, the Sixth in 1788 and the Seventh in 1789.

been proposed in 1759 by Robert Symmer[1] as a development of the ideas published by du Fay in 1733. 'My notion,' said Symmer, 'is that the operations of electricity do not depend upon one single positive power, according to the opinion generally received; but upon two distinct, positive, and active powers, which, by contrasting, and, as it were, counteracting each other produce the various phenomena of electricity; and that, when a body is said to be positively electrified, it is not simply that it is possessed of a larger share of electric matter than in a natural state; nor, when it is said to be negatively electrified, of a less; but that, in the former case, it is possessed of a larger portion of one of those active powers, and in the latter, of a larger portion of the other; while a body in its natural state remains unelectrified, from an equal ballance of those two powers within it.'

Coulomb developed this idea: 'Whatever be the cause of electricity,' he says,[2] 'we can explain all the phenomena by supposing that there are two electric fluids, the parts of the same fluid repelling each other according to the inverse square of the distance, and attracting the parts of the other fluid according to the same inverse square law.' 'The supposition of two fluids,' he adds, 'is moreover in accord with all those discoveries of modern chemists and physicists, which have made known to us various pairs of gases whose elasticity is destroyed by their admixture in certain proportions—an effect which could not take place without something equivalent to a repulsion between the parts of the same gas, which is the cause of its elasticity, and an attraction between the parts of different gases, which accounts for the loss of elasticity on combination.'

According, then, to the two-fluid theory, the 'natural fluid' contained in all matter can be decomposed, under the influence of an electric field, into equal quantities of vitreous and resinous electricity, which, if the matter be conducting, can then fly to the surface of the body. The abeyance of the characteristic properties of the opposite electricities when in combination was sometimes further compared to the neutrality manifested by the compound of an acid and an alkali.

The publication of Coulomb's views led to some controversy between the partisans of the one-fluid and two-fluid theories; the latter was soon generally adopted in France, but was stoutly opposed in Holland by Van Marum and in Italy by Volta. The chief difference between the rival hypotheses is that, in the two-fluid

[1] *Phil. Trans.* li (1759), p. 371 [2] Sixth Memoir, p. 561

theory, both the electric fluids are movable within the substance of a solid conductor ; while in the one-fluid theory the actual electric fluid is mobile, but the particles of the conductor are fixed. The dispute could therefore be settled only by a determination of the actual motion of electricity in discharges ; and this was beyond the reach of experiment.

In his Fourth Memoir Coulomb showed that electricity in equilibrium is confined to the surface of conductors, and does not penetrate to their interior substance ; and in the Sixth Memoir [1] he virtually establishes the result that the electric force near a conductor is proportional to the surface-density of electrification.

Since the overthrow of the doctrine of electric effluvia by Aepinus, the aim of electricians had been to establish their science upon the foundation of a law of action at a distance, resembling that which had led to such triumphs in Celestial Mechanics. When the law first stated by Priestley was at length decisively established by Coulomb, its simplicity and beauty gave rise to a general feeling of complete trust in it as the best attainable conception of electrostatic phenomena. The result was that attention was almost exclusively focused on action-at-a-distance theories, until the time, long afterwards, when Faraday led natural philosophers back to the right path.

Coulomb rendered great services to magnetic theory. It was he who in 1777, by simple mechanical reasoning, completed the overthrow of the hypothesis of vortices.[2] He also, in the second of the Memoirs already quoted,[3] confirmed Michell's law, according to which the particles of the magnetic fluids attract or repel each other with forces proportional to the inverse square of the distance. Coulomb, however, went beyond this, and endeavoured to account for the fact that the two magnetic fluids, unlike the two electric fluids, cannot be obtained separately ; for when a magnet is broken into two pieces, one containing its north and the other its south pole, it is found that each piece is an independent magnet possessing two poles of its own, so that it is impossible to obtain a north or south pole in a state of isolation. Coulomb explained this by supposing [4] that the magnetic fluids are permanently imprisoned within the molecules of magnetic bodies, so as to be incapable of crossing from one molecule to the next ; each molecule therefore under all

[1] p. 677 [2] *Mém. présentés par divers Savans*, ix (1780), p. 165

[3] *Mém. de l'Acad.* (1785), p. 593. Gauss finally established the law by a much more refined method.

[4] In his Seventh Memoir, *Mém. de l'Acad.* (1789), p. 488

circumstances contains as much of the boreal as of the austral fluid, and magnetisation consists simply in a separation of the two fluids to opposite ends of each molecule. Such a hypothesis evidently accounts for the impossibility of separating the two fluids to opposite ends of a body of finite size.

In spite of the advances which have been recounted, the mathematical development of electric and magnetic theory was scarcely begun at the close of the eighteenth century; and many erroneous notions were still widely entertained. In a Report [1] which was presented to the French Academy in 1800, it was assumed that the mutual repulsion of the particles of electricity on the surface of a body is balanced by the resistance of the surrounding air; and for long afterwards the electric force outside a charged conductor was confused with a supposed additional pressure in the atmosphere.

Electrostatical theory was, however, suddenly advanced to quite a mature state of development by Siméon Denis Poisson (1781–1840), in a memoir which was read to the French Academy in 1812.[2] As the opening sentences show, he accepted the conceptions of the two-fluid theory.

'The theory of electricity which is most generally accepted,' he says, 'is that which attributes the phenomena to two different fluids, which are contained in all material bodies. It is supposed that molecules of the same fluid repel each other and attract the molecules of the other fluid; these forces of attraction and repulsion obey the law of the inverse square of the distance; and at the same distance the attractive power is equal to the repellent power; whence it follows that, when all the parts of a body contain equal quantities of the two fluids, the latter do not exert any influence on the fluids contained in neighbouring bodies, and consequently no electrical effects are discernible. This equal and uniform distribution of the two fluids is called the *natural state*; when this state is disturbed in any body, the body is said to be *electrified*, and the various phenomena of electricity begin to take place.

'Material bodies do not all behave in the same way with respect to the electric fluid. Some, such as the metals, do not appear to exert any influence on it, but permit it to move about freely in their substance; for this reason they are called *conductors*; others, on the contrary—very dry air, for example—oppose the passage of the electric fluid in their interior, so that they can prevent the fluid accumulated in conductors from being dissipated throughout space.'

[1] On Volta's discoveries

[2] *Mém. de l'Institut* (1811), pt. i, p. 1; pt. ii, p. 163

When an excess of one of the electric fluids is communicated to a metallic body, this charge distributes itself over the surface of the body, forming a layer whose thickness at any point depends on the shape of the surface. The resultant force due to the repulsion of all the particles of this surface-layer must vanish at any point in the interior of the conductor, since otherwise the natural state existing there would be disturbed ; and Poisson showed that by aid of this principle it is possible in certain cases to determine the distribution of electricity in the surface layer. For example, a well-known proposition of the theory of attractions asserts that a hollow shell whose bounding surfaces are two similar and similarly situated ellipsoids exercises no attractive force at any point within the interior hollow ; and it may thence be inferred that if an electrified metallic conductor has the form of an ellipsoid, the charge will be distributed on it proportionally to the normal distance from the surface to an adjacent similar and similarly situated ellipsoid.

Poisson went on to show that this result was by no means all that might with advantage be borrowed from the theory of attractions. Lagrange, in a memoir on the motion of gravitating bodies, had shown [1] that the components of the attractive force at any point can be simply expressed as the derivates of the function which is obtained by adding together the masses of all the particles of an attracting system, each divided by its distance from the point ; and Laplace had shown [2] that this function V satisfies the equation

$$\frac{\partial^2 V}{\partial x^2} + \frac{\partial^2 V}{\partial y^2} + \frac{\partial^2 V}{\partial z^2} = 0$$

in space free from attracting matter. Poisson himself showed later, in 1813,[3] that when the point (x, y, z) is within the substance of the attracting body, this equation of Laplace must be replaced by

$$\frac{\partial^2 V}{\partial x^2} + \frac{\partial^2 V}{\partial y^2} + \frac{\partial^2 V}{\partial z^2} = -4\pi\rho,$$

where ρ denotes the density of the attracting matter at the point. In the present memoir Poisson called attention to the utility of this function V in electrical investigations, remarking that its value over the surface of any conductor must be constant.

[1] *Mém. de Berlin* (1777). The theorem was afterwards published, and ascribed to Laplace, in a memoir by Legendre on the Attractions of Spheroids, which will be found in the *Mém. par divers Savans*, published in 1785.

[2] *Mém. de l'Acad.* (1782, published in 1785). p. 113

[3] *Bull. de la Soc. Philomathique* iii (1813), p. 388

The known formulae for the attractions of spheroids show that when a charged conductor is spheroidal, the repellent force acting on a small charged body immediately outside it will be directed at right angles to the surface of the spheroid, and will be proportional to the thickness of the surface-layer of electricity at this place. Poisson suspected that this theorem might be true for conductors not having the spheroidal form—a result which, as we have seen, had been already virtually given by Coulomb; and Laplace suggested to Poisson the following proof, applicable to the general case. The force at a point immediately outside the conductor can be divided into a part s due to the part of the charged surface immediately adjacent to the point, and a part S due to the rest of the surface. At a point close to this, but just inside the conductor, the force S will still act; but the force s will evidently be reversed in direction. Since the resultant force at the latter point vanishes, we must have $S = s$; so the resultant force at the exterior point is $2s$. But s is proportional to the charge per unit area of the surface, as is seen by considering the case of an infinite plate; which establishes the theorem.

When several conductors are in presence of each other, the distribution of electricity on their surfaces may be determined by the principle, which Poisson took as the basis of his work, that at any point in the interior of any one of the conductors, the resultant force due to all the surface-layers must be zero. He discussed, in particular, one of the classical problems of electrostatics, namely that of determining the surface-density on two charged conducting spheres placed at any distance from each other. The solution depends on double gamma functions in the general case; when the two spheres are in contact, it depends on ordinary gamma functions. Poisson gave a solution in terms of definite integrals, which is equivalent to that in terms of gamma functions; and after reducing his results to numbers, compared them with Coulomb's experiments.

The rapidity with which in a single memoir Poisson passed from the barest elements of the subject to such recondite problems as those just mentioned may well excite admiration. His success is, no doubt, partly explained by the high state of development to which analysis had been advanced by the great mathematicians of the eighteenth century; but even after allowance has been made for what is due to his predecessors, Poisson's investigation must be accounted a splendid memorial of his genius.

Some years later Poisson turned his attention to magnetism;

and, in a masterly paper [1] presented to the French Academy in 1824, gave a remarkably complete theory of the subject.

His starting-point is Coulomb's doctrine of two imponderable magnetic fluids, arising from the decomposition of a neutral fluid, and confined in their movements to the individual elements of the magnetic body, so as to be incapable of passing from one element to the next.

Suppose that an amount m of the positive magnetic fluid is located at a point (x, y, z) ; the components of the *magnetic intensity*, or force exerted on unit magnetic pole, at a point (ξ, η, ζ) will evidently be

$$-m\frac{\partial}{\partial\xi}\left(\frac{1}{r}\right), \quad -m\frac{\partial}{\partial\eta}\left(\frac{1}{r}\right), \quad -m\frac{\partial}{\partial\zeta}\left(\frac{1}{r}\right),$$

where r denotes $\{(\xi - x)^2 + (\eta - y)^2 + (\zeta - z)^2\}^{\frac{1}{2}}$. Hence if we consider next a magnetic element in which equal quantities of the two magnetic fluids are displaced from each other parallel to the x-axis, the components of the magnetic intensity at (ξ, η, ζ) will be the negative derivates, with respect to ξ, η, ζ respectively, of the function

$$A\frac{\partial}{\partial x}\left(\frac{1}{r}\right),$$

where the quantity A, which does not involve (ξ, η, ζ), may be called the *magnetic moment* of the element ; it may be measured by the couple required to maintain the element in equilibrium at a definite angular distance from the magnetic meridian.

If the displacement of the two fluids from each other in the element is not parallel to the axis of x, it is easily seen that the expression corresponding to the last is

$$A\frac{\partial}{\partial x}\left(\frac{1}{r}\right) + B\frac{\partial}{\partial y}\left(\frac{1}{r}\right) + C\frac{\partial}{\partial z}\left(\frac{1}{r}\right),$$

where the vector (A, B, C) now denotes the magnetic moment of the element.

Thus the magnetic intensity at an external point (ξ, η, ζ) due to any magnetic body has the components

$$\left(-\frac{\partial V}{\partial\xi}, \quad -\frac{\partial V}{\partial\eta}, \quad -\frac{\partial V}{\partial\zeta}\right),$$

where

$$V = \iiint\left(A\frac{\partial}{\partial x} + B\frac{\partial}{\partial y} + C\frac{\partial}{\partial z}\right)\left(\frac{1}{r}\right) dx\, dy\, dz$$

integrated throughout the substance of the magnetic body, and

[1] *Mem. de l'Acad.* v, p. 247

where the vector (A, B, C) or **I** [1] represents the magnetic moment per unit-volume, or, as it is generally called, the *magnetisation*. The function V was afterwards named by Green the *magnetic potential*.

Poisson, by integrating by parts the preceding expression for the magnetic potential, obtained it in the form

$$V = \iint \frac{1}{r} (\mathbf{I} \cdot \mathbf{dS}) - \iiint \frac{1}{r} \operatorname{div} \mathbf{I} \, dx \, dy \, dz, [2]$$

the first integral being taken over the surface S of the magnetic body, and the second integral being taken throughout its volume. This formula shows that the magnetic intensity produced by the body in external space is the same as would be produced by a fictitious distribution of magnetic fluid, consisting of a layer over its surface, of surface-charge (**I** . **dS**) per element dS, together with a volume distribution of density — div **I** throughout its substance. These fictitious magnetisations are generally known as *Poisson's equivalent surface- and volume-distributions of magnetism.*

Poisson, moreover, perceived that at a point in a very small cavity excavated within the magnetic body, the magnetic potential has a limiting value which is independent of the shape of the cavity as the dimensions of the cavity tend to zero; but that this is not true of the magnetic intensity, which in such a small cavity depends on the shape of the cavity. Taking the cavity to be spherical, he showed that the magnetic intensity within it is

$$\operatorname{grad} V + \tfrac{4}{3}\pi \mathbf{I}, [3]$$

where **I** denotes the magnetisation at the place.

This memoir also contains a discussion of the magnetism temporarily induced in soft iron and other magnetisable metals by the approach of a permanent magnet. Poisson accounted for the properties of temporary magnets by assuming that they contain embedded in their substance a great number of small spheres, which are perfect conductors for the magnetic fluids; so that the resultant magnetic intensity in the interior of one of these small spheres must be zero. He showed that such a sphere, when placed in a field of

[1] In the present work, vectors will generally be distinguished by heavy type.

[2] If the components of a vector **a** are denoted by (a_x, a_y, a_z), the quantity $a_x b_x + a_y b_y + a_z b_z$ is called the *scalar product* of two vectors **a** and **b**, and is denoted by (**a** . **b**).
The quantity $\frac{\partial a_x}{\partial x} + \frac{\partial a_y}{\partial y} + \frac{\partial a_z}{\partial z}$ is called the *divergence* of the vector **a**, and is denoted by div **a**.

[3] The vector whose components are $-\frac{\partial V}{\partial x}, -\frac{\partial V}{\partial y}, -\frac{\partial V}{\partial z}$ is denoted by grad V.

magnetic intensity **F**, must acquire a magnetic moment of amount $\frac{3}{4\pi}\mathbf{F}\times$ the volume of the sphere, in order to counteract within the sphere the force **F**. Thus if k_p denote the total volume of these spheres contained within a unit volume of the temporary magnet, the magnetisation will be **I**, where

$$\tfrac{4}{3}\pi\mathbf{I} = k_p\mathbf{F},$$

and **F** denotes the magnetic intensity within a spherical cavity excavated in the body. This is *Poisson's law of induced magnetism.*

It is known that some substances acquire a greater degree of temporary magnetisation than others when placed in the same circumstances. Poisson accounted for this by supposing that the quantity k_p varies from one substance to another. But the experimental data show that for soft iron k_p must have a value very near unity, which would obviously be impossible if k_p is to mean the ratio of the volume of spheres contained within a region to the total volume of the region.[1] The physical interpretation assigned by Poisson to his formulae must therefore be rejected, although the formulae themselves retain their value.

Poisson's electrical and magnetical investigations were generalised and extended in 1828 by George Green[2] (1793–1841). Green's treatment is based on the properties of the function already used by Lagrange, Laplace and Poisson, which represents the sum of all the electric or magnetic charges in the field, divided by their respective distances from some given point. To this function Green gave the name *potential*, by which it has always since been known.[3]

Near the beginning of the memoir is established the celebrated formula connecting surface and volume integrals, which is now generally called *Green's Theorem*, and of which Poisson's result on the equivalent surface- and volume-distributions of magnetisation is a particular application. By using this theorem to investigate the properties of the potential, Green arrived at many results of remarkable beauty and interest. We need only mention, as an example of the power of his method, the following : suppose that there is a hollow conducting shell, bounded by two closed surfaces, and that a number of electrified bodies are placed, some within and some

[1] This objection was advanced by Maxwell in sect. 430 of his *Treatise*. An attempt to overcome it was made by E. Betti ; cf. his *Teoria d. forze Newtoniane* (Pisa, 1879).

[2] *An essay on the application of mathematical analysis to the theories of electricity and magnetism* (Nottingham, 1828) ; reprinted in *The Mathematical Papers of the late George Green*, p. 1.

[3] Euler in 1744 (*De methodis inveniendi* . . .) had spoken of the *vis potentialis*—what would now be called the potential energy—possessed by an elastic body when bent.

without it; and let the inner surface and interior bodies be called the interior system, and the outer surface and exterior bodies be called the exterior system. Then all the electrical phenomena of the interior system, relative to attractions, repulsions and densities, will be the same as if there were no exterior system, and the inner surface were a perfect conductor, put in communication with the earth; and all those of the exterior system will be the same as if the interior system did not exist, and the outer surface were a perfect conductor, containing a quantity of electricity equal to the whole of that originally contained in the shell itself and in all the interior bodies.

It will be evident that electrostatics had by this time attained a state of development in which further progress could be hoped for only in the mathematical superstructure, unless experiment should unexpectedly bring to light phenomena of an entirely new character. This will therefore be a convenient place to pause and consider the rise of another branch of electrical philosophy.

Chapter III

GALVANISM, FROM GALVANI TO OHM.

UNTIL the last decade of the eighteenth century, electricians were occupied solely with statical electricity. Their attention was then turned in a different direction.

In a work entitled *Recherches sur l'origine des sentiments agréables et désagréables*, which was published [1] in 1752, Johann Georg Sulzer (1720–79) had mentioned that if two pieces of metal, the one of lead and the other of silver, be joined together in such a manner that their edges touch, and if they be placed on the tongue, a taste is perceived 'similar to that of vitriol of iron,' although neither of these metals applied separately gives any trace of such a taste. 'It is not probable,' he says, 'that this contact of the two metals causes a solution of either of them, liberating particles which might affect the tongue; and we must therefore conclude that the contact sets up a vibration in their particles, which, by affecting the nerves of the tongue, produces the taste in question.'

This observation was not suspected to have any connection with electrical phenomena, and it played no part in the inception of the next discovery, which indeed was suggested by a mere accident.

Luigi Galvani, born at Bologna in 1737, occupied from 1775 onwards a chair of Anatomy in his native city. For many years before the event which made him famous he had been studying the susceptibility of the nerves to irritation; and, having been formerly a pupil of Beccaria, he was also interested in electrical experiments. One day in the latter part of the year 1780 he had, as he tells us,[2] 'dissected and prepared a frog, and laid it on a table, on which, at some distance from the frog, was an electric machine. It happened by chance that one of my assistants touched the inner crural nerve of the frog, with the point of a scalpel; whereupon at once the muscles of the limbs were violently convulsed.

'Another of those who used to help me in electrical experiments thought he had noticed that at this instant a spark was drawn from the conductor of the machine. I myself was at the time occupied with a totally different matter; but when he drew my attention to

[1] *Mém. de l'Acad. de Berlin* (1752), p. 356

[2] Aloysii Galvani, *De Viribus Electricitatis in Motu Musculari*: Commentarii Bononiensi, vii (1791), p. 363

this, I greatly desired to try it for myself, and discover its hidden principle. So I, too, touched one or other of the crural nerves with the point of the scalpel, at the same time that one of those present drew a spark; and the same phenomenon was repeated exactly as before.'[1]

After this, Galvani had the idea of trying whether the electricity of thunderstorms would induce muscular contractions equally well with the electricity of the machine. Having successfully experimented with lightning, he 'wished,' as he writes,[2] 'to try the effect of atmospheric electricity in calm weather. My reason for this was an observation I had made, that frogs which had been suitably prepared for these experiments and fastened, by brass hooks in the spinal marrow, to the iron lattice round a certain hanging-garden at my house, exhibited convulsions not only during thunderstorms, but sometimes even when the sky was quite serene. I suspected these effects to be due to the changes which take place during the day in the electric state of the atmosphere; and so, with some degree of confidence, I performed experiments to test the point; and at different hours for many days I watched frogs which I had disposed for the purpose; but could not detect any motion in their muscles. At length, weary of waiting in vain, I pressed the brass hooks, which were driven into the spinal marrow, against the iron lattice, in order to see whether contractions could be excited by varying the incidental circumstances of the experiment. I observed contractions tolerably often, but they did not seem to bear any relation to the changes in the electrical state of the atmosphere.

'However, at this time, when as yet I had not tried the experiment except in the open air, I came very near to adopting a theory that the contractions are due to atmospheric electricity, which, having slowly entered the animal and accumulated in it, is suddenly discharged when the hook comes in contact with the iron lattice. For it is easy in experimenting to deceive ourselves, and to imagine we see the things we wish to see.

'But I took the animal into a closed room, and placed it on an iron plate; and when I pressed the hook which was fixed in the spinal marrow against the plate, behold! the same spasmodic contractions as before. I tried other metals at different hours on various days, in several places, and always with the same result,

[1] According to a story which has often been repeated, but which rests on no sufficient evidence, the frog was one of a number which had been procured for the Signora Galvani who, being in poor health, had been recommended to take a soup made of these animals as a restorative.

[2] loc. cit. p. 377

except that the contractions were more violent with some metals than with others. After this I tried various bodies which are not conductors of electricity, such as glass, gums, resins, stones and dry wood ; but nothing happened. This was somewhat surprising, and led me to suspect that electricity is inherent in the animal itself. This suspicion was strengthened by the observation that a kind of circuit of subtle nervous fluid (resembling the electric circuit which is manifested in the Leyden jar experiment) is completed from the nerves to the muscles when the contractions are produced.

'For, while I with one hand held the prepared frog by the hook fixed in its spinal marrow, so that it stood with its feet on a silver box, and with the other hand touched the lid of the box, or its sides, with any metallic body, I was surprised to see the frog become strongly convulsed every time that I applied this artifice.' [1]

Galvani thus ascertained that the limbs of the frog are convulsed whenever a connection is made between the nerves and muscles by a metallic arc, generally formed of more than one kind of metal ; and he advanced the hypothesis that the convulsions are caused by the transport of a peculiar fluid from the nerves to the muscles, the arc acting as a conductor. To this fluid the names *Galvanism* and *Animal Electricity* were soon generally applied. Galvani himself considered it to be the same as the ordinary electric fluid, and, indeed, regarded the entire phenomenon as similar to the discharge of a Leyden jar.

The publication of Galvani's views soon engaged the attention of the learned world, and gave rise to an animated controversy between those who supported Galvani's own view, those who believed galvanism to be a fluid distinct from ordinary electricity, and a third school who altogether refused to attribute the effects to a supposed fluid contained in the nervous system. The leader of the last-named party was Alessandro Volta (1745–1827), Professor of Natural Philosophy in the University of Pavia, who in 1792 put forward the view [2] that the stimulus in Galvani's experiment is derived essentially from the connection of two different metals by a moist body. 'The metals used in the experiments, being applied to the moist bodies of animals, can by themselves, and of their proper virtue, excite and dislodge the electric fluid from its state of rest ; so that the organs of the animal act only passively.' At first he inclined to combine this theory of metallic stimulus with a certain degree of belief in such a fluid as Galvani had supposed ; but

[1] This observation was made in 1786.
[2] *Phil. Trans.* lxxxiii (1793), pp. 10, 27

after the end of 1793 he denied the existence of animal electricity altogether.

From this standpoint Volta continued his experiments and worked out his theory. The following quotation from a letter [1] which he wrote later to Gren, the editor of the *Neues Journal d. Physik*, sets forth his view in a more developed form :

'The contact of different conductors, particularly the metallic, including pyrites and other minerals, as well as charcoal, which I call dry conductors, or of the *first class*, with moist conductors, or conductors of the *second class*, agitates or disturbs the electric fluid, or gives it a certain impulse. Do not ask in what manner : it is enough that it is a principle, and a general principle. This impulse, whether produced by attraction or any other force, is different or unlike, both in regard to the different metals and to the different moist conductors ; so that the direction, or at least the power, with which the electric fluid is impelled or excited, is different when the conductor A is applied to the conductor B, or to another C. In a perfect circle of conductors, where either one of the second class is placed between two different from each other of the first class, or, contrariwise, one of the first class is placed between two of the second class different from each other, an electric stream is occasioned by the predominating force either to the right or to the left—a circulation of this fluid, which ceases only when the circle is broken, and which is renewed when the circle is again rendered complete.'

Another philosopher who, like Volta, denied the existence of a fluid peculiar to animals, but who took a somewhat different view of the origin of the phenomenon, was Giovanni Fabroni (1752–1822) of Florence, who in 1796, having placed two plates of different metals in water, observed [2] that one of them was partially oxidised when they were put in contact ; from which he rightly concluded that some chemical action is inseparably connected with galvanic effects.

The feeble intensity of the phenomena of galvanism, which compared poorly with the striking displays obtained in electrostatics, was responsible for some falling off of interest in them towards the end of the eighteenth century ; and the last years of their illustrious discoverer were clouded by misfortune. Being attached to the old order which was overthrown by the armies of the French Revolution, he refused in 1798 to take the oath of allegiance to the

[1] *Phil. Mag.* iv (1799), pp. 59, 163, 306

[2] *Phil. Journal*, 4to, iii, p. 308 ; iv, p. 120 ; *Journal de Physique*, vi, p. 348

newly constituted Cisalpine Republic, and was deposed from his professorial chair. A profound melancholy, which had been induced by domestic bereavement, was aggravated by poverty and disgrace; and, unable to survive the loss of all he held dear, he died broken-hearted before the end of the year.[1]

Scarcely more than a year after the death of Galvani, the new science suddenly regained the eager attention of philosophers. This renewal of interest was due to the discovery by Volta, in the early spring of 1800, of a means of greatly increasing the intensity of the effects. Hitherto all attempts to magnify the action by enlarging or multiplying the apparatus had ended in failure. If a long chain of different metals was used instead of only two, the convulsions of the frog were no more violent. But Volta now showed [2] that if any number of couples, each consisting of a zinc disk and a copper disk in contact, were taken, and if each couple was separated from the next by a disk of moistened pasteboard (so that the order was copper, zinc, pasteboard, copper, zinc, pasteboard, etc.), the effect of the *pile* thus formed was much greater than that of any galvanic apparatus previously introduced. When the highest and lowest disks were simultaneously touched by the fingers, a distinct shock was felt; and this could be repeated again and again, the pile apparently possessing within itself an indefinite power of recuperation. It thus resembled a Leyden jar endowed with a power of automatically re-establishing its state of tension after each explosion; with, in fact, 'an inexhaustible charge, a perpetual action or impulsion on the electric fluid.'

Volta unhesitatingly pronounced the phenomena of the pile to be in their nature electrical. The circumstances of Galvani's original discovery had prepared the minds of philosophers for this belief, which was powerfully supported by the similarity of the physiological effects of the pile to those of the Leyden jar, and by the observation that the galvanic influence was conducted only by those bodies—e.g. the metals—which were already known to be good conductors of static electricity. But Volta now supplied a still more convincing proof. Taking a disk of copper and one of zinc, he held each by an insulating handle and applied them to each other for an instant. After the disks had been separated, they were brought into contact with a delicate electroscope, which indicated by the divergence of its straws that the disks were now electrified—

[1] A decree of reinstatement had been granted, but had not come into operation at the time of Galvani's death.

[2] *Phil. Trans.* (1800), p. 403

the zinc had, in fact, acquired a positive and the copper a negative electric charge.[1] Thus the mere contact of two different metals, such as those employed in the pile, was shown to be sufficient for the production of effects undoubtedly electrical in character.

On the basis of this result a definite theory of the action of the pile was put forward. It was not originally published by Volta himself, but it is contained in a Report which Biot wrote for a committee appointed to examine Volta's work.[2] Suppose first that a disk of zinc is laid on a disk of copper, which in turn rests on an insulating support. The experiment just described shows that the electric fluid will be driven from the copper to the zinc. We may then, according to Volta, represent the state or 'tension' of the copper by the number $-\frac{1}{2}$, and that of the zinc by the number $+\frac{1}{2}$, the difference being arbitrarily taken as unity, and the sum being (on account of the insulation) zero. It will be seen that Volta's idea of 'tension' was a mingling of two ideas, which in modern electric theory are clearly distinguished from each other—namely, electric charge and electric potential.

Now let a disk of moistened pasteboard be laid on the zinc, and a disk of copper on this again. Since the uppermost copper is not in contact with the zinc, the contact action does not take place between them; but since the moist pasteboard is a conductor, the copper will receive a charge from the zinc. Thus the states will now be represented by $-\frac{2}{3}$ for the lower copper, $+\frac{1}{3}$ for the zinc, and $+\frac{1}{3}$ for the upper copper, giving a zero sum as before.

If, now, another zinc disk is placed on the top, the states will be represented by -1 for the lower copper, 0 for the lower zinc and upper copper, and $+1$ for the upper zinc.

In this way it is evident that the difference between the numbers indicating the tensions of the uppermost and lowest disks in the pile will always be equal to the number of pairs of metallic disks contained in it. If the pile is insulated, the sum of the numbers indicating the states of all the disks must be zero; but if the lowest disk is connected to earth, the tension of this disk will be zero, and the numbers indicating the states of all the other disks will be increased by the same amount, their mutual differences remaining unchanged.

The pile as a whole is thus similar to a Leyden jar; when the experimenter touches the uppermost and lowest disks, he receives

[1] Abraham Bennet (1750–99) had previously shown (*New Experiments in Electricity* [1789], pp. 86–102) that many bodies, when separated after contact, are oppositely electrified; he conceived that different bodies have different attractions or capacities for electricity.

[2] cf. *Opere di Alessander Volta*, ed. Naz. (Milan, 1918–), ii, p. 110

the shock of its discharge, the intensity being proportional to the number of disks.

The moist layers played no part in Volta's theory beyond that of conductors.[1] It was soon found that when the moisture is acidified the pile is more efficient, but this was attributed solely to the superior conducting power of acids.

Volta fully understood and explained the impossibility of constructing a pile from disks of metal alone, without making use of moist substances. As he showed in 1801, if disks of various metals are placed in contact in any order, the extreme metals will be in the same state as if they touched each other directly without the intervention of the others ; so that the whole is equivalent merely to a single pair. When the metals are arranged in the order silver, copper, iron, tin, lead, zinc, each of them becomes positive with respect to that which precedes it, and negative with respect to that which follows it ; but the moving force from the silver to the zinc is equal to the sum of the moving forces of the metals comprehended between them in the series.

When a connection was maintained for some time between the extreme disks of a pile by the human body, sensations were experienced which seemed to indicate a continuous activity in the entire system. Volta inferred that the electric current persists during the whole time that communication by conductors exists all round the circuit, and that the current is suspended only when this communication is interrupted. 'This endless circulation or perpetual motion of the electric fluid,' he says, 'may seem paradoxical, and may prove inexplicable ; but it is none the less real, and we can, so to speak, touch and handle it.'

Volta announced his discovery in a letter to Sir Joseph Banks, dated from Como, 20 March 1800. Sir Joseph, who was then President of the Royal Society, communicated the news to William Nicholson (1753–1815), founder of the *Journal* which is generally known by his name, and his friend Anthony Carlisle (1768–1840), afterwards a distinguished surgeon. On the 30th of the following month, Nicholson and Carlisle set up the first pile made in England. In repeating Volta's experiments, having made the contact more secure at the upper plate of the pile by placing a drop of water there, they noticed [2] a disengagement of gas round the conducting wire at this point ; whereupon they followed up the matter by

[1] Volta had inclined, in his earlier experiments on galvanism, to locate the seat of power at the interfaces of the metals with the moist conductors. Cf. his letter to Gren, *Phil. Mag.* iv (1799), p. 62.

[2] *Nicholson's Journal* (4to), iv, p. 179 (1800) ; *Phil. Mag.* vii, p. 337 (1800)

introducing a tube of water, into which the wires from the terminals of the pile were plunged. Bubbles of an inflammable gas were liberated at one wire, while the other wire became oxidised ; when platinum wires were used, oxygen and hydrogen were evolved in a free state, one at each wire. This effect, which was nothing less than the electric decomposition of water into its constituent gases, was obtained on 2 May 1800.[1]

Although it had long been known that frictional electricity is capable of inducing chemical action,[2] the discovery of Nicholson and Carlisle was of the first magnitude. It was at once extended by William Cruickshank (1745–1800) of Woolwich, who showed [3] that solutions of metallic salts are also decomposed by the current ; and William Hyde Wollaston (1766–1828) seized on it as a test [4] of the identity of the electric currents of Volta with those obtained by the discharge of frictional electricity. He found that water could be decomposed by currents of either type, and inferred that all differences between them could be explained by supposing that voltaic electricity as commonly obtained is ' less intense, but produced in much larger quantity.' Later in the same year (1801), Martin van Marum (1750–1837) and Christian Heinrich Pfaff (1773–1852) arrived at the same conclusion by carrying out on a large scale [5] Volta's plan of using the pile to charge batteries of Leyden jars ; while Johann Wilhelm Ritter (1776–1810) showed [6] that the electricities furnished at the poles of the pile attract each other, and the electricities at the corresponding poles of two similar piles repel.

The discovery of Nicholson and Carlisle made a great impression on the mind of Humphry Davy (1778–1829), a young Cornishman who about this time was appointed Professor of Chemistry at the Royal Institution in London. Davy at once began to experiment with voltaic piles, and in November 1800,[7] showed that it gives no current when the water between the pairs of plates is pure, and that its power of action is ' in great measure proportional to the power of the conducting fluid substance between the double plates to oxydate the zinc.' This result, as he immediately perceived, did not

[1] It was obtained independently four months later by J. W. Ritter.

[2] Beccaria (*Lettere dell' elettricismo* [Bologna, 1758], p. 282) had reduced mercury and other metals from their oxides by discharges of frictional electricity ; and Priestley had obtained an inflammable gas from certain organic liquids in the same way. Cavendish in 1781 had established the constitution of water by electrically exploding hydrogen and oxygen.

[3] *Nicholson's Journal* (4to), iv (1800), pp. 187, 245 ; *Phil. Mag.* vii (1800), p. 337

[4] *Phil. Trans.* xci (1801), p. 427

[5] *Phil. Mag.* xii (1802), p. 161

[6] Gilbert's *Annalen*, viii (1801), p. 385

[7] *Nicholson's Journal* (4to), iv (1800), pp. 275, 326, 337, 380, 394, 527 ; Davy's *Works*, ii, p. 155

harmonise well with Volta's views on the source of electricity in the pile, but was, on the other hand, in agreement with Fabroni's idea that galvanic effects are always accompanied by chemical action. After a series of experiments he definitely concluded that 'the galvanic pile of Volta acts only when the conducting substance between the plates is capable of oxydating the zinc ; and that, in proportion as a greater quantity of oxygen enters into combination with the zinc in a given time, so in proportion is the power of the pile to decompose water and to give the shock greater. It seems therefore reasonable to conclude, though with our present quantity of facts we are unable to explain the exact mode of operation, that the oxydation of the zinc in the pile, and the chemical changes connected with it, are *somehow* the cause of the electrical effects it produces.' This principle of oxidation guided Davy in designing many new types of pile, with elements chosen from the whole range of the known metals.

Davy's chemical theory of the pile was supported by Wollaston [1] and by Nicholson,[2] the latter of whom urged that the existence of piles in which only *one* metal is used (with more than one kind of fluid) is fatal to any theory which places the seat of the activity in the contact of dissimilar metals.

Davy afterwards proposed [3] a theory of the voltaic pile which combines ideas drawn from both the 'contact' and 'chemical' explanations. He supposed that before the circuit is closed, the copper and zinc disks in each contiguous pair assume opposite electrostatic states, in consequence of inherent 'electrical energies' possessed by the metals ; and when a communication is made between the extreme disks by a wire, the opposite electricities annihilate each other, as in the discharge of a Leyden jar. If the liquid (which Davy compared to the glass of a Leyden jar) were incapable of decomposition, the current would cease after this discharge ; but the liquid in the pile is composed of two elements which have inherent attractions for electrified metallic surfaces, hence arises chemical action, which removes from the disks the outermost layers of molecules, whose energy is exhausted, and exposes new metallic surfaces. The electrical energies of the copper and zinc are consequently again exerted, and the process of electromotion continues. Thus the contact of metals is the cause which *disturbs* the equilibrium, while the chemical changes continually *restore* the conditions under which the contact energy can be exerted.

In this and other memoirs Davy asserted that chemical affinity

[1] *Phil. Trans.* xci (1801), p. 427
[2] *Nicholson's Journal,* i (1802), p. 142
[3] *Phil. Trans.* xcvii (1807), p. 1

is essentially of an electrical nature. 'Chemical and electrical attractions,' he declared,[1] 'are produced by the same cause, acting in one case on particles, in the other on masses, of matter ; and the same property, under different modifications, is the cause of all the phenomena exhibited by different voltaic combinations.'

The further elucidation of this matter came chiefly from researches on electro-chemical decomposition, which we must now consider.

Ritter in 1803 [2] constructed a column formed of disks of silver alternating with disks of damp cloth, and sent through this column the current from a voltaic pile. When the connection with the pile was broken, the silver-and-cloth column acted as a voltaic pile on its own account. Ritter himself misunderstood the phenomenon, comparing the action of the original current to that of a current which is used to charge a condenser. The correct explanation was given in 1805 by Volta,[3] who showed that so long as the original current is passing through Ritter's column, the water in the cloth disks is being decomposed, oxygen and hydrogen being collected at the opposite faces of any disk. When the charging current is cut off, these products of decomposition set up a reverse electromotive force, the Ritter column now acting as a pile formed with two different fluids and one metal. This was the discovery of *voltaic polarisation ;* the Ritter column may be regarded as a precursor of the modern 'accumulator' cell.

A phenomenon which had greatly surprised Nicholson and Carlisle in their early experiments was the appearance of the products of galvanic decomposition at places remote from each other. The first attempt to account for this was made in 1806 by Theodor von Grothuss [4] (1785–1822) and by Davy,[5] who advanced a theory that the terminals at which water is decomposed have attractive and repellent powers ; that the pole whence resinous electricity issues has the property of attracting hydrogen and the metals, and of repelling oxygen and acid substances, while the positive terminal has the power of attracting oxygen and repelling hydrogen ; and that these forces are sufficiently energetic to destroy or suspend the usual operation of chemical affinity in the water molecules nearest the terminals. The force due to each terminal was supposed to diminish with the distance from the terminal. When the molecule nearest

[1] *Phil. Trans.* cxvi, pt. 3 (1826), p. 383
[2] Voigt's *Magazin f. Naturk.* vi (1803), p. 181
[3] Gilbert's *Annalen,* xix (1805), p. 490
[4] *Annales de Chimie,* lviii (1806), p. 54
[5] Bakerian lecture for 1806, *Phil. Trans.* xcvii (1807), p. 1. A theory similar to that of Grothuss and Davy was communicated by Peter Mark Roget (1779–1869) in 1807 to the Philosophical Society of Manchester ; cf. Roget's *Galvanism,* § 106.

one of the terminals has been decomposed by the attractive and repellent forces of the terminal, one of its constituents is liberated there, while the other constituent, by virtue of electrical forces (the oxygen and hydrogen being in opposite electrical states), attacks the next molecule, which is then decomposed. The surplus constituent from this attacks the next molecule, and so on. Thus a chain of decompositions and recompositions was supposed to be set up among the molecules intervening between the terminals.

The hypothesis of Grothuss and Davy was attacked in 1825 by Auguste De La Rive [1] (1801–73) of Geneva, on the ground of its failure to explain what happens when different liquids are placed in series in the circuit. If, for example, a solution of zinc sulphate is placed in one compartment, and water in another, and if the positive pole is placed in the solution of zinc sulphate, and the negative pole in the water, De La Rive found that oxide of zinc is developed round the latter; although decomposition and recomposition of zinc sulphate could not take place in the water, which contained none of it. Accordingly, he supposed the constituents of the decomposed liquid to be bodily transported across the liquids, in close union with the moving electricity. In the electrolysis of water, one current of electrified hydrogen was supposed to leave the positive pole, and become decomposed into hydrogen and electricity at the negative pole, the hydrogen being there liberated as a gas. Another current in the same way carried electrified oxygen from the negative to the positive pole. In this scheme the chain of successive decompositions imagined by Grothuss does not take place, the only molecules decomposed being those adjacent to the poles.

The appearance of the products of decomposition at the separate poles could be explained either in Grothuss' fashion by assuming dissociations throughout the mass of liquid, or in De La Rive's by supposing particular dissociated atoms to travel considerable distances. Perhaps a preconceived idea of economy in Nature deterred the workers of that time from accepting the two assumptions together, when either of them separately would meet the case. Yet it is to this apparent redundancy that later researches have pointed as the truth. Nature is what she is, and not what we would make her.

De La Rive was one of the most thoroughgoing opponents of Volta's contact theory of the pile; even in the case when two metals are in contact in air only, without the intervention of any liquid, he attributed the electric effect wholly to the chemical affinity of the air for the metals.

[1] *Annales de Chimie*, xxviii (1825), p. 190

During the long interval between the publication of the rival hypotheses of Grothuss and De La Rive, little real progress was made with the special problems of the cell; but meanwhile electric theory was developing in other directions. One of these, to which our attention will first be turned, was the electro-chemical theory of the celebrated Swedish chemist, Jöns Jacob Berzelius (1779–1848).

Berzelius founded his theory,[1] which had been in one or two of its features anticipated by Davy,[2] on inferences drawn from Volta's contact effects. 'Two bodies,' he remarked, 'which have affinity for each other, and which have been brought into mutual contact, are found upon separation to be in opposite electrical states. That which has the greatest affinity for oxygen usually becomes positively electrified, and the other negatively.'

This seemed to him to indicate that chemical affinity arises from the play of electric forces, which in turn spring from electric charges within the atoms of matter. To be precise, he supposed each atom to possess two poles, which are the seat of opposite electrifications, and whose electrostatic field is the cause of chemical affinity.

By aid of this conception Berzelius drew a simple and vivid picture of chemical combination. Two atoms, which are about to unite, dispose themselves so that the positive pole of one touches the negative pole of the other; the electricities of these two poles then discharge each other, giving rise to the heat and light which are observed to accompany the act of combination.[3] The disappearance of these leaves the compound molecule with the two remaining poles; and it cannot be dissociated into its constituent atoms again until some means is found of restoring to the vanished poles their charges. Such a means is afforded by the action of the galvanic pile in electrolysis: the opposite electricities of the current invade the molecules of the electrolyte, and restore the atoms to their original state of polarisation.

If, as Berzelius taught, all chemical compounds are formed by the mutual neutralisation of pairs of atoms, it is evident that they must have a binary character. Thus he conceived a salt to be compounded of an acid and an oxide, and each of these to be compounded of two other constituents. Moreover, in any compound the electropositive member would be replaceable only by another electropositive member, and the electro-negative member only by another member also electronegative; so that the substitution of,

[1] *Memoirs of the Acad. of Stockholm*, 1812; *Nicholson's Journal*, xxxiv (1813), pp. 142, 153, 240, 319; xxxv, pp. 38, 118, 159

[2] *Phil. Trans.* xcvii (1807), p. 1

[3] This idea was Davy's.

for example, chlorine for hydrogen in a compound would be impossible—an inference which was overthrown by subsequent discoveries in chemistry.

Berzelius succeeded in bringing the most curiously diverse facts within the scope of his theory. Thus ' the combination of polarised atoms requires a motion to turn the opposite poles to each other; and to this circumstance is owing the facility with which combination takes place when one of the two bodies is in the liquid state, or when both are in that state; and the extreme difficulty, or nearly impossibility, of effecting an union between bodies, both of which are solid. And again, since each polarised particle must have an electric atmosphere, and as this atmosphere is the predisposing cause of combination, as we have seen, it follows, that the particles cannot act but at certain distances, proportioned to the intensity of their polarity; and hence it is that bodies, which have affinity for each other, always combine nearly on the instant when mixed in the liquid state, but less easily in the gaseous state, and the union ceases to be possible under a certain degree of dilatation of the gases; as we know by the experiments of Grothuss, that a mixture of oxygen and hydrogen in due proportions, when rarefied to a certain degree, cannot be set on fire at any temperature whatever.' And again: 'Many bodies require an elevation of temperature to enable them to act upon each other. It appears, therefore, that heat possesses the property of augmenting the polarity of these bodies.'

Berzelius accounted for Volta's electromotive series by assuming the electrification at one pole of an atom to be somewhat more or somewhat less than what would be required to neutralise the charge at the other pole. Thus each atom would possess a certain net or residual charge, which might be of either sign; and the order of the elements in Volta's series could be interpreted simply as the order in which they would stand when ranged according to the magnitude of this residual charge. As we shall see, this conception was afterwards overthrown by Faraday.

Berzelius permitted himself to publish some speculations on the nature of heat and electricity, which bring vividly before us the outlook of an able thinker in the first quarter of the nineteenth century. The great question, he says, is whether the electricities and caloric are matter or merely phenomena. If the title of matter is to be granted to such things as are ponderable, then these problematic entities are certainly not matter; but thus to narrow the application of the term is, he believes, a mistake; and he inclines to the opinion that caloric is truly matter, possessing chemical affinities without obeying the law of gravitation, and that light and all radiations consist in modes of propagating such matter. This conclusion makes it easier to decide regarding electricity. 'From the relation which exists between caloric and the electricities,' he remarks, 'it is clear that what may be true with regard to the materiality of one of them must also be true with regard to that of the other. There are, however, a quantity of phenomena produced by electricity which do not admit of explanation without admitting at the same time that electricity is matter. Electricity, for instance, very often detaches everything which covers the surface of those bodies which conduct it. It, indeed, passes through conductors without leaving any trace of its passage; but it penetrates non-conductors which oppose its course, and makes a perforation precisely of the same description as would have been made by something which had need of place for its passage. We often observe this when electric jars are broken by an overcharge, or when the electric shock is passed through a number of cards, etc. We may therefore, at least with some probability, imagine caloric and the electricities to be matter, destitute of gravitation, but possessing affinity to gravitating bodies. When they are not confined by these affinities, they tend to place themselves in equilibrium in the universe. The suns destroy at every moment this equilibrium, and they send the re-united electricities in the form of luminous rays towards the planetary bodies, upon the surface of which the rays, being arrested, manifest themselves as caloric; and this last in its turn, during the time required to replace it in equilibrium in the universe, supports the chemical activity of organic and inorganic nature.'

It was scarcely to be expected that anything so speculative as Berzelius' electric conception of chemical combination would be confirmed in all particulars by subsequent discovery; and, as a matter of fact, it did not as a coherent theory survive the lifetime of its author. But some of its ideas have persisted, and among them the conviction which lies at its foundation, that chemical affinities are, in the last resort, of electrical origin.

While the attention of chemists was for long directed to the theory of Berzelius, the interest of electricians was diverted from it by a discovery of the first magnitude in a different region.

That a relation of some kind subsists between electricity and magnetism had been suspected by the philosophers of the eighteenth century. The suspicion was based in part on some curious effects produced by lightning, of a kind which may be illustrated by a paper published in the *Philosophical Transactions* in 1735.[1] A tradesman of Wakefield, we are told, 'having put up a great number of knives and forks in a large box, and having placed the box in the corner of a large room, there happen'd in July, 1731, a sudden storm of thunder, lightning, etc., by which the corner of the room was damaged, the Box split, and a good many knives and forks melted, the sheaths being untouched. The owner emptying the box upon a Counter where some Nails lay, the Persons who took up the knives, that lay upon the Nails, observed that the knives took up the Nails.'

Lightning thus came to be credited with the power of magnetising steel; and it was doubtless this which led Franklin[2] in 1751 to attempt to magnetise a sewing-needle by means of the discharge of Leyden jars. The attempt was indeed successful; but, as Van Marum afterwards showed, it was doubtful whether the magnetism was due directly to the current.

More experiments followed.[3] In 1805 Jean-Nicolas-Pierre Hachette (1769–1834) and Charles-Bernard Desormes (1777–1862) attempted to determine whether an insulated voltaic pile, freely suspended, is oriented by terrestrial magnetism; but without positive result. In 1807 Hans Christian Oersted (1777–1851), Professor of Natural Philosophy in Copenhagen, announced his intention of examining the action of electricity on the magnetic needle; but it was not for some years that his hopes were realised. If one of his pupils is to be believed,[4] he was 'a man of genius, but a very unhappy experimenter; he could not manipulate instruments. He must always have an assistant, or one of his auditors who had easy hands, to arrange the experiment.'

During a course of lectures which he delivered in the winter of 1819–20 on 'Electricity, Galvanism and Magnetism,' the idea occurred to him that the changes observed with the compass needle during a thunderstorm might give the clue to the effect of which he was in search; and this led him to think that the experiment should be tried with the galvanic circuit closed instead of open, and to inquire whether any effect is produced on a magnetic needle when an electric current is passed through a neighbouring wire. At first he placed the wire at right angles to the needle, but observed no result. After the end of a lecture in which this negative experiment had been shown, the idea occurred to him to place the wire parallel to the needle; on trying it, a pronounced deflection was

[1] *Phil. Trans.* xxxix (1735), p. 74

[2] Letter vi from Franklin to Collinson

[3] In 1774 the Electoral Academy of Bavaria proposed the question, 'Is there a real and physical analogy between electric and magnetic forces?' as the subject of a prize.

[4] cf. a letter from Hansteen inserted in Bence Jones's *Life of Faraday*, ii, p. 395

observed, and the relation between magnetism and the electric current was discovered. After confirmatory experiments with more powerful apparatus, the public announcement was made in July 1820.[1]

Oersted did not determine the quantitative laws of the action, but contented himself with a statement of the qualitative effect and some remarks on its cause, which recall the magnetic speculations of Descartes ; indeed, Oersted's conceptions may be regarded as linking those of the Cartesian school to those which were introduced subsequently by Faraday. 'To the effect which takes place in the conductor and in the surrounding space,' he wrote, 'we shall give the name of the *conflict of electricity*.' 'The electric conflict acts only on the magnetic particles of matter. All non-magnetic bodies appear penetrable by the electric conflict, while magnetic bodies, or rather their magnetic particles, resist the passage of this conflict. Hence they can be moved by the impetus of the contending powers.

'It is sufficiently evident from the preceding facts that the electric conflict is not confined to the conductor, but dispersed pretty widely in the circumjacent space.

'From the preceding facts we may likewise collect, that this conflict performs circles ; for without this condition, it seems impossible that the one part of the uniting wire, when placed below the magnetic pole, should drive it toward the east, and when placed above it toward the west ; for it is the nature of a circle that the motions in opposite parts should have an opposite direction.'

Oersted's discovery was described at the meeting of the French Academy on 11 September 1820 by an academician (Arago) who had just returned from abroad. Several investigators in France repeated and extended his experiments ; and the first precise analysis of the effect was published by two of these, Jean-Baptiste Biot (1774–1862) and Félix Savart (1791–1841), who, at a meeting of the Academy of Sciences on 30 October 1820 announced [2] that the action experienced by a pole of austral or boreal magnetism, when placed at any distance from a straight wire carrying a voltaic current, may be thus expressed : 'Draw from the pole a perpendicular to the wire ; the force on the pole is at right angles to this line and to the wire, and its intensity is proportional to the reciprocal of the distance.' This result was soon further analysed, the attractive force being divided into constituents, each of which was supposed to be

[1] *Experimenta circa effectum conflictus electrici in acum magneticam* (Copenhagen, 1820) ; German trans. in Schweigger's *Journal für Chemie und Physik*, xxix (1820), p. 275 ; English trans. in Thomson's *Annals of Philosophy*, xvi (1820), p. 273

[2] *Annales de Chimie*, xv (1820), p. 222 ; *Journal de Phys.* xci (1820), p. 151

due to some particular element of the current ; in its new form the law may be stated thus : *the magnetic force due to an element* **ds** *of a circuit, in which a current* **i** *is flowing, at a point whose vector distance from* **ds** *is* **r**, *is (in suitable units)*

$$\frac{i}{r^3}[\mathbf{ds} \cdot \mathbf{r}]\ ^{[1]} \quad \text{or} \quad \text{curl}\,\frac{i\mathbf{ds}}{r}.^{[2]}$$

It was now recognised that a magnetic field may be produced as readily by an electric current as by a magnet ; and, as Arago soon showed,[3] this, like any other magnetic field, is capable of inducing magnetisation in iron. The question naturally suggested itself as to whether the similarity of properties between currents and magnets extended still further, e.g. whether conductors carrying currents would, like magnets, experience ponderomotive forces when placed in a magnetic field, and whether such conductors would consequently like magnets, exert ponderomotive forces on each other.

The first step towards answering these inquiries was taken by Oersted [4] himself. 'As,' he said, 'a body cannot put another in motion without being moved in its turn, when it possesses the requisite mobility, it is easy to foresee that the galvanic arc must be moved by the magnet' ; and this he verified experimentally.[5]

The next step came from André-Marie Ampère (1775–1836), who at the meeting of the Academy on 18 September, exactly a week after the news of Oersted's first discovery had arrived, showed that two parallel wires carrying currents attract each other if the currents are in the same direction, and repel each other if the currents are in opposite directions. During the next three years Ampère continued to prosecute the researches thus inaugurated, and in 1825 published his collected results in one of the most celebrated memoirs [6] in the history of natural philosophy.

Ampère introduces his work by proclaiming himself a follower

[1] If **a** and **b** denote two vectors, the vector whose components are $(a_y b_z - a_z b_y,\ a_z b_x - a_x b_z,\ a_x b_y - a_y b_x)$ is called the *vector product* of **a** and **b**, and is denoted by [**a**, **b**]. Its direction is at right angles to those of **a** and **b**, and its magnitude is represented by twice the area of the triangle formed by them.

[2] If **a** denotes any vector, the vector whose components are $\frac{\partial a_z}{\partial y} - \frac{\partial a_y}{\partial z},\ \frac{\partial a_x}{\partial z} - \frac{\partial a_z}{\partial x},\ \frac{\partial a_y}{\partial x} - \frac{\partial a_x}{\partial y}$ is denoted by curl **a**.

[3] *Annales de Chimie*, xv (1820), p. 93

[4] Schweigger's *Journal für Chem. u. Phys.* xxix (1820), p. 364 ; Thomson's *Annals of Philosophy*, xvi (1820), p. 375

[5] Davy found in 1821 (*Phil. Trans.* cxi [1821], p. 425) that the electric arc is deflected by a magnet.

[6] *Mém. de l'Acad.* vi (1825), p. 175

of that school which explained all physical phenomena in terms of equal and oppositely directed forces between pairs of particles ; and he renounces the attempt to seek more speculative, though possibly more fundamental, explanations in terms of the motions of ultimate fluids and aethers. Nevertheless, he indicates two conceptions of this latter character, on which such explanations might be founded.

In the first [1] he suggests that the ponderomotive forces between circuits carrying electric currents may be due to ' the reaction of the elastic fluid which extends throughout all space, whose vibrations produce the phenomena of light,' and which is ' put in motion by electric currents.' This fluid or aether can, he says, ' be no other than that which results from the combination of the two electricities.'

In the second conception,[2] Ampère suggests that the interspaces between the metallic molecules of a wire which carries a current may be occupied by a fluid composed of the two electricities, not in the proportions which form the neutral fluid, but with an excess of that one of them which is opposite to the electricity peculiar to the molecules of the metal, and which consequently masks this latter electricity. In this inter-molecular fluid the opposite electricities are continually being dissociated and recombined ; a dissociation of the fluid within one inter-molecular interval having taken place, the positive electricity thus produced unites with the negative electricity of the interval next to it in the direction of the current, while the negative electricity of the first interval unites with the positive electricity of the next interval in the other direction. Such interchanges, according to this hypothesis, constitute the electric current.

Ampère's memoir is, however, but little occupied with the more speculative side of the subject. His first aim was to investigate thoroughly by experiment the ponderomotive forces on electric currents.

' When,' he remarks, ' M. Oersted discovered the action which a current exercises on a magnet, one might certainly have suspected the existence of a mutual action between two circuits carrying currents ; but this was not a necessary consequence ; for a bar of soft iron also acts on a magnetised needle, although there is no mutual action between two bars of soft iron.'

Ampère, therefore, submitted the matter to the test of the laboratory, and discovered that circuits carrying electric currents exert ponderomotive forces on each other, and that ponderomotive forces are exerted on such currents by magnets. To the science

[1] *Recueil d'observations électro-dynamiques*, p. 215 ; and the memoir just cited, pp. 285, 370

[2] *Recueil d'observations électro-dynamiques*, pp. 297, 300, 371

which deals with the mutual action of currents he gave the name *electrodynamics* [1]; and he showed that the action obeys the following laws :

(1) The effect of a current is reversed when the direction of the current is reversed.

(2) The effect of a current flowing in a circuit twisted into small sinuosities is the same as if the circuit were smoothed out.

(3) The force exerted by a closed circuit on an element of another circuit is at right angles to the latter.

(4) The force between two elements of circuits is unaffected when all linear dimensions are increased proportionately, the current-strengths remaining unaltered.

From these data, together with his assumption that the force between two elements of circuits acts along the line joining them, Ampère obtained an expression of this force : the deduction may be made in the following way :

Let **ds**, **ds'** be the elements, **r** the line joining them and *i*, *i'* the current strengths. From (2) we see that the effect of **ds** on **ds'** is the vector sum of the effects of *dx*, *dy*, *dz* on **ds'**, where these are the three components of **ds** ; so the required force must be of the form :

r × a scalar quantity which is linear and homogeneous in **ds** ; and it must similarly be linear and homogeneous in **ds'** ; so using (1), we see that the force must be of the form

$$\mathbf{F} = ii'\mathbf{r}\,\{(\mathbf{ds}\,.\,\mathbf{ds'})\,\phi\,(r) + (\mathbf{ds}\,.\,\mathbf{r})\,(\mathbf{ds'}\,.\,\mathbf{r})\,\psi\,(r)\},$$

where ϕ and ψ denote undetermined functions of *r*.

From (4) it follows that when *ds*, *ds'*, *r* are all multiplied by the same number, **F** is unaffected : this shows that

$$\phi(r) = \frac{A}{r^3} \text{ and } \psi(r) = \frac{B}{r^5},$$

where A and B denote constants. Thus we have

$$\mathbf{F} = ii'\mathbf{r}\left\{\frac{A(\mathbf{ds}\,.\,\mathbf{ds'})}{r^3} + \frac{B(\mathbf{ds}\,.\,\mathbf{r})\,(\mathbf{ds'}\,.\,\mathbf{r})}{r^5}\right\}.$$

Now, by (3), the resolved part of **F** along **ds'** must vanish when integrated round the circuit **s**, i.e. it must be a complete differential when **dr** is taken to be equal to $-$ **ds**. That is to say,

$$\frac{A(\mathbf{ds}\,.\,\mathbf{ds'})\,(\mathbf{r}\,.\,\mathbf{ds'})}{r^3} + \frac{B(\mathbf{ds}\,.\,\mathbf{r})\,(\mathbf{ds'}\,.\,\mathbf{r})^2}{r^5}$$

[1] Loc. cit. p. 298

must be a complete differential ; or

$$-\frac{A}{2r^3}d(\mathbf{r}\cdot\mathbf{ds}')^2+\frac{B}{r^5}(\mathbf{ds}\cdot\mathbf{r})(\mathbf{r}\cdot\mathbf{ds}')^2$$

must be a complete differential ; and therefore

$$d\,\frac{A}{2r^3}=-\frac{B}{r^5}(\mathbf{ds}\cdot\mathbf{r}),$$

or

$$-\frac{3A}{2r^4}\,dr=\frac{B}{r^4}\,dr,$$

or

$$B=-\tfrac{3}{2}A.$$

Thus finally we have

$$\mathbf{F}=\text{Constant}\times ii'\mathbf{r}\left\{\frac{2}{r^3}(\mathbf{ds}\cdot\mathbf{ds}')-\frac{3}{r^5}(\mathbf{ds}\cdot\mathbf{r})(\mathbf{ds}'\cdot\mathbf{r})\right\}.$$

This is Ampère's formula : the multiplicative constant depends of course on the units chosen and may be taken to be -1.

The weakness of Ampère's work evidently lies in the assumption that the force is directed along the line joining the two elements ; for in the analogous case of the action between two magnetic molecules, we know that the force is *not* directed along the line joining the molecules. It is therefore of interest to find the form of **F** when this restriction is removed.

For this purpose we observe that we can add to the expression already found for **F** any term of the form

$$\phi(r)\,.\,(\mathbf{ds}\cdot\mathbf{r})\,.\,\mathbf{ds}',$$

where $\phi(r)$ denotes any arbitrary function of r ; for since

$$(\mathbf{ds}\cdot\mathbf{r})=-r\,.\,ds\,.\,\frac{dr}{ds},$$

this term vanishes when integrated round the circuit **s** ; and it contains **ds** and **ds′** linearly and homogeneously, as it should. We can also add any terms of the form

$$d\{\mathbf{r}\,.\,(\mathbf{ds}'\cdot\mathbf{r})\,.\,\chi(r)\},$$

where $\chi(r)$ denotes any arbitrary function of r, and d denotes differentiation along the arc **s**, keeping **ds′** fixed (so that $\mathbf{dr}=-\mathbf{ds}$) ; this differential may be written

$$-\mathbf{ds}\,.\,(\mathbf{ds}'\cdot\mathbf{r})\,.\,\chi(r)-\mathbf{r}\chi(r)\,(\mathbf{ds}'\cdot\mathbf{ds})-\frac{1}{r}\chi'(r)\,\mathbf{r}\,(\mathbf{ds}\cdot\mathbf{r})\,(\mathbf{ds}'\cdot\mathbf{r}).$$

In order that the law of Action and Reaction may not be violated, we must combine this with the former additional term so as to obtain an expression symmetrical in **ds** and **ds′**; and hence we see finally *that the general value of* **F** *is given by the equation*

$$\mathbf{F} = -ii'\mathbf{r}\left\{\frac{2}{r^3}(\mathbf{ds}\,.\,\mathbf{ds}') - \frac{3}{r^5}(\mathbf{ds}\,.\,\mathbf{r})(\mathbf{ds}'.\,\mathbf{r})\right\}$$
$$+\chi(r)(\mathbf{ds}'\,.\,\mathbf{r})\,\mathbf{ds} + \chi(r)(\mathbf{ds}\,.\,\mathbf{r})\,\mathbf{ds}' + \chi(r)(\mathbf{ds}\,.\,\mathbf{ds}')\,\mathbf{r}$$
$$+\frac{1}{r}\chi'(r)(\mathbf{ds}\,.\,\mathbf{r})(\mathbf{ds}'\,.\,\mathbf{r})\,\mathbf{r}.$$

The simplest form of this expression is obtained by taking

$$\chi(r) = \frac{ii'}{r^3},$$

when we obtain

$$\mathbf{F} = \frac{ii'}{r^3}\{(\mathbf{ds}\,.\,\mathbf{r})\,\mathbf{ds}' + (\mathbf{ds}'\,.\,\mathbf{r})\mathbf{ds} - (\mathbf{ds}\,.\,\mathbf{ds}')\mathbf{r}\}.$$

The comparatively simple expression in brackets is the vector part of the quaternion product of the three vectors **ds**, **r**, **ds′**.[1]

From any of these values of **F** we can find the ponderomotive force exerted by the whole circuit **s** on the element **ds′**; it is, in fact, from the last expression,

$$ii'\int_s \frac{1}{r^3}\{(\mathbf{ds}'\,.\,\mathbf{r})\,\mathbf{ds} - (\mathbf{ds}\,.\,\mathbf{ds}')\,\mathbf{r}\},$$

or

$$ii'\int_s \left[\mathbf{ds}'.\frac{[\mathbf{ds}\,.\,\mathbf{r}]}{r^3}\right],$$

or

$$i'\,[\mathbf{ds}'\,.\,\mathbf{B}],$$

where

$$\mathbf{B} = i\int_s \frac{1}{r^3}\,[\mathbf{ds}\,.\,\mathbf{r}].$$

Now this value of **B** is precisely the value found by Biot and Savart[2] for the magnetic intensity at **ds′** due to the current i in the circuit s. Thus we see that the ponderomotive force on a current element **ds′** in a magnetic field **B** is i' [**ds′**. **B**].

[1] The simpler form of **F** given in the text is, if the term in **ds′** be omitted, the form given by Grassmann, *Ann. d. Phys.* lxiv (1845), p. 1. For further work on his subject cf. Tait, *Proc. R. S. Edin.* viii (1873), p. 220; Helmholtz, *Berl. Akad. Monatsber.* (1873), p. 91; and Korteweg, *Journal für Math.* xc (1881), p. 49. Helmholtz assumes that the interaction between two current elements is derivable from a potential, and this entails the existence of a couple in addition to a force along the line joining the elements. cf. the discussion in chap. vii of the present work.

[2] See ante, p. 82

Ampère developed to a considerable extent the theory of the equivalence of magnets with circuits carrying currents ; and showed that an electric current is equivalent, in its magnetic effects, to a distribution of magnetism on any surface terminated by the circuit, the axes of the magnetic molecules being everywhere normal to this surface :[1] such a magnetised surface is called a *magnetic shell.* He preferred, however, to regard the current rather than the magnetic fluid as the fundamental entity, and considered magnetism to be really an electrical phenomenon ; each magnetic molecule owes its properties, according to this view, to the presence within it of a small closed circuit in which an electric current is perpetually flowing.

The impression produced by Ampère's memoir was great and lasting. Writing half a century afterwards, Maxwell speaks of it as 'one of the most brilliant achievements in science.' 'The whole,' he says, 'theory and experiment seems as if it had leaped, full-grown and full-armed, from the brain of the "Newton of electricity." It is perfect in form and unassailable in accuracy ; and it is summed up in a formula from which all the phenomena may be deduced, and which must always remain the cardinal formula of electro-dynamics.'

Heaviside, however, in 1888 expressed a different opinion[2]: 'It has been stated, on no less authority than that of the great Maxwell, that Ampère's law of force between a pair of current elements is the cardinal formula of electrodynamics. If so, should we not be always using it? Do we *ever* use it? Did Maxwell in his Treatise? Surely there is some mistake. I do not in the least mean to rob Ampère of the credit of being the father of electro-dynamics ; I would only transfer the name of cardinal formula to another due to him, expressing the mechanical force on an element of a conductor supporting current in any magnetic field—the vector product of current and induction. There is something real about it ; it is not like his force between a pair of unclosed elements ; it is fundamental ; and, as everybody knows, it is in continual use, either actually or virtually (through electromotive force), both by theorists and practicians.'

Not long after the discovery by Oersted of the connection between galvanism and magnetism, a connection was discovered between galvanism and heat. In 1822 Thomas Johann Seebeck

[1] Loc. cit. p. 367
[2] *Electrician* (28 Dec. 1888), p. 229 ; Heaviside's *Electrical Papers*, ii, p. 500

(1770–1831), of Berlin discovered [1] that an electric current can be set up in a circuit of metals, without the interposition of any liquid, merely by disturbing the equilibrium of temperature. Let a ring be formed of copper and bismuth soldered together at the two extremities ; to establish a current it is only necessary to heat the ring at one of these junctions. To this new class of circuits the name *thermo-electric* was given.

It was found that the metals can be arranged as a *thermo-electric* series, in the order of their power of generating currents when thus paired, and that this order is quite different from Volta's order of electromotive potency. Indeed antimony and bismuth, which are near each other in the latter series, are at opposite extremities of the former.

The currents generated by thermo-electric means are generally feeble ; and the mention of this fact brings us to the question, which was about this time engaging attention, of the efficacy of different voltaic arrangements.

Comparisons of a rough kind had been instituted soon after the discovery of the pile. The French chemists Antoine-François de Fourcroy (1755–1809), Louis-Nicolas Vauquelin (1763–1829) and Louis-Jacques Thénard (1777–1857) found [2] in 1801, on varying the size of the metallic disks constituting the pile, that the sensations produced on the human frame were unaffected so long as the number of disks remained the same ; but that the power of burning finely drawn wire was altered ; and that the latter power was proportional to the total surface of the disks employed, whether this were distributed among a small number of large disks, or a large number of small ones. This was explained by supposing that small plates give a small quantity of the electric fluid with a high velocity, while large plates give a larger quantity with no greater velocity. Shocks, which were supposed to depend on the velocity of the fluid alone, would therefore not be intensified by increasing the size of the plates.

The effect of varying the conductors which connect the terminals of the pile was also studied. Nicolas Gautherot (1753–1803) observed [3] that water contained in tubes which have a narrow opening

[1] *Abhandl. d. Berlin Akad.* x (1822–3), p. 265. *Ann. d. Phys.* lxxiii (1823), pp. 115, 430; vi (1826), pp. 1, 133, 253

Volta had previously noticed that a silver plate whose ends were at different temperatures appeared to act like a voltaic cell.

Further experiments were performed by James Cumming (1777–1861), Professor of Chemistry at Cambridge, *Trans. Camb. Phil. Soc.* ii (1823), p. 47 ; and by Antoine César Becquerel (1788–1878), *Annales de Chimie*, xxxi (1826), p. 371.

[2] *Ann. de Chimie*, xxxix (1801), p. 103

[3] *Annales de Chim.* xxxix (1801), p. 203

does not conduct voltaic currents so well as when the opening is more considerable. This experiment is evidently very similar to that which Beccaria had performed half a century previously [1] with electrostatic discharges.

As we have already seen, Cavendish investigated very completely the power of metals to conduct electrostatic discharges; their power of conducting voltaic currents was now examined by Davy.[2] His method was to connect the terminals of a voltaic battery by a path containing water (which it decomposed), and also by an alternative path consisting of the metallic wire under examination. When the length of the wire was less than a certain quantity, the water ceased to be decomposed; Davy measured the lengths and weights of wires of different materials and cross-sections under these limiting circumstances; and, by comparing them, showed that the conducting power of a wire formed of any one metal is inversely proportional to its length and directly proportional to its sectional area, but independent of the shape of the cross-section.[3] The latter fact, as he remarked, showed that voltaic currents pass through the substance of the conductor and not along its surface.

Davy, in the same memoir, compared the conductivities of various metals, and studied the effect of temperature; he found that the conductivity varied with the temperature, being 'lower in some inverse ratio as the temperature was higher.'

He also observed that the same magnetic power is exhibited by every part of the same circuit, even though it be formed of wires of different conducting powers pieced into a chain, so that 'the magnetism seems directly as the quantity of electricity which they transmit.'

The current which flows in a given voltaic circuit evidently depends not only on the conductors which form the circuit, but also on the driving-power of the battery. In order to form a complete theory of voltaic circuits, it was therefore necessary to extend Davy's laws by taking the driving-power into account. This advance was effected in 1826 by Georg Simon Ohm [4] (1787–1854).

Ohm had already carried out a considerable amount of experimental work on the subject, and had, e.g., discovered that if a

[1] See p. 55

[2] *Phil. Trans.* cxi (1821), p. 425. His results were confirmed afterwards by Becquerel, *Annales de Chimie*, xxxii (1825), p. 423.

[3] These results had been known to Cavendish.

[4] *Ann. d. Phys.* vi (1826), p. 459; vii, pp. 45, 117; *Die Galvanische Kette mathematisch bearbeitet*, Berlin, 1827; translated in Taylor's *Scientific Memoirs*, ii (1841), p. 401. cf. also subsequent papers by Ohm in Kastner's *Archiv für d. ges. Naturlehre*, and Schweigger's *Jahrbuch*.

number of voltaic cells are placed in series in a circuit, the current is proportional to their number if the external resistance is very large, but is independent of their number if the external resistance is small. He now essayed the task of combining all the known results into a consistent theory.

For this purpose he adopted the idea of comparing the flow of electricity in a current to the flow of heat along a wire, the theory of which had been familiar to all physicists since the publication of Fourier's *Théorie analytique de la chaleur* in 1822. 'I have proceeded,' he says, 'from the supposition that the communication of the electricity from one particle takes place directly only to the one next to it, so that no immediate transition from that particle to any other situate at a greater distance occurs. The magnitude of the flow between two adjacent particles, under otherwise exactly similar circumstances, I have assumed to be proportional to the difference of the electric forces existing in the two particles; just as, in the theory of heat, the flow of caloric between two particles is regarded as proportional to the difference of their temperatures.'

The comparison between the flow of electricity and the flow of heat suggested the propriety of introducing a quantity whose behaviour in electrical problems should resemble that of temperature in the theory of heat. The difference in the values of such a quantity at two points of a circuit would provide what was so much needed, namely, a measure of the 'driving-power' acting on the electricity between these points. To carry out this idea, Ohm recurred to Volta's theory of the electrostatic condition of the open pile. It was customary to measure the 'tension' of a pile by connecting one terminal to earth and testing the other terminal by an electroscope. Accordingly Ohm says: 'In order to investigate the changes which occur in the electric condition of a body A in a perfectly definite manner, the body is each time brought, under similar circumstances, into relation with a second moveable body of invariable electrical condition, called the *electroscope*; and the force with which the electroscope is repelled or attracted by the body is determined. This force is termed the *electroscopic force* of the body A.

'The same body A may also serve to determine the electroscopic force in various parts of the same body. For this purpose take the body A of very small dimensions, so that when we bring it into contact with the part to be tested of any third body, it may from its smallness be regarded as a substitute for this part; then its electroscopic force, measured in the way described, will, when it happens to be different at the various places, make known the relative differences with regard to electricity between these places.'

Ohm assumed, as was customary at that period, that when two metals are placed in contact, 'they constantly maintain at the point of contact the same difference between their electroscopic forces.' He accordingly supposed that each voltaic cell possesses a definite tension, or discontinuity of electroscopic force, which is to be regarded as its contribution to the driving-force of any circuit in which it may be placed. This assumption confers a definite meaning on his use of the term 'electroscopic force'; the force in question is identical with the electrostatic potential. But Ohm and his contemporaries did not correctly understand the relation of galvanic conceptions to the electrostatic functions of Poisson. The electroscopic force in the open pile was generally identified with the thickness of the electrical stratum at the place tested; while Ohm, recognising that electric currents are not confined to the surface of the conductors, but penetrate their substance, seems to have thought of the electroscopic force at a place in a circuit as being proportional to the volume-density of electricity there—an idea in which he was confirmed by the relation which, in an analogous case, exists between the temperature of a body and the volume-density of heat supposed to be contained in it.

Denoting, then, by S the current which flows in a wire of conductivity γ, when the difference of the electroscopic forces at the terminals is E, Ohm writes

$$S = \gamma E.$$

From this formula it is easy to deduce the laws already given by Davy. Thus, if the area of the cross-section of a wire is A, we can by placing n such wires side by side construct a wire of cross-section nA. If the quantity E is the same for each, equal currents will flow in the wires; and therefore the current in the compound wire will be n times that in the single wire; so when the quantity E is unchanged, the current is proportional to the cross-section; that is, the conductivity of a wire is directly proportional to its cross-section, which is one of Davy's laws.

In spite of the confusion which was attached to the idea of electroscopic force, and which was not dispelled for some years, the publication of Ohm's memoir marked a great advance in electrical philosophy. It was now clearly understood that the current flowing in any conductor depends only on the conductivity inherent in the conductor and on another variable which bears to electricity the same relation that temperature bears to heat; and, moreover, it was realised that this latter variable is the link connecting the theory of currents with the older theory of electrostatics. These principles were a sufficient foundation for future progress; and much of the work which was published in the second quarter of the century was no more than the natural development of the principles laid down by Ohm.[1]

It is painful to relate that the discoverer had long to wait before the merits of his great achievement were officially recognised.[2] He was, however, awarded the Copley Medal of the Royal Society in 1841: and twenty-two years after the publication of the memoir on the galvanic circuit, he was promoted to a university professorship; this he held for the five years which remained until his death in 1854.

[1] Ohm's theory was confirmed experimentally by several investigators, among whom may be mentioned Gustav Theodor Fechner (1801–87), *Maassbestimmungen über die Galvanische Kette* (Leipzig, 1831), and Charles Wheatstone (1802–75), *Phil. Trans.* clxxxiv (1843), p. 303.

[2] An interesting account of the reception of Ohm's electrical researches by his contemporaries is given by H. J. J. Winter in *Phil. Mag.* xxxv (1944), p. 371.

Chapter IV

THE LUMINIFEROUS MEDIUM, FROM BRADLEY TO FRESNEL

ALTHOUGH Newton, as we have seen, refrained from committing himself to any doctrine regarding the ultimate nature of light, the writers of the next generation interpreted his criticism of the wave-theory as equivalent to an acceptance of the corpuscular hypothesis. As it happened, the chief optical discovery of this period tended to support the latter theory, by which it was first and most readily explained. In 1728 James Bradley (1692–1762), at that time Savilian Professor of Astronomy at Oxford, sent to the Astronomer Royal (Halley) an 'Account of a new discovered motion of the Fix'd Stars.' [1] In observing the star γ in the head of the Dragon, with the object of discovering its parallax, he had found that during the winter of 1725–6 the transit across the meridian was continually more southerly, while during the following summer its original position was restored by a motion northwards. Such an effect could not be explained as a result of parallax; and eventually Bradley guessed it to be due to the gradual propagation of light.[2]

Thus, let CA denote a ray of light, falling on the line BA; and suppose that the eye of the observer is travelling along BA, with a velocity which is to the velocity of light as BA is to CA. Then the corpuscle of light, by which the object is discernible to the eye at A, would have been at C when the eye was at B. The tube of a telescope must therefore be pointed in the direction BC, in order to receive the rays from an object whose light is really propagated in the direction CA. The angle BCA measures the difference between the real and apparent positions of the object; and it is evident from the figure that the sine of this angle is to the sine of the visible inclination of the object to the line in which the eye is moving, as the velocity of the eye is to the velocity of light. Observations such as Bradley's will therefore enable us to deduce

[1] *Phil. Trans.* xxxv (1728), p. 637

[2] Roemer, in a letter to Huygens of date 30 Dec. 1677, mentions a suspected displacement of the apparent position of a star, due to the motion of the earth at right angles to the line of sight. cf. *Correspondance de Huygens*, viii, p. 53.

the ratio of the mean orbital velocity of the earth to the velocity of light, or, as it is called, *the constant of aberration* ; from its value Bradley calculated that light is propagated from the sun to the earth in 8 minutes 12 seconds, which, as he remarked, 'is as it were a Mean betwixt what had at different times been determined from the eclipses of Jupiter's satellites.' [1]

With the exception of Bradley's discovery, which was primarily astronomical rather than optical, the eighteenth century was decidedly barren, as regards both the experimental and the theoretical investigation of light ; in curious contrast to the brilliance of its record in respect of electrical researches. But some attention must be given to a suggestive study [2] of the aether, for which the younger John Bernoulli (1710–90) was in 1736 awarded the prize of the French Academy. His ideas seem to have been originally suggested by an attempt [3] which his father, the elder John Bernoulli (1667–1748), had made in 1701 to connect the law of refraction with the mechanical principle of the composition of forces. If two opposed forces whose ratio is μ maintain in equilibrium a particle which is free to move only in a given plane, it follows from the triangle of forces that the directions of the forces must obey the relation

$$\sin i = \mu \sin r,$$

where i and r denote the angles made by these directions with the normals to the plane. This is the same equation as that which expresses the law of refraction, and the elder Bernoulli conjectured that a theory of light might be based on it ; but he gave no satisfactory physical reason for the existence of forces along the incident and refracted rays. This defect his son now proceeded to remove.

All space, according to the young Bernoulli, is permeated by a fluid aether, containing an immense number of excessively small whirlpools. The elasticity which the aether appears to possess, and in virtue of which it is able to transmit vibrations, is really due to the presence of these whirlpools ; for, owing to centrifugal force,

[1] Struve in 1845 found for the constant of aberration the value 20″·445, which he afterwards corrected to 20″·463. This was superseded in 1883 by the value 20″·492, determined by M. Nyrén. The observations of both Struve and Nyrén were made with the transit in the prime vertical. The method now generally used depends on the measurement of differences of meridian zenith distances. In the present century values have been found which range from 20″·523 down to 20″·445. The value calculated from the solar parallax and the velocity of light is 20″·511, if for the parallax we take the value 8″·790 found by H. Spencer Jones in 1941, and for the velocity of light take the value $2{\cdot}99776 \times 10^{10}$ cm/sec.

[2] Printed in 1752, in the *Recueil des pièces qui ont remportés les prix de l'Acad.*, tome iii

[3] *Acta eruditorum* (1701), p. 19

each whirlpool is continually striving to dilate, and so presses against the neighbouring whirlpools. It will be seen that Bernoulli is a thorough Cartesian in spirit ; not only does he reject action at a distance, but he insists that even the elasticity of his aether shall be explicable in terms of matter and motion.

This aggregate of small vortices, or 'fine-grained turbulent motion,' as it came to be called a century and a half later,[1] is interspersed with solid corpuscles, whose dimensions are small compared with their distances apart. These are pushed about by the whirlpools whenever the aether is disturbed, but never travel far from their original positions.

A source of light communicates to its surroundings a disturbance which condenses the nearest whirlpools ; these by their condensation displace the contiguous corpuscles from their equilibrium position ; and these in turn produce condensations in the whirlpools next beyond them, so that vibrations are propagated in every direction from the luminous point. It is curious that Bernoulli speaks of these vibrations as *longitudinal*, and actually contrasts them with those of a stretched cord, which, 'when it is slightly displaced from its rectilinear form, and then let go, performs *transverse* vibrations in a direction at right angles to the direction of the cord.' When it is remembered that the objection to longitudinal vibrations, on the score of polarisation, had already been clearly stated by Newton, and that Bernoulli's aether closely resembles that which Maxwell invented in 1861–2 for the express purpose of securing transversality of vibration, one feels that perhaps no man ever so narrowly missed a great discovery.

Bernoulli explained refraction by combining these ideas with those of his father. Within the pores of ponderable bodies the whirlpools are compressed, so the centrifugal force must vary in intensity from one medium to another. Thus a corpuscle situated in the interface between two media is acted on by a greater elastic force from one medium than from the other ; and by applying the triangle of forces to find the conditions of its equilibrium, the law of Snell and Descartes may be obtained.

Not long after this, the echoes of the old controversy between Descartes and Fermat about the law of refraction were awakened [2] by Pierre-Louis-Moreau de Maupertuis (1698–1759).

It will be remembered that according to Descartes, the velocity of light is greatest in dense media, while according to Fermat the

[1] cf. Lord Kelvin's vortex-sponge aether, described later in this work

[2] *Mém. de l'Acad.* (1744), p. 417

propagation is swiftest in free aether. The arguments of the corpuscular theory convinced Maupertuis that on this particular point Descartes was in the right; but nevertheless he wished to retain for science the beautiful method by which Fermat had derived his result. This he now proposed to do by modifying Fermat's principle so as to make it agree with the corpuscular theory; instead of assuming that light follows the *quickest* path, he supposed that 'the path described is that by which the quantity of action is the least'; and this *action* he defined to be proportional to the sum of the spaces described, each multiplied by the velocity with which it is traversed. Thus instead of Fermat's expression

$$\int dt \quad \text{or} \quad \int \frac{ds}{v}$$

(where t denotes time, v velocity and ds an element of the path) Maupertuis introduced

$$\int v\, ds$$

as the quantity which is to assume its minimum value when the path of integration is the actual path of the light. Since Maupertuis' v, which denotes the velocity according to the corpuscular theory, is proportional to the reciprocal of Fermat's v, which denotes the velocity according to the wave-theory, the two expressions are really equivalent, and lead to the same law of refraction. Maupertuis' memoir is, however, of great interest from the point of view of dynamics; for his suggestion was subsequently developed by himself and by Euler and Lagrange into a general principle which covers the whole range of Nature, so far as Nature is a dynamical system.

The natural philosophers of the eighteenth century for the most part, like Maupertuis, accepted the corpuscular hypothesis; but the wave-theory was not without defenders. Franklin declared [1] for it; and the celebrated mathematician Leonhard Euler (1707–83) ranged himself on the same side, being impressed by the notion that the emission of particles would cause a diminution in the mass of the radiating body, which was not observed, while the emission of waves involved no such consequence. In a work entitled *Nova Theoria Lucis et Colorum*, published [2] while he was living under the patronage of Frederic the Great at Berlin, he insisted strongly on the resemblance between light and sound; the whole of the space through which the heavenly bodies move is filled with a subtle

[1] Letter xxiii, written in 1752

[2] L. Euleri, *Opuscula varii argumenti* (Berlin, 1746), p. 169

matter, the aether, and light consists in vibrations of this aether; 'light is in the aether the same thing as sound in air.' Accepting Newton's doctrine that colour depends on wave-length, he in this memoir supposed the frequency greatest for red light, and least for violet; but a few years later [1] he adopted the opposite opinion.

The chief novelty of Euler's writings on light is his explanation of the manner in which material bodies appear coloured when viewed by white light; and, in particular, of the way in which the colours of thin plates are produced. He denied that such colours are due to a more copious reflection of light of certain particular periods, and supposed that they represent vibrations generated within the body itself under the stimulus of the incident light. A coloured surface, according to this hypothesis, contains large numbers of elastic molecules, which, when agitated, emit light of period depending only on their own structure. The colours of thin plates Euler explained in the same way; the elastic response and free period of the plate at any place would, he conceived, depend on its thickness at that place; and in this way the dependence of the colour on the thickness was accounted for, the phenomena as a whole being analogous to well-known effects observed in experiments on sound.

Euler, who remained at Berlin from 1741 to 1766, numbered among his pupils a niece of Frederic, the Princess of Anhalt-Dessau, to whom in 1760–1 he wrote a series of letters setting forth his views on natural philosophy. He anticipated Maxwell in asserting that the source of all electrical phenomena is the same aether that propagates light: electricity is nothing but a derangement of the equilibrium of the aether. 'A body must become electrical,' he said, 'whenever the aether contained in its pores becomes more or less elastic than that which is lodged in adjacent bodies. This takes place when a greater quantity of aether is introduced into the pores of a body, or when part of the aether which it contained is forced out. In the former case, the aether becomes more compressed, and consequently more elastic; in the other, it becomes rarer, and loses its elasticity. In both cases, it is no longer in equilibrium with that which is external; and the efforts which it makes to recover its equilibrium, produce all the phenomena of electricity.[2]

Not only electrical, but also gravitational phenomena, were explained in terms of the aether.[3] The explanation depended on

[1] *Mém. de l'Acad. de Berlin* (1752), p. 262

[2] Letter of 4 July 1761

[3] *Leonhardi Euleri Opera Omnia*, series tertia i, pp. 4, 149

supposing that the pressure of the aether increases with the distance from the centre of the earth, say as Constant $- 1/r$, so that the force pressing a body towards the earth is stronger than that directed away from it, the balance of these forces being the weight of the body. On this hypothesis, the force on each atom would be proportional to the volume of the atom, and therefore the weight of an atom must be proportional to its volume, that is to say, the densities of all atoms are equal; the fact that the densities of bodies differ from each other is accounted for by assuming that the atoms are not in contact.

An attempt to improve the corpuscular theory in another direction was made in 1752 by the Marquis de Courtivron,[1] and independently in the following year by T. Melvill.[2] These writers suggested, as an explanation of the different refrangibility of different colours, that 'the differently colour'd rays are projected with different velocities from the luminous body: the red with the greatest, violet with the least, and the intermediate colours with intermediate degrees of velocity.' On this supposition, as its authors pointed out, the amount of aberration would be different for every different colour; and the satellites of Jupiter would change colour, from white through green to violet, through an interval of more than half a minute before their immersion into the planet's shadow; while at emersion the contrary succession of colours should be observed, beginning with red and ending in white. The testimony of practical astronomers was soon given that such appearances are not observed; and the hypothesis was accordingly abandoned.

In spite of Euler's assertion that a single aether should serve for all purposes, there was in the latter part of the eighteenth century a tendency in some quarters to add to the number of aethers by postulating one whose function was to transmit radiant heat, e.g. from a fire through the air to surrounding bodies. It was, however, shown in 1800[3] by Sir William Herschel (1738–1822) that rays of ordinary light transmit a certain amount of heat, that this effect is more marked for light at the red end of the spectrum than for light at the violet end, and that beyond the red end there are rays which transmit heat but do not affect the human sense of vision: these rays can be reflected and refracted like ordinary light. An obvious inference (which Herschel at first accepted, but later rejected for reasons which need not be discussed here) is that radiant heat is essentially of the same nature as light, and therefore does not

[1] Courtivron's *Traité d'optique* (1752)

[2] *Phil. Trans.* xlviii (1753), p. 262

[3] *Phil. Trans.* xc (1800), pp. 255, 284, 293, 437

require a separate aether. This identification was for long regarded as doubtful, but in the course of the next half-century it was shown experimentally [1] that radiant heat has all the characteristic properties of light (polarisation, double refraction, interference), and the view originally suggested by Herschel's work came to be accepted universally. It was shown in the following year by J. W. Ritter [2] that invisible rays exist also beyond the violet end of the spectrum, capable of chemical actions such as blackening chloride of silver: and Young showed [3] by projecting Newton's rings on paper covered with silver chloride) that these invisible rays have the property of interference (for which see below).

The fortunes of the wave-theory began to brighten at the end of the century, when a new champion arose. Thomas Young, born at Milverton in Somersetshire in 1773, and trained to the practice of medicine, began to write on optical theory in 1799. In his first paper [4] he remarked that, according to the corpuscular theory, the velocity of emission of a corpuscle must be the same in all cases, whether the projecting force be that of the feeble spark produced by the friction of two pebbles, or the intense heat of the sun itself—a thing almost incredible. This difficulty does not exist in the undulatory theory, since all disturbances are known to be transmitted through an elastic fluid with the same velocity. The reluctance which some philosophers felt to filling all space with an elastic fluid he met with an argument which strangely foreshadows the electric theory of light: 'That a medium resembling in many properties that which has been denominated ether does really exist, is undeniably proved by the phenomena of electricity. The rapid transmission of the electrical shock shows that the electric medium is possessed of an elasticity as great as is necessary to be supposed for the propagation of light. Whether the electric ether is to be considered the same with the luminous ether, if such a fluid exists, may perhaps at some future time be discovered by experiment: hitherto I have not been able to observe that the refractive power of a fluid undergoes any change by electricity.'

Young then proceeds to show the superior power of the wave-

[1] J. E. Bérard, *Ann. chim.* lxxxv (1813), p. 309
A. M. Ampère, *Bibl. univ.* xlviii (1832), p. 225; *Ann. d. phys.* xxvi (1832), p. 161; *Ann. chim.* lviii (1835), p. 432
J. D. Forbes, *Phil. Mag.* v, p. 209; vii (1835), p. 349
M. Melloni, *Ann. chim.* lxv (1837), p. 5, and many other papers in the same journal
H. L. Fizeau and L. Foucault, *Comptes Rendus* xxv (1847), p. 447
F. de la Provostaye and P. Desains, *Ann. chim.* xxvii (1849), p. 109

[2] *Ann. d. Phys.* vii (1801), p. 527

[3] *Phil. Trans.* xciv (1804), p. 1

[4] *Phil. Trans.* xc (1800), p. 106

theory to explain reflection and refraction. In the corpuscular theory it is difficult to see why part of the light should be reflected and another part of the same beam refracted ; but in the undulatory theory there is no trouble, as is shown by analogy with the partial reflection of sound from a cloud or denser stratum of air : 'Nothing more is necessary than to suppose all refracting media to retain, by their attraction, a greater or less quantity of the luminous ether, so as to make its density greater than that which it possesses in a vacuum, without increasing its elasticity.' This is precisely the hypothesis adopted later by Fresnel and Green.

In 1801 Young made a discovery of the first magnitude [1] when attempting to explain Newton's rings on the principles of the wave-theory. Rejecting Euler's hypothesis of induced vibrations, he assumed that the colours observed all exist in the incident light, and showed that they could be derived from it by a process which was now for the first time recognised in optical science.

The idea of this process was not altogether new, for it had been used by Newton in his theory of the tides. 'It may happen,' he wrote,[2] 'that the tide may be propagated from the ocean through different channels towards the same port, and may pass in less time through some channels than through others, in which case the same generating tide, being thus divided into two or more succeeding one another, may produce by composition new types of tide.' Newton applied this principle to explain the anomalous tides at Batsha in Tonkin, which had previously been described by Halley.[3]

Young's own illustration of the principle is evidently suggested by Newton's. 'Suppose,' he says,[4] 'a number of equal waves of water to move upon the surface of a stagnant lake, with a certain constant velocity, and to enter a narrow channel leading out of the lake ; suppose then another similar cause to have excited another equal series of waves, which arrive at the same channel, with the same velocity, and at the same time with the first. Neither series of waves will destroy the other, but their effects will be combined ; if they enter the channel in such a manner that the elevations of one series coincide with those of the other, they must together produce a series of greater joint elevations ; but if the elevations of one series are so situated as to correspond to the depressions of the other, they must exactly fill up those depressions, and the surface of the water must remain smooth. Now I maintain that similar

[1] *Phil. Trans.* xcii (1802), pp. 12, 387
[2] *Principia*, bk. iii, prop. 24
[3] *Phil. Trans.* xiv (1684), p. 681
[4] Young's *Works*, i, p. 202

effects take place whenever two portions of light are thus mixed; and this I call the general law of the *interference* of light.'

Thus, 'whenever two portions of the same light arrive to the eye by different routes, either exactly or very nearly in the same direction, the light becomes most intense when the difference of the routes is any multiple of a certain length, and least intense in the intermediate state of the interfering portions; and this length is different for light of different colours.'

Young's explanation of the colours of thin plates as seen by reflection was, then, that the incident light gives rise to two beams which reach the eye: one of these beams has been reflected at the first surface of the plate, and the other at the second surface; and these two beams produce the colours by their interference.

One difficulty encountered in reconciling this theory with observation arose from the fact that the central spot in Newton's rings (where the thickness of the thin film of air is zero) is black and not white, as it would be if the interfering beams were similar to each other in all respects. To account for this Young showed, by analogy with the impact of elastic bodies, that when light is reflected at the surface of a denser medium, its phase is retarded by half an undulation: so that the interfering beams at the centre of Newton's rings destroy each other. The correctness of this assumption he verified by substituting essence of sassafras (whose refractive index is intermediate between those of crown and flint glass) for air in the space between the lenses; as he anticipated, the centre of the ring system was now white.

Newton had long before observed that the rings are smaller when the medium producing them is optically more dense. Interpreted by Young's theory, this definitely proved that the wave-length of light is shorter in dense media, and therefore that its velocity is less.

The publication of Young's papers occasioned a fierce attack on him in the *Edinburgh Review*, from the pen of Henry Brougham, afterwards Lord Chancellor of England. Young replied in a pamphlet, of which it is said [1] that only a single copy was sold; and there can be no doubt that Brougham for the time being achieved his object of discrediting the wave-theory.[2] Its failure to make much impression in the years immediately following its appearance was strikingly shown in 1807–10, when Sir William Herschel

[1] Peacock's *Life of Young*

[2] 'Strange fellow,' wrote Macaulay, when half a century afterwards he found himself sitting beside Brougham in the House of Lords, 'his powers gone: his spite immortal.'

published three papers [1] on Newton's rings. Herschel ignored Young's theory altogether, and, in place of Newton's fits of easy reflection and easy transmission, proposed an explanation based on 'the principle of a critical separation of the colours which takes place at certain angles of incidence, admitting the reflection of some rays at the same angle of incidence at which others are transmitted.'

Young now turned his attention to the fringes of shadows. In the corpuscular explanation of these, it was supposed that the attractive forces which operate in refraction extend their influence to some distance from the surfaces of bodies, and inflect such rays as pass close by. If this were the case, the amount of inflection should obviously depend on the strength of the attractive forces, and consequently on the refractive indices of the bodies—a proposition which had been refuted by the experiments of 'sGravesande. The cause of diffraction effects was thus wholly unknown, until Young, in the Bakerian lecture for 1803,[2] showed that the principle of interference is concerned in their formation; for when a hair is placed in the cone of rays diverging from a luminous point, the internal fringes (i.e. those within the geometrical shadow) disappear when the light passing on one side of the hair is intercepted. His conjecture as to the origin of the interfering rays was not so fortunate; for he attributed the fringes outside the geometrical shadow to interference between the direct rays and rays reflected at the diffracting edge; and supposed the internal fringes of the shadow of a narrow object to be due to the interference of rays inflected by the two edges of the object.

The success of so many developments of the wave-theory led Young to inquire more closely into its capacity for solving the chief outstanding problem of optics—that of the behaviour of light in crystals. The beautiful construction for the extraordinary ray given by Huygens had lain neglected for a century; and the degree of accuracy with which it represented the observations was unknown. At Young's suggestion Wollaston [3] investigated the matter experimentally, and showed that the agreement between his own measurements and Huygens' rule was remarkably close. 'I think,' he wrote, 'the result must be admitted to be highly favourable to the Huygenian theory; and, although the existence of two refractions at the same time, in the same substance, be not well accounted for, and still less their interchange with each other, when a ray of light

[1] *Phil. Trans.* xcvii (1807), p. 180; xcix (1809), p. 259; c (1810), p. 149
[2] *Phil. Trans.* xciv (1804), p. 1; Young's *Works*, i, p. 179
[3] *Phil. Trans.* xcii (1802), p. 381

is made to pass through a second piece of spar situated transversely to the first, yet the oblique refraction, when considered alone, seems nearly as well explained as any other optical phenomenon.'

Meanwhile the advocates of the corpuscular theory were not idle; and in the next few years a succession of discoveries on their part, both theoretical and experimental, seemed likely to imperil the good position to which Young had advanced the rival hypothesis.

The first of these was a dynamical explanation of the refraction of the extraordinary ray in crystals, which was published in 1808 by Laplace.[1] His method is an extension of that by which Maupertuis had accounted for the refraction of the ordinary ray, and which since Maupertuis' day had been so developed that it was now possible to apply it to problems of all degrees of complexity. Laplace assumes that the crystalline medium acts on the light corpuscles of the extraordinary ray so as to modify their velocity, in a ratio which depends on the inclination of the extraordinary ray to the axis of the crystal; so that, in fact, the difference of the squares of the velocities of the ordinary and extraordinary rays is proportional to the square of the sine of the angle which the latter ray makes with the axis. The principle of least action then leads to a law of refraction identical with that found by Huygens' construction with the spheroid; just as Maupertuis' investigation led to a law of refraction for the ordinary ray identical with that found by Huygens' construction with the sphere.

The law of refraction for the extraordinary ray may also be deduced from Fermat's principle of least time, provided that the velocity is taken inversely proportional to that assumed in the principle of least action; and the velocity appropriate to Fermat's principle agrees with that found by Huygens, being, in fact, proportional to the radius of the spheroid. These results are obvious extensions of those already obtained for ordinary refraction.

Laplace's theory was promptly attacked by Young,[2] who pointed out the improbability of such a system of forces as would be required to impress the requisite change of velocity on the light corpuscles. If the aim of controversial matter is to convince the contemporary world, Young's paper must be counted unsuccessful; but it permanently enriched science by proposing a dynamical foundation for double refraction on the principles of the wave-theory. 'A

[1] *Mém. de l'Inst.* (1809), p. 300; *Journal de Physique*, lxviii (Jan. 1809), p. 107; *Mém. de la Soc. d'Arcueil*, ii

[2] *Quarterly Review*, Nov. 1809; Young's *Works*, i, p. 220

solution,' he says, ' might be deduced upon the Huygenian principles, from the simplest possible supposition, that of a medium more easily compressible in one direction than in any direction perpendicular to it, as if it consisted of an infinite number of parallel plates connected by a substance somewhat less elastic. Such a structure of the elementary atoms of the crystal may be understood by comparing them to a block of wood or of mica. Mr Chladni found that the mere obliquity of the fibres of a rod of Scotch fir reduced the velocity with which it transmitted sound in the proportion of 4 to 5. It is therefore obvious that a block of such wood must transmit every impulse in spheroidal—that is, oval—undulations ; and it may also be demonstrated, as we shall show at the conclusion of this article, that the spheroid will be truly elliptical when the body consists either of plane and parallel strata, or of equidistant fibres, supposing both to be extremely thin, and to be connected by a less highly elastic substance ; the spheroid being in the former case oblate and in the latter oblong.' Young then proceeds to a formal proof that ' an impulse is propagated through every perpendicular section of a lamellar elastic substance in the form of an elliptic undulation.' This must be regarded as the beginning of the dynamical theory of light in crystals. It was confirmed in a striking way not long afterwards by Brewster,[1] who found that compression in one direction causes an isotropic transparent solid to become doubly refracting.

Meanwhile, in January 1808, the French Academy had proposed as the subject for the physical prize in 1810, ' To furnish a mathematical theory of double refraction, and to confirm it by experiment.' Among those who resolved to compete was Etienne-Louis Malus (1775–1812), a colonel of engineers who had seen service with Napoleon's expedition to Egypt. While conducting experiments towards the end of 1808 in a house in the Rue des Enfers in Paris, Malus happened to analyse with a rhomb of Iceland spar the light of the setting sun reflected from the window of the Luxembourg, and was surprised to notice that the two images were of very different intensities. Following up this observation, he found that light which had been reflected from glass acquires thereby a modification similar to that which Huygens had noticed in rays which have experienced double refraction, and which Newton had explained by supposing rays of light to have ' sides.' This discovery appeared so important that without waiting for the prize competition he communicated it to the Academy in December 1808, and published

[1] *Phil. Trans.* cv (1815), p. 60

it in the following month.[1] 'I have found,' he said, 'that this singular disposition, which has hitherto been regarded as one of the peculiar effects of double refraction, can be completely impressed on the luminous molecules by all transparent solids and liquids. For example, light reflected by the surface of water at an angle of 52° 45′ has all the characteristics of one of the beams produced by the double refraction of Iceland spar, whose principal section is parallel to the plane which passes through the incident ray and the reflected ray. If we receive this reflected ray on any doubly refracting crystal, whose principal section is parallel to the plane of reflection, it will not be divided into two beams as a ray of ordinary light would be, but will be refracted according to the ordinary law.'

After this Malus found that light which has been refracted at the surface of any transparent substance likewise possesses in some degree this property, to which he gave the name *polarisation.* The memoir [2] which he finally submitted to the Academy, and which contains a rich store of experimental and analytical work on double refraction, obtained the prize in 1810 ; its immediate effect as regards the rival theories of the ultimate nature of light was to encourage the adherents of the corpuscular doctrine ; for it brought into greater prominence the phenomena of polarisation, of which the wave-theorists, still misled by the analogy of light with sound, were unable to give any account.

The successful discoverer was elected to the Academy of Sciences, and became a member of the celebrated club of Arcueil.[3] But his health, which had been undermined by the Egyptian campaign, now broke down completely ; and he died, at the age of thirty-six, in the following year.

The polarisation of a reflected ray is in general incomplete—i.e. the ray displays only imperfectly the properties of light which has been polarised by double refraction ; but for one particular angle of incidence, which depends on the reflecting body, the polarisation of the reflected ray is complete. Malus measured with considerable accuracy the polarising angles for glass and water, and attempted to connect them with the other optical constants of these substances, the refractive indices and dispersive powers, but without success. The matter was afterwards taken up by David

[1] *Nouveau Bulletin des Sciences, par la Soc. Philomatique,* i (1809), p. 266 ; *Mémoires de la Soc. d'Arcueil,* ii, 1809

[2] *Mém. présentés à l'Inst. par divers Savans,* ii (1811), p. 303

[3] So called from the village near Paris where Laplace and Berthollet had their country houses, and where the meetings took place. The club consisted of a dozen of the most celebrated scientific men in France.

Brewster (1781–1868), who in 1815 [1] showed that there is complete polarisation by reflection when the reflected and refracted rays satisfy the condition of being at right angles to each other.

Almost at the same time Brewster made another discovery which profoundly affected the theory of double refraction. It had till then been believed that double refraction is always of the type occurring in Iceland spar, to which Huygens' construction is applicable. Brewster now found this belief to be erroneous, and showed that in a large class of crystals there are two axes, instead of one, along which there is no double refraction. Such crystals are called *biaxal*, the simpler type to which Iceland spar belongs being called *uniaxal*.

The wave-theory at this time was still encumbered with difficulties. Diffraction was not satisfactorily explained ; for polarisation no explanation of any kind was forthcoming ; the Huygenian construction appeared to require two different luminiferous media within doubly refracting bodies ; and the universality of that construction had been impugned by Brewster's discovery of biaxal crystals.

The upholders of the emission theory, emboldened by the success of Laplace's theory of double refraction, thought the time ripe for their final triumph ; and as a step to this, in March 1817 they proposed Diffraction as the subject of the Academy's prize for 1818. Their expectation was disappointed ; and the successful memoir afforded the first of a series of reverses by which, in the short space of seven years, the corpuscular theory was completely overthrown.

The author was Augustin Fresnel (1788–1827), the son of an architect, and himself a civil engineer in the Government service in Normandy. During the brief dominance of Napoleon after his escape from Elba in 1815, Fresnel fell into trouble for having enlisted in the small army which attempted to bar the exile's return ; and it was during a period of enforced idleness following on his arrest that he commenced to study diffraction. In his earliest memoir [2] he propounded a theory similar to that of Young, which was spoiled like Young's theory by the assumption that the fringes depend on light reflected by the diffracting edge. Observing, however, that the blunt and sharp edges of a knife produce exactly the same fringes, he became dissatisfied with this attempt, and on 15 July 1816, presented to the Academy a supplement to his paper,[3] in which, for the first time, diffraction-effects are referred to their

[1] *Phil. Trans.* cv (1815), p. 125

[2] *Annales de chimie* (2), i (1816), p. 239 ; *Œuvres*, i, p. 89

[3] *Œuvres*, i, p. 129

true cause—namely, the mutual interference of the secondary waves emitted by those portions of the original wave-front which have not been obstructed by the diffracting screen. Fresnel's method of calculation utilised the principles of both Huygens and Young; he summed the effects due to different portions of the same primary wave-front.

The sketch presented to the Academy in 1816 was during the next two years developed into an exhaustive memoir,[1] which was submitted for the Academy's prize.

It so happened that the earliest memoir, which had been presented to the Academy in the autumn of 1815, had been referred to a Commission of which the reporter was François Arago (1786–1853); Arago was so much impressed that he sought the friendship of the author, of whom he was later a strenuous champion.

A champion was indeed needed when the larger memoir was submitted; for Laplace, Poisson and Biot, who constituted a majority of the Commission to which it was referred, were all zealous supporters of the corpuscular theory. During the examination, however, Fresnel was vindicated in a somewhat curious way. He had calculated in the memoir the diffraction-patterns of a straight edge, of a narrow opaque body bounded by parallel sides, and of a narrow opening bounded by parallel edges, and had shown that the results agreed excellently with his experimental measures. Poisson, when reading the manuscript, happened to notice that the analysis could be extended to other cases, and in particular that it would indicate the existence of a bright spot at the centre of the shadow of a circular screen. He suggested to Fresnel that this and some further consequences should be tested experimentally; this was done,[2] and the results were found to confirm the new theory. The concordance of observation and calculation was so admirable in all cases where a comparison was possible that the prize was awarded to Fresnel without further hesitation.

In the same year in which the memoir on diffraction was submitted, Fresnel published an investigation [3] of the influence of the earth's motion on light. We have already seen that aberration was explained by its discoverer in terms of the corpuscular theory; and it was Young who first showed [4] how it may be explained on the wave-hypothesis. 'Upon considering the phenomena of the

[1] *Mém. de l'Acad.* v (1826), p. 339; *Œuvres*, i, p. 247

[2] The bright spot in the centre of the shadow had been noticed in the early part of the eighteenth century by J. N. Delisle.

[3] *Annales de Chimie*, ix (1818), p. 57; *Œuvres*, ii, p. 627

[4] *Phil. Trans.* xciv (1804), p. 1; Young's *Works*, i, p. 188

aberration of the stars,' he wrote, 'I am disposed to believe that the luminiferous aether pervades the substance of all material bodies with little or no resistance, as freely perhaps as the wind passes through a grove of trees.' In fact, if we suppose the aether surrounding the earth to be at rest and unaffected by the earth's motion, the light waves will not partake of the motion of the telescope, which we may suppose directed to the true place of the star, and the image of the star will therefore be displaced from the central spider-line at the focus by a distance equal to that which the earth describes while the light is travelling through the telescope. This agrees with what is actually observed.

But a host of further questions now suggest themselves. Suppose, for instance, that a slab of glass with a plane face is carried along by the motion of the earth, and it is desired to adjust it so that a ray of light coming from a certain star shall not be bent when it enters the glass ; must the surface be placed at right angles to the true direction of the star as freed from aberration, or to its apparent direction as affected by aberration? The question whether rays coming from the stars are refracted differently from rays originating in terrestrial sources had been raised originally by Michell [1]; and Robison and Wilson [2] had asserted that the focal length of an achromatic telescope should be increased when it is directed to a star towards which the earth is moving, owing to the change in the relative velocity of light. Arago [3] submitted the matter to the test of experiment, and concluded that the light coming from any star behaves in all cases of reflection and refraction precisely as it would if the star were situated in the place which it appears to occupy in consequence of aberration, and the earth were at rest; so that the apparent refraction in a moving prism is equal to the absolute refraction in a fixed prism.

Fresnel now set out to provide a theory capable of explaining Arago's result. To this end he adopted Young's suggestion, that the refractive powers of transparent bodies depend on the concentration of aether within them; and made it more precise by assuming that the aethereal density in any body is proportional to the square of the refractive index. Thus, if c denote the velocity of light *in vacuo*, and if c_1 denote its velocity in a given material body at rest, so that $\mu = c/c_1$ is the refractive index, then the densities

[1] *Phil. Trans.* lxxiv (1784), p. 35 [2] *Trans. R. S.* (Edin.), i, Hist., p. 30

[3] Arago communicated his results to the Institute in 1810, but did not publish them until much later, in *Comptes Rendus* viii (1839), 326 and xxxvi (1853), 38. cf. Biot, *Astron. phys.* (3rd edn.), v, p. 364. The accuracy of Arago's experiment can scarcely have been such as to demonstrate absolutely his result.

ρ and ρ_1 of the aether in interplanetary space and in the body respectively will be connected by the relation

$$\rho_1 = \mu^2 \rho.$$

Fresnel further assumed that, when a body is in motion, part of the aether within it is carried along—namely, that part which constitutes the excess of its density over the density of aether *in vacuo*; while the rest of the aether within the space occupied by the body is stationary. Thus the density of aether carried along is $(\rho_1 - \rho)$ or $(\mu^2 - 1)\rho$, while a quantity of aether of density ρ remains at rest. The velocity with which the centre of gravity of the aether within the body moves forward in the direction of propagation is therefore

$$\frac{\mu^2 - 1}{\mu^2} w,$$

where w denotes the component of the velocity of the body in this direction. This is to be added to the velocity of propagation of the light waves within the body; so that in the moving body the absolute velocity of light is

$$c_1 + \frac{\mu^2 - 1}{\mu^2} w.$$

Many years afterwards Stokes[1] put the same supposition in a slightly different form. Suppose the whole of the aether within the body to move together, the aether entering the body in front, and being immediately condensed, and issuing from it behind, where it is immediately rarefied. On this assumption a mass ρw of aether must pass in unit time across a plane of area unity, drawn anywhere within the body in a direction at right angles to the body's motion; and therefore the aether within the body has a drift-velocity $-w\rho/\rho_1$ relative to the body; so the velocity of light relative to the body will be $c_1 - w\rho/\rho_1$, and the absolute velocity of light in the moving body will be

$$c_1 + w - \frac{w\rho}{\rho_1},$$

or

$$c_1 + \frac{\mu^2 - 1}{\mu^2} w, \qquad \text{as before.}$$

This formula was experimentally confirmed in 1851 by H. Fizeau,[2] who measured the displacement of interference-fringes

[1] *Phil. Mag.* xxviii (1846), p. 76

[2] *Annales de Chimie*, lvii (1859), p. 385. Also by A. A. Michelson and E. W. Morley, *Am. Journ. Science*, xxxi (1886), p. 377.

formed by light which had passed through a column of moving water.

The same result may easily be deduced from an experiment performed by Hoek.[1] In this a beam of light was divided into two portions, one of which was made to pass through a tube of water AB and was then reflected at a mirror C, the light being afterwards allowed to return to A without passing through the water; while the other portion of the bifurcated beam was made to describe the same path in the reverse order, i.e. passing through the water on its return journey from C instead of on the outward journey. On causing the two portions of the beam to interfere, Hoek found that no difference of phase was produced between them when the apparatus was oriented in the direction of the terrestrial motion.

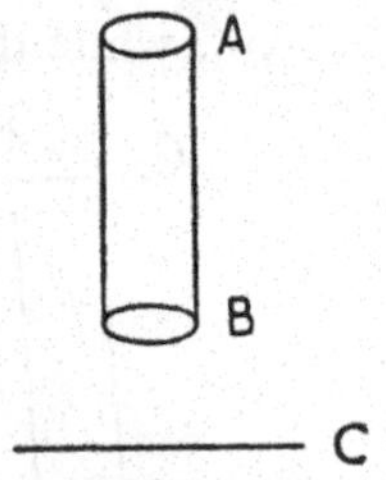

Let w denote the velocity of the earth, supposed to be directed from the tube towards the mirror. Let c/μ denote the velocity of light in the water at rest, and $c/\mu + \phi$ the velocity of light in the water when moving. Let l denote the length of the tube. The magnitude of the distance BC does not affect the experiment, so we may suppose it zero.

The time taken by the first portion of the beam to perform its journey is evidently

$$\frac{l}{c/\mu + \phi - w} + \frac{l}{c + w},$$

while the time for the second portion of the beam is

$$\frac{l}{c - w} + \frac{l}{c/\mu - \phi + w}.$$

The equality of these expressions gives at once, when terms of higher orders than the first in w/c are neglected.

$$\phi = (\mu^2 - 1)\, w/\mu^2,$$

which is Fresnel's expression.[2]

On the basis of this formula, Fresnel proceeded to solve the problem of refraction in moving bodies. Suppose that a prism

[1] *Archives Néerl.* iii (1868), p. 180

[2] Fresnel's law may also be deduced from the principle that the *amount* of light transmitted by a slab of transparent matter must be the same whether the slab is at rest or in motion; otherwise the equilibrium of exchanges of radiation would be vitiated. cf. Larmor, *Phil. Trans.* clxxxv (1893), p. 775.

$A_0C_0B_0$ is carried along by the earth's motion *in vacuo*, its face A_0C_0 being at right angles to the direction of motion ; and that light from a star is incident normally on this face. The rays experience no refraction at incidence ; and we have only to consider the effect produced by the second surface A_0B_0. Suppose that during an interval τ of time the prism travels from the position $A_0C_0B_0$ to the position $A_1C_1B_1$, while the luminous disturbance at C_0 travels to B_1, and the

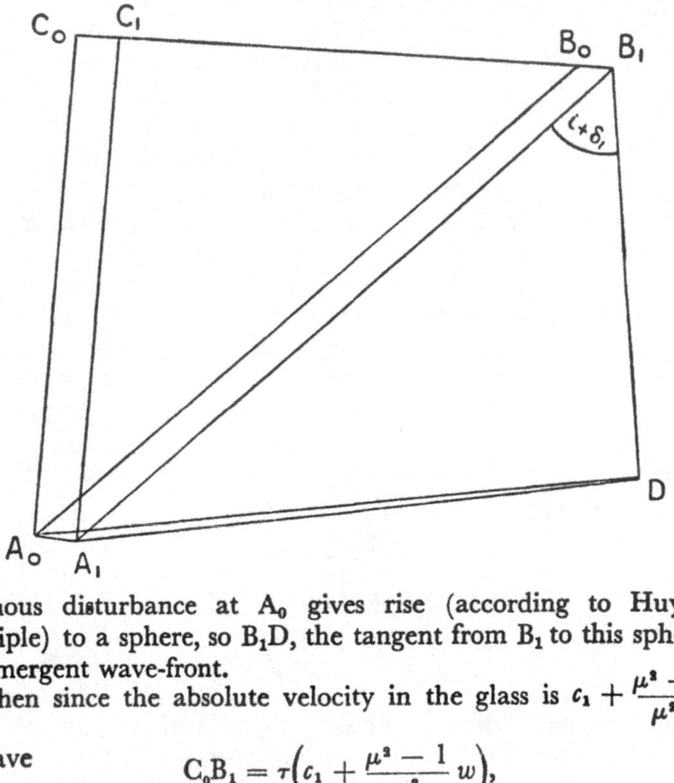

luminous disturbance at A_0 gives rise (according to Huygens' Principle) to a sphere, so B_1D, the tangent from B_1 to this sphere, is the emergent wave-front.

Then since the absolute velocity in the glass is $c_1 + \frac{\mu^2 - 1}{\mu^2} w$,

we have
$$C_0B_1 = \tau\left(c_1 + \frac{\mu^2 - 1}{\mu^2} w\right),$$
$$A_0D = \tau c,$$
$$A_0A_1 = \tau w.$$

If we write $C_1\hat{A}_1B_1 = i$, and denote the total deviation of the wave-front by δ_1, we have

$$A_1D = A_0D - A_1A_0 \cos \delta_1 = \tau c - \tau w \cos \delta_1,$$

$$C_1B_1 = C_0B_1 - C_0C_1 = \tau\left(c_1 + \frac{\mu^2 - 1}{\mu^2} w - w\right) = \tau\left(c_1 - \frac{w c_1^2}{c^2}\right) \text{ since } \mu = \frac{c}{c_1}$$

and therefore (neglecting second-order terms in w/c)

$$\frac{\sin(i+\delta_1)}{\sin i} = \frac{\sin A_1\hat{B}_1D}{\sin i} = \frac{\left(A_1D \big/ A_1B_1\right)}{\left(C_1B_1 \big/ A_1B_1\right)} = \frac{A_1D}{C_1B_1} = \frac{c - w\cos\delta_1}{c_1 - w\dfrac{c_1^2}{c_1^2}}$$

$$= \frac{c}{c_1} + \frac{w}{c} - \frac{w}{c_1}\cos\delta_1$$

Denoting by δ the value of δ_1 when w is zero, we have therefore

$$\frac{\sin(i+\delta)}{\sin i} = \frac{c}{c_1}.$$

Subtracting this equation from the preceding, we have

$$\frac{(\delta-\delta_1)\cos(i+\delta)}{\sin i} = -\frac{w}{c} + \frac{w}{c_1}\cos\delta = -\frac{w}{c} + \frac{w}{c}\frac{\sin(i+\delta)\cos\delta}{\sin i}$$

whence

$$\frac{\delta-\delta_1}{\sin\delta} = \frac{w}{c}.$$

Now the telescope by which the emergent wave-front B_1D is received is itself being carried forward by the earth's motion ; and we must therefore apply the usual correction for aberration in order to find the apparent direction of the emergent ray. But this correction is $w \sin \delta/c$, and precisely counteracts the effect which has been calculated as due to the motion of the prism. So finally we see that the motion of the earth has no first-order influence on the refraction of light from the stars.

Fresnel inferred from his formula that if observations were made with a telescope filled with water, the aberration would be unaffected by the presence of the water—a result which was verified by Airy [1] in 1871. He showed, moreover, that the apparent positions of terrestrial objects, carried along with the observer, are not displaced by the earth's motion ; that experiments in refraction and interference are not influenced by any motion which is common to the source, apparatus and observer ; and that light travels between given points of a moving material system by the path of least time. These predictions have also been confirmed by observation : Respighi [2] in 1861, and Hoek [3] in 1868, experimenting with a telescope filled with water and a terrestrial source of light, found that no effect was produced on the phenomena of reflection and refraction by altering the orientation of the apparatus relative to the direction

[1] *Proc. R. S.* xx (1871), p. 35

[2] *Mem. Accad. Sci. Bologna,* ii (1861), p. 279

[3] *Ast. Nach.* lxxiii (1868), p. 193

of the earth's motion. E. Mascart [1] in 1872 studied experimentally the question of the effect of motion of the source or recipient of light in all its bearings, and showed that the light of the sun and that derived from artificial sources are alike incapable of revealing by diffraction-phenomena the translatory motion of the earth.

The greatest problem now confronting the investigators of light was to reconcile the facts of polarisation with the principles of the wave-theory. Young had long been pondering over this, but had hitherto been baffled by it. In 1816 he received a visit from Arago, who told him of a new experimental result which he and Fresnel had lately obtained [2]—namely, that two pencils of light, polarised in planes at right angles, do not interfere with each other under circumstances in which ordinary light shows interference-phenomena, but always give by their reunion the same intensity of light, whatever be their difference of path.

Arago had not long left him when Young, reflecting on the new experiment, discovered the long-sought key to the mystery: it consisted in the very alternative which Bernoulli had rejected eighty years before, of supposing that the vibrations of light are executed at right angles to the direction of propagation.

Young's ideas were first embodied in a letter to Arago,[3] dated 12 January 1817. 'I have been reflecting,' he wrote, 'on the possibility of giving an imperfect explanation of the affection of light which constitutes polarisation, without departing from the genuine doctrine of undulations. It is a principle in this theory, that all undulations are simply propagated through homogeneous mediums in concentric spherical surfaces like the undulations of sound, consisting simply in the direct and retrograde motions of the particles in the direction of the radius, with their concomitant condensation and rarefactions. And yet it is possible to explain in this theory a transverse vibration, propagated also in the direction of the radius, and with equal velocity, the motions of the particles being in a certain constant direction with respect to that radius; and this is a *polarisation*.'

In an article on 'Chromatics,' which was written in September

[1] *Ann. de l'Ecole Normale* (2), i (1872), p. 157

[2] It was not published until 1819, in *Annales de Chimie*, x; Fresnel's *Œuvres*, i, p. 509. By means of this result, Fresnel was able to give a complete explanation of a class of phenomena which Arago had discovered in 1811, viz. that when polarised light is transmitted through thin plates of sulphate of lime or mica, and afterwards analysed by a prism of Iceland spar, beautiful complementary colours are displayed. Young had shown that these effects are due essentially to interference, but had not made clear the part played by polarisation.

[3] Young's *Works*, i, p. 380

of the same year [1] for the supplement to the *Encyclopaedia Britannica*, he says [2]: 'If we assume as a mathematical postulate, on the undulating theory, without attempting to demonstrate its physical foundation, that a transverse motion may be propagated in a direct line, we may derive from this assumption a tolerable illustration of the subdivision of polarised light by reflection in an oblique plane,' by 'supposing the polar motion to be resolved' into two constituents, which fare differently at reflection.

In a further letter to Arago, dated 29 April 1818 Young recurred to the subject of transverse vibrations, comparing light to the undulations of a cord agitated by one of its extremities.[3] This letter was shown by Arago to Fresnel, who at once saw that it presented the true explanation of the non-interference of beams polarised in perpendicular planes, and that the latter effect could even be made the basis of a proof of the correctness of Young's hypothesis; for if the vibration of each beam be supposed resolved into three components, one along the ray and the other two at right angles to it, it is obvious from the Arago-Fresnel experiment that the components in the direction of the ray must vanish; in other words, that the vibrations which constitute light are executed in the wave-front.

It must be remembered that the theory of the propagation of waves in an elastic solid was as yet unknown, and light was still always interpreted by the analogy with the vibrations of sound in air, for which the direction of vibration is the same as that of propagation. It was therefore necessary to give some justification for the new departure. With wonderful insight Fresnel indicated [4] the precise direction in which the theory of vibrations in ponderable bodies needed to be extended in order to allow of waves similar to those of light: 'the geometers,' he wrote, 'who have discussed the vibrations of elastic fluids hitherto have taken account of no accelerating forces except those arising from the difference of condensation or dilatation between consecutive layers.' He pointed out that if we also suppose the medium to possess a rigidity, or power of resisting distortion, such as is manifested by all actual solid bodies, it will be capable of transverse vibration. The absence

[1] Peacock's *Life of Young*, p. 391

[2] Young's *Works*, i, p. 279

[3] This analogy had been given by Hooke in a communication to the Royal Society on 15 Feb. 1671–2. But there seems no reason to suppose that Hooke appreciated the point now advanced by Young.

[4] *Annales de Chimie*, xvii (1821), p. 180; *Œuvres*, i, p. 629. Young had already drawn attention to this point. 'It is difficult,' he says in his *Lectures on Natural Philosophy* (ed. 1807), vol. i, p. 138, 'to compare the lateral adhesion, or the force which resists the detrusion of the parts of a solid, with any form of direct cohesion. This force constitutes the rigidity or hardness of a solid body, and is wholly absent from liquids.'

of longitudinal waves in the aether he accounted for by supposing that the forces which oppose condensation are far more powerful than those which oppose distortion, and that the velocity with which condensations are propagated is so great compared with the speed of the oscillations of light, that a practical equilibrium of pressure is maintained perpetually.

The nature of ordinary non-polarised light was next discussed. 'If then,' Fresnel wrote,[1] 'the polarisation of a ray of light consists in this, that all its vibrations are executed in the same direction, it results from any hypothesis on the generation of light waves, that a ray emanating from a single centre of disturbance will always be polarised in a definite plane at any instant. But an instant afterwards, the direction of the motion changes, and with it the plane of polarisation; and these variations follow each other as quickly as the perturbations of the vibrations of the luminous particle; so that even if we could isolate the light of this particular particle from that of other luminous particles, we should doubtless not recognise in it any appearance of polarisation. If we consider now the effect produced by the union of all the waves which emanate from the different points of a luminous body, we see that at each instant, at a definite point of the aether, the general resultant of all the motions which commingle there will have a determinate direction, but this direction will vary from one instant to the next. So direct light can be considered as the union, or more exactly as the rapid succession, of systems of waves polarised in all directions. According to this way of looking at the matter, the act of polarisation consists not in creating these transverse motions, but in decomposing them in two invariable directions, and separating the components from each other; for then, in each of them, the oscillatory motions take place always in the same plane.'

He then proceeded to consider the relation of the direction of vibration to the plane of polarisation. 'Apply these ideas to double refraction, and regard a uniaxal crystal as an elastic medium in which the accelerating force which results from the displacement of a row of molecules perpendicular to the axis, relative to contiguous rows, is the same all round the axis; while the displacements parallel to the axis produce accelerating forces of a different intensity, stronger if the crystal is 'repulsive,' and weaker if it is 'attractive.' The distinctive character of the rays which are ordinarily refracted being that of propagating themselves with the same velocity in all directions, we must admit that their oscillatory motions are executed

[1] loc. cit. p. 185

at right angles to the plane drawn through these rays and the axis of the crystal; for then the displacements which they occasion, always taking place along directions perpendicular to the axis, will, by hypothesis, always give rise to the same accelerating forces. But, with the conventional meaning which is attached to the expression *plane of polarisation*, the plane of polarisation of the ordinary rays is the plane through the axis; thus, *in a pencil of polarised light, the oscillatory motion is executed at right angles to the plane of polarisation.*'

This result afforded Fresnel a foothold in dealing with the problem which occupied the rest of his life; henceforth his aim was to base the theory of light on the dynamical properties of the luminiferous medium.

The first topic which he attacked from this point of view was the propagation of light in crystalline bodies. Since Brewster's discovery that many crystals do not conform to the type to which Huygens' construction is applicable, the wave-theory had to some extent lost credit in this region. Fresnel, now, by what was perhaps the most brilliant of all his efforts,[1] not only reconquered the lost territory, but added a new domain to science.

He had, as he tells us himself, never believed the doctrine that in crystals there are two different luminiferous media, one to transmit the ordinary and the other the extraordinary waves. The alternative to which he inclined was that the two velocities of propagation were really the two roots of a quadratic equation, derivable in some way from the theory of a single aether. Could this equation be obtained, he was confident of finding the explanation, not only of double refraction, but also of the polarisation by which it is always accompanied.

The first step was to take the case of uniaxal crystals, which had been discussed by Huygens, and to see whether Huygens' sphere and spheroid could be replaced by, or made to depend on, a *single* surface.[2]

Now a wave propagated in any direction through a uniaxal crystal can be resolved into two plane-polarised components; one of these, the 'ordinary ray,' is polarised in the principal section,

[1] His first memoir on double refraction was presented to the Academy on 19 Nov. 1821, but has not been published except in his collected works, *Œuvres*, ii, p. 261. It was followed by other papers in 1822; and the results were finally collected in a memoir which was printed in 1827, *Mém. de l'Acad.* vii, p. 45; *Œuvres*, ii, p. 479.

[2] In attempting to reconstruct Fresnel's course of thought at this period, the present writer has derived much help from the Life prefixed to the *Œuvres de Fresnel*. Both Fresnel and Young were singularly fortunate in their biographers: Peacock's *Life of Young*, and this notice of Fresnel, which was the last work of Verdet, are excellent reading.

and has a velocity v_1, which may be represented by the radius of Huygens' sphere—say,

$$v_1 = b\,;$$

while the other, the 'extraordinary ray,' is polarised in a plane at right angles to the principal section, and has a wave-velocity v_2, which may be represented by the perpendicular drawn from the centre of Huygens' spheroid on the tangent-plane parallel to the plane of the wave. If the spheroid be represented by the equation

$$\frac{y^2 + z^2}{a^2} + \frac{x^2}{b^2} = 1,$$

and if (l, m, n) denote the direction cosines of the normal to the plane of the wave, we have therefore

$$v_2^2 = a^2(m^2 + n^2) + b^2 l^2.$$

But the quantities $1/v_1$ and $1/v_2$, as given by these equations, are easily seen to be the lengths of the semi-axes of the ellipse in which the spheroid

$$b^2(y^2 + z^2) + a^2x^2 = 1$$

is intersected by the plane

$$lx + my + nz = 0\,;$$

and thus the construction in terms of Huygens' sphere and spheroid can be replaced by one which depends only on a single surface, namely the spheroid

$$b^2(y^2 + z^2) + a^2x^2 = 1.$$

Having achieved this reduction, Fresnel guessed that the case of biaxal crystals could be covered by substituting for the latter spheroid an ellipsoid with three unequal axes—say,

$$\frac{x^2}{\epsilon_1} + \frac{y^2}{\epsilon_2} + \frac{z^2}{\epsilon_3} = 1.$$

If $1/v_1$ and $1/v_2$ denote the lengths of the semi-axes of the ellipse in which this ellipsoid is intersected by the plane

$$lx + my + nz = 0,$$

it is well known that v_1 and v_2 are the roots of the equation in v

$$\frac{l^2}{\dfrac{1}{\epsilon_1} - v^2} + \frac{m^2}{\dfrac{1}{\epsilon_2} - v^2} + \frac{n^2}{\dfrac{1}{\epsilon_3} - v^2} = 0\,;$$

and accordingly Fresnel conjectured that the roots of this equation represent the velocities, in a biaxal crystal, of the two plane-polarised waves whose normals are in the direction (l, m, n).

Having thus arrived at his result by reasoning of a purely geometrical character, he now devised a dynamical scheme to suit it.

The vibrating medium within a crystal he supposed to be ultimately constituted of particles subjected to mutual forces; and on this assumption he showed that the elastic force of restitution when the system is disturbed must depend linearly on the displacement. In this first proposition a difference is apparent between Fresnel's and a true elastic-solid theory; for in actual elastic solids the forces of restitution depend not on the absolute displacement, but on the strains, i.e. the *relative* displacements.

In any crystal there will exist three directions at right angles to each other, for which the force of restitution acts in the same line as the displacement: the directions which possess this property are named *axes of elasticity*. Let these be taken as axes, and suppose that the elastic forces of restitution for unit displacements in these three directions are $1/\epsilon_1$ $1/\epsilon_2$ $1/\epsilon_3$ respectively. That the elasticity should vary with the direction of the molecular displacement seemed to Fresnel to suggest that the molecules of the material body either take part in the luminous vibration, or at any rate influence in some way the elasticity of the aether.

A unit displacement in any arbitrary direction (α, β, γ) can be resolved into component displacements ($\cos\alpha$, $\cos\beta$, $\cos\gamma$) parallel to the axes, and each of these produces its own effect independently; so the components of the force of restitution are

$$\frac{\cos\alpha}{\epsilon_1}, \quad \frac{\cos\beta}{\epsilon_2}, \quad \frac{\cos\gamma}{\epsilon_3}.$$

This resultant force has not in general the same direction as the displacement which produced it; but it may always be decomposed into two other forces, one parallel and the other perpendicular to the direction of the displacement; and the former of these is evidently

$$\frac{\cos^2\alpha}{\epsilon_1} + \frac{\cos^2\beta}{\epsilon_2} + \frac{\cos^2\gamma}{\epsilon_3}.$$

The surface

$$r^4 = \frac{x^2}{\epsilon_1} + \frac{y^2}{\epsilon_2} + \frac{z^2}{\epsilon_3}$$

will therefore have the property that the square of its radius vector in any direction is proportional to the component in that direction of the elastic force due to a unit displacement in that direction: it is called the *surface of elasticity*.

Consider now a displacement along one of the axes of the section on which the surface of elasticity is intersected by the plane of the wave. It is easily seen that in this case the component of the elastic force at right angles to the displacement acts along the normal to the wave-front; and Fresnel assumes that it will be without influence on the propagation of the vibrations, on the ground of his fundamental hypothesis that the vibrations of light are performed solely in the wave-front. This step is evidently open to criticism; for in a dynamical theory everything should be deduced from the laws of motion without special assumptions. But granting his contention, it follows that such a displacement will retain its direction, and will be propagated as a plane-polarised wave with a definite velocity.

Now, in order that a stretched cord may vibrate with unchanged period, when its tension is varied, its length must be increased proportionally to the square root of its tension; and similarly the wave-length of a luminous vibration of given period is proportional to the square root of the elastic force (per unit displacement), which urges the molecules of the medium parallel to the wave-front. Hence the velocity of propagation of a wave, measured at right angles to its front, is proportional to the square root of the component, along the direction of displacement, of the elastic force per unit displacement; and the velocity of propagation of such a plane-polarised wave as we have considered is proportional to the radius vector of the surface of elasticity in the direction of displacement.

Moreover, any displacement in the given wave-front can be resolved into two, which are respectively parallel to the two axes of the diametral section of the surface of elasticity by a plane parallel to this wave-front; and it follows from what has been said that each of these component displacements will be propagated as an independent plane-polarised wave, the velocities of propagation being proportional to the axes of the section,[1] and therefore inversely proportional to the axes of the section of the inverse surface of this with respect to the origin, which is the ellipsoid

$$\frac{x^2}{\epsilon_1}+\frac{y^2}{\epsilon_2}+\frac{z^2}{\epsilon_3}=1.$$

But this is precisely the result to which, as we have seen, Fresnel had been led by purely geometrical considerations; and thus his geometrical conjecture could now be regarded as substantiated by a study of the dynamics of the medium.

[1] It is evident from this that the *optic axes*, or lines of single wave-velocity, along which there is no double refraction, will be perpendicular to the two circular sections of the surface of elasticity.

It is easy to determine the wave-surface or locus at any instant —say, $t = 1$—of a disturbance originated at some previous instant —say, $t = 0$—at some particular point—say, the origin. For this wave-surface will evidently be the envelope of plane waves emitted from the origin at the instant $t = 0$—that is, it will be the envelope of planes

$$lx + my + nz - v = 0,$$

where the constants l, m, n, v are connected by the identical equation

$$l^2 + m^2 + n^2 = 1,$$

and by the relation previously found—namely,

$$\frac{l^2}{\frac{1}{\epsilon_1} - v^2} + \frac{m^2}{\frac{1}{\epsilon_2} - v^2} + \frac{n^2}{\frac{1}{\epsilon_3} - v^2} = 0.$$

By the usual procedure for determining envelopes, it may be shown that the locus in question is the surface of the fourth degree

$$\frac{x^2}{\epsilon_1 r^2 - 1} + \frac{y^2}{\epsilon_2 r^2 - 1} + \frac{z^2}{\epsilon_3 r^2 - 1} = 0,$$

which is called *Fresnel's wave-surface.*[1] It is a two-sheeted surface, as must evidently be the case from physical considerations. In uniaxal crystals, for which ϵ_2 and ϵ_3 are equal, it degenerates into the sphere

$$r^2 = 1/\epsilon_2,$$

and the spheroid

$$\epsilon_2 x^2 + \epsilon_1 (y^2 + z^2) = 1.$$

It is to these two surfaces that tangent planes are drawn in the construction given by Huygens for the ordinary and extraordinary refracted rays in Iceland spar. As Fresnel observed, exactly the same construction applies to biaxal crystals, when the two sheets of the wave-surface are substituted for Huygens' sphere and spheroid.

'The theory which I have adopted,' says Fresnel at the end of this memorable paper, 'and the simple constructions which I have deduced from it, have this remarkable character, that all the unknown quantities are determined together by the solution of the problem. We find at the same time the velocities of the ordinary ray and of the extraordinary ray, and their planes of polarisation. Physicists

[1] Another construction for the wave-surface is the following, which is due to MacCullagh, *Coll. Works*, p. 1. Let the ellipsoid

$$\epsilon_1 x^2 + \epsilon_2 y^2 + \epsilon_3 z^2 = 1$$

be intersected by a plane through its centre, and on the perpendicular to that plane take lengths equal to the semi-axes of the section. The locus of these extremities is the wave surface.

who have studied attentively the laws of Nature will feel that such simplicity and such close relations between the different elements of the phenomenon are conclusive in favour of the hypothesis on which they are based.'

The question as to the correctness of Fresnel's construction was discussed for many years afterwards. A striking consequence of it was pointed out in 1832 by William Rowan Hamilton (1805–65), Royal Astronomer of Ireland, who remarked [1] that the surface defined by Fresnel's equation has four conical points, at each of which there is an infinite number of tangent planes; consequently, a single ray, proceeding from a point within the crystal in the direction of one of these points, must be divided on emergence into an infinite number of rays, constituting a conical surface. Hamilton also showed that there are four planes, each of which touches the wave-surface in an infinite number of points, constituting a circle of contact: so that a corresponding ray incident externally should be divided within the crystal into an infinite number of refracted rays, again constituting a conical surface.

These singular and unexpected consequences were shortly afterwards verified experimentally by Humphrey Lloyd,[2] and helped greatly to confirm belief in Fresnel's theory. It should, however, be observed that conical refraction only shows his form of the wave-surface to be correct in its general features, and is no test of its accuracy in all details. But it was shown experimentally by Stokes in 1872,[3] Glazebrook in 1879 [4] and Hastings in 1887,[5] that the construction of Huygens and Fresnel is certainly correct to a very high degree of approximation; and Fresnel's final formulae have since been regarded as unassailable. The dynamical substructure on which he based them is, as we have seen, open to objection; but, as Stokes observed [6]: 'If we reflect on the state of the subject as Fresnel found it, and as he left it, the wonder is, not that he failed to give a rigorous dynamical theory, but that a single mind was capable of effecting so much.'

In a second supplement to his first memoir on double refraction, presented to the Academy on 26 November 1821,[7] Fresnel indicated the lines on which his theory might be extended so as to take account

[1] *Trans. Roy. Irish Acad.* xvii (1833), p. 1; Hamilton's *Math. Papers*, i, p. 164

[2] *Trans. Roy. Irish Acad.* xvii (1833), p. 145. Strictly speaking, the bright cone which is usually observed arises from rays *adjacent* to the singular ray; the latter can, however, be observed, its enfeeblement by dispersion into the conical form causing it to appear dark.

[3] *Proc. R. S.* xx (1872), p. 443

[4] *Phil. Trans.* clxxi (1879), p. 421

[5] *Am. Jour. Sci.* (3), xxxv (1887), p. 60

[6] *Brit. Assoc. Rep.* (1862), p. 254

[7] *Œuvres*, ii, p. 438

of dispersion. 'The molecular groups, or the particles of bodies,' he wrote, 'may be separated by intervals which, though small, are certainly not altogether insensible relatively to the length of a wave.' Such a coarse-grainedness of the medium would, as he foresaw, introduce into the equations terms by which dispersion might be explained; indeed, the theory of dispersion which was afterwards given by Cauchy was actually based on this principle. It seems likely that, towards the close of his life, Fresnel was contemplating a great memoir on dispersion,[1] which was never completed.

Fresnel had reason at first to be pleased with the reception of his work on the optics of crystals, for in August 1822 Laplace spoke highly of it in public; and when at the end of the year a seat in the Academy became vacant, he was encouraged to hope that the choice would fall on him. In this he was disappointed.[2] Meanwhile his researches were steadily continued; and in January 1823 the very month of his rejection, he presented to the Academy a theory in which reflection and refraction [3] are referred to the dynamical properties of the luminiferous media.

As in his previous investigations, he assumes that the vibrations which constitute light are executed at right angles to the plane of polarisation. He adopts Young's principle, that reflection and refraction are due to differences in the inertia of the aether in different material bodies, and supposes (as in his memoir on aberration) that the inertia is proportional to the inverse square of the velocity of propagation of light in the medium. The conditions which he proposes to satisfy at the interface between two media are that the displacements of the adjacent molecules, resolved parallel to this interface, shall be equal in the two media; and that the energy of the reflected and refracted waves together shall be equal to that of the incident wave.

On these assumptions the intensity of the reflected and refracted light may be obtained in the following way:

Consider first the case in which the incident light is polarised in the plane of incidence, so that the displacement is at right angles to the plane of incidencc; let the amplitude of the displacement

[1] cf. the biography in *Œuvres de Fresnel*, i, p. xcvi

[2] Writing to Young in the spring of 1823, he says: 'Tous ces mémoires, que dernièrement j'ai présentés coup sur coup à l'Académie des Sciences, ne m'en ont pas cependant ouvert la porte. C'est M. Dulong qui a été nommé pour remplir la place vacante dans la section de physique. . . . Vous voyez, Monsieur, que la théorie des ondulations ne m'a point porté bonheur: mais cela ne m'en dégoûte pas: et je me console de ce malheur en m'occupant d'optique avec une nouvelle ardeur.'

[3] The MSS was for some time believed to be lost, but was ultimately found among the papers of Fourier, and printed in *Mém. de l'Acad.* xi (1832), p. 393; *Œuvres*, i, p. 767.

at a given point of the interface be f for the incident ray, g for the reflected ray and h for the refracted ray.

The quantities of energy propagated per second across unit cross-section of the incident, reflected and refracted beams are proportional respectively to

$$c_1\rho_1 f^2, \quad c_1\rho_1 g^2, \quad c_2\rho_2 h^2,$$

where c_1, c_2, denote the velocities of light, and ρ_1, ρ_2 the densities of aether, in the two media; and the cross-sections of the beams which meet the interface in unit area are

$$\cos i, \quad \cos i, \quad \cos r$$

respectively. The principle of conservation of energy therefore gives

$$c_1\rho_1 \cos i \,.\, f^2 = c_1\rho_1 \cos i \,.\, g^2 + c_2\rho_2 \cos r \,.\, h^2.$$

The equation of continuity of displacement at the interface is

$$f + g = h.$$

Eliminating h between these two equations, and using the formulae

$$\frac{\sin^2 r}{\sin^2 i} = \frac{c_2^2}{c_1^2} = \frac{\rho_1}{\rho_2},$$

we obtain the equation

$$\frac{f}{g} = -\frac{\sin (i - r)}{\sin (i + r)}.$$

Thus when the light is polarised in the plane of reflection the amplitude of the reflected wave is

$$\frac{\sin (i - r)}{\sin (i + r)} \times \text{the amplitude of the incident vibration.}$$

Fresnel shows in a similar way that when the light is polarised at right angles to the plane of reflection, the ratio of the amplitudes of the reflected and incident waves is

$$\frac{\tan (i - r)}{\tan (i + r)}.$$

These formulae are generally known as *Fresnel's sine law* and *Fresnel's tangent law* respectively. They had, however, been discovered experimentally by Brewster some years previously. When the incidence is perpendicular, so that i and r are very small, the ratio of the amplitudes becomes

$$\text{Limit}\ \frac{i - r}{i + r},$$

or

$$\frac{\mu_2 - \mu_1}{\mu_2 + \mu_1},$$

where μ_2 and μ_1 denote the refractive indices of the media. This formula had been given previously by Young [1] and Poisson,[2] on the supposition that the elasticity of the aether is of the same kind as that of air in sound.

When $i + r = 90°$, tan $(i + r)$ becomes infinite; and thus a theoretical explanation is obtained for Brewster's law, that if the incidence is such as to make the reflected and refracted rays perpendicular to each other, the reflected light will be wholly polarised in the plane of reflection.

Fresnel's investigation can scarcely be called a dynamical theory in the strict sense, as the qualities of the medium are not defined. His method was to work backwards from the known properties of light, in the hope of arriving at a mechanism to which they could be attributed; he succeeded in accounting for the phenomena in terms of a few simple principles, but was not able to specify an aether which would in turn account for these principles. The 'displacement' of Fresnel could not be a displacement in an elastic solid of the usual type, since its normal component is not continuous across the interface between two media.[3]

The theory of ordinary reflection was completed by a discussion of the case in which light is reflected totally. This had formed the subject of some of Fresnel's experimental researches several years before; and in two papers [4] presented to the Academy in November 1817 and January 1818, he had shown that light polarised in any plane inclined to the plane of reflection is partly 'depolarised' by total reflection, and that this is due to differences of phase which are introduced between the components polarised in and perpendicular to the plane of reflection. 'When the reflection is total,' he said, 'rays polarised in the plane of reflection are reflected nearer

[1] Article 'Chromatics,' *Encycl. Britt. Suppl.* [2] *Mém. de l'Inst.* ii, 1817, p. 305

[3] Fresnel's theory of reflection can, however, be reconciled with the electromagnetic theory of light, by identifying his 'displacement' with the electric force.

[4] *Œuvres de Fresnel*, i, pp. 441, 487

the surface of the glass than those polarised at right angles to the same plane, so that there is a difference in the paths described.'

This change of phase he now deduced from the formulae already obtained for ordinary reflection. Considering light polarised in the plane of reflection, the ratio of the amplitudes of the reflected and incident light is, as we have seen,

$$-\frac{\sin(i-r)}{\sin(i+r)};$$

when the sine of the angle of incidence is greater than μ_2/μ_1, so that total reflection takes place, this ratio may be written in the form

$$e^{\theta\sqrt{-1}}$$

where θ denotes a real quantity defined by the equation

$$\tan\tfrac{1}{2}\theta = \frac{(\mu_1^2\sin^2 i - \mu_2^2)^{\frac{1}{2}}}{\mu_1\cos i}.$$

Fresnel interpreted this expression to mean [1] that the amplitude of the reflected light is equal to that of the incident, but that the two waves differ in phase by an amount θ. The case of light polarised at right angles to the plane of reflection may be treated in the same way, and the resulting formulae are completely confirmed by experiment.

A few months after the memoir on reflection had been presented, Fresnel was elected to a seat in the Academy; and during the rest of his short life honours came to him both from France and abroad. In 1827 the Royal Society awarded him the Rumford medal; but Arago, to whom Young had confided the mission of conveying the medal, found him dying; and eight days afterwards he breathed his last.

By the genius of Young and Fresnel the wave-theory of light was established in a position so strong that henceforth the corpuscular hypothesis was unable to recruit any adherents among the younger men.[2] Two striking experiments which were devised later seemed to be decisive in the controversy. The first was due to G. B. Airy (1801–92), who remarked [3] that according to the corpuscular theory the colours of thin plates are produced solely by the light reflected from the second surface of the plate, whereas in the wave-theory they depend on interference between light reflected

[1] On this interpretation cf. De Morgan's *Trigonometry and Double Algebra*, pp. 119, 139

[2] Brewster (*b.* 1781) still inclined to the corpuscular theory in the 1830's.

[3] *Trans. Camb. Phil. Soc.* iv (1833), p. 279 (read 14 Nov. 1831)

at the first and second surfaces. Hence if, in the production of Newton's rings, we can by any means prevent reflection at the first surface, the rings ought not to appear according to the wave-theory, but ought still to be seen if the other hypothesis is true. Airy arranged an experiment in which a lens was laid upon a polished metallic surface. By using light polarised at right angles to the plane of reflection, and incident at the polarising angle, he secured that the light coming to the eye should be that reflected from the metal, unmixed with other light: the rings were not seen, so the verdict was in favour of the wave-theory.

The second experiment was performed in 1850, when Foucault [1] and Fizeau,[2] carrying out a plan long before imagined by Arago, directly measured the velocity of light in air and in water, and found that on the question so long debated between the rival schools the adherents of the undulatory theory had been in the right.

[1] *Comptes Rendus*, xxx (1850), p. 551 [2] ibid. p. 562

Chapter V

THE AETHER AS AN ELASTIC SOLID

When Young and Fresnel put forward the view that the vibrations of light are performed at right angles to its direction of propagation, they at the same time pointed out that this peculiarity might be explained by making a new hypothesis regarding the nature of the luminiferous medium ; namely, that it possesses the power of resisting attempts to distort its shape. It is by the possession of such a power that solid bodies are distinguished from fluids, which offer no resistance to distortion ; the idea of Young and Fresnel may therefore be expressed by the simple statement that *the aether behaves as an elastic solid.* After the death of Fresnel this conception was developed in a brilliant series of memoirs to which our attention must now be directed.

The elastic-solid theory meets with one obvious difficulty at the outset. If the aether has the qualities of a solid, how is it that the planets in their orbital motions are able to journey through it at immense speeds without encountering any perceptible resistance? This objection was first satisfactorily answered by Sir George Gabriel Stokes [1] (1819–1903), who remarked that such substances as pitch and shoemaker's wax, though so rigid as to be capable of elastic vibration, are yet sufficiently plastic to permit other bodies to pass slowly through them. The aether, he suggested, may have this combination of qualities in an extreme degree, behaving like an elastic solid for vibrations so rapid as those of light, but yielding like a fluid to the much slower progressive motions of the planets.

Stokes's explanation harmonises in a curious way with Fresnel's hypothesis that the velocity of longitudinal waves in the aether is indefinitely great compared with that of the transverse waves ; for it is found by experiment with actual substances that the ratio of the velocity of propagation of longitudinal waves to that of transverse waves increases rapidly as the medium becomes softer and more plastic.

[1] *Trans. Camb. Phil. Soc.* viii (1845), p. 287

In attempting to set forth a parallel between light and the vibrations of an elastic substance, the investigator is compelled more than once to make a choice between alternatives. He may, for instance, suppose that the vibrations of the aether are executed either parallel to the plane of polarisation of the light or at right angles to it ; and he may suppose that the different refractive powers of different media are due either to differences in the inertia of the aether within the media, or to differences in its power of resisting distortion, or to both these causes combined. There are, moreover, several distinct methods for avoiding the difficulties caused by the presence of longitudinal vibrations ; and as, alas ! we shall see, a further source of diversity is to be found in that liability to error from which no man is free. It is therefore not surprising that the list of elastic-solid theories is a long one.

At the time when the transversality of light was discovered, no general method had been developed for investigating mathematically the properties of elastic bodies ; but under the stimulus of Fresnel's discoveries, some of the best intellects of the age were attracted to the subject. The volume of *Memoirs of the Academy* which contains Fresnel's theory of crystal-optics contains also a memoir by Claud Louis-Marie-Henri Navier [1] (1785-1836), at that time Professor of Mechanics in Paris, in which the correct equations of vibratory motion for a particular type of elastic solid were for the first time given. Navier supposed the medium to be ultimately constituted of an immense number of particles, which act on each other with forces directed along the lines joining them, and depending on their distances apart ; and showed that if $\mathbf{e}$ denote the (vector) displacement of the particle whose undisturbed position is (x, y, z), and if ρ denote the density of the medium, the equation of motion is

$$\rho\frac{\partial^2\mathbf{e}}{\partial t^2} = -\, 3n \text{ grad div } \mathbf{e} - n \text{ curl curl } \mathbf{e},$$

where n denotes a constant which measures the *rigidity*, or power of resisting distortion, of the medium. All such elastic properties of the body as the velocity of propagation of waves in it must evidently depend on the ratio n/ρ.

Among the referees of one of Navier's papers was Augustine-Louis Cauchy (1789–1857), one of the greatest analysts of the

[1] *Mém. de l'Acad.* vii, p. 375. The memoir was presented in 1821, and published in 1827.

nineteenth century,[1] who, becoming interested in the question, published in 1828 [2] a discussion of it from an entirely different point of view. Instead of assuming, as Navier had done, that the medium is an aggregate of point-centres of force, and thus involving himself in doubtful molecular hypotheses, he devised a method of directly studying the elastic properties of matter in bulk, and by its means showed that the vibrations of an isotropic solid are determined by the equation

$$\rho\frac{\partial^2 \mathbf{e}}{\partial t^2} = -(k + \tfrac{4}{3}n)\text{ grad div }\mathbf{e} - n\text{ curl curl }\mathbf{e}\,;$$

here n denotes, as before, the constant of rigidity; and the constant k, which is called the *modulus of compression*,[3] denotes the ratio of a pressure to the cubical compression produced by it. Cauchy's equation evidently differs from Navier's in that two constants, k and n, appear instead of one. The reason for this is that a body constituted from point-centres of force in Navier's fashion has its moduli of rigidity and compression connected by the relation [4]

$$k = \frac{5}{3}n.$$

Actual bodies do not necessarily obey this condition; e.g. for

[1] Hamilton's opinion, written in 1833, is worth repeating: 'The principal theories of algebraical analysis (under which I include Calculi) require to be entirely remodelled; and Cauchy has done much already for this great object. Poisson also has done much; but he does not seem to me to have nearly so logical a mind as Cauchy, great as his talents and clearness are; and both are in my judgment very far inferior to Fourier, whom I place at the head of the French School of Mathematical Philosophy, even above Lagrange and Laplace, though I rank *their* talents above those of Cauchy and Poisson.' (*Life of Sir W. R. Hamilton*, ii, p. 58.)

William Thomson (Lord Kelvin) and Heaviside were warm admirers of Fourier. In May 1840 Thomson, being then aged fifteen, borrowed Fourier's *Théorie de la Chaleur* from the Glasgow University library, and read it through in a fortnight. It was the strongest of all the influences that shaped his mind.

'No-one admires Fourier more than I do,' said Heaviside (*Electromagnetic Theory*, ii, p. 32).

[2] Cauchy, *Exercices de Mathématiques*, iii (1828), p. 160

[3] This notation was introduced at a later period, but is used here in order to avoid subsequent changes.

[4] In order to construct a body whose elastic properties are not limited by this equation, William John Macquorn Rankine (1820-72) considered a continuous fluid in which a number of point-centres of force are situated; the fluid is supposed to be partially condensed round these centres, the elastic atmosphere of each nucleus being retained round it by attraction. An additional volume elasticity due to the fluid is thus acquired; and no relation between k and n is now necessary. cf. Rankine's *Miscellaneous Scientific Papers*, pp. 81 sqq.

Sir William Thomson (Lord Kelvin), in 1889, formed a solid not obeying Navier's condition by using pairs of dissimilar atoms. cf. Thomson's *Papers*, iii, p. 395; cf. also *Baltimore Lectures*, pp. 123 sqq.

india-rubber k is much larger than $5n/3$,[1] and there seems to be no reason why we should impose it on the aether.

In the same year Poisson [2] succeeded in solving the differential equation which had thus been shown to determine the wave-motions possible in an elastic solid. The solution, which is both simple and elegant, may be derived as follows: let the displacement vector $\mathbf{e}$ be resolved into two components, of which one $\mathbf{c}$ is *circuital*, or satisfies the condition

$$\operatorname{div} \mathbf{c} = 0,$$

while the other $\mathbf{b}$ is *irrotational*, or satisfies the condition

$$\operatorname{curl} \mathbf{b} = 0.$$

The equation takes the form

$$\rho \frac{\partial^2}{\partial t^2}(\mathbf{b} + \mathbf{c}) - n\nabla^2\mathbf{c} - \left(k + \frac{4}{3}n\right)\nabla^2\mathbf{b} = 0.$$

The terms which involve $\mathbf{b}$ and those which involve $\mathbf{c}$ must be separately zero, since they represent respectively the irrotational and the circuital parts of the equation. Thus, $\mathbf{c}$ satisfies the pair of equations

$$\rho \frac{\partial^2 \mathbf{c}}{\partial t^2} = n\nabla^2\mathbf{c}, \quad \operatorname{div} \mathbf{c} = 0;$$

while $\mathbf{b}$ is to be determined from

$$\rho \frac{\partial^2 \mathbf{b}}{\partial t^2} = \left(k + \frac{4}{3}n\right)\nabla^2\mathbf{b}, \quad \operatorname{curl} \mathbf{b} = 0.$$

A particular solution of the equations for $\mathbf{c}$ is easily seen to be

$$c_x = \mathrm{A} \sin \lambda \left(z - t\sqrt{\frac{n}{\rho}}\right), \quad c_y = \mathrm{B} \sin \lambda \left(z - t\sqrt{\frac{n}{\rho}}\right), \quad c_z = 0,$$

which represents a transverse plane wave propagated with velocity $\sqrt{(n/\rho)}$. It can be shown that the general solution of the differential equations for $\mathbf{c}$ is formed of such waves as this, travelling in all directions, superposed on each other.

A particular solution of the equations for $\mathbf{b}$ is

$$b_x = 0, \quad b_y = 0, \quad b_z = \mathrm{C} \sin \lambda \left(z - t\sqrt{\frac{k + \frac{4}{3}n}{\rho}}\right),$$

[1] It may, however, be objected that india-rubber and other bodies which fail to fulfil Navier's relation are not true solids. On this historic controversy, cf. Todhunter and Pearson's *History of Elasticity*, i, p. 496.

[2] *Mém. de l'Acad.* viii (1828), p. 623. Poisson takes the equation in the restricted form given by Navier; but this does not affect the question of wave-propagation.

which represents a longitudinal wave propagated with velocity

$$\sqrt{(k + \tfrac{4}{3}n)/\rho}\ ;$$

the general solution of the differential equation for **b** is formed by the superposition of such waves as this, travelling in all directions.

Poisson thus discovered that the waves in an elastic solid are of two kinds : those in **c** are transverse, and are propagated with velocity $(n/\rho)^{\frac{1}{2}}$; while those in **b** are longitudinal, and are propagated with velocity $\{(k + \tfrac{4}{3}n)/\rho\}^{\frac{1}{2}}$. The latter are[1] waves of dilatation and condensation, like sound waves ; in the **c** waves, on the other hand, the medium is not dilated or condensed, but only distorted in a manner consistent with the preservation of a constant density.[2]

The researches which have been mentioned hitherto have all been concerned with isotropic bodies. Cauchy in 1828 [3] extended the equations to the case of crystalline substances. This, however, he accomplished only by reverting to Navier's plan of conceiving an elastic body as a cluster of particles which attract each other with forces depending on their distances apart ; the aelotropy he accounted for by supposing the particles to be packed more closely in some directions than in others.

The general equations thus obtained for the vibrations of an elastic solid contain twenty-one constants ; six of these depend on the initial stress, so that if the body is initially without stress, only fifteen constants are involved. If, the initial stress being retained, the medium is supposed to be symmetrical with respect to three mutually orthogonal planes, the twenty-one constants reduce to nine, and the equations which determine the vibrations may be written in the form [4]

$$\frac{\partial^2 e_x}{\partial t^2} = (a + \mathrm{G})\frac{\partial^2 e_x}{\partial x^2} + (h + \mathrm{H})\frac{\partial^2 e_x}{\partial y^2} + (g + \mathrm{I})\frac{\partial^2 e_x}{\partial z^2}$$
$$+ 2\frac{\partial}{\partial x}\left(a\frac{\partial e_x}{\partial x} + h\frac{\partial e_y}{\partial y} + g\frac{\partial e_z}{\partial z}\right),$$

[1] cf. Stokes, 'On the Dynamical Problem of Diffraction,' *Camb. Phil. Trans.* ix, 1849, p. 1 : *Math. and Phys. Papers* ii, p. 243

[2] It may easily be shown that any disturbance, in either isotropic or crystalline media, for which the direction of vibration of the molecules lies in the wave-front or surface of constant phase, must satisfy the equation

$$\operatorname{div} \mathbf{e} = 0,$$

where **e** denotes the displacement ; if, on the other hand, the direction of vibration of the molecules is perpendicular to the wave-front, the disturbance must satisfy the equation

$$\operatorname{curl} \mathbf{e} = 0.$$

These results were proved by M. O'Brien, *Trans. Camb. Phil. Soc.* vii (1842), p. 397

[3] *Exercices de Math.* iii (1828), p. 188

[4] These are substantially equations (68) on p. 208 of the third volume of the *Exercices.*

and two similar equations. The three constants G, H, I represent the stresses across planes parallel to the co-ordinate planes in the undisturbed state of the aether.[1]

On the basis of these equations, Cauchy worked out a theory of light, of which an instalment relating to crystal-optics was presented to the Academy in 1830.[2] Its characteristic features will now be sketched.

By substitution in the equations last given, it is found that when the wave-front of the vibration is parallel to the plane of yz, the velocity of propagation must be $(h + \mathrm{G})^{\frac{1}{2}}$ if the vibration takes place parallel to the axis of y, and $(g + \mathrm{G})^{\frac{1}{2}}$ if it takes place parallel to the axis of z. Similarly when the wave-front is parallel to the plane of zx, the velocity must be $(h + \mathrm{H})^{\frac{1}{2}}$ if the vibration is parallel to the axis of x, and $(f + \mathrm{H})^{\frac{1}{2}}$ if it is parallel to the axis of z; and when the wave-front is parallel to the plane of xy, the velocity must be $(g + \mathrm{I})^{\frac{1}{2}}$ if the vibration is parallel to the axis of x, and $(f + \mathrm{I})^{\frac{1}{2}}$ if it is parallel to the axis of y.

Now it is known from experiment that the velocity of a ray polarised parallel to one of the planes in question is the same, whether its direction of propagation is along one or the other of the axes in that plane; so, if we assume that the vibrations which constitute light are executed parallel to the plane of polarisation, we must have

$$f + \mathrm{H} = f + \mathrm{I}, \quad g + \mathrm{I} = g + \mathrm{G}, \quad h + \mathrm{H} = h + \mathrm{G};$$

or,

$$\mathrm{G} = \mathrm{H} = \mathrm{I}.$$

This is the assumption made in the memoir of 1830; the theory based on it is generally known as *Cauchy's First Theory*[3]; the equilibrium pressures G, H, I being all equal, are taken to be zero.

If, on the other hand, we make the alternative assumption that the vibrations of the aether are executed at right angles to the plane of polarisation, we must have

$$h + \mathrm{H} = g + \mathrm{I}, \quad f + \mathrm{I} = h + \mathrm{G}, \quad g + \mathrm{G} = f + \mathrm{H};$$

[1] G, H, I are tensions when they are positive and pressures when they are negative.

[2] *Mém. de l'Acad.* x, p. 293.

In the previous year (*Mém. de l'Acad.* ix, p. 114) Cauchy had stated that the equations of elasticity lead in the case of uniaxal crystals to a wave-surface of which two sheets are a sphere and spheroid as in Huygens' theory.

[3] The equations and results of Cauchy's First Theory of crystal-optics were independently obtained shortly afterwards by Franz Ernst Neumann (1798–1895); cf. *Ann. d. Phys.* xxv (1832), p. 418, reprinted as no. 76 of Ostwald's *Klassiker der exakten Wissenschaften*, with notes by A. Wangerin.

the theory based on this supposition is known as *Cauchy's Second Theory*; it was published in 1836.[1]

In both theories Cauchy imposes the condition that the section of two of the sheets of the wave-surface made by any one of the co-ordinate planes is to be formed of a circle and an ellipse, as in Fresnel's theory; this yields the three conditions

$$3bc = f(b + c + f)\ ; \quad 3ca = g(c + a + g)\ ; \quad 3ab = h(a + b + h).$$

Thus in the first theory we have these together with the equations

$$\mathrm{G} = 0, \quad \mathrm{H} = 0, \quad \mathrm{I} = 0,$$

which express the condition that the undisturbed state of the aether is unstressed; and the aethereal vibrations are executed parallel to the plane of polarisation. In the second theory we have the three first equations, together with

$$f - \mathrm{G} = g - \mathrm{H} = h - \mathrm{I}\ ;$$

and the plane of polarisation is interpreted to be the plane at right angles to the direction of vibration of the aether.

Either of Cauchy's theories accounts tolerably well for the phenomena of crystal-optics; but the wave-surface (or rather the two sheets of it which correspond to nearly transverse waves) is not exactly Fresnel's. In both theories the existence of a third wave, formed of nearly longitudinal vibrations, is a formidable difficulty. Cauchy himself anticipated that the existence of these vibrations would ultimately be demonstrated by experiment, and in one place[2] conjectured that they might be of a calorific nature. A further objection to Cauchy's theories is that the relations between the constants do not appear to admit of any simple physical interpretation, being evidently assumed for the sole purpose of forcing the formulae into some degree of conformity with the results of experiment. And further difficulties will appear when we proceed subsequently to compare the properties which are assigned to the aether in crystal-optics with those which must be postulated in order to account for reflection and refraction.

To the latter problem Cauchy soon addressed himself, his investigations being in fact published[3] in the same year (1830) as the first of his theories of crystal-optics.

At the outset of any work on refraction, it is necessary to assign a cause for the existence of refractive indices, i.e. for the variation

[1] *Comptes Rendus*, ii (1836), p. 341; *Mém. de l'Acad.* xviii (1839), p. 153

[2] *Mém. de l'Acad.* xviii (1839, published 1842), p. 153

[3] *Bull. des Sciences Math.* xiv (1830), p. 6

in the velocity of light from one body to another. Huygens, as we have seen, suggested that transparent bodies consist of hard particles which interact with the aethereal matter, modifying its elasticity. Cauchy in his earlier papers [1] followed this lead more or less closely, assuming that the density ρ of the aether is the same in all media, but that its rigidity n varies from one medium to another.

Let the axis of x be taken at right angles to the surface of separation of the media, and the axis of z parallel to the intersection of this interface with the incident wave-front; and suppose, first, that the incident vibration is executed at right angles to the plane of incidence, so that it may be represented by

$$e_z = f\left(-x\cos i - y\sin i + \sqrt{\frac{n}{\rho}}\,t\right),$$

where i denotes the angle of incidence; the reflected wave may be represented by

$$e_z = \mathrm{F}\left(x\cos i - y\sin i + \sqrt{\frac{n}{\rho}}\,t\right),$$

and the refracted wave by

$$e_z = f_1\left(-x\cos r - y\sin r + \sqrt{\frac{n'}{\rho}}\,t\right),$$

where r denotes the angle of refraction, and n' the rigidity of the second medium.

To obtain the conditions satisfied at the reflecting surface, Cauchy assumed (without assigning reasons) that the x and y components of the stress across the xy plane are equal in the media on either side of the interface. This implies in the present case that the quantities

$$n\frac{\partial e_z}{\partial x} \quad \text{and} \quad n\frac{\partial e_z}{\partial y}$$

are to be continuous across the interface: so we have

$$n\cos i\,.\,(f' - \mathrm{F}') = n'\cos r\,.\,f_1'\,; \quad n\sin i\,.\,(f' + \mathrm{F}') = n'\sin r\,.\,f_1'.$$

Eliminating f_1', we have

$$\frac{\mathrm{F}'}{f'} = \frac{\sin(r-i)}{\sin(r+i)}.$$

Now this is Fresnel's sine law for the ratio of the intensity of the reflected ray to that of the incident ray; and it is known that the light to which it applies is that which is polarised parallel to the

[1] As will appear, his views on this subject subsequently changed.

plane of incidence. Thus Cauchy was driven to the conclusion that, in order to satisfy the known facts of reflection and refraction, the vibrations of the aether must be supposed executed at right angles to the plane of polarisation of the light.

The case of a vibration performed in the plane of incidence he discussed in the same way. It was found that Fresnel's tangent law could be obtained by assuming that e_x and the normal pressure across the interface have equal values in the two contiguous media.

The theory thus advanced was encumbered with many difficulties. In the first place, the identification of the plane of polarisation with the plane at right angles to the direction of vibration was contrary to the only theory of crystal-optics which Cauchy had as yet published. In the second place, no reasons were given for the choice of the conditions at the interface. Cauchy's motive in selecting these particular conditions was evidently to secure the fulfilment of Fresnel's sine law and tangent law; but the results are inconsistent with the true boundary conditions, which were given later by Green.

It is probable that the results of the theory of reflection had much to do with the decision, which Cauchy now made,[1] to reject the first theory of crystal-optics in favour of the second. After 1836 he consistently adhered to the view that the vibrations of the aether are performed at right angles to the plane of polarisation. In that year he made another attempt to frame a satisfactory theory of reflection,[2] based on the assumption just mentioned, and on the following boundary conditions: at the interface between two media curl $\mathbf{e}$ is to be continuous, and (taking the axis of x normal to the interface) $\partial e_x/\partial x$ is also to be continuous.

Again we find no very satisfactory reasons assigned for the choice of the boundary conditions; and as the continuity of $\mathbf{e}$ itself across the interface is not included amongst the conditions chosen, they are obviously open to criticism; but they lead to Fresnel's sine and tangent equations, which correctly express the actual behaviour of light.[3] Cauchy remarks that in order to justify them it is necessary to abandon the assumption of his earlier theory, that the density of the aether is the same in all material bodies.

[1] *Comptes Rendus*, ii (1836), p. 341

[2] *Comptes Rendus*, ii (1836), p. 341: 'Mémoire sur la dispersion de la lumière' (*Nouveaux exercices de Math.* [1836], p. 203)

[3] These boundary conditions of Cauchy's are, as a matter of fact, satisfied by the electric force in the electromagnetic theory of light. The continuity of curl $\mathbf{e}$ is equivalent to the continuity of the magnetic vector across the interface, and the continuity of $\partial e_x/\partial x$ leads to the same equation as the continuity of the component of electric force in the direction of the intersection of the interface with the plane of incidence.

It may be remarked that neither in this nor in Cauchy's earlier theory of reflection is any trouble caused by the appearance of longitudinal waves when a transverse wave is reflected, for the simple reason that he assumes the boundary conditions to be only four in number ; and these can all be satisfied without the necessity for introducing any but transverse vibrations.

These features bring out the weakness of Cauchy's method of attacking the problem. His object was to derive the properties of light from a theory of the vibrations of elastic solids. At the outset he had already in his possession the differential equations of motion of the solid, which were to be his starting-point, and the equations of Fresnel, which were to be his goal. It only remained to supply the boundary conditions at an interface, which are required in the discussion of reflection, and the relations between the elastic constants of the solid, which are required in the optics of crystals. Cauchy seems to have considered the question from the purely analytical point of view. Given certain differential equations, what supplementary conditions must be adjoined to them in order to produce a given analytical result ? The problem when stated in this form admits of more than one solution ; and hence it is not surprising that within the space of ten years the great French mathematician produced two distinct theories of crystal-optics and three distinct theories of reflection,[1] almost all yielding correct or nearly correct final formulae, and yet mostly irreconcilable with each other, and involving incorrect boundary conditions and improbable relations between elastic constants.

Cauchy's theories, then, resemble Fresnel's in postulating types of elastic solid which do not exist, and for whose assumed properties no dynamical justification is offered. The same objection applies, though in a less degree, to the original form of a theory of reflection and refraction which was discovered about this time[2] almost simultaneously by James MacCullagh (1809–47) of Trinity College, Dublin, and Franz Neumann (1798–1895) of Königsberg. To these authors is due the merit of having extended the laws of reflection to crystalline media ; but the principles of the theory were

[1] One yet remains to be mentioned.

[2] The outlines of the theory were published by MacCullagh in *Brit. Assoc. Rep.*, 1835 ; and his results were given in *Phil. Mag.* x (Jan. 1837) and in *Proc. Roy. Irish Acad.* xviii, Jan. 1837 ; cf. Hamilton's memoir of MacCullagh in *Proc. R. S.*, v (1847), p. 712. Neumann's memoir was presented to the Berlin Academy towards the end of 1835, and published in 1837 in *Abh. Berl. Ak. aus dem Jahre 1835*, *Math. Klasse*, p. 1. So far as publication is concerned, the priority would seem to belong to MacCullagh ; but there are reasons for believing that the priority of discovery really rests with Neumann, who had arrived at his equations a year before they were communicated to the Berlin Academy.

originally derived in connection with the simpler case of isotropic media, to which our attention will for the present be confined.

MacCullagh and Neumann felt that the great objection to Fresnel's theory of reflection was its failure to provide for the continuity of the normal component of displacement at the interface between two media ; it is obvious that a discontinuity in this component could not exist in any true elastic-solid theory, since it would imply that the two media do not remain in contact. Accordingly, they made it a fundamental condition that all three components of the displacement must be continuous at the interface, and found that the sine law and tangent law can be reconciled with this condition only by supposing that the aether vibrations are parallel to the plane of polarisation : which supposition they accordingly adopted. In place of the remaining three true boundary conditions, however, they used only a single equation, derived by assuming that transverse incident waves give rise only to transverse reflected and refracted waves, and that the conservation of energy holds for these —i.e. that the masses of aether put in motion, multiplied by the squares of the amplitudes of vibration, are the same before and after incidence. This is, of course, the same device as had been used previously by Fresnel ; it must, however, be remarked that the principle is unsound as applied to an ordinary elastic solid ; for in such a body the refracted and reflected energy would in part be carried away by longitudinal waves.

In order to obtain the sine and tangent laws, MacCullagh and Neumann found it necessary to assume that the inertia of the luminiferous medium is everywhere the same, and that the differences in behaviour of this medium in different substances are due to differences in its elasticity. The two laws may then be deduced in much the same way as in the previous investigations of Fresnel and Cauchy.

Although to insist on continuity of displacement at the interface was a decided advance, the theory of MacCullagh and Neumann scarcely showed as yet much superiority over the quasi-mechanical theories of their predecessors. Indeed, MacCullagh himself expressly disavowed any claim to regard his theory, in the form to which it had then been brought, as a final explanation of the properties of light. ' If we are asked,' he wrote, ' what reasons can be assigned for the hypotheses on which the preceding theory is founded, we are far from being able to give a satisfactory answer. We are obliged to confess that, with the exception of the law of *vis viva*, the hypotheses are nothing more than fortunate conjectures. These conjectures are very probably right, since they have led to elegant laws which are fully borne out by experiments ; but this is all we can assert respecting them. We cannot attempt to deduce them from first principles ; because, in the theory of light, such principles are still to be sought for. It is certain, indeed, that light is produced by undulations, propagated, with transversal vibrations, through a highly elastic aether ; but the constitution of this aether, and the laws of its connection (if it has any connection) with the particles of bodies, are utterly unknown.'

The needful reformation of the elastic-solid theory of reflection was effected by Green, in a paper [1] read to the Cambridge Philosophical Society in December 1837. Green, though inferior to Cauchy as an analyst, was his superior in physical insight. Instead of designing boundary equations for the express purpose of yielding Fresnel's sine and tangent formulae, he set to work to determine the conditions which are actually satisfied at the interfaces of real elastic solids.

These he obtained by means of general dynamical principles. In an isotropic medium which is strained, the potential energy per unit volume due to the state of stress is

$$\phi = \frac{1}{2}\left(k + \frac{4}{3}n\right)\left(\frac{\partial e_x}{\partial x} + \frac{\partial e_y}{\partial y} + \frac{\partial e_z}{\partial z}\right)^2 + \frac{1}{2}n\left\{\left(\frac{\partial e_z}{\partial y} + \frac{\partial e_y}{\partial z}\right)^2 + \left(\frac{\partial e_x}{\partial z} + \frac{\partial e_z}{\partial x}\right)^2 + \left(\frac{\partial e_y}{\partial x} + \frac{\partial e_x}{\partial y}\right)^2 - 4\frac{\partial e_y}{\partial y}\frac{\partial e_z}{\partial z} - 4\frac{\partial e_z}{\partial z}\frac{\partial e_x}{\partial x} - 4\frac{\partial e_x}{\partial x}\frac{\partial e_y}{\partial y}\right\},$$

where $\mathbf{e}$ denotes the displacement, and k and n denote the two elastic constants already introduced ; by substituting this value of ϕ in the general variational equation

$$\iiint \rho\left\{\frac{\partial^2 e_x}{\partial t^2}\delta e_x + \frac{\partial^2 e_y}{\partial t^2}\delta e_y + \frac{\partial^2 e_z}{\partial t^2}\delta e_z\right\} dxdydz = -\iiint \delta\phi\, dxdydz$$

(where ρ denotes the density), the equation of motion may be deduced.

But this method does more than merely furnish the equation of motion

$$\rho\frac{\partial^2 \mathbf{e}}{\partial t^2} = -\left(k + \frac{4}{3}n\right)\text{grad div }\mathbf{e} - n\text{ curl curl }\mathbf{e}\ ;$$

or,

$$\rho\frac{\partial^2 \mathbf{e}}{\partial t^2} = -\left(k + \frac{1}{3}n\right)\text{grad div }\mathbf{e} + n\nabla^2\mathbf{e},$$

which had already been obtained by Cauchy ; for it also yields

[1] *Trans. Camb. Phil. Soc.* vii (1838), pp. 1, 113 ; Green's *Math. Papers*, p. 245

the boundary conditions which must be satisfied at the interface between two elastic media in contact; these are, as might be guessed by physical intuition, that the three components of the displacement [1] and the three components of stress across the interface are to be equal in the two media. If the axis of x be taken normal to the interface, the latter three quantities are

$$\left(k - \frac{2}{3}n\right) \operatorname{div} \mathbf{e} + 2n\frac{\partial e_x}{\partial x}, \quad n\left(\frac{\partial e_z}{\partial x} + \frac{\partial e_x}{\partial z}\right), \quad \text{and} \quad n\left(\frac{\partial e_x}{\partial y} + \frac{\partial e_y}{\partial x}\right).$$

The correct boundary conditions being thus obtained, it was a simple matter to discuss the reflection and refraction of an incident wave by the procedure of Fresnel and Cauchy. The result found by Green was that if the vibration of the aethereal molecules is executed at right angles to the plane of incidence, the intensity of the reflected light obeys Fresnel's sine law, provided the rigidity n is assumed to be the same for all media, but the inertia ρ to vary from one medium to another. Since the sine law is known to be true for light polarised in the plane of incidence, Green's conclusion confirmed the hypotheses of Fresnel, that the vibrations are executed at right angles to the plane of polarisation, and that the optical differences between media are due to the different densities of aether within them.

It now remained for Green to discuss the case in which the incident light is polarised at right angles to the plane of incidence, so that the motion of the aethereal particles is parallel to the intersection of the plane of incidence with the front of the wave. In this case it is impossible to satisfy all the six boundary conditions without assuming that longitudinal vibrations are generated by the act of reflection. Taking the plane of incidence to be the plane of yz, and the interface to be the plane of xy, the incident wave may be represented by the equations

$$e_y = \mathrm{A}\frac{\partial}{\partial z}f(t + lz + my); \quad e_z = -\mathrm{A}\frac{\partial}{\partial y}f(t + lz + my);$$

where, if i denote the angle of incidence, we have

$$l = \sqrt{\frac{\rho_1}{n}}\cos i, \quad m = -\sqrt{\frac{\rho_1}{n}}\sin i.$$

There will be a transverse reflected wave,

$$e_y = \mathrm{B}\frac{\partial}{\partial z}f(t - lz + my); \quad e_z = -\mathrm{B}\frac{\partial}{\partial y}f(t - lz + my);$$

[1] These first three conditions are of course not dynamical but geometrical.

and a transverse refracted wave,

$$e_y = C\frac{\partial}{\partial z}f(t + l_1 z + my); \quad e_z = -C\frac{\partial}{\partial y}f(t + l_1 z + my),$$

where, since the velocity of transverse waves in the second medium is $\sqrt{n/\rho_2}$, we can determine l_1 from the equation

$$l_1^2 + m^2 = \frac{\rho_2}{n};$$

there will also be a longitudinal reflected wave,

$$e_y = D\frac{\partial}{\partial y}f(t - \lambda z + my); \quad e_z = D\frac{\partial}{\partial z}f(t - \lambda z + my),$$

where λ is determined by the equation

$$\lambda^2 + m^2 = \frac{\rho_1}{k_1 + \frac{4}{3}n};$$

and a longitudinal refracted wave,

$$e_y = E\frac{\partial}{\partial y}f(t + \lambda_1 z + my); \quad e_z = E\frac{\partial}{\partial z}f(t + \lambda_1 z + my),$$

where λ_1 is determined by

$$\lambda_1^2 + m^2 = \frac{\rho_2}{k_2 + \frac{4}{3}n}.$$

Substituting these values for the displacement in the boundary conditions which have been already formulated, we obtain the equations which determine the intensities of the reflected and refracted waves; in particular, it appears that the amplitude of the reflected transverse wave is given by the equation

$$\frac{A - B}{A + B} = \frac{l_1\rho_1}{l\rho_2} + \frac{m^2}{l}\frac{(\rho_1 - \rho_2)^2}{\rho_2(\lambda\rho_2 + \lambda_1\rho_1)}.$$

Now if the elastic constants of the media are such that the velocities of propagation of the longitudinal waves are of the same order of magnitude as those of the transverse waves, the direction cosines of the longitudinal reflected and refracted rays will in general have real values, and these rays will carry away some of the energy which is brought to the interface by the incident wave. Green avoided this difficulty by adopting Fresnel's suggestion that the resistance of the aether to compression may be very large in comparison with the resistance to distortion, as is actually the case with such substances as jelly and caoutchouc; in this case the longitudinal waves are degraded in much the same way as the transverse refracted ray is degraded when there is total reflection, and so do not carry away energy. Making this supposition, so that k_1 and k_2 are very large, the quantities λ and λ_1 have the value $m\sqrt{-1}$, and we have

$$\frac{A-B}{A+B}=\frac{l_1\rho_1}{l\rho_2}-\frac{m}{l}\frac{(\rho_1-\rho_2)^2}{\rho_2(\rho_1+\rho_2)}\sqrt{-1}.$$

Thus we have

$$\left|\frac{B}{A}\right|^2=\frac{\left(\frac{\rho_2}{\rho_1}+1\right)^2\left(\frac{\rho_2}{\rho_1}-\frac{l_1}{l}\right)^2+\left(\frac{\rho_2}{\rho_1}-1\right)^4\frac{m^2}{l^2}}{\left(\frac{\rho_2}{\rho_1}+1\right)^2\left(\frac{\rho_2}{\rho_1}+\frac{l_1}{l}\right)^2+\left(\frac{\rho_2}{\rho_1}-1\right)^4\frac{m^2}{l^2}}.$$

This expression represents the ratio of the intensity of the transverse reflected wave to that of the incident wave. It does not agree with Fresnel's tangent formula ; and both on this account and also because (as we shall see) this theory of reflection does not harmonise well with the elastic-solid theory of crystal-optics, it must be concluded that the vibrations of a Greenian solid do not furnish an exact parallel to the vibrations which constitute light.

The success of Green's investigation from the standpoint of dynamics, set off by its failure in the details last mentioned, stimulated MacCullagh to fresh exertions. At length he succeeded in placing his own theory, which had all along been free from reproach so far as agreement with optical experiments was concerned, on a sound dynamical basis ; thereby effecting that reconciliation of the theories of light and dynamics which had been the dream of every physicist since the days of Descartes.

The central feature of MacCullagh's investigation,[1] which was presented to the Royal Irish Academy in 1839, is the introduction of a new type of elastic solid. He had, in fact, concluded from Green's results that it was impossible to explain optical phenomena satisfactorily by comparing the aether to an elastic solid of the ordinary type, which resists compression and distortion ; and he saw that the only hope of the situation was to devise a medium which should be as strictly conformable to dynamical laws as Green's elastic solid, and yet should have its properties specially designed to fulfil the requirements of the theory of light. Such a medium he now described.

If as before we denote by **e** the vector displacement of a point of the medium from its equilibrium position, it is well known that

[1] *Trans. Roy. Irish Acad.* xxi (1848) p. 17 ; MacCullagh's *Coll. Works*, p. 145

the vector curl $\mathbf{e}$ denotes twice the rotation of the part of the solid in the neighbourhood of the point (x, y, z) from its equilibrium orientation. In an ordinary elastic solid, the potential energy of strain depends only on the change of size and shape of the volume elements—on their compression and distortion, in fact. For MacCullagh's new medium, on the other hand, the potential energy depends only on the *rotation* of the volume elements.

Since the medium is not supposed to be in a state of stress in its undisturbed condition, the potential energy per unit volume must be a quadratic function of the derivates of $\mathbf{e}$; so that in an isotropic medium this quantity ϕ must be formed from the only invariant which depends solely on the rotation and is quadratic in the derivates, that is from $(\text{curl }\mathbf{e})^2$; thus we may write

$$\phi = \tfrac{1}{2}\mu\left\{\left(\frac{\partial e_z}{\partial y}-\frac{\partial e_y}{\partial z}\right)^2 + \left(\frac{\partial e_x}{\partial z}-\frac{\partial e_z}{\partial x}\right)^2 + \left(\frac{\partial e_y}{\partial x}-\frac{\partial e_x}{\partial y}\right)^2\right\},$$

where μ is a constant.

The equation of motion is now to be determined, as in the case of Green's aether, from the variational equation

$$\iiint \rho\left\{\frac{\partial^2 e_x}{\partial t^2}\delta e_x + \frac{\partial^2 e_y}{\partial t^2}\delta e_y + \frac{\partial^2 e_z}{\partial t^2}\delta e_z\right\} dx\, dy\, dz = -\iiint \delta\phi\, dx\, dy\, dz;$$

the result is

$$\rho\frac{\partial^2 \mathbf{e}}{\partial t^2} = -\mu\,\text{curl curl }\mathbf{e}.$$

It is evident from this equation that if div $\mathbf{e}$ is initially zero it will always be zero; we shall suppose this to be the case, so that no longitudinal waves exist at any time in the medium. One of the greatest difficulties which beset elastic-solid theories is thus completely removed.

The equation of motion may now be written

$$\rho\frac{\partial^2 \mathbf{e}}{\partial t^2} = \mu\nabla^2\mathbf{e},$$

which shows that transverse waves are propagated with velocity $\sqrt{(\mu/\rho)}$.

From the variational equation we may also determine the boundary conditions which must be satisfied at the interface between two media; these are, that the three components of $\mathbf{e}$ are to be continous across the interface, and that the two components of μ curl $\mathbf{e}$ parallel to the interface are also to be continuous across it. One of these five conditions, namely, the continuity of the normal component of **e**, is really dependent on the other four ; for if we take the axis of x normal to the interface, the equation of motion gives

$$\rho \frac{\partial^2 e_x}{\partial t^2} = -\frac{\partial}{\partial y}(\mu \operatorname{curl} \mathbf{e})_z + \frac{\partial}{\partial z}(\mu \operatorname{curl} \mathbf{e})_y,$$

and as the quantities ρ, $(\mu \operatorname{curl} \mathbf{e})_z$, and $(\mu \operatorname{curl} \mathbf{e})_y$ are continuous across the interface, the continuity of $\partial^2 e_x/\partial t^2$ follows. Thus the only independent boundary conditions in MacCullagh's theory are the continuity of the tangential components of **e** and of μ curl **e**.[1] It is easily seen that these are equivalent to the boundary conditions used in MacCullagh's earlier paper, namely, the equation of *vis viva* and the continuity of the three components of **e** ; and thus the 'rotationally elastic' aether of this memoir furnishes a dynamical foundation for the memoir of 1837.

The extension to crystalline media is made by assuming the potential energy per unit volume to have, when referred to the principal axes, the form

$$A\left(\frac{\partial e_z}{\partial y} - \frac{\partial e_y}{\partial z}\right)^2 + B\left(\frac{\partial e_x}{\partial z} - \frac{\partial e_z}{\partial x}\right)^2 + C\left(\frac{\partial e_y}{\partial x} - \frac{\partial e_x}{\partial y}\right)^2,$$

where A, B, C denote three constants which determine the optical behaviour of the medium ; it is readily seen that the wave-surface is Fresnel's, and that the plane of polarisation contains the displacement, and is at right angles to the rotation.

MacCullagh's work was regarded with doubt by his own and the succeeding generation of mathematical physicists, and can scarcely be said to have been properly appreciated until FitzGerald drew attention to it forty years afterwards. But there can be no doubt that MacCullagh really solved the problem of devising a medium whose vibrations, calculated in accordance with the correct laws of dynamics, should have the same properties as the vibrations of light.

[1] MacCullagh's equations may readily be interpreted in the electromagnetic theory of light (see chap. viiii and ix of this book) ; **e** corresponds to the magnetic force, μ curl **e** to the electric force, and curl **e** to the electric displacement. This interpretation is due to Heaviside, *Electrician*, xxvi (1891), p. 360 : Lord Kelvin had regarded rotation as the analogue of magnetic force. Heaviside's discussion is as follows : let **E** be the torque. The force is $-\operatorname{curl} \mathbf{E}$, so we have the equation of motion $-\operatorname{curl} \mathbf{E} = \mu \frac{\partial \mathbf{H}}{\partial t}$, if **H** is the velocity and μ the density. Also the torque is proportional to the rotation, so curl **H** $= c\frac{\partial \mathbf{E}}{\partial t}$, where c is the reciprocal of the quasi-rigidity. These are the electromagnetic equations.

The hesitation which was felt in accepting the rotationally elastic aether arose mainly from the want of any readily conceived example of a body endowed with such a property. This difficulty was removed in 1889 by Sir William Thomson (Lord Kelvin), who designed mechanical models possessed of rotational elasticity.[1] Suppose, for example,[2] that a structure is formed of spheres, each sphere being in the centre of the tetrahedron formed by its four nearest neighbours. Let each sphere be joined to these four neighbours by rigid bars, which have spherical caps at their ends so as to slide freely on the spheres. Such a structure would, for small deformations, behave like an incompressible perfect fluid. Now attach to each bar a pair of gyroscopically mounted flywheels, rotating with equal and opposite angular velocities, and having their axes in the line of the bar; a bar thus equipped will require a couple to hold it at rest in any position inclined to its original position, and the structure as a whole will possess that kind of quasi-elasticity which was first imagined by MacCullagh.

This particular representation is not perfect, since a system of forces would be required to hold the model in equilibrium if it were irrotationally distorted. Lord Kelvin subsequently invented another structure free from this defect.[3]

The work of Green proved a stimulus not only to MacCullagh but to Cauchy, who now (1839) published yet a third theory of reflection.[4] This appears to have owed its origin to a remark of Green's,[5] that the longitudinal wave might be avoided in either of two ways—namely, by supposing its velocity to be indefinitely great or indefinitely small. Green curtly dismissed the latter alternative and adopted the former, on the ground that the equilibrium of the medium would be unstable if its compressibility were negative (as it must be if the velocity of longitudinal waves is to vanish). Cauchy, without attempting to meet Green's objection, took up the study of a medium whose elastic constants are connected by the equation

$$k + \tfrac{4}{3}n = 0,$$

so that the longitudinal vibrations have zero velocity; and showed that if the aethereal vibrations are supposed to be executed at right angles to the plane of polarisation, and if the rigidity of the

[1] W. Thomson was, said Heaviside, 'most intensely mechanical, and could not accept any ether unless he could make a model of it' (Heaviside's *Electromagnetic Theory*, iii, p. 479).

[2] *Comptes Rendus*, cix (16 Sept. 1889), p. 453: Kelvin's *Math. and Phys. Papers*, iii, p. 466

[3] *Proc. R. S. Edin.* xvii (17 Mar. 1890), p. 127: Kelvin's *Math. and Phys. Papers*, iii, p. 468

[4] *Comptes Rendus*, ix (25 Nov. 1839), p. 676; and (2 Dec. 1839), p. 726

[5] Green's *Math. Papers*, p. 246

aether is assumed to be the same in all media, a ray which is reflected will obey the sine law and tangent law of Fresnel. The boundary conditions which he adopted in order to obtain this result were the continuity of the displacement $\mathbf{e}$ and of its derivate $\partial\mathbf{e}/\partial x$, where the axis of x is taken at right angles to the interface.[1] These are not the true boundary conditions for general elastic solids; but in the particular case now under discussion, where the rigidity is the same in the two media, they yield the same equations as the conditions correctly given by Green.

The aether of Cauchy's third theory of reflection is well worthy of some further study. It is generally known as the *contractile* or *labile*[2] aether, the names being due to William Thomson (Lord Kelvin, 1824–1907), who discussed it long afterwards.[3] It may be defined as an elastic medium of (negative) compressibility such as to make the velocity of the longitudinal wave zero; this implies that no work is required to be done in order to give the medium any small irrotational disturbance. An example is furnished by homogeneous foam free from air and held from collapse by adhesion to a containing vessel.

Cauchy, as we have seen, did not attempt to refute Green's objection that such a medium would be unstable; but, as Thomson remarked, every possible infinitesimal motion of the medium is, in the elementary dynamics of the subject, proved to be resolvable into coexistent wave-motions. If, then, the velocity of propagation for each of the two kinds of wave-motion is real, the equilibrium must be stable, provided the medium either extends through boundless space or has a fixed containing vessel as its boundary.

When the rigidity of the luminiferous medium is supposed to have the same value in all bodies, the conditions to be satisfied at an interface reduce to the continuity of the displacement $\mathbf{e}$, of the tangential components of curl $\mathbf{e}$, and of the scalar $(k+\frac{4}{3}n)$ div $\mathbf{e}$ across the interface.

Now we have seen that when a transverse wave is incident on an interface, it gives rise in general to reflected and refracted waves of both the transverse and the longitudinal species. In the case of the contractile aether, for which the velocity of propagation of the longitudinal waves is very small, the ordinary construction for refracted waves shows that the directions of propagation of the

[1] *Comptes Rendus*, x (2 Mar. 1840), p. 347; xxvii (1848), p. 621; xxviii (1849), p. 25. *Mém. de l'Acad.* xxii (1848), pp. 17, 29

[2] *Labile* or *neutral* is a term used of such equilibrium as that of a rigid body on a perfectly smooth horizontal plane.

[3] *Phil. Mag.* xxvi (1888), p. 414

reflected and refracted longitudinal waves will be almost normal to the interface. The longitudinal waves will therefore contribute only to the component of displacement normal to the interface, not to the tangential components ; in other words, the only tangential components of displacement at the interface are those due to the three transverse waves—the incident, reflected and refracted. Moreover, the longitudinal waves do not contribute at all to curl **e** ; and, therefore, in the contractile aether, the conditions that the tangential components of **e** and of n curl **e** shall be continuous across an interface are satisfied by the distortional part of the disturbance taken alone. The condition that the component of **e** normal to the interface is to be continuous is not satisfied by the distortional part of the disturbance taken alone, but is satisfied when the distortional and compressional parts are taken together.

The energy carried away by the longitudinal waves is infinitesimal, as might be expected, since no work is required in order to generate an irrotational displacement. Hence, with this aether, the behaviour of the transverse waves at an interface may be specified without considering the irrotational part of the disturbance at all, by the conditions that the conservation of energy is to hold and that the tangential components of **e** and of n curl **e** are to be continuous. But if we identify these transverse waves with light, assuming that the displacement **e** is at right angles to the plane of polarisation of the light, and assuming moreover that the rigidity n is the same in all media [1] (the differences between media depending on differences in the inertia ρ), we have exactly the assumptions of Fresnel's theory of light ; whence it follows that transverse waves in the labile aether must obey in reflection the sine law and tangent law of Fresnel.

The great advantage of the labile aether is that it overcomes the difficulty about securing continuity of the normal component of displacement at an interface between two media ; the light-waves taken alone do not satisfy this condition of continuity ; but the total disturbance consisting of light waves and irrotational disturbance taken together does satisfy it ; and this is ensured without allowing the irrotational disturbance to carry off any of the energy.[2]

William Thomson (Lord Kelvin), who devoted much attention to the labile aether, was at one time led to doubt the validity

[1] This condition is in any case necessary for stability, as was shown by R. T. Glazebrook ; cf. Thomson, *Phil. Mag.* xxvi, (1888), p. 500.

[2] The labile-aether theory of light may be compared with the electromagnetic theory, by interpreting the displacement **e** as the electric force, and ρ**e** as the electric displacement.

of this explanation of light[1]; for when investigating the radiation of energy from a vibrating rigid globe embedded in an infinite elastic-solid aether, he found that in some cases the irrotational waves would carry away a considerable part of the energy if the aether were of the labile type. This difficulty, however, was removed by the observation[2] that it is sufficient for the fulfilment of Fresnel's laws if the velocity of the irrotational waves in one of the two media is very small, without regard to the other medium. Following up this idea, Thomson assumed that in space void of ponderable matter the aether is practically incompressible by the forces concerned in light waves, but that in the space occupied by liquids and solids it has a negative compressibility, so as to give zero velocity for longitudinal aether waves in these bodies. This assumption was based on the conception that material atoms move through space without displacing the aether; a conception which, as Thomson remarked, contradicts the old scholastic axiom that two different portions of matter cannot simultaneously occupy the same space.[3] He supposed the aether to be attracted and repelled by the atoms, and thereby to be condensed or rarefied.[4]

The year 1839, which saw the publication of MacCullagh's dynamical theory of light and Cauchy's theory of the labile aether, was memorable also for the appearance of a memoir by Green on crystal-optics.[5] This really contains two distinct theories, which respectively resemble Cauchy's First and Second Theories: in one of them, the stresses in the undisturbed state of the aether are supposed to vanish, and the vibrations of the aether are supposed to be executed parallel to the plane of polarisation of the light; in the other theory, the initial stresses are not supposed to vanish, and the aether vibrations are at right angles to the plane of polarisation. The two investigations are generally known as Green's First and Second Theories of crystal-optics.

The foundations of both theories are, however, the same. Green first of all determined the potential energy of a strained crystalline solid; this in the most general case involves 27 constants, or 21 if

[1] *Baltimore Lectures* (edn. 1904), p. 214 [2] ibid. p. 411

[3] Michell and Boscovich in the eighteenth century had taught the doctrine of the mutual penetration of matter, i.e. that two substances may be in the same place at the same time without excluding each other; cf. Priestley's *History*, i, p. 392.

[4] cf. *Baltimore Lectures*, pp. 413–14, 463 and Appendices A and E. 'I am afraid nothing except the complete overthrow of his whole notion of how the functions of the ether are produced will cure Sir W. T.' (FitzGerald to Heaviside, 4 Feb. 1889). 'I had a long correspondence with Lord Kelvin, and his last letter says that he gives up everything he ever wrote about the ether. I hope he is not still quite so down in the mouth about it.' (FitzGerald to Heaviside, 8 June 1896).

[5] *Trans. Cambridge Phil. Soc.* VII (1839), p. 121; Green's *Math. Papers*, p. 293

there is no initial stress.[1] If, however, as is here assumed, the medium possesses three planes of symmetry at right angles to each other, the number of constants reduces to 12, or to 9 if there is no initial stress; if $\mathbf{e}$ denote the displacement, the potential energy per unit volume may be written

$$\phi = G\frac{\partial e_x}{\partial x} + H\frac{\partial e_y}{\partial y} + I\frac{\partial e_z}{\partial z}$$

$$+\tfrac{1}{2}G\left\{\left(\frac{\partial e_x}{\partial x}\right)^2 + \left(\frac{\partial e_y}{\partial x}\right)^2 + \left(\frac{\partial e_z}{\partial x}\right)^2\right\} + \tfrac{1}{2}H\left\{\left(\frac{\partial e_x}{\partial y}\right)^2 + \left(\frac{\partial e_y}{\partial y}\right)^2 + \left(\frac{\partial e_z}{\partial y}\right)^2\right\}$$

$$+ \tfrac{1}{2}I\left\{\left(\frac{\partial e_x}{\partial z}\right)^2 + \left(\frac{\partial e_y}{\partial z}\right)^2 + \left(\frac{\partial e_z}{\partial z}\right)^2\right\}$$

$$+ \tfrac{3}{2}a\left(\frac{\partial e_x}{\partial x}\right)^2 + \tfrac{3}{2}b\left(\frac{\partial e_y}{\partial y}\right)^2 + \tfrac{3}{2}c\left(\frac{\partial e_z}{\partial z}\right)^2$$

$$+f'\frac{\partial e_y}{\partial y}\frac{\partial e_z}{\partial z} + g'\frac{\partial e_x}{\partial x}\frac{\partial e_z}{\partial z} + h'\frac{\partial e_x}{\partial x}\frac{\partial e_y}{\partial y}$$

$$+ \tfrac{1}{2}f\left(\frac{\partial e_y}{\partial z} + \frac{\partial e_z}{\partial y}\right)^2 + \tfrac{1}{2}g\left(\frac{\partial e_z}{\partial x} + \frac{\partial e_x}{\partial z}\right)^2 + \tfrac{1}{2}h\left(\frac{\partial e_x}{\partial y} + \frac{\partial e_y}{\partial x}\right)^2.$$

The usual variational equation

$$\iiint \rho\left\{\frac{\partial^2 e_x}{\partial t^2}\delta e_x + \frac{\partial^2 e_y}{\partial t^2}\delta e_y + \frac{\partial^2 e_z}{\partial t^2}\delta e_z\right\} dx\,dy\,dz = -\iiint \delta\phi\, dx\,dy\,dz,$$

then yields the differential equations of motion, namely:

$$\rho\frac{\partial^2 e_x}{\partial t^2} = (a + G)\frac{\partial^2 e_x}{\partial x^2} + (h + H)\frac{\partial^2 e_x}{\partial y^2} + (g + I)\frac{\partial^2 e_x}{\partial z^2}$$

$$+\frac{\partial}{\partial x}\left(a\frac{\partial e_x}{\partial x} + h\frac{\partial e_y}{\partial y} + g\frac{\partial e_z}{\partial z}\right) + \frac{\partial}{\partial x}\left(a\frac{\partial e_x}{\partial x} + h'\frac{\partial e_y}{\partial y} + g'\frac{\partial e_z}{\partial z}\right),$$

and two similar equations.

These differ from Cauchy's fundamental equations in having greater generality; for Cauchy's medium was supposed to be built up of point-centres of force attracting each other according to some function of the distance; and, as we have seen, there are limitations in this method of construction, which render it incompetent to represent the most general type of elastic solid. Cauchy's equations for crystalline media are, in fact, exactly analogous to the equations originally found by Navier for isotropic media, which contain only one elastic constant instead of two.

[1] For there are 21 terms in a homogeneous function of the second degree in six variables.

The number of constants in the above equations still exceeds the three which are required to specify the properties of a biaxal crystal; and Green now proceeds to consider how the number may be reduced. The condition which he imposes for this purpose is that for two of the three waves whose front is parallel to a given plane, the vibration of the aethereal molecules shall be accurately in the plane of the wave: in other words, that two of the three waves shall be purely distortional, the remaining one being consequently a normal vibration. This condition gives five relations,[1] which may be written:

$$a=b=c=\tfrac{1}{3}\mu;$$
$$f' = \mu - 2f; \qquad g' = \mu - 2g; \qquad h' = \mu - 2h;$$

where μ denotes a new constant.[2]

Thus the potential energy per unit volume may be written

$$\begin{aligned}\phi = {} & G\frac{\partial e_x}{\partial x} + H\frac{\partial e_y}{\partial y} + I\frac{\partial e_z}{\partial z} \\ & + \tfrac{1}{2}G\left\{\left(\frac{\partial e_x}{\partial x}\right)^2 + \left(\frac{\partial e_y}{\partial x}\right)^2 + \left(\frac{\partial e_z}{\partial x}\right)^2\right\} + \tfrac{1}{2}H\left\{\left(\frac{\partial e_x}{\partial y}\right)^2 + \left(\frac{\partial e_y}{\partial y}\right)^2 + \left(\frac{\partial e_z}{\partial y}\right)^2\right\} \\ & \qquad + \tfrac{1}{2}I\left\{\left(\frac{\partial e_x}{\partial z}\right)^2 + \left(\frac{\partial e_y}{\partial z}\right)^2 + \left(\frac{\partial e_z}{\partial z}\right)^2\right\} \\ & + \tfrac{1}{2}\mu\left(\frac{\partial e_x}{\partial x} + \frac{\partial e_y}{\partial y} + \frac{\partial e_z}{\partial z}\right)^2 \\ & + \tfrac{1}{2}f\left\{\left(\frac{\partial e_y}{\partial z} + \frac{\partial e_z}{\partial y}\right)^2 - 4\frac{\partial e_y}{\partial y}\frac{\partial e_z}{\partial z}\right\} + \tfrac{1}{2}g\left\{\left(\frac{\partial e_z}{\partial x} + \frac{\partial e_x}{\partial z}\right)^2 - 4\frac{\partial e_z}{\partial z}\frac{\partial e_x}{\partial x}\right\} \\ & \qquad + \tfrac{1}{2}h\left\{\left(\frac{\partial e_x}{\partial y} + \frac{\partial e_y}{\partial x}\right)^2 - 4\frac{\partial e_x}{\partial x}\frac{\partial e_y}{\partial y}\right\}.\end{aligned}$$

At this point Green's two theories of crystal-optics diverge from

[1] As Green showed, the hypothesis of transversality really involves the existence of planes of symmetry, so that it alone is capable of giving 14 relations between the 21 constants; and 3 of the remaining 7 constants may be removed by change of axes, leaving only four.

[2] It was afterwards shown by Barré de Saint-Venant (1797–1886), *Journal de Math.* vii (1863), p. 399, that if the initial stresses be supposed to vanish, the conditions which must be satisfied among the remaining nine constants $a, b, c, f, g, h, f' g', h'$, in order that the wave-surface may be Fresnel's, are the following:

$$\begin{cases} (3b-f)(3c-f) = (f+f')^2 \\ (3c-g)(3a-g) = (g+g')^2 \\ (3a-h)(3b-h) = (h+h')^2 \\ (3a-g)(3b-h)(3c-f) + (3a-h)(3b-f)(3c-g) = 2(f+f')(g+g')(h+h'). \end{cases}$$

These reduce to Green's relations when the additional equation $b=c$ is assumed.

Saint-Venant disputed the validity of Green's relations, asserting that they are compatible only with isotropy. On this controversy cf. R. T. Glazebrook, *Brit. Assoc. Report* (1885), p. 171, and Karl Pearson in Todhunter & Pearson's *History of Elasticity*, ii, § 147.

each other. According to the first theory, the initial stresses G, H, I are zero, so that

$$\phi = \tfrac{1}{2}\mu\left(\frac{\partial e_x}{\partial x} + \frac{\partial e_y}{\partial y} + \frac{\partial e_z}{\partial z}\right)^2$$
$$+ \tfrac{1}{2}f\left\{\left(\frac{\partial e_y}{\partial z} + \frac{\partial e_z}{\partial y}\right)^2 - 4\frac{\partial e_y}{\partial y}\frac{\partial e_z}{\partial z}\right\} + \tfrac{1}{2}g\left\{\left(\frac{\partial e_z}{\partial x} + \frac{\partial e_x}{\partial z}\right)^2 - 4\frac{\partial e_z}{\partial z}\frac{\partial e_x}{\partial x}\right\}$$
$$+ \tfrac{1}{2}h\left\{\left(\frac{\partial e_x}{\partial y} + \frac{\partial e_y}{\partial x}\right)^2 - 4\frac{\partial e_x}{\partial x}\frac{\partial e_y}{\partial y}\right\}.$$

This expression contains the correct number of constants, namely, four: three of them represent the optical constants of a biaxal crystal, and one (namely, μ) represents the square of the velocity of propagation of longitudinal waves. It is found that the two sheets of the wave-surface which correspond to the two distortional waves form a Fresnel's wave-surface; the third sheet, which corresponds to the longitudinal wave, being an ellipsoid. The directions of polarisation and the wave-velocities of the distortional waves are identical with those assigned by Fresnel, provided it is assumed that the direction of vibration of the aether particles is parallel to the plane of polarisation; but this last assumption is of course inconsistent with Green's theory of reflection and refraction.

In his Second Theory, Green, like Cauchy, used the condition that for the waves whose fronts are parallel to the co-ordinate planes, the wave-velocity depends only on the plane of polarisation, and not on the direction of propagation. He thus obtained the equations already found by Cauchy

$$\mathrm{G} - f = \mathrm{H} - g = \mathrm{I} - h.$$

The wave-surface in this case also is Fresnel's, provided it is assumed that the vibrations of the aether are executed at right angles to the plane of polarisation.

The principle which underlies the Second Theories of Green and Cauchy is that the aether in a crystal resembles an elastic solid which is unequally pressed or pulled in different directions by the unmoved ponderable matter. This idea appealed strongly to W. Thomson (Kelvin), who long afterwards developed it further,[1] arriving at the following interesting result: let an incompressible solid, isotropic when unstrained, be such that its potential energy per unit volume is

$$\tfrac{1}{2}q\left(\frac{1}{a} + \frac{1}{\beta} + \frac{1}{\gamma} - 3\right),$$

[1] *Proc. R. S. Edin.* xv (1887), p. 21; *Phil. Mag.* xxv (1888), p. 116; *Baltimore Lectures*, pp. 228–59

where q denotes its modulus of rigidity when unstrained, and $\alpha^{\frac{1}{2}}$, $\beta^{\frac{1}{2}}$, $\gamma^{\frac{1}{2}}$ denote the proportions in which lines parallel to the axes of strain are altered; then if the solid be initially strained in a way defined by given values of α, β, γ by forces applied to its surface, and if waves of distortion be superposed on this initial strain, the transmission of these waves will follow exactly the laws of Fresnel's theory of crystal-optics, the wave-surface being

$$\frac{x^2}{\frac{\alpha}{q}r^2 - 1} + \frac{y^2}{\frac{\beta}{q}r^2 - 1} + \frac{z^2}{\frac{\gamma}{q}r^2 - 1} = 0.$$

There is some difficulty in picturing the manner in which the molecules of ponderable matter act upon the aether so as to produce the initial strain required by this theory. Lord Kelvin utilised [1] the suggestion to which we have already referred, namely, that the aether may pervade the atoms of matter so as to occupy space jointly with them, and that its interaction with them may consist in attractions and repulsions exercised throughout the regions interior to the atoms. These forces may be supposed to be so large in comparison with those called into play in free aether that the resistance to compression may be overcome, and the aether may be (say) condensed in the central region of an isolated atom, and rarefied in its outer parts. A crystal may be supposed to consist of a group of spherical atoms in which neighbouring spheres overlap each other; in the central regions of the spheres the aether will be condensed, and within the lens-shaped regions of overlapping it will be still more rarefied than in the outer parts of a solitary atom, while in the interstices between the atoms its density will be unaffected. In consequence of these rarefactions and condensations, the reaction of the aether on the atoms tends to draw inwards the outermost atoms of the group, which, however, will be maintained in position by repulsions between the atoms themselves; and thus we can account for the pull which, according to the present hypothesis, is exerted on the aether by the ponderable molecules of crystals.

Analysis similar to that of Cauchy's and Green's Second Theory of crystal-optics may be applied to explain the doubly refracting property which is possessed by strained glass; but in this case the formulae derived are found to conflict with the results of experiment. The discordance led Kelvin to doubt the truth of the whole theory. 'After earnest and hopeful consideration of the stress theory of double refraction during fourteen years,' he said,[2] 'I am unable

[1] *Baltimore Lectures*, p. 253

[2] *Baltimore Lectures*, p. 258

to see how it can give the true explanation either of the double refraction of natural crystals, or of double refraction induced in isotropic solids by the application of unequal pressures in different directions.'

It is impossible to avoid noticing throughout all Kelvin's work evidences of the deep impression which was made upon him by the writings of Green. The same may be said of Kelvin's friend and contemporary Stokes ; and, indeed, it is no exaggeration to describe Green as the real founder of that 'Cambridge school' of natural philosophers, of which Kelvin, Stokes, Rayleigh, Clerk Maxwell, Lamb, J. J. Thomson, Larmor and Love were the most illustrious members in the latter half of the nineteenth century. In order to understand the peculiar position occupied by Green, it is necessary to recall something of the history of mathematical studies at Cambridge.

The century which elapsed between the death of Newton and the scientific activity of Green was the darkest in the history of the University. It is true that Cavendish and Young were educated at Cambridge ; but they, after taking undergraduate courses, removed to London. In the entire period the only natural philosopher of distinction who lived and taught at Cambridge was Michell ; and for some reason which at this distance of time it is difficult to understand fully, Michell's researches seem to have attracted little or no attention among his collegiate contemporaries and successors, who silently acquiesced when his discoveries were attributed to others, and allowed his name to perish entirely from Cambridge tradition.

A few years before Green published his first paper a notable revival of mathematical learning swept over the University ; the fluxional symbolism, which since the time of Newton had isolated Cambridge from the continental schools, was abandoned in favour of the differential notation, and the works of the great French analysts were introduced and eagerly read. Green undoubtedly received his own early inspiration from this source, chiefly from Poisson ; but in clearness of physical insight and conciseness of exposition he far excelled his masters ; and the slight volume of his collected papers has to this day a charm which is wanting in their voluminous writings.

The neglect of his work during his lifetime (he died in 1841) has often been remarked. It is, however, to be remembered that in 1840 the Lucasian professor at Cambridge was a man who never wrote anything, that the Plumian professor was Challis, whose attention was engaged in his hydrodynamical researches, that the

Lowndean professor was Dean of Ely and lived there, that Airy had removed to Greenwich and was almost altogether occupied with astronomy, that Stokes was an undergraduate and that William Thomson had not yet entered the University. Green's great memoirs of 1837 and 1839 on the elastic-solid theory were referred to in the *Trans. Camb. Phil. Soc.* for 1842 by M. O'Brien, who was evidently well acquainted with them; but when Stokes was writing on hydrodynamics in 1843 he did not know that Green had solved the problem of the motion of an ellipsoid in a fluid; while Thomson knew nothing of Green (except from a reference in a paper by Murphy to the electrical memoir published at Nottingham in 1828), until January 1845, when Hopkins gave him a copy of the Nottingham essay. Before a year was over Thomson's enthusiasm caused everyone to know about Green.

In spite of the advances which were made in the great memoirs of the year 1839, the fundamental question as to whether the aether particles vibrate parallel or at right angles to the plane of polarisation was still unanswered. More light was thrown on this problem ten years later by Stokes's investigation of diffraction.[1] Stokes showed that on almost any conceivable hypothesis regarding the aether, a disturbance in which the vibrations are executed at right angles to the plane of diffraction must be transmitted round the edge of an opaque body with less diminution of intensity than a disturbance whose vibrations are executed parallel to that plane. It follows that when light, of which the vibrations are oblique to the plane of diffraction, is so transmitted, the plane of vibration will be more nearly at right angles to the plane of diffraction in the diffracted than in the incident light. Stokes himself performed experiments to test the matter, using a grating in order to obtain strong light diffracted at a large angle, and found that when the plane of polarisation of the incident light was oblique to the plane of diffraction, the plane of polarisation of the diffracted light was more nearly parallel to the plane of diffraction. This result, which was afterwards confirmed by L. Lorenz,[2] appeared to confirm decisively the hypothesis of Fresnel, that the vibrations of the aethereal particles are executed at right angles to the plane of polarisation.

Three years afterwards Stokes indicated [3] a second line of proof leading to the same conclusion. It had long been known that the blue light of the sky, which is due to the scattering of the sun's direct

[1] *Trans. Camb. Phil. Soc.* ix (1849), p. 1; Stokes's *Math. and Phys. Papers*, ii, p. 243

[2] *Ann. d. Phys.* cxi (1860), p. 315; *Phil. Mag.* xxi (1861), p. 321

[3] *Phil. Trans.* (1852), p. 463; Stokes's *Math. and Phys. Papers*, iii, p. 267. cf. the footnote added on p. 361 of the *Math. and Phys. Papers*

rays by small particles or molecules in the atmosphere, is partly polarised. The polarisation is most marked when the light comes from a part of the sky distant 90° from the sun, in which case it must have been scattered in a direction perpendicular to that of the direct sunlight incident on the small particles ; and the polarisation is in the plane through the sun.

If, then, the axis of y be taken parallel to the light incident on a small particle at the origin, and the scattered light be observed along the axis of x, this scattered light is found to be polarised in the plane xy. Considering the matter from the dynamical point of view, we may suppose the material particle to possess so much inertia (compared to the aether) that it is practically at rest. Its motion relative to the aether, which is the cause of the disturbance it creates in the aether, will therefore be in the same line as the incident aethereal vibration, but in the opposite direction. The disturbance must be transversal, and must therefore be zero in a polar direction and a maximum in an equatorial direction, its amplitude being, in fact, proportional to the sine of the polar distance. The polar line must, by considerations of symmetry, be the line of the incident vibration. Thus we see that none of the light scattered in the x-direction can come from that constituent of the incident light which vibrates parallel to the x axis ; so the light observed in this direction must consist of vibrations parallel to the z axis. But we have seen that the plane of polarisation of the scattered light is the plane of xy ; and therefore the vibration is at right angles to the plane of polarisation.[1]

The phenomena of diffraction and of polarisation by scattering thus agreed in confirming the result arrived at in Fresnel's and Green's theory of reflection. The chief difficulty in accepting it arose in connection with the optics of crystals. As we have seen, Green and Cauchy were unable to reconcile the hypothesis of aethereal vibrations at right angles to the plane of polarisation with the correct formulae of crystal-optics, at any rate so long as the aether within crystals was supposed to be free from initial stress. The underlying reason for this can be readily seen. In a crystal, where the elasticity is different in different directions, the resistance to distortion depends solely on the orientation of the plane of distortion, which in the case of light is the plane through the directions of propagation and vibration. Now it is known that for light propagated parallel to one of the axes of elasticity of a crystal, the velocity of propagation depends only on the plane of polarisation of the light, being the

[1] The theory of polarisation by small particles was afterwards investigated by Lord Rayleigh, *Phil. Mag.* xli (1871), p. 447.

same whichever of the two axes lying in that plane is the direction of propagation. Comparing these results, we see that the plane of polarisation must be the plane of distortion, and therefore the vibrations of the aether particles must be executed parallel to the plane of polarisation.[1]

A way of escape from this conclusion suggested itself to Stokes,[2] and later to Rankine[3] and Lord Rayleigh.[4] What if the aether in a crystal, instead of having its elasticity different in different directions, were to have its rigidity invariable and its *inertia* different in different directions? This would bring the theory of crystal-optics into complete agreement with Fresnel's and Green's theory of reflection, in which the optical differences between media are attributed to differences of inertia of the aether contained within them. The only difficulty lies in conceiving how aelotropy of inertia can exist; and all three writers overcame this obstacle by pointing out that a solid which is immersed in a fluid may have its effective inertia different in different directions. For instance, a coin immersed in water moves much more readily in its own plane than in the direction at right angles to this.

Suppose then that twice the kinetic energy per unit volume of the aether within a crystal is represented by the expression

$$\rho_1\left(\frac{\partial e_x}{\partial t}\right)^2 + \rho_2\left(\frac{\partial e_y}{\partial t}\right)^2 + \rho_3\left(\frac{\partial e_z}{\partial t}\right)^2,$$

and that the potential energy per unit volume has the same value as in space void of ordinary matter. The aether is assumed to be incompressible, so that div $\mathbf{e}$ is zero: the potential energy per unit volume is therefore

$$\phi = \tfrac{1}{2} n \left\{\left(\frac{\partial e_z}{\partial y} + \frac{\partial e_y}{\partial z}\right)^2 + \left(\frac{\partial e_x}{\partial z} + \frac{\partial e_z}{\partial x}\right)^2 + \left(\frac{\partial e_y}{\partial x} + \frac{\partial e_x}{\partial y}\right)^2 - 4\frac{\partial e_y}{\partial y}\frac{\partial e_z}{\partial z} - 4\frac{\partial e_z}{\partial z}\frac{\partial e_x}{\partial x} - 4\frac{\partial e_x}{\partial x}\frac{\partial e_y}{\partial y}\right\},$$

where n denotes as usual the rigidity.

[1] In Fresnel's theory of crystal-optics, in which the aether vibrations are at right angles to the plane of polarisation, the velocity of propagation depends only on the direction of vibration, not on the plane through this and the direction of transmission.

[2] Stokes, in a letter to Lord Rayleigh, inserted in his *Memoir and Scientific Correspondence*, ii, p. 99, explains that the idea presented itself to him while he was writing the paper on fluid motion which appeared in *Trans. Camb. Phil. Soc.* viii (1843), p. 105. He suggested the wave-surface to which this theory leads in *Brit. Assoc. Rep.* (1862), p. 269.

[3] *Phil. Mag.* (4), i (1851), p. 441

[4] *Phil. Mag.* (4), xli (1871), p. 519

The variational equation of motion is

$$\iiint \left\{ \rho_1 \frac{\partial^2 e_x}{\partial t^2} \delta e_x + \rho_2 \frac{\partial^2 e_y}{\partial t^2} \delta e_y + \rho_3 \frac{\partial^2 e_z}{\partial t^2} \delta e_z \right\} dx\, dy\, dz$$

$$= - \iiint \left\{ \delta\phi - p\delta \left(\frac{\partial e_x}{\partial x} + \frac{\partial e_y}{\partial y} + \frac{\partial e_z}{\partial z} \right) \right\} dx\, dy\, dz,$$

where p denotes an undetermined function of (x, y, z) : the term in p being introduced on account of the kinematical constraint expressed by the equation

$$\text{div}\, \mathbf{e} = 0.$$

The equations of motion which result from this variational equation are

$$\rho_1 \frac{\partial^2 e_x}{\partial t^2} = - \frac{\partial p}{\partial x} + n\nabla^2 e_x,$$

and two similar equations. It is evident that p resembles a hydrostatic pressure.

Substituting in these equations the analytical expression for a plane wave, we readily find that the velocity V of the wave is connected with the direction cosines (λ, μ, ν) of its normal by the equation

$$\frac{\lambda^2}{n - \rho_1 V^2} + \frac{\mu^2}{n - \rho_2 V^2} + \frac{\nu^2}{n - \rho_3 V^2} = 0.$$

When this is compared with Fresnel's relation between the velocity and direction of a wave, it is seen that the new formula differs from his only in having the reciprocal of the velocity in place of the velocity. About 1867 Stokes carried out a series of experiments in order to determine which of the two theories was most nearly conformable to the facts. He found the construction of Huygens and Fresnel to be decidedly the more correct, the difference between the results of it and the rival construction being about a hundred times the probable error of observation.[1]

The hypothesis that in crystals the inertia depends on direction seemed therefore to be discredited when the theory based on it was compared with the results of observation. But when, in 1888, W. Thomson (Kelvin) revived Cauchy's theory of the labile aether, the question naturally arose as to whether that theory could be extended so as to account for the optical properties of crystals : and it was shown by R. T. Glazebrook [2] that the correct formulae

[1] *Proc. R. S.* xx (1872), p. 443. After these experiments Stokes gave it as his opinion (*Phil. Mag.* xli [1871], p. 521) that the true theory of crystal-optics was yet to be found. On the accuracy of Fresnel's construction cf. Glazebrook, *Phil. Trans.* clxxi (1879), p. 421, and Hastings, *Am. Journ. Sci.* (3), xxxv (1887), p. 60.

[2] *Phil. Mag.* xxvi (1888), p. 521 ; xxviii (1889), p. 110

of crystal-optics are obtained when the Cauchy-Thomson hypothesis of zero velocity for the longitudinal wave is combined with the Stokes-Rankine-Rayleigh hypothesis of aelotropic inertia.

For, on reference to the formulae which have been already given, it is obvious that the equation of motion of an aether having these properties must be

$$(\rho_1 \ddot{e}_x, \rho_2 \ddot{e}_y, \rho_3 \ddot{e}_z) = - n \text{ curl curl } \mathbf{e},$$

where $\mathbf{e}$ denotes the displacement, n the rigidity, and (ρ_1, ρ_2, ρ_3) the inertia ; and this equation leads by the usual analysis to Fresnel's wave-surface. The displacement $\mathbf{e}$ of the aethereal particles is not, however, accurately in the wave-front, as in Fresnel's theory, but is at right angles to the direction of the ray, in the plane passing through the ray and the wave-normal.[1]

Having now traced the progress of the elastic-solid theory so far as it is concerned with the propagation of light in ordinary isotropic media and in crystals, we must consider the attempts which were made about this time to account for the optical properties of a more peculiar class of substances.

It was found by Arago in 1811 [2] that the state of polarisation of a beam of light is altered when the beam is passed through a plate of quartz along the optic axis. The phenomenon was studied shortly afterwards by Biot, who showed [3] that the alteration consists in a rotation of the plane of polarisation about the direction of propagation : the angle of rotation is proportional to the thickness of the plate and inversely proportional to the square of the wave-length.

In some specimens of quartz the rotation is from left to right, in others from right to left. This distinction was shown by Sir John Herschel [4] (1792–1871) in 1820 to be associated with differences in the crystalline form of the specimens, the two types bearing the same relation to each other as a right-handed and left-handed helix respectively. Fresnel [5] and W. Thomson (Kelvin) [6] proposed the term *helical* to denote the property of rotating the plane of

[1] This theory of crystal-optics may be assimilated to the electromagnetic theory by interpreting the elastic displacement $\mathbf{e}$ as electric force, and the vector $(\rho_1 e_x, \rho_2 e_y \rho_3 e_z)$ as electric displacement.

[2] *Mém. de l'Institut* (1811), pt. i, p. 115 sqq.

[3] *Mém. de l'Institut* (1812), pt. i, p. 218 sqq. ; *Annales de Chim.* ix (1818), p. 372 ; x (1819), p. 63

[4] *Trans. Camb. Phil. Soc.* i (1820), p. 43

[5] *Mém. de l'Institut*, vii (1827), p. 45, at p. 73. The memoir was presented to the Academy in three parts, on 26 Nov. 1821, 22 Jan. 1822 and 22 Apr. 1822 respectively.

[6] *Baltimore Lectures*, p. 31

polarisation, exhibited by such bodies as quartz ; the less appropriate term *natural rotatory polarisation* is, however, generally used.[1]

Biot showed that many liquid organic bodies, e.g. turpentine and sugar solutions, possess the natural rotatory property. We might be led to infer the presence of a helical structure in the molecules of such substances ; and this inference is supported by the study of their chemical constitution, for they are invariably of the ' mirror-image ' or ' enantiomorphous ' type, in which one of the atoms (generally carbon) is asymmetrically linked to other atoms.

The next advance in the subject was due to Fresnel,[2] who showed that in naturally active bodies the velocity of propagation of circularly polarised light is different according as the polarisation is right-handed or left-handed. From this property the rotation of the plane of polarisation of a plane-polarised ray may be immediately deduced ; for the plane-polarised ray may be resolved into two rays circularly polarised in opposite senses, and these advance in phase by different amounts in passing through a given thickness of the substance : at any stage they may be recompounded into a plane-polarised ray, the azimuth of whose plane of polarisation varies with the length of path traversed.

It is readily seen from this that a ray of light incident on a crystal of quartz will in general bifurcate into two refracted rays, each of which will be elliptically polarised, i.e. will be capable of resolution into two plane-polarised components which differ in phase by a definite amount. The directions of these refracted rays may be determined by Huygens' construction, provided the wave-surface is supposed to consist of a sphere and spheroid which *do not touch.*

The first attempt to frame a theory of naturally active bodies was made by MacCullagh in 1836.[3] Suppose a plane wave of light to be propagated within a crystal of quartz. Let (x, y, z) denote the co-ordinates of a vibrating molecule, when the axis of x is taken at right angles to the plane of the wave, and the axis of z at right angles to the axis of the crystal. Using Y and Z to denote the displacements parallel to the axes of y and z respectively at any time t, MacCullagh assumed that the differential equations which determine Y and Z are

$$\frac{\partial^2 Y}{\partial t^2} = c_1^2 \frac{\partial^2 Y}{\partial x^2} + \mu \frac{\partial^3 Z}{\partial x^3}$$

$$\frac{\partial^2 Z}{\partial t^2} = c_2^2 \frac{\partial^2 Z}{\partial x^2} - \mu \frac{\partial^3 Y}{\partial x^3}$$

[1] The term *rotatory* may be applied with propriety to the property discovered by Faraday, which will be discussed later.

[2] *Annales de Chim.* xxviii (1825), p. 147

[3] *Trans. Roy. Irish Acad.* xvii (1837), p. 461 ; MacCullagh's *Coll. Works,* p. 63

where μ denotes a constant on which the natural rotatory property of the crystal depends. In order to avoid complications arising from the ordinary crystalline properties of quartz, we shall suppose that the light is propagated parallel to the optic axis, so that we can take c_1 equal to c_2.

Assuming first that the beam is circularly polarised, let it be represented by

$$Y = A \sin\left\{\frac{2\pi}{\tau}(lx - t)\right\}, \quad Z = \pm A \cos\left\{\frac{2\pi}{\tau}(lx - t)\right\},$$

the ambiguous sign being determined according as the circular polarisation is right-handed or left-handed.

Substituting in the above differential equations, we have

$$1 = c_1^2 l^2 \mp \mu . \frac{2\pi}{\tau} . l^3,$$

or

$$l = \frac{1}{c_1} \pm \frac{\pi\mu}{\tau c_1^4}.$$

Since $1/l$ denotes the velocity of propagation, it is evident that the reciprocals of the velocities of propagation of a right-handed and left-handed beam differ by the quantity

$$\frac{2\pi\mu}{\tau c_1^4}.$$

from which it is easily shown that the angle through which the plane of polarisation of a plane-polarised beam rotates in unit length of path is

$$\frac{2\pi^2\mu}{\tau^2 c_1^4}.$$

If we neglect the variation of c_1 with the period of the light, this expression satisfies Biot's law that the angle of rotation in unit length of path is proportional to the inverse square of the wave-length.

MacCullagh's investigation can be scarcely called a theory, for it amounts only to a reduction of the phenomena to empirical, though mathematical, laws ; but it was on this foundation that later workers built the theory which is now accepted.[1]

[1] The later developments of this theory will be discussed in a subsequent chapter ; but mention may here be made of an attempt which was made in 1856 by Carl Neumann, then a very young man, to provide a rational basis for MacCullagh's equations. Neumann showed that the equations may be derived from the hypothesis that the relative displacement of one aethereal particle with respect to another acts on the latter according to the same law as an element of an electric current acts on a magnetic pole. cf. the preface to C. Neumann's *Die magnetische Drehung der Polarisationsebene des Lichtes* (Halle, 1863).

The great investigators who developed the theory of light after the death of Fresnel devoted considerable attention to the optical properties of metals. Their researches in this direction must now be reviewed.

The most striking properties of metals are the power of brilliantly reflecting light at all angles of incidence, which is so well shown by the mirrors of reflecting telescopes, and the opacity, which causes a train of waves to be extinguished before it has proceeded many wave-lengths into a metallic medium. That these two attributes are connected appears probable from the fact that certain non-metallic bodies—e.g. aniline dyes—which strongly absorb the rays in certain parts of the spectrum, reflect those rays with almost metallic brilliance. A third quality in which metals differ from transparent bodies, and which, as we shall see, is again closely related to the other two, is in regard to the polarisation of the light reflected from them. This was first noticed by Malus; and in 1830 Sir David Brewster [1] showed that plane-polarised light incident on a metallic surface remains polarised in the same plane after reflection if its polarisation is either parallel or perpendicular to the plane of reflection, but that in other cases the reflected light is polarised elliptically.

It was this discovery of Brewster's which suggested to the mathematicians a theory of metallic reflection. For, as we have seen, elliptic polarisation is obtained when plane-polarised light is totally reflected at the surface of a transparent body; and this analogy between the effects of total reflection and metallic reflection led to the surmise that the latter phenomenon might be treated in the same way as Fresnel had treated the former, namely, by introducing imaginary quantities into the formulae of ordinary reflection. On these principles mathematical formulae were devised by MacCullagh[2] and Cauchy.[3]

To explain their method, we shall suppose the incident light to be polarised in the plane of incidence. According to Fresnel's sine law, the amplitude of the light (polarised in this way) reflected from a transparent body is to the amplitude of the incident light in the ratio

$$J = \frac{\sin(i - r)}{\sin(i + r)},$$

where i denotes the angle of incidence and r is determined from the equation

$$\sin i = \mu \sin r.$$

[1] *Phil. Trans*, 1830, p. 287

[2] *Proc. Roy. Irish Acad.* i (1836), p. 2; ii (1843), p. 376: *Trans. Roy. Irish Acad.* xviii (1837), p. 71: MacCullagh's *Coll. Works*, pp. 58, 132, 230

[3] *Comptes Rendus*, vii (1838), p. 953; viii (1839), pp. 553, 658, 961; xxvi (1848), p. 86

MacCullagh and Cauchy assumed that these equations hold good also for reflection at a metallic surface, provided the refractive index μ is replaced by a complex quantity

$$\mu = \nu(1 - \kappa\sqrt{-1}) \qquad \text{say,}$$

where ν and κ are to be regarded as two constants characteristic of the metal. We have therefore

$$J = \frac{\tan i - \tan r}{\tan i + \tan r} = \frac{(\mu^2 - \sin^2 i)^{\frac{1}{2}} - \cos i}{(\mu^2 - \sin^2 i)^{\frac{1}{2}} + \cos i}$$

If then we write

$$\nu^2(1 - \kappa\sqrt{-1})^2 - \sin^2 i = U^2 e^{2v\sqrt{-1}},$$

so that equations defining U and v are obtained by equating separately the real and the imaginary parts of this equation, we have

$$J = \frac{Ue^{v\sqrt{-1}} - \cos i}{Ue^{v\sqrt{-1}} + \cos i},$$

and this may be written in the form

$$|J|e^{\delta\sqrt{-1}}$$

where

$$\begin{cases} |J|^2 = \dfrac{U^2 + \cos^2 i - 2U\cos v\cos i}{U^2 + \cos^2 i + 2U\cos v\cos i} \\[2ex] \tan\delta = \dfrac{2U\cos i\sin v}{U^2 - \cos^2 i}. \end{cases}$$

The quantities $|J|$ and δ are interpreted in the same way as in Fresnel's theory of total reflection: that is, we take $|J|^2$ to mean the ratio of the intensities of the reflected and incident light, while δ measures the change of phase experienced by the light in reflection.

The case of light polarised at right angles to the plane of incidence may be treated in the same way.

When the incidence is perpendicular, U evidently reduces to $\nu(1 + \kappa^2)^{\frac{1}{2}}$, and v reduces to $-\arctan^{-1}\kappa$. For silver at perpendicular incidence almost all the light is reflected, so $|J|^2$ is nearly unity; this requires $\cos v$ to be small, and κ to be very large. The extreme case in which κ is indefinitely great but ν indefinitely small, so that the quasi-index of refraction is a pure imaginary, is generally known as the case of *ideal silver*.

The physical significance of the two constants ν and κ was more or less distinctly indicated by Cauchy; in fact, as the difference between metals and transparent bodies depends on the constant κ, it is evident that κ must in some way measure the opacity of the substance. This will be more clearly seen if we inquire how the elastic-solid theory of light can be extended so as to provide a physical basis for the formulae of MacCullagh and Cauchy.[1] The sine formula of Fresnel, which was the starting-point of our investigation of metallic reflection, is a consequence of Green's elastic-solid theory ; and the differences between Green's results and those which we have derived arise solely from the complex value which we have assumed for μ. We have therefore to modify Green's theory in such a way as to obtain a complex value for the index of refraction.

Take the plane of incidence as plane of xy, and the metallic surface as plane of yz. If the light is polarised in the plane of incidence, so that the light vector is parallel to the axis of z, the incident light may be taken to be a function of the argument

$$ax + by + ct,$$

where

$$\frac{a}{c} = -\left(\frac{\rho}{n}\right)^{\frac{1}{2}} \cos i, \qquad \frac{b}{c} = -\left(\frac{\rho}{n}\right)^{\frac{1}{2}} \sin i\;;$$

here i denotes the angle of incidence, ρ the inertia of the aether, and n its rigidity.

Let the reflected light be a function of the argument

$$a_1 x + by + ct,$$

where, in order to secure continuity at the boundary, b and c must have the same values as before. Since Green's formulae are to be still applicable, we must have

$$\frac{a_1}{b} = \cot r,$$

where $\sin i = \mu \sin r$, but μ has now a complex value. This equation may be written in the form

$$a_1^2 + b^2 = \frac{\mu^2 \rho c^2}{n}.$$

Let the complex value of μ^2 be written

$$\mu^2 = \frac{\rho_1}{\rho} - \mathrm{A}\sqrt{-1},$$

the real part being written ρ_1/ρ in order to exhibit the analogy with Green's theory of transparent media ; then we have

$$a_1^2 + b^2 = \frac{\rho_1}{n} c^2 - \frac{\rho c^2}{n} \mathrm{A}\sqrt{-1}.$$

[1] This was done by Lord Rayleigh, *Phil. Mag.* xliii (1872), p. 321

But an equation of this kind must (as in Green's theory) represent the condition to be satisfied in order that the quantity

$$e^{(a_1 x + by + ct)\sqrt{-1}}$$

may satisfy the differential equation of motion of the aether; from which we see that the equation of motion of the aether in the metallic medium is probably of the form

$$\rho_1 \frac{\partial^2 e_z}{\partial t^2} + \rho c A \frac{\partial e_z}{\partial t} = n \left(\frac{\partial^2 e_z}{\partial x^2} + \frac{\partial^2 e_z}{\partial y^2} \right).$$

This equation of motion differs from that of a Greenian elastic solid by reason of the occurrence of the term in $\partial e_z / \partial t$. But this is evidently a 'viscous' term, representing something like a frictional dissipation of the energy of luminous vibrations; a dissipation which, in fact, occasions the opacity of the metal. Thus the term which expresses opacity in the equation of motion of the luminiferous medium appears as the origin of the peculiarities of metallic reflection.[1] It is curious to notice how closely this accords with the idea of Huygens, that metals are characterised by the presence of soft particles which damp the vibrations of light.

There is, however, one great difficulty attending this explanation of metallic reflection, which was first pointed out by Lord Rayleigh.[2] We have seen that for ideal silver μ^2 is real and negative; and therefore A must be zero and ρ_1 negative; that is to say, the inertia of the luminiferous medium in the metal must be negative. This seems to destroy entirely the physical intelligibility of the theory as applied to the case of ideal silver.

The difficulty is a deep-seated one, and was not overcome for many years. The direction in which the true solution lies will suggest itself when we consider the resemblance which has already been noticed between metals and those substances which show 'surface colour'—e.g. the aniline dyes. In the case of the latter substances, the light which is so copiously reflected from them lies within a restricted part of the spectrum; and it therefore seems probable that the phenomenon is not to be attributed to the existence of dissipative terms, but that it belongs rather to the same class of effects as dispersion, and is to be referred to the same causes. In fact, dispersion means that the value of the refractive index of a substance with respect to any kind of light depends on the period

[1] It is easily seen that the amplitude is reduced by the factor $\epsilon^{-2\pi\kappa}$ when light travels one wavelength in the metal; κ is generally called the *coefficient of absorption*.

[2] loc. cit.

of the light; and we have only to suppose that the physical causes which operate in dispersion cause the refractive index to become imaginary for certain kinds of light, in order to explain satisfactorily both the surface colours of the aniline dyes and the strong reflecting powers of the metals.

Dispersion was the subject of several memoirs by the founders of the elastic-solid theory. So early as 1830 Cauchy's attention was directed [1] to the possibility of constructing a mathematical theory of this phenomenon on the basis of Fresnel's 'Hypothesis of Finite Impacts' [2]—i.e. the assumption that the radius of action of one particle of the luminiferous medium on its neighbours is so large as to be comparable with the wave-length of light. Cauchy supposed the medium to be formed, as in Navier's theory of elastic solids, of a system of point-centres of force: the force between two of these point-centres, m at (x, y, z), and μ at $(x + \Delta x, y + \Delta y, z + \Delta z)$, may be denoted by $m\mu f(r)$, where r denotes the distance between m and μ. When this medium is disturbed by light waves propagated parallel to the z axis, the displacement being parallel to the x axis, the equation of motion of m is evidently

$$\frac{\partial^2 \xi}{\partial t^2} = \sum_{\mu} \mu f(r + \rho) \frac{\Delta x + \Delta \xi}{r + \rho},$$

where ξ denotes the displacement of m, $(\xi + \Delta\xi)$ the displacement of μ, and $(r + \rho)$ the new value of r. Substituting for ρ its value, and retaining only terms of the first degree in $\Delta\xi$, this equation becomes

$$\frac{\partial^2 \xi}{\partial t^2} = \Sigma\mu \frac{f(r)}{r} \Delta\xi + \Sigma\mu \frac{d}{dr}\left\{\frac{f(r)}{r}\right\}\frac{(\Delta x)^2}{r}\Delta\xi.$$

Now, by Taylor's theorem, since ξ depends only on z, we have

$$\Delta\xi = \frac{\partial \xi}{\partial z}\Delta z + \frac{1}{2\,!}\frac{\partial^2 \xi}{\partial z^2}(\Delta z)^2 + \frac{1}{3\,!}\frac{\partial^3 \xi}{\partial z^3}(\Delta z)^3 \ldots$$

Substituting, and remembering that summations which involve odd powers of Δz must vanish when taken over all the point-centres within the sphere of influence of m, we obtain an equation of the form

$$\frac{\partial^2 \xi}{\partial t^2} = a\frac{\partial^2 \xi}{\partial z^2} + \beta\frac{\partial^4 \xi}{\partial z^4} + \gamma\frac{\partial^6 \xi}{\partial z^6} + \ldots,$$

where $a, \beta, \gamma \ldots$ denote constants.

Each successive term on the right-hand side of this equation involves an additional factor $(\Delta z)^2/\lambda^2$ as compared with the pre-

[1] *Bull. des Sc. Math.* xiv (1830), p. 9; 'Sur la dispersion de la lumière,' *Nouv. Exercices de Math.*, 1836 [2] cf. p. 123

ceding term, where λ denotes the wave-length of the light; so if the radii of influence of the point-centres were indefinitely small in comparison with the wave-length of the light, the equation would reduce to

$$\frac{\partial^2 \xi}{\partial t^2} = a \frac{\partial^2 \xi}{\partial z^2},$$

which is the ordinary equation of wave-propagation in one dimension in non-dispersive media. But if the medium is so coarse-grained that λ is not large compared with the radii of influence, we must retain the higher derivates of ξ. Substituting

$$\xi = e^{2\pi i(z - c_1 t)/\lambda}$$

in the differential equation with these higher derivates retained, we have

$$c_1^2 = a - \beta\left(\frac{2\pi}{\lambda}\right)^2 + \gamma\left(\frac{2\pi}{\lambda}\right)^4 \ldots,$$

which shows that c_1, the velocity of the light in the medium, depends on the wave-length λ; as it should do in order to explain dispersion.

Dispersion is, then, according to the view of Fresnel and Cauchy, a consequence of the coarse-grainedness of the medium. Since the luminiferous medium was found to be dispersive only within material bodies, it seemed natural to suppose that in these bodies the aether is *loaded* by the molecules of matter, and that dispersion depends essentially on the ratio of the wave-length to the distance between adjacent material molecules. This theory, in one modification or another, held its ground until forty years later it was overthrown by the facts of anomalous dispersion.

The distinction between aether and ponderable matter was more definitely drawn in memoirs which were published independently in 1841–2 by F. E. Neumann [1] and Matthew O'Brien.[2] These authors supposed the ponderable particles to remain sensibly at rest while the aether surges round them, and is acted on by them with forces which are proportional to its displacement. Thus [3] the equation of motion of the aether becomes

$$\rho \frac{\partial^2 \mathbf{e}}{\partial t^2} = -(k + \tfrac{4}{3}n) \text{ grad div } \mathbf{e} - n \text{ curl curl } \mathbf{e} - C\mathbf{e},$$

where C denotes a constant on which the phenomena of dispersion

[1] *Berlin Abhandlungen aus dem* Jahre 1841, Zweiter Teil (Berlin, 1843), p. 1

[2] *Trans. Camb. Phil. Soc.* vii (1842), p. 397

[3] O'Brien, loc. cit. §§ 15, 28

depend. For polarised plane waves propagated parallel to the axis of x, this equation becomes

$$\rho \frac{\partial^2 \mathbf{e}}{\partial t^2} = n \frac{\partial^2 \mathbf{e}}{\partial x^2} - \mathrm{C}\mathbf{e} ;$$

and if $\mathbf{e}$ is proportional to

$$e^{\frac{2\pi\sqrt{-1}}{\tau}\left(t - \frac{x}{\mathrm{V}}\right)},$$

where τ denotes the period and V the velocity of the light, we have

$$\frac{n}{\mathrm{V}^2} = \rho - \frac{\mathrm{C}}{4\pi^2}\tau^2,$$

an equation which expresses the dependence of the velocity on the period.

The attempt to represent the properties of the aether by those of an elastic solid lost some of its interest after the rise of the electromagnetic theory of light. But in 1867, before the electromagnetic hypothesis had attracted much attention, an elastic-solid theory in many respects preferable to its predecessors was presented to the French Academy [1] by Joseph Boussinesq (1842–1929). Until this time, as we have seen, investigators had been divided into two parties, according as they attributed the optical properties of different bodies to variations in the inertia of the luminiferous medium, or to variations in its elastic properties. Boussinesq, taking up a position apart from both these schools, assumed that the aether is exactly the same in all material bodies as in interplanetary space, in regard both to inertia and to rigidity, and that the optical properties of matter are due to interaction between the aether and the material particles, as had been imagined more or less by Neumann and O'Brien. These material particles he supposed to be disseminated in the aether, in much the same way as dust-particles floating in the air.

If $\mathbf{e}$ denote the displacement at the point (x, y, z) in the aether, and $\mathbf{e}'$ the displacement of the ponderable particles at the same place, the equation of motion of the aether is

$$\rho \frac{\partial^2 \mathbf{e}}{\partial t^2} = -(k + \tfrac{1}{3}n) \text{ grad div } \mathbf{e} + n\nabla^2\mathbf{e} - \rho_1 \frac{\partial^2 \mathbf{e}'}{\partial t^2}, \qquad (1)$$

where ρ and ρ_1 denote the densities of the aether and matter

[1] *Journal de Math.* (2), xiii (1868), pp. 313, 425 ; cf. also *Comptes Rendus*, cxvii (1893), pp. 80, 139, 193. Equations kindred to some of those of Boussinesq were afterwards deduced by Karl Pearson, *Proc. Lond. Math. Soc.* xx (1889), p. 297, from the hypothesis that the strain-energy involves the velocities.

respectively, and k and n denote as usual the elastic constants of the aether. This differs from the ordinary Cauchy-Green equation only in the presence of the term $\rho_1 \partial^2 \mathbf{e}' / \partial t^2$, which represents the effect of the inertia of the matter. To this equation we must adjoin another expressing the connection between the displacements of the matter and of the aether; if we assume that these are simply proportional to each other—say,

$$\mathbf{e}' = \mathrm{A}\mathbf{e}, \tag{2}$$

where the constant A depends on the nature of the ponderable body—our equation becomes

$$(\rho + \rho_1 \mathrm{A}) \frac{\partial^2 \mathbf{e}}{\partial t^2} = -(k + \tfrac{1}{3}n) \operatorname{grad} \operatorname{div} \mathbf{e} + n\nabla^2 \mathbf{e},$$

which is essentially the same equation as is obtained in those older theories which suppose the inertia of the luminiferous medium to vary from one medium to another. So far there would seem to be nothing very new in Boussinesq's work. But when we proceed to consider crystal-optics, dispersion and rotatory polarisation, the advantage of his method becomes evident: he retains equation (1) as a formula universally true—at any rate for bodies at rest; while equation (2) is varied to suit the circumstances of the case. Thus dispersion can be explained if, instead of equation (2), we take the relation

$$\mathbf{e}' = \mathrm{A}\mathbf{e} - \mathrm{D}\nabla^2 \mathbf{e},$$

where D is a constant which measures the dispersive power of the substance; the rotation of the plane of polarisation of sugar solutions can be explained if we suppose that in these bodies equation (2) is replaced by

$$\mathbf{e}' = \mathrm{A}\mathbf{e} + \mathrm{B} \operatorname{curl} \mathbf{e},$$

where B is a constant which measures the rotatory power; and the optical properties of crystals can be explained if we suppose that for them equation (2) is to be replaced by the equations

$$e_x' = \mathrm{A}_1 e_x, \qquad e_y' = \mathrm{A}_2 e_y, \qquad e_z' = \mathrm{A}_3 e_z$$

When these values for the components of $\mathbf{e}'$ are substituted in equation (1), we evidently obtain the same formulae as were derived from the Stokes-Rankine-Rayleigh hypothesis of inertia different in different directions in a crystal; to which Boussinesq's theory of crystal-optics is practically equivalent.

The optical properties of bodies in motion may be accounted for by modifying equation (1), so that it takes the form

$$\rho\frac{\partial^2 \mathbf{e}}{\partial t^2} = -(k+\tfrac{1}{3}n)\text{ grad div }\mathbf{e} + n\nabla^2\mathbf{e} - \rho_1\left(\frac{\partial}{\partial t}+w_x\frac{\partial}{\partial x}+w_y\frac{\partial}{\partial y}+w_z\frac{\partial}{\partial z}\right)^2\mathbf{e},'$$

where $\mathbf{w}$ denotes the velocity of the ponderable body. If the body is an ordinary isotropic one, and if we consider light propagated parallel to the axis of z, in a medium moving in that direction, the light-vector being parallel to the axis of x, the equation reduces to

$$\rho\frac{\partial^2 e_x}{\partial t^2} = n\frac{\partial^2 e_x}{\partial z^2} - \rho_1 A\left(\frac{\partial}{\partial t}+w\frac{\partial}{\partial z}\right)^2 e_x;$$

substituting

$$e_x = f(z - Vt),$$

where V denotes the velocity of propagation of light in the medium estimated with reference to the fixed aether, we obtain for V the value

$$\left(\frac{n}{\rho+\rho_1 A}\right)^{\frac{1}{2}} + \frac{\rho_1 A}{\rho+\rho_1 A}w.$$

The absolute velocity of light is therefore increased by the amount $\rho_1 Aw/(\rho+\rho_1 A)$ owing to the motion of the medium; and this may be written $(\mu^2-1)\,w/\mu^2$, where μ denotes the refractive index; so that Boussinesq's theory leads to the same formula as had been given half a century previously by Fresnel.[1]

It is Boussinesq's merit to have clearly asserted that all space, both within and without ponderable bodies, is occupied by one identical aether, the same everywhere both in inertia and elasticity; and that all aethereal processes are to be represented by two kinds of equations, of which one kind expresses the invariable equations of motion of the aether, while the other kind expresses the interaction between aether and matter. Many years afterwards these ideas were revived in connection with the electromagnetic theory, in the modern forms of which they are indeed of fundamental importance.

[1] cf. p. 110

Chapter VI

FARADAY

Towards the end of the year 1812, Davy received a letter in which the writer, a bookbinder's journeyman named Michael Faraday, expressed a desire to escape from trade, and obtain employment in a scientific laboratory. With the letter was enclosed a neatly written copy [1] of notes which the young man—he was twenty-one years of age—had made of Davy's own public lectures. The great chemist replied courteously, and arranged an interview; at which he learnt that his correspondent had educated himself by reading the volumes which came into his hands for binding. 'There were two,' Faraday wrote later, 'that especially helped me, the *Encyclopaedia Britannica*, from which I gained my first notions of electricity, and Mrs Jane Marcet's *Conversations on Chemistry*, which gave me my foundation in that science.' Already, before his application to Davy, he had performed a number of chemical experiments, and had made for himself a voltaic pile, with which he had decomposed several compound bodies.

At Davy's recommendation Faraday was in the following spring appointed to a post in the laboratory of the Royal Institution, which had been established at the close of the eighteenth century under the auspices of Count Rumford; and here he remained for the whole of his active life, first as assistant, then (on the death of Davy in 1829) as director of the laboratories, and from 1833 onwards as the occupant of a chair of chemistry which was founded for his benefit.

For many years Faraday was directly under Davy's influence, and was occupied chiefly in chemical investigations. But in 1821, when the new field of inquiry opened by Oersted's discovery was attracting attention, he wrote an *Historical Sketch of Electro-Magnetism*,[2] as a preparation for which he carefully repeated the experiments described by the writers he was reviewing; and this seems to have been the beginning of the researches to which his fame is chiefly due.

The memoir which stands first in the published volumes of Faraday's electrical work [3] was communicated to the Royal Society on 24 November 1831. The investigation was inspired, as he tells

[1] Still preserved at the Royal Institution in London.
[2] Published in *Annals of Philosophy*, ii (1821), pp. 195, 274; iii (1822), p. 107
[3] *Experimental Researches in Electricity*, by Michael Faraday; 3 vols.

us, by the hope of discovering analogies between the behaviour of electricity as observed in motion in currents, and the behaviour of electricity at rest on conductors. Static electricity was known to possess the power of 'induction'—i.e. of causing an opposite electrical state on bodies in its neighbourhood; was it not possible that electric currents might show a similar property? The idea at first was that if in any circuit a current were made to flow, any adjacent circuit would be traversed by an induced current, which would persist exactly as long as the inducing current. Faraday found that this was not the case; a current was indeed induced, but it lasted only for an instant, being in fact perceived only when the primary current was started or stopped. It depended, as he soon convinced himself, not on the mere existence of the inducing current, but on its variation.[1]

Faraday now set himself to determine the laws of induction of currents, and for this purpose devised a new way of representing the state of a magnetic field. Philosophers had been long accustomed [2] to illustrate magnetic power by strewing iron filings on a sheet of paper, and observing the curves in which they dispose themselves when a magnet is brought underneath. These curves suggested to Faraday [3] the idea of *lines of magnetic force* [4]; or curves whose direction at every point coincides with the direction of the magnetic intensity at that point; the curves in which the iron filings arrange themselves on the paper resemble these curves so far as is possible subject to the condition of not leaving the plane of the paper.

With these lines of magnetic force Faraday conceived all space to be filled. Every line of force is a closed curve, which in some part of its course passes through the magnet to which it belongs.[5] Hence if any small closed curve be taken in space, the lines of force which intersect this curve must form a tubular surface returning into itself;

[1] This great discovery was only narrowly missed by Ampère; cf. S. P. Thompson, *Phil. Mag.* xxxix (1895), p. 534.

[2] The practice goes back at least as far as Niccolo Cabeo; indeed the curves traced by Petrus Peregrinus on his globular lodestone (cf. p. 33) were projections of lines of force. Among eighteenth-century writers La Hire mentions the use of iron filings, *Mém. de l'Acad*, 1717. Faraday had referred to them in his electromagnetic paper of 1821, *Exp. Res.* ii, p. 127.

[3] They were first defined in *Exp. Res.* § 114: 'By magnetic curves, I mean the lines of magnetic forces, however modified by the juxtaposition of poles, which could be depicted by iron filings; or those to which a very small magnetic needle would form a tangent.'

[4] The term *lines of force* (*lineae virtutis*) had been used by Aristotelian-scholastic philosophers in connection with magnetism, e.g. in the *Philosophia magnetica* (1629) of Niccolo Cabeo.

[5] *Exp. Res.* iii, p. 405

such a surface is called a *tube of force*. From a tube of force we may derive information not only regarding the direction of the magnetic intensity, but also regarding its magnitude ; for the product of this magnitude [1] and the cross-section of any tube is constant along the entire length of the tube.[2] On the basis of this result, Faraday conceived the idea of partitioning all space into compartments by tubes, each tube being such that this product has the same definite value. For simplicity, each of these tubes may be called a ' unit line of force ' ; the strength of the field is then indicated by the separation or concentration of the unit lines of force,[3] so that the number of them which intersect a unit area placed at right angles to their direction at any point measures the intensity of the magnetic field at that point.

Faraday constantly thought in terms of lines of force. ' I cannot refrain,' he wrote, in 1851,[4] ' from again expressing my conviction of the truthfulness of the representation, which the idea of lines of force affords in regard to magnetic action. All the points which are experimentally established in regard to that action—i.e. all that is not hypothetical—appear to be well and truly represented by it.' [5]

Faraday found that a current is induced in a circuit either when the strength of an adjacent current is altered, or when a magnet is brought near to the circuit, or when the circuit itself is moved about in presence of another current or a magnet. He saw from the first [6] that in all cases the induction depends on the relative motion of the circuit and the lines of magnetic force in its vicinity. The precise nature of this dependence was the subject of long-continued further experiments. In 1832 he found [7] that the currents produced by induction under the same circumstances in different wires are proportional to the conducting powers of the wires—a result which

[1] Within the substance of magnetised bodies we must in this connection understand the magnetic intensity to be that experienced in a crevice whose sides are perpendicular to the lines of magnetisation ; in other words, we must take it to be what since Maxwell's time has been called the magnetic induction.

[2] *Exp. Res.* § 3073. This theorem was first proved by the French geometer Michel Chasles in his memoir on the attraction of an ellipsoidal sheet, *Journal de l'Ecole Polyt.* xv (1837), p. 266.

[3] ibid. § 3122. ' The relative amount of force, or of lines of force, in a given space is indicated by their concentration or separation—i.e. by their number in that space.'

[4] *Exp. Res.* § 3174

[5] Some of Faraday's most distinguished contemporaries were far from sharing this conviction. ' I declare,' wrote Sir George Airy in 1855, ' that I can hardly imagine anyone who practically and numerically knows this agreement ' between observation and the results of calculation based on action at a distance, ' to hesitate an instant in the choice between this simple and precise action, on the one hand, and anything so vague and varying as lines of force, on the other hand.' cf. Bence Jones's *Life of Faraday*, ii, p. 353.

[6] *Exp. Res.*, § 116 [7] ibid. § 213

showed that the induction consists in the production of a definite electromotive force, independent of the nature of the wire, and dependent only on the intersections of the wire and the magnetic curves. This electromotive force is produced whether the wire forms a closed circuit (so that a current flows) or is open (so that electric tension results).

All that now remained was to inquire in what way the electromotive force depends on the relative motion of the wire and the lines of force. The answer to this inquiry is, in Faraday's own words,[1] that 'whether the wire moves directly or obliquely across the lines of force, in one direction or another, it sums up the amount of the forces represented by the lines it has crossed,' so that 'the quantity of electricity thrown into a current is directly as the number of curves intersected.'[2] The induced electromotive force is, in fact, simply proportional to the number of the unit lines of magnetic force intersected by the wire per second.

This is the fundamental principle of the induction of currents. Faraday is undoubtedly entitled to the full honour of its discovery; but for a right understanding of the progress of electrical theory at this period, it is necessary to remember that many years elapsed before all the conceptions involved in Faraday's principle became clear and familiar to his contemporaries; and that in the meantime the problem of formulating the laws of induced currents was approached with success from other points of view. There were indeed many obstacles to the direct appropriation of Faraday's work by the mathematical physicists of his own generation; not being himself a mathematician, he was unable to address them in their own language; and his favourite mode of representation by moving lines of force repelled analysts who had been trained in the school of Laplace and Poisson. Moreover, the idea of electromotive force itself, which had been applied to currents a few years previously in Ohm's memoir, was, as we have seen, still involved in obscurity and misapprehension.

A curious question which arose out of Faraday's theory was whether a bar magnet which is rotated on its own axis carries its lines of magnetic force in rotation with it. Faraday himself believed that the lines of force do not rotate.[3] On this view a revolving magnet like the earth is to be regarded as moving through its own lines of force, so that it must become charged at the equator and poles with electricity of opposite signs; and if a wire not partaking

[1] *Exp. Res.*, § 3082 [2] ibid. § 3115 [3] ibid. § 3090

in the earth's rotation were to have sliding contact with the earth at a pole and at the equator, a current would steadily flow through it. Experiments confirmatory of these views were made by Faraday himself[1]; but they do not strictly prove his hypothesis that the lines of force remain at rest; for it is easily seen[2] that, if they were to rotate, that part of the electromotive force which would be produced by their rotation would be derivable from a potential, and so would produce no effect in closed circuits such as Faraday used.

Three years after the commencement of Faraday's researches on induced currents he was led to an important extension of them by an observation which was communicated to him by another worker. William Jenkin had noticed that an electric shock may be obtained with no more powerful source of electricity than a single cell, provided the wire through which the current passes is long and coiled; the shock being felt when contact is broken.[3] As Jenkin did not choose to investigate the matter further, Faraday took it up, and showed[4] that the powerful momentary current, which was observed when the circuit was interrupted, was really an induced current governed by the same laws as all other induced currents, but with this peculiarity, that the induced and inducing currents now flowed in the same circuit. In fact, the current in its steady state establishes in the surrounding region a magnetic field, whose lines of force are linked with the circuit; and the removal of these lines of force when the circuit is broken originates an induced current, which greatly reinforces the primary current just before its final extinction. To this phenomenon the name of *self-induction* has been given.

The circumstances attending the discovery of self-induction occasioned a comment from Faraday on the number of suggestions which were continually being laid before him. He remarked that although at different times a large number of authors had presented him with their ideas, this case of Jenkin was the only one in which any result had followed. 'The volunteers are serious embarrassments generally to the experienced philosopher.'[5]

The discoveries of Oersted, Ampère and Faraday had shown the

[1] *Exp. Res.*, §§ 218, 3109, etc.

[2] cf. W. Weber, *Ann. d. Phys.* lii (1841), p. 353; S. Tolver Preston, *Phil. Mag.* xix (1885), p. 131. In 1891 S. T. Preston (*Phil. Mag.* xxxi, p. 100) designed a crucial experiment to test the question, but it was not tried for want of a sufficiently delicate electrometer.

[3] A similar observation had been made by Joseph Henry (1797–1878), and published in the *Amer. Jour. Sci.* xxii (1832), p. 408. The spark at the rupture of a spirally-wound circuit had been often observed, e.g. by Pouillet and Nobili.

[4] *Exp. Res.*, § 1048

[5] Bence Jones's *Life of Faraday*, ii, p. 45

close connection of magnetic with electric science. But the connection of the different branches of electric science with each other was still not altogether clear. Although Wollaston's experiments of 1801 had in effect proved the identity in kind of the currents derived from frictional and voltaic sources, the question was still regarded as open thirty years afterwards,[1] no satisfactory explanation being forthcoming of the fact that frictional electricity appeared to be a surface-phenomenon, whereas voltaic electricity was conducted within the interior substance of bodies. To this question Faraday now applied himself; and in 1833 he succeeded [2] in showing that every known effect of electricity—physiological, magnetic, luminous, calorific, chemical and mechanical—may be obtained indifferently either with the electricity which is obtained by friction or with that obtained from a voltaic battery. Henceforth the identity of the two was beyond dispute.

Some misapprehension, however, has existed among later writers as to the conclusions which may be drawn from this identification. What Faraday proved is that the process which goes on in a wire connecting the terminals of a voltaic cell is of the same nature as the process which for a short time goes on in a wire by which a condenser is discharged. He did not prove, and did not profess to have proved, that this process consists in the actual movement of a quasi-substance, electricity, from one plate of the condenser to the other, or of two quasi-substances, the resinous and vitreous electricities, in opposite directions. The process had been pictured in this way by many of his predecessors, notably by Volta; and it has since been so pictured by most of his successors; but from such assumptions Faraday himself carefully abstained.

What is common to all theories, and is universally conceded, is that the rate of increase in the total quantity of electrostatic charge within any volume-element is equal to the excess of the influx over the efflux of current from it.[3] This statement may be represented by the equation

$$\frac{\partial \rho}{\partial t} + \operatorname{div} \mathbf{s} = 0, \tag{1}$$

where ρ denotes the volume-density of electrostatic charge, and

[1] cf. John Davy, *Phil. Trans.* (1832), p. 259; W. Ritchie, ibid., p. 279. Davy suggested that the electrical power, 'according to the analogy of the solar ray,' might be 'not a simple power, but a combination of powers, which may occur variously associated, and produce all the varieties of electricity with which we are acquainted.'

[2] *Exp. Res.*, series iii

[3] The first satisfactory experimental proof of the conservation of electric charge was given by Faraday in *Phil. Mag.* xxii (1843), p. 200; *Exp. Res.*, ii, p. 279.

$\mathbf{s}$ the current, at the place (x, y, z) at the time t. Volta's assumption is really one way of interpreting this equation physically : it presents itself when we compare equation (1) with the equation

$$\frac{\partial \rho}{\partial t} + \text{div}\,(\rho \mathbf{v}) = 0,$$

which is the equation of continuity for a fluid of density ρ and velocity $\mathbf{v}$. We may identify the two equations by supposing $\mathbf{s}$ to be of the same physical nature as the product $\rho\mathbf{v}$; and this is precisely what is done by those who accept Volta's assumption.

But other assumptions might be made which would equally well furnish physical interpretations to equation (1). For instance, if we suppose ρ to be the convergence of *any* vector of which $\mathbf{s}$ is the time-flux,[1] equation (1) is satisfied automatically ; we can picture this vector as being of the nature of a displacement. By such an assumption we should avoid altogether the necessity for regarding the conduction current as an actual flow of electric charges, or for speculating whether the drifting charges are positive or negative ; and there would be no longer anything surprising in the production of a null effect by the coalescence of electric charges of opposite signs.

Faraday himself wished to leave the matter open, and to avoid any definite assumption.[2] Perhaps the best indication of his views is afforded by a laboratory note [3] of date 1837 :

'After much consideration of the manner in which the electric forces are arranged in the various phenomena generally, I have come to certain conclusions which I will endeavour to note down without committing myself to any opinion as to the cause of electricity, i.e., as to the nature of the power. If electricity exist independently of matter, then I think that the hypothesis of one fluid will not stand against that of two fluids. There are, I think, evidently what I may call two elements of power, of equal force and acting toward each other. But these powers may be distinguished only by *direction*, and may be no more separate than the north and south forces in the

[1] In symbols,

$$\text{div}\,\mathbf{q} = -\rho$$
$$\frac{\partial \mathbf{q}}{\partial t} = \mathbf{s}$$

where $\mathbf{q}$ denotes the vector in question.

[2] 'His principal aim,' said Helmholtz in the Faraday Lecture of 1881, 'was to express in his new conceptions only facts, with the least possible use of hypothetical substances and forces. This was really a progress in general scientific method, destined to purify science from the last remains of metaphysics.'

[3] Bence Jones's *Life of Faraday*, ii, p. 77

elements of a magnetic needle. They may be the polar points of the forces originally placed in the particles of matter.'

It may be remarked that since the rise of the mathematical theory of electrostatics, the controversy between the supporters of the one-fluid and the two-fluid theories had become manifestly barren. The analytical equations, in which interest was now largely centred, could be interpreted equally well on either hypothesis; and there seemed to be little prospect of discriminating between them by any new experimental discovery. But a problem does not lose its fascination because it appears insoluble. 'I said once to Faraday,' wrote Stokes to his father-in-law in 1879, 'as I sat beside him at a British Association dinner, that I thought a great step would be made when we should be able to say of electricity that which we say of light, in saying that it consists of undulations. He said to me he thought we were a long way off that yet.' [1]

For his next series of researches,[2] Faraday reverted to subjects which had been among the first to attract him as an apprentice attending Davy's lectures: the voltaic pile, and the relations of electricity to chemistry.

It was at this time generally supposed that the decomposition of a solution, through which an electric current is passed, is due primarily to attractive and repellent forces exercised on its molecules by the metallic terminals at which the current enters and leaves the solution. Such forces had been assumed both in the hypothesis of Grothuss and Davy, and in the rival hypothesis of De La Rive [3]; the chief difference between these being that whereas Grothuss and Davy supposed a chain of decompositions and recompositions in the liquid, De La Rive supposed the molecules adjacent to the terminals to be the only ones decomposed, and attributed to their fragments the power of travelling through the liquid from one terminal to the other.

To test this doctrine of the influence of terminals, Faraday moistened a piece of paper in a saline solution, and supported it in the air on wax, so as to occupy part of the interval between two needle points which were connected with an electric machine. When the machine was worked the current was conveyed between the needle points by way of the moistened paper and the two air-intervals on either side of it; and under these circumstances it was found that the salt underwent decomposition. Since in this case no metallic terminals of any kind were in contact with the solution,

[1] Stokes's *Scientific Correspondence*, i, p. 353
[2] *Exp. Res.* § 450 (1833)
[3] cf. pp. 76, 77

it was evident that all hypotheses which attributed decomposition to the action of the terminals were untenable.

The ground being thus cleared by the demolition of previous theories, Faraday was at liberty to construct a theory of his own. He retained one of the ideas of Grothuss' and Davy's doctrine, namely that a chain of decompositions and recombinations takes place in the liquid ; but these molecular processes he attributed not to any action of the terminals, but to a power possessed by the electric current itself, at all places in its course through the solution. If as an example we consider neighbouring molecules A, B, C, D, . . . of the compound—say water, which was at that time believed to be directly decomposed by the current—Faraday supposed that before the passage of the current the hydrogen of A would be in close union with the oxygen of A, and also in a less close relation with the oxygen atoms of B, C, D, . . . : these latter relations being conjectured to be the cause of the attraction of aggregation in solids and fluids.[1] When an electric current is sent through the liquid, the affinity of the hydrogen of A for the oxygen of B is strengthened if A and B lie along the direction of the current ; while the hydrogen of A withdraws some of its bonds from the oxygen of A, with which it is at the moment combined. So long as the hydrogen and oxygen of A remain in association, the state thus induced is merely one of polarisation ; but the compound molecule is unable to stand the strain thus imposed on it, and the hydrogen and oxygen of A part company from each other. Thus decompositions take place, followed by recombinations : with the result that after each exchange an oxygen atom associates itself with a partner nearer to the positive terminal, while a hydrogen atom associates with a partner nearer to the negative terminal.

This theory explains why, in all ordinary cases, the evolved substances appear only at the terminals ; for the terminals are the limiting surfaces of the decomposing substance ; and, except at them, every particle finds other particles having a contrary tendency with which it can combine. It also explains why, in numerous cases, the atoms of the evolved substances are not retained by the terminals (an obvious difficulty in the way of all theories which suppose the terminals to attract the atoms) : for the evolved substances are expelled from the liquid, not drawn out by an attraction.

Many of the perplexities which had harassed the older theories were at once removed when the phenomena were regarded from Faraday's point of view. Thus, mere mixtures (as opposed to chemical

[1] *Exp. Res.*, § 523

compounds) are not separated into their constituents by the electric current; although there would seem to be no reason why the Grothuss-Davy polar attraction should not operate as well on elements contained in mixtures as on elements contained in compounds.

In the latter part of the same year (1833) Faraday took up the subject again.[1] It was at this time that he introduced the terms which have ever since been generally used to describe the phenomena of electrochemical decomposition. To the terminals by which the electric current passes into or out of the decomposing body he gave the name *electrodes*. The electrode of high potential, at which oxygen, chlorine, acids, etc., are evolved, he called the *anode*, and the electrode of low potential, at which metals, alkalis and bases are evolved, the *cathode*. Those bodies which are decomposed directly by the current he named *electrolytes*; the parts into which they are decomposed, *ions*; the acid ions, which travel to the anode, he named *anions*; and the metallic ions, which pass to the cathode, *cations*.

Faraday now proceeded to test the truth of a supposition which he had published rather more than a year previously,[2] and which indeed had apparently been suspected by Gay-Lussac and Thénard [3] so early as 1811; namely, that the rate at which an electrolyte is decomposed depends solely on the intensity of the electric current passing through it, and not at all on the size of the electrodes or the strength of the solution. Having established the accuracy of this law,[4] he found by a comparison of different electrolytes that the mass of any ion liberated by a given quantity of electricity is proportional to its *chemical equivalent*, i.e. to the amount of it required to combine with some standard mass of some standard element. If an element is n-valent, so that one of its atoms can hold in combination n atoms of hydrogen, the chemical equivalent of this element may be taken to be $1/n$ of its atomic weight; and therefore Faraday's result may be expressed by saying that an electric current will liberate exactly one atom of the element in question in the time which it would take to liberate n atoms of hydrogen.[5]

The quantitative law seemed to Faraday [6] to indicate that 'the atoms of matter are in some way endowed or associated with electrical powers, to which they owe their most striking qualities, and amongst

[1] *Exp. Res.*, § 661 [2] ibid. § 377, Dec. 1832
[3] *Recherches physico-chimiques faites sur la pile* (Paris, 1811), p. 12
[4] *Exp. Res.*, §§ 713–821
[5] In the modern units, 96,580 coulombs of electricity must pass round the circuit in order to liberate of each ion a number of grams equal to the quotient of the atomic weight by the valency.
[6] *Exp. Res.*, § 852

them their mutual chemical affinity.' Looking at the facts of electrolytic decomposition from this point of view, he showed how natural it is to suppose that the electricity which passes through the electrolyte is the exact equivalent of that which is *possessed* by the atoms separated at the electrodes; which implies that *there is a certain absolute quantity of the electric power associated with each atom of matter.*

The claims of this splendid speculation he advocated with conviction. 'The harmony,' he wrote,[1] 'which it introduces into the associated theories of definite proportions and electrochemical affinity is very great. According to it, the equivalent weights of bodies are simply those quantities of them which contain equal quantities of electricity, or have naturally equal electric powers; it being the ELECTRICITY which *determines* the equivalent number, *because* it determines the combining force. Or, if we adopt the atomic theory or phraseology, then the atoms of bodies which are equivalent to each other in their ordinary chemical action, have equal quantities of electricity naturally associated with them. But,' he added, 'I must confess I am jealous of the term *atom*: for though it is very easy to talk of atoms, it is very difficult to form a clear idea of their nature, especially when compound bodies are under consideration.'

These discoveries and ideas tended to confirm Faraday in preferring, among the rival theories of the voltaic cell, that one to which all his antecedents and connections predisposed him. The controversy between the supporters of Volta's contact hypothesis on the one hand, and the chemical hypothesis of Davy and Wollaston on the other, had now been carried on for a generation without any very decisive result. In Germany and Italy the contact explanation was generally accepted, under the influence of Christian Heinrich Pfaff (1773–1852) of Kiel, and of Ohm, and, among the younger men, of Gustav Theodor Fechner (1801–87) of Leipzig,[2] and Stefano Marianini (1790–1866) of Modena. Among French writers De La Rive, of Geneva, was, as we have seen, active in support of

[1] ibid. § 869

[2] Johann Christian Poggendorff (1796–1877) of Berlin, for long the editor of the *Annalen der Physik*, leaned originally to the chemical side, but in 1838 became convinced of the truth of the contact theory, which he afterwards actively defended. Moritz Hermann Jacobi (1801–74) of Dorpat, is also to be mentioned among its advocates.

Faraday's first series of investigations on this subject were made in 1834 (*Exp. Res.*, series viii). In 1836 De La Rive followed on the same side with his *Recherches sur la Cause de l'Electr. Voltaique*. The views of Faraday and De La Rive were criticised by Pfaff, *Revision der Lehre vom Galvanismus* (Kiel, 1837), and by Fechner, *Ann. d. Phys.* xlii (1837), p. 481, and xliii (1838), p. 433; translated *Phil. Mag.* xiii (1838), pp. 205, 367. Faraday returned to the question in 1840, *Exp. Res.*, series xvi and xvii.

the chemical hypothesis; and this side in the dispute had always been favoured by the English philosophers.

There is no doubt that when two different metals are put in contact, a difference of potential is set up between them without any apparent chemical action; but while the contact party regarded this as a direct manifestation of a 'contact force' distinct in kind from all other known forces of nature, the chemical party explained it as a consequence of chemical affinity or incipient chemical action between the metals and the surrounding air or moisture. There is also no doubt that the continued activity of a voltaic cell is always accompanied by chemical unions or decompositions; but while the chemical party asserted that these constitute the efficient source of the current, the contact party regarded them as secondary actions, and attributed the continual circulation of electricity to the perpetual tendency of the electromotive force of contact to transfer charge from one substance to another.

One of the most active supporters of the chemical theory among the English physicists immediately preceding Faraday was Peter Mark Roget (1779–1869), to whom are due two of the strongest arguments in its favour. In the first place, carefully distinguishing between the *quantity* of electricity put into circulation by a cell and the *tension* at which this electricity is furnished, he showed that the latter quantity depends on the 'energy of the chemical action'[1]—a fact which, when taken together with Faraday's discovery that the quantity of electricity put into circulation depends on the amount of chemicals consumed, places the origin of voltaic activity beyond all question. Roget's principle was afterwards verified by Faraday[2] and by De La Rive[3]; 'the electricity of the voltaic pile is proportionate in its intensity to the intensity of the affinities concerned in its production,' said the former in 1834; while De La Rive wrote in 1836, 'The intensity of the currents developed in combinations and in decompositions is exactly proportional to the degree of affinity which subsists between the atoms whose combination or separation has given rise to these currents.'

Not resting here, however, Roget brought up another argument of far-reaching significance. 'If,' he wrote,[4] 'there could exist a

[1] 'The absolute quantity of electricity which is thus developed, and made to circulate, will depend upon a variety of circumstances, such as the extent of the surfaces in chemical action, the facilities afforded to its transmission, etc. But its degree of intensity, or *tension*, as it is often termed, will be regulated by other causes, and more especially by the energy of the chemical action.' (Roget's *Galvanism* [1832], § 70.)

[2] *Exp. Res.*, §§ 908, 909, 916, 988, 1958

[3] *Annales de chimie* lxi (1836), p. 38

[4] Roget's *Galvanism* (1832), § 113

power having the property ascribed to it by the [contact] hypothesis, namely, that of giving continual impulse to a fluid in one constant direction, without being exhausted by its own action, it would differ essentially from all the other known powers in nature. All the powers and sources of motion, with the operation of which we are acquainted, when producing their peculiar effects, are expended in the same proportion as those effects are produced ; and hence arises the impossibility of obtaining by their agency a perpetual effect ; or, in other words, a perpetual motion. But the electromotive force ascribed by Volta to the metals when in contact is a force which, as long as a free course is allowed to the electricity it sets in motion, is never expended, and continues to be exerted with undiminished power, in the production of a never-ceasing effect. Against the truth of such a supposition the probabilities are all but infinite.'

This principle, which is little less than the doctrine of conservation of energy applied to a voltaic cell, was reasserted by Faraday. The process imagined by the contact school 'would,' he wrote, 'indeed be a *creation of power*, like no other force in nature.' In all known cases energy is not generated, but only transformed. There is no such thing in the world as 'a pure creation of force ; a production of power without a corresponding exhaustion of something to supply it.' [1]

As time went on, each of the rival theories of the cell became modified in the direction of the other. The contact party admitted the importance of the surfaces at which the metals are in contact with the liquid, where of course the chief chemical action takes place ; and the chemical party confessed their inability to explain the state of tension which subsists before the circuit is closed, without introducing hypotheses just as uncertain as that of contact force.

Faraday's own view on this point [2] was that a plate of amalgamated zinc, when placed in dilute sulphuric acid, 'has power so far to act, by its attraction for the oxygen of the particles in contact with it, as to place the similar forces already active between these and the other particles of oxygen and the particles of hydrogen in the water, in a peculiar state of tension or polarity, and probably also at the same time to throw those of its own particles which are in contact with the water into a similar but opposed state. Whilst this state is retained, no further change occurs : but when it is relieved by completion of the circuit, in which case the forces determined in opposite directions, with respect to the zinc and the electrolyte, are found exactly competent to neutralise each other,

[1] *Exp. Res.*, § 2071 (1840)
[2] *Exp. Res.*, § 949

then a series of decompositions and recompositions takes place amongst the particles of oxygen and hydrogen which constitute the water, between the place of contact with the platina and the place where the zinc is active : these intervening particles being evidently in close dependence upon and relation to each other. The zinc forms a direct compound with those particles of oxygen which were, previously, in divided relation to both it and the hydrogen : the oxide is removed by the acid, and a fresh surface of zinc is presented to the water, to renew and repeat the action.'

These ideas were developed further by the later adherents of the chemical theory, especially by Faraday's friend Christian Friedrich Schönbein [1] of Basle (1799–1868), the discoverer of ozone. Schönbein made the hypothesis more definite by assuming that when the circuit is open, the molecules of water adjacent to the zinc plate are electrically polarised, the oxygen side of each molecule being turned towards the zinc and being negatively charged, while the hydrogen side is turned away from the zinc and is positively charged. In the third quarter of the nineteenth century, the general opinion was in favour of some such conception as this. Helmholtz [2] attempted to grasp the molecular processes more intimately by assuming that the different chemical elements have different attractive powers (exerted only at small distances) for the vitreous and resinous electricities ; thus potassium and zinc have strong attractions for positive charges, while oxygen, chlorine and bromine have strong attractions for negative electricity. This differs from Volta's original hypothesis in little else but in assuming two electric fluids where Volta assumed only one. It is evident that the contact difference of potential between two metals may be at once explained by Helmholtz's hypothesis, as it was by Volta's ; and the activity of the voltaic cell may be referred to the same principles ; for the two ions of which the liquid molecules are composed will also possess different attractive powers for the electricities, and may be supposed to be united respectively with vitreous and resinous charges. Thus when two metals are immersed in the liquid, the circuit being open, the positive ions are attracted to the negative metal and the negative ions to the positive metal, thereby causing a polarised arrangement of the liquid molecules near the metals. When the circuit is closed, the positively charged surface of the positive metal is dissolved into the fluid ; and as the atoms carry their charge with them, the

[1] *Ann. d. Phys.* lxxviii (1849), p. 289, translated *Archives des sc. phys.* xiii (1850), p. 192. Faraday and Schönbein for many years carried on a correspondence, which has been edited by G. W. A. Kahlbaum and F. V. Darbishire (London, Williams and Norgate).

[2] In his celebrated memoir of 1847 on the Conservation of Energy

positive charge on the immersed surface of this metal must be perpetually renewed by a current flowing in the outer circuit.

It will be seen that Helmholtz did not adhere to Davy's doctrine of the electrical nature of chemical affinity quite as simply or closely as Faraday, who preferred it in its most direct and uncompromising form. 'All the facts show us,' he wrote,[1] 'that that power commonly called chemical affinity can be communicated to a distance through the metals and certain forms of carbon; that the electric current is only another form of the forces of chemical affinity; that its power is in proportion to the chemical affinities producing it; that when it is deficient in force it may be helped by calling in chemical aid, the want in the former being made up by an equivalent of the latter; that, in other words, *the forces termed chemical affinity and electricity are one and the same.*'

In the interval between Faraday's earlier and later papers on the cell, some important results on the same subject were published by Frederic Daniell (1790–1845), Professor of Chemistry in King's College, London.[2] Daniell showed that when a current is passed through a solution of a salt in water, the ions which carry the current are those derived from the salt, and not the oxygen and hydrogen ions derived from the water; this follows since a current divides itself between different mixed electrolytes according to the difficulty of decomposing each, and it is known that pure water can be electrolysed only with great difficulty. Daniell further showed that the ions arising from, say, sodium sulphate are not Na_2O and SO_3, but Na and SO_4; and that in such a case as this, sulphuric acid is formed at the anode and soda at the cathode by secondary action, giving rise to the observed evolution of oxygen and hydrogen respectively at these terminals.

The researches of Faraday on the decomposition of chemical compounds placed between electrodes maintained at different potentials led him in 1837 to reflect on the behaviour of such substances as oil of turpentine or sulphur, when placed in the same situation. These bodies do not conduct electricity, and are not decomposed; but if the metallic faces of a condenser are maintained at a definite potential difference, and if the space between them is occupied by one of these insulating substances, it is found that the charge on either face depends on the nature of the insulating substance. If for any particular insulator the charge has a value ϵ times the value which it would have if the intervening body were air, the number ϵ may be regarded as a measure of the influence

[1] *Exp. Res.*, § 918 [2] *Phil. Trans.* (1839), p. 97

which the insulator exerts on the propagation of electrostatic action through it; it was called by Faraday the *specific inductive capacity* of the insulator.[1]

The discovery of this property of insulating substances or *dielectrics* raised the question as to whether it could be harmonised with the old ideas of electrostatic action. Consider, for example, the force of attraction or repulsion between two small electrically charged bodies. So long as they are in air, the force is proportional to the inverse square of the distance; but if the medium in which they are immersed be partly changed—e.g. if a globe of sulphur be inserted in the intervening space—this law is no longer valid; the change in the dielectric affects the distribution of electric intensity throughout the entire field.

The problem could be satisfactorily solved only by forming a physical conception of the action of dielectrics; and such a conception Faraday now put forward.

The original idea had been promulgated long before by his master Davy. Davy, it will be remembered,[2] in his explanation of the voltaic pile, had supposed that at first, before chemical decompositions take place, the liquid plays a part analogous to that of the glass in a Leyden jar, and that in this is involved an electric polarisation of the liquid molecules.[3] This hypothesis was now developed by Faraday. Referring first to his own work on electrolysis, he asserted [4] that the behaviour of a dielectric is exactly the same as that of an electrolyte, up to the point at which the electrolyte breaks down under the electric stress; a dielectric being, in fact, a body which is capable of sustaining the stress without suffering decomposition.

'For,' he argued,[5] 'let the electrolyte be water, a plate of ice being coated with platina foil on its two surfaces, and these coatings connected with any continued source of the two electrical powers, the ice will charge like a Leyden arrangement, presenting a case of common induction, but no current will pass. If the ice be liquefied, the induction will now fall to a certain degree, because a current can now pass; but its passing is dependent upon a *peculiar molecular arrangement* of the particles consistent with the transfer of the elements of the electrolyte in opposite directions. . . . As, therefore, in the

[1] *Exp. Res.*, § 1252 (1837). Cavendish had discovered specific inductive capacity long before, but his papers were still unpublished.

[2] cf. p. 75

[3] This is expressly stated in Davy's *Elements of Chemical Philosophy* (1812), div. i, § 7, where he lays it down that an essential 'property of non-conductors' is 'to receive electrical polarities.'

[4] *Exp. Res.*, §§ 1164, 1338, 1343, 1621

[5] *Exp. Res.*, § 1164

electrolytic action, *induction* appeared to be the *first* step, and *decomposition* the *second* (the power of separating these steps from each other by giving the solid or fluid condition to the electrolyte being in our hands) ; as the induction was the same in its nature as that through air, glass, wax, etc., produced by any of the ordinary means ; and as the whole effect in the electrolyte appeared to be an action of the particles thrown into a peculiar or polarised state, I was glad to suspect that common induction itself was in all cases an *action of contiguous particles*, and that electrical action at a distance (i.e. ordinary inductive action) never occurred except through the influence of the intervening matter.'

Thus at the root of Faraday's conception of electrostatic induction lay this idea that the whole of the insulating medium through which the action takes place is in a state of polarisation similar to that which precedes decomposition in an electrolyte. 'Insulators,' he wrote,[1] 'may be said to be bodies whose particles can retain the polarised state, whilst conductors are those whose particles cannot be permanently polarised.'

The conception which he at this time entertained of the polarisation may be reconstructed from what he had already written concerning electrolytes. He supposed [2] that in the ordinary or unpolarised condition of a body, the molecules consist of atoms which are bound to each other by the forces of chemical affinity, these forces being really electrical in their nature ; and that the same forces are exerted, though to a less degree, between atoms which belong to different molecules, thus producing the phenomena of cohesion. When an electric field is set up, a change takes place in the distribution of these forces ; some are strengthened and some are weakened, the effect being symmetrical about the direction of the applied electric force.

Such a polarised condition acquired by a dielectric when placed in an electric field presents an evident analogy to the condition of magnetic polarisation which is acquired by a mass of soft iron when placed in a magnetic field ; and it was therefore natural that in discussing the matter Faraday should introduce *lines of electric force*, similar to the lines of magnetic force which he had employed so successfully in his previous researches. A line of electric force he defined to be a curve whose tangent at every point has the same direction as the electric intensity.

[1] *Exp. Res.*, § 1338

[2] This must not be taken to be more than an idea which Faraday mentions as present to his mind. He declined as yet to formulate a definite hypothesis.

The changes which take place in an electric field when the dielectric is varied may be very simply described in terms of lines of force. Thus if a mass of sulphur, or other substance of high specific inductive capacity, is introduced into the field, the effect is as if the lines of force tend to crowd into it; as W. Thomson (Kelvin) showed later, they are altered in the same way as the lines of flow of heat, in a case of steady conduction of heat, would be altered by introducing a body of greater conducting power for heat. By studying the figures of the lines of force in a great number of individual cases, Faraday was led to notice that they always dispose themselves as if they were subject to a mutual repulsion, or as if the tubes of force had an inherent tendency to dilate.[1]

It is interesting to interpret by aid of these conceptions the law of Priestley and Coulomb regarding the attraction between two oppositely charged spheres. In Faraday's view, the medium intervening between the spheres is the seat of a system of stresses, which may be represented by an attraction or tension along the lines of electric force at every point, together with a mutual repulsion of these lines, or pressure laterally. Where a line of force ends on one of the spheres, its tension is exercised on the sphere; in this way, every surface-element of each sphere is pulled outwards. If the spheres were entirely removed from each other's influence, the state of stress would be uniform round each sphere, and the pulls on its surface-elements would balance, giving no resultant force on the sphere. But when the two spheres are brought into each other's presence, the unit lines of force become somewhat more crowded together on the sides of the spheres which face than on the remote sides, and thus the resultant pull on either sphere tends to draw it toward the other. When the spheres are at distances great compared with their radii, the attraction is nearly proportional to the inverse square of the distance, which is Priestley's law.

In the following year (1838) Faraday amplified [2] his theory of electrostatic induction, by making further use of the analogy with the induction of magnetism. Fourteen years previously Poisson had imagined [3] an admirable model of the molecular processes which accompany magnetisation; and this was now applied with very little change by Faraday to the case of induction in dielectrics. 'The particles of an insulating dielectric,' he suggested,[4] 'whilst under induction may be compared to a series of small magnetic needles, or, more correctly still, to a series of small insulated

[1] *Exp. Res.*, §§ 1224, 1297 (1837)
[2] *Exp. Res.*, series xiv
[3] cf. p. 65
[4] *Exp. Res.*, § 1679

conductors. If the space round a charged globe were filled with a mixture of an insulating dialectric, as oil of turpentine or air, and small globular conductors, as shot, the latter being at a little distance from each other so as to be insulated, then these would in their condition and action exactly resemble what I consider to be the condition and action of the particles of the insulating dielectric itself. If the globe were charged, these little conductors would all be polar; if the globe were discharged, they would all return to their normal state, to be polarised again upon the recharging of the globe.'

That this explanation accounts for the phenomena of specific inductive capacity may be seen by what follows, which is substantially a translation into electrostatical language of Poisson's theory of induced magnetism.[1]

Let ρ denote volume-density of electric charge. For each of Faraday's 'small shot' the integral

$$\iiint \rho dx\, dy\, dz,$$

integrated throughout the shot, will vanish, since the total charge of the shot is zero; but if $\mathbf{r}$ denote the vector (x, y, z), the integral

$$\iiint \rho\, \mathbf{r}\, dx\, dy\, dz$$

will not be zero, since it represents the electric polarisation of the shot; if there are N shot per unit volume, the quantity

$$\mathbf{P} = \mathrm{N} \iiint \rho\, \mathbf{r}\, dx\, dy\, dz$$

will represent the total polarisation per unit volume. If $\mathbf{d}$ denote the electric force, and $\mathbf{E}$ the average value of $\mathbf{d}$, then $\mathbf{P}$ will be proportional to $\mathbf{E}$, say

$$\mathbf{P} = (\epsilon - 1)\, \mathbf{E}.$$

By integration by parts, assuming all the quantities concerned to vary continuously and to vanish at infinity, we have

$$\iiint \left(\mathrm{P}_x \frac{\partial}{\partial x} + \mathrm{P}_y \frac{\partial}{\partial y} + \mathrm{P}_z \frac{\partial}{\partial z} \right) \phi\,(x, y, z)\, dx\, dy\, dz = -\iiint \phi \,\mathrm{div}\, \mathbf{P}\, dx\, dy\, dz,$$

where ϕ denotes an arbitrary function, and the volume-integrals are taken throughout infinite space. This equation shows that the polar distribution of electric charge on the shot is equivalent to a volume distribution throughout space, of density

$$\bar{\rho} = -\,\mathrm{div}\, \mathbf{P}.$$

[1] W. Thomson (Kelvin), *Camb. and Dub. Math. Journal*, i (Nov. 1845), p. 75; reprinted in his *Papers on Electrostatics and Magnetism*, § 43 sqq.; F. O. Mossotti, *Arch. des sc. phys.* vi (Geneva, 1847), p. 193; *Mem. della Soc. Ital. Modena* (2), xiv (1850), p. 49

Now the fundamental equation of electrostatics may in suitable units be written,

$$\text{div}\ \mathbf{d} = \rho\ ;$$

and this gives on averaging

$$\text{div}\ \mathbf{E} = \rho_1 + \bar{\rho},$$

where ρ_1 denotes the volume-density of free electric charge, i.e. excluding that in the doublets ; or

$$\text{div}\ (\mathbf{E} + \mathbf{P}) = \rho_1,$$

or

$$\text{div}\ (\epsilon\ \mathbf{E}) = \rho_1.$$

This is the fundamental equation of electrostatics, as modified in order to take into account the effect of the specific inductive capacity ϵ.

The conception of action propagated step by step through a medium by the influence of contiguous particles had a firm hold on Faraday's mind, and was applied by him in almost every part of physics. 'It appears to me possible,' he wrote in 1838,[1] 'and even probable, that magnetic action may be communicated to a distance by the action of the intervening particles, in a manner having a relation to the way in which the inductive forces of static electricity are transferred to a distance ; the intervening particles assuming for the time more or less of a peculiar condition, which (though with a very imperfect idea) I have several times expressed by the term *electrotonic state*.' [2]

The same set of ideas sufficed to explain electric currents. Conduction, Faraday suggested,[3] might be 'an action of contiguous particles, dependent on the forces developed in electrical excitement ; these forces bring the particles into a state of tension or polarity ; [4] and being in this state the contiguous particles have a power or capability of communicating these forces, one to the other, by which they are lowered and discharge occurs.'

After working strenuously for the ten years which followed the discovery of induced currents, Faraday found in 1841 that his health was affected [5] ; and for four years he rested. A second period of brilliant discoveries began in 1845.

[1] *Exp. Res.*, § 1729

[2] This name had been devised in 1831 to express the state of matter subject to magneto-electric induction ; cf. *Exp. Res.*, § 60.

[3] *Exp. Res.* iii, p. 513

[4] As in electrostatic induction in dielectrics.

[5] The description of his symptoms has given rise to the conjecture that he was suffering *inter alia* from mercury poisoning : in electric connections he made great use of cups of mercury, some of which would no doubt be spilt occasionally on the floor of the laboratory and ultimately vaporised.

Many experiments had been made at different times by various investigators [1] with the purpose of discovering a connection between magnetism and light. These had generally taken the form of attempts to magnetise bodies by exposure in particular ways to particular kinds of radiation; and a successful issue had been more than once reported, only to be negatived on re-examination.

The true path was first indicated by Sir John Herschel. After his discovery of the connection between the outward form of quartz crystals and their property of rotating the plane of polarisation of light, Herschel remarked that a rectilinear electric current, deflecting a needle to right and left all round it, possesses a helicoidal dissymmetry similar to that displayed by the crystals. The analogy, therefore, 'led me,' he wrote,[2] 'to conclude that a similar connection exists, and must turn up somehow or other, between the electric current and polarised light, and that the plane of polarisation would be deflected by magneto-electricity.'

Herschel never submitted his prediction to the test of experiment; this was done by Faraday, who so far back as 1834 [3] had transmitted polarised light through an electrolytic solution during the passage of the current, in the hope of observing a change of polarisation. This early attempt failed; but in September, 1845, he varied the experiment by placing a piece of heavy glass between the poles of an excited electro-magnet; and found that the plane of polarisation of a beam of light was rotated when the beam travelled through the glass parallel to the lines of force of the magnetic field.[4]

In the following year a Swiss physicist, E. F. Wartmann, described experiments indicating that a phenomenon exactly similar to Faraday's magnetic rotation of the plane of polarisation of light can be observed when rays of radiant heat are made to pass through rock-salt in a strong magnetic field; a complete confirmation of this result was obtained a little later by F. de la Provostaye and P. Q. Desains.[5]

In the year following Faraday's discovery, Airy [6] suggested a way of representing the effect analytically; as might have been expected, this was by modifying the equations which had been

[1] e.g. by Morichini, of Rome, in 1813, *Quart. Journ. Sci.* xix, p. 338; by Samuel Hunter Christie, of Cambridge, in 1825, *Phil. Trans.* (1826), p. 219; and by Mary Somerville in the same year, *Phil. Trans.* (1826), p. 132

[2] Sir J. Herschel in Bence Jones's *Life of Faraday*, p. 205

[3] *Exp. Res.*, § 951

[4] *Ib.*, § 2152

[5] *Ann. chim.* (3), xxvii (1849), p. 232

[6] *Phil. Mag.* xxviii (1846), p. 469

already introduced by MacCullagh for the case of naturally active bodies. In MacCullagh's equations

$$\begin{cases} \dfrac{\partial^2 Y}{\partial t^2} = c_1^2 \dfrac{\partial^2 Y}{\partial x^2} + \mu \dfrac{\partial^3 Z}{\partial x^3} \\ \dfrac{\partial^2 Z}{\partial t^2} = c_1^2 \dfrac{\partial^2 Z}{\partial x^2} - \mu \dfrac{\partial^3 Y}{\partial x^3}, \end{cases}$$

the terms $\partial^3 Z/\partial x^3$ and $\partial^3 Y/\partial x^3$ change sign with x, so that the rotation of the plane of polarisation is always right-handed or always left-handed with respect to the direction of the beam. This is the case in naturally active bodies; but the rotation due to a magnetic field is in the same absolute direction whichever way the light is travelling, so that the derivations with respect to x must be of even order. Airy proposed the equations

$$\begin{cases} \dfrac{\partial^2 Y}{\partial t^2} = c_1^2 \dfrac{\partial^2 Y}{\partial x^2} + \mu \dfrac{\partial Z}{\partial t} \\ \dfrac{\partial^2 Z}{\partial t^2} = c_1^2 \dfrac{\partial^2 Z}{\partial x^2} - \mu \dfrac{\partial Y}{\partial t}, \end{cases}$$

where μ denotes a constant, proportional to the strength of the magnetic field which is used to produce the effect. He remarked, however, that instead of taking $\mu\partial Z/\partial t$ and $\mu\partial Y/\partial t$ as the additional terms, it would be possible to take $\mu\partial^3 Z/\partial t^3$ and $\mu\partial^3 Y/\partial t^3$, or $\mu\partial^3 Z/\partial x^2\partial t$ and $\mu\partial^3 Y/\partial x^2\partial t$, or any other derivates in which the number of differentiations is odd with respect to t and even with respect to x. It may, in fact, be shown by the method previously applied to MacCullagh's formulae that, if the equations are

$$\begin{cases} \dfrac{\partial^2 Y}{\partial t^2} = c_1^2 \dfrac{\partial^2 Y}{\partial x^2} + \mu \dfrac{\partial^{r+s} Z}{\partial x^r \partial t^s} \\ \dfrac{\partial^2 Z}{\partial t^2} = c_1^2 \dfrac{\partial^2 Z}{\partial x^2} - \mu \dfrac{\partial^{r+s} Y}{\partial x^r \partial t^s}, \end{cases}$$

where $(r + s)$ is an odd number, the angle through which the plane of polarisation rotates in unit length of path is a numerical multiple of

$$\frac{\mu}{\tau^{r+s-1} c_1^{\,r+1}},$$

where τ denotes the period of the light. Now it was shown by Verdet [1] that the magnetic rotation is approximately proportional to the inverse square of the wave-length; and hence we must have

$$r + s = 3;$$

[1] *Comptes Rendus*, lvi (1863), p. 630

so that the only equations capable of correctly representing Faraday's effect are either

$$\begin{cases} \dfrac{\partial^2 Y}{\partial t^2} = c_1^2 \dfrac{\partial^2 Y}{\partial x^2} + \mu \dfrac{\partial^3 Z}{\partial x^2 \partial t} \\ \dfrac{\partial^2 Z}{\partial t^2} = c_1^2 \dfrac{\partial^2 Z}{\partial x^2} - \mu \dfrac{\partial^3 Y}{\partial x^2 \partial t} \end{cases}$$

or

$$\begin{cases} \dfrac{\partial^2 Y}{\partial t^2} = c_1^2 \dfrac{\partial^2 Y}{\partial x^2} + \mu \dfrac{\partial^3 Z}{\partial t^3} \\ \dfrac{\partial^2 Z}{\partial t^2} = c_1^2 \dfrac{\partial^2 Z}{\partial x^2} - \mu \dfrac{\partial^3 Y}{\partial t^3}. \end{cases}$$

The former pair arise, as will appear later, in Maxwell's theory of rotatory polarisation; the latter pair, which were suggested in 1868 by Boussinesq,[1] follow from that physical theory of the phenomenon which was more generally accepted by later workers in the electromagnetic theory.[2] There is practically no difference between them.

Airy's work on the magnetic rotation of light was limited in the same way as MacCullagh's work on the rotatory power of quartz; it furnished only an analytical representation of the effect, without attempting to justify the equations. The earliest endeavour to provide a physical theory seems to have been made [3] in 1858, in the inaugural dissertation of Carl Neumann, of Halle.[4] Neumann assumed that every element of an electric current exerts force on the particles of the aether; and in particular that this is true of the molecular currents which constitute magnetisation, although in this case the force vanishes except when the aethereal particle is already in motion. If $\mathbf{e}$ denote the displacement of the aethereal particle m, the force in question may be represented by the term

$$km \left[\frac{\partial \mathbf{e}}{\partial t} \cdot \mathbf{K} \right]$$

where $\mathbf{K}$ denotes the imposed magnetic field, and k denotes a magneto-optic constant characteristic of the body. When this term is introduced into the equations of motion of the aether, they take the form which had been suggested by Airy; whence Neumann's

[1] *Journal de Math.* xiii (1868), p. 430

[2] Y and Z being interpreted as components of electric force

[3] But see two letters from Stokes to W. Thomson of dates 12 Dec. and 13 Dec. 1848, printed by Larmor in *Proc. Camb. Phil. Soc.* xxii (1924), p. 76

[4] *Explicare tentatur, quomodo fiat, ut lucis planum polarisationis per vires el. vel mag. declinetur. Halis Saxonum* (1858). The results were republished in a tract *Die magnetische Drehung der Polarisationsebene des Lichtes* (Halle, 1863).

hypothesis is seen to lead to the incorrect conclusion that the rotation is independent of the wave-length.

The rotation of plane-polarised light depends, as Fresnel had shown,[1] on a difference between the velocities of propagation of the right-handed and left-handed circularly polarised waves into which plane-polarised light may be resolved. In the case of magnetic rotation, this difference was shown by Verdet to be proportional to the component of the magnetic force in the direction of propagation of the light; and Cornu [2] showed further that the mean of the velocities of the right-handed and left-handed waves is equal to the velocity of light in the medium when there is no magnetic field. From these data, by Fresnel's geometrical method, the wave-surface in the medium may be obtained; it is found to consist of two spheres (one relating to the right-handed and one to the left-handed light), each identical with the spherical wave-surface of the unmagnetised medium, displaced from each other along the lines of magnetic force.[3]

The discovery of the connection between magnetism and light gave interest to a short paper of a speculative character which Faraday published [4] in 1846, under the title *Thoughts of Ray-Vibrations*. In this it is possible to trace the progress of Faraday's thought towards something like an electromagnetic theory of light.

Considering first the nature of ponderable matter, he suggests that an ultimate atom may be nothing else than a field of force—electric, magnetic and gravitational—surrounding a point-centre; on this view, which is substantially that of Michell [5] and Boscovich, an atom would have no definite size, but ought rather to be conceived of as completely penetrable, and extending throughout all space; and the molecule of a chemical compound would consist not of atoms side by side, but of 'spheres of power mutually penetrated, and the centres even coinciding.' [6]

All space being thus permeated by lines of force, Faraday suggested that light and radiant heat might be transverse vibrations propagated along these lines of force. In this way he proposed to 'dismiss the aether,' or rather to replace it by lines of force between centres, the centres together with their lines of force constituting the particles of material substances.

[1] cf. p. 159 [2] *Comptes Rendus*, xcii (1881), p. 1368
[3] Cornu, *Comptes Rendus*, xcix (1884), p. 1045
[4] *Phil. Mag.* (3), xxviii (1846), p. 345; *Exp. Res.* iii, p. 447
[5] Michell was led to this opinion by reflecting on Baxter's doctrine of the immateriality of the soul; cf. one of the papers written by Sir William Herschel at Bath and not published until 1911, in Dreyer's edition of Herschel's works.
[6] cf. Bence Jones's *Life of Faraday*, ii, p. 178

If the existence of a luminiferous aether were to be admitted, Faraday suggested that it might be the vehicle of magnetic force; 'for,' he wrote in 1851,[1] 'it is not at all unlikely that if there be an aether, it should have other uses than simply the conveyance of radiations.' This sentence may be regarded as the origin of the electromagnetic theory of light.

At the time when the *Thoughts on Ray-Vibrations* were published, Faraday was evidently trying to comprehend everything in terms of lines of force; his confidence in which had been recently justified by another discovery. A few weeks after the first observation of the magnetic rotation of light, he noticed [2] that a bar of the heavy glass which had been used in this investigation, when suspended between the poles of an electromagnet, set itself *across* the line joining the poles: thus behaving in the contrary way to a bar of an ordinary magnetic substance, which would tend to set itself *along* this line. A simpler manifestation of the effect was obtained when a cube or sphere of the substance was used; in such forms it showed a disposition to move from the stronger to the weaker places of the magnetic field. The pointing of the bar was then seen to be merely the resultant of the tendencies of each of its particles to move outwards into the positions of weakest magnetic action.

Many other bodies besides heavy glass were found to display the same property; in particular, bismuth.[3] The name *diamagnetic* was given to them.

'Theoretically,' remarked Faraday, 'an explanation of the movements of the diamagnetic bodies might be offered in the supposition that magnetic induction caused in them a contrary state to that which it produced in magnetic matter; i.e. that if a particle of each kind of matter were placed in the magnetic field, both would become magnetic, and each would have its axis parallel to the resultant of magnetic force passing through it; but the particle of magnetic matter would have its north and south poles opposite, or facing toward the contrary poles of the inducing magnet, whereas with the diamagnetic particles the reverse would be the case; and hence would result approximation in the one substance, recession in the other. Upon Ampère's theory, this view would be equivalent to the supposition that, as currents are induced in iron and magnetics parallel to those existing in the inducing magnet or battery wire, so

[1] *Exp. Res.*, § 3075 [2] *Phil. Trans.* (1846), p. 21; *Exp. Res.*, § 2253

[3] The repulsion of bismuth in the magnetic field had been previously observed by A. Brugmans in 1778; *Antonii Brugmans Magnetismus*, Leyden, 1778.

in bismuth, heavy glass and diamagnetic bodies the currents induced are in the contrary direction.' [1]

This explanation became generally known as the 'hypothesis of diamagnetic polarity'; it represents diamagnetism as similar to ordinary induced magnetism in all respects, except that the direction of the induced polarity is reversed. It was accepted by other investigators, notably by W. Weber, Plücker, Reich and Tyndall; but was afterwards displaced from the favour of its inventor by another conception, more agreeable to his peculiar views on the nature of the magnetic field. In this second hypothesis, Faraday supposed an ordinary magnetic or *paramagnetic* [2] body to be one which offers a specially easy passage to lines of magnetic force, so that they tend to crowd into it in preference to other bodies; while he supposed a diamagnetic body to have a low degree of conducting power for the lines of force, so that they tend to avoid it. 'If, then,' he reasoned,[3] 'a medium having a certain conducting power occupy the magnetic field, and then a portion of another medium or substance be placed in the field having a greater conducting power, the latter will tend to draw up towards the place of greatest force, displacing the former.' There is an electrostatic effect to which this is quite analogous; a charged body attracts a body whose specific inductive capacity is greater than that of the surrounding medium, and repels a body whose specific inductive capacity is less; in either case the tendency is to afford the path of best conductance to the lines of force.[4]

For some time the advocates of the 'polarity' and 'conduction' theories of diamagnetism carried on a controversy which, indeed, like the controversy between the adherents of the one-fluid and two-fluid theories of electricity, persisted after it had been shown that the rival hypotheses were mathematically equivalent, and that no experiment could be suggested which would distinguish between them.

Meanwhile new properties of magnetisable bodies were being discovered. In 1847 Julius Plücker (1801–68), Professor of Natural Philosophy in the University of Bonn, while repeating and extending Faraday's magnetic experiments, observed [5] that certain uniaxal crystals, when placed between the two poles of a magnet,

[1] *Exp. Res.*, § 2429–30
[2] This term was introduced by Faraday, *Exp. Res.*, § 2790.
[3] *Exp. Res.*, § 2798
[4] The mathematical theory of the motion of a magnetisable body in a non-uniform field of force was discussed by W. Thomson (Kelvin) in 1847.
[5] *Ann. d. Phys.* lxxii (1847), p. 315; Taylor's *Scientific Memoirs*, v, p. 353

tend to set themselves so that the optic axis has the equatorial position. At this time Faraday was continuing his researches ; and, while investigating the diamagnetic properties of bismuth, was frequently embarrassed by the occurrence of anomalous results. In 1848 he ascertained that these were in some way connected with the crystalline form of the substance, and showed [1] that when a crystal of bismuth is placed in a field of uniform magnetic force (so that no tendency to motion arises from its diamagnetism) it sets itself so as to have one of its crystalline axes directed along the lines of force.

At first he supposed this effect to be distinct from that which had been discovered shortly before by Plücker. 'The results,' he wrote,[2] 'are altogether very different from those produced by diamagnetic action. They are equally distinct from those discovered and described by Plücker, in his beautiful researches into the relation of the optic axis to magnetic action ; for there the force is equatorial, whereas here it is axial. So they appear to present to us a new force, or a new form of force, in the molecules of matter, which, for convenience sake, I will conventionally designate by a new word, as the *magnecrystallic* force.' Later in the same year, however, he recognised [3] that 'the phaenomena discovered by Plücker and those of which I have given an account have one common origin and cause.'

The idea of the 'conduction' of lines of magnetic force by different substances, by which Faraday had so successfully explained the phenomena of diamagnetism, he now applied to the study of the magnetic behaviour of crystals. 'If,' he wrote,[4] 'the idea of conduction be applied to these magnecrystallic bodies, it would seem to satisfy all that requires explanation in their special results. A magnecrystallic substance would then be one which in the crystallised state could conduct onwards, or permit the exertion of the magnetic force with more facility in one direction than another ; and that direction would be the magnecrystallic axis. Hence, when in the magnetic field, the magnecrystallic axis would be urged into a position coincident with the magnetic axis, by a force correspondent to that difference, just as if two different bodies were taken, when the one with the greater conducting power displaces that which is weaker.'

This hypothesis led Faraday to predict the existence of another type of magnecrystallic effect, as yet unobserved. 'If such a view were correct,' he wrote,[5] 'it would appear to follow that a

[1] *Phil. Trans.* (1849), p. 1 ; *Exp. Res.*, § 2454
[2] *Exp. Res.*, § 2469
[3] ibid. § 2605
[4] ibid. § 2837
[5] *Exp. Res.*, § 2839

diamagnetic body like bismuth ought to be less diamagnetic when its magnecrystallic axis is parallel to the magnetic axis than when it is perpendicular to it. In the two positions it should be equivalent to two substances having different conducting powers for magnetism, and therefore if submitted to the differential balance ought to present differential phaenomena.' This expectation was realised when the matter was subjected to the test of experiment.[1]

The series of Faraday's *Experimental Researches in Electricity* end in the year 1855. The closing period of his life was quietly spent at Hampton Court, in a house placed at his disposal by the kindness of the Queen ; and here on 25 August 1867 he passed away.

Among experimental philosophers Faraday holds by universal consent the foremost place. The memoirs in which his discoveries are enshrined will never cease to be read with admiration and delight ; and future generations will preserve with an affection not less enduring the personal records and familiar letters, which recall the memory of his humble and unselfish spirit.

[1] ibid. § 2841

Chapter VII

THE MATHEMATICAL ELECTRICIANS OF THE MIDDLE OF THE NINETEENTH CENTURY

WHILE Faraday was engaged in discovering the laws of induced currents in his own way, by use of the conception of lines of force, his contemporary Franz Neumann was attacking the same problem from a different point of view. Neumann preferred to take Ampère as his model; and in 1845 published a memoir,[1] in which the laws of induction of currents were deduced by the help of Ampère's analysis.

Among the assumptions on which Neumann based his work was a rule which had been formulated, not long after Faraday's original discovery, by Emil Lenz,[2] and which may be enunciated as follows: when a conducting circuit is moved in a magnetic field, the induced current flows in such a direction that the ponderomotive forces on it tend to oppose the motion.

Let ds denote an element of the circuit which is in motion, and let Cds denote the component, taken in the direction of motion, of the ponderomotive force exerted by the inducing current on ds, when the latter is carrying unit current; so that the value of C is known from Ampère's theory. Then Lenz's rule requires that the product of C into the strength of the induced current should be negative. Neumann assumed that this is because it consists of a negative coefficient multiplying the square of C; that is, he assumed the induced electromotive force to be proportional to C. He further assumed it to be proportional to the velocity v of the motion; and thus obtained for the electromotive force induced in ds the expression

$$- \epsilon v C ds,$$

where ϵ denotes a constant coefficient. By aid of this formula, in the earlier part[3] of the memoir, he calculated the induced currents in various particular cases.

But having arrived at the formulae in this way, Neumann

[1] *Berlin Abhandlungen* (1845), p. 1; (1848) p. 1; reprinted as no. 10 and no. 36 of Ostwald's *Klassiker*; translated *Journal de Math.* xiii (1848), p. 113

[2] *Ann. d. Phys.* xxxi (1834), p. 483

[3] §§ 1–8. It may be remarked that Neumann, in making use of Ohm's law, was (like everyone else at this time) unaware of the identity of electroscopic force with electrostatic potential.

noticed [1] a peculiarity in them which suggested a totally different method of treating the subject. In fact, on examining the expression for the current induced in a circuit which is in motion in the field due to a magnet, it appeared that this induced current depends only on the alteration caused by the motion in the value of a certain function; and, moreover, that this function is no other than the potential of the ponderomotive forces which, according to Ampère's theory, act between the circuit, supposed traversed by unit current, and the magnet.

Accordingly, Neumann now proposed to reconstruct his theory by taking this potential function as the foundation.

The nature of Neumann's potential, and its connection with Faraday's theory, will be understood from the following considerations:

The potential energy of a magnetic molecule **M** in a field of magnetic intensity **B** is (**B**.**M**); and therefore the potential energy of a current i flowing in a circuit s in this field is

$$i \iint_S (\mathbf{B} \cdot d\mathbf{S}),$$

where S denotes a diaphragm bounded by the circuit s; as is seen at once on replacing the circuit by its equivalent magnetic shell S. If the field **B** be produced by a current i' flowing in a circuit s', we have, by the formula of Biot and Savart,

$$\mathbf{B} = i' \int_{s'} \frac{[d\mathbf{s}' \cdot \mathbf{r}]}{r^3}$$

$$= i' \int_{s'} \operatorname{curl} \frac{d\mathbf{s}'}{r}.$$

Hence, the mutual potential energy of the two currents is

$$ii' \iiint_{s'\ S} \left(\operatorname{curl} \frac{d\mathbf{s}'}{r} \cdot d\mathbf{S}\right),$$

which by Stokes's transformation may be written in the form

$$ii' \iint_{s\ s'} \frac{(d\mathbf{s} \cdot d\mathbf{s}')}{r}.$$

This expression represents the amount of mechanical work which must be performed against the electrodynamic ponderomotive

[1] § 9

forces, in order to separate the two circuits to an infinite distance apart, when the current strengths are maintained unaltered.

The above potential function has been obtained by considering the ponderomotive forces; but it can now be connected with Faraday's theory of induction of currents. For by interpreting the expression

$$\iint_S (\mathbf{B} \cdot \mathbf{dS})$$

in terms of lines of force, we see that the potential function represents the product of i into the number of unit lines of magnetic force due to s', which pass through the gap formed by the circuit s; and since by Faraday's law the currents induced in s depend entirely on the variation in the number of these lines, it is evident that the potential function supplies all that is needed for the analytical treatment of the induced currents. This was Neumann's discovery.

The electromotive force induced in a circuit s by the motion of other circuits s', carrying currents i', is thus proportional to the time-rate of variation of the potential

$$i' \iint_{s\,s'} \frac{(\mathbf{ds} \cdot \mathbf{ds}')}{r};$$

so that if we denote by $\mathbf{a}$ the vector

$$i' \int_{s'} \frac{\mathbf{ds}'}{r},$$

which, of course, is a function of the position of the element ds from which r is measured, then the electromotive force induced in any circuit element ds by any alteration in the currents which give rise to $\mathbf{a}$ is

$$\left(\frac{\partial \mathbf{a}}{\partial t} \cdot \mathbf{ds}\right).$$

The induction of currents is therefore governed by the vector $\mathbf{a}$; this, which is generally known as the *vector-potential*, has from Neumann's time onwards played a great part in electrical theory. It may be readily interpreted in terms of Faraday's conceptions; for $(\mathbf{a} . \mathbf{ds})$ represents the total number of unit lines of magnetic force which have passed across the line-element $\mathbf{ds}$ prior to the instant t. The vector-potential may in fact be regarded as the analytical measure of Faraday's *electrotonic state*.[1]

While Neumann was endeavouring to comprehend the laws of induced currents in an extended form of Ampère's theory, another

[1] cf. p. 189

investigator was attempting a still more ambitious project: no less than that of uniting electrodynamics into a coherent whole with electrostatics.

Wilhelm Weber (1804–90) was in the earlier part of his scientific career a friend and colleague of Gauss at Göttingen. In 1837, however, he became involved in political trouble. The union of Hanover with the British Empire, which had subsisted since the accession of the Hanoverian dynasty to the British throne, was in that year dissolved by the operation of the Salic law; the Princess Victoria succeeded to the crown of England, and her uncle Ernest-Augustus to that of Hanover. The new king revoked the free constitution which the Hanoverians had for some time enjoyed; and Weber, who took a prominent part in opposing this action, was deprived of his professorship. From 1843 to 1849, when his principal theoretical researches in electricity were made, he occupied a chair in the University of Leipzig.

The theory of Weber was in its origin closely connected with the work of another Leipzig professor, Gustav Theodor Fechner, who in 1845 [1] introduced certain assumptions regarding the nature of electric currents. Fechner supposed every current to consist in a streaming of electric charges, the vitreous charges travelling in one direction, and the resinous charges, equal to them in magnitude and number, travelling in the opposite direction with equal velocity. He further supposed that like charges attract each other when they are moving parallel to the same direction, while unlike charges attract when they are moving in opposite directions. On these assumptions he succeeded in bringing Faraday's induction effects into connection with Ampère's laws of electrodynamics.

In 1846 Weber,[2] adopting the same assumptions as Fechner, analysed the phenomena in the following way.

The formula of Ampère for the ponderomotive force between two elements ds, ds' of currents i, i', may be written

$$F = ii'\,ds\,ds'\left(\frac{2}{r}\frac{d^2r}{ds\,ds'} - \frac{1}{r^2}\frac{dr}{ds}\frac{dr}{ds'}\right).$$

Suppose now that λ units of vitreous electricity are contained in unit length of the wire s, and are moving with velocity u; and that an equal quantity of resinous electricity is moving with velocity u in the opposite direction; so that $i = 2\lambda u$.

Let λ', u', denote the corresponding quantities for the other current;

[1] *Ann. d. Phys.* lxiv (1845), p. 337

[2] *Leipzig Abhandl.*, 1846, p. 209; *Ann. d. Phys.* lxxiii (1848), p. 193; English translation in Taylor's *Scientific Memoirs*, v (1852), p. 489

and let the suffix $_1$ be taken to refer to the action between the positive charges in the two wires, the suffix $_2$ to the action between the positive charge in s and the negative charge in s', the suffix $_3$ to the action between the negative charge in s and the positive charge in s', and the suffix $_4$ to the action between the negative charges in the two wires. Then we have

$$\left(\frac{dr}{dt}\right)_1 = u\frac{dr}{ds} + u'\frac{dr}{ds'},$$

and

$$\left(\frac{d^2r}{dt^2}\right)_1 = u^2\frac{d^2r}{ds^2} + 2uu'\frac{d^2r}{ds\,ds'} + u'^2\frac{d^2r}{ds'^2}.$$

By aid of these and the similar equations with the suffixes $_2$, $_3$, $_4$, the equation for the ponderomotive force may be transformed into the equation

$$\mathrm{F} = \frac{\lambda\lambda'\,ds\,ds'}{r^2}\left\{\begin{aligned}&\left(r\frac{d^2r}{dt^2}\right)_1 - \left(r\frac{d^2r}{dt^2}\right)_2 - \left(r\frac{d^2r}{dt^2}\right)_3 + \left(r\frac{d^2r}{dt^2}\right)_4\\ &-\tfrac{1}{2}\left(\frac{dr}{dt}\right)_1^2 + \tfrac{1}{2}\left(\frac{dr}{dt}\right)_2^2 + \tfrac{1}{2}\left(\frac{dr}{dt}\right)_3^2 - \tfrac{1}{2}\left(\frac{dr}{dt}\right)_4^2\end{aligned}\right\}.$$

But this is the equation which we should have obtained had we set out from the following assumptions: that the ponderomotive force between two current-elements is the resultant of the force between the positive charge in ds and the positive charge in ds', of the force between the positive charge in ds and the negative charge in ds', etc.; and that any two electrified particles of charges e and e', whose distance apart is r, repel each other with a force of magnitude

$$\frac{ee'}{r^2}\left\{r\frac{d^2r}{dt^2} - \frac{1}{2}\left(\frac{dr}{dt}\right)^2\right\}.$$

Two such charges would, of course, also exert on each other an electrostatic repulsion, whose magnitude in these units would be $ee'c^2/r^2$, where c denotes a constant [1] of the dimensions of a velocity,

[1] Two different systems of units are used in electromagnetic theory. In the first, the *electrostatic* system of units, a unit electric charge is defined to be such that two unit charges at unit distance apart repel each other with unit ponderomotive force. In the other, the *electromagnetic* system of units, a unit magnetic pole is defined to be such that two unit poles at unit distance apart repel each other with unit ponderomotive force. In this system a unit of electric current is defined to satisfy the condition that the mutual electrodynamic potential of two closed circuits carrying unit currents is equal to the mutual potential of two magnetic shells of strength unity whose boundaries are the two circuits respectively; and the unit of electric charge is then defined to be the quantity of electricity conveyed in unit time by a unit current. This electromagnetic unit of electric charge is c times the unit electrostatic charge.

In the investigation in the test above, electromagnetic units are used; in the electrostatic system of units, the ponderomotive force of repulsion between two moving electrified charges is

$$\frac{ee'}{r^2}\left\{1 + \frac{r}{c^2}\frac{d^2r}{dt^2} - \frac{1}{2c^2}\left(\frac{dr}{dt}\right)^2\right\}$$

whose value is approximately 3×10^{10} cm./sec. So that on these assumptions the total repellent force would be

$$\frac{ee'c^2}{r^2}\left\{1 + \frac{r}{c^2}\frac{d^2r}{dt^2} - \frac{1}{2c^2}\left(\frac{dr}{dt}\right)^2\right\}.$$

This expression for the force between two electric charges was taken by Weber as the basis of his theory. Weber's is the first of the *electron theories*—a name given to any theory which attributes the phenomena of electrodynamics to the agency of moving electric charges, the forces on which depend not only on the position of the charges (as in electrostatics), but also on their velocities.

The latter feature of Weber's theory led its earliest critics to deny that his law of force could be reconciled with the principle of conservation of energy. They were, however, mistaken on this point, as may be seen from the following considerations. The above expression for the force between two charges may be written in the form

$$-\frac{\partial U}{\partial r} + \frac{d}{dt}\left(\frac{\partial U}{\partial\left(\frac{\partial r}{\partial t}\right)}\right),$$

where U denotes the expression

$$\frac{ee'c^2}{r}\left\{1 + \frac{1}{2c^2}\left(\frac{\partial r}{\partial t}\right)^2\right\}.$$

Consider now two material particles at distance r apart, whose mechanical kinetic energy is T, and whose mechanical potential energy is V, and which carry charges e and e'. The equations of motion of these particles will be exactly the same as the equations of motion of a dynamical system for which the kinetic energy is

$$T - \frac{ee'}{2r}\left(\frac{\partial r}{\partial t}\right)^2,$$

and the potential energy is

$$V + \frac{ee'c^2}{r}.$$

To such a system the principle of conservation of energy may be applied : the equation of energy is, in fact,

$$T + V - \frac{ee'}{r^2}\left(\frac{\partial r}{\partial t}\right)^2 + \frac{ee'c^2}{r} = \text{constant}.$$

The first objection made to Weber's theory is thus disposed of; but another and more serious one now presents itself. The occurrence of the negative sign with the term $-\frac{ee'}{2r}\left(\frac{\partial r}{\partial t}\right)^2$ implies that a charge behaves somewhat as if its mass were negative, so that in certain circumstances its velocity might increase indefinitely under the action of a force opposed to the motion. This is one of the vulnerable points of Weber's theory, and has been the object of much criticism. In fact,[1] suppose that one charged particle of mass μ is free to move, and that the other charges are spread uniformly over the surface of a hollow spherical insulator in which the particle is enclosed. The equation of conservation of energy is

$$\tfrac{1}{2}(\mu - ep)v^2 + V = \text{constant},$$

where e denotes the charge of the particle, v its velocity, V its potential energy with respect to the mechanical forces which act on it, and p denotes the quantity

$$\iint \frac{\sigma}{r} \cos^2 (v \hat{.} r)\, dS,$$

where the integration is taken over the sphere, and where σ denotes the surface-density ; p is independent of the position of the particle μ within the sphere. If now the electric charge on the sphere is so great that ep is greater than μ, then v^2 and V must increase and diminish together, which is evidently absurd.

Leaving this objection unanswered, we proceed to show how Weber's law of force between electrons leads to the formulae for the induction of currents.

The mutual energy of two moving charges is

$$\frac{ee'c^2}{r}\left\{1 - \frac{1}{2c^2}\left(\frac{\partial r}{\partial t}\right)^2\right\},$$

or

$$\frac{ee'c^2}{r}\left[1 - \frac{\{(\mathbf{r}\cdot\mathbf{v}') - (\mathbf{r}\cdot\mathbf{v})\}^2}{2c^2r^2}\right],$$

where $\mathbf{v}$ and $\mathbf{v}'$ denote the velocities of the charges ; so that the mutual energy of two current-elements containing charges e, e' respectively of each kind of electricity, is

$$\frac{ee'}{2r^3}[-\{(\mathbf{r}\cdot\mathbf{v}') - (\mathbf{r}\cdot\mathbf{v})\}^2 + \{(\mathbf{r}\cdot\mathbf{v}') + (\mathbf{r}\cdot\mathbf{v})\}^2$$
$$+ \{-(\mathbf{r}\cdot\mathbf{v}') - (\mathbf{r}\cdot\mathbf{v})\}^2 - \{(-\mathbf{r}\cdot\mathbf{v}') + (\mathbf{r}\cdot\mathbf{v})\}^2],$$

or

$$\frac{4ee'(\mathbf{r}\cdot\mathbf{v}')(\mathbf{r}\cdot\mathbf{v})}{r^3}.$$

[1] This example was given by Helmholtz, *Journal für Math.* lxxv (1873), p. 35 ; *Phil. Mag.* xliv (1872), p. 530.

If ds, ds' denote the lengths of the elements, and i, i' the currents in them, we have

$$i\mathbf{ds} = 2e\mathbf{v}, \quad i'\mathbf{ds}' = 2e'\mathbf{v}' ;$$

so the mutual energy of two current-elements is

$$\frac{ii'}{r^3} (\mathbf{r} \cdot \mathbf{ds}') (\mathbf{r} \cdot \mathbf{ds}).$$

The mutual energy of $i\mathbf{ds}$ with all the other currents is therefore

$$i (\mathbf{ds} \cdot \mathbf{a}),$$

where $\mathbf{a}$ denotes a vector-potential

$$\int_{s'} i' \frac{(\mathbf{r} \cdot \mathbf{ds}')\,\mathbf{r}}{r^3}.$$

By reasoning similar to Neumann's, it may be shown that the electromotive force induced in $\mathbf{ds}$ by any alteration in the rest of the field is

$$-\left(\mathbf{ds} \cdot \frac{\partial \mathbf{a}}{\partial t}\right);$$

and thus a complete theory of induced currents may be constructed.

The necessity for induced currents may be inferred by general reasoning from the first principles of Weber's theory. When a circuit s moves in the field due to currents, the velocity of the vitreous charges in s is, owing to the motion of s, not equal and opposite to that of the resinous charges ; this gives rise to a difference in the forces acting on the vitreous and resinous charges in s ; and hence the charges of opposite sign separate from each other and move in opposite directions.

The assumption that positive and negative charges move with equal and opposite velocities relative to the matter of the conductor is one to which, for various reasons which will appear later, objection may be taken ; but it is an integral part of Weber's theory, and cannot be excised from it. In fact, if this condition were not satisfied, and if the law of force were Weber's, electric currents would exert forces on electrostatic charges at rest [1] ; as may be seen by the following example. Let a current flow in a closed circuit formed by arcs of two concentric circles and the portions of the radii connecting their extremities ; then, if Weber's law were true, and if only one kind of electricity were in motion, the current would evidently exert an electrostatic force on a charge placed at the centre of the circles.

[1] This remark was first made by Clausius, *Journal für Math.* lxxxii (1877), p. 86 ; the simple proof given above is due to Grassmann, *Journal für Math.* lxxxiii (1877), p. 57.

It has been shown,[1] indeed, that the assumption of opposite electricities moving with equal and opposite velocities in a circuit is almost inevitable in any theory of the type of Weber's, so long as the mutual action of two charges is assumed to depend only on their relative (as opposed to their absolute) motion.

The law of Weber is not the only one of its kind ; an alternative to it was suggested by Bernhard Riemann (1826–66), in a course of lectures which were delivered [2] at Göttingen in 1861, and which were published after his death by K. Hattendorff. Riemann proposed as the electrokinetic energy of two electrons $e(x, y, z)$ and $e'(x', y', z')$ the expression

$$-\tfrac{1}{2}\frac{ee'}{r}\left\{\left(\frac{\partial x}{\partial t}-\frac{\partial x'}{\partial t}\right)^2+\left(\frac{\partial y}{\partial t}-\frac{\partial y'}{\partial t}\right)^2+\left(\frac{\partial z}{\partial t}-\frac{\partial z'}{\partial t}\right)^2\right\};$$

or in the electrostatic system of units

$$-\frac{ee'}{2rc^2}\left\{\left(\frac{\partial x}{\partial t}-\frac{\partial x'}{\partial t}\right)^2+\left(\frac{\partial y}{\partial t}-\frac{\partial y'}{\partial t}\right)^2+\left(\frac{\partial z}{\partial t}-\frac{\partial z'}{\partial t}\right)^2\right\};$$

this differs from the corresponding expression given by Weber only in that the relative velocity of the two electrons is substituted in place of the component of this velocity along the radius vector. Eventually, as will be seen later, the laws of Riemann and Weber were both abandoned in favour of a third alternative.

At the time, however, Weber's discovery was felt to be a great advance ; and indeed it had, perhaps, the greatest share in awakening mathematical physicists to a sense of the possibilities latent in the theory of electricity. Beyond this, its influence was felt in general dynamics ; for Weber's electrokinetic energy, which resembled kinetic energy in some respects and potential energy in others, could not be precisely classified under either head ; and its introduction, by helping to break down the distinction which had hitherto sub-

[1] H. Lorberg, *Journal für Math.* lxxxiv (1878), p. 305

[2] *Schwere, Elektricität und Magnetismus, nach den Vorlesungen von B. Riemann* (Hannover, 1875), p. 326. Another alternative to Weber's law had been discovered by Gauss so far back as 1835, but was not published until after his death ; cf. Gauss' *Werke*, v, p. 616. It may be remarked that Riemann's form of the Lagrangian function might be written, when the electrostatic term is included,

$$L = -\frac{ee'c^2}{r\sqrt{\{1-(v/c)^2\}}}$$

where v denotes the relative velocity ; or in the electrostatic system of units

$$L = -\frac{ee'}{r\sqrt{\{1-(v/c)^2\}}}$$

sisted between the two parts of the kinetic potential, prepared the way for the modern transformation theory of dynamics.[1]

Another subject whose development was stimulated by the work of Weber was the theory of gravitation. That gravitation is propagated by the action of a medium, and consequently is a process requiring time for its accomplishment, had been an article of faith with many generations of physicists. Indeed, the dependence of the force on the distance between the attracting bodies seemed to suggest this idea ; for a propagation which is truly instantaneous would, perhaps, be more naturally conceived to be effected by some kind of rigid connection between the bodies, which would be more likely to give a force independent of the mutual distance.

It is obvious that, if the simple law of Newton is abandoned, there is a wide field of rival hypotheses from which to choose its successor. The first notable attempt to discuss the question was made by Laplace.[2] Laplace supposed gravity to be produced by the impulsion on the attracted body of a 'gravific fluid,' which flows with a definite velocity toward the centre of attraction—say, the sun. If the attracted body or planet is in motion, the velocity of the fluid relative to it will be compounded of the absolute velocity of the fluid and the reversed velocity of the planet, and the force of gravity will act in the direction thus determined, its magnitude being unaltered by the planet's motion. This amounts to supposing that gravity is subject to an aberrational effect similar to that observed in the case of light. It is easily seen that the modification thus introduced into Newton's law may be represented by an additional perturbing force, directed along the tangent to the orbit in the opposite sense to the motion, and proportional to the planet's velocity and to the inverse square of the distance from the sun. By considering the influence of this force on the secular equation of the moon's motion, Laplace found that the velocity of the gravific fluid must be at least a hundred million times greater than that of light.

The assumptions made by Laplace are evidently in the highest degree questionable ; but the generation immediately succeeding, overawed by his fame, seems to have found no way of improving on them. Under the influence of Weber's ideas, however, astronomers began to think of modifying Newton's law by adding a term involving the velocities of the bodies. Tisserand [3] in 1872 discussed

[1] cf. Whittaker, *Analytical Dynamics*, chaps. ii, iii, xi

[2] *Mécanique Céleste*, livre x, chap. vii, § 22

[3] *Comptes Rendus*, lxxv (1872), p. 760. cf. also *Comptes Rendus*, cx (1890), p. 313, and Holzmüller, *Zeitschrift für Math. u. Phys.* (1870), p. 69.

the motion of the planets round the sun on the supposition that the law of gravitation is the same as Weber's law of electrodynamic action, so that the force is

$$F = \frac{fm\mu}{r^2}\left\{1 - \frac{1}{h^2}\left(\frac{dr}{dt}\right)^2 + \frac{2}{h^2} r \frac{d^2r}{dt^2}\right\},$$

where f denotes the constant of gravitation, m the mass of the planet, μ the mass of the sun, r the distance of the planet from the sun, and h the velocity of propagation of gravitation. The equations of motion may be rigorously integrated by the aid of elliptic functions [1]; but the simplest procedure is to write

$$F = \frac{fm\mu}{r^2} + F_1,$$

and, regarding F_1 as a perturbing function, to find the variation of the constants of elliptic motion. Tisserand showed that the perturbations of all the elements are zero or periodic, and quite insensible, except that of the longitude of perihelion, which has a secular part. If h be assumed equal to the velocity of light, the effect would be to rotate the major axis of the orbit of Mercury in the direct sense 14″ in a century.

Now, as it happened, a discordance between theory and observation was known to exist in regard to the motion of Mercury's perihelion; for Le Verrier had found that the attraction of the planets might be expected to turn the perihelion 527″ in the direct sense in a century, whereas the motion actually observed was greater than this by 38″. It is evident, however, that only $\frac{3}{8}$ of the excess is explained by Tisserand's adoption of Weber's law; and it seemed therefore that this suggestion would prove as unprofitable as Le Verrier's own hypothesis of an intra-mercurial planet. But it was found later [2] that $\frac{3}{4}$ of the excess could be explained by substituting Riemann's electrodynamic law for Weber's, and that a combination of the laws of Riemann and Weber would give exactly the amount desired.[3]

After the publication of his memoir on the law of force between electrons, Weber turned his attention to the question of diamagnetism, and developed Faraday's idea regarding the explanation of diamagnetic phenomena by the effects of electric currents induced

[1] This had been done in an inaugural dissertation by Seegers at Göttingen in 1864.

[2] By Maurice Lévy, *Comptes Rendus*, cx (1890), p. 545

[3] The consequences of adopting the electrodynamic law of Clausius (for which see later) were discussed by Oppenheim, *Zur Frage nach der Fortpflanzungsgeschwindigkeit der Gravitation*, Wien, 1895.

in the diamagnetic bodies.[1] Weber remarked that if, with Ampère, we assume the existence of molecular circuits in which there is no ohmic resistance, so that currents can flow without dissipation of energy, it is quite natural to suppose that currents would be induced in these molecular circuits if they were situated in a varying magnetic field; and he pointed out that such induced molecular currents would confer upon the substance the properties characteristic of diamagnetism.

The difficulty with this hypothesis is to avoid explaining too much; for, if it be accepted, the inference seems to be that all bodies, without exception, should be diamagnetic. Weber escaped from this conclusion by supposing that in iron and other magnetic substances there exist permanent molecular currents, which do not owe their origin to induction, and which, under the influence of the impressed magnetic force, set themselves in definite orientations. Since a magnetic field tends to give such a direction to a pre-existing current that its course becomes opposed to that of the current which would be induced by the increase of the magnetic force, it follows that a substance stored with such pre-existing currents would display the phenomena of paramagnetism. The bodies ordinarily called paramagnetic are, according to this hypothesis, those bodies in which the paramagnetism is strong enough to mask the diamagnetism.

The radical distinction which Weber postulated between the natures of paramagnetism and diamagnetism seemed to be confirmed by many facts which were discovered subsequently. Thus in 1895 P. Curie showed[2] that the magnetic susceptibility per gramme-molecule is connected with the temperature by laws which are different for paramagnetic and diamagnetic bodies. For the former it varies in inverse proportion to the absolute temperature, whereas for diamagnetic bodies it is independent of the temperature.

The conclusions which followed from the work of Faraday and Weber were adverse to the hypothesis of magnetic fluids; for according to that hypothesis the induced polarity would be in the same direction whether due to a change of orientation of pre-existing molecular magnets, or to a fresh separation of magnetic fluids in the molecules. 'Through the discovery of diamagnetism,' wrote Weber[3] in 1852, 'the hypothesis of electric molecular currents

[1] *Leipzig Berichte*, i (1847), p. 346; *Ann. d. Phys.* lxxiii (1848), p. 241; translated Taylor's *Scientific Memoirs*, v, p. 477; *Abhandl. der K. Sächs. Ges.* i (1852), p. 483; *Ann. d. Phys.* lxxxvii (1852), p. 145; trans. Tyndall and Francis' *Scientific Memoirs*, p. 163

[2] *Annales de Chimie* (7), v (1845), p. 289

[3] *Ann. d. Phys.* lxxxvii (1852), p. 145; Tyndall and Francis's *Sci. Mem.*, p. 163

in the interior of bodies is corroborated, and the hypothesis of magnetic fluids in the interior of bodies is refuted.' The latter hypothesis was, moreover, unable to account for the phenomena shown by bodies which are strongly magnetic, like iron ; for it is found that when the magnetising force is gradually increased to a very large value, the magnetisation induced in such bodies does not increase in proportion, but tends to a saturation value. This effect cannot be explained on the assumptions of Poisson, but is easily deducible from those of Weber ; for, according to Weber's theory, the magnetising force merely orients existing magnets ; and when it has attained such a value that all of them are oriented in the same direction, there is nothing further to be done.

If the elementary magnets were supposed to be free to orient themselves without encountering any resistance, it is evident that a very small magnetising force would suffice to turn them all parallel to each other, and thus would produce immediately the greatest possible intensity of induced magnetism. To overcome this difficulty Weber assumed that every displacement of a molecular circuit is resisted by a couple. The form of the hypothesis which ultimately gained acceptance was that the orientation is resisted by couples which arise from the mutual action of the molecular magnets themselves. In the unmagnetised condition the molecules 'arrange themselves so as to satisfy their mutual attraction by the shortest path, and thus form a complete closed circuit of attraction,' as D. E. Hughes wrote [1] in 1883 ; when an external magnetising force is applied, these small circuits are broken up ; and at any stage of the process a molecular magnet is in equilibrium under the joint influence of the external force and the forces due to the other molecules.[2]

This hypothesis, due originally to Weber, was advocated by Maxwell,[3] and was later developed by J. A. Ewing ;[4] its consequences may be illustrated by the following simple example [5] :

Consider two magnetic molecules, each of magnetic moment m, whose centres are fixed at a distance c apart. When undisturbed, they dispose themselves in the position of stable equilibrium, in which they point in the same direction along the line c. Now let an increasing magnetic force H be made to act on them in a direction

[1] *Proc. R. S.* xxxv (1883), p. 178

[2] On the history of this, see W. Peddie, *Nature*, cxx (1927), p. 80

[3] *Treatise on Elect. & Mag.*, § 443

[4] *Phil. Mag.* xxx (1890), p. 205 ; *Magnetic Induction in Iron and other Metals* (London, 1892) ; cf. also E. Warburg, *Phil. Mag.* xv (1883), p. 246

[5] E. G. Gallop, *Messenger of Math.* xxvii (1897), p. 6

at right angles to the line *c*. The magnets turn towards the direction of H; and when H attains the value $3m/c^3$, they become perpendicular to the line *c*, after which they remain in this position, when H is increased further. Thus they display the phenomena of induction initially proportional to the magnetising force, and of saturation. If the magnetising force H be supposed to act parallel to the line *c*, in the direction in which the axes originally pointed, the magnets will remain at rest. But if H acts in the opposite direction, the equilibrium will be stable only so long as H is less than m/c^3; when H increases beyond this limit, the equilibrium becomes unstable, and the magnets turn over so as to point in the direction of H; when H is gradually decreased to zero, they remain in their new positions, thus illustrating the phenomenon of residual magnetism. By taking a large number of such pairs of magnetic molecules, originally oriented in all directions, and at such distances that the pairs do not sensibly influence each other, we may construct a model whose behaviour under the influence of an external magnetic field will closely resemble the actual behaviour of ferromagnetic bodies.

In order that the magnets in the model may come to rest in their new positions after reversal, it will be necessary to suppose that they experience some kind of dissipative force which damps the oscillations; to this would correspond in actual magnetic substances the electric currents which would be set up in the neighbouring mass when the molecular magnets are suddenly reversed; in either case, the sudden reversals are attended by a transformation of magnetic energy into heat.

In 1871 Weber restated his theory of magnetism [1] in terms of a theory of electric particle-charges: an Ampèrean molecular current was now supposed to be constituted by an electric charge moving in an orbit around a fixed electric charge of opposite sign, a picture which reappeared forty years later in the electron orbits of the Rutherford-Bohr atom. In the twentieth century, however, many theoretical physicists were at first inclined to doubt whether the magnetic theory of Ampère and Weber could be successfully expressed in terms of the new electron theory; these doubts were set at rest by P. Langevin,[2] who gave a complete formulation. In this the diamagnetic phenomenon is regarded in all cases as the sole initial effect of applying an external magnetic field; paramagnetism is attributed to subsequent mutual actions between molecules.

The transformation of energy from one form to another is a

[1] *Leipzig Abhandl. Math. Phys.* x (1873), p. 1; English translation in *Phil. Mag.* xliii (1872), pp. 1, 119

[2] *Annales chim. phys.* v (1905), p. 70

subject which was first treated in a general fashion shortly before the middle of the nineteenth century. It had long been known that the energy of motion and the energy of position of a dynamical system are convertible into each other, and that the amount of their sum remains invariable when the system is self-contained. This principal of conservation of dynamical energy had been extended to optics by Fresnel, who had assumed [1] that the energy brought to to an interface by incident light is equal to the energy carried away from the interface by the reflected and refracted beams. A similar conception was involved in Roget's and Faraday's defence [2] of the chemical theory of the voltaic cell ; they argued that the work done by the current in the outer circuit must be provided at the expense of the chemical energy stored in the cell, and showed that the quantity of electricity sent round the circuit is proportional to the quantity of chemicals consumed, while its tension is proportional to the strength of the chemical affinities concerned in the reaction. This theory was extended and completed by James Prescott Joule, of Manchester, in 1841. Joule accepted the principle which had been maintained, and virtually established by Rumford and Davy in 1798–9, that heat is producible from mechanical work and convertible into it ; and he saw that a closer investigation of this principle was necessary in order to achieve a complete theory of the voltaic cell. He therefore measured [3] the amount of heat evolved in unit time in a metallic wire, through which a current of known strength was passed ; he found the amount to be proportional to the resistance of the wire multiplied by the square of the current strength ; or (as follows from Ohm's law) to the current strength multiplied by the difference of electric tensions at the extremities of the wire.

The quantity of energy yielded up as heat in the outer circuit being thus known, it became possible to consider the transference of energy in the circuit as a whole. ' When,' wrote Joule, ' any voltaic arrangement, whether simple or compound, passes a current of electricity through any substance, whether an electrolyte or not, the total voltaic heat which is generated in any time is proportional to the number of atoms which are electrolysed in each cell of the circuit, multiplied by the virtual intensity of the battery ; if a decomposing cell be in the circuit, the virtual intensity of the battery is reduced in proportion to its resistance to electrolysation.' In the same year he [4] enhanced the significance of this by showing that the quantities

[1] cf. p. 182 [2] cf. p. 124 [3] *Phil. Mag.* xix (1841), p. 260 ; Joule's *Scientific Papers*, i, p. 60
[4] *Phil. Mag.* xx (1841), p. 98 ; cf. also *Phil. Mag.* xxii (1843), p. 204

of heat which are evolved by the combustion of the equivalents of bodies are proportional to the intensities of their affinities for oxygen, as measured by the electromotive force of a battery required to decompose the oxide electrolytically.

The theory of Roget and Faraday, thus perfected by Joule, enables us to trace quantitatively the transformations of energy in the voltaic cell and circuit. The primary source of energy is the chemical reaction: in a Daniell cell, $Zn|ZnSO_4|CuSO_4|Cu$, for instance, it is the substitution of zinc for copper as the partner of the sulphion. The strength of the chemical affinities concerned is in this case measured by the difference of the heats of formation of zinc sulphate and copper sulphate; and it is this which determines the electromotive force of the cell.[1] The amount of energy which is changed from the chemical to the electrical form in a given interval of time is measured by the product of the strength of the chemical affinity into the quantity of chemicals decomposed in that time, or (what is the same thing) by the product of the electromotive force of the cell into the quantity of electricity which is circulated. This energy may be either dissipated as heat in conformity to Joule's law, or otherwise utilised in the outer circuit.

Joule was the first to determine the mechanical equivalent of heat by methods which fulfil the essential condition that no ultimate change of state is produced in the matter operated on. The notions of Rumford and Davy regarding a mechanical equivalent of heat were however in the minds of many of his contemporaries; there was a general tendency to return to the doctrine of Boyle, Hooke, Newton and Euler, that heat is not a particular quasi-material entity, but is an oscillatory motion of the smallest parts of bodies; and the 'material' theory decayed until by about 1860 it had completely disappeared. Among the writers who contributed to this result may be mentioned C. F. Mohr of Coblenz (1806–79)[2] and Julius Robert Mayer (1814–78),[3] who was a medical man in Heilbronn. Mayer does not refer to Mohr's paper, although it was published in the same journal and contains almost all that was of any value in his own contribution; and neither seems to have had any influence on contemporary thought, since Helmholtz does not

[1] The heat of formation of a gramme-molecule of $ZnSO_4$ is greater than the heat of formation of a gramme-molecule of $CuSO_4$ by about 50,000 calories; and with divalent metals, 46,000 calories per gramme-molecule corresponds to an e.m.f. of one volt; so the e.m.f. of a Daniell cell should be 50/46 volts, which is nearly the case.

[2] *Annal. der Chimie* xxiv (1837), p. 141; translated *Phil. Mag.* (5) ii (1876), p. 110

[3] *Annal. der Chimie*, xlii (1842), p. 233; translated *Phil. Mag.* xxiv (1862), p. 371. Mayer's paper was first declined by the editors of the *Ann. d. Phys.*

refer to them in the work now to be mentioned. The movement culminated in a memoir published in 1847 by Hermann von Helmholtz [1] (1821–94), under the title *On the Conservation of Force.* It was read originally to the Physical Society of Berlin [2]; but though the younger physicists of the Society received it with enthusiasm, the prejudices of the older generation [3] prevented its acceptance for the *Annalen der Physik*; and it was eventually published as a separate treatise.[4]

In the memoir it was asserted [5] that the conservation of energy is a universal principle of nature; that the kinetic and potential [6] energy of dynamical systems may be converted into heat according to definite quantitative laws, as taught by Rumford, Davy and Joule; and that any of these forms of energy may be converted into the chemical, electrostatic, voltaic and magnetic forms.

A paper "*On the Mechanical Equivalent of Heat*" was read by Joule to a meeting of the British Association [7] at Oxford in 1847. Among the audience was William Thomson, who had taken his degree as Second Wrangler at Cambridge two years previously, and who now saw that in order to bring the entire subject of heat and heat-engines into a form which should be adapted to mathematical treatment and should reveal scientific laws, it was necessary first of all to discover a method of measuring temperature absolutely, which should be altogether independent of the properties of any particular substance. He further saw that this could be founded on a paper which had been published more than forty years previously by Nicolas Léonard Sadi Carnot (1796–1832), elder son of the 'organiser of victory' of the French Revolution. Carnot's work, which

[1] Helmholtz, like Mayer, was a medical man. The origin and function of animal heat had long been a matter of discussion in their profession.

[2] On 23 July 1847

[3] Based largely on the dread of a revival of something like Hegel's natural philosophy

[4] Berlin, G. A. Reimer. English translation in Tyndall and Francis's *Scientific Memoirs*, p. 114. The publisher, to Helmholtz' 'great surprise,' gave him an honorarium. cf. *Hermann von Helmholtz*, by Leo Koenigsberger; English translation by F. A. Welby.

[5] Helmholtz had been partly anticipated by W. R. Grove, in his lectures on the *Correlation of Physical Forces*, which were delivered in 1843 and published in 1846. Grove, after asserting that heat is 'purely dynamical' in its nature, and that the various 'physical forces' may be transformed into each other, remarked: 'The great problem which remains to be solved, in regard to the correlation of physical forces, is the establishment of their equivalent of power, or their measurable relation to a given standard.'

[6] These terms appeared later. *Kinetic energy* was first used in an article by W. Thomson and P. G. Tait in *Good Words* (then edited by Charles Dickens), October 1862. The term *potential energy* was first used by W. J. M. Rankine in 1853 [*Proc. Glasgow Phil. Soc.* iii (1853), p. 276 = Rankine's *Misc. Sci. Papers* (1881), p. 203]. There was however little or nothing new in this, for the thing signified was what Lazare Carnot had called *force vive virtuelle* in 1803, and it had been a familiar concept in dynamics for generations past, while the word *potential* had been in common use since Green had introduced it in 1828.

[7] *Brit. Ass. Rep.* (1847), pt. 2, p. 55

appeared at Paris in 1824 under the title *Réflexions sur la Puissance Motrice du Feu*, was a theoretical study of the machines which produce mechanical work from heat. In such a machine there must be constituent parts which serve the same purposes as the steam, the boiler and the condenser of a steam-engine : that is to say, a *working substance* (analogous to the steam), a *source* of heat (analogous to the boiler) and a *sink* of heat (analogous to the condenser). Carnot considered a process in which the working substance and all parts of the engine are in the same state at the end of the process as at the beginning—what may be called a *cyclical process*, or *cycle*. He showed that in order to obtain the maximum efficiency in the production of mechanical work, it is necessary that there should not be any direct flow of heat between bodies whose temperatures differ by a finite amount ; when this condition is satisfied, and when moreover the engine is supposed to work without friction, the cyclical process is reversible, and the engine may be said to be 'perfect.'

Carnot used for heat the term *calorique*, which had been introduced by Lavoisier [1] ; and he assumed that the amount of caloric taken from the source in a reversible cycle is equal to the amount given up to the sink, just as the amount of water going over the top of a waterfall is equal to the amount that falls on the bottom. He showed that with this assumption the mechanical work obtained depends only on the amount of caloric and the temperatures of the source and sink ; a result which is quite correct when his 'caloric' is identified with the quantity known later as 'entropy.' [2]

Thomson's new idea was thus expressed in his first paper [3] : 'The characteristic property of the scale which I now propose is, that all degrees have the same value ; that is, that a unit of heat descending from a body A at the temperature T° of this scale, to a body B at the temperature (T − 1)°, would give out the same mechanical effect, whatever be the number T. This may justly be termed an absolute scale, since its characteristic is quite independent of the physical properties of any specific substance.'

The 'heat' referred to by Thomson in this sentence was Carnot's 'caloric,' and he was not as yet clear regarding its relation to Joule's 'heat.' The key to this problem was supplied by Clausius in a communication to the Berlin Academy of February 1850 [4] ; and Thomson's principle could now take the form [5] : 'In a Carnot

[1] *Traité élémentaire de chimie*, 1789

[2] The term *entropy* was introduced by Clausius, *Ann. d. Phys.* cxxv (1865), p. 390

[3] *Proc. Camb. Phil. Soc.* i (5 June 1848), p. 66 ; *Phil. Mag.* xxxiii (Oct. 1848), p. 313 ; Thomson's *Math. and Phys. Papers*, i, p. 100

[4] *Ann. d. Phys.* lxxix (1850), pp. 368, 500

[5] *Trans. R. S. Edin.* xx (Mar. 1851), p. 261 ; Thomson's *Math. and Phys. Papers*, i, p. 174

cycle of a reversible engine, the quantity of heat-energy taken from the source is to the quantity of heat-energy given up to the sink, in the ratio of the absolute temperatures of the source and sink. The difference of these amounts of heat-energy is given up to the external world in the form of mechanical work.'

The zero of absolute temperature could therefore be defined by the property that if the sink were at zero absolute temperature, the whole of the heat-energy supplied by the source could be converted into mechanical work. For convenience, the marks 0° and 100° on the scale were taken to be defined as in the Centigrade system: and it was found that the zero of absolute temperature would then be at − 273°·1.

Thomson's principle as stated above may be generalised into the statement that in any reversible process which leaves the working substance in the same state as it was originally,

$$\int \frac{dQ}{T} = 0$$

where T is the absolute temperature at which the element of heat dQ is absorbed by the working substance. Thus the integral $\int dQ/T$ has the same value for every process which leads from a state A to a state B of the working substance, and which is reversible. The value of this integral is called the *increase of entropy* of the working substance in passing from the state A to the state B.

The importance of the principle of energy in connection with the voltaic cell was emphasised by Helmholtz in his memoir of 1847, and by W. Thomson in 1851 [1]: the equations have subsequently received only one important modification, which is due to Helmholtz.[2] Helmholtz pointed out that the electrical energy furnished by a voltaic cell need not be derived exclusively from the energy of the chemical reactions: for the cell may also operate by abstracting heat-energy from neighbouring bodies, and converting this into electrical energy. The extent to which this takes place is determined by a law which was discovered in 1855 by Thomson.[3] Thomson showed that if E denotes the 'available energy,' i.e. possible output of mechanical work, of a system maintained at the absolute temperature T, then a fraction

$$\frac{T}{E}\frac{dE}{dT}$$

of this work is obtained, not at the expense of the thermal or chemical

[1] *Phil. Mag.* ii (1851), pp. 429, 551; Kelvin's *Math. and Phys. Papers*, i, pp. 472, 490

[2] *Berlin Sitzungsber.* (1882), pp. 22, 825; (1883), p. 647

[3] *Quart. Jour. Math.* i (April 1855), p. 57; Kelvin's *Math. and Phys. Papers*, i, p. 297, eqn. (7)

energy of the system itself, but at the expense of the thermal energy of neighbouring bodies. Now in the case of the voltaic cell, the principle of Roget, Faraday and Joule is expressed by the equation

$$E = \lambda,$$

where E denotes the available or electrical energy, which is measured by the electromotive force of the cell, and where λ denotes the heat of the chemical reaction which supplies this energy. In accordance with Thomson's principle, we must replace this equation by

$$E = \lambda + T\frac{dE}{dT},$$

which is the correct relation between the electromotive force of a cell and the energy of the chemical reactions which occur in it. In general the term λ is much larger than the term $T dE/dT$; but in certain classes of cells—e.g. concentration cells—λ is zero; in which case the whole of the electrical energy is procured at the expense of the thermal energy of the cells' surroundings.[1]

To establish Thomson's equation, we use the principle that if dQ is the quantity of heat communicated to a working substance at the absolute temperature T in a series of reversible changes, then dQ/T, being the differential of the entropy of the working substance, is a perfect differential. In the case of the cell, if a quantity e of electricity passes in the circuit, and if $\mho$ is the internal energy of the cell, so that $\lambda = -\frac{\partial \mho}{\partial e}$, then

$$dQ = E\,de + d\mho = (E - \lambda)\,de + \frac{\partial \mho}{\partial T}\,dT.$$

Therefore $\frac{E-\lambda}{T}\,de + \frac{1}{T}\frac{\partial \mho}{\partial T}\,dT$ is a perfect differential,

so $$\frac{\partial}{\partial T}\left(\frac{E-\lambda}{T}\right) = \frac{\partial}{\partial e}\left(\frac{1}{T}\frac{\partial \mho}{\partial T}\right) = \frac{1}{T}\frac{\partial^2 \mho}{\partial T\,\partial e} = -\frac{1}{T}\frac{\partial \lambda}{\partial T}$$

or $$\frac{1}{T}\frac{\partial E}{\partial T} - \frac{E-\lambda}{T^2} = 0$$

or $$E = \lambda + T\frac{\partial E}{\partial T}$$

which is Thomson's equation.

Now returning to Helmholtz' memoir of 1847, let us consider his discussion of the energy of an electrostatic field. It will be

[1] It is at the junctions that this interchange of thermal energy with the outside world generally takes place: it is related to the Peltier effect, described later.

convenient to suppose that the system has been formed by continually bringing from a very great distance infinitesimal quantities of electricity, proportional to the quantities already present at the various points of the system ; so that the charge is always distributed proportionally to the final distribution. Let e typify the final charge at any point of space, and V the final potential at this point. Then at any stage of the process the charge and potential at this point will have the values λe and λV, where λ denotes a proper fraction. At this stage let charges $ed\lambda$ be brought from a great distance and added to the charges λe. The work required for this is

$$\Sigma\, e\, d\lambda \,.\, \lambda\, V,$$

so the total work required in order to bring the system from infinite dispersion to its final state is

$$\Sigma e\, V \,.\int_0^1 \lambda d\lambda, \quad \text{or} \quad \tfrac{1}{2}\Sigma e\, V.$$

By reasoning similar to that used in the case of electrostatic distributions, it may be shown that the energy of a magnetic field, which is due to permanent magnets and which also contains bodies susceptible to magnetic induction, is

$$\tfrac{1}{2}\iiint \rho_0\, \phi\, dx\, dy\, dz,$$

where ρ_0 denotes the density of Poisson's equivalent magnetisation, for the permanent magnets only, and ϕ denotes the magnetic potential.[1]

Helmholtz, moreover, applied the principle of energy to systems containing electric currents. For instance, when a magnet is moved in the vicinity of a current, the energy taken from the battery may be equated to the sum of that expended as Joulean heat and that communicated to the magnet by the electromagnetic force : and this equation shows that the current is not proportional to the electromotive force of the battery, i.e. it reveals the existence of Faraday's magneto-electric induction. As, however, Helmholtz was at the time unacquainted with the conception of the electrokinetic energy stored in connection with a current, his equations were for the most part defective. But in the case of the mutual action of a current and a permanent magnet, he obtained the correct result that the time-integral of the induced electromotive force in the circuit is equal to the increase which takes place in the potential of the magnet towards a current of a certain strength in the circuit.

[1] We suppose all transitions to be continuous, so as to avoid the necessity for writing surface-integrals separately.

The correct theory of the energy of magnetic and electromagnetic fields is due mainly to W. Thomson (Kelvin). Thomson's researches on this subject commenced with one or two short investigations regarding the ponderomotive forces which act on temporary magnets. In 1847 he discussed [1] the case of a small iron sphere placed in a magnetic field, showing that it is acted on by a ponderomotive force represented by $-\text{grad}\, cR^2$, where c denotes a constant, and R denotes the magnetic force of the field; such a sphere must evidently tend to move towards the places where R^2 is greatest. The same analysis may be applied to explain why diamagnetic bodies tend to move, as in Faraday's experiments, from the stronger to the weaker parts of the field.

Two years later Thomson presented to the Royal Society a memoir [2] in which the results of Poisson's theory of magnetism were derived from experimental data, without making use of the hypothesis of magnetic fluids; and this was followed in 1850 by a second memoir,[3] in which Thomson drew attention to the fact previously noticed by Poisson,[4] that the magnetic intensity at a point within a magnetised body depends on the shape of the small cavity in which the exploring magnet is placed. Thomson distinguished two vectors [5]; one of these, by later writers generally denoted by **B**, represents the magnetic intensity at a point situated in a small crevice in the magnetised body, when the faces of the crevice are at right angles to the direction of magnetisation; the vector **B** is always circuital, that is, $\text{div}\, \mathbf{B} = 0$. The other vector, generally denoted by **H**, represents the magnetic intensity in a narrow tubular cavity tangential to the direction of magnetisation; it is an irrotational vector, that is, $\text{curl}\, \mathbf{H} = 0$. The magnetic potential tends at any point to a limit which is independent of the shape of the cavity in which the point is situated; and the space-gradient of this limit is identical with **H**. Thomson called **B** the 'magnetic force according to the electromagnetic definition,' and **H** the 'magnetic force according to the polar definition'; but the names *magnetic induction* and *magnetic force*, proposed by Maxwell, have been generally used by later writers.

It may be remarked that the vector to which Faraday applied the term 'magnetic force,' and which he represented by lines of force, is not **H**, but **B**; for the number of unit lines of force passing

[1] *Camb. and Dub. Math. Journal*, ii (1847), p. 230; W. Thomson's *Papers on Electrostatics and Magnetism*, p. 499; cf. also *Phil. Mag.* xxxvii (1850), p. 241

[2] *Phil. Trans.* (1851), p. 243; Thomson's *Papers on Elect. and Mag.*, p. 345

[3] *Phil. Trans.* (1851), p. 269; *Papers on Elect. and Mag.*, p. 382

[4] cf. p. 64

[5] loc. cit. § 78 of the original paper, and § 517 of the reprint

through any gap must depend only on the gap, and not on the particular diaphragm filling up the gap, across which the flux is estimated ; and this can be the case only if the vector which is represented by the lines of force is a circuital vector.

Thomson introduced a number of new terms into magnetic science—as indeed he did into every science in which he was interested. The ratio of the measure of the induced magnetisation $\mathbf{I}_i$, in a temporary magnet, to the magnetising force $\mathbf{H}$, he named the *susceptibility* ; it is positive for paramagnetic and negative for diamagnetic bodies, and is connected with Poisson's constant k_p[1] by the relation

$$\kappa = \frac{3}{4\pi}\frac{k_p}{1-k_p},$$

where κ denotes the susceptibility. By an easy extension of Poisson's analysis it is seen that the magnetic induction and magnetic force are connected by the equation

$$\mathbf{B} = \mathbf{H} + 4\pi\mathbf{I},$$

where $\mathbf{I}$ denotes the total intensity of magnetisation : so if $\mathbf{I}_0$ denote the permanent magnetisation, we have

$$\begin{aligned}\mathbf{B} &= \mathbf{H} + 4\pi\mathbf{I}_i + 4\pi\mathbf{I}_0,\\ &= \mu\mathbf{H} + 4\pi\mathbf{I}_0,\end{aligned}$$

where μ denotes $(1 + 4\pi\kappa)$; μ was called by Thomson the *permeability*.

In 1851 Thomson extended his magnetic theory so as to include magnecrystallic phenomena. The mathematical foundations of the theory of magnecrystallic action had been laid by anticipation, long before the experimental discovery of the phenomenon, in a memoir read by Poisson to the Paris Academy in February 1824. Poisson, as will be remembered, had supposed temporary magnetism to be due to 'magnetic fluids,' movable within the infinitely small 'magnetic elements' of which he assumed magnetisable matter to be constituted. He had not overlooked the possibility that in crystals these magnetic elements might be non-spherical (e.g. ellipsoidal), and symmetrically arranged ; and had remarked that a portion of such a crystal, when placed in a magnetic field, would act in a manner depending on its orientation. The relations connecting the induced magnetisation $\mathbf{I}$ with the magnetising force $\mathbf{H}$ he had given in a form equivalent to

$$\begin{cases} I_x = aH_x + b'H_y + c''H_z,\\ I_y = a''H_x + bH_y + c'H_z,\\ I_z = a'H_x + b''H_y + cH_z.\end{cases}$$

[1] cf. p. 65

Thomson now showed[1] that the nine coefficients a, b', c'' . . ., introduced by Poisson, are not independent of each other. For a sphere composed of the magnecrystalline substance, if placed in a uniform field of force, would be acted on by a couple : and the work done by this couple when the sphere, supposed of unit volume, performs a complete revolution round the axis of x may be easily shown to be $\pi H\ (1 - H_x^2/H^2)\ (- b'' + c')$. But this work must be zero, since the system is restored to its primitive condition ; and hence b'' and c' must be equal. Similarly $c'' = a'$, and $a'' = b'$. By change of axes three more coefficients may be removed, so that the equations may be brought to the form

$$I_x = \kappa_1 H_x, \quad I_y = \kappa_2 H_y, \quad I_z = \kappa_3 H_z,$$

where κ_1, κ_2, κ_3 may be called the *principal magnetic susceptibilities.*

In the same year (1851) Thomson investigated the energy which, as was evident from Faraday's work on self-induction, must be stored in connection with every electric current. He showed that, in his own words,[2] ' the value of a current in a closed conductor, left without electromotive force, is the quantity of work that would be got by letting all the infinitely small currents into which it may be divided along the lines of motion of the electricity come together from an infinite distance, and make it up. Each of these "infinitely small currents" is of course in a circuit which is generally of finite length ; it is the section of each partial conductor and the strength of the current in it that must be infinitely small.'

Discussing next the mutual energy due to the approach of a permanent magnet and a circuit carrying a current, he arrived at the remarkable conclusion that in this case there is no electrokinetic energy which depends on the mutual action ; the energy is simply the sum of that due to the permanent magnets and that due to the currents. If a permanent magnet is caused to approach a circuit carrying a current, the electromotive force acting in the circuit is thereby temporarily increased ; the amount of energy dissipated as Joulean heat, and the speed of the chemical reactions in the cells, are temporarily increased also. But the increase in the Joulean heat is exactly equal to the increase in the energy derived from consumption of chemicals, together with the mechanical work done on the magnet by the operator who moves it ; so that the balance of energy is perfect, and none needs to be added to or taken from the electrokinetic form. It will now be evident why it was

[1] *Phil. Mag.* (4), i (1851), p. 177 ; *Papers on Electrostatics and Magnetism,* p. 471
[2] *Papers on Electrostatics and Magnetism,* p. 446

that Helmholtz escaped in this case the errors into which he was led in other cases by his neglect of electrokinetic energy ; for in this case there was no electrokinetic energy to neglect.

Two years later, in 1853, Thomson [1] gave a new form to the expression for the energy of a system of permanent and temporary magnets.

We have seen that the energy of such a system is represented by

$$\tfrac{1}{2}\iiint \rho_0\phi \, dx\, dy\, dz,$$

where ρ_0 denotes the density of Poisson's equivalent magnetisation for the permanent magnets, and ϕ denotes the magnetic potential, and where the integration may be extended over the whole of space. Substituting for ρ_0 its value $-$ div $\mathbf{I}_0$,[2] the expression may be written in the form

$$-\tfrac{1}{2}\iiint \phi \operatorname{div} \mathbf{I}_0 \, dx\, dy\, dz\ ;$$

or, integrating by parts,

$$-\tfrac{1}{2}\iiint (\mathbf{I}_0 \cdot \operatorname{grad} \phi)\, dx\, dy\, dz, \quad \text{or} \quad -\tfrac{1}{2}\iiint (\mathbf{H} \cdot \mathbf{I}_0)\, dx\, dy\, dz.$$

Since $\mathbf{B} = \mu\mathbf{H} + 4\pi\mathbf{I}_0$, this expression may be written in the form

$$-\frac{1}{8\pi}\iiint (\mathbf{H} \cdot \mathbf{B})\, dx\, dy\, dz + \frac{1}{8\pi}\iiint \mu\mathbf{H}^2\, dx\, dy\, dz\ ;$$

but the former of these integrals is equivalent to

$$-\frac{1}{8\pi}\iiint (\mathbf{B} \cdot \operatorname{grad} \phi)\, dx\, dy\, dz, \quad \text{or} \quad -\frac{1}{8\pi}\iiint \phi \operatorname{div} \mathbf{B}\, dx\, dy\, dz,$$

which vanishes, since **B** is a circuital vector. The energy of the field, therefore, reduces to

$$\frac{1}{8\pi}\iiint \mu\mathbf{H}^2\, dx\, dy\, dz,$$

integrated over all space ; which is equivalent to Thomson's form.[3]

In the same memoir Thomson returned to the question of the energy which is possessed by a circuit in virtue of an electric current

[1] *Proc. Glasgow Phil. Soc.* iii (1853), p. 281 ; Kelvin's *Math. and Phys. Papers,* i, p. 521

[2] cf. p. 64

[3] The form actually given by Thomson was

$$\frac{1}{8\pi}\iiint \left(\mathbf{H}^2 + \frac{4\pi}{\kappa}\mathbf{I}^2\right) dx\, dy\, dz,$$

which reduces to the above when we neglect that part of $\mathbf{I}^2$ which is due to the permanent magnetism, over which we have no control.

circulating in it. As he remarked, the energy may be determined by calculating the amount of work which must be done in and on the circuit in order to double the circuit on itself while the current is sustained in it with constant strength ; for Faraday's experiments show that a circuit doubled on itself has no stored energy. Thomson found that the amount of work required may be expressed in the form $\frac{1}{2}Li^2$, where i denotes the current strength, and L, which is called the *coefficient of self-induction,* depends only on the form of the circuit.

It may be noticed that in the doubling process the inherent electrodynamic energy is being given up, and yet the operator is doing positive work. The explanation of this apparent paradox is that the energy derived from both these sources is being used to save the energy which would otherwise be furnished by the battery, and which is expended in Joulean heat.

Thomson next proceeded [1] to show that the energy which is stored in connection with a circuit in which a current is flowing may be expressed as a volume-integral extended over the whole of space, similar to the integral by which he had already represented the energy of a system of permanent and temporary magnets. The theorem, as originally stated by its author, applied only to the case of a single circuit ; but it may be established for a system formed by any number of circuits in the following way :

If N_s denote the number of unit tubes of magnetic induction which are linked with the sth circuit, in which a curent i_s is flowing, the electrokinetic energy of the system is $\frac{1}{2}\Sigma N_s i_s$; which may be written $\frac{1}{2}\Sigma I_r$, where I_r denotes the total current flowing through the gap formed by the rth unit tube of magnetic induction. But if $\mathbf{H}$ denote the (vector) magnetic force, and H its numerical magnitude, it is known that $(1/4\pi)\int H ds$, integrated along a closed line of magnetic induction, measures the total current flowing through the gap formed by the line. The energy is therefore $(1/8\pi)\ \Sigma \int H ds$, the summation being extended over all the unit tubes of magnetic induction, and the integration being taken along them. But if dS denote the cross-section of one of these tubes, we have $B dS = 1$, where B denotes the numerical magnitude of the magnetic induction $\mathbf{B}$: so the energy is $(1/8\pi)\ \Sigma B dS \int H ds$; and as the tubes fill all space, we may replace $\Sigma dS \int ds$ by $\iiint dx\, dy\, dz$. Thus the energy takes the form

[1] *Nichols' Cyclopaedia*, 2nd edn., 1860, article ' Magnetism, dynamical relations of' ; reprinted in Thomson's *Papers on Elect. and Mag.*, p. 447, and his *Math. and Phys. Papers*, p. 532

$(1/8\pi) \iiint BH\, dx\, dy\, dz$, where the integration is extended over the whole of space; and since in the present case $B = \mu H$, the energy may also be represented by $(1/8\pi) \iiint \mu H^2\, dx\, dy\, dz$.

But this is identical with the form which was obtained for a field due to permanent and temporary magnets. It thus appears that in all cases the stored energy of a system of electric currents and permanent and temporary magnets is

$$\frac{1}{8\pi} \iiint \mu H^2\, dx\, dy\, dz,$$

where the integration is extended over all space.

It must, however, be remembered that this represents only what in thermodynamics is called the 'available energy'; and it must further be remembered that part even of this available energy may not be convertible into mechanical work within the limitations of the system: e.g. the electrokinetic energy of a current flowing in a single closed perfectly conducting circuit cannot be converted into any other form so long as the circuit is absolutely rigid. All that we can say is that the changes in this stored electrokinetic energy correspond to the work furnished by the system in any change.

The above form suggests that the energy may not be localised in the substance of the circuits and magnets, but may be distributed over the whole of space, an amount $(\mu H^2/8\pi)$ of energy being contained in each unit volume. This conception was afterwards adopted by Maxwell, in whose theory it is of fundamental importance.

While Thomson was investigating the energy stored in connection with electric currents, the equations of flow of the currents were being generalised by Gustav Kirchhoff (1824–87). In 1848 Kirchhoff[1] extended Ohm's theory of linear conduction to the case of conduction in three dimensions; this could be done without much difficulty by making use of the analogy with the flow of heat, which had proved so useful to Ohm. In Kirchhoff's memoir a system is supposed to be formed of three-dimensional conductors, through which steady currents are flowing. At any point let V denote the 'tension' or 'electroscopic force'—a quantity the significance of which in electrostatics was not yet correctly known. Then, within the substance of any homogeneous conductor, the function V must satisfy Laplace's equation $\nabla^2 V = 0$; while at the air-surface of each conductor, the derivate of V taken along

[1] *Ann. d. Phys.* lxxv (1848), p. 189; Kirchhoff's *Ges. Abhandl.*, p. 33

the normal must vanish. At the interface between two conductors formed of different materials, the function V has a discontinuity, which is measured by the value of Volta's contact force for the two conductors ; and, moreover, the condition that the current shall be continuous across such an interface requires that $k\partial V/\partial N$ shall be continuous, where k denotes the ohmic specific conductivity of the conductor, and $\partial/\partial N$ denotes differentiation along the normal to the interface. The equations which have now been mentioned suffice to determine the flow of electricity in the system.

Kirchhoff also showed that the currents distribute themselves in the conductors in such a way as to generate the least possible amount of Joulean heat ; as is easily seen, since the quantity of Joulean heat generated in unit time is

$$\iiint k \left\{ \left(\frac{\partial V}{\partial x}\right)^2 + \left(\frac{\partial V}{\partial y}\right)^2 + \left(\frac{\partial V}{\partial z}\right)^2 \right\} dx\, dy\, dz,$$

where k, as before, denotes the specific conductivity ; and this integral has a stationary value when V satisfies the equation

$$\frac{\partial}{\partial x}\left(k\,\frac{\partial V}{\partial x}\right) + \frac{\partial}{\partial y}\left(k\,\frac{\partial V}{\partial y}\right) + \frac{\partial}{\partial z}\left(k\,\frac{V}{\partial z}\right) = 0.$$

Kirchhoff next applied himself to establish harmony between electrostatical conceptions and the theory of Ohm. That theory had now been before the world for twenty years, and had been verified by numerous experimental researches ; in particular, a careful investigation was made at this time (1848) by Rudolph Kohlrausch (1809–58), who showed [1] that the difference of the electric ' tensions ' at the extremities of a voltaic cell, measured electrostatically with the circuit open, was for different cells proportional to the electromotive force measured by the electrodynamic effects of the cell with the circuit closed ; and, further,[2] that when the circuit was closed, the difference of the tensions, measured electrostatically, at any two points of the outer circuit was proportional to the ohmic resistance existing between them. But in spite of all that had been done, it was still uncertain how ' tension,' or ' electroscopic force,' or ' electromotive force ' should be interpreted in the language of theoretical electrostatics ; it will be remembered that Ohm himself, perpetuating a confusion which had originated with Volta, had identified electroscopic force with density of electric charge, and had assumed that the electricity in a

[1] *Ann. d. Phys.* lxxv (1848), p. 220 [2] *Ann. d. Phys.* lxxviii (1849), p. 1

conductor is at rest when it is distributed uniformly throughout the substance of the conductor.

The uncertainty was finally removed in 1849 by Kirchhoff,[1] who identified Ohm's electroscopic force with the electrostatic potential. That this identification is correct may be seen by comparing the different expressions which have been obtained for electric energy; Helmholtz' expression[2] shows that the energy of a unit charge at any place is proportional to the value of the electrostatic potential at that place; while Joule's result[3] shows that the energy liberated by a unit charge in passing from one place in a circuit to another is proportional to the difference of the electric tensions at the two places. It follows that tension and potential are the same thing.

The work of Kirchhoff was followed by several other investigations which belong to the borderland between electrostatics and electrodynamics. One of the first of these was the study of the Leyden jar discharge.

Early in the century Wollaston, in the course of his experiments on the decomposition of water, had observed that when the decomposition is effected by a discharge of static electricity, the hydrogen and oxygen do not appear at separate electrodes; but that at each electrode there is evolved a mixture of the gases, as if the current had passed through the water in both directions. After this F. Savary[4] had noticed that the discharge of a Leyden jar magnetises needles in alternating layers, and had conjectured that 'the electric motion during the discharge consists of a series of oscillations.' A similar remark was made in connection with a similar observation by Joseph Henry of Washington, in 1842.[5] 'The phenomena,' he wrote, 'require us to admit the existence of a principal discharge in one direction, and then several reflex actions backward and forward, each more feeble than the preceding, until equilibrium is restored.' Helmholtz had repeated the same suggestion in his essay on the conservation of energy; and in 1853 W. Thomson[6] verified it, by investigating the mathematical theory of the discharge, as follows:

Let C denote the capacity of the jar, i.e. the measure of the charge when there is unit difference of potential between the coatings; let R denote the ohmic resistance of the discharging circuit,

[1] ibid. lxxviii (1849), p. 506; Kirchhoff's *Ges. Abhandl.*, p. 49; *Phil. Mag.* (3) xxxvii (1850), p. 463
[2] cf. p. 218 [3] cf. p. 212 [4] *Annales de Chimie*, xxxiv (1827), p. 5
[5] *Proc. Am. Phil. Soc.* ii (1842), p. 193
[6] *Phil. Mag.* (4) v (1853), p. 400; Kelvin's *Math. and Phys. Papers*, i, p. 540

and L its coefficient of self-induction. Then if at any instant t the charge of the condenser be Q, and the current in the wire be i, we have $i = dQ/dt$; while Ohm's law, modified by taking self-induction into account, gives the equation

$$Ri + L\frac{di}{dt} = -\frac{Q}{C}.$$

Eliminating i, we have

$$L\frac{d^2Q}{dt^2} + R\frac{dQ}{dt} + \frac{1}{C}Q = 0,$$

an equation which shows that when $R^2C < 4L$, the subsidence of Q to zero is effected by oscillations of period

$$2\pi\left(\frac{1}{LC} - \frac{R^2}{4L^2}\right)^{-\frac{1}{2}}.$$

If $R^2C > 4L$, the discharge is not oscillatory.

This simple result may be regarded as the beginning of the theory of electric oscillations.[1]

Thomson was at this time much engaged in the problems of submarine telegraphy; and thus he was led to examine the vexed question of the 'velocity of electricity' over long insulated wires and cables. Various workers had made experiments on this subject at different times, but with hopelessly discordant results. Their attempts had generally taken the form of measuring the interval of time between the appearance of sparks at two spark gaps in the same circuit, between which a great length of wire intervened, but which were brought near each other in order that the discharges might be seen together. In one series of experiments, performed by Watson at Shooter's Hill in 1747–8,[2] the circuit was four miles in length, two miles through wire and two miles through the ground; but the discharges appeared to be perfectly simultaneous; whence Watson concluded that the velocity of propagation of electric effects is too great to be measurable.

In 1834 Charles Wheatstone,[3] Professor of Experimental Philosophy in King's College, London, by examining in a revolving mirror sparks formed at the extremities of a circuit, found the velocity of electricity in a copper wire to be about one and a half times the velocity of light. In 1850 H. Fizeau and E. Gounelle,[4]

[1] Thomson's theory was verified by W. Feddersen of Leipzig in a series of papers published in 1857–66; these have been reprinted as Ostwald's *Klassiker*, no. 166. Feddersen was the first to investigate the oscillating discharge fully and correctly.

[2] *Phil. Trans.* xlv (1748), pp. 49, 491

[3] *Phil. Trans.* (1834), p. 583

[4] *Comptes Rendus*, xxx (1850), p. 437

experimenting with the telegraph lines from Paris to Rouen and to Amiens, obtained a velocity about one-third that of light for the propagation of electricity in an iron wire, and nearly two-thirds that of light for the propagation in a copper wire.

The reasons for these discrepancies were not understood at the time. The large value of the velocity found by Wheatstone was mainly due to the fact that his wire was not straight, but was coiled in twenty straight windings : and the effect travels in a zig-zag or spiral wire more quickly than in a straight one.[1]

An important advance was made when Faraday,[2] early in 1854, showed experimentally that a submarine cable, formed of copper wire covered with gutta-percha, 'may be assimilated exactly to an immense Leyden battery ; the glass of the jars represents the gutta-percha ; the internal coating is the surface of the copper wire,' while the outer coating corresponds to the sea water. It follows that in all calculations relating to the propagation of electric disturbances along submarine cables, the electrostatic capacity of the cable must be taken into account.

The theory of signalling by cable originated in a correspondence between Stokes and Thomson in 1854.[3] In the case of long submarine lines, the speed of signalling is so much limited by the electrostatic factor that electromagnetic induction has no sensible effect ; and it was accordingly neglected in the investigation. In view of other applications of the analysis, however, we shall suppose that the cable has a self-induction L per unit length, and that R denotes the ohmic resistance, and C the capacity per unit length, V the electric potential at a distance x from one terminal, and i the current at this place. Ohm's law, as modified for inductance, is expressed by the equation

$$-\frac{\partial V}{\partial x} = L\frac{\partial i}{\partial t} + Ri\,;$$

moreover, since the rate of accumulation of charge in unit length at x is $-\partial i/\partial x$, and since this increases the potential at the rate $-(1/C)\,\partial i/\partial x$, we have

$$C\frac{\partial V}{\partial t} = -\frac{\partial i}{\partial x}.$$

Eliminating i between these two equations, we have an equation

[1] cf. J. Stefan, *Wien Anzeiger*, xxviii (1891), p. 106

[2] *Proc. Roy. Inst.*, 20 Jan. 1854 ; *Phil. Mag.* vii (June 1854), p. 396 ; *Exp. Res.* iii, pp. 508, 521

[3] The Dover–Calais cable was laid in 1851 ; the Holyhead–Howth cable in 1853. The first Atlantic cable, which lasted three weeks, was laid in 1858 ; the second and third in 1866.

first obtained by Oliver Heaviside [1] (1850–1925), namely

$$\frac{1}{C}\frac{\partial^2 V}{\partial x^2} = L\frac{\partial^2 V}{\partial t^2} + R\frac{\partial V}{\partial t},$$

which is known as the *equation of telegraphy*.[2]

Thomson, in one of his letters [3] to Stokes, in 1854, obtained this equation in the form which applies to Atlantic cables, i.e. with the term in L neglected. In this form it is the same as Fourier's equation for the linear propagation of heat: so that the known solutions of Fourier's theory may be used in a new interpretation. If we substitute

$$V = e^{2nt\sqrt{-1} + \lambda x},$$

we obtain

$$\lambda = \pm(1 + \sqrt{-1})(nCR)^{\frac{1}{2}};$$

and therefore a typical elementary solution of the equation is

$$V = e^{-(nCR)^{\frac{1}{2}}x}\sin\{2nt - (nCR)^{\frac{1}{2}}x\}.$$

The form of this solution shows that if a regular harmonic variation of potential is applied at one end of a cable, the phase is propagated with a velocity which is proportional to the square root of the frequency of the oscillations: since therefore the different harmonics are propagated with different velocities, it is evident that no definite 'velocity of transmission' is to be expected for ordinary signals. If a potential is suddenly applied at one end of the cable, a certain time elapses before the current at the other end attains a definite percentage of its maximum value; but it may easily be shown [4] that this retardation is proportional to the resistance and to the capacity, and each of these is proportional to the length of the cable: so the retardation is proportional to the square of the length: thus the apparent velocity of propagation would be less, the greater the length of cable used.

The case of a telegraph wire insulated in the air on poles is different from that of a cable; for here the capacity is small, and it is necessary to take into account the inductance. If in the general equation of telegraphy we write

$$V = e^{nx\sqrt{-1} + \mu t},$$

we obtain the equation

$$\mu = -\frac{R}{2L} \pm \left(\frac{R^2}{4L^2} - \frac{n^2}{CL}\right)^{\frac{1}{2}};$$

[1] *Phil. Mag.* ii (Aug. 1876), p. 135

[2] We have neglected leakage, which is beside our present purpose.

[3] *Proc. R. S.* vii (May 1855), p. 382; Kelvin's *Math. and Phys. Papers,* ii, p. 61

[4] This result, indeed, follows at once from the theory of dimensions.

as the capacity is small, we may replace the quantity under the radical by its second term : and thus we see that a typical elementary solution of the equation is

$$V = e^{-\frac{Rt}{2L}} \sin n \{x - (CL)^{-\frac{1}{2}} t\} ;$$

this shows that any harmonic disturbance, and therefore any disturbance whatever, is propagated along the wire with velocity $(CL)^{-\frac{1}{2}}$. The difference between propagation in a telegraph wire and propagation in an oceanic cable is, as Thomson remarked, similar to the difference between the propagation of an impulsive pressure through a long column of fluid in a tube when the tube is rigid (case of the telegraph wire) and when it is elastic, so as to be capable of local distension (case of the cable, the distension corresponding to the effect of capacity) ; in the former case, as is well known, the impulse is propagated with a definite velocity, namely, the velocity of sound in the fluid.

The work of Thomson on signalling along cables was followed in 1857 by a celebrated investigation [1] of Kirchhoff's, on the propagation of electric disturbance along a telegraph wire of circular cross-section.

Kirchhoff assumed that the electric charge is practically all resident on the surface of the wire, and that the current is uniformly distributed over its cross-section ; his idea of the current was the same as that of Fechner and Weber, namely, that it consists of equal streams of vitreous and resinous electricity flowing in opposite directions. Denoting the electric potential by V, the charge per unit length of wire by e, the length of the wire by l, and the radius of its cross-section by a, he showed that V is determined approximately by the equation [2]

$$V = 2e \log (l/a).$$

The next factor to be considered is the mutual induction of the current-elements in different parts of the wire. Assuming with

[1] *Ann. d. Phys.* c (1857), pp. 193, 251 ; Kirchhoff's *Ges. Abhandl.*, p. 131 ; *Phil. Mag.* xiii (1857), p. 393

[2] His method of obtaining this equation was to calculate separately the effects of (1) the portion of the wire within a distance ϵ on either side of the point considered, where ϵ denotes a length small compared with l, but large compared with a, and (2) the rest of the wire. He thus obtained the equation

$$V = 2e \log \frac{2\epsilon}{a} + \int \frac{e' ds'}{r},$$

where the integration is to be taken over all the length of the wire except the portion 2ϵ : the equation given in the text was then derived by an approximation, which, however, is open to some objection.

Weber that the electromotive force induced in an element **ds** due to another element **ds**′ carrying a current i' is derivable from a vector-potential

$$i' \frac{(\mathbf{r}.\mathbf{ds}')\mathbf{r}}{r^3},$$

Kirchhoff found for the vector-potential due to the entire wire the approximate value

$$w = 2i \log (l/a),$$

where i denotes the strength of the current [1]; the vector-potential being directed parallel to the wire. Ohm's law then gives the equation

$$i = -\pi k a^2 \left(\frac{\partial V}{\partial x} + \frac{1}{c^2}\frac{\partial w}{\partial t}\right),$$

where k denotes the specific conductivity of the material of which the wire is composed; and finally the principle of conservation of electricity gives the equation

$$\frac{\partial i}{\partial x} = -\frac{\partial e}{\partial t}.$$

Denoting $\log (l/a)$ by γ, and eliminating e, i, w from these four equations, we have

$$\frac{\partial^2 V}{\partial x^2} = \frac{1}{c^2}\frac{\partial^2 V}{\partial t^2} + \frac{1}{2\gamma\kappa\pi a^2}\frac{\partial V}{\partial t},$$

which is, as might have been expected, the equation of telegraphy. When the term in $\partial V/\partial t$ is ignored, as we have seen is in certain cases permissible, the equation becomes

$$\frac{\partial^2 V}{\partial x^2} = \frac{1}{c^2}\frac{\partial^2 V}{\partial t^2},$$

which shows that *the electric disturbance is propagated along the wire with the velocity c.*[2] Kirchhoff's procedure has, in fact, involved the calculation of the capacity and self-induction of the wire, and is thus able to supply the definite values of the quantities which were left undetermined in the general equation of telegraphy.

[1] This expression was derived in a similar way to that for V, by an intermediate formula

$$w = 2i \log \frac{2\epsilon}{a} + \int \frac{i' ds'}{r} \cos\theta \cos\theta',$$

where θ and θ' denote respectively the angles made with **r** by **ds** and **ds**′.

[2] In referring to the original memoirs of Weber and Kirchhoff, it must be remembered that the quantity which in the present work is denoted by c, and which represents the velocity of light in free aether, was by these writers denoted by $c/\sqrt{2}$. Weber, in fact, denoted by c the relative velocity with which two charges must approach each other in order that the force between them, as calculated by his formula, should vanish.

It must also be remembered that those writers who accepted the hypothesis that currents consist of equal and opposite streams of vitreous and resinous electricity, were accustomed to write $2i$ to denote the current strength.

The velocity c, whose importance was thus demonstrated, has already been noticed in connection with Weber's law of force; it is a factor of proportionality, which must be introduced when electrodynamic phenomena are described in terms of units which have been defined electrostatically,[1] or conversely when units which have been defined electrodynamically are used in the description of electrostatic phenomena. That the factor which is introduced on such occasions must be of the dimensions (length/time), may be easily seen; for the electrostatic repulsion between electric charges is a quantity of the same kind as the electrodynamic repulsion between two definite *lengths of* wire, carrying currents which may be specified by the amount of charge which travels past any point in unit *time*.

Shortly before the publication of Kirchhoff's memoir, the value of c had been determined by Weber and Kohlrausch[2]; their determination rested on a comparison of the measures of the charge of a Leyden jar, as obtained by a method depending on electrostatic attraction, and by a method depending on the effects of the current produced by discharging the jar. The resulting value was nearly

$$c = 3.1 \times 10^{10} \text{ cm./sec.};$$

which was the same, within the limits of the errors of measurement, as the speed with which light travels in interplanetary space. The coincidence was noticed by Kirchhoff, who was thus the first to discover the important fact that the velocity with which an electric disturbance is propagated along a perfectly conducting aerial wire is equal to the velocity of light.

In a second memoir published in the same year, Kirchhoff[3] extended the equations of propagation of electric disturbance to the case of three-dimensional conductors.

As in his earlier investigation, he divided the electromotive force at any point into two parts, of which one is the gradient of the electrostatic potential ϕ, and the other is the derivate with respect to the time (with sign reversed) of a vector-potential $\mathbf{a}$; so that if $\mathbf{s}$ denote the current and k the specific conductivity, Ohm's law is expressed by the equation

$$\mathbf{s} = k\left(c^2 \operatorname{grad} \phi - \frac{\partial \mathbf{a}}{\partial t}\right).$$

[1] cf. p. 202

[2] *Ann. d. Phys.* xcix (1856), p. 10. Weber was really following up the work of absolute measurements begun by himself and Gauss in connection with terrestrial magnetism.

[3] *Ann. d. Phys.* cii (1857), p. 529; *Ges. Abhandl.*, p. 154

Kirchhoff calculated the value of **a** by aid of Weber's formula for the inductive action of one current-element on another ; the result is

$$\mathbf{a} = \iiint \frac{dx'\, dy'\, dz'}{r^3} (\mathbf{r} \cdot \mathbf{s}')\, \mathbf{r},$$

where **r** denotes the vector from the point (x, y, z), at which **a** is measured, to any other point (x', y', z') of the conductor, at which the current is **s** ; and the integration is extended over the whole volume of the conductor. The remaining general equations are the ordinary equation of the electrostatic potential

$$\nabla^2 \phi + 4\pi\rho = 0$$

(where ρ denotes the density of electric charge), and the equation of conservation of electricity

$$\frac{\partial \rho}{\partial t} + \operatorname{div} \mathbf{s} = 0.$$

It will be seen that Kirchhoff's electrical researches were greatly influenced by those of Weber. The latter investigations, however, did not enjoy unquestioned authority ; for there was still a question as to whether the expressions given by Weber for the mutual energy of two current-elements, and for the mutual energy of two electrons, were to be preferred to the rival formulae of Neumann and Riemann. The matter was examined in 1870 by Helmholtz, in a series of memoirs [1] to which reference has already been made.[2] Helmholtz remarked that, for two elements **ds**, **ds**′, carrying currents i, i', the electrodynamic energy is

$$\frac{ii'\,(\mathbf{ds} \cdot \mathbf{ds}')}{r},$$

according to Neumann, and

$$\frac{ii'}{r^3}\,(\mathbf{r} \cdot \mathbf{ds})\,(\mathbf{r} \cdot \mathbf{ds}'),$$

according to Weber ; and that these expressions differ from each other only by the quantity

$$\frac{ii'\, ds ds'}{r} \{- \cos (ds \cdot ds') + \cos (r \cdot ds) \cos (r \cdot ds') \},$$

or

$$ii'\, ds ds' \frac{d^2 r}{ds\, ds'} ;$$

since this vanishes when integrated round either circuit, the two

[1] *Journal für Math.* lxxii (1870), p. 57 ; lxxv (1873), p. 35 ; lxxviii (1874), p. 273
[2] cf. p. 204

formulae give the same result when applied to entire currents. A general formula including both that of Neumann and that of Weber is evidently

$$\frac{ii'\,(\mathbf{ds}\,.\,\mathbf{ds}')}{r} + kii'\,\frac{d^2r}{ds\,ds'}\,ds\,ds',$$

where k denotes an arbitrary constant.[1]

Helmholtz's result suggested to Clausius[2] a new form for the law of force between electrons; namely, that which is obtained by supposing that two electrons of charges e, e', and velocities $\mathbf{v}$, $\mathbf{v}'$, possess electrokinetic energy of amount

$$\frac{ee'\,(\mathbf{v}\,.\,\mathbf{v}')}{r} + kee'\,\frac{d^2r}{ds\,ds'}\,vv'.$$

Subtracting from this the mutual electrostatic potential energy, which is $ee'c^2/r$, we may write the mutual kinetic potential of the two electrons in the form

$$\frac{ee'}{r}\left(\frac{\partial x}{\partial t}\frac{\partial x'}{\partial t} + \frac{\partial y}{\partial t}\frac{\partial y'}{\partial t} + \frac{\partial z}{\partial t}\frac{\partial z'}{\partial t} - c^2\right) + kee'\,\frac{d^2r}{ds\,ds'}\,vv',$$

where (x, y, z) denote the co-ordinates of e, and (x', y', z') those of e'.

The unknown constant k has clearly no influence so long as closed circuits only are considered; if k be replaced by zero, the expression for the kinetic potential becomes

$$\frac{ee'}{r}\left(\frac{\partial x}{\partial t}\frac{\partial x'}{\partial t} + \frac{\partial y}{\partial t}\frac{\partial y'}{\partial t} + \frac{\partial z}{\partial t}\frac{\partial z'}{\partial t} - c^2\right),$$

or, in the electrostatic system of units,

$$\frac{ee'}{r}\left(\frac{1}{c^2}\frac{\partial x}{\partial t}\frac{\partial x'}{\partial t} + \frac{1}{c^2}\frac{\partial y}{\partial t}\frac{\partial y'}{\partial t} + \frac{1}{c^2}\frac{\partial z}{\partial t}\frac{\partial z'}{\partial t} - 1\right),$$

which, as will appear later, closely resembles the corresponding expression in the Lorentz theory of electrons.[3]

Clausius' formula has the great advantage over Weber's, that it does not compel us to assume equal and opposite velocities for

[1] cf. H. Lamb, *Proc. Lond. Math. Soc.* xiv (1883), p. 301

[2] *Journal für Math.* lxxxii (1877), p. 85; *Phil. Mag.* x (1880), p. 255

[3] In fact, when any number of electrons are present, the part of the kinetic potential which concerns any one of them—say e—may be written

$$L_e = e\left(\frac{1}{c^2}a_x\frac{\partial x}{\partial t} + \frac{1}{c^2}a_y\frac{\partial y}{\partial t} + \frac{1}{c^2}a_z\frac{\partial z}{\partial t} - \phi\right)$$

where $\mathbf{a}$ and ϕ denote potential-functions defined by the equations

$$\mathbf{a} = \iiint\frac{\rho'\mathbf{v}'}{r}\,dx'\,dy'\,dz', \qquad \phi = \iiint\frac{\rho'}{r}\,dx'\,dy'\,dz'.$$

To convert these into Lorentz's formulae, it is only necessary to retard the potentials

the vitreous and resinous charges in an electric current ; on the other hand, Clausius' expression involves the absolute velocities of the electrons, while Weber's depends only on their relative motion ; and therefore Clausius' theory requires the assumption of a fixed aether in space, to which the velocities **v** and **v**′ may be referred.

When the behaviour of finite electrical systems is predicted from the formulae of Weber, Riemann and Clausius, the three laws do not always lead to concordant results. For instance, if a circular current be rotated with constant angular velocity round its axis, according to Weber's law there would be a development of free electricity on a stationary conductor in the neighbourhood ; whereas, according to Clausius' formula there would be no induction on a stationary body, but electrification would appear on a body turning with the circuit as if rigidly connected with it. Again,[1] let a magnet be suspended within a hollow metallic body, and let the hollow body be suddenly charged or discharged ; then, according to Clausius' theory, the magnet is unaffected ; but according to Weber's and Riemann's theories it experiences an impulsive couple. And again, if an electrified disk be rotated in its own plane, under certain circumstances a steady current will be induced in a neighbouring circuit according to Weber's law, but not according to the other formulae.

An interesting objection to Clausius' theory was brought forward in 1879 by Fröhlich[2]—namely, that when a charge of free electricity and a constant electric current are at rest relatively to each other, but partake together of the translatory motion of the earth in space, a force should act between them if Clausius' law were true. It was, however, shown by Budde[3] that the circuit itself acquires an electrostatic charge, partly as a result of the same action which causes the force on the external conductor, and partly as a result of electrostatic induction by the charge on the external conductor ; and that the total force between the circuit and external conductor is thus reduced to zero.[4]

We have seen that the discrimination between the different laws

[1] The two following crucial experiments, with others, were suggested by E. Budde, *Ann. d. Phys.* xxx (1887), p. 100.

[2] *Ann. d. Phys.* ix (1880), p. 261

[3] *Ann. d. Phys.* x (1880), p. 553

[4] This case of a charge and current moving side by side was afterwards examined by FitzGerald (*Trans. Roy. Dub. Soc.* i [1882], p. 319 : *Scient. Writings of G. F. FitzGerald*, p. 111) without reference to Clausius' formula, from the standpoint of Maxwell's theory. The result obtained was the same—namely, that the electricity induced on the conductor carrying the current neutralises the ponderomotive force between the current and the external charge. The question was also discussed in the light of relativity-theory by Lorentz, *Phys. ZS.* xi (1910), p. 1234.

of electrodynamic force is closely connected with the question whether in an electric current there are two kinds of electricity moving in opposite directions, or only one kind moving in one direction. On the unitary hypothesis, that the current consists in a transport of one kind of electricity with a definite velocity relative to the wire, it might be expected that a coil rotated rapidly about its own axis would generate a magnetic field different from that produced by the same coil at rest. Experiments to determine the matter were performed by A. Föppl [1] and by E. L. Nichols and W. S. Franklin,[2] but with negative results. The latter investigators found that the velocity of electricity must be such that the quantity conveyed past a specified point in a unit of time, when the direction of the current was that in which the coil was travelling, did not differ from that transferred when the current and coil were moving in opposite directions by as much as one part in ten million, even when the velocity of the wire was 9,096 cm./sec. They considered that they would have been able to detect a change of deflection due to the motion of the coil, even though the velocity of the current had been considerably greater than a thousand million metres per second.

During the decades in the middle of the century considerable progress was made in the science of thermo-electricity, whose beginnings we have already described.[3] In Faraday's laboratory notebook, under the date 28 July 1836, we read : [4] 'Surely the converse of thermo-electricity ought to be obtained experimentally. Pass current through a circuit of antimony and bismuth.'

Unknown to Faraday, the experiment here indicated had already been made, although its author had arrived at it by a different train of ideas. In 1834 Jean Charles Peltier [5] (1785–1845) attempted the task, which was afterwards performed with success by Joule,[6] of measuring the heat evolved by the passage of an electric current through a conductor. He found that a current produces in a homogeneous conductor an elevation of temperature, which is the same in all parts of the conductor where the cross-section is the same ; but he did not succeed in connecting the thermal phenomena quantitatively with the strength of the current—a failure which was due chiefly to the circumstance that his attention was fixed on the rise of temperature rather than on the amount of the heat evolved. But incidentally the investigation led to an important discovery

[1] *Ann. d. Phys.* xxvii (1886), p. 410
[2] *Amer. Jour. Sci.* xxxvii (1889), p. 103
[3] cf. pp. 88, 89
[4] Bence Jones's *Life of Faraday*, ii, p. 79
[5] *Annales de Chimie*, lvi (1834), p. 371
[6] cf. p. 212

—namely, that when a current was passed in succession through two conductors made of dissimilar metals, there was an evolution of heat at the junction; and that this depended on the direction of the current; for if the junction was heated when the current flowed in one sense, it was cooled when the current flowed in the opposite sense. This *Peltier effect*, as it is called, is quite distinct from the ordinary Joulean liberation of heat, in which the amount of energy set free in the thermal form is unaffected by a reversal of the current; the Joulean effect is, in fact, proportional to the square of the current strength, while the Peltier effect is proportional to the current strength directly. The Peltier heat which is absorbed from external sources when a current i flows for unit time through a junction from one metal B to another metal A may therefore be denoted by

$$\Pi_B^A(T)i,$$

where T denotes the absolute temperature of the junction. The function $\Pi_B^A(T)$ is found to be expressible as the difference of two parts, of which one depends on the metal A only, and the other on the metal B only; thus we can write

$$\Pi_B^A(T) = \Pi_A(T) - \Pi_B(T).$$

In 1851 a general theory of thermo-electric phenomena was constructed on the foundation of Seebeck's [1] and Peltier's discoveries by W. Thomson.[2] Consider a circuit formed of two metals, A and B, and let one junction be maintained at a slightly higher temperature $(T + \delta T)$ than the temperature T of the other junction. As Seebeck had shown, a thermo-electric current will be set up in the circuit. Thomson saw that such a system might be regarded as a heat engine, which absorbs a certain quantity of heat at the hot junction, and converts part of this into electrical energy, liberating the rest in the form of heat at the cold junction. If the Joulean evolution of heat be neglected, the process is reversible, and must obey the second law of thermodynamics; that is, the sum of the quantities of heat absorbed, each divided by the absolute temperature at which it is absorbed, must vanish. Thus we have

$$\frac{\Pi_B^A(T+\delta T)}{T+\delta T} - \frac{\Pi_B^A(T)}{T} = 0;$$

so the Peltier effect $\Pi_B^A(T)$ must be directly proportional to the

[1] cf. pp. 88, 89

[2] *Proc. R. S. Edin.* iii (1851), p. 91; *Phil. Mag.* iii (1852), p. 529; Kelvin's *Math. and Phys. Papers*, i, p. 316. cf. also *Trans. R. S. Edin.* xxi (1854), p. 123, reprinted in *Papers*, i, p. 232; and *Phil. Trans.* cxlvi (1856), p, 649, reprinted in *Papers*, ii, p. 189.

absolute temperature T. This result, however, as Thomson well knew, was contradicted by the observations of Cumming, who had shown that when the temperature of the hot junction is gradually increased, the electromotive force rises to a maximum value and then decreases. The contradiction led Thomson to predict the existence of a hitherto unrecognised thermo-electric phenomenon—namely, a reversible absorption of heat at places in the circuit other than the junctions. Suppose that a current flows along a wire which is of the same metal throughout, but varies in temperature from point to point. Thomson showed that heat must be liberated at some points and absorbed at others, so as either to accentuate or to diminish the differences of temperature at the different points of the wire. Suppose that the heat absorbed from external sources when unit electric charge passes from the absolute temperature T to the temperature $(T + \delta T)$ in a metal A is denoted by $S_A(T)\delta T$. The thermodynamical equation now takes the corrected form

$$\frac{\Pi_B^A(T + \delta T)}{T + \delta T} - \frac{\Pi_B^A(T)}{T} + \{S_B(T) - S_A(T)\}\frac{\delta T}{T} = 0.$$

Since the metals A and B are quite independent, this gives

$$\frac{\Pi_A(T + \delta T)}{T + \delta T} - \frac{\Pi_A(T)}{T} - S_A(T)\frac{\delta T}{T} = 0,$$

or

$$S_A(T) = T\frac{d}{dT}\left\{\frac{\Pi_A(T)}{T}\right\}.$$

This equation connects Thomson's 'specific heat of electricity' $S_A(T)$ with the Peltier effect.

In 1870 P. G. Tait[1] found experimentally that the specific heat of electricity in pure metals is proportional to the absolute temperature. We may therefore write $S_A(T) = \sigma_A T$, where σ_A denotes a constant characteristic of the metal A. The thermodynamical equation then becomes

$$\frac{d}{dT}\left\{\frac{\Pi_A(T)}{T}\right\} = \sigma_A,$$

or

$$\Pi_A(T) = \pi_A T + \sigma_A T^2,$$

where π_A denotes another constant characteristic of the metal. The chief part of the Peltier effect arises from the term $\pi_A T$.

By the investigations which have been described in the present chapter, the theory of electric currents was considerably advanced

[1] *Proc. R. S. Edin.* vii (1870), p. 308. cf. also Batelli, *Atti della R. Acc. di Torino*, xxii (1886), p. 48, translated *Phil. Mag.* xxiv (1887), p. 295.

in several directions. In all these researches, however, attention was fixed on the conductor carrying the current as the seat of the phenomenon. In the following period, interest was centred not so much on the conductors which carry charges and currents, as on the processes which take place in the dielectric media around them.

Chapter VIII

MAXWELL

SINCE the time of Descartes, natural philosophers have never ceased to speculate on the manner in which electric and magnetic influences are transmitted through space. About the middle of the nineteenth century, speculation assumed a definite form, and issued in a rational theory.

Among those who thought much on the matter was Karl Friedrich Gauss (1777–1855). In a letter [1] to Weber of date 19 March 1845, Gauss remarked that he had long ago proposed to himself to supplement the known forces which act between electric charges by other forces, such as would cause electric actions to be propagated between the charges with a finite velocity. But he expressed himself as determined not to publish his researches until he should have devised a mechanism by which the transmission could be conceived to be effected ; and this he had not succeeded in doing.

More than one attempt to realise Gauss' aspiration was made by his pupil Riemann. In a fragmentary note,[2] which appears to have been written in 1853, but which was not published until after his death, Riemann proposed an aether whose elements should be endowed with the power of resisting compression, and also (like the elements of MacCullagh's aether) of resisting changes of orientation. The former property he conceived to be the cause of gravitational and electrostatic effects, and the latter to be the cause of optical and magnetic phenomena. The theory thus outlined was apparently not developed further by its author ; but in a short investigation [3] which was published posthumously in 1867,[4] he returned to the question of the process by which electric action is propagated through space. In this memoir he proposed to replace Poisson's equation for the electrostatic potential, namely,

$$\nabla^2 V + 4\pi\rho = 0,$$

by the equation

$$\nabla^2 V - \frac{1}{c^2}\frac{\partial^2 V}{\partial t^2} + 4\pi\rho = 0,$$

[1] Gauss' *Werke*, v, p. 629
[2] Riemann's *Werke*, 2e Aufl., p. 526
[3] *Ann. d. Phys.* cxxxi (1867), p. 237 ; Riemann's *Werke*, 2e Aufl., p. 288 ; *Phil. Mag.* xxxiv (1867), p. 368
[4] It had been presented to the Göttingen Academy in 1858, but afterwards withdrawn.

according to which the changes of potential due to changing electrification would be propagated outwards from the charges with a velocity *c*. This, so far as it goes, is in agreement with the view which is now accepted as correct ; but Riemann's hypothesis was too slight to serve as the basis of a complete theory. Success came only when the properties of the intervening medium were taken into account.

In that power to which Gauss attached so much importance, of devising dynamical models and analogies for obscure physical phenomena, perhaps no-one has ever excelled W. Thomson [1] ; and to him, jointly with Faraday, is due the credit of having initiated the theory of the electric medium. In one of his earliest papers, written at the age of seventeen when he was a very young freshman at Cambridge,[2] Thomson compared the distribution of electrostatic force, in a region containing electrified conductors, with the distribution of the flow of heat in an infinite solid ; the equipotential surfaces in the one case correspond to the isothermal surfaces in the other, and an electric charge corresponds to a source of heat.[3]

It may, perhaps, seem as if the value of such an analogy as this consisted merely in the prospect which it offered of comparing, and thereby extending, the mathematical theories of heat and electricity. But to the physicist its chief interest lay rather in the idea that formulae which relate to the electric field, and which had been deduced from laws of action at a distance, were shown to be identical with formulae relating to the theory of heat, which had been deduced from hypotheses of action between contiguous particles. 'This paper,' as Maxwell said long afterwards, 'first introduced into mathematical science that idea of electrical action carried on by means of a continuous medium, which, though it had been announced by Faraday, and used by him as the guiding idea of his researches, had never been appreciated by other men of

[1] As will appear from the present chapter, Maxwell had the same power in a very marked degree. It has always been cultivated by the 'Cambridge school' of natural philosophers, founded by Green, Stokes and W. Thomson, which was dominated by the belief that all physical action is founded on dynamics. The value of a dynamical model is, that it will have properties other than those which suggested its construction ; the question then arises as to whether these properties are found in nature.

[2] *Camb. Math. Jour.* iii (1842), p. 71 ; reprinted in Thomson's *Papers on Electrostatics and Magnetism*, p. 1. Also *Camb. and Dub. Math. Jour.* i (Nov. 1845), p. 75 ; reprinted in *Papers*, p. 15.

[3] As regards this comparison, Thomson had been anticipated by Chasles, *Journal de l'Ec. Polyt.* xv (1837), p. 266, who had shown that attraction according to Newton's law gives rise to the same fields as the steady conduction of heat, both depending on Laplace's equation $\nabla^2 V = 0$.

It will be remembered that Ohm had used an analogy between thermal conduction and galvanic phenomena.

science, and was supposed by mathematicians to be inconsistent with the law of electrical action, as established by Coulomb, and built on by Poisson.'

In 1846—the year after he had taken his degree as second wrangler at Cambridge—Thomson investigated[1] the analogies of electric phenomena with those of elasticity. For this purpose he examined the equations of equilibrium of an incompressible elastic solid which is in a state of strain; and showed that the distribution of the vector which represents the elastic displacement might be assimilated to the distribution of the electric force in an electrostatic system. This, however, as he went on to show, is not the only analogy which may be perceived with the equations of elasticity; for the elastic displacement may equally well be identified with a vector **a**, defined in terms of the magnetic induction **B** by the relation

$$\text{curl}\,\mathbf{a} = \mathbf{B}.$$

The vector **a** is equivalent to the vector-potential which had been used in the memoirs of Neumann, Weber and Kirchhoff, on the induction of currents; but Thomson arrived at it independently by a different process, and without being at the time aware of the identification.

The results of Thomson's memoir seemed to suggest a picture of the propagation of electric or magnetic force: might it not take place in somewhat the same way as changes in the elastic displacement are transmitted through an elastic solid? These suggestions were not at the time pursued further by their author; but they helped to inspire another young Cambridge man to take up the matter a few years later. James Clerk Maxwell, by whom the problem was eventually solved, was born in 1831, the son of a landed proprietor in Kirkcudbrightshire. He was educated at Edinburgh, and at Trinity College, Cambridge, of which society he became in 1855 a Fellow; and in 1855–6, not long after his election to Fellowship, he communicated to the Cambridge Philosophical Society the first of his endeavours[2] to form a mechanical conception of the electromagnetic field.

Maxwell had been reading Faraday's *Experimental Researches*; and, gifted as he was with a physical imagination akin to Faraday's, he had been profoundly impressed by the theory of lines of force. At the same time, he was a trained mathematician; and the distinguishing feature of almost all his researches was the union of the

[1] *Camb. and Dub. Math. Jour.* ii (1847), p. 61; Thomson's *Math. and Phys. Papers*, i, p. 76 [2] *Trans. Camb. Phil. Soc.* x (1864), p. 27; Maxwell's *Scientific Papers*, i, p. 155

imaginative and the analytical faculties to produce results partaking of both natures. This first memoir may be regarded as an attempt to connect the ideas of Faraday with the mathematical analogies which had been devised by Thomson.[1]

Maxwell considered first the illustration of Faraday's lines of force which is afforded by the lines of flow of a liquid. The lines of force represent the direction of a vector; and the magnitude of this vector is everywhere inversely proportional to the cross-section of a narrow tube formed by such lines. This relation between magnitude and direction is possessed by any circuital vector; and in particular by the vector which represents the velocity at any point in a fluid, if the fluid be incompressible. It is therefore possible to represent the magnetic induction **B**, which is the vector represented by Faraday's lines of magnetic force, as the velocity of an incompressible fluid. Such an analogy had been indicated some years previously by Faraday himself,[2] who had suggested that along the lines of magnetic force there may be a 'dynamic condition,' analogous to that of the electric current, and that, in fact, 'the physical lines of magnetic force are currents.'

The comparison with the lines of flow of a liquid is applicable to electric as well as to magnetic lines of force. In this case the vector which corresponds to the velocity of the fluid is, in free aether, the electric force **E**. But when different dielectrics are present in the field, the electric force is not a circuital vector, and therefore cannot be represented by Faraday's type of lines of force; in fact, the equation

$$\operatorname{div} \mathbf{E} = 0$$

is now replaced by the equation

$$\operatorname{div} (\epsilon \mathbf{E}) = 0,$$

where ϵ denotes the specific inductive capacity or dielectric constant at the place (x, y, z). It is, however, evident from this equation that the vector $\epsilon\mathbf{E}$ is circuital; this vector, which will be denoted by **D**, bears to **E** a relation similar to that which the magnetic induction **B** bears to the magnetic force **H**. It is the vector **D** which is represented by Faraday's lines of electric force, and which in the hydrodynamical analogy corresponds to the velocity of the incompressible fluid.

In comparing fluid motion with electric fields it is necessary to introduce sources and sinks into the fluid to correspond to the

[1] Light is thrown on the origins of Clerk Maxwell's electric ideas by a series of letters written by him to Thomson and published in *Proc. Camb. Phil. Soc.* xxxii (1936), p. 695.

[2] *Exp. Res.*, § 3,269 (1852)

electric charges ; for **D** is not circuital at places where there is free charge. The magnetic analogy is therefore somewhat the simpler.

In the latter half of his memoir Maxwell discussed how Faraday's 'electrotonic state' might be represented in mathematical symbols. This problem he solved by borrowing from Thomson's investigation of 1847 the vector **a**, which is defined in terms of the magnetic induction by the equation

$$\text{curl}\,\mathbf{a} = \mathbf{B}\,;$$

if, with Maxwell, we call **a** the *electrotonic intensity*, the equation is equivalent to the statement that 'the entire electrotonic intensity round the boundary of any surface measures the number of lines of magnetic force which pass through that surface.' The electromotive force of induction at the place (x, y, z) is $-\partial\mathbf{a}/\partial t$: as Maxwell said, 'the electromotive force on any element of a conductor is measured by the instantaneous rate of change of the electrotonic intensity on that element.' From this it is evident that **a** is no other than the vector-potential which had been employed by Neumann, Weber and Kirchhoff, in the calculation of induced currents ; and we may take [1] for the electrotonic intensity due to a current i' flowing in a circuit $\mathbf{s}'$ the value which results from Neumann's theory namely,

$$\mathbf{a} = i' \int \frac{\mathbf{ds}'}{r}.$$

It may, however, be remarked that the equation

$$\text{curl}\,\mathbf{a} = \mathbf{B},$$

taken alone, is insufficient to determine **a** uniquely ; for we can choose **a** so as to satisfy this, and also to satisfy the equation

$$\text{div}\,\mathbf{a} = \psi,$$

where ψ denotes any arbitrary scalar. There are, therefore, an infinite number of possible functions **a**. With the particular value of **a** which has been adopted, we have

$$\begin{aligned}\text{div}\,\mathbf{a} &= \frac{\partial}{\partial x} i' \int_{s'} \frac{dx'}{r} + \frac{\partial}{\partial y} i' \int_{s'} \frac{dy'}{r} + \frac{\partial}{\partial z} i' \int_{s'} \frac{dz'}{r} \\ &= -i' \int_{s'} \left(dx' \cdot \frac{\partial}{\partial x'} + dy' \cdot \frac{\partial}{\partial y'} + dz' \cdot \frac{\partial}{\partial z'} \right) \left(\frac{1}{r} \right) \\ &= -i' \int_{s'} d\left(\frac{1}{r} \right) \\ &= 0\,;\end{aligned}$$

so the vector-potential **a** which we have chosen is circuital.

[1] cf. p. 200

In this memoir the physical importance of the operators *curl* and *div* first became evident [1]; for, in addition to those applications which have been mentioned, Maxwell showed that the connection between the strength **s** of a current and the magnetic field **H**, to which it gives rise, may be represented by the equation

$$4\pi\mathbf{s} = \text{curl}\,\mathbf{H}\;;$$

this equation is equivalent to the statement that 'the entire magnetic intensity round the boundary of any surface measures the quantity of electric current which passes through that surface.'

In the same year (1856) in which Maxwell's investigation was published, W. Thomson [2] put forward an alternative interpretation of magnetism. He had now come to the conclusion, from a study of the magnetic rotation of the plane of polarisation of light, that magnetism possesses a rotatory character; and suggested that the resultant angular momentum of the thermal motions of a body [3] might be taken as the measure of the magnetic moment. 'The explanation,' he wrote, 'of all phenomena of electromagnetic attraction or repulsion, or of electromagnetic induction, is to be looked for simply in the inertia or pressure of the matter of which the motions constitute heat. Whether this matter is or is not electricity, whether it is a continuous fluid interpermeating the spaces between molecular nuclei, or is itself molecularly grouped; or whether all matter is continuous, and molecular heterogeneousness consists in finite vortical or other relative motions of contiguous parts of a body; it is impossible to decide, and, perhaps, in vain to speculate, in the present state of science.'

The two interpretations of magnetism, in which the linear and rotatory characters respectively are attributed to it, occur frequently in the subsequent history of the subject. The former was amplified in 1858, when Helmholtz published his researches [4] on vortex motion; for Helmholtz showed that if a magnetic field produced by electric currents is compared to the flow of an incompressible fluid, so that the magnetic vector is represented by the fluid velocity, then the electric currents correspond to the vortex filaments in the fluid. This analogy correlates many theorems in hydrodynamics and

[1] These operators had, however, occurred frequently in the writings of Stokes, especially in his memoir of 1849 on the *Dynamical Theory of Diffraction.*

[2] *Proc. R. S.* viii (1856), p. 150; xi (1861), p. 327, footnote; *Phil Mag.* xiii (1857), p. 198; *Baltimore Lectures*, Appendix F

[3] This was written shortly before the kinetic theory of gases was developed by Clausius and Maxwell.

[4] *Jour. für Math.* lv (1858), p. 25; Helmholtz' *Wiss. Abh.* i, p. 101; translated *Phil. Mag.* xxxiii (1867), p. 485

electricity; for instance, the theorem that a re-entrant vortex-filament is equivalent to a uniform distribution of doublets over any surface bounded by it, corresponds to Ampère's theorem of the equivalence of electric currents and magnetic shells.

In his memoir of 1855, Maxwell had not attempted to construct a mechanical model of electrodynamic actions, but had expressed his intention of doing so. 'By a careful study,' he wrote,[1] 'of the laws of elastic solids, and of the motions of viscous fluids, I hope to discover a method of forming a mechanical conception of this electrotonic state adapted to general reasoning'; and in a foot-note he referred to the effort which Thomson had already made in this direction. Six years elapsed, however, before anything further on the subject was published. In the meantime, Maxwell had become (1856–60) Professor of Natural Philosophy in Marischal College, Aberdeen,[2] and then (1860–5) in King's College, London,[3] In the London chair he had opportunities of personal contact with Faraday, whom he had long reverenced. It is true that in 1857 he wrote to Forbes that he was 'by no means as yet a convert to the views which Faraday maintained,' but in 1858 he wrote of Faraday as 'the nucleus of everything electric since 1830.' Faraday had now concluded the *Experimental Researches*, and was living in retirement at Hampton Court; but his thoughts frequently recurred to the great problem which he had brought so near to solution. It appears from his note book that in 1857[4] he was speculating whether the velocity of propagation of magnetic action is of the same order as that of light, and whether it is affected by the susceptibility to induction of the bodies through which the action is transmitted.

[1] Maxwell's *Scientific Papers*, i, p. 188

[2] His reason for leaving Trinity for the Aberdeen chair was the long summer vacation as it then obtained in the Scottish universities, which enabled him to spend half the year at Glenlair.

One of his students in Aberdeen, David Gill, afterwards Her Majesty's Astronomer at the Cape of Good Hope, wrote long afterwards: 'In those days a professor was little better than a schoolmaster—and Maxwell was not a good schoolmaster; only some four or five of us, in a class of seventy or eighty, got much out of him. We used to remain with him for a couple of hours after lectures, till his terrible wife came and dragged him away to a miserable dinner at three o'clock in the afternoon. By himself he was the most delightful and sympathetic of beings—often dreaming and suddenly awakening—then he spoke of what he had been thinking. Much of it we could not understand at the time, and some of it we afterwards remembered and understood.'

There is good authority for the statement that the 'terrible wife' wanted Maxwell to live as a country gentleman—hunting, shooting and fishing—and that she was rude to his scientific friends.

[3] In 1860 the two Colleges at Aberdeen were united, and Maxwell lost his chair at Marischal College.

[4] Bence Jones's *Life of Faraday*, ii, p. 379

The answer to this question was furnished in 1861–2, when Maxwell fulfilled his promise of devising a mechanical conception of the electromagnetic field.[1]

In the interval since the publication of his previous memoir Maxwell had become convinced by Thomson's arguments that magnetism is in its nature rotatory. 'The transference of electrolytes in fixed directions by the electric current, and the rotation of polarised light in fixed directions by magnetic force, are,' he wrote, 'the facts the consideration of which has induced me to regard magnetism as a phenomenon of rotation, and electric currents as phenomena of translation.' This conception of magnetism he brought into connection with Faraday's idea, that tubes of force tend to contract longitudinally and to expand laterally. Such a tendency may be attributed to centrifugal force, if it be assumed that each tube of force contains fluid which is in rotation about the axis of the tube. Accordingly Maxwell supposed that, in any magnetic field, the medium is in rotation about the lines of magnetic force; each unit tube of force may for the present be pictured as an isolated vortex.

The energy of the motion per unit volume is proportional to $\mu\mathbf{H}^2$, where μ denotes the density of the medium, and $\mathbf{H}$ denotes the linear velocity at the circumference of each vortex. But, as we have seen,[2] Thomson had already shown that the energy of any magnetic field, whether produced by magnets or by electric currents, is

$$\frac{1}{8\pi}\iiint \mu\mathbf{H}^2\, dx\, dy\, dz,$$

where the integration is taken over all space, and where μ denotes the magnetic permeability, and $\mathbf{H}$ the magnetic force. It was therefore natural to identify the density of the medium at any place with the magnetic permeability, and the circumferential velocity of the vortices with the magnetic force.

But an objection to the proposed analogy now presents itself. Since two neighbouring vortices rotate in the same direction, the particles in the circumference of one vortex must be moving in the opposite direction to the particles contiguous to them in the circumference of the adjacent vortex; and it seems, therefore, as if the motion would be discontinuous. Maxwell escaped from this difficulty by imitating a well-known mechanical arrangement. When it is desired that two wheels should revolve in the same sense, an 'idle' wheel is inserted between them so as to be in gear with both.

[1] *Phil. Mag.* xxi (1861), pp. 161, 281, 338; xxiii (1862), pp. 12, 85; Maxwell's *Scientific Papers*, i, p. 451
[2] cf. pp. 222, 224

The model of the electromagnetic field to which Maxwell arrived by the introduction of this device greatly resembles that proposed by Bernoulli in 1736.[1] He supposed a layer of particles, acting as idle wheels, to be interposed between each vortex and the next, and to roll without sliding on the vortices; so that each vortex tends to make the neighbouring vortices revolve in the same direction as itself. The particles were supposed to be not otherwise constrained, so that the velocity of the centre of any particle would be the mean of the circumferential velocities of the vortices between which it is placed. This condition yields (in suitable units) the analytical equation

$$4\pi \mathbf{s} = \operatorname{curl} \mathbf{H},$$

where the vector $\mathbf{s}$ denotes the flux of the particles, so that its x component s_x denotes the quantity of particles transferred in unit time across unit area perpendicular to the x direction. On comparing this equation with that which represents Oersted's discovery, it is seen that the flux $\mathbf{s}$ of the movable particles interposed between neighbouring vortices is the analogue of the electric current.

It will be noticed that in Maxwell's model the relation between electric current and magnetic force is secured by a connection which is not of a dynamical, but of a purely kinematical character. The above equation simply expresses the existence of certain non-holonomic constraints within the system.

If from any cause the rotatory velocity of some of the cellular vortices is altered, the disturbance will be propagated from that part of the model to all other parts, by the mutual action of the particles and vortices. This action is determined, as Maxwell showed, by the relation

$$\mu \frac{\partial \mathbf{H}}{\partial t} = - \operatorname{curl} \mathbf{E}$$

which connects $\mathbf{E}$, the force exerted on a unit quantity of particles at any place in consequence of the tangential action of the vortices, with $\frac{\partial \mathbf{H}}{\partial t}$, the rate of change of velocity of the neighbouring vortices. It will be observed that this equation is not kinematical but dynamical. On comparing it with the electromagnetic equations

$$\left\{ \begin{aligned} \operatorname{curl} \mathbf{a} &= \mu \mathbf{H}, \\ \text{Induced electromotive force} &= - \frac{\partial \mathbf{a}}{\partial t}, \end{aligned} \right.$$

it is seen that $\mathbf{E}$ must be interpreted electromagnetically as the

[1] cf. p. 95

induced electromotive force. Thus the motion of the particles constitutes an electric current, the tangential force with which they are pressed by the matter of the vortex cells constitutes electromotive force, and the pressure of the particles on each other may be taken to correspond to the tension or potential of the electricity.

The mechanism must next be extended so as to take account of the phenomena of electrostatics. For this purpose Maxwell assumed that the particles, when they are displaced from their equilibrium position in any direction, exert a tangential action on the elastic substance of the cells; and that this gives rise to a distortion of the cells, which in turn calls into play a force arising from their elasticity, equal and opposite to the force which urges the particles away from the equilibrium position. When the exciting force is removed, the cells recover their form, and the electricity returns to its former position. The state of the medium, in which the electric particles are displaced in a definite direction, is assumed to represent an electrostatic field. Such a displacement does not itself constitute a current, because when it has attained a certain value it remains constant; but the variations of displacement are to be regarded as currents, in the positive or negative direction according as the displacement is increasing or diminishing.

The conception of the electrostatic state as a displacement of something from its equilibrium position was not altogether new, although it had not been previously presented in this form. Thomson, as we have seen, had compared electric force to the displacement in an elastic solid; and Faraday, who had likened the particles of a ponderable dielectric to small conductors embedded in an insulating medium,[1] had supposed that when the dielectric is subjected to an electrostatic field, there is a displacement of electric charge on each of the small conductors. The motion of these charges, when the field is varied, is equivalent to an electric current; and it was from this precedent that Maxwell derived the principle, which became of cardinal importance in his theory, that variations of displacement are to be counted as currents. But in adopting the idea, he altogether transformed it; for Faraday's conception of displacement was applicable only to ponderable dielectrics, and was in fact introduced solely in order to explain why the specific inductive capacity of such dielectrics is different from that of free aether; whereas according to Maxwell there is displacement wherever there is electric force, whether material bodies are present or not.

The difference between the conceptions of Faraday and Maxwell

[1] cf. p. 187

in this respect may be illustrated by an analogy drawn from the theory of magnetism. When a piece of iron is placed in a magnetic field, there is induced in it a magnetic distribution, say of intensity **I**; this induced magnetisation exists only within the iron, being zero in the free aether outside. The vector **I** may be compared to the polarisation or displacement, which according to Faraday is induced in dielectrics by an electric field; and the electric current constituted by the variation of this polarisation is then analogous to $\partial \mathbf{I}/\partial t$. But the entity which was called by Maxwell the electric displacement in the dielectric is analogous not to **I**, but to the magnetic induction **B**; the Maxwellian displacement current corresponds to $\partial \mathbf{B}/\partial t$, and may therefore have a value different from zero even in free aether.

It may be remarked in passing that the term *displacement*, which was thus introduced, and which has been retained in the later development of the theory, is perhaps not well chosen; what in the early models of the aether was represented as an actual displacement, was in later investigations conceived of as a change of structure rather than of position in the elements of the aether.

Maxwell's model is of such cardinal importance that it may be well to recapitulate its principles, in the form in which he described them in a letter to W. Thomson of date 10 December 1861.

'I suppose that the "magnetic medium" is divided into small portions or cells, the divisions or cell walls being composed of a single stratum of spherical particles, these particles being "electricity." The substance of the cells I suppose to be highly elastic, both with respect to compression and distortion; and I suppose the connection between the cells and the particles in the cell walls to be such that there is perfect rolling without slipping between them and that they act on each other tangentially.

'I then find that if the cells are set in rotation, the medium exerts a stress equivalent to a hydrostatic pressure combined with a longitudinal tension along the lines of axes of rotation.

'If there be two similar systems, the first a system of magnets, electric currents and bodies capable of magnetic induction, and the second composed of cells and cell walls, the density of the cells everywhere proportional to the capacity for magnetic induction at the corresponding point of the other, and the magnitude and direction of the cells proportional to the magnetic force, then

'1 All the mechanical magnetic forces in the one system will be proportional to forces in the other arising from centrifugal force.

'2 All the electric currents in the one system will be proportional to currents of the particles forming the cell walls in the other.

'3 All the electromotive forces in the one system, whether arising from changes of position of magnets or currents, or from motions of conductors, or from changes of intensity of magnets or currents will be proportional to forces urging the particles of the cell walls arising from the tangential action of the rotating cells when their velocity is increasing or diminishing.

'4 If in a non-conducting body the mutual pressure of the particles of the cell walls (which corresponds to electric tension) diminishes in a given direction, the particles will be urged in that direction by their mutual pressure, but will be restrained by their connection with the substance of the cells. They will, therefore, produce strain in the cells till the elasticity called forth balances the tendency of the particles to move. Thus there will be a displacement of particles proportional to the electromotive force, and when this force is removed, the particles will recover from displacement.'

Maxwell supposed the electromotive force acting on the electric particles to be connected with the displacement **D** which accompanies it, by an equation of the form

$$\mathbf{D} = \frac{1}{4\pi c_1^2}\,\mathbf{E},$$

where c_1 denotes a constant which depends on the elastic properties of the cells. The displacement current $\frac{\partial \mathbf{D}}{\partial t}$ must now be inserted in the relation which connects the current with the magnetic force; and thus we obtain the equation

$$\operatorname{curl} \mathbf{H} = 4\pi \mathbf{S},$$

where the vector **S**, which is called the *total current*, is the sum of the convection current **s** and the displacement current $\frac{\partial \mathbf{D}}{\partial t}$. By performing the operation *div* on both sides of this equation, it is seen that the total current is a circuital vector. In the model, the total current is represented by the total motion of the rolling particles; and this is conditioned by the rotations of the vortices in such a way as to impose the kinematic relation

$$\operatorname{div} \mathbf{S} = 0.$$

Having obtained the equation of motion of his system of vortices and particles, Maxwell proceeded to determine the rate of propagation of disturbances through it. He considered in particular the case in which the substance represented is a dielectric, so that the conduction current is zero. If, moreover, the constant μ be supposed to have the value unity, the equations may be written

$$\begin{cases} \operatorname{div} \mathbf{H} = 0. \\ c_1^2 \operatorname{curl} \mathbf{H} = \dfrac{\partial \mathbf{E}}{\partial t}, \\ -\operatorname{curl} \mathbf{E} = \dfrac{\partial \mathbf{H}}{\partial t}. \end{cases}$$

Eliminating **E**, we see[1] that **H** satisfies the equations

$$\begin{cases} \operatorname{div} \mathbf{H} = 0, \\ \dfrac{\partial^2 \mathbf{H}}{\partial t^2} = c_1^2 \nabla^2 \mathbf{H}. \end{cases}$$

But these are precisely the equations which the light vector satisfies in a medium in which the velocity of propagation is c_1: it follows that disturbances are propagated through the model by waves which are similar to waves of light, the magnetic (and similarly the electric) vector being in the wave-front. For a plane-polarised wave propagated parallel to the axis of z, the equations reduce to

$$-c_1^2 \frac{\partial \mathrm{H}_y}{\partial z} = \frac{\partial \mathrm{E}_x}{\partial t}, \quad c_1^2 \frac{\partial \mathrm{H}_x}{\partial z} = \frac{\partial \mathrm{E}_y}{\partial t}, \quad \frac{\partial \mathrm{E}_y}{\partial z} = \frac{\partial \mathrm{H}_x}{\partial t}, \quad -\frac{\partial \mathrm{E}_x}{\partial z} = \frac{\partial \mathrm{H}_y}{\partial t},$$

whence we have

$$c_1 \mathrm{H}_y = \mathrm{E}_x, \quad -c_1 \mathrm{H}_x = \mathrm{E}_y;$$

these equations show that the electric and magnetic vectors are at right angles to each other.

The question now arises as to the magnitude of the constant c_1.[2] This may be determined by comparing different expressions for the energy of an electrostatic field. The work done by an electromotive force **E** in producing a displacement **D** is

$$\int_0^{\mathbf{D}} \mathbf{E} \, . \, d\mathbf{D} \quad \text{or} \quad \tfrac{1}{2}\mathbf{ED}$$

per unit volume, since **E** is proportional to **D**. But if it be assumed that the energy of an electrostatic field is resident in the dielectric, the amount of energy per unit volume may be calculated by con-

[1] For if **a** denote any vector, we have identically

$$\nabla^2 \mathbf{a} + \operatorname{grad} \operatorname{div} \mathbf{a} + \operatorname{curl} \operatorname{curl} \mathbf{a} = 0.$$

[2] For criticisms on the procedure by which Maxwell determined the velocity of propagation of disturbance, cf. P. Duhem, *Les Théories Electriques de J. Clerk Maxwell*, Paris, 1902.

sidering the mechanical force required in order to increase the distance between the plates of a condenser, so as to enlarge the field comprised between them. The result is that the energy per unit volume of the dielectric is $\epsilon \mathbf{E}'^2/8\pi$, where ϵ denotes the specific inductive capacity of the dielectric and $\mathbf{E}'$ denotes the electric force, measured in terms of the electrostatic unit; if $\mathbf{E}$ denotes the electric force expressed in terms of the electrodynamic units used in the present investigation, we have $\mathbf{E} = c\mathbf{E}'$, where c denotes the constant which [1] occurs in transformations of this kind. The energy is therefore $\epsilon \mathbf{E}^2/8\pi c^2$ per unit volume. Comparing this with the expression for the energy in terms of $\mathbf{E}$ and $\mathbf{D}$, we have

$$\mathbf{D} = \epsilon \mathbf{E}/4\pi c^2,$$

and therefore the constant c_1 has the value $c\epsilon^{-\frac{1}{2}}$. Thus the result is obtained that the velocity of propagation of disturbances in Maxwell's medium is $c\epsilon^{-\frac{1}{2}}$, where ϵ denotes the specific inductive capacity and c denotes the velocity for which Kohlrausch and Weber had found [2] the value $3{\cdot}1 \times 10^{10}$ cm./sec.

Now by this time the velocity of light was known, not only from the astronomical observations of aberration and of Jupiter's satellites, but also by direct terrestrial experiments. In 1849 Hippolyte Louis Fizeau [3] had determined it by rotating a toothed wheel so rapidly that a beam of light transmitted through the gap between two teeth and reflected back from a mirror was eclipsed by one of the teeth on its return journey. The velocity of light was calculated from the dimensions and angular velocity of the wheel and the distance of the mirror; the result being $3{\cdot}15 \times 10^{10}$ cm./sec.[4]

[1] cf. pp. 202, 232 [2] cf. p. 232

[3] *Comptes Rendus*, xxix (1849), p. 90. A determination made by Cornu in 1874 was on this principle.

[4] A different experimental method was employed in 1862 by Léon Foucault (*Comptes Rendus*, lv, pp. 501, 792); in this a ray from an origin O was reflected by a revolving mirror M to a fixed mirror, and so reflected back to M, and again to O. It is evident that the returning ray MO must be deviated by twice the angle through which M turns while the light passes from M to the fixed mirror and back. The value thus obtained by Foucault for the velocity of light was $2{\cdot}98 \times 10^{10}$ cm./sec. Subsequent determinations by Michelson in 1879 (*Ast. Papers of the Amer. Ephemeris*, i), and by Newcomb in 1882 (ibid. ii) depended on the same principle.

As was shown afterwards by Lord Rayleigh (*Nature*, xxiv, p. 382; xxv, p. 52) and by Gibbs (*Nature*, xxxiii, p. 582), the value obtained for the velocity of light by the methods of Fizeau and Foucault represents the *group-velocity*, not the *wave-velocity*; the eclipses of Jupiter's satellites also give the group-velocity, and so does the value deduced from the coefficient of aberration; cf. Ehrenfest, *Ann. d. Phys.* xxxiii (1910), p. 1,571, and Rayleigh, *Phil. Mag.* xxii (1911), p. 130. In a non-dispersive medium the group-velocity coincides with the wave-velocity.

The velocity of light in dispersive media was directly investigated by Michelson in 1883–4, with results in accordance with theory.

Maxwell was impressed, as Kirchhoff had been before him, by the close agreement between the electric ratio c and the velocity of light [1]; and having demonstrated that the propagation of electric disturbance resembles that of light, he did not hesitate to assert the identity of the two phenomena. 'We can scarcely avoid the inference,' he said, 'that light consists in the transverse undulations of the same medium which is the cause of electric and magnetic phenomena.' Thus was answered the question which Priestley had asked almost exactly a hundred years before [2]: 'Is there any electric fluid *sui generis* at all, distinct from the aether?'

The presence of the dielectric constant ϵ in the expression $c\epsilon^{-\frac{1}{2}}$, which Maxwell had obtained for the velocity of propagation of electromagnetic disturbances, suggested a further test of the identity of these disturbances with light: for the velocity of light in a medium is known to be inversely proportional to the refractive index of the medium, and therefore the refractive index should be, according to the theory, proportional to the square root of the specific inductive capacity. At the time, however, Maxwell did not examine whether this relation was confirmed by experiment.

In what has preceded, the magnetic permeability μ has been supposed to have the value unity. If this is not the case, the velocity of propagation of disturbance may be shown, by the same analysis, to be $c\epsilon^{-\frac{1}{2}}\mu^{-\frac{1}{2}}$; so that it is diminished when μ is greater than unity, i.e. in paramagnetic bodies. This inference had been anticipated by Faraday: 'Nor is it likely,' he wrote,[3] 'that the paramagnetic body oxygen can exist in the air and not retard the transmission of the magnetism.'

It was inevitable that a theory so novel and so capacious as that of Maxwell should involve conceptions which his contemporaries understood with difficulty and accepted with reluctance. Of these the most difficult and unacceptable was the principle that the total current is always a circuital vector; or, as it is generally expressed, that 'all currents are closed.' According to the older electricians, a current which is employed in charging a condenser is not closed, but terminates at the coatings of the condenser, where charges are accumulating. Maxwell, on the other hand, taught that the dielectric between the coatings is the seat of a process—the *displacement current*—which is proportional to the rate of increase of the

[1] He had 'worked out the formulae in the country, before seeing Weber's result.' cf. Campbell and Garnett's *Life of Maxwell*, p. 244.

[2] Priestley's *History*, p. 488

[3] Faraday's laboratory notebook for 1857; cf. Bence Jones's *Life of Faraday*, ii, p. 380

electric force in the dielectric; and that this process produces the same magnetic effects as a true current, and forms, so to speak, a continuation, through the dielectric, of the charging current, so that the latter may be regarded as flowing in a closed circuit.[1]

Another characteristic feature of Maxwell's theory is the conception—for which, as we have seen, he was largely indebted to Faraday and Thomson—that magnetic energy is the kinetic energy of a medium occupying the whole of space, and that electric energy is the energy of strain of the same medium. By this conception electromagnetic theory was brought into such close parallelism with the elastic-solid theories of the aether, that it was bound to issue in an electromagnetic theory of light.

Maxwell's views were presented in a more developed form in a memoir entitled 'A Dynamical Theory of the Electromagnetic Field,' which was read to the Royal Society in 1864[2]; in this the architecture of his system was displayed, stripped of the scaffolding by aid of which it had been first erected.

As the equations employed were for the most part the same as had been set forth in the previous investigation, they need only be briefly recapitulated. We shall measure all electric scalars and vectors in the electrostatic system of units, and we shall measure all magnetic scalars and vectors in the electromagnetic system.[3] The magnetic induction $\mu\mathbf{H}$, being a circuital vector, may be expressed in terms of a vector-potential $\mathbf{A}$ by the equation

$$\mu\mathbf{H} = \text{curl}\,\mathbf{A}.$$

The electric displacement $\mathbf{D}$ is connected with the volume-density ρ of free electric charge by the electrostatic equation

$$\text{div}\,\mathbf{D} = 4\pi\rho.$$

The principle of conservation of electricity yields the equation

$$\text{div}\,\mathbf{s} = -\,\partial\rho/\partial t,$$

where $\mathbf{s}$ denotes the conduction current.

The law of induction of currents—namely, that the total electromotive force in any circuit is proportional to the rate of decrease

[1] The existence of the displacement current in space free from matter or electric charge seemed very questionable to Maxwell's contemporaries, and its postulation was a stumbling-block to the acceptance of his theory. In the twentieth century it was seen that the term representing the displacement current is required in order that the equations of the electromagnetic field should be relativistically invariant, so Maxwell's innovation was really a relativity-correction.

[2] *Phil. Trans*, clv (1865), p. 459; Maxwell's *Scient. Papers*, i, p. 526

[3] cf. p. 202

of the number of lines of magnetic induction which pass through it—may be written

$$- \text{curl } \mathbf{E} = \frac{1}{c} \mu \frac{\partial \mathbf{H}}{\partial t};$$

from which it follows that the electric force $\mathbf{E}$ must be expressible in the form

$$\mathbf{E} = - \frac{1}{c} \frac{\partial \mathbf{A}}{\partial t} + \text{grad } \psi,$$

where ψ denotes some scalar function. The quantities $\mathbf{A}$ and ψ which occur in this equation are not as yet completely determinate; for the equation by which $\mathbf{A}$ is defined in terms of the magnetic induction specifies only the circuital part of $\mathbf{A}$; and as the irrotational part of $\mathbf{A}$ is thus indeterminate, it is evident that ψ also must be indeterminate. Maxwell decided the matter by assuming[1] $\mathbf{A}$ to be a circuital vector; thus

$$\text{div } \mathbf{A} = 0,$$

and therefore

$$\text{div } \mathbf{E} = - \nabla^2 \psi,$$

from which equation it is evident that ψ represents the electrostatic potential.

The principle which is peculiar to Maxwell's theory must now be introduced. Currents of conduction are not the only kind of currents; even in the older theory of Faraday, Thomson and Mossotti, it had been assumed that electric charges are set in motion in the particles of a dielectric when the dielectric is subjected to an electric field; and the predecessors of Maxwell would not have refused to admit that the motion of these charges is in some sense a current. Suppose, then, that $\mathbf{S}$ denotes the total current which is capable of generating a magnetic field; since the integral of the magnetic force round any curve is proportional to the electric current which flows through the gap enclosed by the curve, we have in the units aready specified

$$c \text{ curl } \mathbf{H} = 4\pi \mathbf{S}.$$

In order to determine $\mathbf{S}$, we may consider the case of a condenser whose coatings are supplied with electricity by a conduction current s per unit area of coating. If $\pm \sigma$ denote the surface-density of electric charge on the coatings, we have

$$s = \partial \sigma / \partial t, \quad \text{and} \quad D = 4\pi\sigma,$$

[1] This is the effect of the introduction of (F′, G′, H′) in § 98 of the memoir; cf. also Maxwell's *Treatise on Electricity and Magnetism*, § 616.

where D denotes the magnitude of the electric displacement **D** in the dielectric between the coatings; so $4\pi s = \frac{\partial D}{\partial t}$. But since the total current is to be circuital, its value in the dielectric must be the same as the value **s** which it has in the rest of the circuit; that is, the current in the dielectric has the value $\frac{1}{4\pi}\frac{\partial \mathbf{D}}{\partial t}$. We shall assume that the current in dielectrics always has this value, so that in the general equations the total current must be understood to be $\mathbf{s} + \frac{1}{4\pi}\frac{\partial \mathbf{D}}{\partial t}$.

The above equations, together with those which express the proportionality of **E** to **D** in insulators, and to **s** in conductors ($\mathbf{D} = \epsilon\mathbf{E}$ when ϵ denotes the specific inductive capacity, and $\mathbf{s} = \kappa\mathbf{E}$ when κ denotes the ohmic conductivity), constituted Maxwell's system for a field formed by isotropic bodies which are not in motion. When the magnetic field is due entirely to currents (including both conduction currents and displacement currents), so that there is no magnetisation, we have

$$\nabla^2\mathbf{A} = -\text{curl curl }\mathbf{A} = -\text{curl }\mathbf{H}$$
$$= -\frac{4\pi\mathbf{S}}{c},$$

so that the vector-potential is connected with the total current by an equation of the same form as that which connects the scalar potential with the density of electric charge. To these potentials Maxwell inclined to attribute a physical significance; he supposed ψ to be analogous to a pressure subsisting in the mass of particles in his model, and **A** to be the measure of the electrotonic state. The two functions are, however, of merely analytical interest, and do not correspond to physical entities. For let two oppositely charged conductors, placed close to each other, give rise to an electrostatic field throughout all space. In such a field the vector-potential **A** is everywhere zero, while the scalar potential ψ has a definite value at every point. Now let these conductors discharge each other; the electrostatic force at any point of space remains unchanged until the point in question is reached by a wave of disturbance, which is propagated outwards from the conductors with the velocity of light, and which annihilates the field as it passes over it. But this order of events is not reflected in the behaviour of Maxwell's functions ψ and **A**; for at the instant of discharge, ψ is everywhere annihilated, and **A** suddenly acquires a finite value throughout all space.

As the potentials do not possess any physical significance, it is desirable to remove them from the equations. This was afterwards done by Maxwell himself, who [1] in 1868 proposed to base the electromagnetic theory of light solely on the equations

$$c \operatorname{curl} \mathbf{H} = 4\pi\mathbf{S},$$

$$-c \operatorname{curl} \mathbf{E} = \frac{\partial \mathbf{B}}{\partial t},$$

together with the equations which define **S** in terms of **E**, and **B** in terms of **H**.

After tracing to the action of the surrounding medium both the magnetic and the electric attractions and repulsions, and showing that they are proportional to the inverse square of the distance, Maxwell inquired [2] whether the attraction of gravitation, which follows the same law of the distance, may not also be traceable to the action of a surrounding medium.

Gravitation differs from magnetism and electricity in this: that the bodies concerned are all of the same kind, instead of being of opposite signs, like magnetic poles or electrified bodies, and that the force between these bodies is an attraction and not a repulsion, as is the case between like electric and magnetic bodies. If, then, gravitational potential energy is regarded as located in the medium, the energy per c.c. at any place must be

$$\alpha - \beta R^2$$

where α and β are positive constants and R is the gravitational force per gram at the place. If we assume (as everybody in Maxwell's day assumed) that energy is essentially positive, then the constant α must have a value greater than βR^2, where R is the greatest value of the gravitational force at any place in the universe: and hence at any place where the gravitational force vanishes, the intrinsic energy in the medium must have an enormously great value. 'As I am unable to understand in what way a medium can possess such properties,' said Maxwell, 'I cannot go any further in this direction in searching for the cause of gravitation.'

The memoir of 1864 contained an extension of the equations to the case of bodies in motion; the consideration of which naturally

[1] *Phil. Trans.* clviii (1868), p. 643; Maxwell's *Scientific Papers*, ii, p. 125
[2] In § 82 of the memoir of 1864

revives the question as to whether the aether is in any degree carried along with a body which moves through it. Maxwell did not formulate any express doctrine on this subject; but his custom was to treat matter as if it were merely a modification of the aether, distinguished only by altered values of such constants as the magnetic permeability and the specific inductive capacity; so that his theory may be said to involve the assumption that matter and aether move together. In deriving the equations which are applicable to moving bodies, he made use of Faraday's principle that the electromotive force induced in a body depends only on the relative motion of the body and the lines of magnetic force, whether one or the other is in motion absolutely. From this principle it may be inferred that the equation which determines the electric force [1] in terms of the potentials, in the case of a body which is moving with velocity **w**, is

$$\mathbf{E} = \frac{1}{c}[\mathbf{w} \cdot \mu\mathbf{H}] - \frac{1}{c}\frac{\partial \mathbf{A}}{\partial t} + \text{grad}\,\psi.$$

Maxwell thought that the scalar quantity ψ in this equation represented the electrostatic potential; but the researches of other investigators [2] have indicated that it represents the sum of the electrostatic potential and the quantity $(1/c)\,(\mathbf{A} \cdot \mathbf{w})$.

The electromagnetic theory of light was moreover extended in this memoir so as to account for the optical properties of crystals. For this purpose Maxwell assumed that in crystals the values of the coefficients of electric and magnetic induction depend on direction, so that the equation

$$\mu\mathbf{H} = \text{curl}\,\mathbf{A}$$

is replaced by

$$(\mu_1 H_x, \mu_2 H_y, \mu_3 H_z) = \text{curl}\,\mathbf{A};$$

and similarly the equation

$$\mathbf{E} = \mathbf{D}/\epsilon$$

is replaced by

$$\mathbf{E} = (c_1^2 D_x, c_2^2 D_y, c_3^2 D_z).$$

The other equations are the same as in isotropic media; so that the propagation of disturbance is readily seen to depend on the equation

$$\left(\mu_1 \frac{\partial^2 H_x}{\partial t^2}, \mu_2 \frac{\partial^2 H_y}{\partial t^2}, \mu_3 \frac{\partial^2 H_z}{\partial t^2}\right)$$
$$= -c^2\,\text{curl}\,\{c_1^2\,(\text{curl}\,\mathbf{H})_x, c_2^2\,(\text{curl}\,\mathbf{H})_y, c_3^2\,(\text{curl}\,\mathbf{H})_z\}.$$

Now, if μ_1, μ_2, μ_3 are supposed equal to each other, this equa-

[1] It may be here remarked that later writers distinguished between the electric force in a moving body and the electric force in the aether through which the body is moving, and that **E** in the present equation corresponds to the former of these vectors.

[2] Helmholtz, *Jour. für Math.* lxxviii (1874), p. 309; H. W. Watson, *Phil. Mag.* (5), xxv (1888), p. 271

tion is the same as the equation of motion of MacCullagh's aether in crystalline media,[1] the magnetic force **H** corresponding to MacCullagh's elastic displacement; and we may therefore immediately infer that Maxwell's electromagnetic equations yield a satisfactory theory of the propagation of light in crystals, provided it is assumed that the magnetic permeability is (for optical purposes) the same in all directions, and provided the plane of polarisation is identified with the plane which contains the magnetic vector. It is readily shown that the direction of the ray is at right angles to the magnetic vector and the electric force, and that the wave-front is the plane of the magnetic vector and the electric displacement.[2]

After this Maxwell proceeded to investigate the propagation of light in metals. The difference between metals and dielectrics, so far as electricity is concerned, is that the former are conductors; and it was therefore natural to seek the cause of the optical properties of metals in their ohmic conductivity. This idea at once suggested a physical reason for the opacity of metals—namely, that within a metal the energy of the light vibrations is converted into Joulean heat in the same way as the energy of ordinary electric currents.

The equations of the electromagnetic field in the metal may be written

$$\begin{cases} c\,\mathrm{curl}\,\mathbf{H} = 4\pi\mathbf{S}, \\ -c\,\mathrm{curl}\,\mathbf{E} = \dfrac{\partial \mathbf{H}}{\partial t}, \\ \mathbf{S} = \mathbf{s} + \dfrac{\partial \mathbf{D}}{\partial t} = \kappa\mathbf{E} + \dfrac{\epsilon}{4\pi c^2}\dfrac{\partial \mathbf{E}}{\partial t}, \end{cases}$$

where κ denotes the ohmic conductivity; whence it is seen that the electric force satisfies the equation

$$\epsilon\frac{\partial^2 \mathbf{E}}{\partial t^2} + 4\pi\kappa\frac{\partial \mathbf{E}}{\partial t} = c^2\nabla^2\mathbf{E}.$$

This is of the same form as the corresponding equation in the elastic-solid theory[3]; and, like it, furnishes a satisfactory general explanation of metallic reflection. It is indeed correct in all details,

[1] cf. p. 144 et sqq.

[2] In the memoir of 1864 Maxwell left open the choice between the above theory and that which is obtained by assuming that in crystals the specific inductive capacity is (for optical purposes) the same in all directions, while the magnetic permeability is aeolotropic. In the latter case the plane of polarisation must be identified with the plane which contains the electric displacement. Nine years later, in his *Treatise* (§ 794), Maxwell definitely adopted the former alternative.

[3] cf. p. 163

so long as the period of the disturbance is not too short—i.e. so long as the light-waves considered belong to the extreme infra-red region of the spectrum ; but if we attempt to apply the theory to the case of ordinary light, we are confronted by the difficulty which Lord Rayleigh indicated in the elastic-solid theory,[1] and which attends all attempts to explain the peculiar properties of metals by inserting a viscous term in the equation. The difficulty is that, in order to account for the properties of ideal silver, we must suppose the coefficient of $\frac{\partial^2 \mathbf{E}}{\partial t^2}$ negative—that is, the dielectric constant of the metal must be negative, which would imply instability of electrical equilibrium in the metal. The problem, as we have already remarked,[2] was solved only when its relation to the theory of dispersion was rightly understood.

At this time important developments were in progress in the last-named subject. Since the time of Fresnel, theories of dispersion had proceeded [3] from the assumption that the radii of action of the particles of luminiferous media are so large as to be comparable with the wave-length of light. It was generally supposed that the aether is loaded by the molecules of ponderable matter, and that the amount of dispersion depends on the ratio of the wave-length to the distance between adjacent molecules. This hypothesis was, however, seen to be inadequate, when, in 1862, F. P. Leroux [4] found that a prism filled with the vapour of iodine refracted the red rays to a greater degree than the blue rays ; for in all theories which depend on the assumption of a coarse-grained luminiferous medium, the refractive index increases with the frequency of the light.

Leroux's phenomenon, to which the name *anomalous dispersion* was given,[5] was shown by later investigators [6] to be generally associated with 'surface colour,' i.e. the property of brilliantly reflecting incident light of some particular frequency. Such an association seemed to indicate that the dispersive property of a substance is intimately connected with a certain frequency of vibration which is peculiar to that substance, and which, when it happens

[1] cf. p. 164. cf. also Rayleigh, *Phil. Mag.* (5) xii (1881), p. 81, and H. A. Lorentz, *Over de Theorie de Terugkaatsing*, Arnhem, 1875.

[2] cf. p. 164 [3] cf. p. 165

[4] *Comptes Rendus*, lv (1862), p. 126. In 1870 C. Christiansen (*Ann. d. Phys.* cxli, p. 479 ; cxliii, p. 250) observed a similar effect in a solution of fuchsin.

[5] Fox Talbot had discovered anomalous dispersion many years previously, but had not published his work.

[6] Especially by Kundt, in a series of papers in the *Annalen d. Phys.*, from vol. cxlii (1871) onwards.

to fall within the limits of the visible spectrum, is apparent in the surface colour. This idea of a frequency of vibration peculiar to each kind of ponderable matter had already appeared in another connection. Sir John Herschel[1] first noticed that sulphate of quinine shows a blue colour when light is incident on it under certain circumstances; later it was found that many other substances, if placed in a dark room and exposed to invisible radiations beyond the violet end of the visible spectrum, emit a bluish or greenish light. To this phenomenon the name *fluorescence* was given by Stokes in 1852,[2] in a famous paper disclosing its true nature. He remarked: 'Nothing seems more natural than to suppose that the incident vibrations of the luminiferous aether produce vibratory movements among the ultimate molecules of sensitive substances, and that the molecules in turn, swinging on their own account, produce vibrations in the luminiferous aether, and thus cause the sensation of light. The periodic times of these vibrations depend on the periods in which the molecules are disposed to swing, not upon the periodic time of the incident vibrations.'

The principle here introduced, of considering the molecules as dynamical systems which possess natural free periods, and which interact with the incident vibrations, lies at the basis of all the later nineteenth-century theories of dispersion. The earliest of these was devised by Maxwell, who, in the Cambridge Mathematical Tripos for 1869,[3] published the results of the following investigation:

A model of a dispersive medium may be constituted by embedding systems which represent the atoms of ponderable matter in a medium which represents the aether. We may picture each atom[4] as composed of a single massive particle supported symmetrically by springs from the interior face of a massless spherical shell; if the shell be fixed, the particle will be capable of executing vibrations about the centre of the sphere, the effect of the springs being equivalent to a force on the particle proportional to its distance from the centre. The atoms thus constituted may be supposed to occupy small spherical cavities in the aether, the outer shell of each atom being in contact with the aether at all points and partaking in its motion. An immense number of atoms is supposed to exist in each unit volume of the dispersive medium, so that the medium as a whole is fine-grained.

[1] *Phil. Trans.* (1845), p. 143; *Proc. R. S.* v (1845), p. 547

[2] *Phil. Trans.* (1852), p. 463; Stokes's *Coll. Papers*, iii, p. 267

[3] *Cambridge Calendar*, 1869; republished by Lord Rayleigh, *Phil. Mag.* xlviii (1899), p. 151. cf. also Rayleigh, *Phil. Mag.* xxxiii (1917), p. 496

[4] This illustration is due to W. Thomson.

Suppose that the potential energy of strain of free aether per unit volume is

$$\tfrac{1}{2}\mathrm{E}\left(\frac{\partial\eta}{\partial x}\right)^2,$$

where η denotes the displacement and E an elastic constant; so that the equation of wave-propagation in free aether is

$$\rho\frac{\partial^2\eta}{\partial t^2} = \mathrm{E}\frac{\partial^2\eta}{\partial x^2},$$

where ρ denotes the aethereal density.

Then if σ denote the mass of the atomic particles in unit volume, $(\eta + \zeta)$ the total displacement of an atomic particle at the place x at time t, $\sigma p^2\zeta$ the attractive force, it is evident that for the compound medium the kinetic energy per unit volume is

$$\tfrac{1}{2}\rho\left(\frac{\partial\eta}{\partial t}\right)^2 + \tfrac{1}{2}\sigma\left(\frac{\partial\eta}{\partial t} + \frac{\partial\zeta}{\partial t}\right)^2,$$

and the potential energy per unit volume is

$$\tfrac{1}{2}\mathrm{E}\,(\partial\eta/\partial x)^2 + \tfrac{1}{2}\sigma p^2\zeta^2.$$

The equations of motion, derived by the process usual in dynamics, are

$$\left\{\begin{aligned} &\rho\frac{\partial^2\eta}{\partial t^2} + \sigma\left(\frac{\partial^2\eta}{\partial t^2} + \frac{\partial^2\zeta}{\partial t^2}\right) - \mathrm{E}\frac{\partial^2\eta}{\partial x^2} = 0,\\ &\sigma\left(\frac{\partial^2\eta}{\partial t^2} + \frac{\partial^2\zeta}{\partial t^2}\right) + \sigma p^2\zeta = 0.\end{aligned}\right.$$

So eliminating ζ by operating with $\left(1 + \frac{1}{p^2}\frac{\partial^2}{\partial t^2}\right)$ on the first equation and substituting from the second, we have

$$\left(1 + \frac{\sigma}{\rho}\right)\frac{\partial^2\eta}{\partial t^2} - c^2\frac{\partial^2\eta}{\partial x^2} + \frac{1}{p^2}\frac{\partial^4\eta}{\partial t^4} - \frac{c^2}{p^2}\frac{\partial^4\eta}{\partial x^2\partial t^2} = 0.$$

Either of the terms $\frac{\partial^4\eta}{\partial t^4}$ or $\frac{\partial^4\eta}{\partial x^2\partial t^2}$ would be sufficient to produce dispersion. Consider the propagation, through the medium thus constituted, of vibrations whose frequency is n, and whose velocity of propagation in the medium is v; so that η and ζ are harmonic functions of $n(t - x/v)$. Substituting these values in the differential equations, we obtain

$$\frac{1}{v^2} = \frac{\rho}{\mathrm{E}} + \frac{\sigma p^2}{\mathrm{E}\,(p^2 - n^2)}.$$

Now, ρ/E has the value $1/c^2$, where c denotes the velocity of light in free aether; and c/v is the refractive index μ of the medium for vibrations of frequency n. So the equation, which may be written

$$\mu^2 = 1 + \frac{\sigma p^2}{\rho (p^2 - n^2)},$$

determines the refractive index of the substance for vibrations of any frequency n. The same formula was independently obtained from similar considerations three years later by W. Sellmeier.[1]

If the oscillations are very slow, the incident light being in the extreme infra-red part of the spectrum, n is small, and the equation gives approximately $\mu^2 = (\rho + \sigma)/\rho$; for such oscillations, each atomic particle and its shell move together as a rigid body, so that the effect is the same as if the aether were simply loaded by the masses of the atomic particles, its rigidity remaining unaltered.

The dispersion of light within the limits of the visible spectrum is for most substances controlled by a natural frequency p which corresponds to a vibration beyond the violet end of the visible spectrum; so that, n being smaller than p, we may expand the fraction in the formula of dispersion, and obtain the equation

$$\mu^2 = 1 + \frac{\sigma}{\rho}\left(1 + \frac{n^2}{p^2} + \frac{n^4}{p^4} + \ldots\right),$$

which resembles the formula of dispersion in Cauchy's theory[2]; indeed, we may say that Cauchy's formula is the expansion of Maxwell's formula in a series which, as it converges only when n has values within a limited range, fails to represent the phenomena outside that range.

The theory as given above is defective in that it becomes meaningless when the frequency n of the incident light is equal to the frequency p of the free vibrations of the atoms. This defect may be remedied by supposing that the motion of an atomic particle relative to the shell in which it is contained is opposed by a dissipative force varying as the relative velocity; such a force suffices to prevent the forced vibration from becoming indefinitely great as the period of the incident light approaches the period of free vibration of the atoms; its introduction is justified by the fact that vibrations in this part of the spectrum suffer absorption in passing through the medium. When the incident vibration is not in the same region of the spectrum as the free vibration, the absorption is not of much importance, and may be neglected.

It is shown by the spectroscope that the atomic systems which

[1] *Ann. d. Phys.* cxlv (1872), pp. 399, 520; cxlvii (1872), pp. 386, 525. cf. also Helmholtz, *Ann. d. Phys.* cliv (1875), p. 582

[2] cf. p. 166.

emit and absorb radiation in actual bodies possess more than one distinct characteristic frequency. The theory already given may, however, readily be extended [1] to the case in which the atoms have several natural frequencies of vibration; we have only to suppose that the external massless rigid shell is connected by springs to an interior massive rigid shell, and that this again is connected by springs to another massive shell inside it, and so on. The corresponding extension of the equation for the refractive index is

$$\mu^2 - 1 = \frac{c_1}{p_1^2 - n^2} + \frac{c_2}{p_2^2 - n^2} + \ldots,$$

where $p_1, p_2 \ldots$ denote the frequencies of the natural periods of vibration of the atom.

The validity of the Maxwell-Sellmeier formula of dispersion was strikingly confirmed by experimental researches in the closing years of the nineteenth century.[2] In 1897 Rubens [3] showed that the formula represents closely the refractive indices of sylvin (potassium chloride) and rock salt, with respect to light and radiant heat of wave-lengths between 4,240 Ångström units and 223,000 Å. The constants in the formula being known from this comparison, it was possible to predict the dispersion for radiations of still lower frequency; and it was found that the square of the refractive index should have a negative value (indicating complete reflection) for wave-lengths 370,000 to 550,000 Å. in the case of rock salt, and for wave-lengths 450,000 to 670,000 Å. in the case of sylvin. This inference was verified experimentally in the following year.

It may seem strange that Maxwell, having successfully employed his electromagnetic theory to explain the propagation of light in isotropic media, in crystals, and in metals, should have omitted to apply it to the problem of reflection and refraction. This is all the more surprising, as the study of the optics of crystals had already revealed a close analogy between the electromagnetic theory and MacCullagh's elastic-solid theory; and in order to explain reflection and refraction electromagnetically, nothing more was necessary than to transcribe MacCullagh's investigation of the same problem, interpreting $\frac{\partial \mathbf{e}}{\partial t}$ (the time-flux of the displacement of MacCullagh's aether) as the magnetic force, and curl $\mathbf{e}$ as the electric displacement. As in MacCullagh's theory the difference between the contiguous media

[1] This subject was developed by Lord Kelvin in the *Baltimore Lectures*.

[2] A discussion of the whole subject in the light of sixty years' subsequent work was given by C. G. Darwin, *Proc. R. S.* cxlvi (1934), p. 17.

[3] *Ann. d. Phys.* lx (1897), p. 454

is represented by a difference of their elastic constants, so in the electromagnetic theory it may be represented by a difference in their specific inductive capacities. From a letter which Maxwell wrote to Stokes in 1864, and which has been preserved,[1] it appears that the problem of reflection and refraction was engaging Maxwell's attention at the time when he was preparing his Royal Society memoir on the electromagnetic field ; but he was not able to satisfy himself regarding the conditions which should be satisfied at the interface between the media. He seems to have been in doubt which of the rival elastic-solid theories to take as a pattern ; and it is not unlikely that he was led astray by relying too much on the analogy between the electric displacement and an elastic displacement.[2] For in the elastic-solid theory all three components of the displacement must be continuous across the interface between two contiguous media ; but Maxwell found that it was impossible to explain reflection and refraction if all three components of the electric displacement were supposed to be continuous across the interface ; and, unwilling to give up the analogy which had hitherto guided him aright, yet unable to disprove [3] the Greenian conditions at bounding surfaces, he seems to have laid aside the problem until some new light should dawn upon it.

This was not the only difficulty which beset the electromagnetic theory. The theoretical conclusion, that the specific inductive capacity of a medium should be equal to the square of its refractive index with respect to waves of long period, was not as yet substantiated by experiment ; and the theory of displacement currents, on which everything else depended, was unfavourably received by the most distinguished of Maxwell's contemporaries. Helmholtz indeed ultimately accepted it, but only after many years ; and W. Thomson (Kelvin) seems never to have thoroughly believed it to the end of his long life. In 1888 he referred to it as a 'curious and ingenious, but not wholly tenable hypothesis,' [4] and proposed [5] to replace it by an extension of the older potential theories. In 1893 he wrote a preface for the English edition of Hertz's *Electric Waves,*

[1] Stokes's *Scientific Correspondence*, ii, pp. 25, 26

[2] It must be remembered that Maxwell pictured the electric displacement as a real displacement of a medium. 'My theory of electrical forces,' he wrote, 'is that they are called into play in insulating media by slight electric displacements, which put certain small portions of the medium into a state of distortion, which, being resisted by the elasticity of the medium, produces an electro-motive force.' (Campbell and Garnett's *Life of Maxwell*, p. 244.)

[3] The letter to Stokes already mentioned appears to indicate that Maxwell for a time doubted the correctness of Green's conditions.

[4] *Nature*, xxxviii (1888), p. 571

[5] *Brit. Assoc. Report* (1888), p. 567

and there appeared to accept 'magnetic waves.' But in 1896 he had some inclination [1] to speculate that alterations of electrostatic force due to rapidly changing electrification are propagated by condensational waves in the luminiferous aether; and in the same year FitzGerald wrote to Heaviside [2]: 'I tried hard to knock down Lord K. with $\Delta^2\epsilon = k\mu \frac{\partial^2 \mathbf{e}}{\partial t^2}$ as a sufficient answer to all his complications as proving that *every* action was propagated with velocity $= (k\mu)^{-\frac{1}{2}}$, but I had to resort to an elaborate discussion of the simple two spheres case to get him to understand the electromagnetic point of view, and though he had nothing to reply, he is not yet convinced, nor does he, I think, even yet understand Maxwell's notion of displacement currents being accompanied by magnetic force. I tried to get him to see that his own investigation of the penetration of alternating currents into conductors was only the viscous motion analogue of light propagation, but he shied at it like a horse at a heap of stones which he is accustomed in another form to use for riding over. He *will* always return to it that these *equations* of light propagation are only those of transverse vibrations in an elastic solid, and his last giving up of everything was, I think, partly due to my reminding him that *that* would require for a continuous current that the parts of the solid should be capable of continuous rotation.' In September 1898 Oliver Lodge wrote to Heaviside: 'Kelvin doesn't even believe in Maxwell's light pressure. He says that part is all wrong.' In 1904 he admitted [3] that a bar magnet rotating about an axis at right angles to its length is equivalent to a lamp emitting light of period equal to the period of the rotation, but gave his final judgment in the sentence [4]: 'The so-called electromagnetic theory of light has not helped us hitherto.'

Thomson appears to have based his ideas of the propagation of electric disturbance on the case which had first become familiar to him—that of the transmission of signals along a wire. He clung to the older view that in such a disturbance the wire is the actual medium of transmission; whereas in Maxwell's theory the function of the wire is merely to guide the disturbance, which is resident in the surrounding dielectric.

This opinion that conductors are the media of propagation of electric disturbance was entertained also by Ludwig Lorenz (1829–1891) of Copenhagen, who independently developed an electro-

[1] cf. Bottomley, in *Nature*, liii (1896), p. 268; Kelvin, ibid., p. 316; J. Willard Gibbs, ibid., p. 509

[2] Letter of 11 June 1896 [3] *Baltimore Lectures*, p. 376 [4] ibid., preface, p. 7

magnetic theory of light[1] a few years after the publication of Maxwell's memoirs. The procedure which Lorenz followed was that which Riemann had suggested[2] in 1858—namely, to modify the accepted formulae of electrodynamics by introducing terms which, though too small to be appreciable in ordinary laboratory experiments, would be capable of accounting for the propagation of electrical effects through space with a finite velocity. We have seen that in Neumann's theory the electric force **E** was determined by the equation

$$\mathbf{E} = \text{grad}\, \phi - \frac{1}{c}\frac{\partial \mathbf{a}}{\partial t}, \tag{1}$$

where ϕ denotes the electrostatic potential defined by the equation

$$\phi = \iiint (\rho'/r)\, dx'dy'dz',$$

ρ' being the density of electric charge at the point (x', y', z'), and where **a** denotes the vector-potential, defined by the equation

$$\mathbf{a} = \frac{1}{c}\iiint (\mathbf{s}'/r)\, dx'dy'dz',$$

$\mathbf{s}'$ being the conduction current at (x', y', z'). We suppose the specific inductive capacity and the magnetic permeability to be everywhere unity.

Lorenz proposed to replace these by the equations

$$\phi = \iiint \{\rho'(t - r/c)/r\}\, dx'dy'dz',$$

$$\mathbf{a} = \frac{1}{c}\iiint \{\mathbf{s}'(t - r/c)/r\}\, dx'dy'dz'\ ;$$

the change consists in replacing the values which ρ' and $\mathbf{s}'$ have at the instant t by those which they have at the instant $(t - r/c)$, which is the instant at which a disturbance travelling with velocity c must leave the place (x', y', z') in order to arrive at the place (x, y, z) at the instant t. Thus the values of the potentials at (x, y, z) at any instant t would, according to Lorenz's theory, depend on the electric state at the point (x', y', z') at the previous instant $(t - r/c)$: as if the potentials were propagated outwards from the charges and currents with velocity c. The functions ϕ and **a** formed in this way are generally known as the *retarded potentials*.

[1] *Oversigt over det K. danske Vid. Selskaps Forhandlinger* (1867), p. 26 ; *Ann. d. Phys.* cxxxi (1867), p. 243 ; *Phil. Mag.* xxxiv (1867), p. 287

[2] cf. p. 240. Riemann's memoir was, however, published only in the same year (1867) as Lorenz's.

The equations by which ϕ and $\mathbf{a}$ have been defined are equivalent to the equations

$$\nabla^2\phi - \frac{1}{c^2}\frac{\partial^2\phi}{\partial t^2} = -4\pi\rho, \tag{2}$$

$$\nabla^2\mathbf{a} - \frac{1}{c^2}\frac{\partial^2\mathbf{a}}{\partial t^2} = -4\pi\mathbf{s}/c, \tag{3}$$

while the equation of conservation of electricity,

$$\operatorname{div}\mathbf{s} + \frac{\partial\rho}{\partial t} = 0$$

gives [1]

$$\operatorname{div}\mathbf{a} + \frac{1}{c}\frac{\partial\phi}{\partial t} = 0. \tag{4}$$

From equations (1), (2), (4), we may readily derive the equation

$$\operatorname{div}\mathbf{E} = 4\pi\rho\ ; \tag{I}$$

and from (1), (3), (4), we have

$$\operatorname{curl}\mathbf{H} = \frac{1}{c}\frac{\partial\mathbf{E}}{\partial t} + \frac{4\pi\mathbf{s}}{c}, \tag{II}$$

where $\mathbf{H}$ or curl $\mathbf{a}$ denotes the magnetic force; while from (1) we have

$$c\operatorname{curl}\mathbf{E} = -\frac{\partial\mathbf{H}}{\partial t}. \tag{III}$$

[1] For we have

$$c\frac{\partial a_x}{\partial x} = \frac{\partial}{\partial x}\iiint\frac{s_x'}{r}\,dx'\,dy'\,dz'$$

$$= \iiint\left\{s_x'\frac{\partial}{\partial x}\left(\frac{1}{r}\right) + \frac{1}{r}\frac{\partial s_x'}{\partial t}\frac{\partial}{\partial x}\left(-\frac{r}{c}\right)\right\}dx'\,dy'\,dz'$$

$$= -\iiint\left\{s_x'\frac{\partial}{\partial x'}\left(\frac{1}{r}\right) + \frac{1}{r}\frac{\partial s_x'}{\partial t}\frac{\partial}{\partial x'}\left(-\frac{r}{c}\right)\right\}dx'\,dy'\,dz'.$$

Integrating the first of these integrals by parts, we have

$$c\frac{\partial a_x}{\partial x} = \iiint\frac{1}{r}\left\{\frac{\partial s_x'}{\partial x'} - \frac{\partial s_x'}{\partial t}\frac{\partial}{\partial x'}\left(-\frac{r}{c}\right)\right\}dx'\,dy'\,dz'$$

$$= \iiint\frac{1}{r}\left[\frac{\partial s_x'}{\partial x'}\right]dx'\,dy'\,dz'$$

where the square brackets imply that the differentiation is to be taken partially with respect to x' only in so far as it occurs explicitly, ignoring its implicit occurrence in the argument $\left(t - \frac{r}{c}\right)$.

We have therefore

$$c\operatorname{div}\mathbf{a} = \iiint\frac{1}{r}\left\{\left[\frac{\partial s_x'}{\partial x'}\right] + \left[\frac{\partial s_y'}{\partial y'}\right] + \left[\frac{\partial s_z'}{\partial z'}\right]\right\}dx'\,dy'\,dz'$$

$$= -\iiint\frac{1}{r}\frac{\partial\rho'}{\partial t}\,dx'\,dy'\,dz',$$ by the equation of conservation of electricity

$$= -\frac{\partial\phi}{\partial t},$$ which is the required equation (4).

The equations (I), (II), (III) are, however, the fundamental equations of Maxwell's theory ; and therefore the theory of L. Lorenz is practically equivalent to that of Maxwell, so far as concerns the propagation of electromagnetic disturbances through free aether. Lorenz himself, however, does not appear to have clearly perceived this ; for in his memoir he postulated the presence of conducting matter throughout space, and was consequently led to equations resembling those which Maxwell had given for the propagation of light in metals. Observing that his equations represented periodic electric currents at right angles to the direction of propagation of the disturbance, he suggested that all luminous vibrations might be constituted by electric currents, and hence that there was 'no longer any reason for maintaining the hypothesis of an aether, since we can admit that space contains sufficient ponderable matter to enable the disturbance to be propagated.'

Lorenz was unable to derive from his equations any explanation of the existence of refractive indices, and his theory lacks the rich physical suggestiveness of Maxwell's ; the value of his memoir lies chiefly in the introduction of the retarded potentials. It may be remarked in passing that Lorenz's retarded potentials are not identical with Maxwell's scalar- and vector-potentials ; for Lorenz's $\mathbf{a}$ is not a circuital vector, and Lorenz's ϕ is not, like Maxwell's, the electrostatic potential, but depends on the positions occupied by the charges at certain previous instants.

For some years no progress was made either with Maxwell's theory or with Lorenz's. Meanwhile, Maxwell had in 1865 resigned his chair at King's College, and had retired to his estate of Glenlair in Kirkcudbrightshire, where he occupied himself in writing a connected account of electrical theory. In 1871 he returned to Cambridge as Professor of Experimental Physics ; and two years later published his *Treatise on Electricity and Magnetism.*

In this celebrated work is comprehended almost every branch of electric and magnetic theory ; but the intention of the writer was to discuss the whole as far as possible from a single point of view, namely, that of Faraday ; so that little or no account was given of the hypotheses which had been propounded in the two preceding decades by the great German electricians. So far as Maxwell's purpose was to disseminate the ideas of Faraday, it was undoubtedly fulfilled ; but the *Treatise* was less successful when considered as the exposition of its author's own views. The doctrines peculiar to Maxwell—the existence of displacement currents, and of electromagnetic vibrations identical with light—were not introduced in the first volume, or in the first half of the second volume; and the account which was given of them was scarcely more complete, and was perhaps less attractive, than that which had been furnished in the original memoirs.

Some matters were, however, discussed more fully in the *Treatise* than in Maxwell's previous writings; and among these was the question of stress in the electromagnetic field.

It will be remembered[1] that Faraday, when studying the curvature of lines of force in electrostatic fields, had noticed an apparent tendency of adjacent lines to repel each other, as if each tube of force were inherently disposed to distend laterally; and that in addition to this repellent or diverging force in the transverse direction, he supposed an attractive or contractile force to be exerted at right angles to it, that is to say, in the direction of the lines of force.

Of the existence of these pressures and tensions Maxwell was fully persuaded; and he determined analytical expressions suitable to represent them. The tension along the lines of force must be supposed to maintain the ponderomotive force which acts on the conductor on which the lines of force terminate; and it may therefore be measured (in the system of units we are now using) by the force which is exerted on unit area of the conductor, i.e. $\epsilon\mathbf{E}^2/8\pi$ or $\mathbf{DE}/8\pi$. The pressure at right angles to the lines of force must then be determined so as to satisfy the condition that the aether is to be in equilibrium.

For this purpose, consider a thin shell of aether included between two equipotential surfaces. The equilibrium of the portion of this shell which is intercepted by a tube of force requires (as in the theory of the equilibrium of liquid films) that the resultant force per unit area due to the above-mentioned normal tensions on its two faces shall have the value $T(1/\rho_1 + 1/\rho_2)$, where ρ_1 and ρ_2 denote the principal radii of curvature of the shell at the place, and where T denotes the lateral stress across unit length of the surface of the shell, T being analogous to the surface-tension of a liquid film.

Now, if t denote the thickness of the shell, the area intercepted on the second face by the tube of force bears to the area intercepted on the first face the ratio $(\rho_1 + t)(\rho_2 + t)/\rho_1\rho_2$; and by the fundamental property of tubes of force, **D** and **E** vary inversely as the cross-section of the tube, so the total force on the second face will bear to that on the first face the ratio

$$\rho_1\rho_2/(\rho_1 + t)(\rho_2 + t),$$

[1] cf. p. 187

or approximately
$$\left(1 - \frac{t}{\rho_1} - \frac{t}{\rho_2}\right);$$

the resultant force per unit area along the outward normal is therefore
$$-\frac{1}{8\pi}\mathbf{DE}\,.\,t\,.(1/\rho_1 + 1/\rho_2),$$

and so we have
$$\mathbf{T} = -\frac{1}{8\pi}\mathbf{DE}\,.\,t;$$

or the pressure at right angles to the lines of force is $1/8\pi\,\mathbf{DE}$ per unit area—that is, it is numerically equal to the tension along the lines of force.

The principal stresses in the medium being thus determined, it readily follows that the stress across any plane, to which the unit vector **N** is normal, is [1]
$$\frac{1}{4\pi}(\mathbf{D}\,.\,\mathbf{N})\mathbf{E} - \frac{1}{8\pi}(\mathbf{D}\,.\,\mathbf{E})\mathbf{N}.$$

Maxwell obtained [2] a similar formula for the case of magnetic fields; the ponderomotive forces on magnetised matter and on conductors carrying currents may be accounted for by assuming a stress in the medium, the stress across the plane **N** being represented by the vector
$$\frac{1}{4\pi}(\mathbf{B}\,.\,\mathbf{N})\mathbf{H} - \frac{1}{8\pi}(\mathbf{B}\,.\,\mathbf{H})\mathbf{N}.$$

This, like the corresponding electrostatic formula, represents a tension across planes perpendicular to the lines of force, and a pressure across planes parallel to them.

It may be remarked that Maxwell made no distinction between stress in the material dielectric and stress in the aether: indeed, so long as it was supposed that material bodies when displaced carry the contained aether along with them, no distinction was possible. In the modifications of Maxwell's theory which were developed many years afterwards by his followers, stresses corresponding to those introduced by Maxwell were assigned to the aether, as distinct from ponderable matter; and it was assumed that the only stresses set up in material bodies by the electromagnetic field are produced indirectly; they may be calculated by the methods of the theory of elasticity, from a knowledge of the ponderomotive forces exerted on the electric charges connected with the bodies.

Another remark suggested by Maxwell's theory of stress in the

[1] cf. W. H. Bragg, *Phil. Mag.* xxxiv (1892), p. 18

[2] Maxwell's *Treatise on Electricity and Magnetism*, § 643

medium is that he considered the question from the purely statical point of view. He determined the stress so that it might produce the required forces on ponderable bodies, and be self-equilibrating in free aether. But [1] if the electric and magnetic phenomena are not really statical, but are kinetic in their nature, the stress or pressure need not be self-equilibrating. This may be illustrated by reference to the hydrodynamical models of the aether shortly to be described, in which perforated solids are immersed in a moving liquid : the ponderomotive forces exerted on the solids by the liquid correspond to those which act on conductors carrying currents in a magnetic field, and yet there is no stress in the medium beyond the pressure of the liquid.

Among the problems to which Maxwell applied his theory of stress in the medium was one which had engaged the attention of many generations of his predecessors. The adherents of the corpuscular theory of light in the eighteenth century believed that their hypothesis would be decisively confirmed if it could be shown that rays of light possess momentum : to determine the matter, several investigators directed powerful beams of light on delicately suspended bodies, and looked for evidences of a pressure due to the impulse of the corpuscles. Such an experiment was performed in 1708 by Homberg,[2] who imagined that he actually obtained the effect in question ; but Mairan and Du Fay in the middle of the century, having repeated his operations, failed to confirm his conclusion.[3]

The subject was afterwards taken up by Michell, who 'some years ago,' wrote Priestley [4] in 1772, 'endeavoured to ascertain the momentum of light in a much more accurate manner than those in which M. Homberg and M. Mairan had attempted it.' He exposed a very thin and delicately suspended copper plate to the rays of the sun concentrated by a mirror, and observed a deflection. He was not satisfied that the effect of the heating of the air had been altogether excluded, but 'there seems to be no doubt,' in Priestley's opinion, 'but that the motion above mentioned is to be ascribed to the impulse of the rays of light.' A similar experiment was made by A. Bennet,[5] who directed the light from the focus of a large lens on writing-paper delicately suspended in an exhausted receiver, but 'could not perceive any motion distinguishable from the effects of heat. Perhaps,' he concluded, 'sensible heat and light may not be caused by the influx or rectilineal projections of fine particles,

[1] cf. V. Bjerknes, *Phil. Mag.* ix (1905), p. 491
[2] *Histoire de l'Acad.* (1708), p. 21
[3] J. J. de Mairan, *Traité de l'Aurore boréale*, p. 370
[4] *History of Vision*, i, p. 387
[5] *Phil. Trans.* lxxxii (1792), p. 81

but by the vibrations made in the universaly diffused *caloric* or matter of heat, or fluid of light.' Thus Bennet, and after him Young,[1] regarded the non-appearance of light repulsion in this experiment as an argument in favour of the undulatory system of light. 'For,' wrote Young, 'granting the utmost imaginable subtility of the corpuscles of light, their effects might naturally be expected to bear some proportion to the effects of the much less rapid motions of the electrical fluid, which are so very easily perceptible, even in their weakest states.'

This attitude is all the more remarkable, because Euler many years before had expressed the opinion that light pressure might be expected just as reasonably on the undulatory as on the corpuscular hypothesis. 'Just as,' he wrote,[2] 'a vehement sound excites not only a vibratory motion in the particles of the air, but there is also observed a real movement of the small particles of dust which are suspended therein, it is not to be doubted but that the vibratory motion set up by the light causes a similar effect.' Euler not only inferred the existence of light pressure, but even (adopting a suggestion of Kepler's) accounted for the tails of comets by supposing that the solar rays, impinging on the atmosphere of a comet, drive off from it the more subtle of its particles.

The question was examined by Maxwell [3] from the point of view of the electromagnetic theory of light, which readily furnishes reasons for the existence of light pressure. For suppose that light falls on a metallic reflecting surface at perpendicular incidence. The light may be regarded as constituted of a rapidly alternating magnetic field; and this must induce electric currents in the surface layers of the metal. But a metal carrying currents in a magnetic field is acted on by a ponderomotive force, which is at right angles to both the magnetic force and the direction of the current, and is therefore, in the present case, normal to the reflecting surface: this ponderomotive force is the light pressure. Thus, according to Maxwell's theory, light pressure is only an extended case of effects which may readily be produced in the laboratory.

The magnitude of the light pressure was deduced by Maxwell from his theory of stresses in the medium. We have seen that the stress across a plane whose unit-normal is **N** is represented by the vector

$$\frac{1}{4\pi}(\mathbf{D}\cdot\mathbf{N})\mathbf{E}-\frac{1}{8\pi}(\mathbf{D}\cdot\mathbf{E})\mathbf{N}+\frac{1}{4\pi}(\mathbf{B}\cdot\mathbf{N})\mathbf{H}-\frac{1}{8\pi}(\mathbf{B}\cdot\mathbf{H})\mathbf{N}.$$

[1] ibid. xcii (1802), p. 12

[2] *Histoire de l'Acad. de Berlin*, ii (1748), p. 117

[3] Maxwell's *Treatise on Electricity and Magnetism*, § 792

Now, suppose that a plane wave is incident perpendicularly on a perfectly reflecting metallic sheet: this sheet must support the mechanical stress which exists at its boundary in the aether. Owing to the presence of the reflected wave, **D** is zero at the surface; and **B** is perpendicular to **N**, so (**B. N**) vanishes. Thus the stress is a pressure of magnitude $1/8\pi$ (**B. H**) normal to the surface: that is, the light pressure is equal to the density of the aethereal energy in the region immediately outside the metal. This was Maxwell's result.

This conclusion has been reached on the assumption that the light is incident normally to the reflecting surface. If, on the other hand, the surface is placed in an enclosure completely surrounded by a radiating shell, so that radiation falls on it from all directions, it may be shown that the light pressure is measured by one-third of the density of aethereal energy.

A different way of inferring the necessity for light pressure was indicated in 1876 by Adolfo Bartoli[1] (1851–96) of Bologna, who showed that, when radiant energy is transported from a cold body to a hot one by means of a moving mirror, the second law of thermodynamics would be violated unless a pressure were exerted on the mirror by the light.

In the year in which Maxwell's treatise was published, Sir William Crookes[2] obtained experimental evidence of a pressure accompanying the incidence of light; but this was soon found to be due to thermal effects; and the existence of a true light pressure was not confirmed experimentally[3] until 1899. Since then the subject has been considerably developed, especially in regard to the part played by the pressure of radiation in cosmical physics. In connection with radiation pressure, it may be mentioned that there is a pressure of radiation against the source—a *recoil* produced by the emission of light. This was investigated in 1909 by J. H. Poynting and Guy Barlow,[4] who employed thin slips of material which were heated by incident radiation. Slips blackened on both sides experience a pressure equal to the energy-density P of the incident radiation, since the emitted radiation from the front and back have equal recoil pressures. Slips blackened on the incident side and brightly silvered

[1] Bartoli, *Sopra i movimenti prodotti dalla luce e dal calore e sopra il radiometro di Crookes*, Firenze, 1876. Also *Nuovo Cimento*(3), xv (1884), p. 193, and Exner's *Rep.* xxi (1885), p. 198.

[2] *Phil. Trans.* clxiv (1874), p. 501. The radiometer was discovered in 1875.

[3] P. Lebedeff, *Archives des Sciences Phys. et Nat.* (4), viii (1899), p. 184; *Ann. d. Phys.* vi (1901), p. 433; E. F. Nichols and G. F. Hull, *Phys. Rev.* xiii (1901), p. 293; *Astrophys. Jour.* xvii (1903), p. 315; W. Gerlach and Alice Golsen, *Zs. f. Phys.* xv (1923), p. 1

[4] *Brit. Ass. Rep.* (1909), p. 385

on the other experience a pressure $\frac{5}{3}$P, the excess being due to the radiation which is emitted according to the cosine law by the front side only of the plate. Plates which are bright on both sides experience a pressure 2P, since they do not become heated, but reflect the radiation.

Another matter which received attention in Maxwell's *Treatise* was the influence of a magnetic field on the propagation of light in material substances. We have already seen [1] that the theory of magnetic vortices had its origin in W. Thomson's speculations on this phenomenon; and Maxwell in his memoir of 1861–2 had attempted by the help of that theory to arrive at some explanation of it. The more complete investigation which is given in the *Treatise* is based on the same general assumptions, namely, that in a medium subjected to a magnetic field there exist concealed vortical motions, the axes of the vortices being in the direction of the lines of magnetic force; and that waves of light passing through the medium disturb the vortices, which thereupon react dynamically on the luminous motion, and so affect its velocity of propagation.

The manner of this dynamical interaction must now be more closely examined. Maxwell supposed that the magnetic vortices are affected by the light waves in the same way as vortex filaments in a liquid would be affected by any other coexisting motion in the liquid. The latter problem had been already discussed in Helmholtz' great memoir on vortex motion; adopting Helmholtz' results, Maxwell assumed for the additional term introduced into the magnetic force by the displacement of the vortices the value $\partial \mathbf{e}/\partial \theta$, where $\mathbf{e}$ denotes the displacement of the medium (i.e. the light vector), and the operator $\partial/\partial\theta$ denotes $H_x\partial/\partial x + H_y\partial/\partial y + H_z\partial/\partial z$, $\mathbf{H}$ denoting the imposed magnetic field. Thus the luminous motion, by disturbing the vortices, gives rise to an electric current in the medium, proportional to curl $\partial\mathbf{e}/\partial\theta$.

Maxwell further assumed that the current thus produced interacts dynamically with the luminous motion in such a manner that the kinetic energy of the medium contains a term proportional to the scalar product of $\dfrac{\partial \mathbf{e}}{\partial t}$ and curl $\partial\mathbf{e}/\partial\theta$. The total kinetic energy of the medium may therefore be written

$$\tfrac{1}{2}\rho\left(\frac{\partial \mathbf{e}}{\partial t}\right)^2 + \tfrac{1}{2}\sigma\left(\frac{\partial \mathbf{e}}{\partial t}\cdot \text{curl } \partial\mathbf{e}/\partial\theta\right),$$

where ρ denotes the density of the medium, and σ denotes a constant

[1] cf. p. 245

which measures the capacity of the medium to rotate the plane of polarisation of light in a magnetic field.

The equation of motion may now be derived as in the elastic-solid theories of light: it is

$$\rho \frac{\partial^2 \mathbf{e}}{\partial t^2} = n\nabla^2 \mathbf{e} - \sigma \frac{\partial^2}{\partial t \partial \theta} \text{curl } \mathbf{e}.$$

When the light is transmitted in the direction of the lines of force, and the axis of x is taken parallel to this direction, the equation reduces to

$$\begin{cases} \rho \dfrac{\partial^2 e_y}{\partial t^2} = n \dfrac{\partial^2 e_y}{\partial x^2} + \sigma \mathrm{H} \dfrac{\partial^3 e_z}{\partial t \partial x^2}, \\ \rho \dfrac{\partial^2 e_z}{\partial t^2} = n \dfrac{\partial^2 e_z}{\partial x^2} - \sigma \mathrm{H} \dfrac{\partial^3 e_y}{\partial t \partial x^2}; \end{cases}$$

and these equations, as we have seen,[1] furnish an explanation of Faraday's phenomenon.

It may be remarked that the term

$$\tfrac{1}{2}\sigma \left(\frac{\partial \mathbf{e}}{\partial t} \cdot \text{curl } \partial \mathbf{e}/\partial \theta \right)$$

in the kinetic energy may by partial integration be transformed into a term

$$\tfrac{1}{2}\sigma \,(\text{curl } \mathbf{e} \cdot \partial \mathbf{e}/\partial \theta),$$ [2]

together with surface terms; or, again, into

$$- \tfrac{1}{2}\sigma \left(\text{curl } \mathbf{e} \cdot \frac{\partial^2 \mathbf{e}}{\partial t \partial \theta} \right),$$

together with surface terms. These different forms all yield the same equation of motion for the medium; but, owing to the differences in the surface terms, they yield different conditions at the boundary of the medium, and consequently give rise to different theories of reflection.

The assumptions involved in Maxwell's treatment of the magnetic rotation of light were such as might scarcely be justified in themselves; but since the discussion as a whole proceeded from sound dynamical principles, and its conclusions were in harmony with experimental results, it was fitted to lead to the more perfect explanations which were afterwards devised by his successors. At the time of Maxwell's death, which happened in 1879, before he had com-

[1] cf. p. 192

[2] This form was suggested by FitzGerald six years later, *Phil. Trans.* clxxi (1880), p. 691; FitzGerald's *Scientific Writings*, p. 45

pleted his forty-ninth year, much yet remained to be done both in this and in the other investigations with which his name is associated ; and the energies of the next generation were largely spent in extending and refining that conception of electrical and optical phenomena whose origin is correctly indicated in its name of *Maxwell's Theory*.

Chapter IX

MODELS OF THE AETHER

THE early attempts of W. Thomson and Maxwell to represent the electric medium by mechanical models opened up a new field of research, to which investigators were attracted as much by its intrinsic fascination as by the importance of the services which it promised to render to physical theory. There seemed to be a possibility of accounting for electric, magnetic and gravitational forces by the action of an intervening aether.

Of the models to which reference has already been made, some —such as those described in Thomson's memoir [1] of 1847 and Maxwell's memoir [2] of 1861–2—attribute a linear character to electric force and electric current, and a rotatory character to magnetism; others—such as that devised by Maxwell in 1855 [3] and afterwards amplified by Helmholtz [4]—regard magnetic force as a linear, and electric current as a rotatory phenomenon. This distinction furnishes a natural classification of models into two principal groups.

Even within the limits of the former group diversity has already become apparent; for in Maxwell's analogy of 1861–2, a continuous vortical motion is supposed to be in progress about the lines of magnetic induction; whereas in Thomson's analogy the vector-potential was likened to the displacement in an elastic solid, so that the magnetic induction at any point would be represented by the twist of an element of volume of the solid from its equilibrium position; or, in symbols,

$$\mathbf{a} = \mathbf{e}, \quad \mathbf{E} = -\frac{1}{c}\frac{\partial \mathbf{e}}{\partial t}, \quad \mathbf{B} = \text{curl } \mathbf{e},$$

where $\mathbf{a}$ denotes the vector-potential, $\mathbf{E}$ the electric force, $\mathbf{B}$ the magnetic induction and $\mathbf{e}$ the elastic displacement.

Thomson's original memoir concluded with a notice of his intention to resume the discussion in another communication. His purpose was fulfilled only in 1890, when [5] he showed that in his model a linear current could be represented by a piece of endless

[1] cf. p. 242 [2] cf. p. 247 [3] cf. p. 242 [4] cf. p. 245

[5] Kelvin's *Math. and Phys. Papers*, iii, p. 436

cord, of the same quality as the solid and embedded in it, if a tangential force were applied to the cord uniformly all round the circuit. The forces so applied tangentially produce a tangential drag on the surrounding solid ; and the rotatory displacement thus caused is everywhere proportional to the magnetic vector.

In order to represent the effect of varying permeability, Thomson abandoned the ordinary type of elastic solid, and replaced it by an aether of MacCullagh's type ; that is to say, an ideal incompressible substance, having no rigidity of the ordinary kind (i.e. elastic resistance to change of shape), but capable of resisting absolute rotation —a property to which the name *gyrostatic rigidity* was given. The rotation of the solid representing the magnetic induction, and the coefficient of gyrostatic rigidity being inversely proportional to the permeability, the normal component of magnetic induction will be continuous across an interface, as it should be.[1]

We have seen above that in models of this kind the electric force is represented by the translatory velocity of the medium. It might therefore be expected that a strong electric field would perceptibly effect the velocity of propagation of light ; and that this does not appear to be the case,[2] is an argument against the validity of the scheme.

We now turn to the alternative conception, in which electric phenomena are regarded as rotatory, and magnetic force is represented by the linear velocity of the medium ; in symbols,

$$\begin{cases} \frac{1}{c}\mathbf{D} = \operatorname{curl} \mathbf{e}, \\ \mathbf{H} = \frac{\partial \mathbf{e}}{\partial t}, \end{cases}$$

where $\mathbf{D}$ denotes the electric displacement, $\mathbf{H}$ the magnetic force and $\mathbf{e}$ the displacement of the medium. In Maxwell's memoir of 1855, and in most of the succeeding writings for many years, attention was directed chiefly to magnetic fields of a steady, or at any rate non-oscillatory, character ; in such fields, the motion of the particles of the medium is continuously progressive ; and it was consequently natural to suppose the medium to be fluid.

Maxwell himself, as we have seen,[3] afterwards abandoned this conception in favour of that which represents magnetic phenomena as rotatory. 'According to Ampère and all his followers,' he wrote

[1] Thomson inclined to believe (*Papers*, iii, p. 465) that light might be correctly represented by the vibratory motion of such a solid.

[2] Wilberforce, *Trans. Camb. Phil. Soc.* xiv (1887), p. 170 ; Lodge, *Phil. Trans.* clxxxix (1897), p. 149 [3] cf. p. 247

in 1870,[1] 'electric currents are regarded as a species of translation, and magnetic force as depending on rotation. I am constrained to agree with this view, because the electric current is associated with electrolysis, and other undoubted instances of translation, while magnetism is associated with the rotation of the plane of polarisation of light.' But the other analogy was felt to be too valuable to be altogether discarded, especially when in 1858 Helmholtz extended it [2] by showing that if magnetic induction is compared to fluid velocity, then electric currents correspond to vortex filaments in the fluid. Two years afterwards Kirchhoff [3] developed it further. If the analogy has any dynamical (as opposed to a merely kinematical) value, it is evident that the ponderomotive forces between metallic rings carrying electric currents should be similar to the ponderomotive forces between the same rings when they are immersed in an infinite incompressible fluid; the motion of the fluid being such that its circulation through the aperture of each ring is proportional to the strength of the electric current in the corresponding ring. In order to decide the question, Kirchhoff attempted, and solved, the hydrodynamical problem of the motion of two thin, rigid rings in an incompressible frictionless fluid, the fluid motion being irrotational; and found that the forces between the rings are numerically equal to those which the rings would exert on each other if they were traversed by electric currents proportional to the circulations.

There is, however, an important difference between the two cases, which was subsequently discussed by W. Thomson, who pursued the analogy in several memoirs.[4] In order to represent the magnetic field by a conservative dynamical system, we shall suppose that it is produced by a number of rings of perfectly conducting material, in which electric currents are circulating; the surrounding medium being free aether. Now any perfectly conducting body acts as an impenetrable barrier to lines of magnetic force; for, as Maxwell showed,[5] when a perfect conductor is placed in a magnetic field, electric currents are induced on its surface in such a way as to make the total magnetic force zero throughout the interior of the conductor.[6] Lines of force are thus deflected by the body in the

[1] *Proc. Lond. Math. Soc.* iii (1870), p. 224; Maxwell's *Scientific Papers*, ii, p. 263

[2] cf. p. 245

[3] *Berlin Monatsb.* (1869), p. 881; *Journal für Math.* lxxi (1869), p. 263; Kirchhoff's *Gesamm. Abhandl.*, p. 404. cf. also C. Neumann, *Leipzig Berichte*, xliv (1892), p. 86.

[4] Thomson's *Papers in Elect. and Mag.*, §§ 573, 733, 751 (1870–2)

[5] Maxwell's *Treatise on Elect. and Mag.*, § 654

[6] For this reason W. Thomson called a perfect conductor an *ideal extreme diamagnetic.*

same way as the lines of flow of an incompressible fluid would be deflected by an obstacle of the same form, or as the lines of flow of electric current in a uniform conducting mass would be deflected by the introduction of a body of this form and of infinite resistance. If, then, for simplicity we consider two perfectly conducting rings carrying currents, those lines of force which are initially linked with a ring cannot escape from their entanglement, and new lines cannot become involved in it. This implies that the total number of lines of magnetic force which pass through the aperture of each ring is invariable. If the coefficients of self and mutual induction of the rings are denoted by L_1, L_2, L_{12}, the electrokinetic energy of the system may be represented by

$$T = \tfrac{1}{2}(L_1 q_1^2 + 2\,L_{12}\,q_1 q_2 + L_2 q_2^2),$$

where q_1, q_2 denote the strengths of the currents ; and the condition that the number of lines of force linked with each circuit is to be invariable gives the equations

$$\begin{aligned} L_1 q_1 + L_{12} q_2 &= \text{constant}, \\ L_{12} q_1 + L_2 q_2 &= \text{constant}. \end{aligned}$$

It is evident that, when the system is considered from the point of view of general dynamics, the electric currents must be regarded as generalised velocities, and the quantities

$$(L_1 q_1 + L_{12} q_2) \quad \text{and} \quad (L_{12} q_1 + L_2 q_2)$$

as momenta. The electromagnetic ponderomotive force on the rings tending to increase any co-ordinate x is $\partial T/\partial x$. In the analogous hydrodynamical system, the fluid velocity corresponds to the magnetic force ; and therefore the circulation through each ring (which is defined to be the integral $\int v\,ds$, taken round a path linked once with the ring) corresponds kinematically to the electric current ; and the flux of fluid through each ring corresponds to the number of lines of magnetic force which pass through the aperture of the ring. But in the hydrodynamical problem the circulations play the part of generalised momenta ; while the fluxes of fluid through the rings play the part of generalised velocities. The kinetic energy may indeed be expressed in the form

$$K = \tfrac{1}{2}(N_1 \kappa_1^2 + 2 N_{12} \kappa_1 \kappa_2 + N_2 \kappa_2^2),$$

where κ_1, κ_2, denote the circulations (so that κ_1 and κ_2 are proportional respectively to q_1 and q_2), and N_1, N_{12}, N_2 depend on the positions of the rings; but this is the Hamiltonian (as opposed to the Lagrangian) form of the energy-function,[1] and the ponderomotive force on the rings tending to increase any co-ordinate x is $-\partial K/\partial x$. Since $\partial K/\partial x$ is equal to $\partial T/\partial x$, we see that the ponderomotive forces on the rings in any position in the hydrodynamical system are equal, *but opposite*, to the ponderomotive forces on the rings in the electric system.

The reason for the difference between the two cases may readily be understood. The rings cannot cut through the lines of magnetic force in the one system, but they can cut through the stream lines in the other; consequently the flux of fluid through the rings is not invariable when the rings are moved, the invariants in the hydrodynamical system being the circulations. If a thin ring, for which the circulation is zero, is introduced into the fluid, it will experience no ponderomotive forces; but if a ring initially carrying no current is introduced into a magnetic field, it will experience ponderomotive forces, owing to the electric currents induced in it by its motion.

Imperfect though the analogy is, it is not without interest. A bar magnet, being equivalent to a current circulating in a wire wound round it, may be compared (as W. Thomson remarked) to a straight tube immersed in a perfect fluid, the fluid entering at one end and flowing out by the other, so that the particles of fluid follow the lines of magnetic force. If two such tubes are presented with like ends to each other, they attract; with unlike ends, they repel. The forces are thus diametrically opposite in direction to those of magnets; but in other respects the laws of mutual action between these tubes and between magnets are precisely the same.[2]

[1] cf. Whittaker, *Analytical Dynamics*, § 109

[2] The mathematical analysis in this case is very simple. A narrow tube through which water is flowing may be regarded as equivalent to a source at one end of the tube and a sink at the other; and the problem may therefore be reduced to the consideration of sinks in an unlimited fluid. If there are two sinks in such a fluid, of strengths m and m', the velocity-potential is

$$m/r + m'/r',$$

where r and r' denote distance from the sinks. The kinetic energy per unit volume of the fluid is

$$\tfrac{1}{2}\rho\left[\left\{\frac{\partial}{\partial x}\left(\frac{m}{r}+\frac{m'}{r'}\right)\right\}^2+\left\{\frac{\partial}{\partial y}\left(\frac{m}{r}+\frac{m'}{r'}\right)\right\}^2+\left\{\frac{\partial}{\partial z}\left(\frac{m}{r}+\frac{m'}{r'}\right)\right\}^2\right],$$

where ρ denotes the density of the fluid; whence it is easily seen that the total energy of the fluid, when the two sinks are at a distance l apart, exceeds the total energy when they are at an infinite distance apart by an amount

$$\rho mm'\iiint\left\{\frac{\partial}{\partial x}\left(\frac{1}{r}\right)\frac{\partial}{\partial x}\left(\frac{1}{r'}\right)+\frac{\partial}{\partial y}\left(\frac{1}{r}\right)\frac{\partial}{\partial y}\left(\frac{1}{r'}\right)+\frac{\partial}{\partial z}\left(\frac{1}{r}\right)\frac{\partial}{\partial z}\left(\frac{1}{r'}\right)\right\}dx\,dy\,dz,$$

Thomson, moreover, investigated[1] the ponderomotive forces which act between two solid bodies immersed in a fluid, when one of the bodies is constrained to perform small oscillations. If, for example, a small sphere immersed in an incompressible fluid is compelled to oscillate along the line which joins its centre to that of a much larger sphere, which is free, the free sphere will be attracted if it is denser than the fluid; while if it is less dense than the fluid, it will be repelled or attracted according as the ratio of its distance from the vibrator to its radius is greater or less than a certain quantity depending on the ratio of its density to the density of the fluid. Systems of this kind were afterwards extensively investigated by C. A. Bjerknes.[2] Bjerknes showed that two spheres which are immersed in an incompressible fluid, and which pulsate (i.e. change in volume) regularly, exert on each other (by the mediation of the fluid) an attraction, determined by the inverse square law, if the pulsations are concordant; and exert on each other a repulsion, determined likewise by the inverse square law, if the phases of the pulsations differ by half a period. If the phases differ by a quarter period, there is no action. It is necessary to suppose that the medium is incompressible, so that all pulsations are propagated instantaneously; otherwise attractions would change to repulsions and vice versa at distances greater than a quarter wave-length.[3] If the

[1] cf. F. Guthrie, *Phil. Mag.* xli (1871), p. 405, specially the letter of Thomson to Guthrie dated 24 Nov. 1870, printed at p. 427.

[2] *Repertorium d. Mathematik* i (Leipzig, 1877), p. 268; *Göttinger Nachrichten* (1876), p. 245; *Comptes Rendus*, lxxxiv (1877), p. 1375; cf. *Nature*, xxiv (1881), p. 360; cf. also H. Witte, *Ann. d. Phys.* xxx (1909), p. 337, with a reply by V. Bjerknes, ibid. xxxi (1910), p. 312, and a further note by Witte, ibid. xxxii (1910), p. 382.

[3] On the mathematical theory of the force between two pulsating spheres in a fluid, cf. W. M. Hicks, *Proc. Camb. Phil. Soc.* iii (1879), p. 276; iv (1880), p. 29.

---

the integration being taken throughout the whole volume of the fluid, except two small spheres s, s', surrounding the sinks. By Green's theorem, this expression reduces at once to

$$\rho m m' \iint \frac{1}{r'} \frac{\partial}{\partial n}\left(\frac{1}{r}\right) dS,$$

where the integration is taken over s and s', and n denotes the interior normal to s or s'. The integral taken over s' vanishes; evaluating the remaining integral, we have

$$-\frac{\rho\, m m'}{l} \iint \frac{\partial}{\partial r}\left(\frac{1}{r}\right) dS, \quad \text{or} \quad 4\pi\rho m m'/l.$$

The energy of the fluid is therefore greater when sinks of strengths m, m' are at a mutual distance l than when sinks of the same strengths are at infinite distance apart by an amount $4\pi\rho m m'/l$. Since, in the case of the tubes, the quantities m correspond to the fluxes of fluid, this expression corresponds to the Lagrangian form of the kinetic energy; and therefore the force tending to increase the co-ordinate x of one of the sinks is $(\partial/\partial x)\,(4\pi\rho m m'/l)$. Whence it is seen that the *like* ends of two tubes *attract*, and the unlike ends *repel*, according to the inverse square law.

spheres, instead of pulsating, oscillate to and fro in straight lines about their mean positions, the forces between them are proportional in magnitude and the same in direction, but opposite in sign, to those which act between two magnets oriented along the directions of oscillation.[1]

The results obtained by Bjerknes were extended by A. H. Leahy [2] to the case of two spheres pulsating in an elastic medium; the wave-length of the disturbance being supposed large in comparison with the distance between the spheres. For this system Bjerknes' results are reversed, the law being now that of attraction in the case of unlike phases, and of repulsion in the case of like phases; the intensity is as before proportional to the inverse square of the distance.

The same author afterwards discussed [3] the oscillations which may be produced in an elastic medium by the displacement, in the direction of the tangent to the cross-section, of the surfaces of tubes of small sectional area: the tubes either forming closed curves, or extending indefinitely in both directions. The direction and circumstances of the motion are in general analogous to ordinary vortex motions in an incompressible fluid; and it was shown by Leahy that, if the period of the oscillation be such that the waves produced are long compared with ordinary finite distances, the displacement due to the tangential disturbances is proportional to the velocity due to vortex rings of the same form as the tubular surfaces. One of these 'oscillatory twists,' as the tubular surfaces may be called, produces a displacement which is analogous to the magnetic force due to a current flowing in a curve coincident with the tube; the strength of the current being proportional to $b^2\omega \sin pt$, where b denotes the radius of the twist, and $\omega \sin pt$ its angular displacement. If the field of vibration is explored by a rectilineal twist of the same period as that of the vibration, the twist will experience a force at right angles to the plane containing the twist and the direction of the displacement which would exist if the twist were removed; if the displacement of the medium be represented by $F \sin pt$, and the angular displacement of the twist by $\omega \sin pt$, the magnitude of the force is proportional to the vector-product of F (in the direction of the displacement) and ω (in the direction of the axis of the twist).

[1] A theory of gravitation has been based by Korn on the assumption that gravitating particles resemble slightly compressible spheres immersed in an incompressible perfect fluid; the spheres execute pulsations, whose intensity corresponds to the mass of the gravitating particles, and thus forces of the Newtonian kind are produced between them. cf. Korn, *Eine Theorie der Gravitation und der elect. Erscheinungen*, Berlin, 1898.

[2] *Trans. Camb. Phil. Soc.* xiv (1884), p. 45

[3] *Trans. Camb. Phil. Soc.* xiv (1885), p. 188

A model of magnetic action may evidently be constructed on the basis of these results. A bar magnet must be regarded as vibrating tangentially, the direction of vibration being parallel to the axis of the body. A cylindrical body carrying a current will have its surface also vibrating tangentially ; but in this case the direction of vibration will be perpendicular to the axis of the cylinder. A statically electrified body, on the other hand, may, as follows from the same author's earlier work, be regarded as analogous to a body whose surface vibrates in the normal direction. Larmor [1] suggested that gravitation between electrons might be explained if the electrons are vacuous and have free periods of vibration in the fluid aether, the explanation following the hydrodynamical pulsatory theory of Bjerknes.

We have now discussed models in which the magnetic force is represented as the velocity in a liquid, and others in which it is represented as the displacement in an elastic solid. Some years before the date of Leahy's memoir, George Francis FitzGerald (1851–1901) [2] had instituted a comparison between magnetic force and the velocity in a quasi-elastic solid of the type first devised by MacCullagh.[3] Of all possible types of aether, this alone could propagate waves having the properties of waves of light, and an analogy with the electromagnetic theory of light is at once evident when it is noticed that the electromagnetic equation

$$\frac{1}{c}\frac{\partial \mathbf{D}}{\partial t} = \text{curl}\,\mathbf{H}$$

is satisfied identically by the values

$$\begin{cases} \dfrac{1}{c}\mathbf{D} = \text{curl}\,\mathbf{e}, \\ \mathbf{H} = \dfrac{\partial \mathbf{e}}{\partial t}, \end{cases}$$

where **e** denotes any vector ; and that, on substituting these values in the other electromagnetic equation,

$$-\,c\,\text{curl}\,(\mathbf{D}/\epsilon) = \frac{\partial \mathbf{H}}{\partial t},$$

we obtain the equation

$$\epsilon\,\frac{\partial^2 \mathbf{e}}{\partial t^2} + c^2\,\text{curl curl}\,\mathbf{e} = 0$$

which is no other than the equation of motion of MacCullagh's

[1] *Phil. Trans.* clxxxvi (1895), p. 697

[2] *Phil. Trans.* clxxi (1880), p. 691 (presented October 1878) ; FitzGerald's *Scientific Writings*, p. 45

[3] cf. p. 142

aether,[1] the specific inductive capacity ϵ corresponding to the reciprocal of MacCullagh's constant of elasticity. In the analogy thus constituted, electric displacement corresponds to the twist of the elements of volume of the aether ; and electric charge must evidently be represented as an intrinsic rotational strain. Thus it appeared possible to extend the capacities of this aether so as to represent not only optical phenomena, but all kinds of magnetic and electric interaction. Mechanical models of the electromagnetic field, based on FitzGerald's analogy, were afterwards studied by A. Sommerfeld,[2] by R. Reiff[3] and by Sir J. Larmor.[4] The last-named author[5] supposed the electric charge to exist in the form of discrete electrons, for the creation of which he suggested the following ideal process[6] : A filament of aether, terminating at two nuclei, is supposed to be removed, and circulatory motion is imparted to the walls of the channel so formed, at each point of its length, so as to produce throughout the medium a rotational strain. When this has been accomplished, the channel is to be filled up again with aether, which is to be made continuous with its walls. When the constraint is removed from the walls of the channel, the circulation imposed on them proceeds to undo itself, until this tendency is balanced by the elastic resistance of the aether with which the channel has been filled up ; thus finally the system assumes a state of equilibrium in which the nuclei, which correspond to a positive and a negative electron, are surrounded by intrinsic rotational strain. On this view, electrons, and hence all material bodies built up from them, are of the nature of structures in the aether, involving an atmosphere of aethereal strain all around them ; action at a distance is abolished. 'All that is needed,' said Larmor,[7] 'is a postulate of free mobility of the nucleus through the aether. This is definitely hypothetical, but it is not an unreasonable postulate, because a rotational aether has the properties of a perfect fluid medium except where differentially rotational motions are concerned, and so would not react on the motion of any structure moving through it except after the manner of an apparent change of inertia.' 'If,' he said elsewhere,[8] 'the electron is a mere passive pole—nucleus of beknottedness in some way—in the aether, conditioned and controlled entirely by the

[1] cf. p. 142 [2] *Ann. d. Phys.* xlvi (1892), p. 139

[3] Reiff, *Elasticität und Elektricität*, Freiburg, 1893

[4] *Phil. Trans.* clxxxv (1893), p. 719

[5] In a supplement, of date August 1894, to his above-cited memoir of 1893

[6] *Phil. Trans.* clxxxv (1894), p. 810 ; cxc (1897), p. 210 ; Larmor, *Aether and Matter* (1900), p. 326

[7] *Brit. Ass. Rep.* (1900), p. 626 [8] *Phil. Mag.* vii (1904), p. 621

aether around it, just as a vortex ring is conditioned by the fluid in which it subsists and is also carried along thereby, then, as in the familiar hydrodynamics of vortices, the motion of the aether determines the motion of the entirely passive electrons, and the idea of force acting between them and the aether is dispensed with.'

Models in which magnetic force is represented by the velocity of an aether are not, however, secure from objection. It is necessary to suppose that the aether is capable of flowing like a perfect fluid in irrotational motion (which would correspond to a steady magnetic field), and that it is at the same time endowed with the power (which is requisite for the explanation of electric phenomena) of resisting the rotation of any element of volume.[1] But when the aether moves irrotationally in the fashion which corresponds to a steady magnetic field, each element of volume acquires after a finite time a rotatory displacement from its original orientation, in consequence of the motion ; and it might therefore be expected that the quasi-elastic power of resisting rotation would be called into play—i.e. that a steady magnetic field would develop electric phenomena.[2]

A further objection to all models in which magnetic force corresponds to velocity is that a strong magnetic field, being in such models represented by a steady drift of the aether, might be expected to influence the velocity of propagation of light. The existence of such an effect appears, however, to be disproved by the experiments of Sir Oliver Lodge[3] ; at any rate, unless it is assumed that the aether has an inertia at least of the same order of magnitude as that of ponderable matter, in which case the motion might be too slow to be measurable.

Again, the evidence in favour of the rotatory as opposed to the linear character of magnetic phenomena had perhaps, on the whole, been strengthened since Thomson originally based his conclusion on the magnetic rotation of light. This brings us to the consideration of an experimental discovery.

[1] Larmor (loc. cit.) suggested the analogy of a liquid filled with magnetic molecules under the action of an external magnetic field.

It has often been objected to the mathematical conception of a perfect fluid that it contains no safeguard against slipping between adjacent layers, so that there is no justification for the usual assumption that the motion of a perfect fluid is continuous. Larmor remarked that a rotational elasticity, such as is attributed to the medium above considered, furnishes precisely such a safeguard ; and that without some property of this kind a continuous frictionless fluid cannot be imagined.

[2] Larmor proposed to avoid this by assuming that the rotation which is resisted by an element of volume of the aether is the vector sum of the series of differential rotations which it has experienced.

[3] *Phil. Trans.* clxxxix (1897), p. 149

In 1879 Edwin H. Hall,[1] at that time a student at Baltimore, repeating an experiment which had been previously suggested by H. A. Rowland, obtained a new action of a magnetic field on electric currents. A strip of gold leaf mounted on glass, forming part of an electric circuit through which a current was passing, was placed between the poles of an electromagnet, the plane of the strip being perpendicular to the lines of magnetic force. The two poles of a sensitive galvanometer were then placed in connection with different parts of the strip, until two points at the same potential were found. When the magnetic field was created or destroyed, a deflection of the galvanometer needle was observed, indicating a change in the relative potential of the two poles. It was thus shown that the magnetic field produces in the strip of gold leaf a new electromotive force, at right angles to the primary electromotive force and to the magnetic force, and proportional to the product of these forces.[2]

From the physical point of view we may therefore regard Hall's effect as an additional electromotive force generated by the action of the magnetic field on the current; or alternatively we may regard it as a modification of the ohmic resistance of the metal, such as would be produced if the molecules of the metal assumed a helicoidal structure about the lines of magnetic force. From the latter point of view, all that is needed is to modify Ohm's law

$$\mathbf{S} = k\mathbf{E}$$

(where **S** denotes electric current, k specific conductivity and **E** electric force) so that it takes the form

$$\mathbf{S} = k\mathbf{E} + h\,[\mathbf{E}\,.\,\mathbf{H}]$$

where **H** denotes the imposed magnetic force, and h denotes a constant on which the magnitude of Hall's phenomenon depends. It is a curious circumstance that the occurrence, in the case of magnetised bodies, of an additional term in Ohm's law, formed from a vector product of **E** and **H**, had been expressly suggested in Maxwell's *Treatise*[3]: although Maxwell had not indicated the possibility of realising it by Hall's experiment.

An interesting application of Hall's discovery was made in the

[1] *Am. Jour. Math.* ii (1879), p. 287; *Am. J. Sci.* **xix** (1880), p. 200, and **xx**, p. 161; *Phil. Mag.* ix (1880), p. 225, and x, p. 301

[2] Oliver Lodge in the 1870's had worked experimentally on the flow of electricity in a metallic sheet, and had come very near to discovering the Hall effect; but he read in Maxwell's *Electricity and Magnetism* (vol. ii, § 501): 'It must be carefully remembered, that the mechanical force which urges a conductor carrying a current across the lines of magnetic force acts, not on the electric current, but on the conductor which carries it'; and this deterred him from making the crucial test.

[3] § 303. cf. Hopkinson, *Phil. Mag.* **x** (1880), p. 430

same year by Boltzmann,[1] who remarked that it offered a prospect of determining the absolute velocity of the electric charges which carry the current in the strip. For if it is supposed that only one kind (vitreous or resinous) of electricity is in motion, the force on one of the charges tending to drive it to one side of the strip will be proportional to the vector product of its velocity and the magnetic intensity. Assuming that Hall's phenomenon is a consequence of this tendency of charges to move to one side of the strip, it is evident that the velocity in question must be proportional to the magnitude of the Hall electromotive force due to a unit magnetic field. On the basis of this reasoning, A. von Ettingshausen [2] found for the current sent by one or two Daniell's cells through a gold strip a velocity of the order of 0.1 cm. per second. It is clear, however, that, if the current consists of both vitreous and resinous charges in motion in opposite directions, Boltzmann's argument fails ; for the two kinds of electricity would give opposite directions to the current in Hall's phenomenon.

In the year following his discovery, Hall [3] extended his researches in another direction, by investigating whether a magnetic field disturbs the distribution of equipotential lines in a dielectric which is in an electric field ; but no effect could be observed.[4] Such an effect, indeed,[5] was not to be expected on theoretical grounds ; for when, in a material system, all the velocities are reversed, the motion is reversed, it being understood that, in the application of this theorem to electrical theory, an electrostatic state is to be regarded as one of rest, and a current as a phenomenon of motion ; and if such a reversal be performed in the present system, the poles of the electromagnet are exchanged, while in the dielectric no change takes place.

We must now consider the bearing of Hall's effect on the question as to whether magnetism is a rotatory or a linear phenomenon.[6] If magnetism be linear, electric currents must be rotatory ; and if Hall's phenomenon be supposed to take place in a horizontal strip of metal, the magnetic force being directed vertically upwards, and the primary current flowing horizontally from north to south, the only geometrical entities involved are the vertical direction and a

[1] *Wien Anz.* xvii (1880), p. 12 ; *Phil. Mag.* ix (1880), p. 307

[2] *Ann. d. Phys.* xi (1880), pp. 432, 1044

[3] *Amer. Jour. Sci.* xx (1880), p. 164

[4] In 1885–6 E. van Aubel (*Bull. de l'Acad. Roy. de Belgique* [3], x, p. 609 ; xii, p. 280) repeated the investigation in an improved form, and confirmed the result that a magnetic field has no influence on the electrostatic polarisation of dielectrics.

[5] H. A. Lorentz, *Arch. Neérl.* xix (1884), p. 123

[6] cf. F. Koláček, *Ann. d. Phys.* lv (1895), p. 503

rotation in the east-and-west vertical plane ; and these are indifferent with respect to a rotation in the north-and-south vertical plane, so that there is nothing in the physical circumstances of the system to determine in which direction the secondary current shall flow. The hypothesis that magnetism is linear appears therefore to be inconsistent with the existence of Hall's effect.[1] There are, however, some considerations which may be urged on the other side. Hall's effect, like the magnetic rotation of light, takes place only in ponderable bodies, not in free aether ; and its direction is sometimes in one sense, sometimes in the other, according to the nature of the substance. It may therefore be doubted whether these phenomena are not of a secondary character, and the argument based on them invalid. Moreover, as FitzGerald remarked,[2] the magnetic lines of force associated with a system of currents are circuital and have no open ends, making it difficult to imagine how alteration of rotation inside them could be produced.

Of the various attempts to represent electric and magnetic phenomena by the motions and strains of a continuous medium, none of those hitherto considered has been found free from objection.[3] Before proceeding to consider models which are not constituted by a continuous medium, mention must be made of a suggestion offered by Riemann in his lectures [4] of 1861. Riemann remarked that the scalar-potential ϕ and vector-potential $\mathbf{a}$, corresponding to his own law of force between electrons, satisfy the equation

$$\frac{\partial \phi}{\partial t} + c \operatorname{div} \mathbf{a} = 0 ;$$

an equation which, as we have seen, is satisfied also by the potentials of L. Lorenz.[5] This appeared to Riemann to indicate that ϕ might represent the density of an aether, of which $c\mathbf{a}$ represents the velocity. It will be observed that on this hypothesis the electric and magnetic forces correspond to *second* derivates of the displacement—a circumstance which makes it somewhat difficult to assimilate the energy possessed by the electromagnetic field to the energy of the model.

We must now proceed to consider those models in which the

[1] Further evidence in favour of the hypothesis that it is the electric phenomena which are linear is furnished by the fact that pyro-electric effects (the production of electric polarisation by warming) occur in acentric crystals, and only in such. cf. M. Abraham, *Encyklopädie der math. Wiss.* iv (2), p. 43.

[2] cf. Larmor, *Phil. Trans.* clxxxv (1894), p. 780

[3] cf. H. Witte, *Ueber den gegenwärtigen Stand der Frage nach einer mechanischen Erklärung der elektrischen Erscheinungen*, Berlin, 1906

[4] Published after his death by K. Hattendorff, under the title *Schwere, Elektricität, und Magnetismus* (1875), p. 330

[5] cf. p. 269

aether is represented as composed of more than one kind of constituent: of these Maxwell's model of 1861–2, formed of vortices and rolling particles, may be taken as the type. Another device of the same class was described in 1885 by FitzGerald[1]; this was constituted of a number of wheels, free to rotate on axes fixed perpendicularly in a plane board; the axes were fixed at the intersections of two systems of perpendicular lines; and each wheel was geared to each of its four neighbours by an indiarubber band. Thus all the wheels could rotate without any straining of the system, provided they all had the same angular velocity; but if some of the wheels were revolving faster than others, the indiarubber bands would become strained. It is evident that the wheels in this model play the same part as the vortices in Maxwell's model of 1861–2: their rotation is the analogue of magnetic force; and a region in which the masses of the wheels are large corresponds to a region of high magnetic permeability. The indiarubber bands of FitzGerald's model correspond to the medium in which Maxwell's vortices were embedded; and a strain on the bands represents dielectric polarisation, the line joining the tight and slack sides of any band being the direction of displacement. A body whose specific inductive capacity is large would be represented by a region in which the elasticity of the bands is feeble. Lastly, conduction may be represented by a slipping of the bands on the wheels.

Such a model is capable of transmitting vibrations analogous to those of light. For if any group of wheels be suddenly set in rotation, those in the neighbourhood will be prevented by their inertia from immediately sharing in the motion; but presently the rotation will be communicated to the adjacent wheels, which will transmit it to their neighbours; and so a wave of motion will be propagated through the medium. The motion constituting the wave is readily seen to be directed in the plane of the wave, i.e. the vibration is transverse. The axes of rotation of the wheels are at right angles to the direction of propagation of the wave, and the direction of polarisation of the bands is at right angles to both these directions.

The elastic bands may be replaced by lines of governor balls[2]: if this be done, the energy of the system is entirely of the kinetic type.[3]

[1] *Scient. Proc. Roy. Dublin Soc.* iv (1885), p. 407; *Phil. Mag.* xix (1885), p. 438; FitzGerald's *Scientific Writings*, pp. 142, 157

[2] FitzGerald's *Scientific Writings*, p. 271

[3] It is of course possible to devise models of this class in which the rotation may be interpreted as having the electric instead of the magnetic character. Such a model was proposed by Boltzmann, *Vorlesungen über Maxwell's Theorie* (Leipzig 1891–93), ii.

Models of types different from the foregoing have been proposed, which involve the idea of *vortex motion.* One of the greatest achievements of Helmholtz was his discovery [1] in 1858 that vortex rings in a perfect fluid are types of motion which possess permanent individuality throughout all changes, and cannot be destroyed, so that they may be regarded as combining and interacting with each other, although each of them consists of a motion pervading the whole of the fluid. The energy of the fluid can be expressed in terms of the positions and strengths of the vortices, and from a knowledge of these characteristics it is possible to determine the future course of the system. As Helmholtz showed, vortex filaments interact mechanically like linear electric circuits (but with attraction replaced by repulsion and vice versa), the strengths of the vortices corresponding to the current strengths, and the velocity of the fluid in the neighbourhood of the filaments corresponding to the magnetic force; magnetic poles would be represented by sources and sinks in the fluid.

The individuality of vortices suggested a connection with the atomic theory of matter. As we have seen,[2] the atomic hypothesis of Leucippus and Democritus had been adopted by Gassendi, whose views in turn profoundly influenced Robert Boyle. To Boyle must be given the chief credit for introducing the notion of chemical element, which became more precise when Antoine-Laurent Lavoisier (1743–94) found that compounds which have a definite individuality contain the chemical elements in constant proportions, and that the total mass concerned in a chemical reaction is the same after it as before it. On the physical side, Newton [3] showed that a gas would obey Boyle's law of compression if it were composed of particles, which repel each other with forces reciprocally proportional to the distances between their centres; and Daniel Bernoulli [4] (1700–82) in 1738 gave a different explanation of Boyle's law, in which the gas was supposed to be constituted of small elastic spheres which collide with each other—a kinetic theory of gases in fact. These researches, chemical and physical, culminated in the work of John Dalton (1766–1844), who [5] proved that to every chemical element could be assigned a definite number, representing the relative weight of an atom of the element.[6]

[1] *Jour. für Math.* lv (1858), p. 25 [2] cf. pp. 12, 13 [3] *Principia*, ii, prop. 23

[4] *Hydrodynamica* (Strasbourg, 1738), § 10. See also Hooke's Lectures *De potentia restitutiva*, London, 1678

[5] In a paper read before the Manchester Philosophical Society in 1803 (*Mem. Manchester Phil. Soc.* i [1805], p. 271), and in his *New System of Chemical Philosophy* (1808)

[6] cf. Kurd Lasswitz, *Geschichte der Atomistik vom Mittelalter bis Newton*, Hamburg, 1890, P. Kirchberger, *La théorie atomique, son histoire et son développement*, Paris, 1930

The earliest attempts to build up a general physical theory on the basis of vortex motion were made in 1867 by William Thomson (Kelvin), and were suggested by a display of smoke rings which he happened to see in the lecture room of his friend, P. G. Tait, in Edinburgh University. He used vortices in the first place to illustrate the properties of ponderable matter rather than of the luminiferous medium, and pointed out that if[1] the atoms of matter are constituted of vortex rings in a perfect fluid, the conservation of matter may be immediately explained. The mutual interactions of atoms may be illustrated by the behaviour of smoke rings, which after approaching each other closely are observed to rebound: and the spectroscopic properties of matter may be referred to the possession by vortex rings of free periods of vibration.[2]

A further recommendation of the vortex-atom hypothesis was that it seemed to throw some light on the mode of propagation of gravity, which since Laplace's investigations on the subject had been believed to be instantaneous. For the actions between two vortex rings in a perfect liquid are not propagated from one to the other: they are due to the fact that each vortex ring is accompanied by motion throughout the liquid, so that each may be regarded as extending over the whole of space, and involving all the others; and the effects are simultaneous everywhere. In the seventeenth century, Michell and Boscovich had suggested that an atom should not be supposed to have a definite size but should be conceived of as completely penetrable, and extending throughout all space; and Faraday had adopted this view in his *Thoughts on Ray Vibrations*[3] of 1846.

In the nineteenth century it was objected to the vortex theory that the virtual inertia of a vortex ring increases as its energy increases;

[1] *Phil. Mag.* xxxiv (1867), p. 15; *Proc. R. S. Edin.* vi, (1869) p. 94

[2] An attempt was made in 1883 by J. J. Thomson (*Phil. Mag.* xv [1883], p. 427) to explain the phenomena of the electric discharge through gases in terms of the theory of vortex atoms. The atoms of a gas were represented as vortex rings: the combination of two atoms to form a molecule would then correspond to the union or pairing of two vortex rings, a phenomenon which was well known experimentally and which Thomson himself had investigated mathematically. Now suppose that there is a quantity of gas in an electric field. The electric field he represented by a distribution of velocity in the medium whose vortex motion constitutes the atoms of the gas: the disturbance due to this distribution of velocity will cause some of the paired vortex rings to dissociate, and so a mathematical theory of the phenomenon of conduction in gases becomes possible. He worked out the conditions for the disruptive discharge, and showed that the intensity of the field necessary to produce discharge should diminish when the gas is rarefied. This investigation is notable inasmuch as it was the beginning of those researches on conduction in rarefied gases which occupied so much of J. J. Thomson's life and led him to his greatest discovery, that of the electron.

[3] *Exp. Res.* iii, p. 447

this fact, however, fits in with later discoveries regarding the connection of mass and energy, and would therefore now be regarded as highly favourable.

It is, however, doubtful whether vortex atoms would be stable. 'It now seems to me certain,' wrote W. Thomson [1] (Kelvin) in 1905, 'that if any motion be given within a finite portion of an infinite incompressible liquid, originally at rest, its fate is necessarily dissipation to infinite distances with infinitely small velocities everywhere; while the total kinetic energy remains constant. After many years of failure to prove that the motion in the ordinary Helmholtz circular ring is stable, I came to the conclusion that it is essentially unstable, and that its fate must be to become dissipated as now described.'

The vortex-atom hypothesis is not the only way in which the theory of vortex motion has been applied to the construction of models of the aether. It was shown in 1880 by W. Thomson [2] that in certain circumstances a mass of fluid can exist in a state in which portions in rotational and irrotational motion are finely mixed together, so that on a large scale the mass is homogeneous, having within any sensible volume an equal amount of vortex motion in all directions. To a fluid having such a type of motion he gave the name *vortex sponge*.

Five years later, FitzGerald [3] discussed the suitability of the vortex sponge as a model of the aether. Since vorticity in a perfect fluid cannot be created or destroyed, the modification of the system which is to be analogous to an electric field must be a polarised state of the vortex motion, and light must be represented by a communication of this polarised motion from one part of the medium to another. Many distinct types of polarisation may readily be imagined: for instance, if the turbulent motion were constituted of vortex rings, these might be in motion parallel to definite lines or planes; or if it were constituted of long vortex filaments, the filaments might be bent spirally about axes parallel to a given direction. The energy of any polarised state of vortex motion would be greater than that of the unpolarised state; so that if the motion of matter had the effect of reducing the polarisation, there would be forces tending to produce that motion. Since the forces due to a small vortex vary inversely as a high power of the distance from it, it seems probable that in the case of two infinite planes, separated by a region of polarised vortex motion, the forces due to the polarisation

[1] *Proc. R. S. Edin.* xxv (1905), p. 565
[2] *Brit. Assoc. Rep.* (1880), p. 473
[3] *Scient. Proc. Roy. Dublin Soc.* iv (1885), p. 407; *Scientific Writings of FitzGerald*, p. 154

between the planes would depend on the polarisation, but not on the mutual distance of the planes—a property which is characteristic of plane distributions whose elements attract according to the Newtonian law.

It is possible to conceive polarised forms of vortex motion which are steady so far as the interior of the medium is concerned, but which tend to yield up their energy in producing motion of its boundary—a, property parallel to that of the aether, which, though itself in equilibrium tends to move objects immersed in it.

In the same year Hicks [1] discussed the possibility of transmitting waves through a medium consisting of an incompressible fluid in which small vortex rings are closely packed together. The wave-length of the disturbance was supposed large in comparison with the dimensions and mutual distances of the rings ; and the translatory motion of the latter was supposed to be so slow that very many waves can pass over any one before it has much changed its position. Such a medium would probably act as a fluid for larger motions. The vibration in the wave-front might be either swinging oscillations of a ring about a diameter, or transverse vibrations of the ring, or apertural vibrations ; vibrations normal to the plane of the ring appear to be impossible. Hicks determined in each case the velocity of translation, in terms of the radius of the rings, the distance of their planes, and their cyclic constant.

The greatest advance in the vortex-sponge theory of the aether was made in 1887, when W. Thomson [2] showed that the equation of propagation of laminar disturbances in a vortex sponge is the same as the equation of propagation of luminous vibrations in the aether. The demonstration, which in the circumstances can scarcely be expected to be either very simple or very rigorous, is as follows :

Let (u, v, w) denote the components of velocity, and p the pressure, at the point (x, y, z) in an incompressible fluid. Let the initial motion be supposed to consist of a laminar motion $\{f(y), 0, 0\}$, superposed on a homogeneous, isotropic and fine-grained distribution (u_0', v_0, w_0) : so that at the origin of time the velocity is $\{f(y) + u_0', v_0, w_0\}$: it is desired to find a function $f(y, t)$ such that at any time t the velocity shall be $\{f(y, t) + u', v, w\}$, where u', v, w, are quantities of which every average taken over a sufficiently large space is zero.

[1] *Brit. Assoc. Rep.* (1885), p. 930

[2] *Phil. Mag.* xxiv (1887), p. 342 ; Kelvin's *Math. and Phys. Papers*, iv, p. 308

Substituting these values of the components of velocity in the equation of motion

$$\frac{\partial u}{\partial t} = - u \frac{\partial u}{\partial x} - v \frac{\partial u}{\partial y} - w \frac{\partial u}{\partial z} - \frac{\partial p}{\partial x},$$

there results

$$\frac{\partial f(y,t)}{\partial t} + \frac{\partial u'}{\partial t} = - f(y,t) \frac{\partial u'}{\partial x} - v \frac{\partial f(y,t)}{\partial y} - u' \frac{\partial u'}{\partial x} - v \frac{\partial u'}{\partial y} - w \frac{\partial u'}{\partial z} - \frac{\partial p}{\partial x}.$$

Take now the xz-averages of both members. The quantities $\partial u'/\partial t$, $\partial u'/\partial x$, v, $\partial p/\partial x$ have zero averages; so the equation takes the form

$$\frac{\partial f(y,t)}{\partial t} = - \mathrm{A}\left(u' \frac{\partial u'}{\partial x} + v \frac{\partial u'}{\partial y} + w \frac{\partial u'}{\partial z}\right),$$

if the symbol A is used to indicate that the xz-average is to be taken of the quantity following. Moreover, the incompressibility of the fluid is expressed by the equation

$$\frac{\partial u'}{\partial x} + \frac{\partial v}{\partial y} + \frac{\partial w}{\partial z} = 0,$$

whence

$$0 = - \mathrm{A}\left(u' \frac{\partial u'}{\partial x} + u' \frac{\partial v}{\partial y} + u' \frac{\partial w}{\partial z}\right).$$

When this is added to the preceding equation, the first and third pairs of terms of the second member vanish, since the x average of any derivate $\partial Q/\partial x$ vanishes if Q is finite for infinitely great values of x; and the equation thus becomes

$$\frac{\partial f(y,t)}{\partial t} = - \mathrm{A}\frac{\partial (u'v)}{\partial y}. \tag{1}$$

From this it is seen that if the turbulent motion were to remain continually isotropic as at the beginning, $f(y, t)$ would constantly retain its critical value $f(y)$. In order to examine the deviation from isotropy, we shall determine $\mathrm{A}\partial(u'v)/\partial t$, which may be done in the following way: multiplying the u- and v-equations of motion by v, u' respectively, and adding, we have

$$v \frac{\partial f(y,t)}{\partial t} + \frac{\partial (u'v)}{\partial t} = - f(y,t) \frac{\partial (u'v)}{\partial x} - v^2 \frac{\partial f(y\,t)}{\partial y} - u' \frac{\partial (u'v)}{\partial x} - v \frac{\partial (u'v)}{\partial y} - w \frac{\partial (u'v)}{\partial z} - v \frac{\partial p}{\partial x} - u' \frac{\partial p}{\partial y}.$$

Taking the xz-average of this, we observe that the first term of the first member disappears, since Av is zero, and the first term of the second member disappears, since $A\partial(u'v)/\partial x$ is zero. Denoting by $\frac{1}{3}R^2$ the average value of u^2, v^2 or w^2, so that R may be called the *average velocity* of the turbulent motion, the equation becomes

$$\frac{\partial}{\partial t}\{Au'v\} = -\tfrac{1}{3}R^2\frac{\partial f(y,t)}{\partial y} - Q,$$

where

$$Q = A\left\{u'\frac{\partial(u'v)}{\partial x} + v\frac{\partial(u'v)}{\partial y} + w\frac{\partial(u'v)}{\partial z} + v\frac{\partial p}{\partial x} + u'\frac{\partial p}{\partial y}\right\}.$$

Let p be written $(p' + \omega)$, where p' denotes the value which p would have if f were zero. The equations of motion immediately give

$$-\nabla^2 p = \left(\frac{\partial u}{\partial x}\right)^2 + \left(\frac{\partial v}{\partial y}\right)^2 + \left(\frac{\partial w}{\partial z}\right)^2 + 2\frac{\partial v}{\partial z}\frac{\partial w}{\partial y} + 2\frac{\partial w}{\partial x}\frac{\partial u}{\partial z} + 2\frac{\partial u}{\partial y}\frac{\partial v}{\partial x};$$

and on subtracting the forms which this equation takes in the two cases, we have

$$-\nabla^2\omega = 2\frac{f\partial(y,t)}{\partial y}\frac{\partial v}{\partial x},$$

which, when the turbulent motion is fine-grained, so that $f(y, t)$ is sensibly constant over ranges within which u', v, w pass through all their values, may be written

$$\omega = -2\frac{\partial f(y,t)}{\partial y}\nabla^{-2}\frac{\partial v}{\partial x}.$$

Moreover, we have

$$0 = A\left\{u'\frac{\partial(u'v)}{\partial x} + v\frac{\partial(u'v)}{\partial y} + w\frac{\partial(u'v)}{\partial z} + v\frac{\partial p'}{\partial x} + u'\frac{\partial p'}{\partial y}\right\};$$

for positive and negative values of u', v, w are equally probable; and therefore the value of the second member of this equation is doubled by adding to itself what it becomes when for u', v, w we substitute $-u'$, $-v$, $-w$; which (as may be seen by inspection of the above equation in $\nabla^2 p$) does not change the value of p'. Comparing this equation with that which determines the value of Q, we have

$$Q = A\left(v\frac{\partial\omega}{\partial x} + u'\frac{\partial\omega}{\partial y}\right),$$

or substituting for ω,

$$Q = -2\frac{\partial f(y,t)}{\partial y}A\left(v\frac{\partial}{\partial x} + u'\frac{\partial}{\partial y}\right)\nabla^{-2}\frac{\partial v}{\partial x}.$$

The isotropy with respect to x and z gives the equation

$$2A\left(v_0\frac{\partial}{\partial x}+u_0'\frac{\partial}{\partial y}\right)\nabla^{-2}\frac{\partial v_0}{\partial x}=A\left\{v_0\left(\frac{\partial^2}{\partial x^2}+\frac{\partial^2}{\partial z^2}\right)\right.$$
$$\left.+\left(u_0'\frac{\partial}{\partial x}+w_0\frac{\partial}{\partial z}\right)\frac{\partial}{\partial y}\right\}\nabla^{-2}v_0.$$

But by integration by parts we obtain the equation

$$A\left(u_0'\frac{\partial}{\partial x}+w_0\frac{\partial}{\partial z}\right)\frac{\partial}{\partial v}\cdot\nabla^{-2}v_0=-A\left(\frac{\partial u_0'}{\partial x}+\frac{\partial w_0}{\partial z}\right)\frac{\partial}{\partial y}\nabla^{-2}v_0\,;$$

and by the condition of incompressibility the second member may be written

$$A\,.\,(\partial v_0/\partial y)\,.\,(\partial/\partial y)\,.\,\nabla^{-2}v_0,\quad\text{or}\quad -Av_0(\partial^2/\partial y^2)\nabla^{-2}v_0\,;$$

so we have

$$Q_0=-\frac{\partial f(y,t)}{\partial y}A\left\{v_0\left(\frac{\partial^2}{\partial x^2}+\frac{\partial^2}{\partial z^2}-\frac{\partial^2}{\partial y^2}\right)\right\}\nabla^{-2}v_0.$$

On account of the isotropy, we may write $\frac{1}{3}$ for

$$\left(\frac{\partial^2}{\partial x^2}-\frac{\partial^2}{\partial y^2}+\frac{\partial^2}{\partial z^2}\right)\nabla^{-2}\,;$$

so

$$Q_0=-\frac{R^2}{9}\frac{\partial f(y,t)}{\partial y}\,;$$

and, therefore,

$$\frac{\partial}{\partial t}_{(t=0)}\{A(u'v)\}=-\frac{2R^2}{9}\left\{\frac{\partial f(y,t)}{\partial y}\right\}_{t=0}.$$

The deviation from isotropy shown by this equation is very small, because of the smallness of $\partial f(y,t)/\partial y$. The equation is therefore not restricted to the initial values of the two members, for we may neglect an infinitesimal deviation from $(2/9)\,R^2$ in the first factor of the second member, in consideration of the smallness of the second factor. Hence for all values of t we have the equation

$$\frac{\partial}{\partial t}A(u'v)=-\tfrac{2}{9}R^2\frac{\partial f(y,t)}{\partial y},$$

which, in combination with (1), yields the result

$$\frac{\partial^2}{\partial t^2}f(y,t)=\frac{2R^2}{9}\frac{\partial^2}{\partial y^2}f(y,t)\,;$$

the form of this equation shows that *laminar disturbances are propagated through the vortex sponge in the same manner as waves of distortion in a homogeneous elastic solid.*

The question of the stability of the turbulent motion remained undecided; and at the time Thomson seems to have thought it likely that the motion would suffer diffusion. But two years later [1] he showed that stability was ensured at any rate when space is filled with a set of approximately straight hollow vortex filaments. FitzGerald [2] subsequently determined the energy per unit volume in a turbulent liquid which is transmitting laminar waves. Writing for brevity

$$(2/9)\,R^2 = V^2, \quad f(y, t) = P, \quad \text{and} \quad A\,(u'v) = \gamma,$$

the equations are

$$\frac{\partial P}{\partial t} = -\frac{\partial \gamma}{\partial y}, \quad \text{and} \quad \frac{\partial \gamma}{\partial t} = -V^2 \frac{\partial P}{\partial y}$$

If the quantity

$$P^2 + \gamma^2/V^2 = 2\Sigma$$

is integrated throughout space, and the variations of the integral with respect to time are determined, it is found that

$$\frac{\partial}{\partial t}\iiint \Sigma\, dx\, dy\, dz = \iiint \left(P\frac{\partial P}{\partial t} + \frac{\gamma}{V^2}\frac{\partial \gamma}{\partial t}\right) dx\, dy\, dz$$

$$= \iiint \left(P\frac{\partial P}{\partial t} - \gamma\frac{\partial P}{\partial v}\right) dx\, dy\, dz.$$

Integrating by parts the second term under the integral, and omitting the superficial terms (which may be at infinity, or wherever energy enters the space under consideration), we have

$$\frac{\partial}{\partial t}\iiint \Sigma\, dx\, dy\, dz = \iiint P\left(\frac{\partial P}{\partial t} + \frac{\partial \gamma}{\partial y}\right) dx\, dy\, dz = 0.$$

Hence it appears that the quantity Σ, which is of the dimensions of energy, must be proportional to the energy per unit volume of the medium. Since P corresponds to one of the two quantities, electric and magnetic force, while γ corresponds to the other, the expression $\frac{1}{2}(P^2 + \gamma^2/V^2)$ for Σ shows that the energy per c.c. of the wave is the sum of the squares of the electric and magnetic forces, in suitable units; a result which shows that there is a pronounced similarity between the dynamics of a vortex sponge and of Maxwell's elastic aether.

A definite vortex-sponge model of the aether was described by

[1] *Proc. Roy. Irish Acad.* (3), i (1889), p. 340; Kelvin's *Math. and Phys. Papers*, iv, p. 202

[2] *Brit. Assoc. Rep.* (1899), p. 632; FitzGerald's *Scientific Writings*, p. 484, see also pp. 254, 472

Hicks in his presidential address to the mathematical section of the British Association in 1895.[1] In this the small motions whose function is to confer the quasi-rigidity were not completely chaotic, but were disposed systematically. The medium was supposed to be constituted of cubical elements of fluid, each containing a rotational circulation complete in itself. In any element, the motion close to the central vertical diameter of the element is vertically upwards; the fluid which is thus carried to the upper part of the element flows outwards over the top, down the sides and up the centre again. In each of the six adjoining elements the motion is similar to this, but in the reverse direction. The rotational motion in the elements confers on them the power of resisting distortion, so that waves may be propagated through the medium as through an elastic solid; but the rotations are without effect on irrotational motions of the fluid, provided the velocities in the irrotational motion are slow compared with the velocity of propagation of distortional vibrations.

A different model was described four years later by FitzGerald.[2] Since the distribution of velocity of a fluid in the neighbourhood of a vortex filament is the same as the distribution of magnetic force around a wire of identical form carrying an electric current, it is evident that the fluid has more energy when the filament has the form of a helix than when it is straight; so if space were filled with vortices, whose axes were all parallel to a given direction, there would be an increase in the energy per unit volume when the vortices were bent into a spiral form; and this could be measured by the square of a vector—say, **E**—which may be supposed parallel to this direction.

If now a single spiral vortex is surrounded by parallel straight ones, the latter will not remain straight, but will be bent by the action of their spiral neighbour. The transference of spirality may be specified by a vector **H**, which will be distributed in circles round the spiral vortex; its magnitude will depend on the rate at which spirality is being lost by the original spiral, and can be taken such that its square is equal to the mean energy of this new motion. The vectors **E** and **H** will then represent the electric and magnetic vectors; the vortex spirals representing tubes of electric force.

FitzGerald's spirality is essentially similar to the laminar motion investigated by Lord Kelvin, since it involves a flow in the direction of the axis of the spiral, and such a flow cannot take place along the

[1] *Brit. Assoc. Rep.* (1895), p. 595

[2] *Proc. Roy. Dublin Soc.*, IX (Dec. 1899), p. 50; FitzGerald's *Scientific Writings*, p. 472

direction of a vortex filament without a spiral deformation of a filament.

Other vortex analogues have been devised for electrostatical systems. One such, which was described in 1888 by W. M. Hicks,[1] depends on the circumstance that if two bodies in contact in an infinite fluid are separated from each other, and if there be a vortex filament which terminates on the bodies, there will be formed at the point where they separate a hollow vortex filament [2] stretching from one to the other, with rotation equal and opposite to that of the original filament. As the bodies are moved apart, the hollow vortex may, through failure of stability, dissociate into a number of smaller ones ; and if the resulting number be very large, they will ultimately take up a position of stable equilibrium. The two sets of filaments —the original filaments and their hollow companions—will be intermingled, and each will distribute itself according to the same law as the lines of force between the two bodies which are equally and oppositely electrified.

Since the pressure inside a hollow vortex is zero, the portion of the surface on which it abuts experiences a diminution of pressure ; the two bodies are therefore attracted. Moreover, as the two bodies separate farther, the distribution of the filaments being the same as that of lines of electric force, the diminution of pressure for each line is the same at all distances, and therefore the force between the two bodies follows the same law as the force between two bodies equally and oppositely electrified. It may be shown that the effect of the original filaments is similar, the diminution of pressure being half as large again as for the hollow vortices.

If another surface were brought into the presence of the others, those of the filaments which encounter it would break off and rearrange themselves so that each part of a broken filament terminates on the new body. This analogy thus gives a complete account of electrostatic actions both quantitatively and qualitatively ; the electric charge on a body corresponds to the number of ends of filaments abutting on it, the sign being determined by the direction of rotation of the filament as viewed from the body.

A magnetic field may be supposed to be produced by the motion of the vortex filaments through the stationary aether, the magnetic force being at right angles to the filament and to its direction of

[1] *Brit. Assoc. Rep.* (1888), p. 577

[2] A hollow vortex is a cyclic motion existing in a fluid without the presence of any actual rotational filaments. On the general theory cf. Hicks, *Phil. Trans.* clxxv (1883), p. 161 ; clxxvi (1885), p. 725 ; cxcii (1898), p. 33.

motion. Electrostatic and magnetic fields thus correspond to states of motion in the medium, in which, however, there is no bodily flow; for the two kinds of filament produce circulation in opposite directions.

It is possible that hollow vortices are better adapted than ordinary vortex filaments for the construction of models of the aether. Such, at any rate, was the opinion of Thomson (Kelvin) in his later years.[1] The analytical difficulties of the subject are formidable, and progress was consequently slow; but among the many mechanical schemes which were devised in the nineteenth century to represent electrical and optical phenomena, none possesses greater interest than that which pictures the aether as a vortex sponge.

Towards the close of the nineteenth century, chiefly under the influence of Larmor, it came to be generally recognised that the aether is an immaterial medium, *sui generis*, not composed of identifiable elements having definite locations in absolute space. The older view had supposed ' the pressures and thrusts of the engineer, and the strains and stresses in the material structures by which he transmits them from one place to another, to be the archetype of the processes by which all mechanical effect is transmitted in nature. This doctrine implies an expectation that we may ultimately discover something analogous to structure in the celestial spaces, by means of which the transmission of physical effect will be brought into line with the transmission of mechanical effect by material framework.' [2] Larmor urged on the contrary that ' we should not be tempted towards explaining the simple group of relations which have been found to define the activity of the aether by treating them as mechanical consequences of concealed structure in that medium; we should rather rest satisfied with having attained to their exact dynamical correlation, just as geometry explores or correlates, without explaining, the descriptive and metric properties of space.' This point of view enabled Larmor's theory to withstand subsequent criticisms based on the principle of relativity, which shattered practically all rival concepts of the aether.[3]

[1] *Proc. Roy. Irish Acad*, i (30 Nov. 1889), p. 340; Kelvin's *Math. and Phys. Papers*, iv, p. 202. ' Rotational vortex cores,' he wrote, ' must be absolutely discarded; and we must have nothing but irrotational revolution and vacuous cores.'

[2] Larmor, *Brit. Ass. Rep.* (1900), p. 618

[3] Models of the aether were discussed, in the light of the knowledge available in 1908, by H. Witte, *Ann. d. Phys.* xxvi (1908), p. 235.

Chapter X

THE FOLLOWERS OF MAXWELL

THE most notable imperfection in the electromagnetic theory of light, as presented in Maxwell's original memoirs, was the absence of any explanation of reflection and refraction. Before the publication of Maxwell's *Treatise*, however, a method of supplying the omission was indicated by Helmholtz.[1] The principles on which the explanation depends are that the normal components of the electric displacement **D** and the magnetic induction **B**, and the tangential components of the electric force **E** and the magnetic force **H**, are to be continuous across the interface at which the reflection takes place ; the optical difference between the contiguous bodies being represented by a difference in their dielectric constants, and the electric vector being assumed to be at right angles to the plane of polarisation.[2] The analysis required is a mere transcription of MacCullagh's theory of reflection,[3] if the derivate of MacCullagh's displacement **e** with respect to the time be interpreted as the magnetic force, μ curl **e** as the electric force, and curl **e** as the electric displacement. The mathematical details of the solution were not given by Helmholtz himself, but were supplied a few years later in the inaugural dissertation of H. A. Lorentz.[4]

In the years immediately following the publication of Maxwell's *Treatise*, a certain amount of evidence in favour of his theory was furnished by experiment. That an electric field is closely concerned with the propagation of light was demonstrated in 1875, when John Kerr[5] showed that dielectrics subjected to powerful electrostatic force acquire the property of double refraction, their optical behaviour being similar to that of uniaxal crystals whose axes are directed along the lines of force.

Other researches undertaken at this time had a more direct

[1] *Jour. für Math.* lxxii (1870), p. 68*n*

[2] Helmholtz (loc. cit.) pointed out that if the optical difference between the media were assumed to be due to a difference in their magnetic permeabilities, it would be necessary to suppose the magnetic vector at right angles to the plane of polarisation in order to obtain Fresnel's sine and tangent formulae of reflection.

[3] cf. pp. 137, 138, 142–4

[4] *Zeitschrift für Math. u. Phys.* xxii (1877), pp. 1, 205 : 'Over de theorie der terugkaatsing en breking van het licht,' Arnhem, 1875. Lorentz' work was based on Helmholtz' equations, but remains substantially unchanged when Maxwell's formulae are substituted.

[5] *Phil. Mag.* i (1875), pp. 337, 446 ; viii (1879), pp. 85, 229 ; xiii (1882), pp. 153, 248

bearing on the questions at issue between the hypothesis of Maxwell and the older potential theories. In 1875–6 Helmholtz [1] and his pupil Schiller [2] attempted to discriminate between the various doctrines and formulae relative to unclosed circuits by performing a crucial experiment.

It was agreed in all theories that a ring-shaped magnet, which returns into itself so as to have no poles, can exert no ponderomotive force on other magnets or on closed electric currents. Helmholtz [3] had, however, shown in 1873 that according to the potential theories such a magnet would exert a ponderomotive force on an unclosed current. The matter was tested by suspending a magnetised steel ring by a long fibre in a closed metallic case, near which was placed a terminal of a Holtz machine. No ponderomotive force could be observed when the machine was put in action so as to produce a brush discharge from the terminal; from which it was inferred that the potential theories do not correctly represent the phenomena, at least when displacement currents and convection currents (such as that of the electricity carried by the electrically repelled air from the terminal) are not taken into account.

Helmholtz also found that when a conductor was rotated in a magnetic field of force symmetrical about the axis of rotation, a difference of potential was induced between the axial and circumferential parts, which was shown by the resulting electrification.

The researches of Helmholtz and Schiller brought into prominence the question as to the effects produced by the translatory motion of electric charges. That the convection of electricity is equivalent to a current had been suggested long before by Faraday.[4] 'If,' he wrote in 1838, 'a ball be electrified positively in the middle of a room and be then moved in any direction, effects will be produced as if a current in the same direction had existed.' Maxwell in his *Treatise* [5] endorsed the 'supposition' that 'a moving electrified body is equivalent to an electric current.' To decide the matter a new experiment inspired by Helmholtz was performed by H. A. Rowland [6] in 1876. The electrified body in Rowland's disposition was a disk of ebonite, coated with gold leaf and capable of turning rapidly round a vertical axis between two fixed plates of

[1] *Monatsberichte d. Acad. d. Berlin* (1875), p. 400; *Ann. d. Phys.* clviii (1876), p. 87

[2] *Ann. d. Phys.* clix (1876), pp. 456, 537; clx (1877), p. 333

[3] The valuable memoirs by Helmholtz in *Jour. für Math.* lxxii (1870), p. 57; lxxv (1873), p. 35; lxxviii (1874), p. 273, to which reference has already been made, contain a full discussion of the various possibilities of the potential theories.

[4] *Exper. Res.*, § 1644

[5] §§ 768–70

[6] *Monatsberichte d. Akad. d. Berlin* (1876), p. 211; *Ann. d. Phys.* clviii (1876), p. 487; *Annales de Chim. et de Phys.* xii (1877), p. 119

glass, each gilt on one side. The gilt faces of the plates could be earthed, while the ebonite disk received electricity from a point placed near its edge ; each coating of the disk thus formed a condenser with the plate nearest to it. An astatic needle was placed above the upper condenser plate, nearly over the edge of the disk ; and when the disk was rotated a magnetic field was found to be produced. This experiment, which has since been repeated under improved conditions by Rowland and Hutchinson,[1] H. Pender,[2] Eichenwald,[3] E. P. Adams [4] and H. Pender and V. Crémieu [5] shows that the 'convection current' produced by the rotation of a charged disk, when the other ends of the lines of force are on an earthed stationary plate parallel to it, produces the same magnetic field as an ordinary conduction current flowing in a circuit which coincides with the path of the convection current. When two disks forming a condenser are rotated together, the magnetic action is the sum of the magnetic actions of each of the disks separately. It appears, therefore, that electric charges cling to the matter of a conductor and move with it, so far as Rowland's phenomenon is concerned.

The first examination of the matter from the point of view of Maxwell's theory was undertaken by J. J. Thomson (1856-1940),[6] in 1881. If an electrostatically charged body is in motion, the change in the location of the charge must produce a continuous alteration of the electric field at any point in the surrounding medium; or, in the language of Maxwell's theory, there must be displacement currents in the medium. It was to these displacement currents that Thomson, in his original investigation, attributed the magnetic effects of moving charges. The particular system which he considered was that formed by a charged spherical conductor, moving uniformly in a straight line. It was assumed that the distribution of electricity remains uniform over the surface during the motion, and that the electric field in any position of the sphere is the same as if the sphere were at rest ; these assumptions are true so long as quantities of order $(v/c)^2$ are neglected, where v denotes the velocity of the sphere and c the velocity of light.

Thomson's method was to determine the displacement currents in the space outside the sphere from the known values of the electric field, and then to calculate the vector potential due to these displacement currents by means of the formula

$$\mathbf{A} = \iiint (\mathbf{S}'/r)\, dx'\, dy'\, dz',$$

[1] *Phil. Mag.* xxvii (1889), p. 445
[2] *Ann. d. Phys.* xi (1901), p. 1
[3] ibid. ii (1901), p. 179 ; v (1903), p. 34
[4] *Amer. Jour. Sci.* xii (1901), p. 155
[5] *Jour. de Phys.* ii (1903), p. 641
[6] *Phil. Mag.* xi (1881), p. 229

where $\mathbf{S}'$ denotes the displacement current at $(x', y' z')$. The magnetic field was then determined by the equation

$$\mathbf{H} = \text{curl}\,\mathbf{A}.$$

A defect in this investigation was pointed out by FitzGerald, who, in a short but most valuable note,[1] published a few months afterwards, observed that the displacement currents of Thomson do not satisfy the circuital condition. This is most simply seen by considering the case in which the system consists of two parallel plates forming a condenser; if one of the plates is fixed, and the other plate is moved towards it, the electric field is annihilated in the space over which the moving plate travels; this destruction of electric displacement constitutes a displacement current, which, considered alone, is evidently not a closed current. The defect, as FitzGerald showed, may be immediately removed by assuming that a moving charge itself is to be counted as a current element; the total current, thus composed of the displacement currents and the convection current, is circuital. Making this correction, FitzGerald found that the magnetic force due to a sphere of charge e moving with velocity v along the axis of z is curl $(0, 0, ev/r)$—a formula which shows that the displacement currents have no resultant magnetic effect, since the term ev/r would be obtained from the convection current alone.

The expressions obtained by Thomson and FitzGerald were correct only to the first order of the small quantity v/c. The effect of including terms of higher order was considered in 1888–9 by Oliver Heaviside,[2] whose solution may be derived in the following manner:

Suppose that a charged system is in motion with uniform velocity v parallel to the axis of z; the total current consists of the displacement current $\frac{1}{4\pi}\frac{\partial \mathbf{E}}{\partial t}$ where $\mathbf{E}$ denotes the electric force, and the convection current $\rho\mathbf{v}$ where ρ denotes the volume-density of electricity. So the equation which connects magnetic force with electric current may be written (taking, as usual, electrostatic units for $\mathbf{E}$ and electromagnetic units for $\mathbf{H}$)

$$\frac{\partial \mathbf{E}}{\partial t} = c\,\text{curl}\,\mathbf{H} - 4\pi\rho\mathbf{v}.$$

[1] *Proc. Roy. Dublin Soc.* iii (Nov. 1881), p. 250; FitzGerald's *Scientific Writings*, p. 102

[2] *Electrician*, 23 Nov. 1888; *Phil. Mag.* xxvii (1889), p. 324

Eliminating **E** between this and the equation

$$c \operatorname{curl} \mathbf{E} = -\frac{\partial \mathbf{H}}{\partial t},$$

and remembering that **H** is here circuital, we have

$$\frac{1}{c^2}\frac{\partial^2 \mathbf{H}}{\partial t^2} - \nabla^2 \mathbf{H} = \frac{4\pi}{c} \operatorname{curl} \rho \mathbf{v}.$$

If, therefore, a vector-potential **a** be defined by the equation

$$\frac{1}{c^2}\frac{\partial^2 \mathbf{a}}{\partial t^2} - \nabla^2 \mathbf{a} = \frac{4\pi}{c} \rho \mathbf{v},$$

the magnetic force will be the curl of **a**; and from the equation for **a** it is evident that the components a_x and a_y are zero, and that a_z is to be determined from the equation

$$\ddot{a}_z/c^2 - \nabla^2 a_z = \frac{4\pi}{c} \rho v.$$

Now, let (x, y, ζ) denote co-ordinates relative to axes which are parallel to the axes (x, y, z), and which move with the charged bodies; then a_z is a function of (x, y, ζ) only; so we have

$$\frac{\partial}{\partial z} = \frac{\partial}{\partial \zeta}, \quad \text{and} \quad \frac{\partial}{\partial t} = -v\frac{\partial}{\partial \zeta};$$

and the preceding equation is readily seen to be equivalent to

$$\frac{\partial^2 a_z}{\partial x^2} + \frac{\partial^2 a_z}{\partial y^2} + \frac{\partial^2 a_z}{\partial \zeta_1^2} = -\frac{4\pi}{c} \rho v,$$

where ζ_1 denotes $(1 - v^2/c^2)^{-\frac{1}{2}}\zeta$. But this is simply Poisson's equation, with ζ_1 substituted for z; so the solution may be transcribed from the known solution of Poisson's equation: it is

$$ca_z = \iiint \frac{\rho' v' \, dx' \, dy' \, d\zeta_1'}{\{(\zeta_1 - \zeta_1')^2 + (x - x')^2 + (y - y')^2\}^{\frac{1}{2}}},$$

the integrations being taken over all the space in which there are moving charges; or

$$ca_z = \iiint \frac{\rho' v' \, dx' \, dy' \, d\zeta'}{\{(\zeta - \zeta')^2 + (1 - v^2/c^2)(x - x')^2 + (1 - v^2/c^2)(y - y')^2\}^{\frac{1}{2}}}.$$

If the moving system consists of a single charge e at the point $\zeta = 0$, this gives

$$ca_z = \frac{ev}{r(1 - v^2 \sin^2 \theta / c^2)^{\frac{1}{2}}},$$

where $\sin^2 \theta = (x^2 + y^2)/r^2$.

It is readily seen that the lines of magnetic force due to the moving point-charge are circles whose centres are on the line of motion, the magnitude of the magnetic force being

$$\frac{ev\,(1 - v^2/c^2)\sin\theta}{cr^2\,(1 - v^2\sin^2\theta/c^2)^{\frac{3}{2}}}.$$

The electric force is radial, its magnitude being

$$\frac{e\,(1 - v^2/c^2)}{r^2\,(1 - v^2\sin^2\theta/c^2)^{\frac{3}{2}}}.$$

The fact that the electric vector due to a moving point-charge is everywhere radial led Heaviside to conclude that the same solution is applicable when the charge is distributed over a perfectly conducting sphere whose centre is at the point, the only change being that **E** and **H** would now vanish inside the sphere. This inference was subsequently found by G. F. C. Searle to be incorrect; a distribution of electric charge on a moving sphere could in fact not be in equilibrium if the electric force were radial, since there would then be nothing to balance the mechanical force exerted on the moving charge (which is equivalent to a current) by the magnetic field. The moving system which gives rise to the same field as a moving point-charge is not a sphere, but an oblate spheroid whose polar axis (which is in the direction of motion) bears to its equatorial axis the ratio $(1 - v^2/c^2)^{\frac{1}{2}} : 1$.[1]

It was moreover shown by W. B. Morton [2] that in the case of the moving electrified sphere, the surface-density is unaltered by the motion, but the lines of force no longer leave the surface perpendicularly.

The energy of the field surrounding a charged sphere is greater when the sphere is in motion than when it is at rest. To determine the additional energy quantitatively (retaining only the lowest significant powers of v/c), we have only to integrate, throughout the space outside the sphere, the expression $\mathbf{H}^2/8\pi$, which represents the electrokinetic energy per unit volume: the result is $e^2v^2/3ac^2$, where e denotes the charge, v the velocity and a the radius of the sphere.

It is evident from this result that the work required to be done in order to communicate a given velocity to the sphere is greater when the sphere is charged than when it is uncharged; that is to

[1] cf. Searle, *Phil. Trans.* clxxxvii (1896), p. 675; and *Phil. Mag.* xliv (1897), p. 329. On the theory of the moving electrified sphere, cf. also J. J. Thomson, *Recent Researches in Elect. and Mag.*, p. 16; O. Heaviside, *Electrical Papers*, i, p. 446, and ii, p. 514; *Electromag. Theory*, i, p. 269; A. Schuster, *Phil. Mag.* xliii (1897), p. 1.

[2] *Phil. Mag.* xli (1896), p. 488

say, the virtual mass of the sphere is increased by an amount $2e^2/3ac^2$, owing to the presence of the charge. This may be regarded as arising from the self-induction of the convection current which is formed when the charge is set in motion. It was suggested by J. Larmor[1] and by W. Wien[2] that the inertia of ordinary ponderable matter may ultimately prove to be of this nature, the atoms being constituted of systems of electrons.[3] It may, however, be remarked that this view of the origin of mass is not altogether consistent with the principle that the electron is an indivisible entity. For the so-called self-induction of the spherical electron is really the *mutual* induction of the convection-currents produced by the elements of electric charge which are distributed over its surface; and the calculation of this quantity presupposes the divisibility of the total charge into elements capable of acting severally in all respects as ordinary electric charges; a property which appears scarcely consistent with the supposed fundamental nature of the electron.

In his (1889) paper in the *Philosophical Magazine* on the moving charge, Heaviside gave for the first time the formula for the mechanical force on an electric charge which is moving in a magnetic field, namely—

charge × the vector product of the velocity and the magnetic induction.

J. J. Thomson had given it as half this amount in 1881.[4]

After the first attempt of J. J. Thomson to determine the field produced by a moving electrified sphere, the mathematical development of Maxwell's theory proceeded rapidly. The problems which admit of solution in terms of known functions are naturally those in which the conducting surfaces involved have simple geometrical forms—planes, spheres and cylinders.[5]

A result which was obtained by Horace Lamb,[6] when investigating electrical motions in a spherical conductor, led to interesting consequences. Lamb found that if a spherical conductor is placed in a rapidly alternating field, the induced currents are almost entirely

[1] *Phil. Trans.* clxxxvi (1895), p. 697

[2] *Arch. Néerl* (3), v (1900), p. 96

[3] Experimental evidence which was at the time thought to support the view that the inertia of electrons is purely electromagnetic was afterwards furnished by W. Kaufmann, *Gött. Nach.* (1901), p. 143; (1902), p. 291.

[4] *Phil. Mag.* xi (1881), p. 227

[5] cf., e.g., C. Niven, *Phil. Trans.* clxxii (1881), p. 307; H. Lamb, *Phil. Trans.* clxxiv (1883), p. 519; J. J. Thomson, *Proc. Lond. Math. Soc.* xv (1884), p. 197; H. A. Rowland, *Phil. Mag.* xvii (1884), p. 413; J. J. Thomson, *Proc. Lond. Math. Soc.* xvii (1886), p. 310; xix (1888), p. 520; and many investigations of Oliver Heaviside, collected in his *Electrical Papers*.

[6] Loc. cit.

confined to a superficial layer; and his result was shortly afterwards generalised by Oliver Heaviside,[1] who showed that whatever be the form of a conductor rapidly alternating currents do not penetrate far into its substance.[2] The reason for this may be readily understood: it is virtually an application of the principle [3] that a perfect conductor is impenetrable to magnetic lines of force. No perfect conductor is known to exist; but [4] if the alternations of magnetic force to which a good conductor such as copper is exposed are very rapid, the conductor has not time (so to speak) to display the imperfection of its conductivity, and the magnetic field is therefore unable to extend far below the surface.

The same conclusion may be reached by different reasoning.[5] When the alternations of the current are very rapid, the ohmic resistance ceases to play a dominant part, and the ordinary equations connecting electromotive force, induction and current are equivalent to the conditions that the currents shall be so distributed as to make the electrokinetic or magnetic energy a minimum. Consider now the case of a single straight wire of circular cross-section. The magnetic energy in the space outside the wire is the same whatever be the distribution of current in the cross-section (so long as it is symmetrical about the centre), since it is the same as if the current were flowing along the central axis; so the condition is that the magnetic energy in the wire shall be a minimum; and this is obviously satisfied when the current is concentrated in the superficial layer, since then the magnetic force is zero in the substance of the wire.

In spite of the advances which were effected by Maxwell and his earliest followers in the theory of electric oscillations, the gulf between the classical electrodynamics and the theory of light was not yet completely bridged. For in all the cases considered in the former science, energy is merely exchanged between one body and another, remaining within the limits of a given system; while in optics the energy travels freely through space, unattached to any material body. The first discovery of a more complete connection between the two theories was made by FitzGerald, who argued that if the unification which had been indicated by Maxwell is valid, it ought to be possible to generate radiant energy by purely electrical

[1] *Electrician*, xiv (10 Jan. 1885) p. 178

[2] The mathematical theory was given by Lord Rayleigh, *Phil. Mag.* xxi (1886), p. 381. cf. Maxwell's *Treatise*, § 689.

[3] cf. p. 281

[4] As was first remarked by Lord Rayleigh, *Phil. Mag.* xiii (1882), p. 344

[5] cf. J. Stefan, *Wiener Sitzungsber.* xcix (1890), p. 319; *Ann. d. Phys.* xli (1890), p. 400

means; 'it seems highly probable,' he said on 5 May 1882,[1] 'that the energy of varying currents is in part radiated into space and so lost to us.' In 1883 [2] he described methods by which the radiant energy could be produced.[3]

FitzGerald's system is what has since become known as the *magnetic oscillator*: it consists of a small circuit, in which the strength of the current is varied according to the simple periodic law. The circuit will be supposed to be a circle of small area S, whose centre is the origin and whose plane is the plane of xy; and the surrounding medium will be supposed to be free aether. The current may be taken to be (in e.-s. units) of strength $cA\cos(2\pi t/T)$, so that the moment of the equivalent magnet (in e.-m. units) is $SA\cos(2\pi t/T)$. Now in the older electrodynamics, the vector-potential due to a magnetic molecule of (vector) moment **M** placed at the origin is $(1/4\pi)$ curl $(\mathbf{M}/r)$, where r denotes distance from the origin. The vector-potential due to FitzGerald's magnetic oscillator would therefore be $(1/4\pi)$ curl **K**, where **K** denotes a vector parallel to the axis of z, and of magnitude $(1/r)\,SA\cos(2\pi t/T)$. The change which is involved in replacing the assumptions of the older electrodynamics by those of Maxwell's theory is in the present case equivalent [4] to retarding the potential; so that the vector-potential **a** due to the oscillator is $(1/4\pi)$ curl **K** where **K** is still directed parallel to the axis of z, and is of magnitude

$$K = \frac{SA}{r}\cos\frac{2\pi}{T}\left(t - \frac{r}{c}\right).$$

The electric force **E** at any point of space is $-\dfrac{1}{c}\dfrac{\partial \mathbf{a}}{\partial t}$, and the magnetic force **H** is curl **a**: so that these quantities may be calculated without difficulty. The electric energy per unit volume is $\mathbf{E}^2/8\pi$; performing the calculations, it is found that the value of this quantity averaged over a period of the oscillation and also averaged over the surface of a sphere of radius r is

$$\frac{\pi A^2S^2}{6c^2r^4T^2}\left(1 + \frac{4\pi^2r^2}{c^2T^2}\right).$$

The part of this which is radiated is evidently that which is pro-

[1] *Trans. Roy. Dublin Soc.* i (1883), p. 325

[2] *Trans. Roy. Dublin Soc.* iii (1883) p. 57; FitzGerald's *Scientific Writings*, p. 122

[3] According to O. J. Lodge, *Journal of the Wireless Society of London*, iii (1922), p. 77, FitzGerald in 1879 read a paper on the 'Impossibility' of producing electric waves, but he struck out the 'Im' afterwards.

[4] cf. pp. 268–9

portional to the inverse square of the distance,[1] so the average value of the radiant energy of electric type at distance r from the oscillator is $2\pi^3A^2S^2/3c^4r^2T^4$ per unit volume. The radiant energy of magnetic type may be calculated in a similar way, and is found to have the same value; so the total radiant energy at distance r is $4\pi^3A^2S^2/3c^4r^2T^4$ per unit volume; and therefore the energy radiated in unit time is $16\pi^4A^2S^2/3c^3T^4$. This is small, unless the frequency is very high; so that ordinary alternating currents would give no appreciable radiation. FitzGerald, however, at the Southport meeting of the British Association, in the same year[2] indicated a method by which the difficulty of obtaining currents of sufficiently high frequency might be overcome: this was, to employ the alternating currents which are produced when a condenser is discharged; in fact, the oscillatory discharge of a Leyden jar. He remarked, however, that the difficulty would be in detecting such waves when they were produced.

The FitzGerald oscillator constructed on this principle is closely akin to the radiator afterwards developed with such success by Hertz; the only difference is that in FitzGerald's arrangement the condenser is used merely as the store of energy (its plates being so close together that the electrostatic field due to the charges is practically confined to the space between them), and the actual source of radiation is the alternating magnetic field due to the circular loop of wire; while in Hertz's arrangement the loop of wire is abolished, the condenser plates are at some distance apart and the source of radiation is the alternating electrostatic field due to their charges.

In the study of electrical radiation, valuable help is afforded by a general theorem on the transfer of energy in the electromagnetic field, which was discovered in 1884 by John Henry Poynting,[3] and independently, almost at the same time, by Heaviside.[4] We have seen that the older writers on electric currents recognised that an electric current is associated with the transport of energy from one place (e.g. the voltaic cell which maintains the current) to another (e.g. an electric motor which is worked by the current); but they supposed the energy to be conveyed by the current itself within the

[1] The other term, which is neglected, is very small compared to the term retained, at great distances from the origin; it is what would be obtained if the effects of induction of the displacement currents were neglected: i.e. it is the energy of the forced displacement currents which are produced directly by the variation of the primary current, and which originate the radiating displacement currents.

[2] *Brit. Assoc. Rep.*, 1883, p. 404; FitzGerald's *Scientific Writings*, p. 129

[3] *Phil. Trans.* clxxv (1884), p. 343

[4] *Electrician* xiv (10 Jan. and 21 Feb. 1885), pp. 178, 306

wire, in much the same way as dynamical energy is carried by water flowing in a pipe; whereas in Maxwell's theory, the storehouse and vehicle of energy is the dielectric medium surrounding the wire. What Poynting achieved was to show that the flux of energy at any place might be expressed by a simple formula in terms of the electric and magnetic forces at the place.

Denoting as usual by **E** the electric force, by **D** the electric displacement, by **H** the magnetic force and by **B** the magnetic induction, the energy stored in unit volume of the medium is [1]

$$(1/8\pi)\,\mathbf{ED} + (1/8\pi)\,\mathbf{BH}\;;$$

so the increase of this in unit time is (since in isotropic media **D** is proportional to **E**, and **B** is proportional to **H**)

$$\frac{1}{4\pi}\left\{\left(\mathbf{E}\,.\,\frac{\partial \mathbf{D}}{\partial t}\right)+\left(\mathbf{H}\,.\,\frac{\partial \mathbf{B}}{\partial t}\right)\right\}$$

or

$$\mathbf{E}\,(\mathbf{S}-\mathbf{s})+\frac{1}{4\pi}\left(\mathbf{H}\,.\,\frac{\partial \mathbf{B}}{\partial t}\right),$$

where **S** denotes the total current, and **s** the current of conduction; or (in virtue of the fundamental electromagnetic equations)

$$-\,(\mathbf{E}\,.\,\mathbf{s}) + (c/4\pi)\,(\mathbf{E}\,.\,\mathrm{curl}\,\mathbf{H}) - (c/4\pi)\,(\mathbf{H}\,.\,\mathrm{curl}\,\mathbf{E}),$$

or

$$-\,(\mathbf{E}\,.\,\mathbf{s}) - (c/4\pi)\,\mathrm{div}\,[\mathbf{E}\,.\,\mathbf{H}].$$

Now $(\mathbf{E}\,.\,\mathbf{s})$ is the amount of electric energy transformed into heat per unit volume per second; and therefore the quantity $-(c/4\pi)$ div $[\mathbf{E}\,.\,\mathbf{H}]$ must represent the deposit of energy in unit volume per second due to the streaming of energy; which shows that the flux of energy is represented by the vector $(c/4\pi)\,[\mathbf{E}\,.\,\mathbf{H}]$.[1] This is Poynting's theorem: *that the flux of energy at any place is represented by the vector product of the electric and magnetic forces, multiplied by $c/4\pi$.*[2]

[1] cf. pp. 222, 224, 253. As usual, we use electrostatic units for **D** and **E**, and electromagnetic units for **B** and **H**.

[1] Of course any circuital vector may be added. A. McAulay (*Phil. Trans.* clxxxiii [1892], p. 685) and H. M. Macdonald (*Electric Waves*, p. 72) propounded forms which differ from Poynting's by a non-circuital vector. K. Birkeland (*Ann. d. Phys.* lii [1894], p. 357), showed that Poynting's expression for the energy flux is the only one which is a function of the electric and magnetic field-strengths only.

[2] The analogue of Poynting's theorem in the theory of the vibrations of an isotropic elastic solid may be easily obtained; for from the equation of motion of an elastic solid,

$$\rho\,\frac{\partial^2 \mathbf{e}}{\partial t^2} = -\,(k + 4n/3)\ \mathrm{grad\ div}\ \mathbf{e} - n\ \mathrm{curl\ curl}\ \mathbf{e},$$

it follows that

$$\frac{\partial}{\partial t}\left\{\tfrac{1}{2}\rho\left(\frac{\partial \mathbf{e}}{\partial t}\right)^2 + \tfrac{1}{2}\,(k + \tfrac{4}{3}n)\,(\mathrm{div}\ \mathbf{e})^2 + \tfrac{1}{2}n\,(\mathrm{curl}\ \mathbf{e})^2\right\} = -\,\mathrm{div}\ \mathbf{W},$$

In the special case of the field which surrounds a straight wire carrying a continuous current, the lines of magnetic force are circles round the axis of the wire, while the lines of electric force are directed along the wire ; hence energy must be flowing in the medium in a direction at right angles to the axis of the wire. A current in any conductor may therefore be regarded as consisting essentially of a convergence of electric and magnetic energy from the medium upon the conductor, and its transformation there into other forms.

This association of a current with motions at right angles to the wire in which it flows doubtless suggested to Poynting the conceptions of a memoir which he published[1] in the following year. When an electric current flowing in a straight wire is gradually increased in strength from zero, the surrounding space becomes filled with lines of magnetic force, which have the form of circles round the axis of the wire. Poynting, adopting Faraday's idea of the physical reality of lines of force, assumed that these lines of force arrive at their places by moving outwards from the wire ; so that the magnetic field grows by a continual emission from the wire of lines of force, which enlarge and spread out like the circular ripples from the place where a stone is dropped into a pond. The electromotive force which is associated with a changing magnetic field was now attributed directly to the motion of the lines of force, so that wherever electromotive force is produced by change in the magnetic field, or by motion of matter through the field, the electric intensity is equal to the number of tubes of magnetic force intersected by unit length in unit time.

A similar conception was introduced in regard to lines of electric force. It was assumed that any change in the total electric induction through a curve is caused by the passage of tubes of force in or out across the boundary ; so that whenever magnetomotive force is

[1] *Phil. Trans.* clxxvi (1885), p. 277

---

where **W** denotes the vector

$$-(k+4n/3)\,\mathrm{div}\,\mathbf{e}\,.\,\frac{\partial \mathbf{e}}{\partial t}+n\left[[\mathrm{curl}\,\mathbf{e}\,.\,\frac{\partial \mathbf{e}}{\partial t}\right];$$

and since the expression which is differentiated with respect to t represents the sum of the kinetic and potential energies per unit volume of the solid (save for terms which give only surface-integrals), it is seen that **W** is the analogue of the Poynting vector. cf. :

K. Pearson, *Mess. of Math.* xix (1889), p. 31
W. Wien, *Ann. d. Phys.* xlv (1892), p. 685
O. Heaviside, *Electrician*, xxvii (3 July 1891) and xxix (29 July 1892), reprinted in *Electromagnetic Theory*, pp. 76, 247
O. Heaviside, *Phil. Trans.* clxxxiii (1892), p. 426
G. Mie, *Wien Sitzunsb.* cvii (1898), Abth. 2a, p. 1113
L. Donati, *Bologna Mem.* (5), vii (1899), p. 633

produced by change in the electric field, or by motion of matter through the field, the magnetomotive force is proportional to the number of tubes of electric force intersected by unit length in unit time.

Poynting, moreover, assumed that when a steady current C flows in a straight wire, C tubes of electric force close in upon the wire in unit time, and are there dissolved, their energy appearing as heat. If E denote the magnitude of the electric force, the energy of each tube per unit length is $\frac{1}{2}$E, so the amount of energy brought to the wire is $\frac{1}{2}$CE per unit length per unit time. This is, however, only half the energy actually transformed into heat in the wire; so Poynting further assumed that E tubes of magnetic force also move in per unit length per unit time, and finally disappear by contraction to infinitely small rings. This motion accounts for the existence of the electric field; and since each tube (which is a closed ring) contains energy of amount $\frac{1}{2}$C, the disappearance of the tubes accounts for the remaining $\frac{1}{2}$CE units of energy dissipated in the wire.

The theory of moving tubes of force was extensively developed by Sir J. J. Thomson.[1] Of the two kinds of tubes—magnetic and electric—which had been introduced by Faraday and used by Poynting, Thomson resolved to discard the former and employ only the latter. This was a distinct departure from Faraday's conceptions, in which, as we have seen, great significance was attached to the physical reality of the magnetic lines; but Thomson justified his choice by inferences drawn from the phenomena of electric conduction in liquids and gases. As will appear subsequently, these phenomena indicate that molecular structure is closely connected with tubes of electrostatic force—perhaps much more closely than with tubes of magnetic force; and Thomson therefore decided to regard magnetism as the secondary effect, and to ascribe magnetic fields, not to the presence of magnetic tubes, but to the motion of electric tubes. In order to account for the fact that magnetic fields may occur without any manifestation of electric force, he assumed that tubes exist in great numbers everywhere in space, either in the form of closed circuits or else terminating on atoms, and that electric force is only perceived when the tubes have a greater tendency to lie in one direction than in another. In a steady magnetic field the positive and negative tubes might be conceived to be moving in opposite directions with equal velocities.

[1] *Phil. Mag.* xxxi (1891), p. 149; Thomson's *Recent Researches in Elect. and Mag.* (1893), chap. i

A beam of light might, from this point of view, be regarded simply as a group of tubes of force which are moving with the velocity of light at right angles to their own length. Such a conception almost amounts to a return to the corpuscular theory; but since the tubes have definite directions perpendicular to the direction of propagation, there would now be no difficulty in explaining polarisation.

The energy accompanying all electric and magnetic phenomena was supposed by Thomson to be ultimately kinetic energy of the aether; the electric part of it being represented by rotation of the aether inside and about the tubes, and the magnetic part being the energy of the additional disturbance set up in the aether by the movement of the tubes. The inertia of this latter motion he regarded as the cause of induced electromotive force.

There was, however, one phenomenon of the electromagnetic field as yet unexplained in terms of these conceptions—namely, the ponderomotive force which is exerted by the field on a conductor carrying an electric current. Now any ponderomotive force consists in a transfer of mechanical momentum from the agent which exerts the force to the body which experiences it; and it occurred to Thomson that the ponderomotive forces of the electromagnetic field might be explained if the moving tubes of force, which enter a conductor carrying a current and are there dissolved, were supposed to possess mechanical momentum, which could be yielded up to the conductor. It is readily seen that such momentum must be directed at right angles to the tube and to the magnetic induction—a result which suggests that the momentum stored in unit volume of the aether may be proportional to the vector product of the electric and magnetic vectors.

For this conjecture reasons of a more definite kind may be given.[1] We have already seen [2] that the ponderomotive forces on material bodies in the electromagnetic field may be accounted for by Maxwell's supposition that across any plane in the aether whose unit normal is **N**, there is a stress represented by

$$\mathbf{P_N} = \frac{1}{4\pi}(\mathbf{D}\,.\,\mathbf{N})\,\mathbf{E} - \frac{1}{8\pi}(\mathbf{D}\,.\,\mathbf{E})\,\mathbf{N} + \frac{1}{4\pi}(\mathbf{B}\,.\,\mathbf{N})\,\mathbf{H} - \frac{1}{8\pi}(\mathbf{B}\,.\,\mathbf{H})\,\mathbf{N}.$$

So long as the field is steady (i.e. electrostatic or magnetostatic) the

[1] The hypothesis that the aether is a storehouse of mechanical momentum, which was first advanced by J. J. Thomson (*Recent Researches in Elect. and Mag.* [1893], p. 13), was afterwards developed by H. Poincaré (*Archives Neérl.* [2], v [1900], p. 252), and by M. Abraham (*Gött, Nach.* [1902], p. 20).

[2] cf. p. 272

resultant of the stresses acting on any element of volume of the aether is zero, so that the element is in equilibrium. But when the field is variable, this is no longer the case. The resultant stress on the aether contained within a surface S is

$$\iint \mathbf{P}_{\mathbf{N}} dS$$

integrated over the surface; transforming this into a volume integral, the term $(1/4\pi)$ $(\mathbf{D} . \mathbf{N})$ $\mathbf{E}$ gives a term $(1/4\pi)$ div $\mathbf{D} . \mathbf{E}$ $+ (1/4\pi)$ $(\mathbf{D} . \nabla)$ $\mathbf{E}$, where ∇ denotes the vector operator $(\partial/\partial x, \partial/\partial y, \partial/\partial z)$; and the first of these terms vanishes, since $\mathbf{D}$ is a circuital vector; the term $-(1/8\pi)$ $(\mathbf{D} . \mathbf{E})$ $\mathbf{N}$ gives in the volume integral a term $(1/8\pi)$ grad $(\mathbf{D} . \mathbf{E})$; and the magnetic terms give similar results. So the resultant force on unit volume of the aether is

$$\frac{1}{4\pi}(\mathbf{D} . \nabla) \mathbf{E} + \frac{1}{8\pi} \text{grad} (\mathbf{D} . \mathbf{E}) + \frac{1}{4\pi}(\mathbf{B} . \nabla) \mathbf{E} + \frac{1}{8\pi} \text{grad} (\mathbf{B} . \mathbf{H})$$

which may be written

$$\frac{1}{4\pi} [\text{curl } \mathbf{E} . \mathbf{D}] + \frac{1}{4\pi} [\text{curl } \mathbf{H} . \mathbf{B}] ;$$

or, by virtue of the fundamental equations for dielectrics,

$$\frac{1}{4\pi c}\left[-\frac{\partial \mathbf{B}}{\partial t} . \mathbf{D}\right] + \frac{1}{4\pi c}\left[\frac{\partial \mathbf{D}}{\partial t} . \mathbf{B}\right], \quad \text{or} \quad \frac{1}{4\pi c}(\partial/\partial t) [\mathbf{D} . \mathbf{B}].$$

This result compels us to adopt one of three alternatives: either to modify the theory so as to reduce to zero the resultant force on an element of free aether—this expedient has not met with general favour;[1] or to assume that the force in question sets the aether in motion—this alternative was chosen by Helmholtz,[2] but is inconsistent with the theory of the aether which was generally received in the closing years of the century; or lastly, with Thomson,[3] to accept the principle that the aether is itself the vehicle of mechanical momentum, of amount $(1/4\pi c)$ $[\mathbf{D} . \mathbf{B}]$ per unit volume.[4]

Maxwell's theory was now being developed in ways which could scarcely have been anticipated by its author. But although every year added something to the superstructure, the foundations remained much as Maxwell had laid them; the doubtful argument by which he had sought to justify the introduction of displacement

[1] It was, however, adopted by G. T. Walker, *Aberration and the Electromagnetic Field*, Camb., 1900.

[2] *Berlin Sitzungsberichte* (1893), p. 649; *Ann. d. Phys.* liii (1894), p. 135. Helmholtz supposed the aether to behave as a frictionless incompressible fluid.

[3] Loc. cit.

[4] As usual, we use e.-s. units for **D** and **E**, and e.-m. units for **B** and **H**.

currents was still all that was offered in their defence. In 1884, however, the theory was established [1] on a different basis by Heinrich Hertz (1857–94).

Hertz, the son of a senator, had originally been intended for the profession of an architect ; but when studying engineering as part of his course, he felt the attraction of pure science, and in 1880 he took a doctorate in physics at Berlin. He was so fortunate as to attract the attention of Helmholtz, who in the same year appointed him his assistant ; in 1883 he became a *privat-docent* at Kiel, and it was here that he began the researches which will now be described.

The train of Hertz's ideas resembles that by which Ampère, on hearing of Oersted's discovery of the magnetic field produced by electric currents, inferred that electric currents should exert ponderomotive forces on each other. Ampère argued that a current, being competent to originate a magnetic field, must be equivalent to a magnet in other respects ; and therefore that currents, like magnets, should exhibit forces of mutual attraction and repulsion.

Ampère's reasoning rests on the assumption that the magnetic field produced by a current is in all respects of the same nature as that produced by a magnet ; in other words, that only one kind of magnetic force exists. This principle of the 'unity of magnetic force' Hertz now proposed to supplement by asserting that the electric force generated by a changing magnetic field is identical in nature with the electric force due to electrostatic charges ; this second principle he called the 'unity of electric force.' [2]

Suppose, then, that a system of electric currents $\mathbf{s}$ exists in otherwise empty space. According to the older theory, these currents give rise to a vector-potential $\mathbf{a}_1$, equal to Pot $(1/c)\mathbf{s}$ [3] ; and the magnetic force $\mathbf{H}_1$ is the curl of $\mathbf{a}_1$; while the electric force $\mathbf{E}_1$, at any point in the field, produced by the variation of the currents, is $-\dfrac{1}{c}\dfrac{\partial \mathbf{a}_1}{\partial t}$.

It is now assumed that the electric force so produced is indistinguishable from the electric force which would be set up by electrostatic charges, and therefore that the system of varying currents exerts ponderomotive forces on electrostatic charges ; the principle of action and reaction then requires that electrostatic

[1] *Ann. d. Phys.* xxiii (1884), p. 84 : English version in Hertz's *Miscellaneous Papers*, translated by D. E. Jones and G. A. Schott, p. 273

[2] That an electric field is produced by a varying magnetic field was shown experimentally by Lodge in 1889, by observing the motion of a gold leaf in a varying magnetic field ; and by A. Righi, *Nuovo Cimento*, v (1901), p. 233.

[3] $a = \text{Pot}\,\beta$ is used to denote the solution of the equation $\nabla^2 a + 4\pi\beta = 0$.

charges should exert ponderomotive forces on a system of varying currents, and consequently (again appealing to the principle of the unity of electric force) that two systems of varying currents should exert on each other ponderomotive forces due to the variations.

But just as Helmholtz,[1] by aid of the principle of conservation of energy, deduced the existence of an electromotive force of induction from the existence of the ponderomotive forces between electric currents (i.e. variable electric systems), so from the existence of ponderomotive forces between variable systems of currents (i.e. variable magnetic systems) we may infer that variations in the rate of change of a variable magnetic system give rise to induced magnetic forces in the surrounding space. The analytical formulae which determine these forces will be of the same kind as in the electric case; so that the induced magnetic force $\mathbf{H}'$ is given by an equation of the form

$$\mathbf{H}' = (1/c^2)\frac{\partial \mathbf{b}_1}{\partial t},$$

where $\mathbf{b}_1$, which is analogous to the vector-potential in the electric case, is a circuital vector whose curl is the electric force $\mathbf{E}_1$ of the variable magnetic system. The value of $\mathbf{b}_1$ is therefore $(1/4\pi)$ curl Pot $\mathbf{E}_1$: so we have

$$\mathbf{H}' = -\frac{1}{4\pi c^2}\frac{\partial^2}{\partial t^2}\operatorname{curl}\operatorname{Pot}\mathbf{a}_1.$$

This must be added to $\mathbf{H}_1$. Writing $\mathbf{H}_2$ for the sum, $\mathbf{H}_1 + \mathbf{H}'$, we see that $\mathbf{H}_2$ is the curl of $\mathbf{a}_2$, where

$$\mathbf{a}_2 = \mathbf{a}_1 - \frac{1}{4\pi c^2}\frac{\partial^2}{\partial t^2}\operatorname{Pot}\mathbf{a}_1;$$

and the electric force $\mathbf{E}_2$ will then be $-\frac{1}{c}\frac{\partial \mathbf{a}_2}{\partial t}$.

This system is not, however, final; for we must now perform the process again with these improved values of the electric and magnetic forces and the vector-potential; and so we obtain for the magnetic force the value curl $\mathbf{a}_3$, and for the electric force the value $-\frac{1}{c}\frac{\partial \mathbf{a}_3}{\partial t}$, where

$$\begin{aligned}\mathbf{a}_3 &= \mathbf{a}_1 - \frac{1}{4\pi c^2}\frac{\partial^2}{\partial t^2}\operatorname{Pot}\mathbf{a}_2 \\ &= \mathbf{a}_1 - \frac{1}{4\pi c^2}\frac{\partial^2}{\partial t^2}\operatorname{Pot}\mathbf{a}_1 + \frac{1}{(4\pi c^2)^2}\frac{\partial^4}{\partial t^4}\operatorname{Pot}\operatorname{Pot}\mathbf{a}_1.\end{aligned}$$

[1] cf. p. 218

This process must again be repeated indefinitely; so finally we obtain for the magnetic force **H** the value curl **a**, and for the electric force **E** the value $-\frac{1}{c}\frac{\partial \mathbf{a}}{\partial t}$, where

$$\mathbf{a} = \mathbf{a}_1 - \frac{1}{4\pi c^2}\frac{\partial^2}{\partial t^2}\text{Pot } \mathbf{a}_1 + \frac{1}{(4\pi c^2)^2}\frac{\partial^4}{\partial t^4}\text{ Pot Pot } \mathbf{a}_1 - \frac{1}{(4\pi c^2)^3}\frac{\partial^6}{\partial t^6}\text{Pot Pot Pot } \mathbf{a}_1 + \ldots$$

It is evident that the quantity **a** thus defined satisfies the equation

$$\nabla^2\mathbf{a} = \nabla^2\mathbf{a}_1 + \frac{1}{c^2}\frac{\partial^2}{\partial t^2}\mathbf{a},$$

or

$$\nabla^2\mathbf{a} - \frac{1}{c^2}\frac{\partial^2}{\partial t^2}\mathbf{a} = -\frac{4\pi}{c}\mathbf{s}.$$

This equation may be written

$$\text{curl } \mathbf{H} = \frac{1}{c}\frac{\partial \mathbf{E}}{\partial t} + \frac{4\pi}{c}\mathbf{s}$$

while the equations $\mathbf{H} = \text{curl } \mathbf{a}$, $\mathbf{E} = -\frac{1}{c}\frac{\partial \mathbf{a}}{\partial t}$ give

$$\text{curl } \mathbf{E} = -\frac{1}{c}\frac{\partial \mathbf{H}}{\partial t}.$$

These are, however, the fundamental equations of Maxwell's theory in the form given in his memoir of 1868.[1]

That Hertz's deduction is ingenious and interesting will readily be admitted. That it is conclusive may scarcely be claimed, for the argument of Helmholtz regarding the induction of currents is not altogether satisfactory; and Hertz, in following his master, is on no surer ground.

In the course of a discussion [2] on the validity of Hertz's assumptions, which followed the publication of his paper, E. Aulinger [3] brought to light a contradiction between the principles of the unity of electric and of magnetic force and the electrodynamics of Weber. Consider an electrostatically charged hollow sphere, in the interior of which is a wire carrying a variable current. According to Weber's theory, the sphere would exert a turning couple on the wire; but according to Hertz's principles, no action would be exerted, since charging the sphere makes no difference to either the electric or the

[1] cf. p. 258

[2] Lorberg, *Ann. d. Phys.* xxvii (1886), p. 666; xxxi (1887), p. 131. Boltzmann, ibid. xxix (1886), p. 598

[3] *Ann. d. Phys.* xxvii (1886), p. 119

magnetic force in its interior. The experiment thus suggested would be a crucial test of the correctness of Weber's theory ; it has the advantage of requiring nothing but closed currents and electrostatic charges at rest ; but the quantities to be observed would be on the limits of observational accuracy.

After his attempt to justify the Maxwellian equations on theoretical grounds, Hertz (who in 1885 had become Professor of Physics at Karlsruhe) turned his attention to the possibility of verifying them by direct experiment. His interest in the matter had first been aroused some years previously, when the Berlin Academy proposed as a prize subject 'To establish experimentally a relation between electromagnetic actions and the polarisation of dielectrics.' Helmholtz suggested to Hertz that he should attempt the solution; but at the time Hertz saw no way of bringing phenomena of this kind within the limits of observation. From this time forward, however, the idea of electric oscillations was continually present to his mind ; and in the spring of 1886 he noticed an effect [1] which formed the starting-point of his later researches. When an open circuit was formed of a piece of copper wire, bent into the form of a rectangle, so that the ends of the wire were separated only by a short air-gap, and when this open circuit was connected by a wire with any point of a circuit through which the spark discharge of an induction coil was taking place, it was found that a spark passed in the air-gap of the open circuit. This was explained by supposing that the change of potential, which is propagated along the connecting wire from the induction coil, reaches one end of the open circuit before it reaches the other, so that a spark passes between them ; and the phenomenon therefore was regarded as indicating a finite velocity of propagation of electric potential along wires.[2]

Continuing his experiments, Hertz [3] found that a spark could be induced in the open or secondary circuit even when it was not in metallic connection with the primary circuit in which the electric

[1] *Ann. d. Phys.* xxxi (1887), p. 421 ; Hertz's *Electric Waves*, translated by D. E. Jones, p. 29

[2] Unknown to Hertz, the transmission of electric waves along wires had been observed in 1870 by Wilhelm von Bezold, *München Sitzungsberichte*, i (1870), p. 113 ; *Phil. Mag.* xl (1870), p. 42. 'If,' he wrote at the conclusion of a series of experiments, 'electrical waves be sent into a wire insulated at the end, they will be reflected at that end. The phenomena which accompany this process in alternating discharges appear to owe their origin to the interference of the advancing and reflected waves,' and, 'an electric discharge travels with the same rapidity in wires of equal length, without reference to the materials of which these wires are made.'

The subject was investigated by O. J. Lodge and A. P. Chattock at almost the same time as Hertz's experiments were being carried out ; mention was made of their researches at the meeting of the British Association in 1888.

[3] loc. cit.

oscillations were generated; and he rightly interpreted the phenomenon by showing that the secondary circuit was of such dimensions as to make the free period of electric oscillations in it nearly equal to the period of the oscillations in the primary circuit; the disturbance which passed from one circuit to the other by induction would consequently be greatly intensified in the secondary circuit by the effect of resonance.

The discovery that sparks may be produced in the air-gap of a secondary circuit, provided it has the dimensions proper for resonance, was of great importance: for it supplied a method of detecting electrical effects in air at a distance from the primary disturbance; a suitable detector was in fact all that was needed in order to observe the propagation of electric waves in free space, and thereby decisively test the Maxwellian theory. To this work Hertz now addressed himself.[1]

Unknown to him, he had been anticipated, about seven years earlier, by David Edward Hughes (1830–1900), who had shown that signals sent from a spark transmitter could be detected at distances up to five hundred yards by a microphone contact (essentially what was later called a *coherer*), to which was connected a telephone in which the signals were heard. He claimed (rightly) that the signals were transmitted by electric waves in the air. In 1879 and 1880 he demonstrated these experiments before the President of the Royal Society (Spottiswoode), Sir George Stokes and Mr W. H. Preece, the Electrician to the Post Office; unfortunately they inclined to the opinion that the effects might be explained by ordinary electromagnetic induction; and Hughes, discouraged, did not print any account of his work until long afterwards,[2] so that the priority of publication belongs to Hertz.

The *oscillator* or primary source of the disturbances studied by Hertz may be constructed of two sheets of metal in the same plane, each sheet carrying a stiff wire which projects towards the other sheet and terminates in a knob; the sheets are to be excited by connecting them to the terminals of an induction coil. The sheets may be regarded as the two coatings of a modified Leyden jar, with air as the dielectric between them; the electric field is extended throughout the air, instead of being confined to the narrow space

[1] Sir Oliver Lodge was about this time independently studying electric oscillations in air in connection with the theory of lightning-conductors; cf. Lodge, *Phil. Mag.* xxvi (1888), p. 217.

[2] In *The Electrician*, xliii (5 May 1899), p. 40

So long before as 1842, Joseph Henry, of Washington, had noticed that the 'inductive' effects of the Leyden jar discharge could be observed at considerable distances, and had even suggested a comparison with 'a spark from flint and steel in the case of light.'

between the coatings, as in the ordinary Leyden jar. Such a disposition ensures that the system shall lose a large part of its energy by radiation at each oscillation.

As in the jar discharge,[1] the electricity surges from one sheet to the other, with a period proportional to $(CL)^{\frac{1}{2}}$, where C denotes the electrostatic capacity of the system formed by the two sheets, and L denotes the self-induction of the connection. The capacity and induction should be made as small as possible in order to make the period small. The detector used by Hertz was that already described, namely, a wire bent into an incompletely closed curve, and of such dimensions that its free period of oscillation was the same as that of the primary oscillation, so that resonance might take place.

Towards the end of the year 1887, when studying the sparks induced in the resonating circuit by the primary disturbance, Hertz noticed [2] that the phenomena were distinctly modified when a large mass of an insulating substance was brought into the neighbourhood of the apparatus; thus confirming the principle that the changing electric polarisation which is produced when an alternating electric force acts on a dielectric is capable of displaying electromagnetic effects.

Early in the following year (1888) Hertz determined to verify Maxwell's theory directly by showing that electromagnetic actions are propagated in air with a finite velocity.[3] For this purpose he transmitted the disturbance from the primary oscillator by two different paths, viz. through the air and along a wire; and having exposed the detector to the joint influence of the two partial disturbances, he observed interference between them. In this way he found the ratio of the velocity of electric waves in air to their velocity when conducted by wires; and the latter velocity he determined by observing the distance between the nodes of stationary waves in the wire, and calculating the period of the primary oscillation. The velocity of propagation of electric disturbances in air was in this way shown to be finite and of the same order as the velocity of light.[4]

[1] cf. p. 227 [2] *Ann. d. Phys.* xxxiv, p. 373; *Electric Waves* (English edn.), p. 95

[3] *Ann. d. Phys.* xxxiv (1888), p. 551; *Electric Waves* (English edn.), p. 107

[4] Hertz's experiments gave the value 45/28 for the ratio of the velocity of electric waves in air to the velocity of electric waves conducted by the wires, and 2×10^{10} cm. per sec. for the latter velocity. These numbers were afterwards found to be open to objection; Poincaré (*Comptes Rendus*, cxi [1890], p. 322) showed that the period calculated by Hertz was $\sqrt{2} \times$ the true period, which would make the velocity of propagation in air equal to that of light $\times \sqrt{2}$. Ernst Lecher (*Wiener Berichte*, 8 May 1890; *Phil. Mag.* xxx [1890], p. 128), experimenting on the velocity of propagation of electric vibrations in wires, found instead of Hertz's 2×10^{10} cm. per sec. a value within 2 per cent of the velocity of light. E. Sarasin and L. de la Rive at Geneva (*Archives des Sc. Phys.* xxix, 1893, pp. 358, 441) finally proved that the velocities of propagation in air and along wires are equal.

Later in 1888 Hertz [1] showed that electric waves in air are reflected at the surface of a wall ; standing waves may thus be produced, and interference may be obtained between direct and reflected beams travelling in the same direction.

The theoretical analysis of the disturbance emitted by a Hertzian radiator according to Maxwell's theory was given by Hertz in the following year.[2]

The effects of the oscillator are chiefly determined by the free electric charges which, alternately appearing at the two sides, generate an electric field by their presence and a magnetic field by their motion. In each oscillation, as the charges on the poles of the radiator increase from zero, lines of electric force, having their ends on these poles, move outwards into the surrounding space. When the charges on the poles attain their greatest values, the lines cease to issue outwards, and the existing lines begin to retreat inwards towards the poles ; but the outer lines of force contract in such a way that their upper and lower parts touch each other at some distance from the radiator, and the remoter portion of each of these lines thus takes the form of a loop ; and when the rest of the line of force retreats inwards towards the radiator, this loop becomes detached and is propagated outwards as radiation. In this way the radiator emits a series of whirl rings, which as they move grow thinner and wider ; at a distance, the disturbance is approximately a plane wave, the opposite sides of the ring representing the two phases of the wave. When one of these rings has become detached from the radiator, the energy contained may subsequently be regarded as travelling outwards with it.

To discuss the problem analytically [3] we take the axis of the radiator as axis of z, and the centre of the spark gap as origin. The field may be regarded as due to an electric doublet formed of a positive and an equal negative charge, displaced from each other along the axis of the vibrator, and of moment

$$Ae^{-pt} \sin (2\pi ct/\lambda),$$

the factor e^{-pt} being inserted to represent the damping.

The simplest method of proceeding, which was suggested by FitzGerald,[4] is to form the retarded potentials ϕ and $\mathbf{a}$ of L. Lorenz.[5]

[1] *Ann. d. Phys.* xxxiv (1888), p. 610 ; *Electric Waves* (English edn.), p. 124

[2] ibid. xxxvi (1889), p. 1 ; *Electric Waves* (English edn.), p. 137

[3] cf. Karl Pearson and A. Lee, *Phil. Trans.* cxciii (1899), p. 165

[4] *Brit. Assoc. Rep., Leeds* (1890), p. 755

[5] cf. p. 268. They had been employed in the meantime by Helmholtz, and by Lord Rayleigh in his *Theory of Sound*. The use of retarded potentials was also recommended in the following year by Poincaré, *Comptes Rendus*, cxiii (1891), p. 515.

These are determined in terms of the charges and their velocities by the equations

$$\phi = \Sigma \frac{(e)_{t-r/c}}{r}, \qquad a_z = \Sigma \frac{1}{cr}\left(e\frac{\partial z}{\partial t}\right)_{t-r/c}$$

whence it is readily shown that in the present case

$$\phi = -\,\partial F/\partial z, \qquad c\mathbf{a} = (0, 0, \partial F/\partial t),$$

where

$$F = \frac{Ae^{-p(t-r/c)}}{r}\sin\frac{2\pi}{\lambda}(ct - r).$$

The electric and magnetic forces are then determined by the equations

$$\mathbf{E} = \text{grad}\,\phi - \frac{1}{c}\frac{\partial \mathbf{a}}{\partial t}, \qquad \mathbf{H} = \text{curl}\,\mathbf{a}.$$

It is found that the electric force may be regarded as compounded of a force ϕ_2, parallel to the axis of the vibrator and depending at any instant only on the distance from the vibrator, together with a force $\phi_1 \sin\theta$ acting in the meridian plane perpendicular to the radius from the centre, where ϕ_1 depends at any instant only on the distance from the vibrator, and θ denotes the angle which the radius makes with the axis of the oscillator. At points on the axis, and in the equatorial plane, the electric force is parallel to the axis. At a great distance from the oscillator, ϕ_2 is small compared with ϕ_1, so the wave is purely transverse. The magnetic force is directed along circles whose centres are on the axis of the radiator; and its magnitude may be represented in the form $\phi_3 \sin\theta$, where ϕ_3 depends only on r and t; at great distances from the radiator, $c\phi_3$ is approximately equal to ϕ_1.

If the activity of the oscillator be supposed to be continually maintained, so that there is no damping, we may replace p by zero, and may proceed as in the case of the magnetic oscillator[1] to determine the amount of energy radiated. The mean outward flow of energy per unit time is found to be $\frac{1}{3}cA^2(2\pi/\lambda)^4$; from which it is seen that the rate of loss of energy by radiation increases greatly as the wave-length decreases.

The action of an electrical vibrator may be studied by the aid of mechanical models. In one of these, devised by Larmor,[2] the aether is represented by an incompressible elastic solid, in which are two cavities, corresponding to the conductors of the vibrator, filled with incompressible fluid of negligible inertia. The electric force is represented by the displacement of the solid. For such rapid alternations as are here considered, the metallic poles behave as

[1] cf. p. 312

[2] *Proc. Camb. Phil. Soc.* vii (1891), p. 165

perfect conductors ; and the tangential components of electric force at their surfaces are zero. This condition may be satisfied in the model by supposing the lining of each cavity to be of flexible sheet-metal, so as to be incapable of tangential displacement ; the normal displacement of the lining then corresponds to the surface-density of electric charge on the conductor.

In order to obtain oscillations in the solid resembling those of an electric vibrator, we may suppose that the two cavities have the form of semicircular tubes forming the two halves of a complete circle. Each tube is enlarged at each of its ends, so as to present a front of considerable area to the corresponding front at the end of the other tube. Thus at each end of one diameter of the circle there is a pair of opposing fronts, which are separated from each other by a thin sheet of the elastic solid.

The disturbance may be originated by forcing an excess of liquid into one of the enlarged ends of one of the cavities. This involves displacing the thin sheet of elastic solid, which separates it from the opposing front of the other cavity, and thus causing a corresponding deficiency of liquid in the enlarged end behind this front. The liquid will then surge backwards and forwards in each cavity between its enlarged ends ; and, the motion being communicated to the elastic solid, vibrations will be generated resembling those which are produced in the aether by a Hertzian oscillator.

In the latter part of the year 1888 the researches of Hertz [1] yielded more complete evidence of the similarity of electric waves to light. It was shown that the part of the radiation from an oscillator which was transmitted through an opening in a screen was propagated in a straight line, with diffraction effects. Of the other properties of light, polarisation existed in the original radiation, as was evident from the manner in which it was produced ; and polarisation in other directions was obtained by passing the waves through a grating of parallel metallic wires ; the component of the electric force parallel to the wires was absorbed, so that in the transmitted beam the electric vibration was at right angles to the wires. This effect obviously resembled the polarisation of ordinary light by a plate of tourmaline. Refraction was obtained by passing the radiation through prisms of hard pitch.[2]

[1] *Ann. d. Phys.* xxxvi (1889), p. 769 ; *Electric Waves* (English edn.), p. 172

[2] O. J. Lodge and J. L. Howard in the same year showed that electric radiation might be refracted and concentrated by means of large lenses ; cf. *Phil. Mag.* xxvii (1889), p. 48. Other demonstrations of optical properties possessed by electromagnetic waves were given by A. Righi (*Bologna Acc. Sci. Mem.* iv [1894], p. 487), and A. Garbasso (*Annalen d. Phys.* liii [1894], p. 534, and *Atti Acc. Torino*, xxx [1895], p. 442).

As Larmor remarked,[1] 'the discoveries of Hertz left no further room for doubt that the physical scheme of Maxwell . . . constituted a real formulation of the underlying unity in physical dynamics.'

The old question as to whether the light vector is in, or at right angles to, the plane of polarisation [2] now presented itself in a new aspect. The wave-front of an electric wave contains two vectors, the electric and magnetic, which are at right angles to each other. Which of these is in the plane of polarisation? The answer was first given theoretically by Heaviside,[3] by working out the formulae for the ratio of the reflected to the incident wave, and then experimentally by FitzGerald and Trouton,[4] who found on reflecting Hertzian waves from a wall of masonry that no reflection was obtained at the polarising angle when the vibrator was in the plane of reflection. The inference from this is that the magnetic vector is in the plane of polarisation of the electric wave, and the electric vector is at right angles to the plane of polarisation. An interesting development followed in 1890, when O. Wiener [5] succeeded in photographing stationary waves of light. The stationary waves were obtained by the composition of a beam incident on a mirror with the reflected beam, and were photographed on a thin film of transparent collodion, placed close to the mirror and slightly inclined to it. If the beam used in such an experiment is plane-polarised, and is incident at an angle of 45°, the stationary vector is evidently that perpendicular to the plane of incidence; but Wiener found that under these conditions the effect was obtained only when the light was polarised in the plane of incidence; so that the chemical activity must be associated with the vector perpendicular to the plane of polarisation—i.e. the electric vector.

In 1890 and the following years appeared several memoirs on the fundamental equations of electromagnetic theory. Hertz, after presenting [6] the general content of Maxwell's theory for bodies at rest, extended [7] the equations to material bodies in motion in the field.

[1] *Brit. Ass. Rep.* (1900), p. 617

[2] cf. p. 154 et sqq.

[3] *Phil. Mag.* xxv (Feb. 1888), p. 130 (in Note A)

[4] *Nature*, xxxix (1889), p. 391

[5] *Ann. d. Phys.* xl (1890), p. 203. cf. a controversy regarding the results: *Comptes Rendus*, cxii (1891), pp. 186, 325, 329, 365, 383, 456; and *Ann. d. Phys.* xli (1890), p. 154; xliii (1891), p. 177; xlviii (1893), p. 119

[6] *Gött. Nach.* (1890), p. 106; *Ann. d. Phys.* xl (1890), p. 577; *Electric Waves* (English edn.), p. 195. In this memoir Hertz advocated the form of the equations which Maxwell had used in his paper of 1868 (cf. *supra*, p. 258) in preference to the earlier form, which involved the scalar and vector potentials.

[7] *Ann. d. Phys.* xli (1890), p. 369; *Electric Waves* (English edn.), p. 241

The propagation of light through a moving dielectric had been discussed previously, on the basis of Maxwell's equations for moving bodies, by J. J. Thomson (*Phil. Mag.* ix [1880], p. 284; *Proc. Camb. Phil. Soc.* v [1885], p. 250).

In a really comprehensive and correct theory, as Hertz remarked, a distinction should be drawn between the quantities which specify the state of the aether at every point, and those which specify the state of the ponderable matter entangled with it. This anticipation has been fulfilled by later investigators ; but Hertz considered that the time was not ripe for such a complete theory, and preferred, like Maxwell, to assume that the state of the compound system—matter plus aether—can be specified in the same way when the matter moves as when it is at rest ; or, as Hertz himself expressed it, that 'the aether contained within ponderable bodies moves with them.'

Maxwell's own hypothesis with regard to moving systems [1] amounted merely to a modification in the equation

$$\frac{1}{c}\frac{\partial \mathbf{B}}{\partial t} = -\operatorname{curl}\mathbf{E},$$

which represents the law that the electromotive force in a closed circuit is measured by the rate of decrease in the number of lines of magnetic induction which pass through the circuit. This law is true whether the circuit is at rest or in motion ; but in the latter case, the **E** in the equation must be taken to be the electromotive force in a stationary circuit whose position momentarily coincides with that of the moving circuit ; and since an electromotive force $(1/c)\,[\mathbf{w}\,.\,\mathbf{B}]$ is generated in matter by its motion with velocity **w** in a magnetic field **B**, we see that **E** is connected with the electromotive force **E′** in the moving ponderable body by the equation

$$\mathbf{E}' = \mathbf{E} + \frac{1}{c}\,[\mathbf{w}\,.\,\mathbf{B}],$$

so that the equation of electromagnetic induction in the moving body is

$$\frac{\partial \mathbf{B}}{\partial t} = -\,c\operatorname{curl}\mathbf{E}' + \operatorname{curl}\,[\mathbf{w}\,.\,\mathbf{B}].$$

Maxwell made no change in the other electromagnetic equations, which therefore retained the customary forms

$$\mathbf{D} = \epsilon\mathbf{E}', \quad \operatorname{div}\mathbf{D} = 4\pi\rho, \quad c\operatorname{curl}\mathbf{H} = \frac{\partial \mathbf{D}}{\partial t} + 4\pi\mathbf{s},$$

Hertz, however, impressed by the duality of electric and magnetic phenomena, modified the last of these equations by assuming that

[1] cf. p. 258–9

a magnetic force $(1/c)$ $[\mathbf{D} . \mathbf{w}]$ is generated in a dielectric which moves with velocity $\mathbf{w}$ in an electric field; such a force would be the magnetic analogue of the electromotive force of induction. A term involving curl $[\mathbf{D} . \mathbf{w}]$ is then introduced into the last equation.

The theory of Hertz resembles in many respects that of Heaviside,[1] who likewise insisted much on the duplex nature of the electromagnetic field, and was in consequence disposed to accept the term involving curl $[\mathbf{D} . \mathbf{w}]$ in the equations of moving media. Heaviside recognised more clearly than his predecessors the distinction between the force $\mathbf{E}'$, which determines the flux $\mathbf{D}$, and the force $\mathbf{E}$, whose curl represents the electric current; and, in conformity with his principle of duality, he made a similar distinction between the magnetic force $\mathbf{H}'$, which determines the flux $\mathbf{B}$, and the force $\mathbf{H}$, whose curl represents the 'magnetic current.' This distinction, as Heaviside showed, is of importance when the system is acted on by 'impressed forces,' such as voltaic electromotive forces or permanent magnetisation; these latter must be included in $\mathbf{E}'$ and $\mathbf{H}'$, since they help to give rise to the fluxes $\mathbf{D}$ and $\mathbf{B}$; but they must not be included in $\mathbf{E}$ and $\mathbf{H}$, since their curls are not electric or magnetic currents; so that in general we have

$$\mathbf{E}' = \mathbf{E} + \mathbf{e}, \quad \mathbf{H}' = \mathbf{H} + \mathbf{h},$$

where $\mathbf{e}$ and $\mathbf{h}$ denote the impressed forces.

Developing the theory by the aid of these conceptions, Heaviside was led to make a further modification. An *impressed force* is best defined in terms of the energy which it communicates to the system; thus, if $\mathbf{e}$ be an impressed electric force, the energy communicated to unit volume of the electromagnetic system in unit time is the scalar product of $\mathbf{e}$ and the electric current. In order that this equation may be true, it is necessary to regard the electric current in a moving medium as composed of the conduction current, displacement current, convection current, and also of the term in curl $[\mathbf{D} . \mathbf{w}]$, whose presence in the equation we have already noticed. This may be called the *current of dielectric convection.* Thus the total current is

$$\mathbf{S} = \frac{1}{4\pi}\frac{\partial \mathbf{D}}{\partial t} + \mathbf{s} + \rho\mathbf{w} + \frac{1}{4\pi}\operatorname{curl}\,[\mathbf{D} . \mathbf{w}],$$

[1] Heaviside's general theory was published in a series of papers in the *Electrician*, from 1885 onwards. His earlier work was republished in his *Electrical Papers* (2 vols., 1892), and his *Electromagnetic Theory* (2 vols, 1894). Mention may be specially made of a memoir in *Phil. Trans.* clxxxiii (1892), p. 423.

where $\rho\mathbf{w}$ denotes the convection current; and the equation connecting current with magnetic force is

$$c \text{ curl } (\mathbf{H}' - \mathbf{h}_0) = 4\pi\mathbf{S},$$

where $\mathbf{h}_0$ denotes the impressed magnetic forces other than that induced by motion of the medium.

We must now consider the advances which were effected during the period following the publication of Maxwell's *Treatise* in some of the special problems of electricity and optics.

We have seen [1] that Maxwell accounted for the rotation of the plane of polarisation of light in a medium subjected to a magnetic field $\mathbf{K}$ by adding to the kinetic energy of the aether, which is represented by $\frac{1}{2}\rho\,(\partial\mathbf{e}/\partial t)^2$, a term $\frac{1}{2}\sigma\left(\frac{\partial\mathbf{e}}{\partial t}\,.\,\text{curl}\,\partial\mathbf{e}/\partial\theta\right)$, where σ is a magneto-optic constant characteristic of the substance through which the light is transmitted, and $\partial/\partial\theta$ stands for $K_x\partial/\partial x + K_y\partial/\partial y + K_z\partial/\partial z$. This theory was developed further in 1879 by FitzGerald,[2] who brought it into closer connection with the electromagnetic theory of light by identifying the curl of the displacement $\mathbf{e}$ of the aethereal particles with the electric displacement; the derivate of $\mathbf{e}$ with respect to the time then corresponds to the magnetic force. Being thus in possession of a definitely electromagnetic theory of the magnetic rotation of light, FitzGerald proceeded to extend it so as to take account of a closely related phenomenon. In 1876 J. Kerr [3] had shown experimentally that when plane-polarised light is regularly reflected from either pole of an iron electromagnet, the reflected ray has a component polarised in a plane at right angles to the ordinary reflected ray. Shortly after this discovery had been made known, FitzGerald [4] had proposed to explain it by means of the same term in the equations as that which accounts for the magnetic rotation of light in transparent bodies. His argument was that if the incident plane-polarised ray be resolved into two rays circularly polarised in opposite senses, the refractive index will have different values for these two rays, and hence the intensities after reflection will be different; so that on recompounding them, two plane-polarised rays will be obtained—one polarised in the plane of incidence, and the other polarised at right angles to it.

The analytical discussion of Kerr's phenomenon, which was given by FitzGerald in his memoir of 1879, was based on these

[1] cf. p. 276
[2] *Phil. Trans.* clxxi (1879), p. 691; FitzGerald's *Scientific Writings*, p. 45
[3] *Phil. Mag.* (5), iii (1877), p. 321
[4] *Proc. R. S.* xxv (1877), p. 447; FitzGerald's *Scientific Writings*, p. 9

ideas ; the most essential features of the phenomenon were explained, but the investigation was in some respects imperfect.[1]

A new and fruitful conception was introduced in 1879–80, when H. A. Rowland [2] suggested a connection between the magnetic rotation of light and the phenomenon which had been discovered by his pupil Hall.[3] Hall's effect may be regarded as a rotation of conduction currents under the influence of a magnetic field ; and if it be assumed that displacement currents in dielectrics are rotated in the same way, the Faraday effect may evidently be explained. Considering the matter from the analytical point of view, the Hall effect may be represented by the addition of a term $k\,[\mathbf{K}\,.\,\mathbf{S}]$ to the electromotive force, where $\mathbf{K}$ denotes the impressed magnetic force, and $\mathbf{S}$ denotes the current ; so Rowland assumed that in dielectrics there is an additional term in the electric force, proportional to $\left[\mathbf{K}\,.\,\frac{\partial \mathbf{D}}{\partial t}\right]$, i.e. proportional to the rate of increase of $[\mathbf{K}\,.\,\mathbf{D}]$. Now it is universally true that the total electric force round a circuit is proportional to the rate of decrease of the total magnetic induction through the circuit ; so the total magnetic induction through the circuit must contain a term proportional to the integral of $[\mathbf{K}\,.\,\mathbf{D}]$ taken round the circuit ; and therefore the magnetic induction at any point must contain a term proportional to curl $[\mathbf{K}\,.\mathbf{D}\,]$. We may therefore write

$$\mathbf{B} = \mathbf{H} + \sigma \operatorname{curl} [\mathbf{K}\,.\,\mathbf{D}],$$

where σ denotes a constant. But if this be combined with the customary electromagnetic equations

$$\operatorname{curl} \mathbf{H} = \frac{1}{c}\frac{\partial \mathbf{D}}{\partial t}, \quad \operatorname{curl} \mathbf{E} = -\frac{1}{c}\frac{\partial \mathbf{B}}{\partial t}, \quad \mathbf{D} = \epsilon \mathbf{E},$$

and all the vectors except $\mathbf{B}$ be eliminated ($\mathbf{K}$ being treated as a constant), we obtain the equation

$$\frac{\partial^2 \mathbf{B}}{\partial t^2} = (c^2/\epsilon)\,\nabla^2 \mathbf{B} + c\sigma \operatorname{curl} (\partial^2 \mathbf{B}/\partial t \partial \theta),$$

[1] cf. Larmor's remarks in his 'Report on the Action of Magnetism on Light,' *Brit. Assoc. Rep.*, 1893 ; and his editorial comments in FitzGerald's *Scientific Writings*. Larmor traced to its source an inconsistency in the equations by which FitzGerald had represented the boundary conditions at an interface between the media. FitzGerald had indeed made the mistake, similar to that which was so often made by the earlier writers on the elastic-solid theory of light, of forgetting that when a medium is assumed to be incompressible, the condition of incompressibility must be introduced into the variational equation of motion (as was done, *supra*, p. 157). Larmor showed that when this correction was made, new terms (resembling the terms in p, *supra*, p. 157) made their appearance ; and the inconsistency in the equations was thus removed.

[2] *Amer. Jour. Math.* ii, p. 354 ; iii, p. 89 ; *Phil. Mag.* xi (1881), p. 254

[3] cf. p. 289

where $\partial/\partial\theta$ stands for $(K_x\partial/\partial_x + K_y\partial/\partial y + K_z\partial/\partial z)$; and this is identical with the equation which Maxwell had given [1] for the motion of the aether in magnetised media. It follows that the assumptions of Maxwell and of Rowland, different though they are physically, lead to the same analytical equations—at any rate so far as concerns propagation through a homogeneous medium.

The connections of Hall's phenomenon with the magnetic rotation of light, and with the reflection of light from magnetised metals, were extensively studied [2] in the years following the publication of Rowland's memoir; but it was not until the modern theory of electrons had been developed that a satisfactory representation of the molecular processes involved in magneto-optic phenomena was attained.

[1] cf. p. 277

[2] The theory of Basset (*Phil. Trans.* clxxxii [1891], p. 371) was, like Rowland's, based on the idea of extending Hall's phenomenon to dielectric media. An objection to this theory was that the tangential component of the electromotive force was not continuous across the interface between a magnetised and an unmagnetised medium; but Basset subsequently overcame this difficulty (*Nature*, lii [1895], p. 618; liii [1895], p. 130; *Amer. Jour. Math.* xix [1897], p. 60)—the effect analogous to Hall's being introduced into the equation connecting electric displacement with electric force, so that the equation took the form

$$\mathbf{E} = (1/\epsilon)\,\mathbf{D} + \sigma\left[\mathbf{K}\,.\,\frac{\partial \mathbf{D}}{\partial t}\right].$$

Basset, in 1893 (*Proc. Camb. Phil. Soc.* viii, p. 68), derived analytical expressions which represent Kerr's magneto-optic phenomenon by substituting a complex quantity for the refractive index in the formulae applicable to transparent magnetised substances.

The magnetic rotation of light and Kerr's phenomenon have been investigated also by R. T. Glazebrook, *Phil. Mag.* xi (1881), p. 397; by J. J. Thomson, *Recent Researches*, p. 482; by D. A. Goldhammer, *Ann. d. Phys.* xlvi (1892), p. 71; xlvii (1892), p. 345; xlviii (1893), p. 740; l (1893), p. 772; by P. Drude, *Ann. d. Phys.* xlvi (1892), p. 353; xlviii (1893), p. 122; xlix (1893), p. 690; lii (1894), p. 496; by C. H. Wind, *Verslagen Kon. Akad. Amsterdam*, 29 Sept 1894; by Reiff, *Ann. d. Phys.* lvii (1896), p. 281; by J. G. Leathem, *Phil. Trans.* cxc (1897), p. 89; *Trans. Camb. Phil. Soc.* xvii (1898), p. 16; and by W. Voigt in many memoirs, and in his treatise, *Magneto- und Elektro-optik*. Larmor's report presented to the British Association in 1893 has been already mentioned.

In most of the later theories the equations of propagation of light in magnetised metals are derived from the two fundamental electro-magnetic equations

$$c \operatorname{curl} \mathbf{H} = 4\pi\mathbf{S}, \quad -c \operatorname{curl} \mathbf{E} = \frac{\partial \mathbf{H}}{\partial t};$$

the total current S being assumed to consist of a part (the displacement-current) proportional to $\frac{\partial \mathbf{E}}{\partial t}$, a part (the conduction current) proportional to **E**, and a part proportional to the vector product of **E** and the magnetisation.

Various mechanical models of media in which magneto-optic phenomena take place have been devised at different times. W. Thomson (*Proc. Lond. Math. Soc.* vi [1875], p.190) investigated the propagation of waves of displacement along a stretched chain whose links contain rotating fly-wheels: cf. also Larmor, *Proc. Lond. Math. Soc.* xxi (1890), p. 423; xxiii (1891), p. 127; F. Hasenöhrl, *Wien Sitzungsberichte*, cvii, 2a (1898), p. 1015; W. Thomson (Kelvin), *Phil. Mag.* xlviii (1899), p. 236, and *Baltimore Lectures*; and FitzGerald, *Electrician*, 4 Aug. 1899 = FitzGerald's *Scientific Writings*, p. 481.

The allied phenomenon of rotary polarisation in naturally active bodies was investigated in 1892 by Goldhammer.[1] It will be remembered [2] that in the elastic-solid theory of Boussinesq, the rotation of the plane of polarisation of sugar solutions had been represented by substituting the equation

$$\mathbf{e}' = \mathbf{A}\mathbf{e} + \mathbf{B}\,\text{curl}\,\mathbf{e}$$

in place of the usual equation

$$\mathbf{e}' = \mathbf{A}\mathbf{e}.$$

Goldhammer now proposed to represent rotatory power in the electromagnetic theory by substituting the equation

$$\mathbf{E} = (1/\epsilon)\,\mathbf{D} + k\,\text{curl}\,\mathbf{D},$$

in place of the customary equation

$$\mathbf{E} = (1/\epsilon)\,\mathbf{D}:$$

the constant k being a measure of the natural rotatory power of the substance concerned. The remaining equations are as usual,

$$c\,\text{curl}\,\mathbf{H} = \frac{\partial \mathbf{D}}{\partial t}, \quad -c\,\text{curl}\,\mathbf{E} = \frac{\partial \mathbf{H}}{\partial t}$$

Eliminating **H** and **E**, we have

$$\frac{\partial^2 \mathbf{D}}{\partial t^2} = (c^2/\epsilon)\,\nabla^2\mathbf{D} + kc^2\,\nabla^2\,\text{curl}\,\mathbf{D}.$$

For a plane wave which is propagated parallel to the axis of x, this equation reduces to

$$\begin{cases} \dfrac{\partial^2 \mathrm{D}_y}{\partial t^2} = \dfrac{c^2}{\epsilon}\dfrac{\partial^2 \mathrm{D}_y}{\partial x^2} - kc^2\,\dfrac{\partial^3 \mathrm{D}_z}{\partial x^3}, \\[2ex] \dfrac{\partial^2 \mathrm{D}_z}{\partial t^2} = \dfrac{c^2}{\epsilon}\dfrac{\partial^2 \mathrm{D}_z}{\partial x^2} + kc^2\,\dfrac{\partial^3 \mathrm{D}_y}{\partial x^3}; \end{cases}$$

and, as MacCullagh had shown in 1836,[3] these equations are competent to represent the rotation of the plane of polarisation.

In the closing years of the nineteenth century, the general theory of aether and electricity assumed a new form. But before discussing the memoirs in which the new conception was unfolded, we shall consider the progress which had been made since the middle of the century in the study of conduction in liquid and gaseous media.

[1] *Jour. de Phys.* (3), i, pp. 205, 345 [2] cf. p. 168 [3] cf. p. 159

Chapter XI

CONDUCTION IN SOLUTIONS AND GASES, FROM FARADAY TO THE DISCOVERY OF THE ELECTRON

The hypothesis which Grothuss and Davy had advanced [1] to explain the decomposition of electrolytes was open to serious objection in more than one respect. Since the electric force was supposed first to dissociate the molecules of the electrolyte into ions, and afterwards to set them in motion toward the electrodes, it would seem reasonable to expect that doubling the electric force would double both the dissociation of the molecules and the velocity of the ions, and would therefore quadruple the electrolysis—an inference which is not verified by observation. Moreover it might be expected, on Grothuss' theory, that some definite magnitude of electromotive force would be requisite for the dissociation, and that no electrolysis at all would take place when the electromotive force was below this value, which again is contrary to experience.

A way of escape from these difficulties was first indicated, in 1850, by Alex. Williamson,[2] who suggested that in compound liquids decompositions and recombinations of the molecules are continually taking place throughout the whole mass of the liquid, quite independently of the application of an external electric force. An atom of one element in the compound is thus paired now with one and now with another atom of another element, and in the intervals between these alliances the atom may be regarded as entirely free. In 1857 this idea was made by R. Clausius,[3] of Zurich, the basis of a theory of electrolysis. According to it, the electromotive force emanating from the electrodes does not effect the dissociation of the electrolyte into ions, since a degree of dissociation sufficient for the purpose already exists in consequence of the perpetual mutability of the molecules of the electrolyte. Clausius assumed that these ions are in opposite electric conditions ; the applied electric force therefore causes a general drift of all the ions of one kind towards the anode, and of all the ions of the other kind towards the cathode. These opposite motions of the two kinds of ions constitute the galvanic current in the liquid.

[1] cf. p. 76

[2] *Phil. Mag.* xxxvii (1850), p. 350 ; Liebig's *Annalen d. Chem. u. Pharm.* lxxvii (1851), p. 37

[3] *Ann. d. Phys.* ci (1857), p. 338 ; *Phil. Mag.* xv (1858), p. 94

The merits of the Williamson-Clausius hypothesis were not fully recognised for many years ; but it became the foundation of that theory of electrolysis which was generally accepted at the end of the century.

Meanwhile another aspect of electrolysis was receiving attention. It had long been known that the passage of a current through an electrolytic solution is attended not only by the appearance of the products of decomposition at the electrodes, but also by changes of relative strength in different parts of the solution itself. Thus in the electrolysis of a solution of copper sulphate, with copper electrodes, in which copper is dissolved off the anode and deposited on the cathode, it is found that the concentration of the solution diminishes near the cathode, and increases near the anode. Some experiments on the subject were made by Faraday[1] in 1835 ; and in 1844 it was further investigated by J. F. Daniell and W. A. Miller,[2] who explained it by asserting that the cation and anion have not (as had previously been supposed) the same facility of moving to their respective electrodes ; but that in many cases the cation appears to move but little, while the transport is effected chiefly by the anion.

This idea was adopted by W. Hittorf (1824–1914), of Münster, who, in the years 1853 to 1859, published[3] a series of memoirs on the migration of the ions. Let the velocity of the anions in the solution be to the velocity of the cations in the ratio $v : u$. Then it is easily seen that if $(u + v)$ molecules of the electrolyte are decomposed by the current, and yielded up as ions at the electrodes, v of these molecules will have been taken from the fluid on the side of the cathode, and u of them from the fluid on the side of the anode. By measuring the concentration of the liquid round the electrodes after the passage of a current, Hittorf determined the ratio v/u in a large number of cases of electrolysis.[4]

The theory of ionic movements was advanced a further stage by F. W. Kohlrausch[5] (1840–1910), of Würzburg. Kohlrausch showed that although the ohmic specific conductivity k of a solution diminishes indefinitely as the strength of the solution is reduced, yet the ratio k/m, where m denotes the number of gramme-equivalents[6]

[1] *Exp. Res.* §§ 525–30

[2] *Phil. Trans.* (1844), p. 1 ; cf. also Pouillet, *Comptes Rendus*, xx (1845), p. 1544

[3] *Ann. d. Phys.* lxxxix (1853), p. 177 ; xcviii (1856), p. 1 ; ciii (1858), p. 1 ; cvi (1859), pp. 337, 513

[4] The ratio $v/(u + v)$ was termed by Hittorf the *transport number* of the anion.

[5] *Ann. d. Phys.* vi (1879), pp. 1, 145. The chief results had been communicated to the Academy of Göttingen in 1876 and 1877.

[6] A gramme-equivalent means a mass of the salt whose weight in grammes is the molecular weight divided by the valency of the ions.

of salt per unit volume, tends to a definite limit when the solution is indefinitely dilute. This limiting value may be denoted by λ. He further showed that λ may be expressed as the sum of two parts, one of which depends on the cation, but is independent of the nature of the anion ; while the other depends on the anion, but not on the cation—a fact which may be explained by supposing that, in very dilute solutions, the two ions move independently under the influence of the electric force. Let u and v denote the velocities of the cation and anion respectively, when the potential difference per cm. in the solution is unity ; then the total current carried through a cube of unit volume is $mE(u + v)$, where E denotes the electric charge carried by one gramme-equivalent of ion.[1] Thus $mE(u + v) =$ total current $= k = m\lambda$, or $\lambda = E(u + v)$. The determination of v/u by the method of Hittorf, and of $(u + v)$ by the method of Kohlrausch, made it possible to calculate the absolute velocities of drift of the ions from experimental data.

Meanwhile, important advances in voltaic theory were being effected in connection with a different class of investigations.

Suppose that two mercury electrodes are placed in a solution of acidulated water, and that a difference of potential, insufficient to produce continuous decomposition of the water, is set up between the electrodes by an external agency. Initially a slight electric current—the polarising current,[2] is observed ; but after a short time it ceases ; and after its cessation the state of the system is one of electrical equilibrium. It is evident that the polarising current must in some way have set up in the cell an electromotive force equal and opposite to the external difference of potential ; and it is also evident that the seat of this electromotive force must be at the electrodes, which are now said to be *polarised*.

An abrupt fall of electric potential at an interface between two media, such as the mercury and the solution in the present case, requires that there should be a field of electric force, of considerable intensity, within a thin stratum at the interface ; and this must owe its existence to the presence of electric charges. Since there is no electric field outside the thin stratum, there must be as much vitreous as resinous electricity present ; but the vitreous charges must preponderate on one side of the stratum, and the resinous charges on the other side ; so that the system as a whole resembles the two coatings of a condenser with the intervening dielectric. In the case of the polarised mercury cathode in acidulated water, there must be on the electrode itself a negative charge ; the surface of this electrode

[1] i.e. E is 96,580 coulombs

[2] cf. p. 76

in the polarised state may be supposed to be either mercury, or mercury covered with a layer of hydrogen. In the solution adjacent to the electrode, there must be an excess of cations and a deficiency of anions, so as to constitute the other layer of the condenser ; these cations may be either mercury cations dissolved from the electrode, or the hydrogen cations of the solution.

It was shown in 1870 by Cromwell Fleetwood Varley (1828–1883)[1] that a mercury cathode, thus polarised in acidulated water, shows a tendency to adopt a definite superficial form, as if the surface-tension at the interface between the mercury and the solution were in some way dependent on the electric conditions. The matter was more fully investigated in 1873 by a young French physicist, then preparing for his inaugural thesis, Gabriel Lippmann.[2] In Lippmann's instrumental disposition, which is called a *capillary electrometer*, mercury electrodes are immersed in acidulated water ; the anode H_0 has a large surface, while the cathode H has a variable surface S small in comparison. When the external electromotive force is applied, it is easily seen that the fall of potential at the large electrode is only slightly affected, while the fall of potential at the small electrode is altered by polarisation by an amount practically equal to the external electromotive force. Lippmann found that the constant of capillarity of the interface at the small electrode was a function of the external electromotive force, and therefore of the difference of potential between the mercury and the electrolyte.

Let V denote the external electromotive force ; we may without loss of generality assume the potential of H_0 to be zero, so that the potential of H is $-$ V. The state of the system may be varied by altering either V or S ; we assume that these alterations may be performed independently, reversibly and isothermally and that the state of the large electrode H_0 is not altered thereby. Let de denote the quantity of electricity which passes through the cell from H_0 to H, when the state of the system is thus varied ; then if E denote the available energy of the system, and γ the surface-tension at H, we have

$$dE = \gamma dS + V de,$$

γ being measured by the work required to increase the surface when no electricity flows through the circuit.

In order that equilibrium may be re-established between the electrode and the solution when the fall of potential at the cathode

[1] *Phil. Trans.* clxi (1871), p. 129

[2] *Comptes Rendus,* lxxvi (1873), p. 1407 ; *Phil. Mag.* xlvii (1874), p. 281 ; *Ann. de Chim. et de Phys.* v (1875), p. 494 ; xii (1877), p. 265

is altered, it will be necessary not only that some hydrogen cations should come out of the solution and be deposited on the electrode, yielding up their charges, but also that there should be changes in the clustering of the charged ions of hydrogen, mercury and sulphion in the layer of the solution immediately adjacent to the electrode. Each of these circumstances necessitates a flow of electricity in the outer circuit; in the one case to neutralise the charges of the cations deposited, and in the other case to increase the surface density of electric charge on the electrode, which forms the opposite sheet of the quasi-condenser. Let $Sf(V)$ denote the total quantity of electricity which has thus flowed in the circuit when the external electromotive force has attained the value V. Then evidently

$$de = d\{ Sf(V) \};$$

so

$$dE = \{ \gamma + V f(V) \} dS + VSf'(V)\, dV.$$

Since this expression must be an exact differential, we have

$$\frac{d\gamma}{dV} + f(V) = 0;$$

so that $-d\gamma/dV$ is equal to that flux of electricity per unit of new surface formed, which will maintain the surface in a constant condition (V being constant) when it is extended. Integrating the previous equation, we have

$$E = S\left\{ \gamma - V\frac{d\gamma}{dV} \right\}.$$

Lippmann found that when the external electromotive force was applied, the surface tension increased at first, until, when the external electromotive force amounted to about one volt, the surface tension attained a maximum value, after which it diminished. He found that $d^2\gamma/dV^2$ was sensibly independent of V, so that the curve which represents the relation between γ and V is a parabola.[1]

The theory so far is more or less independent of assumptions as to what actually takes place at the electrode; on this latter question many conflicting views have been put forward. In 1878 Josiah Willard Gibbs,[2] of Yale (1839–1903), discussed the problem on the supposition that the polarising current is simply an ordinary electrolytic conduction current, which causes a liberation of hydrogen from the ionic form at the cathode. If this be so, the amount of electricity which passes through the cell in any displacement must

[1] Lippman, *Comptes Rendus*, xcv (1882), p. 686

[2] *Trans. Conn. Acad.* iii (1876–8), pp. 108, 343; Gibbs's *Scientific Papers*, i, p. 55

be proportional to the quantity of hydrogen which is yielded up to the electrode in the displacement; so that $d\gamma/dV$ must be proportional to the amount of hydrogen deposited per unit area of the electrode.[1]

A different view of the physical conditions at the polarised electrode was taken by Helmholtz[2] who assumed that the ions of hydrogen which are brought to the cathode by the polarising current do not give up their charges there, but remain in the vicinity of the electrode, and form one face of a quasi-condenser of which the other face is the electrode itself.[3] If σ denote the surface density of electricity on either face of this quasi-condenser, we have, therefore,

$$de = -\,d(S\sigma)\ ; \quad \text{so} \quad \sigma = d\gamma/dV$$

This equation shows that when $d\gamma/dV$ is zero—i.e. when the surface-tension is a maximum—σ must be zero; that is to say, there must be no difference of potential between the mercury and the electrolyte. The external electromotive force is then balanced entirely by the discontinuity of potential at the other electrode H_0; and thus a method is suggested of measuring the latter discontinuity of potential. All previous measurements of differences of potential had involved the employment of more than one interface; and it was not known how the measured difference of potential should be distributed among these interfaces; so that the suggestion of a means of measuring single differences of potential was a distinct advance, even though the hypotheses on which the method was based were somewhat insecure.

A further consequence deduced by Helmholtz from this theory leads to a second method of determining the difference of potential between mercury and an electrolyte. If a mercury surface is rapidly extending, and electricity is not rapidly transferred through the electrolyte, the electric surface-density in the double layer must rapidly decrease, since the same quantity of electricity is being distributed over an increasing area. Thus it may be inferred that a rapidly extending mercury surface in an electrolyte is at the same potential as the electrolyte.

This conception is realised in the *dropping electrode*, in which a

[1] This is embodied in equation (690) of Gibbs's memoir.

[2] *Berlin Monatsber.* (1881), p. 945; *Wiss. Abh.* i, p. 925; *Ann. d. Phys.* xvi. (1882), p. 31. cf. also Planck, *Ann. d. Phys.* xliv (1891), p. 385

[3] The conception of double layers of electricity at the surface of separation of two bodies had been already applied by Helmholtz to explain various other phenomena—e.g. the Volta contact difference of potential of two metals, frictional electricity and 'electric endosmose,' or the transport of fluid which occurs when an electric current is passed through two conducting liquids separated by a porous barrier. cf. Helmholtz, *Berlin Monatsberichte*, 27 Feb. 1879; *Ann. d. Phys.* vii (1879), p. 337; Helmholtz, *Wiss. Abh.* i, p. 855

jet of mercury, falling from a reservoir into an electrolytic solution, is so adjusted that it breaks into drops when the jet touches the solution. According to Helmholtz' conclusion there is no difference of potential between the drops and the electrolyte ; and therefore the difference of potential between the electrolyte and a layer of mercury underlying it in the same vessel is equal to the difference of potential between this layer of mercury and the mercury in the upper reservoir, which difference is a measurable quantity.

It will be seen that according to the theories both of Gibbs and of Helmholtz, and indeed according to all other theories on the subject,[1] $d\gamma/dV$ is zero for an electrode whose surface is rapidly increasing—e.g. a dropping electrode ; that is to say, the difference of potential between an ordinary mercury electrode and the electrolyte, when the surface-tension has its maximum value, is equal to the difference of potential between a dropping electrode and the same electrolyte. This result has been experimentally verified by various investigators, who have shown that the applied electro-motive force when the surface-tension has its maximum value in the capillary electrometer, is equal to the electromotive force of a cell having as electrodes a large mercury electrode and a dropping electrode.

Another memoir which belongs to the same period of Helmholtz'

[1] e.g. that of E. Warburg, *Ann. d. Phys.* xli (1890), p. 1. In this it is assumed that the electrolytic solution near the electrodes originally contains a salt of mercury in solution. When the external electromotive force is applied, a conduction current passes through the electrolyte, which in the body of the electrolyte is carried by the acid and hydrogen ions. At the cathode, Warburg supposed that the hydrogen ions react with the salt of mercury, reducing it to metallic mercury, which is deposited on the electrode. Thus a considerable change in concentration of the salt of mercury is caused at the cathode. At the anode, the acid ions carrying the current attack the mercury of the electrode, and thus increase the local concentration of mercuric salt ; but on account of the size of the anode this increase is trivial and may be neglected.

Warburg thus supposed that the electromotive force of the polarised cell is really that of a concentration cell, depending on the different concentrations of mercuric salt at the electrodes. He found $d\gamma/dV$ to be equal to the amount of mercuric salt at the cathode per unit area of cathode, divided by the electrochemical equivalent of mercury. The equation previously obtained is thus presented in a new physical interpretation.

Warburg connected the increase of the surface-tension with the fact that the surface-tension between mercury and a solution always increases when the concentration of the solution is diminished. His theory, of course, leads to no conclusion regarding the *absolute* potential difference between the mercury and the solution, as Helmholtz' does.

At an electrode whose surface is rapidly increasing—e.g. a dropping electrode—Warburg supposed that the surface-density of mercuric salt tends to zero, so $d\gamma/dV$ is zero.

The explanation of dropping electrodes favoured by Nernst, *Beilage zu den Ann. d. Phys.* lviii (1896), p. 1, was that the difference of potential corresponding to the equilibrium between the mercury and the electrolyte is instantaneously established ; but that ions are withdrawn from the solution in order to form the double layer necessary for this, and that these ions are carried down with the drops of mercury, until the upper layer of the solution is so much impoverished that the double layer can no longer be formed. The impoverishment of the upper layer of the solution was actually observed by W. Palmaer, *Zeitsch. Phys. Chem.* xxv (1898), p. 265 ; xxviii (1899), p. 257 ; xxxvi (1901), p. 664.

career, and which has led to important developments, was concerned with a special class of voltaic cells. The most usual type of cell is that in which the positive electrode is composed of a different metal from the negative electrode, and the evolution of energy depends on the difference in the chemical affinities of these metals for the liquids in the cell. But in the class of cells now considered [1] by Helmholtz, the two electrodes are composed of the same metal (say, copper) ; and the liquid (say, solution of copper sulphate) is more concentrated in the neighbourhood of one electrode than in the neighbourhood of the other. When the cell is in operation, the salt passes from the places of high concentration to the places of low concentration, so as to equalise its distribution ; and this process is accompanied by the flow of a current in the outer circuit between the electrodes. Such cells had been studied experimentally by James Moser a short time previously [2] to Helmholtz' investigation.

The activity of the cell is due to the fact that the available energy of a solution depends on its concentration ; the molecules of salt, in passing from a high to a low concentration, are therefore capable of supplying energy, just as a compressed gas is capable of supplying energy when its degree of compression is reduced. To examine the matter quantitatively, let $nf(n/V)$ denote the term in the available energy of a solution, which is due to the dissolution of n gramme-molecules of salt in a volume V of pure solvent ; the function f will of course depend also on the temperature. Then when dn gramme-molecules of solvent are evaporated from the solution, the decrease in the available energy of the system is evidently equal to the available energy of dn gramme-molecules of liquid solvent, less the available energy of dn gramme-molecules of the vapour of the solvent, together with $nf(n/V)$ less $nf\{n/(V - vdn)\}$, where v denotes the volume of one gramme-molecule of the liquid. But this decrease in available energy must be equal to the mechanical work supplied to the external world, which is $dn \, . \, p_1 \, (v' - v)$, if p_1 denote the vapour pressure of the solution at the temperature in question, and v' denote the volume of one gramme-molecule of vapour. We have therefore

$$\begin{aligned} dn \, . \, p_1 \, (v' - v) = & - \text{available energy of } dn \text{ gramme-molecules of solvent vapour} \\ & + \text{available energy of } dn \text{ gramme-molecules of liquid solvent} \\ & + nf(n/V) - nf\{n/(V - v\, dn)\}. \end{aligned}$$

[1] *Berlin Monatsber.* (1877), p. 713 ; *Phil. Mag.* (5), v (1878), p. 348 ; reprinted with additions in *Ann. d. Phys.* iii (1878), p. 201

[2] *Ann. d. Phys.* iii (1878), p. 216

Subtracting from this the equation obtained by making n zero, we have

$$dn\,.\,(p_1 - p_0)\,(v' - v) = nf\,(n/V) - nf\,\{\,n/(V - vdn)\,\},$$

where p_0 denotes the vapour pressure of the pure solvent at the temperature in question ; so that

$$(p_1 - p_0)\,(v' - v) = -\,(n^2/V^2)\,f'(n/V)\,v.$$

Now, it is known that when a salt is dissolved in water, the vapour pressure is lowered in proportional to the concentration of the salt—at any rate when the concentration is small : in fact, by a law due to Raoult, $(p_0 - p_1)/p_0$ is approximately equal to nv/V ; so that the previous equation becomes

$$p_0V\,(v' - v) = nf'(n/V).$$

Neglecting v in comparison with v', and making use of the equation of state of perfect gases (namely,

$$p_0v' = RT$$

where T denotes the absolute temperature, and R denotes the constant of the equation of state), we have

$$f'(n/V) = RTV/n,$$

and therefore

$$f\,(n/V) = RT\log\,(n/V).$$

Thus in the available energy of one gramme-molecule of a dissolved salt, the term which depends on the concentration is proportional to the logarithm of the concentration ; and hence, if in a concentration cell one gramme-molecule of the salt passes from a high concentration c_2 at one electrode to a low concentration c_1 at the other electrode, its available energy is thereby diminished by an amount proportional to $\log\,(c_2/c_1)$. The energy which thus disappears is given up by the system in the form of electrical work ; and therefore the electromotive force of the concentration cell must be proportional to $\log\,(c_2/c_1)$. The theory of solutions and their vapour pressure was not at the time sufficiently developed to enable Helmholtz to determine precisely the coefficient of $\log\,(c_2/c_1)$ in the expression.[1]

An important advance in the theory of solutions was effected in

[1] The formula given by Helmholtz was that the electromotive force of the cell is equal to $b\,(1 - n)\,v\log\,(c_2/c_1)$, where c_2 and c_1 denote the concentrations of the solution at the electrodes, v denotes the volume of one gramme of vapour in equilibrium with the water at the temperature in question, n denotes the transport number for the cation (Hittorf's $1/n$), and b denotes $q \times$ the lowering of vapour pressure when one gramme-equivalent of salt is dissolved in q grammes of water, where q denotes a large number.

1887, by a young Swedish physicist, Svante Arrhenius (1859–1927).[1] Interpreting the properties discovered by Kohlrausch [2] in the light of the ideas of Williamson and Clausius regarding the spontaneous dissociation of electrolytes, Arrhenius inferred that in very dilute solutions the electrolyte is completely dissociated into ions, but that in more concentrated solutions the salt is less completely dissociated; and that as in all solutions the transport of electricity in the solution is effected solely by the movement of ions, the equivalent conductivity [3] must be proportional to the fraction which expresses the degree of ionisation. By aid of these conceptions it became possible to estimate the dissociation quantitatively, and to construct a general theory of electrolytes.

Contemporary physicists and chemists found it difficult at first to believe that a salt exists in dilute solution only in the form of ions, e.g. that the sodium and chlorine exist separately and independently in a solution of common salt. But there is a certain amount of chemical evidence in favour of Arrhenius' conception. For instance, the tests in chemical analysis are really tests for the ions; iron in the form of a ferrocyanide, and chlorine in the form of a chlorate, do not respond to the characteristic tests for iron and chlorine respectively, which are really the tests for the iron and chlorine ions.

The general acceptance of Arrhenius' views was hastened by the advocacy of W. Ostwald, who brought to light further evidence in their favour. For instance, all permanganates in dilute solution show the same purple colour; and Ostwald considered their absorption spectra to be identical [4]; this identity is easily accounted for on Arrhenius' theory, by supposing that the spectrum in question is that of the anion which corresponds to the acid radicle. The blue colour which is observed in dilute solutions of copper salts, even when the strong solution is not blue, may in the same way be ascribed to a blue copper cation. A striking instance of the same kind is afforded by ferric sulphocyanide; here the strong solution shows a deep red colour, due to the salt itself; but on dilution the colour disappears, the ions being colourless.

If it be granted that ions can have any kind of permanent existence in a salt solution, it may be shown from thermodynamical

[1] *Zeitschrift für phys. Chem.* i (1887), p. 631. Previous investigations, in which the theory was to some extent foreshadowed, were published in *Bihang till Svenska Vet. Ak. Förh.* viii (1884), nos. 13 and 14.

[2] cf. p. 336

[3] i.e. the ohmic specific conductivity of the solution divided by the number of gramme-equivalents of salt per unit volume

[4] Examination of the spectra with higher dispersion does not altogether confirm this conclusion.

considerations that the degree of dissociation must increase as the dilution increases, and that at infinite dilution there must be complete dissociation. For the available energy of a dilute solution of volume V, containing n_1 gramme-molecules of one substance, n_2 gramme-molecules of another, and so on, is (as may be shown by an obvious extension of the reasoning already employed in connection with concentration cells) [1]

$$\sum_r n_r \phi_r (T) + RT \sum_r n_r \log (n_r/V) + \text{the available energy}$$

possessed by the solvent before the introduction of the solutes, where ϕ_r (T) depends on T and on the nature of the rth solute, but not on V, and R denotes the constant which occurs in the equation of state of perfect gases. When the system is in equilibrium, the proportions of the reacting substances will be so adjusted that the available energy has a stationary value for small virtual alterations $\delta n_1, \delta n_2, \ldots$ of the proportions; and therefore

$$0 = \Sigma \delta n_r \,.\, \phi_r (T) + RT\Sigma \delta n_r \,.\, \log (n_r/V) + RT\Sigma \delta n_r.$$

Applying this to the case of an electrolyte in which the disappearance of one molecule of salt (indicated by the suffix $_1$) gives rise to one cation (indicated by the suffix $_2$) and one anion (indicated by the suffix $_3$), we have $\delta n_1 = -\delta n_2 = -\delta n_3$; so the equation becomes

$$0 = \phi_1 (T) - \phi_2 (T) - \phi_3 (T) + RT \log (n_1 V/n_2 n_3) - RT,$$

or

$$n_1 V/n_2 n_3 = \text{a function of T only.}$$

Since in a neutral solution the number of anions is equal to the number of cations, this equation may be written

$$n_2^2 = V n_1 \times \text{a function of T only};$$

it shows that when V is very large (so that the solution is very dilute), n_2 is very large compared with n_1; that is to say, the salt tends towards a state of complete dissociation.

The ideas of Arrhenius contributed to the success of Walther Nernst [2] (1864–1941) in perfecting Helmholtz' theory of concentration cells, and representing their mechanism in a much more definite fashion than had been done heretofore.

In an electrolytic solution let the drift velocity of the cations under unit electric force be u, and that of the anions be v, so that the fraction $u/(u + v)$ of the current is transported by the cations,

[1] cf. pp. 342–3

[2] *Zeitschr. für phys. Chem.* ii (1888), p. 613; iv (1889), p. 129; *Berlin Sitzungsberichte* (1889), p. 83; *Ann. d. Phys.* xlv (1892), p. 360. cf. also Max Planck, *Ann. d. Phys.* xxxix (1890), p. 161; xl (1890), p. 561

and the fraction $v/(u+v)$ by the anions. If the concentration of the solution be c_1 at one electrode, and c_2 at the other, it follows from the formula previously found for the available energy that one gramme-ion of cations, in moving from one electrode to the other, is capable of yielding up an amount[1] RT log (c_2/c_1) of energy; while one gramme-ion of anions going in the opposite direction must absorb the same amount of energy. The total quantity of work furnished when one gramme-molecule of salt is transferred from concentration c_2 to concentration c_1 is therefore

$$\frac{u-v}{u+v}\,\mathrm{RT}\log\frac{c_2}{c_1}.$$

The quantity of electric charge which passes in the circuit when one gramme-molecule of the salt is transferred is proportional to the valency ν of the ions, and the work furnished is proportional to the product of this charge and the electromotive force E of the cell; so that in suitable units we have

$$\mathrm{E}=\frac{\mathrm{RT}}{\nu}\,\frac{u-v}{u+v}\log\frac{c_2}{c_1}.$$

A typical concentration cell to which this formula may be applied may be constituted in the following way: let a quantity of zinc amalgam, in which the concentration of zinc is c_1, be in contact with a dilute solution of zinc sulphate, and let this in turn be in contact with a quantity of zinc amalgam of concentration c_2. When the two masses of amalgam are connected by a conducting wire outside the cell, an electric current flows in the wire from the weak to the strong amalgam,[2] while zinc cations pass through the solution from the strong amalgam to the weak. The electromotive force of such a cell, in which the current may be supposed to be carried solely by cations, is

$$\frac{\mathrm{RT}}{\nu}\log\frac{c_2}{c_1}$$

Not content with the derivation of the electromotive force from considerations of energy, Nernst proceeded to supply a definite mechanical conception of the process of conduction in electrolytes. The ions are impelled by the electric force associated with the gradient of potential in the electrolyte. But this is not the only

[1] The correct law of dependence of the available energy on the temperature was by this time known.

[2] It will hardly be necessary to remark that this supposed direction of the current is purely conventional.

force which acts on them ; for, since their available energy decreases as the concentration decreases, there must be a force assisting every process by which the concentration is decreased. The matter may be illustrated by the analogy of a gas compressed in a cylinder fitted with a piston ; the available energy of the gas decreases as its degree of compression decreases ; and therefore that movement of the piston which tends to decrease the compression is assisted by a force—the 'pressure' of the gas on the piston. Similarly, if a solution were contained within a cylinder fitted with a piston which is permeable to the pure solvent but not to the solute, and if the whole were immersed in pure solvent, the available energy of the system would be decreased if the piston were to move outwards so as to admit more solvent into the solution ; and therefore this movement of the piston would be assisted by a force—the 'osmotic pressure of the solution' as it is called.[1]

Consider, then, the case of a single electrolyte supposed to be perfectly dissociated ; its state will be supposed to be the same at all points of any plane at right angles to the axis of x. Let ν denote the valency of the ions, and V the electric potential at any point. Since [2] the available energy of a given quantity of a substance in very dilute solution depends on the concentration in exactly the same way as the available energy of a given quantity of a perfect gas depends on its density, it follows that the osmotic pressure p for each ion is determined in terms of the concentration and temperature by the equation of state of perfect gases

$$\mathrm{M}p = \mathrm{RT}c,$$

where M denotes the molecular weight of the salt, and c the mass of salt per unit volume.

Consider the cations contained in a parallelepiped at the place x, whose cross-section is of unit area and whose length is dx. The mechanical force acting on them due to the electric field is $-(\nu c/\mathrm{M})\, d\mathrm{V}/dx \,.\, dx$, and the mechanical force on them due to the osmotic pressure is $-dp/dx \,.\, dx$. If u denote the velocity of drift of the cations in a field of unit electric force, the total amount of charge which would be transferred by cations across unit area in unit time under the influence of the electric forces alone would be $-(u\nu c/\mathrm{M})\, d\mathrm{V}/dx$; so, under the influence of both forces, it is

$$-\frac{u\nu c}{\mathrm{M}}\left(\frac{d\mathrm{V}}{dx} + \frac{\mathrm{RT}}{c\nu}\frac{dc}{dx}\right).$$

[1] cf. van't Hoff, *Svenska Vet.-Ak. Handlingar* xxi (1886), no. 17 ; *Zeitschrift für Phys. Chem.* i (1887), p. 481

[2] As follows from the expression obtained, *supra*, p. 343

Similarly, if v denote the velocity of drift of the anions in a unit electric field, the charge transferred across unit area in unit time by the anions is

$$\frac{v\nu c}{\mathrm{M}}\left(-\frac{d\mathrm{V}}{dx}+\frac{\mathrm{RT}}{c\nu}\frac{dc}{dx}\right).$$

We have therefore, if the total current be denoted by i,

$$i=-(u+v)\frac{\nu c}{\mathrm{M}}\frac{d\mathrm{V}}{dx}-(u-v)\frac{\mathrm{RT}}{\mathrm{M}}\frac{dc}{dx},$$

or

$$-\frac{d\mathrm{V}}{dx}dx=\frac{\mathrm{M}dx}{(u+v)\nu c}i+\frac{u-v}{u+v}\frac{\mathrm{RT}}{\nu c}\frac{dc}{dx}dx.$$

The first term on the right evidently represents the product of the current into the ohmic resistance of the parallelepiped dx, while the second term represents the internal electromotive force of the parallelepiped. It follows that if r denote the specific resistance, we must have

$$u+v=\mathrm{M}/r\nu c,$$

in agreement with Kohlrausch's equation [1]; while by integrating the expression for the internal electromotive force of the parallelepiped dx, we obtain for the electromotive force of a cell whose activity depends on the transference of electrolyte between the concentrations c_1 and c_2, the value

$$\frac{u-v}{u+v}\frac{\mathrm{RT}}{\nu}\int\frac{1}{c}\frac{dc}{dx}dx,$$

or

$$\frac{u-v}{u+v}\frac{\mathrm{RT}}{\nu}\log\frac{c_2}{c_1},$$

in agreement with the result already obtained.

It may be remarked that although the current arising from a concentration cell which is kept at a constant temperature is capable of performing work, yet this work is provided, not by any diminution in the total internal energy of the cell, but by the abstraction of thermal energy from neighbouring bodies. This indeed (as may be seen by reference to W. Thomson's general equation of available energy) [2] must be the case with any system whose available energy is exactly proportional to the absolute temperature.

The advances which were effected in the last quarter of the nineteenth century in regard to the conduction of electricity through liquids, considerable though these advances were, may be regarded as the natural development of a theory which had long been before the world. It was otherwise with the kindred problem of the con-

[1] cf. p. 337 [2] cf. p. 217

duction of electricity through gases; for although many generations of philosophers had studied the remarkable effects which are presented by the passage of a current through a rarefied gas, it was not until recent times that a satisfactory theory of the phenomena was discovered.

Some of the electricians of the earlier part of the eighteenth century performed experiments in vacuous spaces; in particular, Hauksbee [1] in 1705 observed a luminosity when glass is rubbed in rarefied air. But the first investigators of the continuous discharge through a rarefied gas seem to have been Gottfried Heinrich Grummert (1719–76?) in 1744,[2] and Watson,[3] who, by means of an electrical machine, sent a current through an exhausted glass tube three feet long and three inches in diameter. 'It was,' he wrote, 'a most delightful spectacle, when the room was darkened, to see the electricity in its passage: to be able to observe not, as in the open air, its brushes or pencils of rays an inch or two in length, but here the coruscations were of the whole length of the tube between the plates, that is to say, thirty-two inches.' Its appearance he described as being on different occasions 'of a bright silver hue; resembling very much the most lively coruscations of the *aurora borealis*'; and 'forming a continued arch of lambent flame.' His theoretical explanation was that the electricity 'is seen, without any preternatural force, pushing itself on through the vacuum by its own elasticity, in order to maintain the equilibrium in the machine'—a conception which follows naturally from the combination of Watson's one-fluid theory with the prevalent doctrine of electrical atmospheres.[4]

A different explanation was put forward by Nollet, who performed electrical experiments in rarefied air at about the same time as Watson,[5] and saw in them a striking confirmation of his own hypothesis of efflux and afflux of electric matter.[6] According to Nollet, the particles of the effluent stream collide with those of the affluent stream which is moving in the opposite direction; and being thus violently shaken, are excited to the point of emitting light.

Almost a century elapsed before anything more was discovered regarding the discharge in vacuous spaces. But in 1838 Faraday,[7] while passing a current from the electrical machine between two brass rods in rarefied air, noticed that the purple haze or stream of

[1] *Phil. Trans.* xxiv (1705), p. 2165; Fra. Hauksbee, *Physico-Mechanical Experiments*, London, 1709

[2] cf. J. H. Winkler, *die Stärke d. elektr. Kraft d. Wassers in gläsernen Gefässen*, Lips., 1746

[3] *Phil. Trans.* xlv (1748), p. 93; xlvii (1752), p. 362

[4] cf. ch. ii

[5] Nollet, *Recherches sur l'Electricité*, 1749, troisième discours

[6] cf. p. 44

[7] *Phil. Trans.* (1838), p. 125; *Exp. Res.* i, § 1526

light which proceeded from the positive pole stopped short before it arrived at the negative rod. The negative rod, which was itself covered with a continuous glow, was thus separated from the purple column by a narrow dark space : to this, in honour of its discoverer, the name *Faraday's dark space* has generally been given by subsequent writers.

That vitreous and resinous electricity give rise to different types of discharge had long been known ; and indeed, as we have seen,[1] it was the study of these differences that led Franklin to identify the electricity of glass with the superfluity of fluid, and the electricity of amber with the deficiency of it. But phenomena of this class are in general much more complex than might be supposed from the appearance which they present at a first examination ; and the value of Faraday's discovery of the negative glow and dark space lay chiefly in the simple and definite character of these features of the discharge, which indicated them as promising subjects for further research. Faraday himself felt the importance of investigations in this direction. 'The results connected with the different conditions of positive and negative discharge,' he wrote,[2] 'will have a far greater influence on the philosophy of electrical science than we at present imagine.'

Twenty more years, however, passed before another notable advance was made. That a subject so full of promise should progress so slowly may appear strange ; but one reason at any rate is to be found in the incapacity of the air-pumps then in use to rarefy gases to the degree required for effective study of the negative glow. The invention of Geissler's mercurial air-pump in 1855 did much to remove this difficulty ; and it was in Geissler's exhausted tubes that Julius Plücker,[3] of Bonn, studied the discharge three years later.

It had been shown by Sir Humphry Davy in 1821 [4] that one form of electric discharge—namely, the arc between carbon poles—is deflected when a magnet is brought near to it. Plücker now performed a similar experiment with the vacuum discharge, and observed a similar deflection. But the most interesting of his results were obtained by examining the behaviour of the negative glow in the magnetic field ; when the negative electrode was reduced to a single point, the whole of the negative light became concentrated along the line of magnetic force passing through this point. In other

[1] cf. p. 47 [2] *Exp. Res.* i, § 1523

[3] *Ann. d. Phys.* ciii (1858), pp. 88, 151 ; civ (1858), pp. 113, 622 ; cv (1858), p. 67 ; cvii (1859), p. 77 ; *Phil. Mag.* xvi (1858), pp. 119, 408 ; xviii (1859), pp. 1, 7

[4] *Phil. Trans.* cxi (1821), p. 425

words, the negative glow disposed itself as if it were constituted of flexible chains of iron filings attached at one end to the cathode.

Plücker noticed that when the cathode was of platinum, small particles were torn off it and deposited on the walls of the glass bulb. ' It is most natural,' he wrote, ' to imagine that the magnetic light is formed by the incandescence of these platinum particles as they are torn from the negative electrode.' He likewise observed that during the discharge the walls of the tube, near the cathode, glowed with a phosphorescent light, and remarked that the position of this light was altered when the magnetic field was changed. This led to another discovery ; for in 1869 Plücker's pupil, W. Hittorf,[1] having placed a solid body between a point-cathode and the phosphorescent light, was surprised to find that a shadow was cast. He rightly inferred from this that the negative glow is formed of rays (glow rays, *Glimmstrahlen,* as he called them) which proceed from the cathode in straight lines, and which cause the phosphorescence when they strike the walls of the tube.

Hittorf's observation was amplified in 1876 by Eugen Goldstein[2] (1850–1930), who found that distinct shadows were cast, not only when the cathode was a single point, but also when it formed an extended surface, provided the shadow-throwing object was placed close to it. This clearly showed that the *cathode rays* (a term now for the first time introduced) are not emitted indiscriminately in all directions, but that each portion of the cathode surface emits rays which are practically confined to a single direction ; and Goldstein found this direction to be normal to the surface. In this respect his discovery established an important distinction between the manner in which cathode rays are emitted from an electrode and that in which light is emitted from an incandescent surface.

The question as to the nature of the cathode rays attracted much attention during the next two decades. In the year following Hittorf's investigation, Cromwell Varley[3] put forward the hypothesis that the rays are composed of ' attenuated particles of matter, projected from the negative pole by electricity ' ; and that it is in virtue of their negative charges that these particles are influenced by a magnetic field.[4]

During some years following this, the properties of highly rarefied

[1] *Ann. d. Phys.* cxxxvi (1869), pp. 1, 197 ; translated, *Annales de Chimie,* xvi (1869), p. 487

[2] *Berlin Monatsberichte* (1876), p. 279

[3] *Proc. R. S.* xix (1871), p. 236

[4] Priestley in 1766 had shown that a current of electrified air flows from the points of bodies which are electrified either vitreously or resinously ; cf. Priestley's *History of Electricity,* p. 591.

gases were investigated by Sir William Crookes (1832–1919). Influenced, doubtless, by the ideas which were developed in connection with his discovery of the radiometer, Crookes,[1] like Varley, proposed to regard the cathode rays as a molecular torrent: he supposed the molecules of the residual gas, coming into contact with the cathode, to acquire from it a resinous charge, and immediately to fly off normally to the surface, by reason of the mutual repulsion exerted by similarly electrified bodies. Carrying the exhaustion to a higher degree, Crookes was enabled to study a dark space which under such circumstances appears between the cathode and the cathode glow; and to show that at the highest rarefactions this dark space (which has since been generally known by his name, and which must not be confused with the dark space discovered by Faraday) enlarges until the whole tube is occupied by it. He suggested that the thickness of the dark space may be a measure of the mean length of free path of the molecules. 'The extra velocity,' he wrote, 'with which the molecules rebound from the excited negative pole keeps back the more slowly moving molecules which are advancing towards that pole. The conflict occurs at the boundary of the dark space, where the luminous margin bears witness to the energy of the collisions.'[2] Thus according to Crookes the dark space is dark and the glow bright because there are collisions in the latter and not in the former. The fluorescence or phosphorescence on the walls of the tube he attributed to the impact of the particles on the glass.

Crookes spoke of the cathode rays as an 'ultra-gaseous' or 'fourth state' of matter. These expressions have led some later writers to ascribe to him the enunciation or prediction of a hypothesis regarding the nature of the particles projected from the cathode, which arose some years afterwards, and which we shall presently describe; but it is clear from Crookes's memoirs that he conceived the particles of the cathode rays to be ordinary gaseous molecules, carrying electric charges; and by 'a new state of matter' he understood simply a state in which the free path is so long that collisions may be disregarded.

Crookes found that two adjacent pencils of cathode rays appeared to repel each other. At the time this was regarded as a direct confirmation of the hypothesis that the rays are streams of electrically charged particles; but it was shown later that the deflection of the rays must be assigned to causes other than mutual repulsion.

[1] *Phil. Trans.* clxx (1879), pp. 135, 641; *Phil. Mag.* vii (1879), p. 57
[2] *Phil. Mag.* vii (1879), p. 57

How admirably the molecular-torrent theory accounts for the deviation of the cathode rays by a magnetic field was shown by the calculations of Eduard Riecke (1845–1915) in 1881.[1] If the axis of z be taken parallel to the magnetic force H, the equations of motion of a particle of mass m, charge e, and velocity (u, v, w) are

$$cm\,du/dt = ev\mathrm{H}, \quad cm\,dv/dt = -eu\mathrm{H}, \quad cm\,dw/dt = 0.$$

The last equation shows that the component of velocity of the particle parallel to the magnetic force is constant; the other equations give

$$u = \mathrm{A} \sin (e\mathrm{H}t/cm), \qquad v = \mathrm{A} \cos (e\mathrm{H}t/cm),$$

showing that the projection of the path on a plane at right angles to the magnetic force is a circle. Thus, in a magnetic field the particles of the molecular torrent describe spiral paths whose axes are the lines of magnetic force.

But the hypothesis of Varley and Crookes was before long involved in difficulties. Tait[2] in 1880 remarked that if the particles are moving with great velocities, the periods of the luminous vibrations received from them should be affected to a measurable extent in accordance with Doppler's principle (for which cf. p. 368). Tait tried to obtain this effect, but without success. It could, however, be argued that if, as Crookes supposed, the particles become luminous only when they have collided with other particles, and have thereby lost part of their velocity, the phenomenon in question is not to be expected.

The alternative to the molecular-torrent theory was to suppose that the cathode radiation is a disturbance of the aether. This view was maintained by several physicists,[3] and notably by Hertz,[4] who rejected Varley's hypothesis when he found experimentally that the rays did not appear to produce any external electric or magnetic force, and were apparently not affected by an electrostatic field. It was, however, pointed out by FitzGerald[5] (in a review of Hertz's *Miscellaneous Papers*) that external space is probably screened from the effects of the rays by other electric actions which take place in the discharge tube.

FitzGerald has left on record that at the British Association meeting in 1896, Lenard and Bjerknes argued in favour of the idea

[1] *Gött. Nach.*, 2 Feb. 1881; reprinted, *Ann. d. Phys.* xiii (1881), p. 191
[2] *Proc. R. S. Edin.* x (1880), p. 430
[3] e.g. E. Wiedemann, *Ann. d. Phys.* x (1880), p. 202; translated, *Phil. Mag.* x (1880), p. 357. E. Goldstein, *Ann. d. Phys.* xii (1881), p. 249 [4] *Ann. d. Phys.* xix (1883), p. 782
[5] *Nature*, lv (5 Nov. 1896), p. 6; FitzGerald's *Scientific Writings*, p. 433

that cathode rays are an aether propagation of some kind independent of matter, while all the English physicists supported the projectile hypothesis. The great difficulty in the former theory was that the rays are deflected by a magnet, and in the latter that they are capable of passing through films of metal which are so thick as to be quite opaque to ordinary light; [1] it seemed inconceivable that particles of matter should not be stopped by even the thinnest gold leaf. At the time of Hertz's experiments on the subject, an attempt to obviate this difficulty was made by J. J. Thomson,[2] who suggested that the metallic film when bombarded by the rays might itself acquire the property of emitting charged particles, so that the rays which were observed on the further side need not have passed through the film. It was Thomson who ultimately found the true explanation; but this depended in part on another order of ideas, whose introduction and development must now be traced.

The tendency, which was now general, to abandon the electron theory of Weber in favour of Maxwell's theory involved certain changes in the conceptions of electric charge. In the theory of Weber, electric phenomena were attributed to the agency of stationary or moving charges, which could most readily be pictured as having a discrete and atom-like existence. The conception of displacement, on the other hand, which lay at the root of the Maxwellian theory, was more in harmony with the representation of electricity as something of a continuous nature; and as Maxwell's views met with increasing acceptance, the atomistic hypothesis seemed to have entered on a period of decay. Its revival was due largely to the advocacy of Helmholtz,[3] who, in a lecture delivered to the Chemical Society of London in 1881, pointed out [4] that it was thoroughly in accord with the ideas of Faraday,[5] on which Maxwell's theory was founded. 'If,' he said, 'we accept the hypothesis that the elementary substances are composed of atoms, we cannot avoid concluding that electricity also, positive as well as negative, is divided into definite elementary portions which behave like atoms of electricity.'

When the conduction of electricity is considered in the light of this hypothesis, it seems almost inevitable to conclude that the

[1] The penetrating power of the rays had been noticed by Hittorf, and by E. Wiedemann and H. Ebert (*Sitzber. d. phys.-med. Soc. zu Erlangen*, 11 Dec. 1891). It was investigated more thoroughly by Hertz (*Ann. d. Phys.* xlv [1892], p. 28) and by his pupil Philipp Lenard of Bonn (*Ann. d. Phys.* li [1894], p. 225; lii [1894], p. 23) who conducted a series of experiments on cathode rays which had passed out of the discharge tube through a thin window of aluminium.

[2] J. J. Thomson, *Recent Researches*, p. 126

[3] cf. also G. Johnstone Stoney, *Phil. Mag.* xi (May 1881), p. 381

[4] *Jour. Chem. Soc.* xxxix (1881), p. 277

[5] cf. p. 180

process is of much the same character in gases as in electrolytes; and before long this view was actively maintained. It had indeed long been known that a compound gas might be decomposed by the electric discharge; and that in some cases the constituents are liberated at the electrodes in such a way as to suggest an analogy with electrolysis. The question had been studied in 1861 by Adolphe Perrot, who examined [1] the gases liberated by the passage of the electric spark through steam. He found that while the product of this action was a detonating mixture of hydrogen and oxygen, there was a decided preponderance of hydrogen at one pole and of oxygen at the other.

The analogy of gaseous conduction to electrolysis was applied by W. Giese,[2] of Berlin, in 1882, in order to explain the conductivity of the hot gases of flames, which had been discovered in the course of the experiments of the Academia del Cimento of Florence in 1667. 'It is assumed,' he wrote, 'that in electrolytes, even before the application of an external electromotive force, there are present atoms or atomic groups—the ions, as they are called—which originate when the molecules dissociate; by these the passage of electricity through the liquid is effected, for they are set in motion by the electric field and carry their charges with them. We shall now extend this hypothesis by assuming that in gases also the property of conductivity is due to the presence of ions. Such ions may be supposed to exist in small numbers in all gases at the ordinary temperature and pressure; and as the temperature rises their numbers will increase.'

Ideas similar to this were presented in a general theory of the discharge in rarefied gases, which was devised two years later by Arthur Schuster (1851–1934), of Manchester.[3] Schuster remarked that when hot liquids are maintained at a high potential, the vapours which rise from them are found to be entirely free from electrification; from which he inferred that a molecule striking an electrified surface in its rapid motion cannot carry away any part of the charge, and that one molecule cannot communicate electricity to another in an encounter in which both molecules remain intact. Thus he was led to the conclusion that dissociation, or *ionisation*, as it came to be called later, of the gaseous molecules is necessary for the passage of electricity through gases.[4]

[1] *Annales de Chimie* (3), lxi, p. 161

[2] *Ann. d. Phys.* xvii (1882), pp. 1, 236, 519

[3] *Proc. R. S.* xxxvii (1884), p. 317

[4] In the case of an elementary gas, this would imply dissociation of the molecule into two atoms chemically alike, but oppositely charged; in electrolysis the dissociation is into two chemically unlike ions.

Schuster advocated the charged-particle theory of cathode rays, and by extending and interpreting an experiment of Hittorf's was able to adduce strong evidence in its favour. He placed the positive and negative electrodes so close to each other that at very low pressures the Crookes dark space extended from the cathode to beyond the anode. In these circumstances it was found that the discharge from the positive electrode always passed to the nearest point of the inner boundary of the Crookes dark space—which, of course, was in the opposite direction to the cathode. Thus, in the neighbourhood of the positive discharge, the current was flowing in two opposite directions at closely adjoining places ; which could scarcely happen unless the current in one direction were carried by particles moving against the lines of force by virtue of their inertia.

Continuing his researches, Schuster[1] showed in 1887 that a steady electric current may be obtained in air between electrodes whose difference of potential is but small, provided that an independent current is maintained in the same vessel ; that is to say, a continuous discharge produces in the air such a condition that conduction occurs with the smallest electromotive forces. This effect he explained by aid of the hypothesis previously advanced ; the ions produced by the main discharge become diffused throughout the vessel, and, coming under the influence of the field set up by the auxiliary electrodes, drift so as to carry a current between the latter.

A discovery related to this was made in the same year by Hertz,[2] in the course of the celebrated researches[3] which have been already mentioned. Happening to notice that the passage of one spark is facilitated by the passage of another spark in its neighbourhood, he followed up the observation, and found the phenomenon to be due to the agency of ultra-violet light emitted by the latter spark. It appeared in fact that the distance across which an electric spark can pass in air is greatly increased when light of very short wave-length is allowed to fall on the spark gap. It was soon found[4] that the effective light is that which falls on the negative electrode of the gap ; and Wilhelm Hallwachs[5] extended the discovery by showing that when a sheet of metal is negatively electrified and exposed to

[1] *Proc. R. S.* xlii (1887), p. 371. Hittorf had discovered that very small electromotive forces are sufficient to cause a discharge across a space through which the cathode radiation is passing.

[2] *Berlin Ber.* (1887), p. 487 ; *Ann. d. Phys.* xxxi (1887), p. 983 ; *Electric Waves* (English edn.), p. 63

[3] cf. p. 322

[4] By E. Wiedemann and H. Ebert, *Ann. d. Phys.* xxxiii (1888), p. 241

[5] *Ann. d. Phys.* xxxiii (1888), p. 301

ultra-violet light, the adjacent air is thrown into a state which permits the charge to leak rapidly away.[1]

E. Rutherford showed [2] that when ultra-violet light falls on a metal plate, the ions produced in the air around are all negative; and he devised an ingenious arrangement for finding their velocities. The great advances in the theory of this *photo-electric effect*, as it came to be called, was however made only in the twentieth century.

Interest was now thoroughly aroused in the problem of conductivity in gases; and it was generally felt that the best hope of divining the nature of the process lay in studying the discharge at high rarefactions. 'If a first step towards understanding the relations between aether and ponderable matter is to be made,' said Lord Kelvin in 1893,[3] 'it seems to me that the most hopeful foundation for it is knowledge derived from experiments on electricity in high vacuum.'

Within the two following years considerable progress was effected in this direction. J. J. Thomson,[4] by a rotating-mirror method, succeeded in measuring the velocity of the cathode rays, finding it to be [5] $1 \cdot 9 \times 10^7$ cm./sec.; a value so much smaller than that of the velocity of light that it was scarcely possible to conceive of the rays as vibrations of the aether. A further blow was dealt at the latter hypothesis when Jean Perrin,[6] having received the rays in a metallic cylinder, found that the cylinder became charged with resinous electricity. Thomson showed that when the rays were deviated by a magnet in such a way that they could no longer enter the cylinder, it no longer acquired a charge. This appeared to demonstrate that the rays transport negative electricity.

With cathode rays is closely connected another type of radiation, which was discovered on 8 November 1895, by Wilhelm Konrad Röntgen (1845–1923) of Würzburg.[7] The discovery seems to have originated in an accident: Röntgen was experimenting with a Crookes tube, which was completely enclosed in an opaque shield of black cardboard. He noticed however that when a current was passed through the tube, a piece of paper painted with barium

[1] Elster and Geitel published a number of papers on this phenomenon in 1889–95, *Ann. d. Phys.* xxxviii (1889), pp. 40, 497; xxxix (1890), p. 332; xli (1890), pp. 161, 166; lii (1894), p. 433; lv (1895), p. 684

[2] *Proc. Camb. Phil. Soc.* ix (1898), p. 401

[3] *Proc. R. S.* liv (1893), p. 389

[4] *Phil. Mag.* xxxviii (1894), p. 358

[5] The value found by the same investigator in 1897 was much larger than this.

[6] *Comptes Rendus*, cxxi (1895), p. 1130. Hertz had previously tried an experiment of the same kind, but without success.

[7] *Sitzungsber. der Würzburger Physikal.-Medic. Gesellschaft*, 28 Dec. 1895; reprinted, *Ann. d. Phys.* lxiv (1898), pp. 1, 12; translated, *Nature*, liii (23 Jan. 1896), p. 274

platino-cyanide, which happened to be lying on the bench, showed fluorescence. Experiments suggested by this proved that radiation, capable of affecting photographic plates and of causing fluorescence in certain substances, is emitted by tubes in which the electric discharge is passing ; and that the radiation proceeds from the place where the cathode rays strike the glass walls of the tube. The *X-rays*, as they were called by their discoverer, were propagated in straight lines, and could neither be refracted by any of the substances which refract light, not deviated from their course by a magnetic field ; they were moreover able to pass with little absorptions through many substances which are opaque to ordinary and ultra-violet light—a property of which considerable use was at once made in surgery ; Schuster has recorded that for some time after the discovery his own laboratory at Manchester was crowded with medical men bringing patients who were believed to have needles in various parts of their bodies.[1]

The nature of the new radiation was the subject of much speculation. Its discoverer suggested that it might prove to represent the long-sought-for longitudinal vibrations of the aether ; while other writers advocated the rival claims of aethereal vortices, infra-red light and ' sifted ' cathode rays. The hypothesis which subsequently obtained general acceptance was first propounded by Schuster [2] in the month following the publication of Röntgen's researches. It is, that the X-rays are transverse vibrations of the aether, of exceedingly small wave-length ; ultra-ultra-violet light, in fact. A suggestion which was put forward later in the year by E. Wiechert [3] and Sir George Stokes,[4] to the effect that the rays are pulses generated in the aether when the glass of the discharge tube is bombarded by the cathode particles, is not really distinct from Schuster's hypothesis ; for ordinary white light likewise consists of pulses, as Gouy [5] had shown ; and the essential feature which distinguishes the Röntgen pulses is that the harmonic vibrations into which they can be resolved by Fourier's analysis are of very short period.

[1] The discovery of X-rays was narrowly missed by several physicists ; in the case of Hertz and Lenard, because of failure to distinguish them from cathode rays. An Oxford physicist, Frederick Smith, having found that photographic plates kept in a box near a Crookes tube were liable to be fogged, told his assistant to keep them in another place !

[2] *Nature*, liii (23 Jan. 1896), p. 268. FitzGerald independently made the same suggestion in a letter to O. J. Lodge, printed in the *Electrician*, xxxvii, p. 372.

[3] *Ann. d. Phys.* lix (1896), p. 321

[4] *Nature*, liv (3 Sept. 1896), p. 427 ; *Proc. Camb. Phil. Soc.* ix (1896), p. 215 ; *Mem. Manchester Lit. & Phil. Soc.* xli (1896–7), No. 15

[5] *Jour. de Phys.* v (1886), p. 354

The rapidity of the vibrations explains the failure of all attempts to refract the X-rays. For in the formula

$$\mu^2 = 1 + \frac{\sigma p^2}{\rho\,(p^2 - n^2)}$$

of the Maxwell-Sellmeier theory,[1] n denotes the frequency, and so is in this case extremely large; whence we have very nearly

$$\mu^2 = 1,$$

i.e. the refractive index of all substances for the X-rays is unity. In fact, the vibrations alternate too rapidly to have an effect on the sluggish systems which are concerned in refraction.

One of the most important properties of X-rays was discovered, shortly after the rays themselves had become known, by J. J. Thomson,[2] who announced that when they pass through a gas (or, as he thought at that time, even a liquid or a solid) they render it conducting. This he attributed, in accordance with the ionic theory of conduction, to 'a kind of electrolysis, the molecule being split up, or nearly split up, by the Röntgen rays'; and this led him to speak of the gas as being *ionised.*

The conductivity produced in gases by this means was at once investigated[3] more closely. It was found that a gas which had acquired conducting power by exposure to X-rays lost this quality when forced through a plug of glass-wool; whence it was inferred that the structure in virtue of which the gas conducts is of so coarse a character that it is unable to survive the passage through the fine pores of the plug. The conductivity was also found to be destroyed when an electric current was passed through the gas—a phenomenon for which a parallel may be found in electrolysis. For if the ions were removed from an electrolytic solution by the passage of a current, the solution would cease to conduct as soon as sufficient electricity had passed to remove them all; and it may be supposed that the conducting agents which are produced in a gas by exposure to X-rays are likewise abstracted from it when they are employed to transport charges.

The same idea may be applied to explain another property of gases exposed to X-rays. The strength of the current through the gas depends both on the intensity of the radiation and also on the electromotive force; but if the former factor be constant, and the electromotive force be increased, the current does not increase

[1] cf. p. 264 [2] *Nature* liii (27 Feb. 1896), p. 391
[3] J. J. Thomson and E. Rutherford, *Phil. Mag.* xlii (1896), p. 392

indefinitely, but tends to attain a certain 'saturation' value. The existence of this saturation value is evidently due to the inability of the electromotive force to do more than to remove the ions as fast as they are produced by the rays.

Meanwhile other evidence was accumulating to show that the conductivity produced in gases by X-rays is of the same nature as the conductivity of the gases from flames and from the path of a discharge, to which the theory of Giese and Schuster had already been applied. One proof of this identity was supplied by observations of the condensation of water vapour into clouds. It had been noticed long before by John Aitken [1] that gases rising from flames cause precipitation of the aqueous vapour from a saturated gas; and Robert von Helmholtz [2] had found that gases through which an electric discharge has been passed possess the same property. It was now shown by C. T. R. Wilson (*b.* 1869),[3] working in the Cavendish Laboratory at Cambridge, that the same is true of gases which have been exposed to X-rays. The explanation furnished by the ionic theory is that in all three cases the gas contains ions which act as centres of condensation for the vapour.

During the year which followed their discovery, the X-rays were so thoroughly examined that at the end of that period they were almost better understood than the cathode rays from which they derived their origin. But the obscurity in which this subject had been so long involved was now to be dispelled.

Lecturing at the Royal Institution on 30 April 1897, J. J. Thomson [4] advanced a new suggestion to reconcile the molecular-torrent hypothesis with Lenard's observations of the passage of cathode rays through material bodies. 'We see from Lenard's table,' he said, 'that a cathode ray can travel through air at atmospheric pressure a distance of about half a centimetre before the brightness of the phosphorescence falls to about half its original value. Now the mean free path of the molecule of air at this pressure is about 10^{-5} cm., and if a molecule of air were projected it would lose half its momentum in a space comparable with the mean free path. Even if we suppose that it is not the same molecule that is carried, the effect of the obliquity of the collisions would reduce the momentum to half in a short multiple of that path.

'Thus, from Lenard's experiments on the absorption of the rays outside the tube, it follows on the hypothesis that the cathode rays

[1] *Trans. R. S. Edin.* xxx (1880), p. 337
[2] *Ann. d. Phys.* xxxii (1887), p. 1
[3] *Proc. R. S.* lix (19 Mar. 1896), p. 338; *Phil. Trans.* clxxxix (1897), p. 265
[4] *Royal Inst. Proc.* xv (1897), p. 419

are charged particles moving with high velocities that the size of the carriers must be small compared with the dimensions of ordinary atoms or molecules.[1] We see on this hypothesis why the magnetic deflection is the same inside the tube whatever be the nature of the gas, for the carriers are the same whatever gas be used. The assumption of a state of matter more finely subdivided than the atom of an element is a somewhat startling one ; but a hypothesis that would involve somewhat similar consequences—viz. that the so-called elements are compounds of some primordial element—has been put forward from time to time by various chemists.'

In order to explain this last remark, it must be recalled that from the very beginnings of physics, amongst the Milesian philosophers in the sixth century B.C., the doctrine had been maintained that all things are but different forms of one single primary and universal substance. Thales conjectured that it might be water, Anaximenes that it might be air, and Anaximander that it was a boundless indefinite something—*τὸ ἄπειρον*—out of which everything is generated and to which everything returns. Modern science has entertained the same idea, in many different forms : as we have seen, Lord Kelvin suggested that all atoms might be vortex rings in an infinite liquid, which would be precisely Anaximander's *τὸ ἄπειρον* ; but when J. J. Thomson discovered that the cathode-ray particles were the same, whatever atoms they were torn out of, his mind recurred chiefly to the hypothesis put forward in 1815–16 by William Prout (1785–1850), namely, that the atomic weights of the elements are whole-number multiples of the atomic weight of hydrogen, and that, in fact, hydrogen is the universal primordial substance.[2] Thomson now conjectured that Prout had been wrong in selecting hydrogen, and that the fundamental place in nature belonged rightly to the cathode-ray particles, or 'corpuscles' as he called them, which were very much smaller than the hydrogen atom and were now proved to be contained in the atoms of all elements.

[1] In 1896 Lorentz, in commenting on Zeeman's discovery (see p. 411), had suggested that the electric atoms concerned in it had masses about $\frac{1}{1000}$ of the hydrogen atom. Thomson himself said that the idea of the particles in cathode rays being much less massive than atoms first arose in his mind when he noticed how great was the deflection of the rays by a magnet. A similar suggestion was made, and a value of the ratio of charge to mass obtained, by E. Wiechert, *Schriften d. physik.-öcon. Gesellsch. zu Königsberg* xxxviii (Jan. 1897.), p. [3]

[2] Prout's papers were published in *Annals of Philosophy*, vi (1815), p. 321, and vii (1816), p. 111. The author's name was not attached to the papers, but was revealed a few months later by the editor (*Annals of Philosophy*, vii [1816], p. 343.) Prout's own words are, 'If the views we have ventured to advance be correct, we may almost consider the *πρώτη ὕλη* of the ancients to be realised in hydrogen.' On the subsequent fortunes of Prout's hypothesis, cf. J. Kendall, *Proc. R. S. Edin.* lxiii (1950), p. 1.

This idea seems to have fascinated him at the time, as well it might.

Later researches—most of them carried out or inspired by Thomson himself—showed the necessity of modifying this opinion as to the place of the cathode-ray corpuscles in the scheme of the universe. Since they carry charges of negative electricity, while the atom as a whole is neutral in respect of charge, it follows that some part of the atom must be positively charged. At this time Thomson seems to have inclined to the idea that positive electricity was, so to speak, disembodied, existing as a kind of cloud within the atom, and not attached to any material corpuscles, so that the mass of the atom was derived entirely from the negative corpuscles of which it was composed. As we shall see, this conception was later discarded.

Thomson's Royal Institution lecture drew from FitzGerald [1] the suggestion that ' we are dealing with free electrons in these cathode rays '—a remark the point of which will become more evident when we come to consider the direction in which the Maxwellian theory was being developed at this time.

Shortly afterwards Thomson himself published an account [2] of experiments in which the only outstanding objections to the charged-particle theory were removed. The chief of these was Hertz's failure to deflect the cathode rays by an electrostatic field. Hertz had caused the rays to travel between parallel plates of metal maintained at different potentials ; but Thomson now showed that in these circumstances the rays generate ions in the rarefied gas, which settle on the plates, and annul the electric force in the intervening space. By carrying the exhaustion to a much higher degree, he removed this source of confusion, and obtained the expected deflection of the rays.

The electrostatic and magnetic deflections taken together suffice to determine the ratio of the mass of a cathode particle to the charge which it carries. For the equation of motion of the particle is

$$m\frac{d^2\mathbf{r}}{dt^2} = e\mathbf{E} + \frac{e}{c}[\mathbf{v} \cdot \mathbf{H}],$$

where $\mathbf{r}$ denotes the vector from the origin to the position of the particle ; $\mathbf{E}$ and $\mathbf{H}$ denote the electric and magnetic forces measured in e.-s. and e.-m. units respectively ; e the charge in e.-s. units, m the mass, and $\mathbf{v}$ the velocity of the particle. By observing the circumstances in which the force $e\mathbf{E}$, due to the electric field, exactly balances the force $e[\mathbf{v} \cdot \mathbf{H}]/c$, due to the magnetic field, it is possible

[1] *Electrician*, xxxix (21 May 1897), p. 103

[2] *Phil. Mag.* xliv (1897), p. 298

to determine **v**; **v** was found to be 15,000 kilometres per second; and it is readily seen from the above equation that a measurement of the deflection in the magnetic field supplies a relation between **v** and *m/e*; so both **v** and *m/e* may be determined. Thomson found the value of *m/e* to be independent of the nature of the rarefied gas; its amount was 10^{-7} (grammes/electromagnetic units of charge) which is only about the thousandth part of the value of *m/e* for the hydrogen atom in electrolysis.[1] If the charge were supposed to be of the same order of magnitude as that on an electrolytic ion, it would be necessary to conclude that the particle whose mass was thus measured is much smaller than the atom, and as we have seen, the conjecture was entertained that it is the primordial unit or corpuscle of which all atoms are ultimately composed.[2]

The nature of the resinously charged corpuscles which constitute cathode rays being thus far determined, it became of interest to inquire whether corresponding bodies existed carrying charges of vitreous electricity—a question to which at any rate a provisional answer was given by W. Wien (1864–1928),[3] of Aachen, in the same year. More than a decade previously E. Goldstein [4] had shown that when the cathode of a discharge tube is perforated, radiation of a certain type passes outward through the perforations into the part of the tube behind the cathode. To this radiation he had given the name *canal rays*. Wien now showed that the canal rays are formed of positively charged particles, obtaining a value of *m/e* immensely larger than Thomson had obtained for the cathode rays, and indeed of the same order of magnitude as the corresponding ratio in electrolysis. It was found moreover that whereas the negatively charged particles are the same, from whatever source they are derived, the positively charged particles are different from each other.

The disparity thus revealed between the corpuscles of cathode rays and the positive ions of Goldstein's rays excited great interest; it seemed to offer a prospect of explaining the curious differences between the relations of vitreous and of resinous electricity to ponderable matter. These phenomena had been studied by many previous investigators; in particular Schuster,[5] in the Bakerian lecture of 1890, had remarked that 'if the law of impact is different

[1] A later and more accurate value for *m/e* is $0{\cdot}568 \times 10^{-7}$ (grammes/electromagnetic units of charge), which is about $\frac{1}{1842}$ the corresponding ratio for the hydrogen atom.

[2] The value of *m/e* for cathode rays was determined also in the same year by W. Kaufmann, *Ann. d. Phys.* lxi, p. 544.

[3] *Verhandl. der physik. Gesells. zu Berlin*, xvi (1897), p. 165; *Ann. d. Phys.* lxv (1898), p. 440. See also P. Ewers, *Ann. d. Phys.* lxix (1899), p. 167

[4] *Berlin Sitzungsber.* (1886), p. 691

[5] *Proc. R. S.* xlvii (1890), p. 526

between the molecules of the gas and the positive and negative ions respectively, it follows that the rate of diffusion of the two sets of ions will in general be different,' and had inferred from his theory of the discharge that 'the negative ions diffuse more rapidly.' This inference was confirmed in 1898 by John Zeleny,[1] who showed that of the ions produced in air by exposure to X-rays, the positive are decidedly less mobile than the negative.

The magnitude of the electric charge on the ions of gases was not known with any degree of certainty until 1897–8, when a plan for determining it was successfully executed by J. J. Thomson and his pupil J. S. E. Townsend (*b.* 1868).[2] The principles on which this celebrated investigation was based are very ingenious. By measuring the current in a gas which is exposed to Röntgen rays and subjected to a known electromotive force, it is possible to determine the value of the product nev, where n denotes the number of ions in unit volume of the gas, e the charge on an ion, and v the mean velocity of the positive and negative ions under the electromotive force. As v had been already determined,[3] the experiment led to a determination of ne; so if n could be found, the value of e might be deduced.

The method employed by Thomson to determine n was founded on the discovery, to which we have already referred, that when X-rays pass through dust-free air, saturated with aqueous vapour, the ions act as nuclei around which the water condenses, so that a cloud is produced by such a degree of saturation as would ordinarily be incapable of producing condensation. The size of the drops was calculated from measurements of the rate at which the cloud sank; and, by comparing this estimate with the measurement of the mass of water deposited, the number of drops was determined, and hence the number n of ions. The value of e consequently deduced was found to be independent of the nature of the gas in which the ions were produced, being approximately the same in hydrogen as in air, and being in both cases the same as for the charge carried by the hydrogen ion in electrolysis.[4]

Even in the latter part of 1899 most physicists did not fully believe in Thomson's ideas. At the British Association meeting at Dover in September of that year, he read a paper *On the existence of masses smaller than the atoms* [5] to a deeply interested but not altogether

[1] Townsend, *Proc. Camb. Phil. Soc.* ix (Feb. 1897), p. 244; *Phil. Mag.* xlv (Feb. 1898), p. 125; J. J. Thomson, *Phil. Mag.* xlvi (1898), p. 120

[2] *Phil. Mag.* xlvi (1898), p. 528 [3] By E. Rutherford, *Phil. Mag.* xliv (1897), p. 422

[4] Townsend, *Phil. Trans.* cxciii (May 1899), p. 129

[5] Printed under a different title in *Phil. Mag.* (5), xlviii (1899), p. 547

convinced audience. He first referred to his 1897 determination of the ratio m/e for the corpuscles which constitute cathode rays ; this he now supplemented by measurements of m/e and e for the negative electrification discharged by ultra-violet light,[1] and also of m/e for the negative electrification produced by an incandescent carbon filament in an atmosphere of hydrogen.[2] It was found that the value of m/e in the case of the ultra-violet light,[3] and also in that of the carbon filament, is the same as for the cathode rays ; and that in the case of the ultra-violet light, e is the same in magnitude as the charge carried by the hydrogen atom in the electrolysis of solutions. He concluded that negative electrification, though it may be produced by very different means, is made up of units each having an electric charge of about 6×10^{-10} electrostatic units, equal to the positive charge carried by the hydrogen atom in electrolysis. In gases at low pressure, these units of negative electricity are always associated with carriers or corpuscles of a definite mass, which is exceedingly small, being only about $1{\cdot}4 \times 10^{-3}$ of that of the hydrogen ion. A certain number of these corpuscles are normally present in every atom, and the ionisation of a gas consists in the detachment of one corpuscle from the gas atom.

Since the publication of Thomson's papers, these general conclusions have been abundantly confirmed. It is now certain that electric charge exists in discrete units, vitreous and resinous, each of magnitude $4{\cdot}80 \times 10^{-10}$ electrostatic units or $1{\cdot}6 \times 10^{-19}$ coulombs. For the corpuscles of negative charge, the name *electrons*, which had been coined some years earlier by G. Johnstone Stoney,[4] soon became universal. Each ion, whether in an electrolytic liquid or in a gas, carries one (or an integral number) of these charges. An electrolytic ion or a positive gaseous ion also contains one or more atoms of matter ; but among the negative ions in a gas, some are not attached to atoms, but are, in fact, free electrons.

That electric currents cannot be continuous, but must consist of discrete electric charges separated by free aether, had been insisted on in earlier papers by Larmor.[5] He inferred this from the experiments of Rowland [1] (the creation of a magnetic field by a rotating electrified conductor) and of Helmholtz [2] (the creation of an electric

[1] cf. p. 356 [2] cf. p. 426

[3] This was announced in a letter to Rutherford of date 23 July 1899. Values of m/e and v for the electrons emitted by a conductor under the influence of ultra-violet light were also found by P. Lenard, *Wien Sitzungsb.* cviii, abth. 2a (Oct. 1899), p. 1649 ; and *Ann. d. Phys.* ii (1900), p. 359.

[4] *Phil. Mag.* [5] xxxviii (1894), p. 418

[5] *Phil. Trans.* (A), clxxxv (1894), p. 719 ; clxxxvi (1895), p. 695. See also *Aether and Matter*, p. 337, and *Phil. Mag.* vii (1904), p. 621

potential-difference by the rotation of a conductor in a magnetic field) and the phenomenon of unipolar induction (in which an electromotive force is induced when a magnet revolves round its axis of symmetry through its own field of force). For example, in the last-named example, he asserted that the electrons in the magnet, as they are moved across the magnetic field of the aether, are subject to forces proportional to the component magnetic intensity in the meridian plane, which produce electrical separation by drifting the positive ions towards the axis and in the direction of the length of the magnet one way, and the negative ions the opposite way.

The discovery of the electron, and the innumerable researches associated with it, established Thomson's reputation as the first of living experimental physicists; and it was in this aspect that he was generally regarded during the latter half of his life. But his contributions to theoretical or mathematical physics were scarcely less important than the fruits of his researches in the laboratory. Perhaps his greatest achievement in pure theory was the discovery (to which we have already referred [3]) of electromagnetic momentum: namely, that in an electromagnetic field there is stored in every unit volume an amount of mechanical momentum, proportional to the vector product of the electric and magnetic vectors. This principle is necessary for the construction of a quantity which will be described later under the name of the *energy tensor*; and without it, neither relativity theory nor quantum-electrodynamics could have developed.

During the whole of his tenure of the Cavendish chair, Thomson was the active head and inspirer of a great research school. The institution of the status of 'advanced student' at Cambridge in 1895 led to a great increase in the numbers of young graduates of other universities who came to work in his laboratory; with the consequence that in a few years nearly all the important chairs of physics in the British Empire were filled by his disciples.

[1] See pp. 305–6 [2] See p. 305 [3] cf. p. 317

Chapter XII

CLASSICAL RADIATION-THEORY

THE fact that a flame becomes intensely yellow when common salt is added to the burning substance has doubtless been known from the earliest times. A somewhat vague suggestion that the phenomenon might be amenable to a quantitative treatment, and so brought within the domain of science, was made in the eighteenth century by Thomas Melvill,[1] who remarked that the yellow colour appeared to be of a definite 'degree of refrangibility.' The problem of radiation was approached from another direction in 1802 by William Hyde Wollaston[2] (1766–1828), who found that the spectrum of sunlight was crossed by seven dark lines, perpendicular to its length. This discovery, however, passed without notice until it was made again in a much ampler fashion by the Bavarian, Joseph von Fraunhofer[3] (1787–1826), who presented to the Munich Academy in 1814–15 a map of the solar spectrum showing a multitude of dark lines, the chief of which he distinguished by letters of the alphabet. Later, by use of a diffraction grating, he measured[4] the wave-lengths corresponding to the strongest lines: the wave-length of the D line for instance was found to be (in modern units[5]) 5887·7 Å[6]: and the wave-length of the yellow light of many flames (which was not yet definitely known to be due to the presence of sodium) was found to be equal to this.

In 1826 W. H. Fox Talbot[7] (1800–77) examined with a prism the light from flames placed before a slit and showed that, as he said, 'a glance at the prismatic spectrum of a flame may show it to contain substances, which it would otherwise require a laborious chemical

[1] *Physical and Literary Essays*, ii (Edinburgh, 1752)

[2] *Phil. Trans.* xcii (1802), p. 365. Wollaston's lines A, B, *f*, *g*, D, E seem to correspond respectively to Fraunhofer's B, D, *b*, F, G, H.

[3] *Ann. d. Phys.* lvi (1817), p. 264

[4] *Denkschr. Akad. d. Wiss. zu München*, viii (1821–2), p. 1

[5] The unit of wave-length now generally used is the *Ångström* or *tenth-metre*, named after the Swedish physicist, Anders Jöns Ångström (1814–74): it is 10^{-10} metres, or one ten-millionth of a millimetre. The visible spectrum is from about 8,000 to 3,900 Ångströms; the ultra-violet rays which can be studied in air extend to about 1,850 Å. Other units occasionally used are the $\mu\mu$, which is 10 Å., and the μ or *micron*, which is 1,000 $\mu\mu$, or $\frac{1}{1000}$ millimetre.

[6] The D line is really double, its components having wave-lengths 5,896·2 Å. and 5,890·2 Å.

[7] *Edin. Jour. Sci.* v (1826), p. 77

analysis to detect.' In particular, the red given to a flame by a salt of strontium and the red due to a salt of lithium may be at once distinguished from each other in this way. Fox Talbot's paper may be regarded as beginning the application of spectroscopy to chemical analysis.[1]

In 1833 William Hallows Miller[2] (1801–80), continuing a study of the absorption of light by coloured substances which had been begun by Sir John Herschel[3] (1792–1871), examined the spectrum of sunlight which had been transmitted through various gases, and found dark lines in it, evidently due to absorption: whence the opinion became general that the Fraunhofer lines were due to absorption in the outer layers of the sun.[4]

The next important contribution to spectroscopy was the enunciation by Christian Doppler[5] (1803–53) of a principle now generally known by his name, namely, that if a source of light be in motion relative to an observer, the period of the waves as received by the observer will be different from the period of the waves as emitted by the source. To prove this, let τ be the original period of the waves emitted by a source travelling towards the observer with a relative velocity v, supposed small compared with the light velocity V in the intervening medium. Then in one second the source describes a distance v, and emits $1/\tau$ waves. The first of the waves will pass the terminus of this distance v at a time v/V after the beginning of the second, and the last of them will pass the terminus at the end of the second. So $1/\tau$ waves pass the terminus in $(1 - v/V)$ seconds, or the period of the waves as they pass the terminus is $\tau(1 - v/V)$.

A corresponding alteration of period is produced if light is reflected from a moving surface. Suppose that a plane wave of light, in a medium in which its velocity is c_1, is incident at an angle i on a surface parallel to the plane of xy, which is moving parallel to the axis of z with velocity v. Then it is easily shown that the increment of period caused by the reflection is $\delta\tau$, where

$$\frac{\delta\tau}{\tau} = -\frac{2v\cos i}{c_1}.$$

From 1849 onwards much attention was paid to the interdepen-

[1] cf. his later work, *Phil. Mag.* (3) iv (1834), p. 112; (3) vii (1835), p. 113; (3) ix (1836), p. 1

[2] *Phil. Mag.* (3), ii (1833), p. 381

[3] *Trans. R. S. Edin.* ix (1823), p. 445

[4] This is asserted explicitly in Herschel's *Treatise on Astronomy*, 1833

[5] *Abh. Köngl. böhm. Gesellsch.* ii (1842), p. 465

dence between the light emitted by a radiating body, and the absorption of light by the same body. Léon Foucault,[1] working with the electric arc, had found in its light a bright yellow radiation whose wave-length was exactly that of Fraunhofer's dark D line. He discovered that if a brighter radiator giving a continuous spectrum was placed behind the arc, so that the spectrum of this radiator was seen *through* the arc, then the D line appeared dark, just as in the solar spectrum. Thus, as he pointed out, the arc emits the D line, but absorbs it when it is emitted by the other radiator. Foucault did not however attribute the D line to sodium, or extend his observation to other lines, or draw any theoretical conclusions.

Some time prior to the summer of 1852,[2] at Cambridge, Stokes, in a conversation with William Thomson (Kelvin), referred to Fraunhofer's discovery of the coincidence in wave-length between the double dark line D of the solar spectrum and the double bright line which was seen in the spectra of many flames, and to a very rigorous experimental test of this coincidence which had been made by Professor W. A. Miller,[3] which showed it to be accurate to an astounding degree of minuteness : and to the fact that the strong yellow light given out when salt is thrown on burning spirit consists almost solely of light corresponding to this double bright line. In a letter to Thomson of date 24 Feb. 1854[4] Stokes mentioned observations of his own, that the bright line D was absent from a candle-flame when the wick was snuffed clean, so as not to project into the luminous envelope, and from an alcohol flame, when the spirit was burned in a watch glass ; which indicated that the substance responsible for the D line was contained not in the alcohol, but in the wick. In 1855 Foucault went to London to receive the Copley Medal of the Royal Society, and described his experiments with the arc to Stokes, who thereupon communicated them to Thomson. 'In conversation with Thomson,' said Stokes,[5] 'I explained the connection of the bright and dark lines by the analogy of a set of piano strings tuned to the same note, which if struck would give out that note, and would also be ready to sound out, to take it up in fact, if it were sounded in air. This would imply absorption of

[1] Foucault's results were communicated to the Société Philomathique on 7 Jan. 1849, and printed in *L'Institut* of 7 Feb. 1849, but they became known generally only when they were reprinted in *Annales de Chimie et de Physique* [3] lviii (1860), p. 476.

[2] The date is fixed by the fact that Thomson never was in Cambridge from the summer of 1852 until 1866 ; and the conversation was certainly in Cambridge : Stokes's *Math. and Phys. Papers*, iv, p. 374.

[3] This work does not appear to have been published by Miller, but see *Phil. Mag.* xxvii (1845), p 81

[4] Printed in Stokes's *Math. and Phys. Papers*, iv, p. 368

[5] Stokes's *Memoir and Sci. Correspondence*, ii (Cambridge, 1907), p. 75

the aerial vibrations. I told Thomson,' he added,[1] 'I believed there was vapour of sodium in the sun's atmosphere.'

Thomson seized eagerly on these two ideas—the dynamical explanation of the dark lines and the application to solar chemistry—and introduced them into his professorial lectures in the University of Glasgow.[2] In a letter to Stokes of 2 March 1854,[3] he suggested that other vapours than sodium might be found in the atmospheres of sun and stars by searching for substances producing, in the spectra of flames, bright lines coinciding with dark lines of the solar and stellar spectra other than the D line.[4]

In 1853 Ångström presented to the Swedish Academy a memoir [5] in which he asserted that an incandescent gas emits luminous rays of the same refrangibility as those which it absorbs at the ordinary temperature—which, though not quite accurate, indicates the extent to which the mutual relations of radiation and absorption were now being studied.

In 1858 the question was approached by a Scottish physicist, Balfour Stewart (1828–87), from a wholly different standpoint, which must now be explained.

In the present chapter, we have spoken as if rays of monochromatic light existed. This, however, is only an approximation to the truth : a ray of light may be nearly monochromatic, but, strictly speaking, it must always include constituents whose wave-lengths differ from each other, however narrow may be the range of the spectrum within which they are comprised. In this respect, optics differs from acoustics, since there is nothing impossible in the notion of a sound of a single definite frequency. When we are considering the radiation emitted by, say, a red-hot poker, we must recognise that it is composed of a continuous aggregate of constituents, whose wave-lengths extend over a wide range of the spectrum.

At the end of the eighteenth and beginning of the nineteenth century, Pierre Prevost [6] (1751–1839), of Geneva, remarked that a

[1] ibid., p. 83

[2] This must have been soon after Stokes's first conversation with him, for he says, 'I had begun teaching it to my class in the session 1852–3, as an old student's notebook which I have shows' (Stokes, *Math. and Phys. Papers*, iv, p. 374).

[3] Printed in Stokes's *Math. and Phys. Papers*, iv, p. 369

[4] cf. also Thomson's presidential address to the British Association in 1871 (*B. A. Rep. 1871*, p. lxxxiv) ; and Rayleigh's obituary of Stokes (*Royal Society Year Book, 1904* ; Rayleigh's *Scientific Papers*, v, p. 173)

[5] *Stockh. Akad. Handl. 1852–3*, p. 229 ; *Phil. Mag.* (4) ix (1855), p. 327 ; *Ann. d. Phys.* xciv (1855), p. 141

[6] *Recherches physico-mécaniques sur la chaleur*, Geneva, 1792 ; *Essai sur le calorique rayonnant*, Geneva, 1809

red-hot body, by emitting radiation, tends to become cooler: it is, however, possible to imagine that a number of such bodies, so placed that they intercept the whole of each other's radiations, exist in a steady state, in which they all remain permanently at some definite temperature. In such a system, each body radiates just as if the other bodies were not present; but it receives radiation from them, and, as Prevost pointed out, it must receive exactly as much heat as it radiates. This became known as Prevost's *Law of Exchanges*.

Prevost's line of thought was now followed up by Balfour Stewart,[1] who observed that a plate of rock salt is much less diathermanous for rays of radiant heat emitted by a mass of the same substance heated to 100° C., than for rays emitted by any other body at the same temperature; considerations based on this led him to the conclusion that *the radiating power of every kind of substance is equal to its absorbing power, for every kind of ray of radiant heat.* The extension to rays of light was obvious. Stokes, who had not thought of extending Prevost's law of exchanges from radiation as a whole to radiation of each particular refrangibility by itself, had not realised that a soda flame which emits the bright D line must *on that very account* absorb light of the same refrangibility.[2]

A year and a half after Balfour Stewart's paper, Kirchhoff[3] of Heidelberg, by an independent investigation, was led to the same discovery, and showed how it could be applied to the chemistry of the sun and stars.[4] His treatment was more complete than that of any of his predecessors, and made a much greater impression at the time on the scientific world.

We can define the *coefficient of absorption* A of a body for radiation of a given wave-length propagated in a given direction by sending a beam of rays of approximately this wave-length in this direction on to the body: then A is defined to be the ratio of the radiation absorbed by the body (measured as energy) to the incident radiation. We define a perfect absorber, or *black body*, as a body whose coefficient of absorption is unity for all types of radiation.

[1] *Trans. R. S. Edin.* xxii (Mar. 1858), p. 1; cf. Rayleigh, *Phil. Mag.* i (1901), p. 98

[2] Stokes, *Mem. and Sci. Correspondence*, ii, p. 84

[3] *Berlin Monatsber.* (Oct. 1859), p. 662; *Phil. Mag.* xix (1860), p. 193; *Ann. d. Phys.* cix (1860), p. 148

Berlin Monatsber. (Dec. 1859), p. 783; *Ann. d. Phys.* cix (1860), p. 275

Phil. Mag. xxi (1861), p. 240; *Chem. News*, iii (1861), p. 115

Berlin Abhandl. (1861 [Phys.]), p. 63; (1862 [Phys.]), p. 227; *Ann. d. Phys.* cxviii (1863), p. 94; cf. W. Ritchie, *Ann. d. Phys.* xxviii (1933), p. 378

A proof of the Stewart-Kirchhoff theorem simpler than Kirchhoff's was given by E. Pringsheim, *Verh. deutsch. Phys. Ges.* iii (1901), p. 81

[4] To him belongs the credit of having been the first to seek for, and find, metals other than sodium in the sun by spectroscopic methods.

Consider a body C inside the cavity formed by a hollow perfectly absorbing shell B, the whole being at the same temperature. Since heat-energy is lost as fast as it is gained by every part of the system in this equilibrium state, the total energy radiated by C must be equal to that part of the radiant energy proceeding from B which is absorbed by C. Now let B be moved about relative to C, or even replaced by another shell at the same temperature. The amount of energy radiated by C is unaffected, and therefore the part of the radiant energy proceeding from B which is absorbed by C remains constant when B varies. But the absorbing coefficient may vary in any manner from point to point of C and from one wave-length to another, so this constancy requires that the energy of radiation of any particular wave-length incident on a surface-element exposed anywhere within a hollow black shell must be independent of the nature and position of the shell, and must depend only on its temperature.

Now consider the interchange of radiation, of wave-lengths in the range from λ to $\lambda + d\lambda$, between surface-elements $d\gamma$ on C and $d\beta$ on B. The energy passing from $d\beta$ to $d\gamma$ is e_0, where this is the energy of radiation of this type which would be exchanged between $d\gamma$ and $d\beta$ if both C and B were black. The energy passing from $d\gamma$ to $d\beta$ consists of two parts, namely, first, the radiation of this type which is actually emitted by $d\gamma$ (which is not now supposed to be necessarily black) and incident on $d\beta$, and which we shall denote by e, and secondly, the radiation originally proceeding from all parts of B which is reflected or scattered by $d\gamma$ in such a way as to fall on $d\beta$. We shall denote this by f, so

$$e_0 = e + f.$$

But a system is possible which differs from this system only in that every ray takes the same path but in the opposite direction: if we consider this system, we see that f is equal to the part of the radiation emitted by $d\beta$ which is reflected or scattered at $d\gamma$ so as to fall somewhere on B: that is,

$$f = e_0 - e_0 a.$$

Hence

$$e_0 = e + e_0 - e_0 a$$

or

$$\frac{e}{a} = e_0.$$

This is the law of Balfour Stewart and Kirchhoff, that *for a given type of radiation* (specified by its wave-length, direction of propagation and state of polarisation) *the emissive power of any body at a given temperature, divided by its coefficient of absorption, is the same for all bodies, and is equal to the emissive power of a black body at that temperature.*[1]

We can now proceed to a definition of the intrinsic brightness, or specific intensity of radiation of a surface. Consider the energy radiated within the range of wave-lengths λ to $\lambda + d\lambda$, in the interval dt of time, and contained within a narrow cone of solid angle $d\omega$, by an area dS of surface, when the direction of the axis of the cone makes an angle i with the normal to the surface. It will evidently be proportional to $d\lambda$, dt, dS, $\cos i$, and $d\omega$, and we may therefore write it

$$H\, dt\, dS \cos i\, d\omega\, d\lambda$$

where H will be called the *intrinsic brightness* of the surface with respect to radiations in the range λ to $\lambda + d\lambda$. A simple integration shows that the total energy radiated per second by unit area in this range of wave-lengths is πH.

If we imagine a closed cavity, with walls at a uniform temperature T, and occupied by a medium of refractive index unity, then the cavity will be filled with radiation : the energy per c.c. represented by radiations in the range λ to $\lambda + d\lambda$ can be represented by

$$F(\lambda, T)\, d\lambda$$

where $F(\lambda, T)$ is independent of the nature of the substance composing the walls. We can now connect $F(\lambda, T)$ with the intrinsic brightness of the walls, supposed black. Within the range λ to $\lambda + d\lambda$, the element dS of surface emits energy $H \cos i\, dS\, d\omega\, d\lambda$ per second into a cone of solid angle $d\omega$ whose direction makes an angle i with the to dS. This travels with velocity c, so the energy due to dS per normal unit length of the cone is $(1/c)\, H \cos i\, dS\, d\omega\, d\lambda$. Thus this energy per c.c. at distance r along the cone is

$$\frac{1}{cr^2} H \cos i\, dS\, d\lambda.$$

The energy per c.c. at this point due to the entire radiating surface is therefore

$$\frac{1}{c} H\, d\lambda \iint \frac{\cos i}{r^2}\, dS \quad \text{or} \quad \frac{1}{c} H\, d\lambda \iint d\Omega$$

(where $d\Omega$ is the element of solid angle subtended at the point)

or

$$\frac{4\pi}{c} H\, d\lambda.$$

Thus we have

$$F(\lambda, T) = \frac{4\pi}{c} H.$$

[1] For more rigorous proofs of the law, see G. C. Evans (*Proc. Amer. Acad.* xlvi [1910], p. 97), and D. Hilbert (*Phys. Zs.* xiii [1912], p. 1056 ; xiv [1913], p. 592).

Kirchhoff showed that the state of radiation, thus described as existing within a hollow chamber whose walls are at temperature T, can be excited if the walls are perfectly reflecting (so that they do not radiate on their own account), provided that a body (which may be of any kind, so long as it has some emissive power for every wave-length) is placed somewhere inside the chamber and maintained at temperature T.

The problem confronting physicists after the discoveries of Balfour Stewart and Kirchhoff was to find the function $F(\lambda, T)$.[1] The first step towards this was made in 1879 by the Austrian Josef Stefan [2] (1835–93), who found empirically that the energy per c.c. of the total radiation, that is $\int F(\lambda, T)\, d\lambda$, was proportional to the fourth power of the absolute temperature. Five years later Boltzmann [3] proved this theoretically, by combining the Second Law of Thermodynamics with the theory of the pressure of radiation. Let $u = \int F(\lambda, T)\, d\lambda$ be the total aethereal energy per c.c. in a hollow chamber whose walls are at temperature T. Then as we have seen,[4] the pressure of the radiation on a surface-element dS exposed to this radiation is PdS where $P = \frac{1}{3}u$. Now regard the aether within a hollow extensible vessel at temperature T as the 'working substance' of a heat engine. The total internal energy is $U = uv$, where v is the volume of the vessel; and the element of mechanical work performed on the external world (by radiation pressure) is $dA = Pdv$ or $\frac{1}{3}udv$. Now if Q be the heat absorbed from the external world,

$$\frac{dQ}{T} \quad \text{or} \quad \frac{dU + dA}{T} \quad \text{is a perfect differential.}$$

so

$$\frac{1}{T}\left\{ d\,(uv) + \frac{1}{3}\,udv \right\} \quad \text{or} \quad \frac{1}{T}\left\{ v\,\frac{du}{dT}\,dT + \frac{4}{3}\,udv \right\}$$

is a perfect differential. Therefore

$$\frac{\partial}{\partial v}\left(\frac{v}{T}\frac{du}{dT}\right) = \frac{\partial}{\partial T}\left(\frac{4}{3}\frac{u}{T}\right), \quad \text{or} \quad \frac{1}{T}\frac{du}{dT} = \frac{4}{3T}\frac{du}{dT} - \frac{4}{3}\frac{u}{T^2}$$

or

$$\frac{du}{dT} = \frac{4u}{T}$$

whence

$$u = aT^4 \quad \text{where } a \text{ is a constant,}$$

which is Stefan's formula.

[1] The assertion that 'all bodies become red-hot at the same temperature,' which may be regarded as the germ from which the recognition of the function F (λ, T) ultimately developed, was first made by Thomas Wedgwood, *Phil. Trans.* lxxxii (1792), p. 270.

[2] *Wien Ber.* lxxix, *abt.* ii (1879), p. 391

[3] *Ann. d. Phys.* xxii (1884), pp. 31, 291 ; cf. also B. Galitzine, *Ann. d. Phys.* xlvii (1892), p. 479

[4] cf. p. 275

Meanwhile, progress was being made in understanding the line-spectra of the chemical elements. In 1871 G. Johnstone Stoney [1] of Dublin expressed the view that the best preparation for studying the structure of these spectra would be to map them on a scale of wave-numbers, i.e. reciprocals of wave-lenths.[2] This suggestion was carried out by another Dublin worker, Walter Noel Hartley [3] (1846–1913), who showed that in the case of, for example, the triplets which occur in the spectrum of zinc, the intervals between the three components of a triplet are the same for all the triplets, when these intervals are expressed in terms of wave-numbers. This was the first discovery of a connection between lines in a spectrum that could be expressed by a numerical law; it pointed to the addition and subtraction of wave-numbers as being the key to the analysis of spectra. In the same year two Cambridge men, George Downing Liveing (1827–1924) and James Dewar (1842–1923) noticed [4] that certain sets of lines in the spectra of the alkalis and alkaline earths evidently formed series. 'This relation,' they said [of the relations between the lines], 'manifests itself in three ways: first, by the repetition of similar groups of lines; secondly, by a law of sequence in distance, producing a diminishing distance between successive repetitions of the same group as they decrease in wave-length; and thirdly, a law of sequence as regards quality, an alternation of sharper and more diffuse groups, with a gradually increasing diffuseness and diminishing intensity of all the related groups as the wave-length diminishes. The first relationship has long since been noticed in the case of the sodium lines, which recur in pairs.'

In 1885 Alfred Cornu [5] (1841–1902) observed sequences of doublets in the ultra-violet spectra of thallium and aluminium, which, from the similarity of their behaviour with regard to reversibility, he concluded to be the successive members of a series.

The time was now ripe for the discovery of a numerical law connecting the different lines of a series. This was made in 1885 by

[1] *Rep. Brit. Ass.* (1871), notes and abstracts, p. 42

[2] A wave-number is generally defined as $10^8\lambda^{-1}$ where λ is a wavelength expressed in Ångströms; so that it represents the number of waves in one cm. in air.

[3] *Jour. Chem. Soc.* xliii (1883), p. 390

[4] *Phil. Trans.* clxxiv (1883), p. 187 (at pp. 213, 214)

[5] *Comptes rendus*, c (1885), p. 1181

the Swiss physicist Johann Jakob Balmer [1] (1825–98), who gave the formula

$$\lambda = \lambda_0 \frac{m^2}{m^2 - 4} \qquad (m = 3, 4, 5, 6, \ldots)$$

for the wave-lengths of the well-known hydrogen lines

$$H_\alpha = \lambda 6563, \quad H_\beta = \lambda 4861, \quad H_\gamma = \lambda 4340, \quad H_\delta = \lambda 4102, \ldots$$

where λ_0 is a constant. If this formula is expressed in wave-numbers it becomes

$$n = n_0\left(1 - \frac{4}{m^2}\right) \qquad (m = 3, 4, 5, \ldots)$$

In 1890 Heinrich Kayser (1853–1940) and Carl Runge (1856–1927) gave [2] the names *Principal Series*, *First Subordinate Series* and *Second Subordinate Series*, to the series of doublets in the spectra of the alkali metals which had been recognised by Liveing and Dewar. In the case of sodium, the principal series consists of the D lines, a doublet at λ3303 and λ3302, and further lines in the ultra-violet. These doublets are sharp and easily reversed. The first subordinate series consists of strong and diffuse doublets, while the second subordinate series consists of fainter but sharper doublets. Each of these series converges to a limit or 'head' at its ultra-violet end.

In the same year Johannes Robert Rydberg [3] of Lund (1854–1919) gave a new general formula for series, which was capable of representing all the series in the spectra of the alkalis, as well as the hydrogen series, namely

$$n = n_0 - \frac{R}{(m + p)^2}$$

where n is the wave-number, n_0 and p are constants, m takes in succession the values of the positive whole numbers, and R is a constant *common to all series and to all elements*, having a value which he gave as

$$R = 109721{\cdot}6\ cm^{-1}.$$

R is called *Rydberg's constant.* Balmer's formula for the hydrogen series thus becomes

$$n = R\left(\frac{1}{4} - \frac{1}{m^2}\right).$$

In the Principal Series of the alkali metals, the n_0 of Rydberg's formula has the same value for the two lines of a doublet, but p has different values, say p_1 and p_2, for the less and more refrangible lines

[1] *Basel Verh.* vii (1885), pp. 548, 750

[2] *Ann. d. Phys.* xli (Sept. 1890), p. 302 ; *Abhandl. Berlin Akad.*, 1890 (*Anh.*), 66 pp.

[3] *Phil. Mag.* [(5)], xxix (Apr. 1890), p. 331 ; *Kongl. Sv. Vetensk. Ak. Handl.* xxxii (1890), no. 11

respectively. For the first or diffuse subordinate series, p has the same value (say d) for the two lines of any doublet; and similarly for the second or sharp subordinate series, p has the same value (say s) for the two lines of any doublet. The convergence wave-number n_0 has the same value for the more refrangible members of the doublets in the two subordinate series. Thus we can write:

Principal series:

less refrangible lines of doublets, $$n = n_0 - \frac{R}{(m + p_1)^2}$$

more refrangible lines of doublets, $$n = n_0 - \frac{R}{(m + p_2)^2}.$$

First or diffuse subordinate series:

less refrangible lines of doublets, $$n = n_0' - \frac{R}{(m + d)^2}$$

more refrangible lines of doublets, $$n = n_0'' - \frac{R}{(m + d)^2}.$$

Second or sharp subordinate series:

less refrangible lines of doublets, $$n = n_0' - \frac{R}{(m + s)^2},$$

more refrangible lines of doublets, $$n = n_0'' - \frac{R}{(m + s)^2}.$$

In 1896 Rydberg discovered [1] that, with this notation,

$$n_0' = \frac{R}{(2 + p_2)^2}, \qquad n_0'' = \frac{R}{(2 + p_1)^2};$$

and with a similar evaluation of n_0 which followed, we have finally:

Principal series:

less refrangible lines of doublets,

$$n = \frac{R}{(1 + s)^2} - \frac{R}{(m + p_1)^2}, \quad m = 2, 3, 4, 5, \ldots$$

more refrangible lines of doublets,

$$n = \frac{R}{(1 + s)^2} - \frac{R}{(m + p_2)^2}, \quad m = 2, 3, 4, 5, \ldots$$

First subordinate series:

less refrangible lines of doublets,

$$n = \frac{R}{(2 + p_2)^2} - \frac{R}{(m + d)^2}, \quad m = 3, 4, 5, \ldots$$

more refrangible lines of doublets,

$$n = \frac{R}{(2 + p_1)^2} - \frac{R}{(m + d)^2}, \quad m = 3, 4, 5, \ldots$$

[1] *Ann. d. Phys.* lviii (1896), p. 674. The same law was found independently by A. Schuster, *Nature* lv (Jan. 1897), pp. 200, 223

Second subordinate series :

less refrangible lines of doublets,

$$n = \frac{R}{(2+p_2)^2} - \frac{R}{(m+s)^2}, \quad m = 2, 3, 4, 5, \ldots$$

more refrangible lines of doublets,

$$n = \frac{R}{(2+p_1)^2} - \frac{R}{(m+s)^2}, \quad m = 2, 3, 4, 5, \ldots$$

It was perceived by Rydberg [1] and by Walther Ritz [2] (1878–1909) that the wave-numbers of all the lines of the principal series and the two subordinate series are *differences* of numbers taken from the sets

$$\frac{1}{(1+s)^2}, \quad \frac{1}{(2+s)^2}, \quad \frac{1}{(3+s)^2}, \quad \frac{1}{(4+s)^2}, \ldots$$

$$\frac{1}{(2+p_1)^2}, \quad \frac{1}{(3+p_1)^2}, \quad \frac{1}{(4+p_1)^2}, \ldots$$

$$\frac{1}{(2+p_2)^2}, \quad \frac{1}{(3+p_2)^2}, \quad \frac{1}{(4+p_2)^2}, \ldots$$

$$\frac{1}{(3+d)^2}, \quad \frac{1}{(4+d)^2}, \ldots$$

and they accordingly asserted the principle (generally known as the *combination-principle*) that *to the spectrum of any element there corresponds a set of ' terms,' such that the wave-numbers of the lines of the spectrum are the differences of these terms taken in pairs.* It was soon found that in the sodium spectrum there are lines not belonging to any of the three series above-mentioned, whose wave-numbers are differences of terms taken from the above set : such as the lines (in the infra-red)

$$\frac{R}{(2+s)^2} - \frac{R}{(3+p_1)^2} \quad \text{and} \quad \frac{R}{(2+s)^2} - \frac{R}{(3+p_2)^2}$$

and the lines (in the infra-red)

$$\frac{R}{(3+p_1)^2} - \frac{R}{(4+d)^2} \quad \text{and} \quad \frac{R}{(3+p_2)^2} - \frac{R}{(4+d)^2}$$

Meanwhile another series had been discovered in the spectra of the alkalis. In 1904 F. A. Saunders [3] found five new lines in the infra-

[1] *Rapp. prés. au congrès int.* (Paris, 1900), ii, p. 200

[2] *Phys. Zs*, ix (1908), p. 521 ; *Ges. Werke*, p. 141

Astrophys. Jour. xxviii (1908), p. 237 ; *Ges. Werke*, p. 163. Many of the predictions made possible by Ritz's theory were verified by F. Paschen, *Ann. d. Phys.* xxvii (1908), p. 537 ; xxix (1909), p. 625.

[3] *Astrophys. Jour.* xx (1904), p. 188 ; xxviii (1908), p. 71

red spectrum of caesium; and in 1907 it was shown by A. Bergmann[1] that these belong to a series which has been called the 'F' or fundamental series, and which exists in the spectra of all the alkalis (where it is in the infra-red) and many other elements (where it is not necessarily in the infra-red). Its head or limit has the wave-number $R/(3 + d)^2$, and its lines involve a new series of terms

$$\frac{1}{(4+f)^2}, \quad \frac{1}{(5+f)^2}, \quad \frac{1}{(6+f)^2}, \ldots$$

It was noticed that under normal circumstances the terms involving s combine only with the p_1 or p_2 terms, that the p_1 and p_2 terms combine only with s and d terms, the d terms only with p terms and f terms. On this account, the sets of terms are usually arranged in the order $s, p, d, f \ldots$ With the aid of powerful spectroscopes, it was found that the d terms and f terms are doubled like the p terms.

Let us now return to consider the radiation emitted by a black body at the absolute temperature T; as before, we denote by $F(\lambda,T)d\lambda$ the energy per c.c.[3] represented by radiations of wave-lengths between λ and $\lambda + d\lambda$.

In 1893 W. Wien,[2] in studying the form of the function $F(\lambda,T)$ made an important discovery, which will now be described.

Suppose that a perfectly reflecting sphere is contracting, its radius at time t being r. Consider a beam of light incident on the inner surface of the sphere at an angle i. At each reflection we have,[3] if τ be the period of the light,

$$\frac{\delta\tau}{\tau} = \frac{2\cos i}{c}\frac{dr}{dt}$$

where c is the velocity of light.

The length of path between two impacts is $2r \cos i$, so the interval of time between two impacts is $2r \cos i/c$. So if $\frac{d\tau}{dt}$ denotes the change in τ in one second, we have

$$\delta\tau = \frac{2r}{c}\cos i \frac{d\tau}{dt},$$

whence

$$\frac{1}{\tau}\frac{d\tau}{dt} = \frac{1}{r}\frac{dr}{dt}, \quad \text{or} \quad \tau = \text{Constant} \times r.$$

Thus as the sphere contracts, the period of the light also contracts, so as always to remain proportional to the radius of the sphere.

[1] *Dissertation*, Jena, 1907. cf. E. Riecke, *Phys. Zs.* ix (1908), p. 241

[2] *Berlin Sitz.* (9 Feb. 1893), p. 55; *Ann. d. Phys.* lii (1894), p. 132. A more careful proof was given by E. Buckingham, *Phil. Mag.* xxiii (1912), p. 920.

[3] cf. p. 368

Moreover if ϵ denote the amount of energy in one wave-length for this particular radiation, we have in the same way

$$\frac{\delta\epsilon}{\epsilon} = -\frac{2\cos i}{c}\frac{dr}{dt} \quad \text{and} \quad \delta\epsilon = \frac{2r\cos i}{c}\frac{d\epsilon}{dt},$$

so

$$\frac{1}{\epsilon}\frac{d\epsilon}{dt} = -\frac{1}{r}\frac{dr}{dt}, \quad \text{or} \quad \epsilon r = \text{Constant}.$$

Thus the total amount of radiant energy in the sphere varies as $1/r$, so the radiant energy per c.c. varies as $1/r^4$ for the wave whose shrinkage we are following, and therefore for all the waves. Now suppose that the sphere is initially filled with radiation of all wave-lengths so as to be in equilibrium with a black body at some definite temperature T. Since the total energy per c.c. under these circumstances varies as T^4, we see that the total radiant energy within the sphere at any instant is that appropriate to equilibrium with a radiating enclosure at temperature varying as $1/r$. We have not yet shown that the partition of energy among the different wave-lengths which is reached by the compression is one which would be in equilibrium with a radiating enclosure. This however follows from the fact that if it were not so, the radiation of certain wave-lengths would have an energy corresponding to a higher temperature than the radiation in certain other wave-lengths; and by introducing bodies capable of absorbing these respective wave-lengths only, we should be able to obtain differences of temperature from which mechanical work could be obtained in contravention of the Second Law of Thermodynamics.

It follows that if, in a distribution of radiant energy which is in equilibrium with a radiating enclosure at temperature T, we multiply the wave-lengths by k ($= r_2/r_1$) and multiply the energy per c.c. in each wave-length by k^{-4}, we shall obtain a distribution of radiant energy which will be in equilibrium with a radiating enclosure at temperature $k^{-1}T$.

Thus $$k^{-4}\,F(\lambda, T)\,d\lambda = F(k\lambda, k^{-1}T)\,d(k\lambda)$$

whence we see that $F(\lambda, T)\,d\lambda$ *must be of the form* $T^5\phi(T\lambda)\,d\lambda$, which in view of later developments we may write $T\lambda^{-4}\phi(T\lambda)\,d\lambda$, *where* ϕ *is a function of its argument* $T\lambda$ *only*. This is *Wien's theorem*.

Hence it follows that if λ_m denotes the wave-length corresponding to the maximum ordinate of a graph in which energy of radiation is plotted against wave-length, we must have

$$\lambda_m T = \text{Constant},$$

so λ_m is inversely proportional to the temperature. This is known as *Wien's displacement law*.

Three years later Wien[1] investigated the form of the function $\phi(T\lambda)$. He remarked that the law, that in an exhausted vessel the radiation is the same as that from a black body at the same temperature as the walls of the vessel, holds also if the radiating body be a gas which is shut off from the vacuous space by a transparent window, and from the exterior by reflecting walls. We postulate, therefore, a gas which possesses a finite absorptive power for radiation of all wave-lengths, and which is radiating under purely thermal excitement. We suppose that Maxwell's law of the distribution of velocities holds,[2] so the number of molecules whose velocity lies between v and $v + dv$ is proportional to

$$v^2 e^{-v^2/a^2} dv,$$

where a denotes a constant, which is connected with the mean square of velocity by the equation

$$\bar{v^2} = \frac{3}{2} a^2.$$

The absolute temperature is therefore proportional to a^2. We assume that each molecule sends out vibrations whose wave-length and intensity depend only on the velocity of the molecule.

As the wave-length λ of the radiation emitted by any molecule is a function of v, it follows that v is a function of λ. The intensity $F(\lambda, T)$ of the radiation whose wave-length lies between λ and $\lambda + d\lambda$ is therefore proportional to the number of molecules that send out radiations of this period, and also to a function of the velocity v (i.e. a function of λ). Thus

$$F(\lambda, T) = g(\lambda)\, e^{-\frac{f(\lambda)}{T}}$$

where g and f denote two unknown functions. But we know from Wien's previous work that

$$F(\lambda, T) = \lambda^{-5}\phi(T\lambda).$$

Combining these results, we have

$$F(\lambda, T) = C\,\lambda^{-5}\, e^{-\frac{b}{\lambda T}},$$

which is *Wien's law of radiation*. Experimental determinations were found to agree with this formula in the region of high frequencies, but not for long waves. At the other end of the graph of $F(\lambda, T)$—

[1] *Ann. d. Phys.* lviii (1896), p. 662; *Phil. Mag.* xliii (1897), p. 214

[2] It may be mentioned that an attempt to determine the law of radiation of a black body by assuming that Maxwell's law of the distribution of velocities holds for its molecules, had been made by Vladimir A. Michelson, *Journal de Physique* (2), vi (1887), p. 467.

the region of low frequencies—a different formula was found to hold, the derivation of which must now be explained.

When several different gases (which may be initially at different temperatures) are mixed, the mixture becomes ultimately of the same temperature throughout. Since the temperature is proportional to the average kinetic energy of a molecule, we see that the system tends to assume a state in which the average kinetic energy of a molecule is the same whether the molecule be light or massive : the inferiority in mass of the lighter molecules is compensated by their moving with greater velocities. This is a particular instance of what is known as the *theorem of equipartition of energy*, which asserts that *in a state of statistical equilibrium at absolute temperature* T, *the total energy of any system which obeys the laws of dynamics, is partitioned in such a way that every degree of freedom possesses on the average the same kinetic energy, namely* $\frac{1}{2}k$T, *where k is known as Boltzmann's constant.* Thus for an atom of a monatomic gas, which has three degrees of freedom, the mean kinetic energy at temperature T is $\frac{3}{2}k$T. The theorem is due to Maxwell [1] and Boltzmann.[2]

Some very familiar phenomena can be explained by reference to the law of equipartition of energy. For instance, if a hollow rigid vessel is filled with air, and an arbitrary disturbance is set up in the air, this disturbance will in time become dissipated into random motions of the molecules, such as are studied in the kinetic theory of gases. The reason is that any motion in the air can be regarded as a superposition of its various types of free vibration. Now the number of types of free vibration is very great, and those types which are of extremely short wave-length vastly outnumber the types of longer wave-lengths. Each type of vibration counts as a degree of freedom in the equipartition theorem, and the theorem therefore predicts that after a time the energy of the disturbance will be practically all in the vibrations of very short wave-length, i.e. it will have become degraded into small irregular disturbances. The random motion of the molecules can be analysed by Fourier's theorem into waves, and the energy represented by waves of lengths λ to $\lambda + d\lambda$ is found to be $4\pi k\mathrm{T}\lambda^{-4}d\lambda$ per unit volume.[3] The thermal motion of the molecules of a gas is, in fact, essentially sound of very short wave-lengths.

[1] *Phil. Mag.* xix (Jan. 1860), p. 19 ; xx (July 1860), p. 21 ; *Trans. Camb. Phil. Soc.* xii (1878), p. 547

[2] *Wien Sitzungsber.* lviii (1868 [Abth. 2]), p. 517 ; lxiii (1871 [Abth. 2]), p. 397

[3] This is half the value found below for the aether, the difference being due to the fact that vibrations in a gas are longitudinal, and in the aether are transverse, each state of polarisation of a wave counting as a separate degree of freedom.

There has however been much difference of opinion regarding the range of validity of the doctrine of equipartition of energy. Lord Kelvin [1] believed that he had disproved the theorem altogether by a 'decisive test case.' The question was taken up by Lord Rayleigh,[2] who showed that Kelvin's reasoning was inconclusive; although he recognised that there were difficulties in applying the theorem to actual gases. 'In the case of argon and helium and mercury vapour' he said, 'the ratio of specific heats (1·67) limits the degrees of freedom of each molecule to the three required for translatory motion. The value (1·4) applicable to the principal diatomic gases gives room for the three kinds of translation and for two kinds of rotation. Nothing is left for rotation round the line joining the atoms, nor for relative motion of the atoms in this line. Even if we regard the atoms as mere points, whose rotation means nothing, there must still exist energy of the last-mentioned kind, and its amount (according to the law) should not be inferior.' The discussion was continued by Poincaré,[3] G. H. Bryan [4] and others.[5]

In 1900 Lord Rayleigh,[6] assuming the truth of the theorem of equipartition of energy, derived from it a formula for the energy of black-body radiation. It was obtained as follows:

In the free vibrations of the aether inside a cubical enclosure whose edge is of length l, each mode of vibration corresponds to values of the components of the electric and magnetic forces involving a factor of the form

$$\begin{matrix}\cos\\ \sin\end{matrix}\left(\frac{p\pi x}{l}\right)\begin{matrix}\cos\\ \sin\end{matrix}\left(\frac{q\pi y}{l}\right)\begin{matrix}\cos\\ \sin\end{matrix}\left(\frac{r\pi z}{l}\right)$$

where p, q, r are whole numbers. The frequency ν of this vibration is

$$\nu = \frac{c}{2l}(p^2 + q^2 + r^2)^{\frac{1}{2}},$$

where c denotes the velocity of light. We have to find how many modes of vibration correspond to the range ν to $\nu + d\nu$ of frequency. Let whole-number values of (p, q, r) be regarded as the co-ordinates of a point; the whole system of points constitutes a cubic array of volume-density unity. Let R denote the distance of the point

[1] *Proc. R. S.* vol. L (1891), p. 85; *Phil. Mag.* xxxiii (May 1892), p. 466

[2] *Phil. Mag.* xxxiii (April 1892), p. 356; *Phil. Mag.* xlix (Jan. 1900), p. 98

[3] *Revue générale des Sciences*, July 1894

[4] 'Report on Thermodynamics,' *Brit. Ass. Rep.* (1894), p. 64; *Proc. Camb. Phil. Soc.* viii (1895), p. 250

[5] W. Peddie, *Proc. R. S. Edin.* xxvi (1905–6), p. 130; J. H. Jeans, *Phil. Mag.* xvii (1909), p. 229; J. Larmor, *Brit. Ass. Rep.* (1913), p. 385

[6] *Phil. Mag.* xlix (1900), p. 539; *Nature*, lxxii (May 1905), p. 54 and (July 1905) p. 243. cf. J. H. Jeans, *Phil. Mag.* (6), x (July 1905), p. 91

(p, q, r) from the origin, so $R^2 = p^2 + q^2 + r^2$. The number of points in the shell for which R ranges from R to R + dR is equal to the included volume (since p, q, r are positive, we take only the positive octant), which is

$$\tfrac{1}{2}\pi R^2 dR.$$

Hence the number of sets of values of p, q, r corresponding to the range ν to $\nu + d\nu$ is

$$\tfrac{1}{2}\pi\left(\frac{2l}{c}\right)^3 \nu^2\, d\nu$$

or in terms of λ (where $\nu = c/\lambda$)

$$4\pi l^3 \lambda^{-4} d\lambda.$$

To obtain the number of modes that are to be counted in applying the theorem of the equipartition of energy, we must double this number, since the vibrations are transverse, and double it again, since the whole energy is (on the average) double the electric or magnetic energy taken alone. Thus the number of modes to be counted is

$$16\pi l^3 \lambda^{-4} d\lambda.$$

At temperature T, each mode possesses the average energy $\tfrac{1}{2}kT$. This gives for the total energy

$$8\pi l^3 kT\lambda^{-4} d\lambda.$$

so *the energy of radiation per c.c. in the range of wave-lengths from* λ *to* $\lambda + d\lambda$ *is*

$$F(\lambda, T) = 8\pi kT\lambda^{-4} d\lambda$$

where k is Boltzmann's constant. *This is Rayleigh's formula.* k has the value $1{\cdot}38 \times 10^{-16}$ erg. (degree)$^{-1}$.

It is obvious that according to this formula the radiant energy would tend to run entirely into waves of the highest frequency, i.e. of infinitesimal wave-length. This is due fundamentally to the fact that the aether is regarded as a continuum, and has therefore an infinite number of degrees of freedom. Rayleigh's formula therefore cannot represent the true law of black-body radiation. It is, however, asymptotically correct for the long waves, as was shown experimentally by H. Rubens and F. Kurlbaum in 1900–1,[1] and it was deduced from the electron theory by Lorentz,[2] who calculated by help of that theory the emissive power and absorption coefficient of a thin sheet of metal on which radiation of great wave-length is incident.

The subsequent history of the radiation problem must be post-

[1] *Berlin Sitzungsb.* xli (Oct. 1900), p. 929 ; *Ann. d. Phys.* iv (Apr. 1901), p. 649

[2] *Proc. Amst. Acad.* v (1902–3), p. 666

poned to a later chapter. It may be mentioned that Lord Kelvin refused to the last to accept the equipartition theorem. Referring to a remark of Rayleigh's—'What would appear to be wanted is some escape from the destructive simplicity of the general conclusion'—he said: 'The simplest way of arriving at this desired result is to deny the conclusion; and so, in the beginning of the twentieth century, to lose sight of a cloud which has obscured the brilliance of the molecular theory of heat and light during the last quarter of the nineteenth century.' [1]

[1] Kelvin, *Baltimore Lectures* (1904), p. 527

Chapter XIII

CLASSICAL THEORY IN THE AGE OF LORENTZ

THE attempts of Maxwell[1] and of Hertz[2] to extend the theory of the electromagnetic field to the case in which ponderable bodies are in motion had not been altogether successful. Neither writer had taken account of any motion of the material particles relative to the aether entangled with them, so that in both investigations the moving bodies were regarded simply as homogeneous portions of the medium which fills all space, distinguished only by special values of the electric and magnetic constants. Such an assumption is evidently inconsistent with the admirable theory by which Fresnel[3] had explained the optical behaviour of moving transparent bodies; it was therefore not surprising that writers subsequent to Hertz should have proposed to replace his equations by others designed to agree with Fresnel's formulae. Before discussing these, however, it may be well to review briefly the evidence for and against the motion of the aether in and adjacent to moving ponderable bodies, as it appeared in the last decade of the nineteenth century.

The phenomena of aberration had been explained by Young[4] on the assumption that the aether around bodies is unaffected by their motion. But it was shown by Stokes[5] in 1845 that this is not the only possible explanation. For suppose that the motion of the earth communicates motion to the neighbouring portions of the aether; this may be regarded as superposed on the vibratory motion which the aethereal particles have when transmitting light: the orientation of the wave-fronts of the light will consequently in general be altered; and the direction in which a heavenly body is seen, being normal to the wave-fronts will thereby be affected. But if the aethereal motion is irrotational, so that the elements of the aether do not rotate, it is easily seen that the direction of propagation of the light in space is unaffected; the luminous disturbance is still propagated in straight lines from the star, while the normal to the wave-front at any point deviates from this line of propagation by the small angle u/c, where u denotes the component of the aethereal velocity at the point, resolved at right angles to the line of propagation, and c denotes the velocity of light. If it be supposed that the aether

[1] cf. p. 259 [2] cf. p. 329 [3] cf. p. 108 [4] cf. p. 108
[5] *Phil. Mag.* xxvii (1845), p. 9; xxviii (1846), p. 76; xxix (1846), p. 6

near the earth is at rest relatively to the earth's surface, the star will appear to be displaced towards the direction in which the earth is moving, through an angle measured by the ratio of the velocity of the earth to the velocity of light, multiplied by the sine of the angle between the direction of the earth's motion and the line joining the earth and star. This is precisely the law of aberration.

An objection to Stokes's theory was pointed out by several writers, amongst others by H. A. Lorentz.[1] This is, that the irrotational motion of an incompressible fluid is completely determinate when the normal component of the velocity at its boundary is given; so that if the aether were supposed to have the same normal component of velocity as the earth, it would not have the same tangential component of velocity. It follows that no motion will in general exist which satisfies Stokes's conditions; and the difficulty is not solved in any very satisfactory fashion by either of the suggestions which have been proposed to meet it. One of these is to suppose that the moving earth does generate a rotational disturbance, which however, being radiated away with the velocity of light, does not affect the steadier irrotational motion; the other, which was advanced by Planck[2] in 1899, is that the two conditions of Stokes's theory—namely, that the motion of the aether is to be irrotational and that at the earth's surface its velocity is to be the same as that of the earth—may both be satisfied if the aether is supposed to be compressible in accordance with Boyle's law, and subject to gravity, so that round the earth it is compressed like the atmosphere; the velocity of light being supposed independent of the condensation of the aether.

Lorentz,[3] in calling attention to the defects of Stokes's theory, proposed to combine the ideas of Stokes and Fresnel, by assuming that the aether near the earth is moving irrotationally (as in Stokes's theory), but that at the surface of the earth the aethereal velocity is not necessarily the same as that of ponderable matter, and that (as in Fresnel's theory) a material body imparts the fraction $(\mu^2 - 1)/\mu^2$ of its own motion to the aether within it. Fresnel's theory is a particular case of this new theory, being derived from it by supposing the velocity potential to be zero.

Aberration is by no means the only astronomical phenomenon which depends on the velocity of propagation of light; we have indeed seen[4] that this velocity was originally determined by

[1] *Archives Néerl*, xxi (1896), p. 103

[2] cf. Lorentz, *Proc. Amst. Acad.* (English edn.), i (1899), p. 443

[3] *Archives Néerl.* xxi (1886), p. 103; cf. also *Zittinsgsversl. Kon. Ak. Amsterdam* (1897–8), p. 266

[4] cf. p. 22

observing the retardation of the eclipses of Jupiter's satellites. It was remarked by Maxwell [1] in 1879 that these eclipses furnish, theoretically at least, a means of determining the velocity of the solar system relative to the aether. For if the distance from the eclipsed satellite to the earth be divided by the observed retardation in time of the eclipse, the quotient represents the velocity of propagation of light in this direction, relative to the solar system; and this will differ from the velocity of propagation of light relative to the aether by the component, in this direction, of the sun's velocity relative to the aether. By taking observations when Jupiter is in different signs of the zodiac, it should therefore be possible to determine the sun's velocity relative to the aether, or at least that component of it which lies in the ecliptic.

The same principles may be applied to the discussion of other astronomical phenomena. Thus the minimum of a variable star of the Algol type will be retarded or accelerated by an interval of time which is found by dividing the projection of the radius from the sun to the earth on the direction from the sun to the Algol variable by the velocity, relative to the solar system, of propagation of light from the variable; and thus the latter quantity may be deduced from observations of the retardation.[2]

Another instance in which the time taken by light to cross an orbit influences an observable quantity is afforded by the astronomy of double stars. Savary [3] long ago remarked that when the plane of the orbit of a double star is not at right angles to the line of sight, an inequality in the apparent motion must be caused by the circumstance that the light from the remoter star has the longer journey to make. Yvon Villarceau [4] showed that the effect might be represented by a constant alteration of the elliptic elements of the orbit (which alteration is of course beyond detection), together with a periodic inequality, which may be completely specified by the following statement: the apparent co-ordinates of one star relative to the other have the values which in the absence of this effect they would have at an earlier or later instant, differing from the actual time by the amount

$$\frac{m_1 - m_2}{m_1 + m_2} \cdot \frac{z}{c},$$

where m_1 and m_1 denote the masses of the stars, c the velocity of

[1] *Proc. R. S.* xxx (1880), p. 108

[2] The velocity of light was found from observations of Algol, by C. V. L. Charlier, *Öfversigt af K. Vet.-Ak. Förhandl.* xlvi (1889), p. 523.

[3] *Conn. des Temps*, 1830

[4] *Additions à la Connaissance des Temps*, 1878; an improved deduction was given by H. Seeliger, *Sitzungsberichte d. K. Ak. zu München*, xix (1889), p. 19.

light, and z the actual distance of the two stars from each other at the time when the light was emitted, resolved along the line of sight. In the existing state of double-star astronomy, this effect would be masked by errors of observation.

Villarceau also examined the consequences of supposing that the velocity of light depends on the velocity of the source by which it is emitted. If, for instance, the velocity of light from a star occulted by the moon were less than the velocity of light reflected by the moon, then the apparent position of the lunar disk would be more advanced in its movement than that of the star, so that at emersion the star would first appear at some distance outside the lunar disk, and at immersion the star would be projected on the interior of the disk at the instant of its disappearance. The amount by which the image of the star could encroach on that of the disk on this account could not be so much as 0″·71 ; encroachment to the extent of more than 1″ has been observed, but is evidently to be attributed for the most part to other causes.[1]

Among the consequences of the finite velocity of propagation of light which are of importance in astronomy, a leading place must be assigned to the principle enunciated by Doppler,[2] that the motion of a source of light relative to an observer modifies the period of the disturbance which is received by him. The phenomenon resembles the depression of the pitch of a note when the source of sound is receding from the observer. In either case, the period of the vibrations perceived by the observer is $(+ v)/c \times$ the natural period, where v denotes the velocity of separation of the source and observer, and c denotes the velocity of propagation of the disturbance. If, for example, the velocity of separation is equal to the orbital velocity of the earth, the D lines of sodium in the spectrum of the source will be displaced towards the red, as compared with lines derived from a terrestrial sodium flame, by about one-tenth of the distance between them. The application of this principle to the determination of the relative velocity of stars in the line of sight, which has proved of great service in astrophysical research, was suggested by Fizeau in 1848.[3]

Passing now from the astronomical observatory, we must examine the information which has been gained in the physical laboratory regarding the effect of the earth's motion on optical phenomena. We have already [4] referred to the investigations by which the truth

[1] The question of the possible dependence of the velocity of light on the velocity of the source will be discussed in Volume II of this work, in connection with Ritz's ' ballistic ' hypothesis. [2] cf. p. 368

[3] An apparatus for demonstrating the Doppler-Fizeau effect in the laboratory was constructed by Belopolsky, *Astrophys. Jour.*, xiii (1901), p. 15. [4] cf. pp. 110–11

of Fresnel's formula was tested. An experiment of a different type was suggested in 1852 by Fizeau,[1] who remarked that, unless the aether is carried along by the earth, the radiation emitted by a terrestrial source should have different intensities in different directions. It was, however, shown long afterwards by Lorentz [2] that such an experiment would not be expected on theoretical grounds to yield a positive result; the amount of radiant energy imparted to an absorbing body is independent of the earth's motion. A few years later Fizeau investigated [3] another possible effect. If a beam of polarised light is sent obliquely through a glass plate, the azimuth of polarisation is altered to an extent which depends, amongst other things, on the refractive index of the glass. Fizeau performed this experiment with sunlight, the light being sent through the glass in the direction of the terrestrial motion, and in the opposite direction; the readings seemed to differ in the two cases, but on account of experimental difficulties the result was indecisive.

Some years later, the effect of the earth's motion on the rotation of the plane of polarisation of light propagated along the axis of a quartz crystal was investigated by Mascart.[4] The result was negative, Mascart stating that the rotation could not have been altered by more than the 1/40,000th part when the orientation of the apparatus was reversed from that of the terrestrial motion to the opposite direction. This was afterwards confirmed by Lord Rayleigh,[5] who found that the alteration, if it existed, could not amount to 1/100,000th part.

P. G. Tait in 1882 suggested [6] that 'if the aether be in motion relative to the earth, the absolute deviations of lines in the diffraction spectrum should be different in different azimuths.'

As Maxwell pointed out in the letter of 1879 already referred to,[7] in terrestrial methods of determining the velocity of light the ray is made to retrace its path, so that any velocity which the earth might possess with respect to the luminiferous medium would affect the time of the double passage only by an amount proportional to the square of the constant of aberration.[8] In 1881, however, Albert Abraham Michelson [9] (1852–1931) remarked that the effect, though

[1] *Ann. d. Phys.* xcii (1854), p. 652 [2] *Proc. Amst. Acad.* (English edn.), iv (1902), p. 678

[3] *Annales de Chim.* (3), lxviii (1860), p. 129; *Ann. d. Phys.* cxiv (1861), p. 554

[4] *Annales de l'Ec. Norm.* (2), i (1872), p. 157 [5] *Phil. Mag.* iv (1902), p. 215

[6] Tait's *Life and Scientific Work* (ed. C. G. Knott), p. 92

[7] *Proc. R. S.* xxx (1880), p. 108

[8] The constant of aberration is the ratio of the earth's orbital velocity to the velocity of light; cf. *supra*, p. 95.

[9] *Amer. Jour. Sci.* xxii (1881), p. 20. His method was afterwards improved; cf. Michelson and E. W. Morley, *Amer. Jour. Sci.* xxxiv (1887), p. 333; Phil. *Mag.* xxiv (1887), p. 449. See also E. W. Morley and D. C. Miller, *Phil. Mag.* viii (1904), p. 753, and ix (1905), p. 680.

of the second order, should be manifested by a measurable difference between the times for rays describing equal paths parallel and perpendicular respectively to the direction of the earth's motion. He produced interference fringes between two pencils of light which had traversed paths perpendicular to each other; but when the apparatus was rotated through a right angle, so that the difference would be reversed, the expected displacement of the fringes could not be perceived. This result was regarded by Michelson himself as a vindication of Stokes's theory,[1] in which the aether in the neighbourhood of the earth is supposed to be set in motion. Lorentz,[2] however, showed that the quantity to be measured had only half the value supposed by Michelson, and suggested that the negative result of the experiment might be explained by that combination of Fresnel's and Stokes's theories which was developed in his own memoir[3]; since, if the velocity of the aether near the earth were, say, half the earth's velocity, the displacement of Michelson's fringes would be insensible.

A sequel to the experiment of Michelson and Morley was performed in 1897, when Michelson[4] attempted to determine by experiment whether the relative motion of earth and aether varies with the vertical height above the terrestrial surface. No result, however, could be obtained to indicate that the velocity of light depends on the distance from the centre of the earth; and Michelson concluded that if there were no choice but between the theories of Fresnel and Stokes, it would be necessary to adopt the latter, and to suppose that the earth's influence on the aether extends to many thousand kilometres above its surface. By this time, however, as will subsequently appear, a different explanation was at hand.

Meanwhile the perplexity of the subject was increased by experimental results which pointed in the opposite direction to that of Michelson. In 1892 Sir Oliver Lodge[5] observed the interference between the two portions of a bifurcated beam of light, which were made to travel in opposite directions round a closed path in the space between two rapidly rotating steel disks. The observations showed that the velocity of light is not affected by the motion of adjacent matter to the extent of 1/200th part of the velocity of the matter. Continuing his investigations Lodge[6] strongly magnetised the

[1] cf. p. 386

[2] *Arch. Néerl.* xxi (1886), p. 103. On the Michelson-Morley experiment cf. also Hicks, *Phil. Mag.* iii (1902), p. 9

[3] cf. p. 387

[4] *Amer. Jour. Sci.* (4), iii (1897), p. 475

[5] *Phil. Trans.* clxxxiv (1893), p. 727

[6] ibid. clxxxix (1897), p. 149. See also Lodge's 'Silvanus Thompson Memorial Lecture,' *Journal of the Röntgen Society*, xviii, July 1922.

moving matter (iron in this experiment), so that the light was propagated across a moving magnetic field ; and electrified it so that the path of the beams lay in a moving electrostatic field ; but in no case was the velocity of the light appreciably affected.

We must now trace the steps by which theoretical physicists not only arrived at a solution of the apparent contradictions furnished by experiments with moving bodies, but so extended the domain of electrical science that it became necessary to enlarge the conceptions of space and time to contain it.

The first memoir in which the new conceptions were unfolded was published by Hendrik Antoon Lorentz [1] (1853–1928) in 1892. The theory of Lorentz was, like those of Weber, Riemann and Clausius,[2] a theory of electrons ; that is to say, all electrodynamical phenomena were ascribed to the agency of moving electric charges, which were supposed in a magnetic field to experience forces proportional to their velocities, and to communicate these forces to the ponderable matter with which they might be associated.[3]

In spite of the fact that the earlier theories of electrons had failed to fulfil the expectations of their authors, the assumption that all electric and magnetic phenomena are due to the presence or motion of individual electric charges was one to which physicists were at this time disposed to give a favourable consideration ; for, as we have seen,[4] evidence of the atomic nature of electricity was now contributed by the study of the conduction of electricity through liquids and gases. Moreover, the discoveries of Hertz [5] had shown that a molecule which is emitting light must contain some system resembling a Hertzian vibrator ; and the essential process in a Hertzian vibrator is the oscillation of electricity to and fro. Lorentz himself from the outset of his career [6] had supposed the interaction of ponderable

[1] *Archives Néerl.* xxv (1892), p. 363 : the theory is given in ch. iv, pp. 432 et sqq.

[2] cf. pp. 201, 206, 234

[3] Some writers have inclined to use the term ' electron theory ' as if it were specially connected with Sir Joseph Thomson's justly celebrated discovery (cf. p. 364, *supra*) that all negative electrons have equal charges. But Thomson's discovery, though undoubtedly of the greatest importance as a guide to the structure of the universe, has hitherto exercised but little influence on general electromagnetic theory. The reason for this is that in theoretical investigations it is customary to denote the changes of electrons by symbols, e_1, e_2, . . . ; and the equality or non-equality of these makes no difference to the equations. To take an illustration from celestial mechanics, it would clearly make no difference in the general equations of the planetary theory if the masses of the planets happened to be all equal.

[4] cf. ch. xi

[5] cf. pp. 322–8

[6] *Verh. d. Ak. v. Wetenschappen, Amsterdam, Deel* xviii, 1878

matter with the electric field to be effected by the agency of electric charges associated with the material atoms.

The principal difference by which the theory now advanced by Lorentz is distinguished from the theories of Weber, Riemann, and Clausius and from Lorentz' own earlier work, lies in the conception which is entertained of the propagation of influence from one electron to another. In the older writings, the electrons were assumed to be capable of acting on each other at a distance, with forces depending on their charges, mutual distances and velocities ; in the present memoir, on the other hand, the electrons were supposed to interact not directly with each other, but with the medium in which they were embedded. To this medium were ascribed the properties characteristic of the aether in Maxwell's theory.

The only respect in which Lorentz' medium differed from Maxwell's was in regard to the effects of the motion of bodies. Impressed by the success of Fresnel's beautiful theory of the propagation of light in moving transparent substances,[1] Lorentz designed his equations so as to accord with that theory, and showed that this might be done by drawing a distinction between matter and aether, and assuming that a moving ponderable body cannot communicate its motion to the aether which surrounds it, or even to the aether which is entangled in its own particles ; so that no part of the aether can be in motion relative to any other part. Such an aether is simply space endowed with certain dynamical properties ; its introduction was the most characteristic and most valuable feature of Lorentz' theory, which differs completely from, for example, the theory of Hertz so far as concerns the electrodynamics of bodies in motion.

The general plan of Lorentz' investigation was, therefore, to reduce all the complicated cases of electromagnetic action to one simple and fundamental case, in which the field contains only free aether with solitary electrons dispersed in it ; the theory which he adopted in this fundamental case was a combination of Clausius' theory of electricity with Maxwell's theory of the aether.

Suppose that $e\ (x, y, z)$ and $e'(x', y', z')$ are two electrons. In the theory of Clausius,[2] the kinetic potential of their mutual action is (in the electrostatic system of units)

$$\frac{ee'}{r}\left(\frac{1}{c^2}\frac{\partial x}{\partial t}\frac{\partial x'}{\partial t}+\frac{1}{c^2}\frac{\partial y}{\partial t}\frac{\partial y'}{\partial t}+\frac{1}{c^2}\frac{\partial z}{\partial t}\frac{\partial z'}{\partial t}-1\right);$$

[1] cf. pp. 108 et sqq. [2] cf. p. 234

so when any number of electrons are present, the part of the kinetic potential which concerns any one of them—say, e—may be written

$$L_e = e\left(\frac{1}{c} a_x \frac{\partial x}{\partial t} + \frac{1}{c} a_y \frac{\partial y}{\partial t} + \frac{1}{c} a_z \frac{\partial z}{\partial t} - \phi\right)$$

where **a** and ϕ denote potential functions, defined by the equations

$$\mathbf{a} = \iiint \frac{\rho' \mathbf{v}'}{cr}\, dx'dy'dz', \qquad \phi = \iiint \frac{\rho'}{r}\, dx'dy'dz' ;$$

ρ denoting the volume-density of electric charge in electrostatic units, and **v** its velocity, and the integration being taken over all space.

We shall now reject Clausius' assumption that electrons act instantaneously at a distance, and replace it by the assumption that they act on each other only through the mediation of an aether which fills all space, and satisfies Maxwell's equations. This modification may be effected in Clausius' theory without difficulty; for, as we have seen,[1] if the state of Maxwell's aether at any point is defined by the electric vector **d** and magnetic vector **h**,[2] these vectors may be expressed in terms of potentials **a** and ϕ by the equations

$$\mathbf{d} = \text{grad}\,\phi - \frac{1}{c}\frac{\partial \mathbf{a}}{\partial t}, \quad \mathbf{h} = \text{curl}\,\mathbf{a} ;$$

and the functions **a** and ϕ may in turn be expressed in terms of the electric charges by the equations

$$\mathbf{a} = \iiint \{(\overline{\rho \mathbf{v}_x})'/cr\}\, dx'dy'dz', \qquad \phi = \iiint \{(\bar{\rho})'/r\}\, dx'dy'dz',$$

where the bars indicate that the values of $(\rho \mathbf{v}_x)'$ and $(\rho)'$ refer to the instant $(t - r/c)$.[3] Comparing these formulae with those given above for Clausius' potentials, we see that the only change which it is necessary to make in Clausius' theory is that of retarding the potentials in the way indicated by L. Lorenz.[4]

It was shown by T. Levi-Civita [5] that when **a** and ϕ are defined by these integrals they satisfy the equation

$$\frac{\partial a_x}{\partial x} + \frac{\partial a_y}{\partial y} + \frac{\partial a_z}{\partial z} + \frac{\partial \phi}{c\partial t} = 0.$$

[1] cf. pp. 255–6

[2] We shall use the small letters **d** and **h** in place of **E** and **H**, when we are concerned with Lorentz' fundamental case, in which the system consists solely of free aether and isolated electrons.

[3] These integrals for **a** and ϕ— had been given in the year before Lorentz' work of 1892, by H. Poincaré, *Comptes Rendus*, cxiii (1891), p. 515.

[4] cf. p. 268

[5] Nuovo Cimento [(4)], vi (1897), p. 93

It had further been shown so long ago as 1858 by Riemann [1] that a function ϕ defined by the above integral satisfies the equation

$$\frac{\partial^2\phi}{\partial x^2} + \frac{\partial^2\phi}{\partial y^2} + \frac{\partial^2\phi}{\partial z^2} - \frac{1}{c^2}\frac{\partial^2\phi}{\partial t^2} = -4\pi\rho\,;$$

but Riemann considered only electric forces, and therefore did not give the equation

$$\frac{\partial^2\mathbf{a}}{\partial x^2} + \frac{\partial^2\mathbf{a}}{\partial y^2} + \frac{\partial^2\mathbf{a}}{\partial z^2} - \frac{1}{c^2}\frac{\partial^2\mathbf{a}}{\partial t^2} = -4\frac{\pi}{c}\rho\mathbf{v}.$$

The electric and magnetic forces, thus defined in terms of the position and motion of the charges, satisfy the Maxwellian equations [2]

$$\begin{cases} \text{div}\,\mathbf{d} = 4\pi\rho, \\ \text{div}\,\mathbf{h} = 0, \\ \text{curl}\,\mathbf{d} = -\dfrac{1}{c}\dfrac{\partial\mathbf{h}}{\partial t}, \\ \text{curl}\,\mathbf{h} = \dfrac{1}{c}\dfrac{\partial\mathbf{d}}{\partial t} + \dfrac{4\pi}{c}\rho\mathbf{v}. \end{cases}$$

The theory of Lorentz is based on these four aethereal equations of Maxwell, together with the equation which determines the ponderomotive force on a charged particle ; this, which we shall now derive, is the contribution furnished by Clausius' theory.

The Lagrangian equations of motion of the electron e are

$$\frac{d}{dt}\left(\frac{\partial L}{\partial\left(\dfrac{\partial x}{\partial t}\right)}\right) - \frac{\partial L}{\partial x} = 0,$$

and two similar equations, where L denotes the total kinetic potential due to all causes, electric and mechanical. The ponderomotive force exerted on the electron by the electromagnetic field has for its x-component

$$\frac{\partial L_e}{\partial x} - \frac{d}{dt}\left(\frac{\partial L_e}{\partial\left(\dfrac{\partial x}{\partial t}\right)}\right),$$

where

$$L_e = e\left(\frac{a_x}{c}\frac{\partial x}{\partial t} + \frac{a_y}{c}\frac{\partial y}{\partial t} + \frac{a_z}{c}\frac{\partial z}{\partial t} - \phi\right)$$

or

$$e\left(\frac{1}{c}\frac{\partial a_x}{\partial x}\frac{\partial x}{\partial t} + \frac{1}{c}\frac{\partial a_y}{\partial x}\frac{\partial y}{\partial t} + \frac{1}{c}\frac{\partial a_z}{\partial x}\frac{\partial z}{\partial t} - \frac{\partial\phi}{\partial x}\right) - \frac{e}{c}\frac{da_x}{dt}\,;$$

[1] In a paper published after his death, *Ann. d. Phys.* [(5)], xi (1867), p. 237

[2] These equations, with the terms in $\rho\mathbf{v}$, were first proposed by FitzGerald at the Southport meeting of the British Association in 1883. cf. Heaviside, *Elect. Papers*, ii, p. 508

which, since

$$\frac{da_x}{dt} = \frac{\partial a_x}{\partial t} + \frac{\partial a_x}{\partial x}\frac{\partial x}{\partial t} + \frac{\partial a_x}{\partial y}\frac{\partial y}{\partial t} + \frac{\partial a_x}{\partial z}\frac{\partial z}{\partial t},$$

reduces to

$$e\left(-\frac{\partial \phi}{\partial x} - \frac{1}{c}\frac{\partial a_x}{\partial t}\right) + \frac{e}{c}\frac{\partial z}{\partial t}\left(\frac{\partial a_z}{\partial x} - \frac{\partial a_x}{\partial z}\right) + \frac{e}{c}\frac{\partial y}{\partial t}\left(\frac{\partial a_y}{\partial x} - \frac{\partial a_x}{\partial y}\right),$$

or

$$ed_x + \frac{e}{c}\left(\frac{\partial y}{\partial t}h_z - \frac{\partial z}{\partial t}h_y\right),$$

so that the force in question is

$$e\mathbf{d} + \frac{e}{c}[\mathbf{v}\,.\,\mathbf{h}],$$

in agreement with the formula obtained by Heaviside[1] in 1889 for the ponderomotive force on an electrified corpuscle of charge e moving with velocity $\mathbf{v}$ in a field defined by the electric force $\mathbf{d}$ and magnetic force $\mathbf{h}$.[2]

The rate of loss of energy by radiation from a charge e having an acceleration Γ was discovered by Larmor[3] in 1897 : it is

$$\frac{2}{3}\frac{e^2}{c^3}\Gamma^2$$

(the charge being measured as usual in electrostatic units). This formula has been very widely applied in physics and astrophysics.

In Lorentz' fundamental case, which has been examined, account was taken only of the ultimate constituents of which the universe was supposed to be composed, namely, corpuscles and the aether. We must now see how to build up from these the more complex systems which are directly presented to our experience.

The electromagnetic field in ponderable bodies, which to our senses appears in general to vary continuously, would present a different aspect if we were able to discern molecular structure ; we should then perceive the individual electrons by which the field is produced, and the rapid fluctuations of electric and magnetic force between them. As it is, the values furnished by our instruments represent averages taken over volumes which, though they appear small to us, are large compared with molecular dimensions.[4] We

[1] cf. p. 310

[2] The derivation of the complete set of equations for the electromagnetic field, including this ponderomotive equation, from the Principle of Least Action, was given by K. Schwarzschild, *Gött. Nach.* (1903), p. 126.

[3] *Phil. Mag.* xliv (1897), p. 503

[4] These principles had been enunciated, and to some extent developed, by J. Willard Gibbs in 1882–3 : *Amer. Jour. Sci.* xxiii, pp. 262, 460 ; xxv, p. 107 ; Gibbs's *Scientific Papers*, ii, pp. 182, 195, 211.

shall denote an average value of this kind by a bar placed over the corresponding symbol.

Lorentz supposed that the phenomena of electrostatic charge and of conduction currents are due to the presence or motion of simple electrons such as have been considered above. The part of $\bar{\rho}$ arising from these is the measurable density of electrostatic charge; this we shall denote by ρ_1. If $\mathbf{w}$ denote the velocity of the ponderable matter, and if the velocity $\mathbf{v}$ of the electrons be written $\mathbf{w} + \mathbf{u}$, then the quantity $\overline{\rho\mathbf{v}}$, so far as it arises from electrons of this type, may be written $\rho_1\mathbf{w} + \overline{\rho\mathbf{u}}$. The former of these terms represents the convection current, and the latter the conduction current.

Consider next the phenomena of dielectrics. Following Faraday, Thomson and Mossotti,[1] Lorentz supposed that each dielectric molecule contains corpuscles charged vitreously and also corpuscles charged resinously. These in the absence of an external field are so arranged as to neutralise each other's electric fields outside the molecule. For simplicity we may suppose that in each molecule only one corpuscle, of charge e, is capable of being displaced from its position; it follows from what has been assumed that the other corpuscles in the molecule exert the same electrostatic action as a charge $-e$ situated at the original position of this corpuscle. Thus if e is displaced to an adjacent position, the entire molecule becomes equivalent to an electric doublet, whose moment is measured by the product of e and the displacement of e. The molecules in unit volume, taken together, will in this way give rise to a (vector) electric moment per unit volume, $\mathbf{P}$, which may be compared to the (vector) intensity of magnetisation in Poisson's theory of magnetism.[2] As in that theory we may replace the doublet distribution $\mathbf{P}$ of the scalar quantity ρ by a volume distribution of ρ, determined by the equation[3]

$$\bar{\rho} = -\operatorname{div}\mathbf{P}.$$

This represents the part of $\bar{\rho}$ due to the dielectric molecules.

Moreover, the scalar quantity ρw_x has also a doublet distribution, to which the same theorem may be applied; the average value of the part of ρw_x, due to dielectric molecules, is therefore determined by the equation

$$\overline{\rho w_x} = -\operatorname{div}(w_x\mathbf{P}) = -w_x \operatorname{div}\mathbf{P} - (\mathbf{P}\cdot\nabla)\,w_x,$$

or

$$\overline{\rho\mathbf{w}} = -\operatorname{div}\mathbf{P}\cdot\mathbf{w} - (\mathbf{P}\cdot\nabla)\,\mathbf{w}.$$

[1] cf. pp. 187, 188 [2] cf. p. 62–5

[3] We assume all transitions gradual, so as to avoid surface distributions.

We have now to find that part of $\overline{\rho \mathbf{u}}$ which is due to dielectric molecules. For a single doublet of moment $\mathbf{p}$ we have, by differentiation,

$$\iiint \rho \mathbf{u} \, dx \, dy \, dz = d\mathbf{p}/dt,$$

where the integration is taken throughout the molecule ; so that

$$\iiint \rho \mathbf{u} \, dx \, dy \, dz = (d/dt)\,(\mathrm{V}\mathbf{P}),$$

where the integration is taken throughout a volume V, which encloses a large number of molecules, but which is small compared with measurable quantities ; and this equation may be written

$$\overline{\rho \mathbf{u}} = \frac{1}{\mathrm{V}} \frac{d}{dt} (\mathrm{V}\mathbf{P}).$$

Now, if $\dfrac{\partial \mathbf{P}}{\partial t}$ refers to differentiation at a fixed point of space (as opposed to a differentiation which accompanies the moving body), we have

$$(d/dt)\,\mathbf{P} = (\partial/\partial t)\,\mathbf{P} + (\mathbf{w} \,.\, \nabla)\,\mathbf{P},$$

and

$$(d/dt)\,\mathrm{V} = \mathrm{V}\,\mathrm{div}\,\mathbf{w}\,;$$

so that

$$\begin{aligned} \overline{\rho \mathbf{u}} &= \frac{\partial \mathbf{P}}{\partial t} + (\mathbf{w} \,.\, \nabla)\,\mathbf{P} + \mathrm{div}\,\mathbf{w} \,.\, \mathbf{P} \\ &= \frac{\partial \mathbf{P}}{\partial t} + \mathrm{curl}\,[\mathbf{P} \,.\, \mathbf{w}] + \mathrm{div}\,\mathbf{P} \,.\, \mathbf{w} + (\mathbf{P} \,.\, \nabla)\,\mathbf{w}, \end{aligned}$$

and therefore

$$\overline{\rho \mathbf{u}} + \overline{\rho \mathbf{w}} = \frac{\partial \mathbf{P}}{\partial t} + \mathrm{curl}\,[\mathbf{P} \,.\, \mathbf{w}].$$

This equation determines the part of $\overline{\rho \mathbf{v}}$ which arises from the dielectric molecules.

The general equations of the aether thus become, when the averaging process is performed,

$$\mathrm{div}\,\bar{\mathbf{d}} = 4\pi\rho_1 - 4\pi\,\mathrm{div}\,\mathbf{P}, \qquad \mathrm{div}\,\bar{\mathbf{h}} = 0,$$

$$\mathrm{curl}\,\bar{\mathbf{d}} = -\frac{1}{c}\frac{\partial \bar{\mathbf{h}}}{\partial t},$$

$$\mathrm{curl}\,\bar{\mathbf{h}} = (1/c)\frac{\partial \bar{\mathbf{d}}}{\partial t} + \frac{4\pi}{c}\left\{ \begin{matrix} \text{convection current + conduction current} \\ + \dfrac{\partial \mathbf{P}}{\partial t} + \mathrm{curl}\,[\mathbf{P} \,.\, \mathbf{w}] \end{matrix} \right\}$$

In order to assimilate these to the ordinary electromagnetic equations, we must evidently write

$$\bar{\mathbf{d}} = \mathbf{E}, \quad \text{the electric force ;}$$

$$\mathbf{E} + 4\pi\,\mathbf{P} = \mathbf{D}, \quad \text{the electric induction ;}$$

$$\bar{\mathbf{h}} = \mathbf{H}, \quad \text{the magnetic vector.}$$

The equations then become (writing ρ for ρ_1, as there is no longer any need to use the subscript),

$$\operatorname{div}\mathbf{D} = 4\pi\rho, \qquad -\operatorname{curl}\mathbf{E} = \frac{1}{c}\frac{\partial\mathbf{H}}{\partial t},$$

$$\operatorname{div}\mathbf{H} = 0, \qquad \operatorname{curl}\mathbf{H} = \frac{4\pi}{c}\mathbf{S},$$

where

$$\mathbf{S} = \text{conduction current} + \text{convection current} + \frac{1}{4\pi}\frac{\partial\mathbf{D}}{\partial t} + \operatorname{curl}[\mathbf{P}\,.\,\mathbf{w}].$$

The term $\frac{1}{4\pi}\cdot\frac{\partial\mathbf{D}}{\partial t}$ in $\mathbf{S}$ evidently represents the displacement-current of Maxwell ; and the term curl $[\mathbf{P}\,.\,\mathbf{w}]$ will be recognised as a modified form of the term curl $[\mathbf{D}\,.\,\mathbf{w}]$, which was first introduced into the equations by Hertz.[1] It will be remembered that Hertz supposed this term to represent the generation of a magnetic force within a dielectric which is in motion in an electric field ; and that Heaviside,[2] by adducing considerations relative to the energy, showed that the term ought to be regarded as part of the total current, and inferred from its existence that a dielectric which moves in an electric field is the seat of an electric current, which produces a magnetic field in the surrounding space. The modification introduced by Lorentz consisted in replacing **D** by $4\pi\mathbf{P}$ in the vector product ; this implied that the moving dielectric does not carry along the aethereal displacement, which is represented by the term **E** in **D**, but only carries along the charges which exist at opposite ends of the molecules of the ponderable dielectric, and which are represented by the term **P**. The part of the total current represented by the term curl $[\mathbf{P}\,.\,\mathbf{w}]$ is generally called the *current of dielectric convection.*

That a magnetic field is produced when an uncharged dielectric is in motion at right angles to the lines of force of a constant electrostatic field had been shown experimentally in 1888 by Röntgen.[3] His experiment consisted in rotating a dielectric disk between the plates of a condenser ; a magnetic field was produced, equivalent to that which would be produced by the rotation of the 'fictitious charges' on the two faces of the dielectric, i.e., charges which bear the same relation to the dielectric polarisation that Poisson's equivalent surface-density of magnetism[4] bears to magnetic polarisation. If U denote the difference of potential between the opposite coatings of the condenser, and ϵ the specific inductive capacity of the dielectric, the surface-density of electric charge on the coatings is proportional

[1] cf. p.329
[2] cf. p. 330
[3] *Ann. d. Phys.* xxxv (1888), p. 264 ; xl (1890), p. 93
[4] cf. p. 64

to $\pm \epsilon U$, and the fictitious charge on the surfaces of the dielectric is proportional to $\mp (\epsilon - 1) U$. It is evident from this that if a plane condenser is charged to a given difference of potential, and is rotated in its own plane, the magnetic field produced is proportional to ϵ if (as in Rowland's experiment [1]) the coatings are rotated while the dielectric remains at rest, but is in the opposite direction, and is proportional to $(\epsilon - 1)$ if (as in Röntgen's experiment) the dielectric is rotated while the coatings remain at rest. If the coatings and dielectric are rotated together, the magnetic action (being the sum of these) should be independent of ϵ—a conclusion which was verified later by Eichenwald.[2]

Hitherto we have taken no account of the possible magnetisation of the ponderable body. This would modify the equations in the usual manner,[3] so that they finally take the form

$$\operatorname{div} \mathbf{D} = 4\pi\rho, \tag{I}$$

$$\operatorname{div} \mathbf{B} = 0, \tag{II}$$

$$\operatorname{curl} \mathbf{H} = \frac{4\pi}{c} \mathbf{S}, \tag{III}$$

$$-\operatorname{curl} \mathbf{E} = \frac{1}{c}\frac{\partial \mathbf{B}}{\partial t}, \tag{IV}$$

where $\mathbf{S}$ denotes the total current formed of the displacement current, the convection current, the conduction current and the current of dielectric convection. Moreover, since

$$\mathbf{S} = \overline{\rho \mathbf{v}} + \frac{1}{4\pi}\frac{\overline{\partial \mathbf{d}}}{\partial t},$$

we have

$$\begin{aligned} \operatorname{div} \mathbf{S} &= \operatorname{div} \overline{\rho \mathbf{v}} + (1/4\pi) \operatorname{div} (\overline{\partial \mathbf{d}}/\partial t) \\ &= \operatorname{div} \overline{\rho \mathbf{v}} + \partial \rho/\partial t, \end{aligned}$$

which vanishes by virtue of the principle of conservation of electricity. Thus

$$\operatorname{div} \mathbf{S} = 0, \tag{V}$$

or the total current is a circuital vector. Equations (I) to (V) are the fundamental equations of Lorentz' theory of electrons.

[1] cf. p. 305

[2] *Ann. d. Phys.* xi (1903), p. 421 ; xiii (1904), p. 919. Eichenwald performed other experiments of a similar character, e.g. he observed the magnetic field due to the changes of polarisation in a dielectric which was moved in a non-homogeneous electric field.

[3] It is possible to construct a purely electronic theory of magnetisation, a magnetic molecule being supposed to contain electrons in orbital revolution. It then appears that the vector which represents the average value of **h** is not **H**, but **B**.

We have now to consider the relation by which the polarisation **P** of dielectrics is determined. If the dielectric is moving with velocity **w**, the ponderomotive force on unit electric charge moving with it is (as in all theories.)[1]

$$\mathbf{E}' = \mathbf{E} + \frac{1}{c}[\mathbf{w}\,.\,\mathbf{B}]. \qquad (1)$$

In order to connect **P** with **E**′, it is necessary to consider the motion of the corpuscles. Let e denote the charge and m the mass of a corpuscle, (ξ, η, ζ) its displacement from its position of equilibrium, $k^2(\xi, \eta, \zeta)$ the restitutive force which retains it in the vicinity of this point; then the equations of motion of the corpuscle are

$$m\frac{d^2\xi}{dt^2} + k^2\xi = e\mathrm{E}_x',$$

and similar equations in η and ζ. When the corpuscle is set in motion by light of frequency n passing through the medium, the displacements and forces will be periodic functions of nt—say,

$$\xi = \mathrm{A}e^{nt\sqrt{-1}}, \qquad \mathrm{E}_x' = \mathrm{E}_0 e^{nt\sqrt{-1}}.$$

Substituting these values in the equations of motion, we obtain

$$\mathrm{A}(k^2 - mn^2) = e\mathrm{E}_0, \quad \text{and therefore} \quad \xi(k^2 - mn^2) = e\mathrm{E}_x'.$$

Thus, if N denote the number of polarisable molecules per unit volume, the polarisation is determined by the equation

$$\mathbf{P} = \mathrm{N}e(\xi, \eta, \zeta) = \mathrm{N}e^2\,\mathbf{E}'/(k^2 - mn^2).$$

In the particular case in which the dielectric is at rest, this equation gives

$$\mathbf{D} = \mathbf{E} + 4\pi\mathbf{P} = \mathbf{E} + 4\pi\mathrm{N}e^2\mathbf{E}/(k^2 - mn^2).$$

But, as we have seen,[2] **D** bears to **E** the ratio μ^2, where μ denotes the refractive index of the dielectric; and therefore the refractive index is determined in terms of the frequency by the equation

$$\mu^2 = 1 + 4\pi e^2\mathrm{N}/(k^2 - mn^2).$$

This formula is equivalent to that which Maxwell and Sellmeier[3] had derived from the elastic-solid theory. Though superficially different, the derivations are alike in their essential feature, which is the assumption that the molecules of the dielectric contain systems which possess free periods of vibration, and which respond to the oscillations of the incident light. The formula may be derived on electromagnetic principles without any explicit reference to electrons;

[1] cf. p. 329 [2] cf. p. 253 [3] cf. p. 264

all that is necessary is to assume that the dielectric polarisation has a free period of vibration.[1]

When the luminous vibrations are very slow, so that n is small, μ^2 reduces to the dielectric constant ϵ [2]; so that the theory of Lorentz leads to the expression

$$\epsilon = 1 + 4\pi N e^2/k^2$$

for the specific inductive capacity in terms of the number and circumstances of the electrons.[3]

Returning now to the case in which the dielectric is supposed to be in motion, the equation for the polarisation may be written

$$4\pi \mathbf{P} = (\mu^2 - 1)\, \mathbf{E}', \text{ where } \mathbf{E}' \text{ is given by (1) ;} \qquad (2)$$

from this equation, Fresnel's formula for the velocity of light in a moving dielectric may be deduced. For, let the axis of z be taken parallel to the direction of motion of the dielectric, which is supposed to be also the direction of propagation of the light ; and, considering a plane-polarised wave, take the axis of x parallel to the electric vector, so that the magnetic vector must be parallel to the axis of y. Then equation (III) above becomes

$$-\frac{\partial H_y}{\partial z} = \frac{1}{c}\frac{\partial D_x}{\partial t} + 4\pi \frac{w}{c}\frac{\partial P_x}{\partial z};$$

equation (IV) becomes (assuming B equal to H, as is always the case in optics),

$$-\frac{\partial E_x}{\partial z} = \frac{1}{c}\frac{\partial H_y}{\partial t}.$$

The equation which defines the electric induction gives

$$D_x = E_x + 4\pi P_x ;$$

[1] A theory of dispersion, which, so far as its physical assumptions and results are concerned, resembles that described above, was published in the same year (1892) by Helmholtz (*Berl. Ber.* [1892], p. 1093 ; *Ann. d. Phys.* xlviii [1893], pp. 389, 723). In this, as in Lorentz' theory, the incident light is supposed to excite sympathetic vibrations in the electric doublets which exist in the molecules of transparent bodies. Helmholtz' equations were, however, derived in a different way from those of Lorentz, being deduced from the Principle of Least Action. The final result is, as in Lorentz' theory, represented (when the effect of damping is neglected) by the Maxwell-Sellmeier formula. Helmholtz' theory was developed further by Reiff (*Ann. d. Phys.* lv [1895], p. 82).

In a theory of dispersion given by Planck (*Berl. Ber.* [1902], p. 470), the damping of the oscillations is assumed to be due to the loss of energy by radiation : so that no new constant is required in order to express it.

Lorentz, in his lectures on the *Theory of Electrons* (Leipzig, 1909, p. 141), suggested that the dissipative term in the equations of motion of dielectric electrons might be ascribed to the destruction of the regular vibrations of the electrons within a molecule by the collisions of the molecule with other molecules.

Some interesting references to the ideas of Hertz on the electromagnetic explanation of dispersion will be found in a memoir by Drude (*Ann. d. Phys.* (6), i [1900], p. 437).

[2] cf. p. 254 [3] cf. p. 188

and equations (1) and (2) give

$$4\pi P_x = (\mu^2 - 1)\left(E_x - \frac{w}{c} H_y\right).$$

Eliminating D_x, P_x and H_y, we have

$$c^2 \frac{\partial^2 E_x}{\partial z^2} = \frac{\partial^2 E_x}{\partial t^2} + \left(\mu^2 - 1\right)\left(\frac{\partial}{\partial t} + w\frac{\partial}{\partial z}\right)^2 E_x\,;$$

or, neglecting w^2/c^2,

$$\frac{\partial^2 E_x}{\partial z^2} = \frac{\mu^2}{c^2}\frac{\partial^2 E_x}{\partial t^2} + \frac{2w(\mu^2 - 1)}{c^2}\frac{\partial^2 E_x}{\partial t\, \partial z}.$$ [1]

Substituting $E_x = e^{n(t - z/V)\sqrt{-1}}$, so that V denotes the velocity of light in the moving dielectric with respect to the fixed aether, we have

$$c^2 = \mu^2 V^2 - 2w(\mu^2 - 1)V,$$

or (neglecting w^2/c^2)

$$V = \frac{c}{\mu} + \frac{\mu^2 - 1}{\mu^2} w,$$

which is the formula of Fresnel.[2] The hypothesis of Fresnel, that a ponderable body in motion carries with it the excess of aether which it contains as compared with space free from matter, is thus seen to be transformed in Lorentz' theory into the supposition that the polarised molecules of the dielectric, like so many small condensers, increase the dielectric constant, and that it is (so to speak) this augmentation of the dielectric constant which travels with the moving matter. One evident objection to Fresnel's theory, namely, that it required the relative velocity of aether and matter to be different for light of different colours, is thus removed; for the theory of Lorentz only requires that the dielectric constant should have different values for light of different colours, and of this a satisfactory explanation is provided by the theory of dispersion.

The correctness of Lorentz' hypothesis, as opposed to that of Hertz (in which the whole of the contained aether was supposed to be transported with the moving body), was afterwards confirmed by various experiments. In 1901 R. Blondlot[3] drove a current of air through a magnetic field, at right angles to the lines of magnetic force. The air current was made to pass between the faces of a condenser, which were connected by a wire, so as to be at the same potential. An electromotive force **E′** would be produced in the air

[1] This equation was first given as a result of the theory of electrons by Lorentz in the last chapter of his memoir of 1892, *Arch. Néerl.* xxv, p. 525. It was also given by Larmor, *Phil. Trans.* clxxxv (1894), p. 821.

[2] cf. p. 110

[3] *Comptes Rendus*, cxxxiii (1901), p. 778

by its motion in the magnetic field ; and, according to the theory of Hertz, this should produce an electric induction **D** of amount $\epsilon\mathbf{E}'$ (where ϵ denotes the specific inductive capacity of the air, which is practically unity) ; so that, according to Hertz, the faces of the condenser should become charged. According to Lorentz' theory, on the other hand, the electric induction **D** is determined by the equation

$$\mathbf{D} = \mathbf{E} + (\epsilon - 1)\,\mathbf{E}'$$

where **E** denotes the electric force on a charge at rest, which is zero in the present case. Thus, according to Lorentz' theory, the charges on the faces would have only $(\epsilon - 1)/\epsilon$ of the values which they would have in Hertz' theory ; that is, they would be practically zero. The result of Blondlot's experiment was in favour of the theory of Lorentz.

An experiment of a similar character was performed in 1905 by H. A. Wilson.[1] In this, the space between the inner and outer coatings of a cylindrical condenser was filled with the dielectric ebonite. When the coatings of such a condenser are maintained at a definite difference of potential, charges are induced on them ; and if the condenser be rotated on its axis in a magnetic field whose lines of force are parallel to the axis, these charges will be altered, owing to the additional polarisation which is produced in the dielectric molecules by their motion in the magnetic field. As before, the value of the additional charge according to the theory of Lorentz is $(\epsilon - 1)/\epsilon$ times its value as calculated by the theory of Hertz. The result of Wilson's experiments was, like that of Blondlot's, in favour of Lorentz.

The reconciliation of the electromagnetic theory with Fresnel's law of the propagation of light in moving bodies was a distinct advance. But the theory of the motionless aether was hampered by one difficulty : it was, in its original form, incompetent to explain the negative result of the experiment of Michelson and Morley.[2] The adjustment of theory to observation in this particular was achieved by means of a remarkable hypothesis which must now be introduced.

In the issue of *Nature* for 16 June 1892,[3] Lodge mentioned that FitzGerald had communicated to him a new suggestion for overcoming the difficulty. This was, to suppose that the dimensions of material bodies are slightly altered when they are in motion relative

[1] *Phil. Trans.* cciv (1905), p. 121

[2] cf. p. 390

[3] *Nature*, xlvi (1892), p. 165

to the aether. Five months afterwards, this hypothesis of FitzGerald's was adopted by Lorentz, in a communication to the Amsterdam Academy [1]; after which it won favour in a gradually widening circle, until eventually it came to be generally taken as the basis of all theoretical investigations on the motion of ponderable bodies through the aether.

Let us first see how it explains Michelson's result. On the supposition that the aether is motionless, one of the two portions into which the original beam of light is divided should accomplish its journey in a time less than the other by w^2l/c^3, where w denotes the velocity of the earth, c the velocity of light and l the length of each arm. This would be exactly compensated if the arm which is pointed in the direction of the terrestrial motion were shorter than the other by an amount $w^2l/2c^2$; as would be the case if the linear dimensions of moving bodies were always contracted in the direction of their motion in the ratio of $(1 - w^2/2c^2)$ to unity. This is FitzGerald's hypothesis of contraction. Since for the earth the ratio w/c is only

$$\frac{30 \text{ km./sec.}}{300{,}000 \text{ km./sec.}},$$

the fraction w^2/c^2 is only one hundred-millionth.

Several further contributions to the theory of electrons in a motionless aether were made in a short treatise [2] which was published by Lorentz in 1895. One of these related to the explanation of an experimental result obtained some years previously by Th. des Coudres,[3] of Leipzig. Des Coudres had observed the mutual inductance of coils in different circumstances of inclination of their common axis to the direction of the earth's motion, but had been unable to detect any effect depending on the orientation. Lorentz now showed that this could be explained by considerations similar to those which Budde and FitzGerald [4] had advanced in a similar case; a conductor carrying a constant electric current and moving with the earth would exert a force on electric charges at relative rest in its vicinity, were it not that this force induces on the surface of the conductor itself a compensating electrostatic charge, whose action annuls the expected effect.

Another result first published in the *Versuch* [5] was that Fresnel's

[1] *Verslagen d. Kon. Ak. van Wetenschappen, I, 1892–3* (26 Nov. 1892), p. 74

[2] *Versuch einer Theorie der electrischen und optischen Erscheinungen in bewegten Körpern,* von H. A. Lorentz; Leiden, E. J. Brill. It was reprinted by Teubner, of Leipzig, in 1906.

[3] *Ann. d. Phys.* xxxviii (1889), p. 73

[4] cf. p. 235

[5] *Versuch,* p. 101: Lorentz, *Theory of Electrons* (1909), pp. 191, 316

formula for the propagation of light in a moving medium requires correction when the medium is dispersive: namely, the formula for the velocity should read,

$$\frac{c}{\mu} + \left(1 - \frac{1}{\mu^2} - \frac{\lambda}{\mu}\frac{d\mu}{d\lambda}\right)w.$$

Subsequent experiments by Zeeman [1] verified the corrected formula with an accuracy that leaves nothing to be desired.

The most satisfactory method of discussing the influence of the terrestrial motion on electrical phenomena is to transform the fundamental equations of the aether and electrons to axes moving with the earth. Taking the axis of x parallel to the direction of the earth's motion, and denoting the velocity of the earth by w, we write

$$x = x_1 + wt, \quad y = y_1, \quad z = z_1,$$

so that (x_1, y_1, z_1) denote co-ordinates referred to axes moving with the earth. Lorentz completed the change of co-ordinates by introducing in place of the variable t a 'local time' t_1, defined by the equation

$$t = t_1 + wx_1/c^2.$$

It is also necessary to introduce, in place of **d** and **h**, the electric and magnetic forces relative to the moving axes: these are [2]

$$\mathbf{d}_1 = \mathbf{d} + \frac{1}{c}[\mathbf{w}\,.\,\mathbf{h}]$$

$$\mathbf{h}_1 = \mathbf{h} + (1/c)\,[\mathbf{d}\,.\,\mathbf{w}]\,;$$

and in place of the velocity **v** of an electron referred to the original fixed axes, we must introduce its velocity $\mathbf{v}_1$ relative to the moving axes, which is given by the equation

$$\mathbf{v}_1 = \mathbf{v} - \mathbf{w}.$$

The fundamental equations of the aether and electrons, referred to the original axes, are

$$\operatorname{div}\mathbf{d} = 4\pi\rho, \qquad \operatorname{curl}\mathbf{d} = -\frac{1}{c}\frac{\partial\mathbf{h}}{\partial t},$$

$$\operatorname{div}\mathbf{h} = 0, \qquad \operatorname{curl}\mathbf{h} = \frac{1}{c}\frac{\partial\mathbf{d}}{\partial t} + \frac{4\pi}{c}\rho\mathbf{v},$$

$$\mathbf{F} = \mathbf{d} + \frac{1}{c}[\mathbf{v}\,.\,\mathbf{h}],$$

where **F** denotes the ponderomotive force on a particle carrying a unit charge.

[1] *Proc. Amst. Ac.* xvii (1914), p. 445 : xvii 1915), p. 398

[2] As usual, we take electrostatic units for **d** and electromagnetic units for **h**; cf. p. 202

By direct transformation from the original to the new variables it is found that, when quantities of order w^2/c^2 and wv/c^2 are neglected, these equations take the form

$$\operatorname{div}_1 \mathbf{d}_1 = 4\pi\rho, \qquad \operatorname{curl}_1 \mathbf{d}_1 = -\frac{1}{c}\,\partial\mathbf{h}_1/\partial t_1,$$

$$\operatorname{div}_1 \mathbf{h}_1 = 0, \qquad \operatorname{curl}_1 \mathbf{h}_1 = (1/c)\,\partial\mathbf{d}_1/\partial t_1 + \frac{4\pi}{c}\,\rho\mathbf{v}_1,$$

$$\mathbf{F} = \mathbf{d}_1 + \frac{1}{c}\,[\mathbf{v}_1 \,.\, \mathbf{h}_1],$$

where $\operatorname{div}_1 \mathbf{d}_1$ stands for

$$\frac{\partial d_{x1}}{\partial x_1} + \frac{\partial d_{y1}}{\partial y_1} + \frac{\partial d_{z1}}{\partial z_1}.$$

Since these have the same form as the original equations, it follows that when terms depending on the square of the constant of aberration are neglected, all electrical phenomena may be expressed with reference to axes moving with the earth by the same equations as if the axes were at rest relative to the aether.

In the last chapter of the *Versuch* Lorentz dicussed those experimental results which were as yet unexplained by the theory of the motionless aether. That the terrestrial motion exerts no influence on the rotation of the plane of polarisation in quartz [1] might be explained by supposing that two independent effects, which are both due to the earth's motion, cancel each other ; but Lorentz left the question undecided. Five years later Larmor [2] criticised this investigation, and arrived at the conclusion that there should be no first order effect ; but Lorentz [3] afterwards maintained his position against Larmor's criticism.

Although the physical conceptions of Lorentz had from the beginning included that of atomic electric charges, the analytical equations had hitherto involved ρ, the volume-density of electric charge ; that is, they had been conformed to the hypothesis of a continuous distribution of electricity in space. It might hastily be supposed that in order to obtain an analytical theory of electrons, nothing more would be required than to modify the formulae by writing e (the charge of an electron) in place of $\rho\, dxdydz$. That this is not the case was shown [4] a few years after the publication of the *Versuch*.

[1] cf. p. 390 [2] Larmor, *Aether and Matter*, 1900

[3] *Proc. Amsterdam Acad.* (English edn.), iv (1902), p. 669

[4] A. Liénard, *L'Eclairage élect.* xvi (1898), pp. 5, 53, 106 ; E. Wiechert, *Arch. Néerl.* (2), v (1900), p. 549. For the solution of many special problems in which the Lienard potentials are involved, cf. G. A. Schott, *Electromagnetic Radiation* (Cambridge University Press, 1912).

Consider, for example, the formula for the scalar potential at any point in the aether.

$$\phi = \iiint \overline{(\rho'/r)}\, dx'dy'dz',$$

where the bar indicates that the quantity underneath it is to have its retarded value.[1]

This integral, in which the integration is extended over all elements of space, must be transformed before the integration can be taken to extend over moving elements of charge. Let de' denote the sum of the electric charges which are accounted for under the heading of the volume-element $dx'dy'dz'$ in the above integral. This quantity de' is not identical with $\bar{\rho}'dx'dy'dz'$. For, to take the simplest case, suppose that it is required to compute the value of the potential-function for the origin at the time t, and that the charge is receding from the origin along the axis of x with velocity u. The charge which is to be ascribed to any position x is the charge which occupies that position at the instant $t - x/c$; so that when the reckoning is made according to intervals of space, it is necessary to reckon within a segment $(x_2 - x_1)$ not the electricity which at any one instant, occupies that segment, but the electricity which at the instant $(t - x_2/c)$ occupies a segment $(x_2 - x'_1)$, where x'_1 denotes the point from which the electricity streams to x_1 in the interval between the instants $(t - x_2/c)$ and $(t - x_1/c)$. We have evidently

$$x_1 - x'_1 = u\,(x_2 - x_1)/c, \quad \text{or} \quad x_2 - x'_1 = (x_2 - x_1)\,(1 + u/c).$$

For this case we should therefore have

$$de' = \frac{x_2 - x'_1}{x_2 - x_1}\bar{\rho}'\, dx'dy'dz' = \left(1 + \frac{u}{c}\right)\bar{\rho}'\, dx'dy'dz'.$$

In the general case, it is only necessary to replace u by the component of velocity of the electric charge in the direction of the radius vector from the point at which the potential is to be computed. This component may be written $v \cos (v\,.\,r)$, where r is measured positively from the point in question to the charge, and v denotes the velocity of the charge. Thus

$$cde' = \{c + v\cos(v \hat{.} r)\}\,\bar{\rho}'dx'dy'dz',$$

and therefore [2]

$$\phi\,(x, y, z, t) = \Sigma\,\frac{ec}{c\bar{r} + \bar{r}\bar{v}\cos(\bar{v}, \bar{r})}$$

where the summation is extended over all the charges in the field,

[1] cf. p. 268

[2] Still using electrostatic units for **d** and electromagnetic units for **h**

and the bars over the letters imply that the position of the charge considered is that which it occupied at the instant $t - \bar{r}/c$. In the same way the vector-potential may be shown to have the value given by

$$a_x(x, y, z, t) = \Sigma \frac{e\bar{v}_x}{c\bar{r} + \bar{r}\,\bar{v}\cos(\bar{v}, \bar{r})}$$

and two similar equations.

Any electromagnetic field is thus expressed in terms of the four functions ϕ, a_x, a_y, a_z (the scalar-potential and the three components of the vector-potential), and these are given by the above formulae in terms of the positions and velocities of the electrons which generate the field. It was however shown in 1904 by E. T. Whittaker[1] that only *two* functions are actually necessary (in place of the four), namely, functions F and G defined by the equations

$$F(x, y, z, t) = \tfrac{1}{2}\Sigma e \log \frac{\bar{r} + \bar{z}' - z}{\bar{r} - (\bar{z}' - z)},$$

$$G(x, y, z, t) = -\tfrac{1}{2} i\,\Sigma e \log \frac{\bar{x}' - x + i(\bar{y}' - y)}{\bar{x}' - x - i(\bar{y}' - y)}$$

where the summation is taken over all the electrons in the field, and where $x'(t)$, $y'(t)$, $z'(t)$ denote the position of the electron at the instant t, and $\bar{x}'(t)$ signifies the value of x' at an instant such that a light-signal sent from the electron at this instant reaches the point (x, y, z) at the instant t, so that $\bar{x}' = x'(t - \bar{r}/c)$. Thus $\bar{x}', \bar{y}', \bar{z}'$ are known functions of (x, y, z, t) when the motions of the electrons are known, and

$$\bar{r}^2 = (\bar{x}' - x)^2 + (\bar{y}' - y)^2 + (\bar{z}' - z)^2.$$

The electric vector (d_x, d_y, d_z) and the magnetic vector (h_x, h_y, h_z) are then given by the formulae

$$d_x = \frac{\partial^2 F}{\partial x \partial z} + \frac{1}{c}\frac{\partial^2 G}{\partial y \partial t}, \qquad h_x = \frac{1}{c}\frac{\partial^2 F}{\partial y \partial t} - \frac{\partial^2 G}{\partial x \partial z}$$

$$d_y = \frac{\partial^2 F}{\partial y \partial z} - \frac{1}{c}\frac{\partial^2 G}{\partial x \partial t}, \qquad h_y = -\frac{1}{c}\frac{\partial^2 F}{\partial x \partial t} - \frac{\partial^2 G}{\partial y \partial z}$$

$$d_z = \frac{\partial^2 F}{\partial z^2} - \frac{1}{c^2}\frac{\partial^2 F}{\partial t^2}, \qquad h_z = \frac{\partial^2 G}{\partial x^2} + \frac{\partial^2 G}{\partial y^2}.$$

It will be noted that F and G are defined in terms of the positions of the electrons alone, and do not explicitly involve their velocities.

Since in the above formulae for **d** and **h** an interchange of electric and magnetic quantities corresponds to a change of G into F

[1] *Proc. Lond. Math. Soc.* (2), i (1904), p. 367

and of F into G, it is clear that the two functions F and G exhibit the duality which is characteristic of electromagnetic theory: thus an electrostatic field can be described by F alone, and a magnetostatic field by G alone; again, if the field consists of a plane wave of light, then the functions F and G correspond respectively to two plane-polarised components into which it can be resolved. Since there are an infinite number of ways of resolving a plane wave of light into two plane-polarised components, it is natural to expect that, corresponding to any given electromagnetic field, there should be an infinite number of pairs of functions F and G capable of describing it, their difference from each other depending on the choice of the axes of co-ordinates—as is in fact the case. Thus there is a physical reason why any particular pair of functions F and G should be specially related to one co-ordinate, and cannot be described by formulae symmetrically related to the three co-ordinates (x, y, z).

From what has been said, it will be evident that, in the closing years of the nineteenth century, electrical investigation was chiefly concerned with systems in motion. The theory of electrons was, however, applied with success in other directions, and notably to the explanation of a new experimental discovery.

The last recorded observation of Faraday[1] was an attempt to detect changes in the period, or in the state of polarisation, of the light emitted by a sodium flame, when the flame was placed in a strong magnetic field. No result was obtained; but the conviction that an effect of this nature remained to be discovered was felt by many of his successors. Tait[2] examined the influence of a magnetic field on the selective absorption of light; impelled thereto, as he explained, by theoretical considerations. For from the phenomenon of magnetic rotation it may be inferred[3] that rays circularly polarised in opposite senses are propagated with different velocities in the magnetised medium; and therefore if only those rays are absorbed which have a certain definite wave-length in the medium, the period of the ray absorbed from a beam of circularly polarised white light will not be the same when the polarisation is right-handed as when it is left-handed. 'Thus,' wrote Tait, 'what was originally a single dark absorption line might become a double line.'

The effect anticipated under different forms by Faraday and Tait was discovered towards the end of 1896, by a young Dutch physicist, who was at the time working in Prof. Kamerlingh Onnes'

[1] Bence Jones's *Life of Faraday*, ii, p. 449
[2] *Proc. R. S. Edin.* ix (1875), p. 118
[3] cf. p. 159

laboratory at Leiden, P. Zeeman (1865–1943).[1] Repeating Faraday's procedure, he placed a sodium flame between the poles of an electromagnet, and observed a widening of the D lines in the spectrum when the magnetising current was applied.

A theoretical explanation of the phenomenon was immediately furnished to Zeeman by Lorentz.[2] The radiation was supposed to be emitted by electrons which describe orbits within the sodium atoms. If e denote the charge (in electrostatic units) of an electron of mass m, the ponderomotive force which acts on it by virtue of the external magnetic field is $(e/c)\left[\frac{d\mathbf{r}}{dt}\,.\,\mathbf{K}\right]$, where **K** denotes the magnetic force in electromagnetic units and $\mathbf{r}$ denotes the displacement of the electron from its position of equilibrium; and therefore, if the force which restrains the electron in its orbit be $\kappa^2\mathbf{r}$, the equation of motion of the electron is

$$m\frac{d^2\mathbf{r}}{dt^2} + \kappa^2\mathbf{r} = \frac{e}{c}[\mathbf{r}\,.\,\mathbf{K}.]$$

The motion of the electron may (as is shown in treatises on dynamics) be represented by the superposition of certain particular solutions called *principal oscillations*, whose distinguishing property is that they are periodic in the time. In order to determine the principal oscillations, we write $\mathbf{r}_0 e^{nt\sqrt{-1}}$ for $\mathbf{r}$, where $\mathbf{r}_0$ denotes a vector which is independent of the time, and n denotes the frequency of the principal oscillation; substituting in the equation, we have

$$(\kappa^2 - mn^2)\,\mathbf{r}_0 = \frac{en}{c}\sqrt{-1}\,[\mathbf{r}_0\,.\,\mathbf{K}.]$$

This equation may be satisfied either (1) if $\mathbf{r}_0$ is parallel to **K**, in which case it reduces to

$$\kappa^2 - mn^2 = 0,$$

so that n has the value $\kappa m^{-\frac{1}{2}}$, or (2) if $\mathbf{r}_0$ is at right angles to **K**, in which case by squaring both sides of the equation we obtain the result

$$(\kappa^2 - mn^2)^2 = \frac{e^2}{c^2}n^2\mathrm{K}^2,$$

which gives for n the approximate values $\kappa m^{-\frac{1}{2}} \pm e\mathrm{K}/2mc$.

[1] *Verslagen der Akad. v. Wet. te Amsterdam*, v (31 Oct. 1896), p. 181, (28 Nov. 1896), p. 242; vi (29 May 1897), p. 13, (26 June 1897) p. 99, (30 Oct. 1897) p. 260; *Phil. Mag.*[(5)], xliii (1897), p. 226

[2] *Phil. Mag.* xliii (1897), p. 232; *Ann. d. Phys.* lxiii (1897), p. 278. It should be mentioned that before the effect was discovered experimentally, Larmor had examined the question theoretically: the electron was not yet discovered, and Larmor did not suppose that anything smaller than an atom could radiate; since the effect depended on the ratio of charge to mass, Larmor found for it a value nearly 2,000 times too small, and concluded that it would be too small to be observed.

When there is no external magnetic field, so that **K** is zero, the three values of n which have been obtained all reduce to $\kappa m^{-\frac{1}{2}}$, which represents the frequency of vibration of the emitted light before the magnetic field is applied. When the field is applied, this single frequency is replaced by the three frequencies $\kappa m^{-\frac{1}{2}}$, $\kappa m^{-\frac{1}{2}} + e\mathrm{K}/2mc$, $\kappa m^{-\frac{1}{2}} - e\mathrm{K}/2mc$; that is to say, the single line in the spectrum is replaced by three lines close together. The apparatus used by Zeeman in his earliest experiments was not of sufficient power to exhibit this triplication distinctly, and the effect was therefore described at first as a widening of the spectral lines.

We have seen above that the principal oscillation of the electron corresponding to the frequency $\kappa m^{-\frac{1}{2}}$ is performed in a direction parallel to the magnetic force **K**. It will therefore give rise to radiation resembling that of a Hertzian vibrator, and the electric vector of the radiation will be parallel to the lines of force of the external magnetic field. It follows that when the light received in the spectroscope is that which has been emitted in a direction at right angles to the magnetic field, this constituent (which is represented by the middle line of the triplet in the spectrum) will appear polarised in a plane at right angles to the field; but when the light received in the spectroscope is that which has been emitted in the direction of the magnetic force, this constituent will be absent.

We have also seen that the principal oscillations of the electron corresponding to the frequencies $\kappa m^{-\frac{1}{2}} \pm e\mathrm{K}/2mc$ are performed in a plane at right angles to the magnetic field **K**. In order to determine the nature of these two principal oscillations, we observe that it is possible for the electron to describe a circular orbit in this plane, if the radius of the orbit be suitably chosen; for in a circular motion the forces $\kappa^2\mathbf{r}$ and $(e/c)\left[\frac{d\mathbf{r}}{dt}\cdot\mathbf{K}\right]$ would be directed towards the centre of the circle; and it would therefore be necessary only to adjust the radius so that these furnish the exact amount of centripetal force required. Such a motion, being periodic, would be a principal oscillation. Moreover, since the force $(e/c)\left[\frac{d\mathbf{r}}{dt}\cdot\mathbf{K}\right]$ changes sign when the sense of the movement in the circle is reversed, it is evident that there are two such circular orbits, corresponding to the two senses in which the electron may circulate; these must, therefore, be no other than the two principal oscillations of frequencies $\kappa m^{-\frac{1}{2}} \pm e\mathrm{K}/2mc$. When the light received in the spectroscope is that which has been emitted in a direction at right angles to the external magnetic field, the circles are seen edgewise, and the light appears polarised in a plane parallel to the field ; but when the light examined is that which has been emitted in a direction parallel to the external magnetic force, the radiations of frequencies $\kappa m^{-\frac{1}{2}} \pm e\text{K}/2mc$ are seen to be circularly polarised in opposite senses. All these theoretical conclusions have been verified by observation.

It was found by Cornu [1] and by C. G. W. König [2] that the more refrangible component (i.e., the one whose period is shorter than that of the original radiation) has its circular vibration in the same sense as the current in the electromagnet. From this it may be inferred that the vibration must be due to a resinously charged electron ; for let the magnetising current and the electron be supposed to circulate round the axis of z in the direction in which a right-handed screw must turn in order to progress along the positive direction of the axis of z ; then the magnetic force is directed positively along the axis of z, and, in order that the force on the electron may be directed inward to the axis of z (so as to shorten the period), the charge on the electron must be negative.

The value of e/m for this negative electron may be determined by measurement of the separation between the components of the triplet in a magnetic field of known strength ; for, as we have seen, the difference of the frequencies of the outer components is $e\text{K}/mc$. The values of e/m thus determined agreed well with the estimations [3] of e/m for the corpuscles of cathode rays.

Towards the end of 1897 the concordance of theory and observation for the Zeeman effect seemed to be perfect, the number of verified predictions, both qualitative and quantitative, being very great. Yet only fourteen months after Zeeman's first publication, a new experimental discovery was made which led ultimately to the abandonment of the Lorentz explanation.

The author was Thomas Preston, who on 22 December 1897 read a paper to the Royal Dublin Society,[4] in which he announced that the spectral lines λ4722 of zinc and λ4800 of cadmium became *quadruplets* in a strong magnetic field.[5] In the following year A. Cornu [6] obtained a quadruplet (obtained by a splitting of the central compoment of the triplet) as the Zeeman pattern for the sodium line D_1, and a sextuplet (obtained by a splitting of all three components of the triplet) for D_2. Similar results were soon obtained by many other workers : in 1900 C. Runge and F. Paschen [7] found

[1] *Comptes Rendus*, cxxv (1897), p. 555

[2] *Ann. d. Phys.* lxii (1897), p. 240

[3] cf. p. 363

[4] *Trans. Roy. Dublin Soc.* vi (1898), p. 385

[5] Later he found that the quadruplets can be resolved into sextuplets by the splitting of the side lines into doublets.

[6] *Comptes rendus*, cxxvi (1898), p. 181

[7] *Phys. Zs.* i (4 Aug. 1900), p. 480

that the green mercury line λ5461 is split into 11 components, and the blue mercury line λ4359 into 8 components. Each component of the Zeeman triplet, in fact, consists of groups formed of several sub-components. Spectral lines of the same substance may be, and generally are, very differently affected ; but [1] lines which belong to the same series have the same type of resolution ; and corresponding lines in the spectra of different elements of the same family (e.g. the alkalis or alkaline earths) are resolved in the same way. It will appear later that the correct explanation of these phenomena depends on considerations which belong to quantum theory.[2]

The phenomenon discovered by Zeeman is closely related to the Faraday magnetic rotation of the plane of polarisation of light.[3] The connection was first indicated by FitzGerald in a discussion at a meeting of the Royal Society on 20 January 1898, at which a paper on the Zeeman effect was read by T. Preston. In the Faraday effect, a beam of polarised light is transmitted through a medium, which is subjected to a magnetic field, parallel to the lines of force of the field, and it is found that the plane of polarisation is rotated. The beam of plane-polarised light may be regarded as the superposition of two beams which are circularly polarised in opposite senses ; and the rotation of the plane of polarisation is then expressed by the statement that the two circularly polarised beams have different velocities of propagation in the medium. Now the Zeeman effect shows that if we consider a radiation which, in the absence of a magnetic field, is represented by a single spectral line, and if a magnetic field is set up and the radiation is observed in a direction parallel to the field, then the spectral line is replaced by two lines, representing radiations circularly polarised in opposite senses, of frequencies respectively a little greater and a little less than the original line. The relation discovered by Balfour Stewart and Kirchhoff [4] then leads us to infer that what is true of emission is true also of absorption, and that therefore if, for example, a sodium flame is placed in a strong magnetic field, the radiations it will absorb will not be of the same frequency as in the absence of the field, but will be rays circularly polarised in opposite senses, of frequences respectively below and above the original frequency. This may be called the *inverse Zeeman effect* ; but according to all theories of dispersion, the velocity of propagation of a ray of light in a transparent medium is controlled in part by the

[1] T. Preston, *Proc. R. S.* lxiii (1898), p. 26

[2] Lorentz attempted to explain them by the classical electron theory in *Phys. Zs.* i (1899), p. 39

[3] cf. p. 190 [4] cf. p. 371

position of the absorption line of the medium.[1] We therefore conclude that for circularly polarised rays of light propagated in a transparent medium parallel to the lines of magnetic force, the velocity of propagation will be slightly different for rays with the two opposite senses of circular polarisation, since they will be controlled by different absorption lines; and this, as we have just seen, is precisely the Faraday effect.

Later in the same year D. A. Goldhammer[2] (1860–1922) of Kasan and Woldemar Voigt[3] (1850–1919) of Göttingen carried FitzGerald's idea further, by considering what happens when, for example, a sodium flame placed in a magnetic field is traversed by a ray of plane-polarised light, of frequency nearly the same as that of the D radiation, and propagated in a direction perpendicular to the lines of magnetic force; they showed that in this case the propagation would depend on the orientation of the plane of polarisation, the ray being doubly refracted, the direction of the lines of force playing the part of optical axis of a uniaxal crystal. This prediction was confirmed experimentally by Voigt and Wiechert.[4] Voigt gave a mathematical treatment covering the combined theories of the Faraday and inverse Zeeman effects, and not depending on the theory of electrons. He remarked that the magnetic rotation should be very large for light whose period is nearly the same as that corresponding to an absorption line; a large rotation is in fact observed when plane-polarised light, whose frequency differs but little from the frequencies of the D lines, is passed through sodium vapour in a direction parallel to the lines of magnetic force.[5]

A different way of connecting the Faraday and Zeeman effects by general reasoning was given by Larmor in 1899.[6] This depended on a theorem which he had proved in 1897[7] to the effect that when in a material molecule there exists an independently vibrating group of ions or electrons, for all of which the ratio e/m of electric charge to inertia is the same, then the influence of a magnetic field H on the

[1] cf. p. 265

[2] Paper read at Kiev, 10 Sept. 1898, printed *Ann. d. Phys.* lxvii (1899), p. 696

[3] *Gött. Nach.* (26 Nov. 1898), p. 355; *Ann. d. Phys.* lxvii (1899), p. 345; *Phys. Zs.* i (1899), p. 138

[4] Goldhammer had conjectured the existence of such a phenomenon as far back as 1887, but no experimental confirmation was at that time obtained. cf. also A. Cotton, *Comptes Rendus*, cxxviii (1899), p. 294; H. Becquerel, *Comptes Rendus*, cxxviii (1899), p. 145; J. Geest, *Arch. Néerl.* (2), x (1905), p. 291

[5] The phenomenon was first observed by D. Macaluso and O. M. Corbino, *Comptes Rendus*, cxxvii (1898), p. 548; *Nuovo Cimento* (4), viii (1898), p. 257; ix (1899), p. 381; *Rend. Lincei* (5), vii (2) (1898), p. 293. cf. P. Zeeman, *Proc. Amsterdam Acad.* v (1902), p. 41; J. J. Hallo, *Arch. Néerl.* (2), x (1905), p. 148

[6] *Proc. Camb. Phil. Soc.* x (6 Mar. 1899), p. 181

[7] *Phil. Mag.* (5), xliv (Dec. 1897), p. 503

motions of the group is precisely the same as that of a rotation with angular velocity ω, equal to $eH/2mc$ (in electrostatic units for e and electromagnetic units for H), imposed on the group around the axis of the field, on the hypothesis that the extraneous forces acting on the ions are symmetrical with respect to this axis. (This is known as the Larmor precession.) Now, Ω being the angular velocity of the displacement vector in a train of circularly polarised waves traversing the medium along the axis, the state of synchronous vibration which it excites in the molecules will have exactly the same formal relation to this train when the magnetic field is off, as it would have to a train with the very slightly different angular velocity $\Omega \pm \omega$ when the magnetic field is on, the sign being different, according as the train is right-handed or left-handed. But change of this angular velocity Ω means change of period of the light; thus the propagation of a circularly polarised wave-train, when the field is on, is identical with that of the same wave-train when the period is altered by its being carried round with angular velocity $\pm\,\omega$ and there is no influencing magnetic field. This is the Faraday effect.[1]

Yet another way of formulating the connection between the Zeeman and Faraday effects is to say that the molecules of material bodies contain electric systems which possess natural periods of vibration, the simplest example of such a system being an electron which is attracted to a fixed centre. Zeeman's effect represents the influence of an external magnetic field on the *free* oscillations of these electric systems, while Faraday's effect represents the influence of the external magnetic field on the *forced* oscillations which the systems perform under the stimulus of incident light.

The optical properties of metals may be explained, according to the theory of electrons, by a slight extension of the analysis which applies to the propagation of light in transparent substances. It is, in fact, only necessary to suppose that some of the electrons in metals are free instead of being bound to the molecules: a supposition which may be embodied in the equations by assuming that an electric force **E** gives rise to a polarisation **P**, where

$$\mathbf{E} = \alpha \frac{\partial^2 \mathbf{P}}{\partial t^2} + \beta \frac{\partial \mathbf{P}}{\partial t} + \gamma \mathbf{P};$$

the term in α represents the effect of the inertia of the electrons; the term in β represents their ohmic drift; and the term in γ represents the effect of the restitutive forces where these exist. This

[1] See also H. Becquerel, *Comptes Rendus*, cxxv (1897), p. 679

equation is to be combined with the customary electromagnetic equations

$$\text{curl } \mathbf{H} = \frac{1}{c}\frac{\partial \mathbf{E}}{\partial t} + \frac{4\pi}{c}\frac{\partial \mathbf{P}}{\partial t}, \qquad - \text{curl } \mathbf{E} = \frac{1}{c}\frac{\partial \mathbf{H}}{\partial t}.$$

In discussing the propagation of light through the metal, we may for convenience suppose that the beam is plane-polarised and propagated parallel to the axis of z, the electric vector being parallel to the axis of x. Thus the equations of motion reduce to

$$\begin{cases} \dfrac{\partial^2 E_x}{\partial z^2} = \dfrac{1}{c^2}\dfrac{\partial^2 E_x}{\partial t^2} + \dfrac{4\pi}{c^2}\dfrac{\partial^2 P_x}{\partial t^2}, \\ E_x = \alpha \dfrac{\partial^2 P_x}{\partial t^2} + \beta \dfrac{\partial P_x}{\partial t} + \gamma P_x. \end{cases}$$

For E_x and P_x we may substitute exponential functions of

$$n\sqrt{-1}\,(t - z\mu/c),$$

where n denotes the frequency of the light, and μ the quasi-index of refraction of the metal : the equations then give at once

$$(\mu^2 - 1)\,(-\alpha n^2 + \beta n\sqrt{-1} + \gamma) = 4\pi.$$

Writing $\nu\,(1 - \kappa\sqrt{-1})$ for μ, so that ν is inversely proportional to the velocity of light in the medium, and κ denotes the coefficient of absorption, and equating separately the real and imaginary parts of the equation, we obtain

$$\begin{cases} \nu^2\,(1 - \kappa^2) = 1 + \dfrac{4\pi\,(\gamma - \alpha n^2)}{\beta^2 n^2 + (\gamma - \alpha n^2)^2}\,; \\ \nu^2\kappa = \dfrac{2\pi\beta n}{\beta^2 n^2 + (\gamma - \alpha n^2)^2}. \end{cases}$$

When the wave-length of the light is very large, the inertia represented by the constant α has but little influence, and the equations reduce to those of Maxwell's original theory [1] of the propagation of light in metals. The formulae were experimentally confirmed for this case by the researches of E. Hagen and H. Rubens [2] with infra-red light ; a relation being thus established between the ohmic conductivity of a metal and its optical properties with respect to light of great wave-length.

[1] cf. p. 260

[2] *Berlin Sitzungsber.* (1903), pp. 269, 410 ; *Ann. d. Phys.* xi (1903), p. 873 ; *Phil. Mag.* vii (1904), p. 157

When, however, the luminous vibrations are performed more rapidly, the effect of the inertia becomes predominant; and if the constants of the metal are such that, for a certain range of values of n, $\nu^2\kappa$ is small, while $\nu^2(1-\kappa^2)$ is negative, it is evident that, for this range of values of n, ν will be small and κ large, i.e. the properties of the metal will approach those of ideal silver.[1] Finally, for indefinitely great values of n, $\nu^2\kappa$ is small and $\nu^2(1-\kappa^2)$ is nearly unity, so that ν tends to unity and κ to zero: an approximation to these conditions is realised in the X-rays.[2]

In the last years of the nineteenth century, attempts were made to form more definite conceptions regarding the behaviour of electrons within metals. It will be remembered that the original theory of electrons had been proposed by Weber[3] for the purpose of explaining the phenomena of electric currents in metallic wires. Weber, however, made but little progress towards an electric theory of metals; for being concerned chiefly with magneto-electric induction and electromagnetic ponderomotive force, he scarcely brought the metal into the discussion at all, except in the assumption that electrons with charges of opposite signs travel with equal and opposite velocities relative to its substance. The more comprehensive scheme of his successors half a century afterwards aimed at connecting in a unified theory all the known electrical properties of metals, such as the conduction of currents according to Ohm's law, the thermo-electric effects of Seebeck, Peltier and W. Thomson, the galvano-magnetic effect of Hall, and other phenomena which will be mentioned subsequently.

The later investigators, indeed, ranged beyond the group of purely electrical properties, and sought by aid of the theory of electrons to explain the conduction of heat. The principal ground on which this extension was justified was an experimental result obtained in 1853 by G. Wiedemann and R. Franz,[4] who found that at any temperature the ratio of the thermal conductivity of a body to its ohmic conductivity is approximately the same for all metals, and that the value of this ratio is proportional to the absolute temperature. In fact, the conductivity of a pure metal for heat is almost independent of the temperature; while the electric conductivity varies in inverse proportion to the absolute temperature,

[1] cf. p. 162

[2] Models illustrating the selective reflection and absorption of light by metallic bodies and by gases were discussed by H. Lamb, *Mem. and Proc. Manchester Lit. and Phil. Soc.* xlii (1898), p. 1; *Proc. Lond. Math. Soc.* xxxii (1900), p. 11; *Trans. Camb. Phil. Soc.* xviii (1900), p. 348.

[3] cf. p. 201

[4] *Ann. d. Phys.* lxxxix (1853), p. 497

so that a pure metal as it approaches the absolute zero of temperature tends to assume the character of a perfect conductor. That the two conductivities are closely related was shown to be highly probable by the experiments of Tait, in which pieces of the same metal were found to exhibit variations in ohmic conductivity exactly parallel to variations in their thermal conductivity.

The attempt to explain the electrical and thermal properties of metals by aid of the theory of electrons rests on the assumption that conduction in metals is more or less similar to conduction in electrolytes ; at any rate, that positive and negative charges drift in opposite directions through the substance of the conductor under the influence of an electric field. It was remarked in 1888 by J. J. Thomson,[1] who must be regarded as the founder of the modern theory, that the differences which are perceived between metallic and electrolytic conduction may be referred to special features in the two cases, which do not affect their general resemblance. In electrolytes the carriers are provided only by the salt, which is dispersed throughout a large inert mass of solvent ; whereas in metals it may be supposed that every molecule is capable of furnishing carriers. Thomson, therefore, proposed to regard the current in metals as a series of intermittent discharges, caused by the rearrangement of the constituents of molecular systems—a conception similar to that by which Grothuss [2] had pictured conduction in electrolytes. This view would, as he showed, lead to a general explanation of the connection between thermal and electrical conductivities.

Most of the writers on metallic conduction in the latter part of the nineteenth century preferred to take the hypothesis of Arrhenius [3] rather than that of Grothuss as a pattern ; and therefore supposed the interstices between the molecules of the metal to be at all times swarming with electric charges in rapid motion. In 1898 E. Riecke [4] effected an important advance by examining the consequences of the assumption that the average velocity of this random motion of the charges is nearly proportional to the square root of the absolute temperature T. P. Drude [5] in 1900 replaced this by the more definite assumption that the kinetic energy of each moving charge is equal to the average kinetic energy of a molecule of a perfect gas at the same temperature, and may therefore be expressed in the form qT, where q denotes a universal constant.

[1] J. J. Thomson, *Applications of Dynamics to Physics and Chemistry* (1888), p. 296. cf. also Giese, *Ann. d. Phys.* xxxvii (1889), p. 576 [2] cf. p. 76 [3] cf. p. 344

[4] *Gött. Nach.* (1898), pp. 48, 137 ; *Ann. d. Phys.* lxvi (1898), pp. 353, 545, 1199 ; ii (1900), p. 835

[5] *Ann. d. Phys.* (4) i (1900), p. 566 ; iii (1900), p. 369 ; vii (1902), p. 687

In the same year J. J. Thomson [1] remarked that it would accord with the conclusions drawn from the study of ionisation in gases to suppose that the vitreous and resinous charges play different parts in the process of conduction. The resinous charges may be conceived of as carried by simple negative corpuscles or electrons, such as constitute the cathode rays ; they may be supposed to move about freely in the interstices between the atoms of the metal. The vitreous charges, on the other hand, may be regarded as more or less fixed in attachment to the metallic atoms. According to this view the transport of electricity is due almost entirely to the motion of the negative charges.

An experiment which was performed at this time by Riecke [2] lent some support to Thomson's hypothesis. A cylinder of aluminium was inserted between two cylinders of copper in a circuit, and a current was passed for such a time that the amount of copper deposited in an electrolytic arrangement would have amounted to over a kilogramme. The weight of each of the three cylinders, however, showed no measurable change ; from which it appeared unlikely that metallic conduction is accompanied by the transport of metallic ions.

Many years later, this view was still more convincingly substantiated when R. C. Tolman and T. D. Stewart [3] gave a fairly direct experimental proof that electric currents in metals are carried by 'free' negative electrons. A coil of wire was rotated at a high speed and suddenly brought to rest : a pulse of electric current was produced, arising from the momentum of the electrons in the rotating coil, which caused them to continue in motion after the rotation of the coil had ceased ; and a quantitative discussion yielded a general confirmation of Thomson's hypothesis.

The ideas of Thomson, Riecke and Drude were combined by Lorentz [4] in an investigation which, as it is the most complete, will here be given as the representative of all of them. More than twenty years later, it was replaced by a theory of metals depending on a new type of statistics ; but at the time it represented a great advance on everything that had preceded it.

It is supposed that the atoms of the metal are fixed, and that in the interstices between them a large number of resinous electrons are in rapid motion. The mutual collisions of the electrons are disregarded, so that their collisions with the fixed atoms alone come

[1] *Rapports prés. au Congrès de Physique*, iii (Paris, 1900), p. 138
[2] *Phys. Zs.* iii (1901), p. 639
[3] *Phys. Rev.* viii (1916), p. 97
[4] *Amsterdam Proc.* (English edn.), vii (1904–5), pp. 438, 585, 684

under consideration ; these are regarded as analogous to collisions between moving and fixed elastic spheres.

The flow of heat and electricity in the metal is supposed to take place in a direction parallel to the axis of x, so that the metal is in the same condition at all points of any plane perpendicular to this direction ; and the flow is supposed to be steady, so that the state of the system is independent of the time.

Consider a slab of thickness dx and of unit area ; and suppose that the number of electrons in this slab whose x-components of velocity lie between u and $u + du$, whose y-components of velocity lie between v and $v + dv$, and whose z-components of velocity lie between w and $w + dw$, is

$$f(u, v, w, x)\, dx\, du\, dv\, dw.$$

One of these electrons, supposing it to escape collision, will in the interval of time dt travel from (x, y, z) to $(x + udt, y + vdt, z + wdt)$; and its x component of velocity will at the end of the interval be increased by an amount $e\mathrm{E}dt/m$, if m and e denote its mass and charge, and E denotes the electric force. Suppose that the number of electrons lost to this group by collisions in the interval dt is $a\, dx\, du\, dv\, dw\, dt$, and that the number added to the group by collisions in the same interval is $b\, dx\, du\, dv\, dw\, dt$. Then we have

$$f(u, v, w, x) + (b - a)\, dt = f(u + e\mathrm{E}\, dt/m, v, w, x + udt),$$

and therefore

$$b - a = \frac{e\mathrm{E}}{m}\frac{\partial f}{\partial u} + u\frac{\partial f}{\partial x}.$$

Now, the law of distribution of velocities which Maxwell postulated for the molecules of a perfect gas at rest is expressed by the equation

$$f = \pi^{-\frac{3}{2}} a^{-3} \mathrm{N} e^{-\frac{r^2}{a^2}},$$

where N denotes the number of moving corpuscles in unit volume, r denotes the resultant velocity of a corpuscle (so that $r^2 = u^2 + v^2 + w^2$), and a denotes a constant which specifies the average intensity of agitation, and consequently the temperature. It is assumed that the law of distribution of velocities among the electrons in a metal is nearly of this form ; but a term must be added in order to represent the general drifting of the electrons parallel to the axis of x. The simplest assumption that can be made regarding this term is that it is of the form

$$u \times \text{a function of } r \text{ only};$$

we shall, therefore, write

$$f = N\pi^{-\frac{3}{2}}\alpha^{-3}e^{-\frac{r^2}{\alpha^2}} + u\chi(r).$$

The value of $\chi(r)$ may now be determined from the equation

$$b - a = \frac{eE}{m}\frac{\partial f}{\partial u} + u\frac{\partial f}{\partial x};$$

for on the left-hand side, the Maxwellian term

$$\pi^{-\frac{3}{2}}\alpha^{-3}Ne^{-\frac{r^2}{\alpha^2}}$$

would give a zero result, since b is equal to a in Maxwell's system; thus $b - a$ must depend solely on the term $u\chi(r)$; and an examination of the circumstances of a collision, in the manner of the kinetic theory of gases, shows that $(b - a)$ must have the form $-ur\chi(r)/l$, where l denotes a constant which is closely related to the mean free path of the electrons. In the terms on the right-hand side of the equation, on the other hand, Maxwell's term gives a result different from zero; and in comparison with this we may neglect the terms which arise from $u\chi(r)$. Thus we have

$$-\frac{ur\chi(r)}{l} = \left(\frac{eE}{m}\frac{\partial}{\partial u} + u\frac{\partial}{\partial x}\right)\frac{N}{\pi^{\frac{3}{2}}\alpha^3}e^{-\frac{r^2}{\alpha^2}},$$

or

$$u\chi(r) = \frac{lu}{\pi^{\frac{3}{2}}r}\cdot e^{-\frac{r^2}{\alpha^2}}\cdot\left\{\frac{2eNE}{m\alpha^5} - \frac{d}{dx}\left(\frac{N}{\alpha^3}\right) - \frac{2Nr^2}{\alpha^6}\frac{d\alpha}{dx}\right\};$$

and thus the law of distribution of velocities is determined.

The electric current i is determined by the equation

$$i = e\iiint uf(u, v, w)\,du\,dv\,dw,$$

where the integration is extended over all possible values of the components of velocity of the electrons. The Maxwellian term in $f(u, v, w)$ furnishes no contribution to this integral, so we have

$$i = e\iiint u^2\chi(r)\,du\,dv\,dw.$$

When the integration is performed, this formula becomes

$$i = \frac{2le}{3\pi^{\frac{1}{2}}}\left(\frac{2e\mathrm{NE}}{ma} - a\frac{d\mathrm{N}}{dx} - \mathrm{N}\frac{da}{dx}\right),$$

or

$$\mathrm{E} = \frac{3\pi^{\frac{1}{2}}m}{4le^2}\frac{a}{\mathrm{N}}i + \frac{m}{2e}\left(\frac{a^2}{\mathrm{N}}\frac{d\mathrm{N}}{dx} + a\frac{da}{dx}\right).$$

The coefficient of i in this equation must evidently represent the ohmic specific resistance of the metal; so if γ denote the specific conductivity, we have

$$\gamma = \frac{4le^2}{3\pi^{\frac{1}{2}}m}\frac{\mathrm{N}}{a}.$$

Let the equation be next applied to the case of two metals A and B in contact at the same temperature T, forming an open circuit in which there is no conduction of heat or electricity (so that i and da/dx are zero). Integrating the equation

$$\mathrm{E} = \frac{m}{2e}\frac{a^2}{\mathrm{N}}\frac{d\mathrm{N}}{dx}$$

across the junction of the metals, we have

$$\text{Discontinuity of potential at junction} = \frac{ma^2}{2e}\log\frac{\mathrm{N_B}}{\mathrm{N_A}};$$

or since $\frac{3}{4}ma^2$, which represents the average kinetic energy of an electron, is by Drude's assumption equal to $q\mathrm{T}$, where q denotes a universal constant, we have

$$\text{Discontinuity of potential at junction} = \frac{2}{3}\frac{q}{e}\mathrm{T}\log\frac{\mathrm{N_B}}{\mathrm{N_A}}.$$

This may be interpreted as the difference of potential connected with the Peltier [1] effect at the junction of two metals; the product of the difference of potential and the current measures the evolution of heat at the junction. The Peltier discontinuity of potential is of the order of a thousandth of a volt, and must be distinguished from Volta's contact difference of potential, which is generally much larger, and which, as it presumably depends on the relation of the metals to the medium in which they are immersed, is beyond the scope of the present investigation.

[1] cf. p. 237

Returning to the general equations, we observe that the flux of energy W is parallel to the axis of x, and is given by the equation

$$W = \tfrac{1}{2}m\iiint ur^2 f(u, v, w)\, du\, dv\, dw,$$

where the integration is again extended over all possible values of the components of velocity; performing the integration, we have

$$W = \frac{2ml}{3\pi^{\frac{1}{2}}}\left(\frac{2ea}{m}\,\mathrm{NE} - a^2\frac{d\mathrm{N}}{dx} - 3\mathrm{N}a^2\frac{da}{dx}\right);$$

or, substituting for E from the equation already found,

$$W = \frac{ma^2}{e}\,i - \frac{4ml}{3\pi^{\frac{1}{2}}}\,\mathrm{N}a^2\frac{da}{dx}.$$

Consider now the case in which there is conduction of heat without conduction of electricity. The flux of energy will in this case be given by the equation

$$W = -\kappa\frac{d\mathrm{T}}{dx},$$

where κ denotes the thermal conductivity of the metal expressed in suitable units; or

$$W = -\kappa\,.\,\frac{3ma}{2q}\,\frac{da}{dx}.$$

If it be assumed that the conduction of heat in metals is effected by motion of the electrons, this expression may be compared with the preceding; thus we have

$$\kappa = \frac{8}{9}\pi^{-\frac{1}{2}}laq\mathrm{N};$$

and comparing this with the formula already found for the electric conductivity, we have

$$\frac{\kappa}{\gamma} = \frac{8}{9}\mathrm{T}\left(\frac{q}{e}\right)^2,$$

an equation which shows that the ratio of the thermal to the electric conductivity is of the form T × a constant which is the same for all metals. This result accords with the law of Wiedemann and Franz.

Moreover, the value of q is known from the kinetic theory of gases; and the value of e had been determined by J. J. Thomson [1] and his followers; substituting these values in the formula for κ/γ, a fair agreement is obtained with the values of κ/γ determined experimentally.

[1] cf. p. 365

It was remarked by J. J. Thomson that if, as is postulated in the above theory, a metal contains a great number of free electrons in temperature equilibrium with the atoms, the specific heat of the metal must depend largely on the energy required in order to raise the temperature of the electrons. Thomson considered that the observed specific heats of metals are smaller than is compatible with the theory, and was thus led to investigate [1] the consequences of his original hypothesis [2] regarding the motion of the electrons, which differs from the one just described in much the same way as Grothuss' theory of electrolysis differs from Arrhenius'. Each electron was now supposed to be free only for a very short time, from the moment when it is liberated by the dissociation of an atom to the moment when it collides with, and is absorbed by, a different atom. The atoms were conceived to be paired in doublets, one pole of each doublet being negatively, and the other positively, electrified. Under the influence of an external electric field the doublets orient themselves parallel to the electric force, and the electrons which are ejected from their negative poles give rise to a current predominantly in this direction. The electric conductivity of the metal may thus be calculated. In order to comprise the conduction of heat in his theory. Thomson assumed that the kinetic energy with which an electron leaves an atom is proportional to the absolute temperature; so that if one part of the metal is hotter than another, the temperature will be equalised by the interchange of corpuscles. This theory, like the other, leads to a rational explanation of the law of Wiedemann and Franz.

The theory of electrons in metals has received support from the study of another phenomenon. It was known to the philosophers of the eighteenth century [3] that the air near an incandescent metal acquires the power of conducting electricity. 'Let the end of a poker,' wrote Canton,[4] 'when red-hot, be brought but for a moment within three or four inches of a small electrified body, and its electrical power will be almost, if not entirely, destroyed.'

The subject continued to attract attention at intervals [5]; and as the process of conduction in gases came to be better understood, the

[1] J. J. Thomson, *The Corpuscular Theory of Matter*, London, 1907 [2] cf. p. 419

[3] Perhaps first to Du Fay in 1733 [4] *Phil. Trans.* lii (1762), p. 457

[5] cf. E. Becquerel, *Annales de Chimie*, xxxix (1853), p. 355; Guthrie, *Phil. Mag.* xlvi (1873), p. 254; also various memoirs by Elster and Geitel in the *Ann. d. Phys.* from 1882 onwards, particularly *Ann. d. Phys.* xxxviii (1889), p. 315. The phenomenon is very noticeable, as Edison showed (*Engineering* [12 Dec. 1884], p. 553), when a filament of carbon is heated to incandescence in a rarefied gas. It was later found that ions are emitted when magnesia, or any of the oxides of the alkaline earth metals, is heated to a dull red heat.

conductivity produced in the neighbourhood of incandescent metals was attributed to the emission of electrically charged particles by the metals. But it was not until the development of J. J. Thomson's theory of ionisation in gases that notable advances were made. In 1899, Thomson [1] determined the ratio of the charge to the mass of the resinously charged ions emitted by a hot filament of carbon in rarefied hydrogen, by observing their deflection in a magnetic field. The value obtained for the ratio was nearly the same as that which he had found for the corpuscles of cathode rays; whence he concluded that the negative ions emitted by the hot carbon were negative electrons.

The corresponding investigation [2] for the positive leak from hot bodies yielded the information that the mass of the positive ions is of the same order of magnitude as the mass of material atoms. There are reasons for believing that these ions are produced from gas which has been absorbed by the superficial layer of the metal.[3]

If, when a hot metal is emitting ions in a rarefied gas, an electromotive force be established between the metal and a neighbouring electrode, either the positive or the negative ions are urged towards the electrode by the electric field, and a current is thus transmitted through the intervening space. When the metal is at a higher potential than the electrode, the current is carried by the vitreously charged ions; when the electrode is at the higher potential, by those with resinous charges. In either case, it is found that when the electromotive force is increased indefinitely, the current does not increase indefinitely likewise, but acquires a certain 'saturation' value. The obvious explanation of this is that the supply of ions available for carrying the current is limited.

When the temperature of the metal is high, the ions emitted are mainly negative; and it is found [4] that in these circumstances, when the surrounding gas is rarefied, the saturation current is almost independent of the nature of the gas or of its pressure. The leak of resinous electricity from a metallic surface in a rarefied gas must therefore depend only on the temperature and on the nature of the metal; and it was shown by O. W. Richardson [5] that the dependence

[1] *Phil. Mag.* xlviii (1899), p. 547

[2] J. J. Thomson, *Proc. Camb. Phil. Soc.* xv (1909), p. 64; O. W. Richardson, *Phil. Mag.* xvi (1908), p. 740

[3] cf. Richardson, *Phil. Trans.* ccvii (1906), p. 1

[4] cf. J. A. McClelland, *Proc. Camb. Phil. Soc.* x (1899), p. 241; xi (1901), p. 296. On the results obtained when the gas is hydrogen, cf. H. A. Wilson, *Phil. Trans.* ccii (1903), p. 243; ccviii (1908), p. 247; and O. W. Richardson, *Phil. Trans.* ccvii (1906), p. 1

[5] *Proc. Camb. Phil. Soc.* xi (1902), p. 286; *Phil. Trans.* cci (1903), p. 497. cf. also H. A. Wilson, *Phil. Trans.* ccii (1903), p. 243

on the temperature might be expressed by an equation of the form

$$i = AT^{\frac{1}{2}}e^{-\frac{b}{T}},$$

where i denotes the saturation current per unit area of surface (which is proportional to the number of ions emitted in unit time), T denotes the absolute temperature and A and b are constants.[1] From 1902 to 1912 this equation was accepted as the basis of the theory of the subject, for which the name *thermionics*, suggested by Richardson, is now in general use.

In order to account for these phenomena, Richardson [2] adopted the hypothesis which had previously been proposed [3] for the explanation of metallic conductivity; namely, that a metal is to be regarded as a sponge-like structure of comparatively large fixed positive ions and molecules, in the interstices of which negative electrons are in rapid motion. He assumed that the free electrons have the same energy as is attributed in the kinetic theory of gases to gas molecules at the same temperature as the metal (an assumption which, as will be explained in the second volume of this work, was subsequently abandoned). Since the electrons do not all escape freely at the surface, he postulated a superficial discontinuity of potential, sufficient to restrain most of them. Thus, let N denote the number of free electrons in unit volume of the metal; then in a parallelepiped whose height measured at right angles to the surface is dx, and whose base is of unit area, the number of electrons whose x-components of velocity are comprised between u and $u + du$ is

$$\pi^{-\frac{1}{2}}\alpha^{-1}Ne^{-\frac{u^2}{\alpha^2}}du\,dx, \quad \text{where} \quad \frac{3}{4}m\alpha^2 = qT,$$

m denoting the mass of an electron, T the absolute temperature, and q the universal constant previously introduced.

Now, an electron whose x-component of velocity is u will arrive at the interface within an interval dt of time, provided that at the beginning of this interval it is within a distance $u\,dt$ of the interface. So the number of electrons whose x-components of velocity are comprised between u and $u + du$ which arrive at unit area of the interface in the interval dt is

$$\pi^{-\frac{1}{2}}\alpha^{-1}Ne^{-\frac{u^2}{\alpha^2}}u\,du\,dt.$$

If the work which an electron must perform in order to escape

[1] The same law applies to the emission from other bodies, e.g. heated alkaline earths, and to the emission of positive ions—at any rate when a steady state of emission has been reached in a gas which is at a definite pressure.

[2] *Phil. Trans.* cci (1903), p. 497

[3] cf. p. 419 et sqq.

through the surface layer be denoted by ϕ, the number of electrons emitted by unit area of metal in unit time is therefore

$$\int_{\frac{1}{2}mu^2=\phi}^{\infty} \pi^{-\frac{1}{2}}a^{-1}Ne^{-\frac{u^2}{a^2}}u\,du, \quad \text{or} \quad \frac{1}{2}\pi^{-\frac{1}{2}}Nae^{-\frac{2\phi}{ma^2}}.$$

The current issuing from unit area of the hot metal is thus

$$\frac{1}{2}\pi^{-\frac{1}{2}}N\epsilon ae^{-\frac{2\phi}{ma^2}}, \quad \text{or} \quad N\epsilon\,.\,(qT/3\pi m)^{\frac{1}{2}}e^{-\frac{3\phi}{2qT}},$$

where ϵ denotes the charge on an electron. This expression, being of the form

$$AT^{\frac{1}{2}}e^{-\frac{b}{T}},$$

agrees with the equation which Richardson had derived from the experimental measures ; and the comparison furnishes the value of the superficial discontinuity of potential which is implied in the existence of ϕ.[1]

The formula of 1902 for the connection between the saturation current and the temperature was subsequently abandoned ; but an account of this development must be postponed to the second volume of the present work.

[1] This discontinuity of potential was found to be 2·45 volts for sodium, 4·1 volts for platinum and 6·1 volts for carbon.

Index of Authors Cited

When many references are made to one author the more important entries are printed in heavy type

VOLUME II:

The Modern Theories

1900–1926

Preface

THE purpose of this volume is to describe the revolution in physics which took place in the first quarter of the twentieth century, and which included the discoveries of Special Relativity, the older Quantum Theory, General Relativity, Matrix Mechanics and Wave Mechanics.

My original intention was to give an account of the history from 1900 to 1950 in a single volume ; but the wealth of material made this undesirable ; and the period from 1926 to 1950 must be reserved for a third book.

I am greatly indebted to Dr E. T. Copson, Regius Professor of Mathematics in the University of St Andrews, and Dr J. M. Whittaker, Vice-Chancellor of the University of Sheffield, for reading the proofs.

E. T. WHITTAKER

48 George Square
Edinburgh, April 1953

Contents

Memorandum on Notation

VECTORS are denoted by letters in black type, as **E**.

The three components of a vector **E** are denoted by E_x, E_y, E_z; and the magnitude of the vector is denoted by E, so that

$$E^2 = E_x^2 + E_y^2 + E_z^2.$$

The *vector product* of two vectors **E** and **H**, which is denoted by [**E** . **H**], is the vector whose components are

$$(E_y H_z - E_z H_y, \quad E_z H_x - E_x H_z, \quad E_x H_y - E_y H_x).$$

Its direction is at right angles to the direction of **E** and **H**, and its magnitude is represented by twice the area of the triangle formed by them.

The *scalar product* of **E** and **H** is $E_x H_x + E_y H_y + E_z H_z$. It is denoted by (**E** . **H**).

The quantity $\frac{\partial E_x}{\partial x} + \frac{\partial E_y}{\partial y} + \frac{\partial E_z}{\partial z}$ is denoted by div **E**.

The vector whose components are

$$\left(\frac{\partial E_z}{\partial y} - \frac{\partial E_y}{\partial z}, \quad \frac{\partial E_x}{\partial z} - \frac{\partial E_z}{\partial x}, \quad \frac{\partial E_y}{\partial x} - \frac{\partial E_x}{\partial y}\right)$$

is denoted by curl **E**.

If V denote a scalar quantity, the vector whose components are $\left(-\frac{\partial V}{\partial x}, \quad -\frac{\partial V}{\partial y}, \quad -\frac{\partial V}{\partial z}\right)$ is denoted by grad V.

The symbol Δ is used to denote the vector operator whose components are $\frac{\partial}{\partial x}, \frac{\partial}{\partial y}, \frac{\partial}{\partial z}$.

Chapter I

THE AGE OF RUTHERFORD

When Röntgen announced his discovery of the X-rays [1] it was natural to suspect some connection between these rays and the fluorescence (or, as it was generally called at that time, phosphorescence) of the part of the vacuum tube from which they were emitted. Accordingly, a number of workers tried to find whether phosphorescent bodies in general emitted radiations which could pass through opaque bodies and then either affect photographic plates or excite phosphorescence in other bodies.

In particular, Henri Becquerel of Paris (1852–1908) resolved to examine the radiations which are emitted, after exposure to the sun, by the double sulphate of uranium and potassium, a substance which had been shown by his father, Edmond Becquerel (1820–91), to have the property of phosphorescence. The result was communicated to the French Academy on 24 February 1896.[2] 'Let a photographic plate,' he said, 'be wrapped in two sheets of very thick black paper, such that the plate is not affected by exposure to the sun for a day. Outside the paper place a quantity of the phosphorescent substance, and expose the whole to the sun for several hours. When the plate is developed, it displays a silhouette of the phosphorescent substance. So the latter must emit radiations which are capable of passing through paper opaque to ordinary light, and of affecting salts of silver.'

At this time Becquerel supposed the radiation to have been excited by the exposure of the phosphorescent substance to the sun ; but a week later he announced [3] that in one experiment the sun had become obscured almost as soon as the exposure was begun, and yet that when the photographic plate was developed, the intensity of the silhouette was as strong as in the other cases : and moreover, he had found that the radiation persisted for an indefinite time after the substance had been removed from the sunlight, and after the luminosity which properly constitutes phosphorescence had died away ; and he was thus led to conclude that the activity was spontaneous and permanent. It was soon found that those salts of uranium which do not phosphoresce—that is, the uranous series of salts—and the metal itself, all emit the rays ; and it became evident that what Becquerel had discovered was a radically new property, possessed by the element uranium in all its chemical compounds.

[1] cf. Vol. I, p. 357 [2] *Comptes Rendus*, cxxii (1896), p. 420
[3] *Comptes Rendus*, cxxii (2 March 1896), p. 501

Very soon he found[1] that the new rays, like the Röntgen and cathode rays, impart conductivity to gases. The conductivity due to X-rays was at that time being investigated at the Cavendish Laboratory, Cambridge, by J. J. Thomson, who had been joined in the summer of 1895 by a young research student from New Zealand named Ernest Rutherford (1871–1937). They found that the conductivity is due to *ions*, or particles carrying electric charges, which are produced in the gas by the radiation, and which are set in motion when an electric field is applied. Rutherford went on to examine the conductivity produced by the rays from uranium (which, as he showed, is likewise due to ionisation), and the absorption of these rays by matter: he found[2] that the rays are not all of the same kind, but that at least two distinct types are present: one of these, to which he gave the name α-rays, is readily absorbed; while another, which he named β-radiation, has a penetrating power a hundred times as great as the α-rays.

Early in 1898 two new workers entered the field. Marya Sklodowska, born in Warsaw in 1867 (*d.* 1934), had studied physics in Paris, and in 1895 had married a young French physicist, Pierre Curie (1859–1906). She now resolved to search for other substances having the properties that Becquerel had found in uranium, and showed in April 1898 that these properties were possessed by compounds of thorium,[3] the element which, of the elements known at that time, stood next to uranium in the order of atomic weights; the same discovery was made simultaneously by G. C. Schmidt[4] in Germany. Madame Curie went on to show that, since the emission of rays by uranium and thorium is unaffected by chemical changes, it must be essentially an *atomic* property.[5] Now the mineral pitchblende, from which the uranium was derived, was found to have an activity much greater than could be accounted for by the uranium contained in it: and from this fact she inferred that the pitchblende must contain yet another 'radio-active' element. Making a systematic chemical analysis, she and her husband in July 1898 discovered a new element which, in honour of her native country, she named *polonium*,[6] and then in December another, having an activity many million times as great as uranium: to this the name *radium* was given.[7] Its spectrum was examined by F. A. Demarçay,[8] and a spectral line was found which was not otherwise identifiable. The next three and a half years were spent chiefly in determining its atomic weight, by a laborious series of successive

[1] *Comptes Rendus*, cxxii (1896), p. 559
[2] This paper was published in *Phil. Mag.*(5) xlvii (1899), p. 109, after Rutherford had left Cambridge for a chair in McGill University.
[3] *Comptes Rendus*, cxxvi (12 April 1898), p. 1101
[4] *Ann. d. Phys.* lxv (19 April 1898), p. 141
[5] Some years later, the Curies described the ideas that had inspired their researches in *Comptes Rendus* cxxxiv (1902), p. 85.
[6] *Comptes Rendus*, cxxvii (1898), p. 175
[7] Ibid. cxxvii (1898), p. 1215
[8] *Comptes Rendus*, cxxvii (1898), p. 1218

fractionations: the value found was 225.[1] Meanwhile another French physicist, André Debierne (*b.* 1874), discovered in the uranium residues yet a further radio-active element,[2] to which he gave the name *actinium*.

Attention was now directed to the α- and β-rays of Rutherford. A few months after their discovery it was shown by Giesel, Becquerel and others, that part of the radiation (the β-rays) was deflected by a magnetic field,[3] while part (the α-rays) was not appreciably deflected.[4] After this Monsieur and Madame Curie [5] found that the deviable rays carry negative electric charges, and Becquerel [6] succeeded in deviating them by an electrostatic field. The deviable or β-rays were thus clearly of the same nature as cathode rays; and when measurements of the electric and magnetic deviations gave for the ratio m/e a value [7] of the order 10^{-7}, the identity of the β-particles with the cathode-ray corpuscles was fully established. They differ only in velocity, the β-rays being very much the swifter.

The α-rays were at this time supposed to be not deviated by a magnetic field: the deviation is in fact small, even when the field is powerful: but in February 1903 Rutherford [8] announced that he had succeeded in deviating them by both magnetic and electrostatic fields. The deviation was in the opposite sense to that of the cathode rays, so the α-radiations must consist of positively charged particles projected with great velocity,[9] and the smallness of the deviation suggested that the expelled particles were massive compared to the electron. A method of observing them was discovered in 1903 by Sir W. Crookes [10] and independently by J. Elster and W. Geitel,[11] who found that when a radio-active substance was brought near a screen of Sidot's hexagonal blende (zinc sulphide), bright scintillations were observed, due to the cleavage of the blende under the bombardment. Rutherford suggested that this property might be used for counting the number of α-particles in the rays.

Meanwhile it had been discovered by P. Villard [12] that in addition to the alpha and beta rays, radium emits a third type of radiation, much more penetrating than either of them, in fact 160 times as penetrating as the beta rays. The thickness of aluminium traversed before the intensity is reduced to one-half is approximately 0·0005 cm. for the α-rays, 0·05 cm. for the β-rays and 8 cm. for the γ-rays, as

[1] Later raised to 226

[2] *Comptes Rendus*, cxxx (2 April 1900), p. 906

[3] F. O. Giesel, *Ann d. Phys.* lxix (1899), p. 834 (working with polonium); Becquerel, *Comptes Rendus*, cxxix (1899), p. 996 (working with radium); S. Meyer and E. v. Schweidler, *Phys. ZS.* (1899), p. 113 (working with polonium and radium)

[4] Becquerel, *Comptes Rendus*, cxxix (1899), p. 1205; cxxx (1900), pp. 206, 372; Curie, ibid., cxxx (1900), p. 73

[5] *Comptes Rendus*, cxxx (1900), p. 647

[6] *Comptes Rendus*, cxxx (1900), p. 809

[7] cf. W. Kaufmann, *Verh. Deutsch Phys. Ges.* ix (1907), p. 667

[8] *Phil. Mag.*(6) v (Feb. 1903), p. 177

[9] This had been conjectured by R. J. Strutt in *Phil. Trans.* cxcvi (1901), p. 507.

[10] *Proc. R.S.* lxxi (30 April 1903), p. 405

[11] *Phys. ZS.* iv (1 May 1903), p. 439

[12] *Comptes Rendus*, cxxx (30 April 1900), p. 1178

Villard's radiation was called. Villard found that the γ-radiation is, like the X-rays, not deviable by magnetic forces.

In 1898 Rutherford was appointed to a chair in McGill University, Montreal, and there with R. B. Owens, the professor of electrical engineering, began an investigation into the radio-activity of the thorium compounds. The conductivity produced by the oxide thoria in the air was found [1] to vary in an unexpected and perplexing manner: it could be altered considerably by slight draughts caused by opening or shutting a door. Eventually Rutherford concluded that thoria emitted [2] very small amounts of some material substance which was itself radio-active, and which could be carried away in an air current: this, to which he gave the name *thorium emanation*, was shown to be a gas belonging to the same chemical family as helium and argon, but of high molecular weight.[3]

Meanwhile in Cambridge C. T. R. Wilson had been developing [4] his cloud-chamber, which was to provide the most powerful of all methods of investigation in atomic physics. In moist air, if a certain degree of supersaturation is exceeded (this can be secured by a sudden expansion of the air) condensation takes place on dust-nuclei, when any are present: if by preliminary operations condensation is made to take place on the dust-nuclei, and the resulting droplets are allowed to settle, the air in the chamber is thereby freed from dust. If now X-rays or radiations from a radio-active substance are passed into the chamber, and if the degree of supersaturation is sufficient, condensation again takes place: this is due to the production of ions by the radiation. Thus the tracks of ionising radiations can be made visible by the sudden expansion of a moist gas, each ion becoming the centre of a visible globule of water. Wilson showed that the ions produced by uranium radiation were identical with those produced by X-rays. J. J. Thomson in July 1899 wrote pointing out the advantages of the Wilson chamber to Rutherford, who henceforth profited immensely by its use. In this way the track of a single atomic projectile or electron could be rendered visible.

An important property, discovered for the first time in connection with thorium emanation, was that the radio-activity connected with it rapidly decreased. This behaviour was found later to be characteristic of all radio-active substances: but in the earliest known cases, uranium and thorium, the half-period (i.e. the time required for the activity to be reduced by one-half) is of the order of millions of years, so the property had not hitherto been noticed. Rutherford found [5] that the intensity of the 'induced radiation' of thorium falls off

[1] Owens, *Phil. Mag.*(5) xlviii (Oct. 1899), p. 360
[2] *Phil. Mag.*(5) xlix (Jan. 1900), p. 1
[3] Soon after this, Friedrich Ernst Dorn of Halle found that radium, like thorium, produced an emanation: *Halle Nat. Ges. Abh.* xxiii (1900).
[4] *Phil. Trans.* clxxxix(A) (1897), p. 265; *Proc. Camb. P.S.* ix (1898), p. 333
[5] *Phil. Mag.* xlix (Feb. 1900), p. 161

exponentially with the time: so that if I_1 is the intensity at any time and I_2 the intensity after the lapse of a time t then

$$I_2 = I_1 e^{-\lambda t}$$

when λ is a constant.

In May 1900 Sir W. Crookes [1] showed that it was possible by chemical means to separate from uranium a small fraction, which he called uranium X, which possessed the whole of the photographic activity of the original substance. He found, moreover, that the activity of the uranium X gradually decayed, while the full activity of the residual uranium was gradually renewed, so that after a sufficient lapse of time it was possible to separate from it a fresh supply of uranium X. These facts had an important share in the formation of the theory.

It was at first supposed that the pure uranium, immediately after the separation, is not radio-active: but F. Soddy (*b.* 1877) observed that though photographically inactive, it is active when tested by the electrical method. Now the α-rays are active electrically but not photographically, whereas the β-rays are active photographically: and the conclusion was drawn [2] that pure uranium emits only α-rays and uranium X only β-rays. Soddy had joined the staff of McGill University in 1900 as Demonstrator in Chemistry, and at once began to assist Rutherford in his work on radio-activity. Further experiments on the thorium emanation involved condensing it by extreme cold, and it was discovered [3] that the emanation was produced not directly by the thorium but by an intermediate substance which, as it had many of the characters of Crookes' uranium X, was named thorium X. This was the first indication that radio-activity involves a chain of transformations of chemical elements.

The work of Rutherford and Soddy on thorium and its radio-active derivatives led them to a general theory of radio-activity, which was published in September 1902–May 1903.[4] The greatest obstacle to a clear understanding of the subject had been, curiously enough, the intense belief of everybody in the principle of conservation of energy: here was an enormous amount of energy being outpoured, and no-one could see where it came from. So long as it was attributed to the absorption of some unknown kind of external radiation, the essence of the matter could not be discovered. Rutherford and Soddy now swept this notion away, and asserted that:

(i) In the radio-active elements radium, thorium and uranium, there is a continuous production of new kinds of matter, which are themselves radio-active.

(ii) When several changes occur together these are not simul-

[1] *Proc. R.S.*(A), lxvi (1900), p. 409

[2] Soddy, *Journ. Chem. Soc.* lxxxi and lxxxii (July 1902), p. 860; Rutherford and A. G. Grier, *Phil. Mag.*(6) iv (Sept. 1902), p. 315

[3] *Phil. Mag.*(6) iv (Sept. 1902), p. 370

[4] *Phil. Mag.*(6) iv (Sept. 1902), p. 370; ibid. (Nov. 1902), p. 569; ibid.(6) v (April 1903), pp. 441, 445; ibid. (May 1903), pp. 561, 576

taneous, but successive; thus thorium produces thorium X, the thorium X produces the thorium emanation and the latter produces an excited activity.

(iii) The phenomenon of radio-activity consists in this, that a certain proportion of the atoms undergo spontaneous transformation into atoms of a different nature: these changes are different in character from any changes that have been dealt with before in chemistry, for the energy comes from intra-atomic sources which are not concerned in chemical reactions.

(iv) The number of atoms that disintegrate in unit time is a definite proportion of the atoms that are present and have not yet disintegrated. The proportion is characteristic of the radio-active body, and is constant for that body. This leads at once to an exponential law of decay with the time: thus if n_0 is the initial number of atoms, and n is the number at time t afterwards, then

$$n = n_0 e^{-\lambda t}$$

where λ is the fraction of the total number which disintegrates in unit time, so the average life of an atom is $1/\lambda$.

(v) The α-rays consist of positively charged particles, whose ratio of mass to charge is over 1,000 times as great as for the electrons in cathode rays. If it is assumed that the value of the charge is the same as for the electron, then the α-ray particles must have a mass of the same order as that of the hydrogen atom.

(vi) The rays emitted are an accompaniment of the change of the atom into the one next produced, and there is every reason to suppose, not merely that the expulsion of a charged particle accompanies the change, but that this expulsion actually *is* the change.

The authors remarked (in the paper of November 1902) that in naturally occurring minerals containing radio-elements, the radio-active changes must have been taking place over a very long period, and it was therefore possible that the ultimate products might have accumulated in sufficient quantity to be detected. As helium is usually found in such minerals, it was suggested that helium might be such a product. Several years passed before this was finally established, but its probability was continually increasing. Soddy left Montreal in 1903 to work with Sir William Ramsay at University College, London, and Rutherford, who was in England in the summer of that year, called on Ramsay and Soddy, and with them detected (by its spectrum) the presence of helium in the emanation of radium. It seemed certain, therefore, that helium occupied some place in the sequence of linear descent which begins with radium, and at first the general expectation was that it would prove to be an end-product. Rutherford, however, entertained the idea that it might be formed from the α-particles,[1] which, as we have seen, were known to be of the same order of mass as hydrogen or helium atoms; that the α-particles, in fact, might be positively charged

[1] *Nature*, lxviii (20 Aug. 1903), p. 366

atoms of helium : and for some years this supposition was debated without a definite conclusion being reached. In 1906 Rutherford determined [1] with greater accuracy the ratio e/m of the α-particles from radium C, and found it to be between $5{\cdot}0 \times 10^3$ and $5{\cdot}2 \times 10^3$, which is only half the value of e/m for the hydrogen atom : this, however, left it undecided whether the α-particle is a hydrogen molecule (molecular weight 2) carrying the ionic charge, or a helium atom (atomic weight 4), carrying twice the ionic charge.

In 1904 William Henry Bragg (1862–1942), at that time professor in the University of Adelaide, South Australia, showed [2] that the α-particle, on account of its mass, has only a small probability of being deviated when passing through matter, and that in general it continues in a fixed direction, gradually losing its energy, until it comes to a stop : the distance traversed may be called the *range* of the α-particle. He found in the case of radium definite ranges for four kinds of α-particles, corresponding to emissions from radium, radium emanation, radium A and radium C : the α-particles from any particular kind of atom are all shot out with the same velocity, but this velocity varies from one kind of atom to another, as might be expected from Rutherford's theory. β-particles, on the other hand, are easily deflected from their paths by collisions with gas molecules, and their tracks in a Wilson cloud-chamber are zig-zag : they are scattered by passing through matter, so that a narrow pencil of β-rays, after passing through a metal plate, emerges as an ill-defined beam.

In 1907 Rutherford was translated to the chair of physics in the University of Manchester. Here he found a young graduate of Erlangen, Hans Geiger, with whom he devised [3] an electrical method of counting the α-particles directly, the *Geiger counter* as it has since been generally called. The α-rays were sent through a gas, exposed to an electric field so strong as to be near the breakdown value at which a discharge must pass. When a single α-particle passed and produced a small ionisation, the ions were accelerated by the electric field and the ionisation was magnified by collisions several thousand times. This made possible the passage of a momentary discharge, which could be registered. This counting of atoms one by one was a great achievement : it was found that the number of α-particles emitted by 1 gram of radium in one second is $3{\cdot}4 \times 10^{10}$: when this was combined with the value of the total charge (found in the second paper), it became clear that an α-particle carries double the electron charge, reversed in sign.

The question as to the possible connection of α-rays with helium was finally settled later in the same year. Rutherford placed a

[1] *Phys. Rev.* xxii (Feb. 1906), p. 122 ; *Phil. Mag.* xii (Oct. 1906), p. 348

[2] *Phil. Mag.* viii (Dec. 1904), p. 719 ; W. H. Bragg and R. Kleeman, *Phil. Mag.* x (Sept. 1905), p. 318 ; Paper read before the Royal Society of South Australia, 6 June 1904.

[3] *Proc. R.S.*(A), lxxxi (27 Aug. 1908), pp. 141, 162

quantity of radium emanation in a glass tube, which was so thin that the α-rays generated by the emanation would pass through its walls : they were received on the walls of a surrounding glass tube, and, after diffusing out, were found to give the spectrum of helium. This proved definitely that the α-particles are helium atoms, carrying two unit positive charges,[1] a conclusion which he had also reached[2] a short time before by a different line of reasoning.

An interesting corroboration of Rutherford's account of the emission of α-rays was obtained somewhat later, when he and Geiger investigated the fluctuations[3] in the recorded numbers of particles emitted by a radio-active substance in successive equal intervals of time. H. Bateman had shown that if the emission is a random one, then the probability that n particles will be observed in unit time is

$$\frac{x^n e^{-x}}{n!}$$

where x is the average number per unit time, and n is a whole number ($n=0, 1, 2, \ldots \infty$). Rutherford and Geiger in 1910 verified this formula experimentally.[4] Yet a further completion of the work on α-particles was a measurement, made with Boltwood,[5] of the volume of helium produced by a large quantity of radium. By combining the result now obtained with that of the counting experiment it was possible to evaluate the number of molecules in a quantity of the substance whose weight in grams is equal to the molecular weight of the substance (the Avogadro number).

The determination of this constant had been the object of many researches in the years immediately preceding, beginning with a notable paper by Einstein.[6] Albert Einstein was born at Ulm in Württemberg on 14 March 1879. The circumstances of his father's business compelled the family to leave Germany ; and after receiving a somewhat irregular education in Switzerland, he became an official in the Patent Office in Berne. It was in this situation that he wrote, in six months, four papers, each of which attracted much attention.[7]

The paper now to be considered was really a sequel to two earlier papers[8] on the statistical-kinetic theory of heat, in which, however, Einstein had only obtained independently certain results which had been published a year or two earlier by Willard Gibbs. He now

[1] Rutherford and T. Royds, *Mem. Manchester Lit. and Phil. Soc.* liii (31 Dec. 1908), p. 1 ; *Phil. Mag.* xvii (Feb. 1909), p. 281

[2] *Nature*, lxxix (5 Nov. 1908), p. 12

[3] That the emission of α-particles is a random process, and so subject to the laws of probability, seems to have been first clearly stated by E. von Schweidler, *Premier Cong. Internat. pour l'Étude de la Radiologie*, Liège, 1905.

[4] *Phil. Mag.* xx (Oct. 1910), p. 698

[5] Rutherford and Boltwood, *Phil. Mag.* xxii (Oct. 1911), p. 586

[6] *Ann. d. Phys.*(4) xvii (1905), p. 549 ; continued in *Ann. d. Phys.* xix (1906), p. 371

[7] Two of these will be referred to in Chapter II and one in Chapter III.

[8] *Ann. d. Phys.* ix (1902), p. 417 ; xi (1903), p. 170

applied these results to the motion of very small particles suspended in a liquid. The particles were supposed to be much larger than a molecule, but it was assumed that as a result of collisions with the molecules of the water, they require a random motion, like that of the molecules of a gas. The average velocity of such a suspended particle, even in the case of particles large enough to be seen with a microscope, might be of observable magnitude : but the direction of its motion would change so rapidly, under the bombardment to which it would be exposed, that it would not be directly measurable. However, as a statistical effect of these transient motions, there would be a resultant motion which might be within the range of visibility. Einstein showed that in a finite interval of time t the mean square of the displacement for a spherical particle of radius a is

$$\frac{RTt}{3\pi a\mu N}$$

where R is the gas-constant, T the temperature, N is Avogadro's number, and μ is the coefficient of viscosity. Thus by this phenomenon the thermal random motion, hitherto a matter of hypothesis, might actually be made a matter of visible demonstration.

The motion of small particles suspended in liquids had been observed as early as 1828 by Robert Brown [1] (1773–1858), a botanist, after whom it was called the *Brownian motion.* Einstein identified the motion studied by him with the Brownian motion, somewhat tentatively in his first paper, but without hesitation in the second.

The theory of the Brownian motion was investigated almost at the same time by M. von Smoluchowski [2] (1872–1917), and it was confirmed experimentally by Th. Svedberg,[3] M. Seddig,[4] and P. Langevin.[5] Particular mention might be made of the experimental studies made in 1908–9 by Jean-Baptiste Perrin (1870–1942) of Paris.[6] These experiments yielded a value of the mean energy of a particle at a definite temperature, and thus enabled him to deduce the value of Avogadro's number, which is $6{\cdot}06 \times 10^{23}$.[7] The direct confirmation of the kinetic theory provided by these researches on the Brownian movement was the means of converting to it some notable former opponents, such as Wilhelm Ostwald and Ernst Mach.

The statistical-kinetic theory of heat was confirmed experimentally in a different way in 1911 by L. Dunoyer,[8] who obtained a parallel beam of sodium molecules by allowing the vapour of

[1] *Phil. Mag.* iv (1828), p. 161

[2] *Bull. Acad. Sci. Cracovie,* vii (1906), p. 577 ; *Ann. d. Phys.* xxi (1906), p. 756

[3] *ZS. Elektrochem,* xii (1906), pp. 853, 909 ; *ZS. phys. chem.* lxv (1909), p. 624 ; lxvi (1909), p. 752 ; lxvii (1909), p. 249 ; lxx (1910), p. 571

[4] *Phys. ZS.* ix (1908), p. 465 [5] *Comptes Rendus,* cxlvi (1908), p. 503

[6] *Comptes Rendus,* cxlvi (1908), p. 967 ; cxlvii (1908), pp. 475, 530 ; cxlix (1909), pp. 477, 549 ; *Ann. Chim. Phys.* xvii (1909), p. 5 ; cf. also R. Fürth, *Ann d. Phys.* liii (1917), p. 117

[7] Another method of determining Avogadro's number is to study the diffusion of ions in a gas under the influence of an electric force.

[8] *Comptes Rendus,* clii (1911), p. 592 ; *Le Radium,* viii (1911), p. 142

heated sodium to pass through two diaphragms pierced with small holes in an exhausted tube: the behaviour of the molecular beams was entirely in agreement with the predictions of the kinetic theory of gases.

In his statistical studies, Einstein also recognised that the thermal motion of the carriers of electric charge in a conductor should give rise to random fluctuations of potential difference between the ends of the conductor. The effect was too small to be detected by the means then available, but many years later, after the development of valve amplification, it was observed by J. B. Johnson,[1] and the theory was studied by H. Nyquist.[2] This phenomenon is one of the causes of the disturbance that is called 'noise' in valve amplifiers.

We must now return to the consideration of the radio-active elements themselves. In 1903–5 Rutherford[3] identified a number of members of the radium sequence later than the emanation: radium A, B and C were known by the summer of 1903, and radium D, E, F were discovered in the next two years. It was suspected that one of these later products was identical with the polonium which had been the first new element found by the Curies, and in fact polonium was shown to be radium F. One of the most remarkable discoveries was that of radium B: for at the time no radiations of any kind could be found accompanying its transformation into radium C, and there was therefore no direct evidence of its existence: the only reason for postulating it was, that to suppose an immediate derivation of radium C from radium A would have violated the laws of radio-active change laid down in 1902–3; and it was therefore necessary to assume the reality of an intermediate body.

As soon as the principle that radio-active elements are derived from each other in series had been established in 1902–3, the suspicion was formed that radium, which is found in nature in uranium ores, might be a descendant of uranium; this conjecture was supported by the facts that uranium is one of the few elements having a higher atomic weight than radium, and that the proportion of radium in pitchblende corresponds roughly with the ratio of activity of radium and uranium. Soddy[4] in 1904 described an experiment which showed that radium is not produced *directly* from uranium: if it is produced at all, it can only be by the agency of intermediate substances. Bertram B. Boltwood (1870–1927), of Yale University,[5] worked on this investigation for several years, and at last in 1907 succeeded in showing that radium is the immediate descendant of

[1] *Nature*, cxix (1927), p. 50; *Phys. Rev.* xxix (1927), p. 367; xxxii (1928), p. 97
The possibility that under certain conditions the thermal motion of electrons in conductors could create a measurable disturbance in amplifiers had been recognised on theoretic grounds by W. Schottky, *Ann. d. Phys.* lvii (1918), p. 541.

[2] *Phys. Rev.* xxxii (1928), p. 110

[3] *Proc. R.S.*(A), lxxiii (22 June 1904), p. 493; *Phil. Trans.* cciv (Nov. 1904), p. 169 (Bakerian lecture); *Phil. Mag.* viii (Nov. 1904), p. 636; *Nature*, lxxi (Feb. 1905), p. 341

[4] *Nature*, lxx (12 May 1904), p. 30

[5] *Nature*, lxx (26 May 1904), p. 80

a new radio-active element which he named *ionium*, and which is itself descended from uranium.[1]

The great number of different radio-active atoms that had now been discovered raised questions concerning their atomic weights, particularly with regard to their position in what was known as the *periodic table* of the chemical elements. In a paper published in 1864, John A. R. Newlands [2] had pointed out that when the chemical elements are arranged according to the numerical values of their atomic weights, the eighth element starting from any given one is, in regard to its properties, closely akin to the first, 'like the eighth note of an octave in music.' This idea he developed in later papers,[3] calling the relationship the 'Law of Octaves.' [4] He read a paper on the subject before the Chemical Society on 1 March 1866; but it was rejected, on the ground that the Society had 'made it a rule not to publish papers of a purely theoretical nature, since it was likely to lead to correspondence of a controversial character.'' [5]

Newland's ideas were adopted and developed a few years later by Dmitri Ivanovich Mendeléev (1834–1907),[6] who arranged the elements in a *periodic table*. From gaps in this he inferred the existence and approximate atomic weights of three hitherto unknown elements, to which he gave the names eka-boron, eka-aluminium and eka-silicon; when these were subsequently discovered (they are now known as scandium, gallium and germanium), the importance of the periodic table became universally recognised; and the inert gases helium etc., when they were discovered still later, were found to fit into it perfectly.

As new members of the radio-active sequences were discovered, it was found in some cases that two or more of the atoms in the series had exactly the same chemical properties, so that they belonged to the same place in Newland's and Mendeléev's periodic table. For instance, in 1905, O. Hahn, working with Sir William Ramsay at University College, London, discovered [7] the parent of thorium X, which he called *radio-thorium*. This was found to be not separable chemically from thorium; and Boltwood found that his ionium also was not separable chemically from thorium. It was shown by A. S. Russell and R. Rossi,[8] working in Rutherford's laboratory, that the optical spectrum of ionium is indistinguishable from the

[1] *Nature*, lxxvi (26 Sept. 1907), p. 544; *Amer. Journ. Sci.* xxiv (Oct. 1907), p. 370; xxv (May 1908), p. 365

[2] *Chem. News*, x (20 Aug. 1864), p. 94

[3] *Chem. News*, xii (18 Aug. 1865), p. 83; xii (25 Aug. 1865), p. 94

[4] The group of elements Helium, Neon, Argon, Krypton, Xenon and Niton was not known at the time; when they are introduced into the table, it is the *ninth* element starting from any given one which is akin to the first. We leave aside the complications associated with the rare earths, etc.

[5] J. A. R. Newlands, *The Periodic Law*; London, E. and F. W. Spon, 1884, p. 23

[6] *ZS. f. Chem.* v (1869), p. 405; *Deutsch. Chem. Gesell. Ber.* iv (1871), p. 348; *Ann. d. Chem.*, Supplementband, viii (1873), p. 133

[7] *Proc. R.S.*(A), lxxvi (24 May 1905), p. 115; *Chem. News*, xcii (1 Dec. 1905), p. 251

[8] *Proc. R.S.*(A), lxxxvii (Dec. 1912), p. 478

spectrum of thorium. The radio-active properties of the three substances are, however, totally different, since the half-value period of thorium is of the order of 10^{10} years, that of ionium is of the order of 10^5 years, and that of radio-thorium is 1·9 years: and the atomic weights are different: but chemically they are different forms of the same element.

Curiously enough, the possibility of such a situation had been suggested so far back as 1886 by Sir William Crookes.[1] 'I conceive, therefore,' he said, 'that when we say the atomic weight of, for instance, calcium is 40, we really express the fact that, while the majority of calcium atoms have an actual atomic weight of 40, there are not a few which are represented by 39 or 41, a less number by 38 or 42, and so on.'

As the investigation of the radio-active atoms progressed still further, many other examples became known of atoms which are inseparable by chemical methods but have different radio-active properties and different atomic weights. Attention was drawn to the matter in 1909 by the Swedish chemists D. Strömholm and Th. Svedberg,[2] and in 1910 by Soddy,[3] and much experimental work relating to it was done by Alexander Fleck.[4]

New light on the problem now came from an unexpected quarter. Sir Joseph Thomson (he had been knighted in 1908) took up work on the canal rays,[5] or *positive rays* as he now called them, and devised a method of 'positive-ray analysis' for finding the values of m/e for the positively charged particles which constitute the rays; the method was to shoot the rays through a narrow tube, so as to obtain a small spot on a phosphorescent screen or a photographic plate, and to subject them between the tube and the screen to an electric field and also a magnetic field, so as to deflect the beam of particles, the electrostatic deflection and the magnetic deflection being perpendicular to each other. He showed that all particles having the same value for m/e would be spread out by the two fields so as to strike the screen in points lying on a parabola; thus, particles of different mass would give different parabolas. Parabolas were found corresponding to the atoms and molecules of various gases in the discharge-tube; and the atomic weights of the particles could be at once inferred from measures of the parabolas. On applying this method of positive-ray analysis to the gas neon, he found[6] in addition to a parabola belonging to atomic weight 20, another corresponding to atomic weight 22. These proved to be, both of them, atoms of neon, but of different masses. Thomson had in fact discovered two ordinary non-radio-active atoms having the same chemical behaviour but different physical characteristics. This result, which was immediately

[1] *Brit. Ass. Rep.*, Birmingham, 1886, p. 569
[2] *ZS. f. Anorg. Chem.* lxi (1909), p. 338 and lxiii (1909), p. 197
[3] *Chem. Soc. Ann. Rep.* (1910), p. 285
[4] cf. Fleck, *Brit. Ass. Rep.*, Birmingham, 1913, p. 447
[5] cf. Vol. I, p. 363
[6] *Proc. R.S.*(A), lxxxix (1 Aug. 1913), p. 1

confirmed by Francis William Aston [1] (1877–1945) showed that the phenomenon of a place in the Newlands-Mendeléev table being occupied by more than one element was not confined to the highest places in the table. Elements which are chemically inseparable, but have different atomic weights, were named by Soddy *isotopes*.

A striking example was furnished when the question as to the end-product of radio-active changes was solved. Having become convinced so early as 1905 that the end-product of the radium series was not helium, Rutherford sought for some other element to fill this position ; and both he and Boltwood suggested that it might be lead,[2] since lead appears persistently as a constituent of uranium-radium minerals: it was possible indeed that lead might be radium G. This proved to be correct, and lead was found to be also the final product of the thorium series. The atomic weights of these two kinds of lead are not, however, equal, that of radium lead being 206 and that of thorium lead being 208. (The atomic weight of ordinary lead is 207·20).[3]

Atoms obtained by radio-active disintegrations occupy places in the periodic table which are determined by what are called the *displacement laws*, first enunciated in 1913 by A. S. Russell,[4] K. Fajans [5] (another pupil of Rutherford's) and F. Soddy,[6] which may be stated as follows: a disintegration with emission of an α-particle causes the atom to descend two places in the Newlands-Mendeléev table (i.e. the atomic weight is diminished) ; a disintegration with emission of a β-particle causes the atom to ascend one place in the table, but does not change the atomic weight.

In 1919–20 it was stated by F. W. Aston [7] that within the limits of experimental accuracy the masses of all the isotopes examined by him were expressed by whole numbers when oxygen was taken as 16 : the only exception was hydrogen, whose mass was 1·008.

Possible methods for separating isotopes were indicated in 1919 by F. A. Lindemann and F. W. Aston,[8] but for long no notable success was attained in practice ; in 1932–3, however, two isotopes of hydrogen were successfully separated by electrolytic methods.[9] The isotopes of neon have been separated by repeated diffusion by Gustave Hertz.[10]

[1] *Brit. Ass. Rep.*, Birmingham, 1913, p. 403

[2] Rutherford, *Radioactivity* (second edn., May 1905), p. 484; Boltwood, *Phil. Mag.*(6) ix (April 1905), p. 599

[3] Soddy, *Ann. Rep. Chem. Soc.*, 1913, p. 269 ; *Chem. News*, cvii (28 Feb. 1913), p. 97 ; *Nature*, xci (20 March 1913), p. 57 ; *Nature*, xcviii (15 Feb. 1917), p. 469

[4] *Chem. News*, cvii (31 Jan. 1913), p. 49

[5] *Phys. ZS.* xiv (15 Feb. 1913), pp. 131 and 136

[6] *Chem. News*, cvii (28 Feb. 1913), p. 97

[7] *Nature*, civ (18 Dec. 1919), p. 393 ; cv (4 March 1920), p. 8 ; *Phil. Mag.* xxxix (April 1920), p. 449 ; ibid. (May 1920), p. 611 ; *Nature*, cv (1 July 1920), p. 547

[8] *Phil. Mag.* xxxvii (May 1919), p. 523

[9] E. W. Washburn and H. C. Urey, *Proc. Nat. Acad. Sci.* xviii (July 1932), p. 496 ; E. W. Washburn, E. R. Smith and M. Frandsen, *Bureau of Standards J. of Research*, xi (Oct. 1933), p. 453 ; G. N. Lewis and R. T. Macdonald, *J. Chem. Phys.* i (June 1933), p. 341

[10] *ZS. f. Phys.* lxxix (1932), p. 108

A new phenomenon in radio-activity was described in 1908 by Rutherford,[1] and confirmed later by Fajans[2] and other workers, namely that in some cases (e.g. radium C, thorium C and actinium C) some of the atoms emitted an α-particle, and in the next transformation a β-particle, while the rest of the atoms reversed the order of the transformations, emitting first a β-particle and afterwards an α-particle. This is known as a *branching* of the series. Rutherford in his original paper expressed the belief that in this way uranium might give rise to the actinium family as well as the radium family; a conjecture which was afterwards generally accepted as correct.

An account must now be given of some notable advances concerned with X-rays. Charles Glover Barkla (1877–1944), when a research student under J. J. Thomson at Cambridge, had become interested in X-rays. In 1902 his work was transferred to Liverpool University, and there in 1904 he discovered that the rays may be partly polarised.[3] In the final dispostion[4] of his experiments, a mass of carbon was subjected to a strong primary beam of X-rays, and so became a source of secondary radiation. A beam of this secondary radiation, propagated in a direction at right angles to that of the primary, was studied. In this second beam was placed a second mass of carbon, and the intensities of tertiary radiation proceeding in directions perpendicular to the direction of propagation of the secondary beam were observed. The X-ray tube was turned round the axis of the secondary beam, while the rest of the apparatus was fixed, and the intensities of the tertiary radiations were observed for different positions of the tube. It was found that the intensity of the tertiary radiation was a maximum when the primary and tertiary beams were parallel, and a minimum when they were at right angles to each other, which showed that the secondary radiation was polarised. This result told decidedly in favour of the hypothesis that X-rays were transverse waves.

Continuing his work on X-rays, Barkla resolved to test a suggestion of J. J. Thomson's, that the number of electrons in an atom might be found by observing the amount of the scattering when X-rays fall on the lighter chemical elements, and comparing it with the scattering produced when they fall on a single electron. In 1903 Thomson had already given, in the first edition of his *Conduction of Electricity through Gases*,[5] a theoretical discussion, based on classical electrodynamics, of the scattering of a pulse of electromagnetic force by an electron on which it is incident. He found

[1] *Nature*, lxxvii (5 March 1908), p. 422

[2] *Phys. ZS.* xii (1911), p. 369; xiii (1912), p. 699

[3] *Nature*, lxix (17 March 1904), p. 463; *Proc. R.S.*(A), lxxiv (1905), p. 474; *Phil. Trans.*(A), cciv (1905), p. 467

[4] *Proc. R.S.*(A), lxxvii (1906), p. 247

[5] J. J. Thomson, *Conduction of Electricity through Gases*, 1st edn. (1903), p. 268; 2nd edn. (1906), p. 321; 3rd edn., Vol. II (1933), p. 256. cf. also J. J. Thomson, *Phil. Mag.* xi (1906), p. 769, where he suggested three different methods of determining the number of electrons in an atom, based respectively on (1) the dispersion of light by gases, (2) the scattering of X-rays by gases, (3) the absorption of β-rays.

that the energy radiated by the electron is $8\pi e^4/3m^2$ times the energy passing through unit area of the wave-front of the primary beam (when the charge e is measured in electromagnetic units). Thus if it is assumed that the electrons, in the chemical element exposed to the X-rays, all scatter independently, the value of the mass-scattering coefficient is

$$\frac{8\pi}{3}\frac{e^4 n}{m^2\rho}$$

where n is the number of electrons per cm^3, and ρ is the density. Now let

N = number of molecules in one gram-molecule
Z = number of free electrons per atom
A = atomic weight.

Then

$$\frac{NZ}{A} = \text{number of electrons in one gram} = \frac{n}{\rho}$$

so the value of the mass-scattering coefficient is

$$\frac{8\pi}{3}\frac{e^4 NZ}{m^2 A}.$$

Barkla [1] found experimentally for the mass-scattering coefficient of the lighter elements (except for hydrogen) a value about 0·2, which would therefore give

$$Z = \frac{3m^2 A}{40\pi e^4 N}.$$

The values accepted at the time for the quantities on the right-hand side of this equation were inaccurate, and the result deduced, namely that there were between 100 and 200 electrons per molecule of air, was replaced by Barkla in 1911 [2] by a much better determination based on Bucherer's value for e/m, Rutherford and Geiger's value for e, and Rutherford's value for N. This gave approximately

$$Z = \tfrac{1}{2}A,$$

i.e. the number of scattering electrons per atom, for the lighter elements, is about half the atomic weight of the element, except in the case of hydrogen, for which Z = 1. These results anticipated later discoveries in a remarkable way.

The secondary X-rays were destined to furnish other contributions to atomic physics. In 1906 Barkla [3] found that in some cases the secondary rays consisted mainly of a radiation which differed altogether in 'hardness,' or penetrating power, from the primary radiation, so that it could not be regarded as the result of 'scattering.'

[1] *Phil. Mag.* vii (May 1904), p. 543. cf. also J. A. Crowther, *Phil. Mag.* xiv (Nov. 1907), p. 653
[2] *Phil. Mag.* xxi (May 1911), p. 648 [3] *Phil. Mag.* xi (June 1906), p. 812

He pursued this matter further, with the help of one of his students, C .A. Sadler, and in 1908 they[1] found that the secondary X-rays emitted by a chemical element exposed to a primary beam of X-rays were of two distinct types :

(i) A scattered radiation, not of great amount, of the same quality as the primary beam.

(ii) A radiation characteristic of the exposed chemical element, and almost, if not quite, *homogeneous*, i.e. all of the same degree of hardness. It was, moreover, emitted uniformly in all directions, unlike the scattered radiation. This characteristic radiation was produced only when the primary X-rays contained a constituent harder than the characteristic radiation that was to be excited. (On this account the characteristic X-rays were often spoken of at the time as 'flourescent.') Barkla found also that the hardness of the characteristic radiation increased as the atomic weight of the emitting chemical element increased.

R. Whiddington[2] found that the primary rays from an X-ray tube can excite the radiation characteristic of an element of atomic weight w only when the velocity of the parent cathode rays exceeds $10^8\,w$ cm/sec ; when the velocity of the primary rays is less than this, only a truly 'scattered' radiation is emitted, resembling the primary.

It was found[3] that the characteristic secondary radiations may be divided into several groups, the radiation belonging to each group becoming more penetrating as the atomic weight of the radiating element increases ; in other words, each chemical element emits a line spectrum of X-rays, each line moving to the more penetrating end of the spectrum as the atomic weight of the element increases. Two groups which were described in 1909 received the notation K and L in 1911, and an M-group was found a little later.[4] The K-series, which is the most penetrating, was found together with the L-series for elements from zirconium (atomic wt. 90·6) to silver (atomic wt. 107·88). For elements heavier than silver, the K-series was difficult to excite, since very great velocities would be required in the exciting cathode rays : and for elements lighter than zirconium, the L-series was difficult to observe because it was so easily absorbed.[5]

It was shown by G. W. C. Kaye[6] that the radiation characteristic of a chemical element can be excited not only by exposing it to a

[1] *Phil. Mag.* xiv (Sept. 1907), p. 408 : xvi (Oct. 1908), p. 550

[2] *Proc. R.S.*(A), lxxxv (April 1911), p. 323

[3] *Proc. Camb. Phil. Soc.* xv (1909), p. 257 ; Barkla and J. Nicol, *Nature*, lxxxiv (Aug. 1910), p. 139

[4] Barkla and V. Collier, *Phil. Mag.* xxiii (June 1912), p. 987

[5] E. H. Kürth, *Phys. Rev.* xviii (1921), p. 461, found for the convergence wave-lengths in Ångströms : K-series of carbon, 42·6, oxygen 23·8 : L-series of carbon, 375, oxygen 248, iron 16·3, copper 12·3 : M-series of iron, 54·3, copper 41·6 : N-series of iron 247, copper 116.

[6] *Phil. Trans.*(A), ccix (Nov. 1908), p. 123. Kaye found that the intensity of general X-radiation was nearly proportional to the atomic weight of the element forming the anticathode ; later, W. Duane and T. Schimizu, *Phys. Rev.* xiv (1919), p. 525, showed that the intensity is proportional to the atomic *number*.

beam of primary X-rays, but also by using it as the anticathode in an X-ray tube, so that it is bombarded by cathode rays. It was suggested that this might be an indirect effect, produced by the mediation of non-characteristic X-rays: but this suggestion was disproved by R. T. Beatty,[1] who proved beyond doubt that the characteristic X-rays are excited directly by the impact of cathode rays. In the following year Beatty [2] verified experimentally a result which had been reached theoretically by J. J. Thomson [3] in 1907, namely that the total intensity of general X-radiation is proportional to the fourth power of the velocity of the exciting electrons.

The question as to whether X-rays were corpuscles or waves was still unsettled in 1910. In that year W. H. Bragg published a paper [4] in which, interpreting his experiments by the light of the corpuscular hypothesis, he arrived at conclusions which were in fact true and of great significance. We have seen [5] that when X-rays are passed through a gas they render it a conductor of electricity, and that this property is due to the production of ions in the gas. Bragg now asserted that the X-rays do not ionise the gas directly; they act by ejecting, from a small proportion of the atoms of the gas, electrons (photo-electrons) of high speed, each of which acts as a β-particle and ionises the gas by detaching electrons in a succession of collisions with molecules along its path. The speed of the ejected electron depends only on the hardness or penetrating power of the X-rays (which was later shown to be, in effect, their frequency), and not at all on their intensity, or on the nature of the atom from which the electron is expelled. What Bragg emphasised as specially remarkable was that the energy of the electron was as great as that of an electron in the beam of cathode rays by which the X-rays had been excited originally: the X-ray pulse seemed to have the property of keeping its energy together in a small bundle, without any of the spreading that might have been expected on the wave theory, and to be able to transfer the whole of this energy to a single electron. He enunciated 'the general principle, that if one radiant entity (α-, β-, γ-, X- or cathode-ray) enters an atom, one and only one entity emerges, carrying with it the energy of the entering entity.' '*One* X-ray provides the energy for *one* β-ray, and similarly in the X-ray bulb, *one* β-ray excites *one* X-ray. No energy is lost in the interchange of forms, β- to X-ray and back again; and the speed of the secondary β-ray is independent of the distance that the X-ray has travelled: so the X-ray cannot diffuse its energy as it goes, that is to say, it is a corpuscle.' It was in fact now established that the X-ray behaves in some ways [6] as a wave and in other ways as a corpuscle.

[1] *Proc. R.S.*(A), lxxxvii (Dec. 1912), p. 511
[2] *Proc. R.S.*(A), lxxxix (1913), p. 314
[3] *Phil. Mag.* xiv (1907), p. 226
[4] *Phil. Mag.* xx (Sept. 1910), p. 385; *Brit. Ass. Rep.* 1911, p. 340; W. H. Bragg and H. L. Porter, *Proc. R.S.*(A), lxxxv (1911), p. 349
[5] cf. Vol. I, p. 359
[6] cf. A. Joffé and N. Dobronrawow, *ZS. f. P.*, xxxiv (1925), p. 889

Bragg's conclusions were fully confirmed in 1911–12 by C. T. R. Wilson,[1] using his method of cloud-chamber photographs. The whole of the region traversed by the primary X-ray beam was seen to be filled with minute streaks and patches of cloud: examining the photographs more closely, the cloudlets were seen to be small thread-like objects, consisting of droplets deposited on ions produced along the paths of the β-particles, which were the actually effective ionising agents.

In the early part of the twentieth century, many attempts were made to test the hypothesis that X-rays are waves, by trying to obtain diffraction-effects with them. In 1899 and 1902 H. Haga and C. H. Wind [2] of Groningen observed a broadening of the image of a wedge-shaped slit, and inferred that the wave-length of the vibrations concerned was of the order of one Ångström.[3] However, in 1908, when B. Walter and R. Pohl [4] repeated the experiments, they found that different times of exposure gave different results as regards the image, and concluded that the effect was not confirmed. In 1912 the question was re-opened when P. P. Koch,[5] making a special study of the blackening of photographic plates in general, re-examined Walter and Pohl's images, and decided that there was evidence of genuine diffraction. Thereupon Arnold J. W. Sommerfeld (1868–1951), Professor at Munich, compared the results of theory with Koch's photometric measurements,[6] and deduced a value of 0·3 Ångströms for the wave-length of the X-rays.

At that time a young student, Peter Paul Ewald (*b.* 1888), who had just taken his doctorate at Munich, was interested in the transmission of light through the atomic lattice of a crystal. Some notion of the dimensions of crystal-lattices could by this time be formed; the Avogadro number (the number of molecules in a number of grams equal to the molecular weight) was known to be approximately 6×10^{23}; this together with a knowledge of the density and molecular weight of a crystal made it possible to estimate that the distance apart of the atoms in a crystal was of the order of 10^{-8} cm. or one Ångström. A junior lecturer in Munich, Max Laue [7] (*b.* 1879), who was in contact with Sommerfeld and Ewald, saw that if the X-rays had a wave-length of the order suggested by Sommerfeld, then the crystal-lattice had the right dimensions for acting as a three-dimensional diffraction grating, so to speak, for the X-rays. He promptly arranged for an experimental test of this idea, which was carried out by W. Friedrich and P. Knipping; and a paper

[1] *Proc. R.S.*(A), lxxxv (April 1911), p. 285; (A), lxxxvii (Sept. 1912), p. 277

[2] *Proc. Amst. Ac.* (25 March 1899) (English edn. i, p. 420) and 27 Sept. 1902 (English edn. v, p. 247). This work was discussed by Sommerfeld, *Phys. ZS.*, i (1899), p. 105 and ii (1900), p. 55.

[3] cf. Vol. I, p. 367, note 5

[4] *Ann. d. Phys.* xxv (1908), p. 715; xxix (1909), p. 331

[5] *Ann. d. Phys.* xxxviii (1912), p. 507

[6] *Ann. d. Phys.* xxxviii (1912), p. 473

[7] About this time Laue's father, who was a general in the German Army, received a title of nobility, so the son was known subsequently as Max von Laue.

was published[1] in June–July 1912 in which it was completely vindicated. A thin pencil of X-rays was allowed to fall on a crystal of zinc sulphide. A photographic plate, placed behind the crystal at right angles to this primary pencil, showed a strong central spot, where it was met by the primary rays, surrounded by a number of other spots, in a regular arrangement: these were situated at the places where the plate was met by diffracted pencils, produced by reflection of the primary X-rays at sets of planes of atoms in the crystal. The positions of the spots in the simplest imaginable case are given by the following rule[2]: Suppose that the atoms are disposed so that the three rectangular co-ordinates of any atom are integral multiples of a length a, every such place being filled, and suppose that the incident rays are parallel to one of the axes. If the distance from an atom A to another atom B is an integral multiple of a, then in the direction AB there will be one of the diffracted pencils that cause the spots. Clearly such directions correspond to all ways of expressing a square as the sum of three squares.

In this case the spots furnish no information regarding the wave-length of the radiation; and indeed the radiation used by Friedrich and Knipping had no definite wave-length, being a heterogeneous mixture of rays whose wave-lengths formed a continuous series.[3] A mathematical theory of the spots in more general cases was given by Laue himself and by other writers.[4]

Laue's discovery was of the first importance, for not only did the diffraction-patterns under suitable conditions serve to determine the wave-length of the X-rays, but the idea was soon developed into a regular method for determining the arrangement of the atoms in crystals; and its merit was fitly recognised by the award to Laue in 1914 of the Nobel Prize for physics. The question as to whether X-rays were corpuscles or waves seemed to be settled in favour of the undulatory hypothesis; it was in fact found that the wave-length of high-frequency X-rays was about one Ångström: but W. H. Bragg wrote[5] 'The problem becomes, it seems to me, not to decide between two theories of X-rays, but to find one theory which possesses the capacity of both'—a remarkable anticipation of the view that was made possible many years later by the discovery of quantum mechanics.

William Lawrence Bragg (*b.* 1890), son of W. H. Bragg, in a paper read before the Cambridge Philosophical Society in the autumn of the same year,[6] introduced considerable simplifications in the theory.

[1] W. Friedrich, P. Knipping and M. Laue, *München Ber.* 8 June, p. 303 and 6 July p. 363, 1912; reprinted *Ann. d. Phys.* xli (1913), p. 971

[2] W. H. Bragg, *Nature*, xc (24 Oct. 1912), p. 219

[3] H. G. J. Moseley and C. G. Darwin, *Phil. Mag.* xxvi (July 1913), p. 210 found a continuous spectrum, with maxima due to characteristic rays from the anticathode.

[4] M. von Laue, loc. cit.: *Ann. d. Phys.* xlii (1913), p. 397; *Phys. ZS.* xiv (1913), p. 1075; H. Moseley and C. G. Darwin, *Nature*, xc (14 Oct. 1912), p. 219; P. P. Ewald, *Phys. ZS.* xiv (1913), p. 1038; xv (1914), p. 399; C. G. Darwin, *Phil. Mag.* xxvii (Feb. 1914), p. 315; ibid. (April 1914), p. 675

[5] *Nature*, xc (28 Nov. 1912), p. 360

[6] *Proc. Camb. Phil. Soc.* xvii (Feb. 1913), p. 43

His leading idea (which replaced Laue's assumption of scattering at the points of a crystal-grating) was that parallel planes in the crystal which are rich in atoms can be regarded, taken together, as a reflecting surface for X-rays; and experiments with a slip of mica about a millimetre thick whose surface was a cleavage plane showed him that the laws of reflection were obeyed when the rays were incident at nearly glancing angles; reflection takes place only when the wave-length λ of the rays, the distance d between the parallel planes in the crystal, and the angle of incidence ϕ, are connected by the relation

$$n\lambda = 2d \cos \phi$$

where n is a small whole number. This is known as the *Bragg law*. His father, continuing the work with him,[1] devised an *X-ray spectrometer*, the principle of which is to allow *monochromatic* X-rays to fall in a fixed direction on a crystal, which is made to turn so that each plane can be examined in detail: and with this instrument the arrangement of the atoms in many different crystals was determined. From this point the study of crystal-structure was developed by the Braggs with great success over an immense range: the Nobel Prize for physics was awarded to them in 1915.

The discoveries regarding X-rays led to a better understanding of the γ-rays from radio-active substance. J. A. Gray[2] established the similarity in nature of γ-rays and X-rays by showing that the γ-rays from RaE excite the characteristic X-radiations (K-series) of several elements, just as very penetrating X-rays would: and that the γ-rays behave similarly to X-rays (both qualitatively and quantitatively) in regard to scattering. In 1914 Rutherford and E. N. da C. Andrade,[3] by methods based on the same principle as those used by the Braggs and by Moseley and Darwin, measured the wave-lengths of the γ-rays from radium B and C. The wave-lengths of γ-rays are usually less than those of X-rays, being generally between 0·01 and 0·1 Ångströms.

In the work of Rutherford and Geiger on counting α-particles by the electric method, carried out in 1908,[4] some of the difficulties that had to be overcome were due to the scattering of α-rays in passing through matter. Geiger made a special study of the scattering for small angles of deflection, and in 1909 Rutherford suggested to one of his research students, E. Marsden, an examination of the possibility of scattering through large angles. As a result of this suggestion, experiments were carried out by Geiger and Marsden, which showed[5] that α-particles fired at a thin plate of matter can be scattered inside the material to such an extent that some of them emerge again on the side of the plate at which they entered: and

[1] *Proc. R.S.*(A), lxxxviii (July 1913), p. 428; lxxxix (Sept. 1913), pp. 246, 248; ibid. (Feb. 1914), p. 468
[2] *Proc. R.S.*(A), lxxxvii (Dec. 1912), p. 489
[3] *Phil. Mag.* xxvii (May 1914), p. 854
[4] cf. p. 7
[5] *Proc. R.S.*(A), lxxxii (July 1909), p. 495; *Phil. Mag.* xxv (1913), p. 604

calculation showed that some of the α-particles must have been deflected at single encounters through angles greater than a right angle.

Now at that time the atom was generally pictured in the form suggested by J. J. Thomson in his Silliman lectures of 1903.[1] He was then working out the consequences of supposing that the negative electrons occupy stationary positions in the atom. In order that the atom as a whole may be electrically neutral, there must be also a positive charge : and he saw that this could not be concentrated in positively charged corpuscles, since a mixed assemblage of negative and positive corpuscular charges could not be in stable equilibrium. He therefore assumed that the positive electrification was uniformly distributed throughout a sphere of radius equal to the radius of the atom as inferred from the kinetic theory of gases (about 10^{-8} cm.) : the negative electrons he supposed to be situated inside this sphere, their total charge being equal and opposite to that of the positive electrification.

In attempting to picture the way in which the negative electrons would dispose themselves, Thomson was guided by some experiments with magnets which had been made many years earlier by Alfred Marshall Mayer of the Stevens Institute of Technology, Hoboken.[2] Mayer magnetised a number of sewing-needles with their points of the same polarity, say south. Each needle was run into a small cork, of such a size that it floated the needle in an upright position, the eye end of the needle just coming through the top of the cork. If three of these vertical magnetic needles are floated in a bowl of water, and the north pole of a large magnet is brought down over them, the mutually repellent needles at once approach each other, and finally arrange themselves at the vertices of an equilateral triangle. With four needles a square is obtained, with five either a regular pentagon or (a less stable configuration) a square with one needle at its centre, and so on. The under-water poles of the floating needles, and the upper pole of the large magnet, were regarded as too far away to exert any appreciable influence, so the problem was practically equivalent to that of a number of south poles in presence of a single large north pole.

Thomson examined theoretically the problem of the configurations assumed by a small number of negative electrons inside a sphere of positive electrification, and found that when the number of electrons was small, they disposed themselves in a regular arrangement, all being at the same distance from the centre ; but when the number of electrons was increased, they tended to arrange themselves in rings or spherical shells, and that the model imitated many of the known properties of atoms, particularly the periodic changes with increase of atomic weight which are set forth in the

[1] Published as *Electricity and Matter* in 1904. cf. J. J. Thomson, *Phil. Mag.* vii (1904), p. 237, and for an earlier model, Lord Kelvin, *Phil. Mag.* iii (1902), p. 257.

[2] *Phil. Mag.*(5) v (1878), p. 397 ; (5) vii (1879), p. 98

Newlands-Mendeléev table. If one of the electrons were displaced slightly from its position of equilibrium, it would be acted on by a restitutive force proportional to the displacement. This was a most desirable property, since it was just what was required for an electronic theory of optical dispersion and absorption; and, moreover, it would explain the monochromatic character of spectral lines: but in no way could Thomson's model be made to give an account of spectral series.[1]

A model atom alternative to Thomson's had been proposed in the same year (1903) by Philipp Lenard[2] (1862–1947) of Kiel, who observed that since cathode-ray particles can penetrate matter, most of the atomic volume must offer no obstacle to their penetration, and who designed his model to exhibit this property. In it there were no electrons and no positive charge separate from the electrons: the atom was constituted entirely of particles which Lenard called *dynamides*, each of which was an electric doublet possessing mass. All the dynamides were supposed to be identical, and an atom contained as many of them as were required to make up its mass. They were distributed throughout the volume of the atom, but their radius was so small ($<0{\cdot}3 \times 10^{-11}$ cm.) compared with the radius of the atom, that most of the atomic volume was actually empty. Lenard's atom, however, never obtained much acceptance, as no evidence could be found for the existence of the dynamides.

The deflection of an α-particle through an angle greater than a right angle was clearly not explicable on the assumption of either Thomson's or Lenard's atom; and Rutherford in December 1910 came to the conclusion that the phenomenon could be explained only by supposing that an α-particle occasionally (but rarely) passed through a very strong electric field, due to a charged nucleus[3] of very small dimensions in the centre of the atom. This was confirmed a year later by C. T. R. Wilson's photographs[4] of cloud-chamber tracks of α-particles which showed violent sudden deflections at encounters with single atoms.

Thus Rutherford was led to what was perhaps the greatest of all his discoveries, that of the structure of the atom; the first account of his theory was published in May 1911.[5] He found that if a model atom were imagined with a central charge concentrated within a sphere of less than 3×10^{-12} cm. radius, surrounded by electricity of the opposite sign distributed throughout the rest of the volume of the atom (about 10^{-8} cm. radius), then this atom would satisfy all the known laws of scattering of α- or β-particles, as found by Geiger and Marsden. The central charge necessary would be Ne,

[1] See, however, an attempt by K. F. Herzfeld, *Wien Ber.* cxxi, 2a (1912), p. 593

[2] *Ann. d. Phys.* xii (1903), p. 714, at p. 736

[3] The term *nucleus* for the central charge seems to have been used first in Rutherford's book *Radioactive Substances and their Radiations*, which was published in 1912.

[4] *Proc. R.S.*(A), lxxxvii (Sept. 1912), p. 279

[5] *Phil. Mag.* xxi (May 1911), p. 669; xxvii (1914), p. 488. For an anticipation that the atom might prove to be of this type, cf. H. Nagaoka, *Phil. Mag.* vii (1904), p. 445.

where e is the electronic charge, and N is a number equal to about half the atomic weight. This fitted in perfectly with the discovery already made by Barkla,[1] that the number of scattering electrons per atom is (for the lighter elements, except hydrogen) about half the atomic weight: for the positive central charge, and the negative charges on the electrons in the space around it, must exactly neutralise each other.

Thus the Rutherford atom is like the solar system, a small positively charged nucleus in the centre, which contains most of the mass of the atom, being surrounded by negative electrons moving around it like planets, at distances of the order of 10^{-8} cm. Occasionally an α-particle passes near enough to unbind and detach an electron and thus ionise the atom: still more infrequently (only about one α-particle in ten thousand, even in the case of heavy elements) the α-particle may come so close to the nucleus as to experience a violent deflection, due to the electrical repulsion between them. The encounters were studied mathematically by C. G. Darwin[2] (*b.* 1887), who found a satisfactory agreement between theory and experiment, and showed that Geiger and Marsden's results could not be reconciled with any law of force except the electrostatic law of the inverse square, which is obeyed to within 3×10^{-12} cm. of the centre of the atom.

Rutherford now laid down the principle[3] that the positive charge on the nucleus (or the number of negative electrons) is the fundamental constant which determines the chemical properties of the atom: this fact explains the existence of isotopes, which have the same nuclear charge but different nuclear masses, and which have the same chemical properties. He pointed out also that gravitation and radio-activity, being unaffected by chemical changes, must depend on the nucleus. His old discovery, that the α-particle is a doubly ionised atom of helium, was now reformulated in the statement, that the α-particle, at the end of its track, captures two electrons (one at a time), and thus becomes a neutral helium atom; he suggested, moreover, that the nucleus of the hydrogen atom might actually be the 'positive electron.' It was seen that the hydrogen nucleus differed from the negative electron not only in the reversal of sign of its charge, but also in having a much greater mass—in fact, almost all the mass of the hydrogen atom[4]; and at the Cardiff meeting of the British Association in 1920, Rutherford proposed for it the name *proton*, which has been universally accepted.

A proposal for removing the uncertainty which still remained as to the precise amount of the nuclear charge was made in 1913 by

[1] cf. p. 15 [2] *Phil. Mag.* xxvii (March 1914), p. 499

[3] *Nature*, xcii (Dec. 1913), p. 423; *Phil. Mag.* xxvi (Oct. 1913), p. 702; xxvii (March 1914), p. 488

[4] Poincaré in his St. Louis lecture of 1904 had said (*Bull. des Sci. Math.* xxviii (1904), p. 302) 'The mass of a body would be the sum of the masses of its positive electrons, the negative electrons not counting.'

A. van der Broek [1] of Utrecht. He remarked that when α-particles are scattered by a nucleus, the amount of scattering per atom, divided by the square of the charge in the nucleus, must be constant. As Geiger and Marsden had shown, this condition is roughly satisfied if the nuclear charge is assumed to be proportional to the atomic weight; but van der Broek now pointed out that it would be satisfied with far greater accuracy if the nuclear charge were assumed to be proportional to the number representing the place of the element in the Newlands-Mendeléev periodic table. He suggested, therefore, that the nuclear charge should be taken to be Ze, where e is the electronic charge (taken positively) and Z is the ordinal number of the element in the periodic table.

This suggestion received a complete confirmation from experiments performed in Rutherford's laboratory at Manchester by Henry Gwyn-Jeffreys Moseley [2] (*b.* 1887, killed at the Suvla Bay landing in the Dardanelles, 10 August 1915) in continuation of the work which he and Darwin had been carrying on together. Moseley exposed the chemical elements, from calcium to nickel, as anticathodes in an X-ray tube, so that under the bombardment of cathode-rays they emitted their characteristic X-ray spectra, consisting essentially of two strong lines (the K- and L-lines) [3]; and the wave-lengths of these lines were determined by the crystal method. Taking either of these lines and following it from element to element, he found that the square root of its frequency increased by a constant quantity as the transition was made from any element to the next higher element in the periodic table; so that the frequency was expressible in the form $k(N-a)^2$, where k was an absolute constant, N was the 'atomic number' or place in the periodic table, and a was a constant which had different values for the K- and L-lines. So there must be in the atom a fundamental number, which increases by unity as we pass from one element to the next in the periodic table; and, having regard to the results of Rutherford, Geiger, Marsden and van der Broek, this quantity can only be the amount of the nuclear charge, expressed in electron-units. Thus *the number of negative electrons which circulate round the nucleus of an atom of a chemical element is equal to the ordinal number of the element in the periodic table.*

Two incidental results of Moseley's work on X-ray spectra must be mentioned. It now became clear that the atomic numbers of iron, cobalt and nickel must be respectively 26, 27, 28, thus confirming the opinion, already suggested by chemical considerations, that cobalt should have a lower place in the periodic table than

[1] *Phys. ZS.* xiv (1913), p. 32; *Nature*, xcii (27 Nov. 1913), p. 372; xcii (25 Dec. 1913), p. 476; *Phil. Mag.* xxvii (March 1914), p. 455

[2] *Phil. Mag.* xxvi (Dec. 1913), p. 1024; xxvii (April 1914), p. 703. Moseley left Manchester for Oxford at the end of 1913, and completed his work there. cf. his obituary notice in *Proc. R.S.*(A), xciii (1917), p. xxii.

[3] cf. p. 16. Actually each of these lines is a multiplet.

nickel, although it has a higher atomic weight; and the vexed questions of the number of elements in the group of the rare earths, and of missing elements in the periodic table, could also be settled, since it was now known what the X-ray spectra of these elements must be.[1] Some predictions which had been made by the Danish chemist, Julius Thomsen (1826–1909) were now verified in a remarkable way.

Thus Rutherford, with the help of the young men in his research school—Geiger, Marsden, Moseley and Darwin [2]—created a definite quantitative theory of the atom, lending itself to mathematical treatment, and satisfying every comparison with experiment. It has been the foundation of all later work.

During the years 1914–18 Rutherford was occupied chiefly with matters connected with the war: but in 1919 he made a contribution [3] of the highest importance to atomic physics. It originated from an observation made by Marsden, who had shown [4] that when an α-particle collides with an atom of hydrogen, the hydrogen atom may be set in such swift motion that it travels (nearly in the direction of the impinging particle) four times as far as the colliding α-particle, and that it may be detected by a scintillation produced on a zinc sulphide screen. Rutherford now showed, by measurements of deflections in magnetic and electric fields, that these scintillations were due to hydrogen atoms carrying unit positive charge, in other words, to hydrogen nuclei, or protons as they soon came to be called.

He next bombarded dry air, and nitrogen, with α-particles, and again found scintillations at long range. The similarity in behaviour of the particles obtained from nitrogen to those previously obtained from the hydrogen led him to suspect that they were identical, i.e. that the long-range particles obtained by bombarding nitrogen with α-particles were actually hydrogen nuclei. The general idea now presented itself, that some of the lighter atoms might be actually disintegrated by a collision with a swift α-particle: going beyond the earlier discovery that an α-particle might be deflected through a large angle by a close collision with a nucleus, he now came to the conclusion that on still more rare occasions (say one α-particle in half a million) it might break up the nucleus. The phenomenon was found to occur markedly with nitrogen, but not with dry oxygen.

In the summer of 1919 Rutherford succeeded J. J. Thomson as the Cavendish Professor of Physics at Cambridge. Continuing his experiments there, he succeeded in proving definitely that the

[1] For individual elements the nuclear charges were found directly by J. Chadwick (*Phil. Mag.* xl (1920), p. 734) by experiments on the scattering of pencils of α-rays. His results agreed with those deduced from Moseley's law of X-ray spectra. Practically all of the elements which have been discovered since Moseley's day, and which fill the gaps that then existed in the periodic table, have been identified by the study of their characteristic X-ray spectra.

[2] And Bohr, whose work will be described in a later chapter

[3] *Phil. Mag.* xxxvii (June 1919), p. 537

[4] *Phil. Mag.* xxvii (May 1914), p. 824

nitrogen atom can be disintegrated by bombarding it with α-particles.[1] As P. M. S. Blackett (*b.* 1897) showed, the tracks of the particles could be seen in the Wilson cloud-chamber. Since the nitrogen nucleus, of charge 7 electronic units, captures the α-particle, of charge 2, and expels the proton, of charge 1, the particle obtained by the transformation must have charge 8, that is, it must be the nucleus of an isotope of oxygen. Since the nitrogen nucleus has mass 14, the captured α-particle has mass 4, and the expelled proton has mass 1, it follows that the oxygen isotope must have mass 17.

In 1921 Rutherford and J. Chadwick (*b.* 1891) found[2] that similar transformations could be produced in boron, fluorine, sodium, aluminium and phosphorus: and other elements were later added to the list. In each case the α-particle was captured and a swift proton was ejected, while a new nucleus of mass three units greater and charge one unit higher was formed. Thus the medieval alchemist's dream of the transmutation of matter was realised at last.

Rutherford died at Cambridge on 19 October 1937, and was buried in Westminster Abbey near the graves of Newton and Kelvin. He was survived by his old teacher J. J. Thomson, who had in 1918 been elected Master of the great foundation of which he had been a member uninterruptedly since 1875.

'How fortunate I have been throughout my life!' Thomson wrote, near the end of it, 'I have had good parents, good teachers, good colleagues, good pupils, good friends, great opportunities, good luck and good health.' He lived to be eighty-three, dying at Trinity Lodge on 30 August 1940, and was buried on 4 September in the Abbey.

[1] *Proc. R.S.*(A), xcvii (July 1920), p. 374; *Engineering*, cx (17 Sept. 1920), p. 382 (a paper read to the British Association at its Cardiff meeting); *Proc. Rhys. Soc.* xxxiii (Aug. 1921), p. 389

[2] *Nature*, cvii (10 March 1921), p. 41

Chapter II

THE RELATIVITY THEORY OF POINCARÉ AND LORENTZ

At the end of the nineteenth century, one of the most perplexing unsolved problems of natural philosophy was that of determining the relative motion of the earth and the aether. Let us try to present the matter as it appeared to the physicists of that time.

According to Newton's First Law of Motion, any particle which is free from the action of impressed forces moves, if it moves at all, with uniform velocity in a straight line. But in order that this statement may have a meaning, it is necessary to define the terms *straight line* and *uniform velocity*; for a particle which is said to be 'moving in a straight line' in a terrestrial laboratory would not appear to be moving in a straight line to an observer on the sun, since he would perceive its motion compounded with the earth's diurnal rotation and her annual revolution in her orbit. We can, however, define a straight line *with reference to a system of axes* $Oxyz$ as the geometrical figure defined by a pair of linear equations between x, y, z; and we can assert as a fact of experience that certain systems of axes $Oxyz$ exist such that free particles move in straight lines with reference to them. Moreover, we can assert that there exist certain ways of measuring time such that the velocity of free particles along their rectilinear paths is uniform. A set of axes in space and a system of time-measurement, which possess these properties, may be called an *inertial system of reference.*

In Newtonian mechanics, if S is an inertial system of reference, and if S′ is another system such that the axes $O'x'y'z'$ of S′ have any uniform motion of pure translation with respect to the axes $Oxyz$ of S, and if the system of time-measurement is the same in the two cases, then S′ is also an inertial system of reference: the Newtonian laws of motion are valid with respect to S′ just as with respect to S. No one inertial system of reference could be regarded as having a privileged status, in the sense that it could properly be said to be fixed while the others were moving. Newtonian mechanics does not involve the notion of the absolute fixity of a point in space.

The laws of Newtonian dynamics thus presuppose the knowledge of a certain set of systems of reference, which is necessary if the laws are to have any meaning. In the nineteenth century many physicists inquired how this set of systems of reference should be described and defined. When Carl Neumann (1832–1925) was appointed professor of mathematics at Leipzig in 1869, he devoted his inaugural lecture [1] to the question, and introduced the name *The Body Alpha* for these systems of reference collectively. W. Thomson (Kelvin) and P. G. Tait in their *Treatise on Natural Philosophy* [2] suggested as a basis for specifying the Body Alpha that the centre of gravity of all matter in the universe might be considered to be *absolutely at rest,* and that the plane in which the angular momentum of the universe round its centre of gravity is the greatest, might be regarded as *fixed in direction in space.* Other writers proposed that the Body Alpha should be based on the system of the fixed stars, or the aggregate of all the bodies in existence.[3]

In the latter part of the nineteenth century the doctrine of the aether, which was justified by the undulatory theory of light, was generally regarded as involving the concepts of rest and motion relative to the aether, and thus to afford a means of specifying absolute position and defining the Body Alpha. Suppose, for instance, that a disturbance is generated at any point in free aether : this disturbance will spread outwards in the form of a sphere : and the centre of this sphere will for all subsequent time occupy an unchanged position relative to the aether. In this way, or in many other ways, we might hope to determine, by electrical or optical experiments, the velocity of the earth's motion relative to the aether.

In the first years of the twentieth century this problem was provoking a fresh series of experimental investigations. The most interesting of these was due to FitzGerald [4] who, shortly before his death in February 1901, commenced to examine the phenomena exhibited by a charged electrical condenser, as it is carried through space by the terrestrial motion. When the plane of the condenser includes the direction of the aether-drift (the ' longitudinal position '), the moving positive and negative charges on its two plates will be equivalent to currents running tangentially in opposite directions in the plates, so that a magnetic field will be set up in the space between them, and magnetic energy must be stored in this space : but when the plane of the condenser is at right angles to the terrestrial motion (the ' transverse position '), the equivalent currents are in the normal direction, and neutralise each other's magnetic action almost completely. FitzGerald's original idea was that, in order to supply the magnetic energy, there must be a mechanical drag on the condenser at the moment of charging, similar to that which would be produced if the mass of a body at the surface of the earth were suddenly to become greater. Moreover, the co-existence of the electric and magnetic fields in the space between the plates would entail [5] the

[1] Afterwards published as a booklet of 32 pages, *Die Principien der Galilei-Newton'schen Theorie* (Leipzig 1870). He returned to the matter in 1904, in the *Festschrift Boltzmann* (Leipzig, 1904), p. 252.

[2] New edition, Cambridge 1890, Vol. I, p. 241

[3] An account of these suggestions is given by G. Giorgi, *Palermo Rend.*, xxxiv (1912), p. 301.

[4] FitzGerald's *Scientific Writings*, p. 557 ; cf. Larmor, ibid., p. 566

[5] cf. Vol. I, p. 318

existence of an electromagnetic momentum proportional to their vector-product. This momentum is easily seen to be (with sufficient approximation) parallel to the plates, and so would not in general have the same direction as the velocity of the condenser relative to the aether: thus the change in the situation in one second might be represented by the annihilation of the momentum existing at the beginning of the second and the creation of the momentum (equal and parallel to it) existing at the end of the second. But two equal and oppositely-parallel momenta at a distance apart constitute an angular momentum: and we may therefore expect that if the condenser is freely suspended, there will in general be a couple acting on it, proportional to the vector-product of the velocity of the condenser and the electromagnetic momentum. This couple would vanish in either the longitudinal or the transverse orientation, but in intermediate positions would tend to rotate the condenser into the longitudinal position; the transverse position would be one of unstable equilibrium.

For both effects a search was made by FitzGerald's pupil F. T. Trouton [1]; in the experiments designed to observe the turning couple, a condenser was suspended in a vertical plane by a fine wire, and charged. The effect to be detected was small: for the magnetic force due to the motion of the charges would be of order (w/c), where w denotes the velocity of the earth: so the magnetic energy of the system, which depends on the square of the force, would be of order $(w/c)^2$: and the couple would likewise be of the second order in (w/c).

No effect of any kind could be detected,[2] a result whose explanation was rightly surmised by P. Langevin [3] to belong to the same order of ideas as FitzGerald's hypothesis of contraction.

It may be remarked that the existence of the couple, had it been observed, would have demonstrated the possibility of drawing on the energy of the earth's motion for purposes of terrestrial utility.

The FitzGerald contraction of matter as it moves through the aether might conceivably be supposed to affect in some way the optical properties of the moving matter: for instance, transparent substances might become doubly refracting. Experiments designed to test this supposition were performed by Lord Rayleigh [4] in 1902 and by D. B. Brace in 1904,[5] but no double refraction comparable with the proportion $(w/c)^2$ of the single refraction could be detected. The FitzGerald contraction of a material body cannot therefore be of the same nature as the contraction which would be produced in the body by pressure, but must be accompanied by such concomitant

[1] *Trans. Roy. Dub. Soc.*, vii (1902), p. 379; F. T. Trouton and H. R. Noble, *Phil. Trans.* ccii (1903), p. 165

[2] This negative result was confirmed in 1926 by R. Tomaschek, *Ann. d. Phys.* lxxviii (1926), p. 743 and lxxx (1926), p. 509; and by C. T. Chase, *Phys. Rev.* xxviii (1926), p. 378.

[3] *Comptes Rendus*, cxl (1905), p. 1171

[4] *Phil. Mag.* iv (1902), p. 678

[5] *Phil. Mag.* vii (1904), p. 317

changes in the relations of the molecules to the aether, that an isotropic substance does not lose its simply refracting character.

Even before the end of the nineteenth century, the failure of so many promising attempts to measure the velocity of the earth relative to the aether had suggested to the penetrating and original mind of Poincaré a new possibility. In his lectures at the Sorbonne in 1899,[1] after describing the experiments so far made, which had yielded no effects involving either the first or the second powers of the coefficient of aberration (i.e. the ratio of the earth's velocity to the velocity of light), he went on to say,[2] 'I regard it as very probable that optical phenomena depend only on the *relative* motions of the material bodies, luminous sources, and optical apparatus concerned, and that this is true not merely as far as quantities of the order of the square of the aberration, but *rigorously*.' In other words, Poincaré believed in 1899 that *absolute motion is indetectible in principle*, whether by dynamical, optical, or electrical means.

In the following year, at an International Congress of Physics held at Paris, he asserted the same doctrine.[3] 'Our aether,' he said, 'does it really exist? I do not believe that more precise observations could ever reveal anything more than *relative* displacements.' After referring to the circumstance that the explanations then current for the negative results regarding terms of the first order in (w/c) were different from the explanations regarding the second order terms, he went on, 'It is necessary to find the *same* explanation for the negative results obtained regarding terms of these two orders: and there is every reason to suppose that this explanation will then apply equally to terms of higher orders, and that the mutual destruction of the terms will be rigorous and absolute.' A new principle would thus be introduced into physics, which would resemble the Second Law of Thermodynamics in as much as it asserted the *impossibility of doing something*: in this case, the impossibility of determining the velocity of the earth relative to the aether.[4]

In a lecture to a Congress of Arts and Science at St Louis, U.S.A., on 24 September 1904, Poincaré gave to a generalised form of this principle the name, *The Principle of Relativity*.[5] 'According to the Principle of Relativity,' he said, 'the laws of physical phenomena must be the same for a "fixed" observer as for an observer who has a uniform motion of translation relative to him: so that we have not, and cannot possibly have, any means of discerning whether we are, or are not, carried along in such a motion.' After examining the records of observation in the light of this principle, he declared,

[1] Edited by E. Néculcéa, and printed in 1901 under the title *Electricité et Optique*, Paris, Carré et Naud. [2] loc. cit., p. 536

[3] *Rapports présentés au Congrès International de Physique réuni à Paris en 1900* (Paris, Gauthier-Villars, 1900), Tome I, p. 1, at pp. 21, 22

[4] In April 1904 Lorentz asserted the same general principle: cf. *Versl. Kon. Akad. v. Wet.*, Amsterdam, Dl. xii (1904), p. 986; English edn. (*Amst. Proc.*), vi (1904), p. 809.

[5] This address appeared in *Bull. des Sc. Math.*(2) xxviii (1904), p. 302; an English translation by G. B. Halsted was published in *The Monist* for January 1905.

'From all these results there must arise an entirely new kind of dynamics, *which will be characterised above all by the rule, that no velocity can exceed the velocity of light.*'

We have now to see how an analytical scheme was devised which enabled the whole science of physics to be reformulated in accordance with Poincaré's Principle of Relativity.

That Principle, as its author had pointed out, required that observers who have uniform motions of translation relative to each other should express the laws of nature in the same form. Let us consider in particular the laws of the electromagnetic field.

Lorentz, as we have seen,[1] had obtained the equations of a moving electric system by applying a transformation to the fundamental equations of the aether. In the original form of this transformation, quantities of order higher than the first in (w/c) were neglected. But in 1900 Larmor [2] extended the analysis so as to include quantities of the second order. Lorentz in 1903 went further still,[3] and obtained the transformation in a form which is exact to all orders of the small quantity (w/c). In this form we shall now consider it.

The fundamental equations of the aether in empty space are

$$\operatorname{div} \mathbf{d} = 0, \qquad c \operatorname{curl} \mathbf{d} = -\frac{\partial \mathbf{h}}{\partial t}$$

$$\operatorname{div} \mathbf{h} = 0, \qquad c \operatorname{curl} \mathbf{h} = \frac{\partial \mathbf{d}}{\partial t}.$$

It is desired to find a transformation from the variables $t, x, y, z, \mathbf{d}, \mathbf{h}$, to new variables $t_1, x_1, y_1, z_1, \mathbf{d}_1, \mathbf{h}_1$, such that the equations in terms of these new variables may take the same form as the original equations, namely

$$\operatorname{div}_1 \mathbf{d}_1 = 0, \qquad c \operatorname{curl}_1 \mathbf{d}_1 = -\frac{\partial \mathbf{h}_1}{\partial t_1}$$

$$\operatorname{div}_1 \mathbf{h}_1 = 0, \qquad c \operatorname{curl}_1 \mathbf{h}_1 = \frac{\partial \mathbf{d}_1}{\partial t_1}.$$

Evidently one particular class of such transformation is that which corresponds to rotations of the axes of co-ordinates about the origin. These may be described as the linear homogeneous transformations of determinant unity which transform the expression $(x^2+y^2+z^2)$ into itself. It had, however, already become clear from Lorentz's earlier work that some of the transformations must involve not only

[1] cf. Vol. I, p. 406. cf. also Lorentz, *Proc. Amst. Acad.* (English edn.), i (1899), p. 427

[2] Larmor, *Aether and Matter* (1900), p. 173

[3] *Proc. Amst. Acad.* (English edn.), vi (1903), p. 809

x, y, z, but also the variable t.[1] So (guided by the approximate formulae already obtained) he now replaced the condition of transforming $(x^2+y^2+z^2)$ into itself, by the condition of transforming the expression $(x^2+y^2+z^2-c^2t^2)$ into itself; and, as we shall now show, he succeeded in proving that the transformations so obtained have the property of transforming the differential equations of the aether in the manner required.

We shall first consider a transformation of this class in which the variables y and z are unchanged. The equations of this transformation may easily be derived by considering that the equation of the rectangular hyperbola

$$x^2-(ct)^2=1$$

(in the plane of the variable x, ct) is unaltered when any pair of conjugate diameters are taken as new axes, and a new unit of length is taken proportional to the length of either of these diameters. The equations of transformation thus obtained are

$$\begin{aligned} ct &= ct_1 \cosh \alpha + x_1 \sinh \alpha \\ x &= x_1 \cosh \alpha + ct_1 \sinh \alpha \\ y &= y_1 \\ z &= z_1 \end{aligned} \qquad (1)$$

where α denotes a constant parameter. The simpler equations previously given by Lorentz[2] may evidently be derived from these by writing $w=c \tanh \alpha$, and neglecting powers of (w/c) above the first. It will be observed that not only is the system of measuring the abscissa x changed, but also the system of measuring the time t: the necessity for this had been recognised in Lorentz's original memoir by his introduction of 'local time.'

Let us find the physical interpretation of this transformation (1). If we consider the point in the (t_1, x_1, y_1, z_1) system for which x_1, y_1, z_1 are all zero, its co-ordinates in the other system are given by the equations

$$t=t_1 \cosh \alpha, \qquad x=ct_1 \sinh \alpha, \qquad y=0, \qquad z=0,$$

so

$$x=ct \tanh \alpha, \qquad y=0, \qquad z=0.$$

Thus if we regard the axes of (x_1, y_1, z_1) and the axes of (x, y, z) as two rectangular co-ordinate systems in space, then the origin of the (x_1, y_1, z_1) system has the co-ordinates $(ct \tanh \alpha, 0, 0)$, that is to

[1] Larmor, *Aether and Matter* (1900), in commenting on the FitzGerald contraction, had recognised that clocks, as well as rods, are affected by motion: a clock moving with velocity v relative to the aether must run slower, in the ratio

$$\sqrt{\left(1-\frac{v^2}{c^2}\right)} : 1.$$

[2] cf. Vol. I, p. 407

say, the origin of the (x_1, y_1, z_1) system moves with a uniform velocity $c \tanh \alpha$ along the x-axis of the (x, y, z) system. Thus if w is the relative velocity of the two systems, we have

$$\cosh \alpha = \left(1 - \frac{w^2}{c^2}\right)^{-\frac{1}{2}}, \qquad \sinh \alpha = \frac{w}{c}\left(1 - \frac{w^2}{c^2}\right)^{-\frac{1}{2}},$$

and Lorentz's transformation between their co-ordinates may be written

$$t = \frac{t_1 - \frac{wx_1}{c^2}}{\sqrt{\left(1 - \frac{w^2}{c^2}\right)}}, \qquad x = \frac{x_1 + wt_1}{\sqrt{\left(1 - \frac{w^2}{c^2}\right)}}, \qquad y = y_1, \qquad z = z_1.$$

In this transformation the variable x plays a privileged part, as compared with y or z. We can of course at once write down similar transformations in which y or z plays the privileged part; and we can combine any number of these transformations by performing them in succession. The aggregate of all the transformations so obtained, combined with the aggregate of all the rotations in ordinary space, constitutes a *group*, to which Poincaré [1] gave the name the group of *Lorentz transformations*.

By a natural extension of the equations formerly given by Lorentz for the electric and magnetic forces, it is seen that the equations for transforming these, when (t, x, v, z) are transformed by equations (1), are

$$\begin{aligned} d_x &= d_{x_1} & h_x &= h_{x_1} \\ d_y &= d_{y_1} \cosh \alpha + h_{z_1} \sinh \alpha & h_y &= h_{y_1} \cosh \alpha - d_{z_1} \sinh \alpha \\ d_z &= d_{z_1} \cosh \alpha - h_{y_1} \sinh \alpha & h_z &= h_{z_1} \cosh \alpha + d_{y_1} \sinh \alpha. \end{aligned} \tag{2}$$

When the original variables are by direct substitution replaced by the new variables defined by (1) and (2) in the fundamental differential equations of the aether, the latter take the form

$$\operatorname{div}_1 \mathbf{d}_1 = 0, \qquad c \operatorname{curl}_1 \mathbf{d}_1 = -\frac{\partial \mathbf{h}_1}{\partial t_1}$$

$$\operatorname{div}_1 \mathbf{h}_1 = 0, \qquad c \operatorname{curl}_1 \mathbf{h}_1 = \frac{\partial \mathbf{d}_1}{\partial t_1}$$

that is to say, *the fundamental equations of the aether retain their form unaltered, when the variables (t, x, y, z) are subjected to the Lorentz*

[1] *Comptes Rendus*, cxl (5 June 1905), p. 1504. It should be added that these transformations had been applied to the equation of vibratory motions many years before by W. Voigt, *Gött. Nach.* (1887), p. 41.

transformation (1), *and at the same time the electric and magnetic intensities are subjected ot the transformation* (2).

The fact that the electric and magnetic intensities undergo the transformation (2) when the co-ordinates undergo the transformation (1), raises the question as to whether the transformation (2) is familiar to us in other connections. That this is so may be seen as follows.

In 1868–9 J. Plücker and A. Cayley introduced into geometry the notion of *line co-ordinates*; if (x_0, x_1, x_2, x_3) and (y_0, y_1, y_2, y_3) are the tetrahedral co-ordinates of two points of a straight line p, and if we write

$$x_m y_n - x_n y_m = p_{mn},$$

then the six quantities

$$p_{01}, \quad p_{02}, \quad p_{03}, \quad p_{23}, \quad p_{31}, \quad p_{12}$$

are called the *line-co-ordinates* of p.

Now suppose that the transformation

$$
\begin{aligned}
x_0 &= x'_0 \cosh \alpha + x'_1 \sinh \alpha \\
x_1 &= x'_1 \cosh \alpha + x'_0 \sinh \alpha \\
x_2 &= x'_2 \\
x_3 &= x'_3
\end{aligned}
\tag{3}
$$

is performed on the co-ordinates (x_0, x_1, x_2, x_3), and the same transformation is performed on the co-ordinates (y_0, y_1, y_2, y_3). Then we have

$$
\begin{aligned}
p_{01} &= x_0 y_1 - x_1 y_0 \\
&= (x'_0 \cosh \alpha + x'_1 \sinh \alpha)(y'_1 \cosh \alpha + y'_0 \sinh \alpha) \\
&\quad - (x'_1 \cosh \alpha + x'_0 \sinh \alpha)(y'_0 \cosh \alpha + y'_1 \sinh \alpha) \\
&= x'_0 y'_1 - x'_1 y'_0 \\
&= p'_{01}
\end{aligned}
$$

and in the same way we find

$$
\begin{aligned}
p_{02} &= p'_{02} \cosh \alpha + p'_{12} \sinh \alpha \\
p_{03} &= p'_{03} \cosh \alpha - p'_{31} \sinh \alpha \\
p_{23} &= p'_{23} \\
p_{31} &= p'_{31} \cosh \alpha - p'_{03} \sinh \alpha \\
p_{12} &= p'_{12} \cosh \alpha + p'_{02} \sinh \alpha.
\end{aligned}
$$

But these equations of transformation of the p's are precisely the same as the equations of transformation (2) of the electric and magnetic intensities, provided we write

$$p_{01} = d_x, \quad p_{02} = d_y, \quad p_{03} = d_z, \quad p_{23} = h_x, \quad p_{31} = h_y, \quad p_{12} = h_z.$$

The line-co-ordinates of a line have this property of transforming like the six components of the electric and magnetic intensities not only for the particular Lorentz transformation (1) but for the *most general* Lorentz transformation. A set of six quantities which transform like the line-co-ordinates of a line when the co-ordinates are subjected to any Lorentz transformation whatever, is called a *six-vector*. Thus we may say that the *quantities* $(d_x, d_y, d_z, h_x, h_y, h_z)$ *constitute a six-vector.*[1] In the older physics, **d** was regarded as a vector, and **h** as a distinct vector : but if an electrostatic system (in which **d** exists but **h** is zero) is referred to axes which are in motion with respect to it, then the magnetic force with respect to these axes will not be zero. The six-vector transformation takes account of this fact, and furnishes the value of the magnetic force which thus appears.

We see, therefore, that in electromagnetic theory, as in Newtonian dynamics, there are *inertial systems* of co-ordinate axes with associated systems of measurement of time, such that the path of a free material particle relative to an inertial system is a straight line described with uniform velocity, and *also* that the equations of the electromagnetic field relative to the inertial system are Maxwell's equations, and *any system of axes which moves with a uniform motion of translation, relative to any given inertial system of axes, is itself an inertial system of axes, the measurement of time and distance in the two systems being connected by a Lorentz transformation. All the laws of nature have the same form in the co-ordinates belonging to one inertial system as in the co-ordinates belonging to any other inertial system.* No inertial system of reference can be regarded as having a privileged status, in the sense that it should be regarded as fixed while the others are moving : the notion of absolute fixity in space, which in the latter part of the nineteenth century was thought to be required by the theory of aether and electrons was shown in 1900–4 by the Poincaré-Lorentz theory of relativity to be without foundation.

Suppose that an inertial system of reference (t, x, y, z) is known on earth : and imagine a distant star which is moving with a uniform velocity relative to this framework (t, x, y, z). The theorem of relativity shows that there exists another framework (t_1, x_1, y_1, z_1) with respect to which the star is at rest, and in which, moreover, a luminous disturbance generated at time t_1 at any point (x_1, y_1, z_1) will spread outwards in the form of a sphere

$$(X_1 - x_1)^2 + (Y_1 - y_1)^2 + (Z_1 - z_1)^2 = c^2(T_1 - t_1)^2,$$

the centre of this sphere occupying for all subsequent time an unchanged position in the co-ordinate system (x_1, y_1, z_1). This framework is peculiarly fitted for the representation of phenomena which happen on the star, whose inhabitants would therefore naturally adopt it as their system of space and time. Beings, on the other hand, who dwell on a body which is at rest with respect to the axes (t, x, y, z), would prefer to use the latter system ; and from the point of view of the universe at large, either of these systems is as good as the other. The electromagnetic equations are the same with respect to both sets of co-ordinates, and therefore neither can

[1] *Raum-Zeit-Vektor II Art* of H. Minkowski, *Gött. Nach.* 1908, p. 53

claim to possess the only property which could confer a primacy—namely, a special relation to the aether.

Some of the consequences of the new theory seemed to contemporary physicists very strange. Suppose, for example, that two inertial sets of axes A and B are in motion relative to each other, and that at a certain instant their origins coincide: and suppose that at this instant a flash of light is generated at the common origin. Then, by what has been said in the subsequent propagation, the wave-fronts of the light, as observed in A and in B, are spheres whose centres are the origins of A and B respectively, and therefore *different* spheres. How can this be?

The paradox is explained when it is remembered that a wave-front is defined to be the locus of points which are *simultaneously* in the same phase of disturbance. Now events taking place at different points, which are simultaneous according to A's system of measuring time, are not in general simultaneous according to B's way of measuring: and therefore what A calls a wave-front is not the same thing as what B calls a wave-front. Moreover, since the system of measuring space is different in the two inertial systems, what A calls a sphere is not the same thing as what B calls a sphere. Thus there is no contradiction in the statement that the wave-fronts for A are spheres with A's origin as centre, while the wave-fronts for B are spheres with B's origin as centre.

In common language we speak of events which happen at different points of space as happening 'at the same instant of time,' and we also speak of events which happen at different instants of time as happening 'at the same point of space.' We now see that such expressions can have a meaning only by virtue of artificial conventions; they do not correspond to any essential physical realities.

It is usual to regard Poincaré as primarily a mathematician, and Lorentz as primarily a theoretical physicist: but as regards their contributions to relativity theory, the positions were reversed: it was Poincaré who proposed the general physical principle, and Lorentz who supplied much of the mathematical embodiment. Indeed, Lorentz was for many years doubtful about the physical theory: in a lecture which he gave in October 1910 [1] he spoke of 'die Vorstellung (die auch Redner nur ungern aufgeben würde), dass Raum und Zeit etwas vollig Verschiedenes seien und dass es eine "wahre Zeit" gebe (die Gleichzeitigkeit würde denn unabhängig vom Orte bestehen).' [2]

A distinguished physicist who visited Lorentz in Holland shortly before his death found that his opinions on this question were unchanged.

We are now in a position to show the connection between the

[1] Printed in *Phys. ZS.* xi (1910), p. 1234

[2] 'The concept (which the present author would dislike to abandon) that space and time are something completely distinct and that a "true time" exists (simultaneity would then have a meaning independent of position).'

Lorentz transformation and FitzGerald's hypothesis of contraction; this connection was first established by Larmor[1] for his approximate form of the Lorentz transformation, which is accurate only to the second order in (w/c), but the extension to the full Lorentz transformation is easy.

Suppose that a rod is moving along the axis of x with uniform velocity w; let the co-ordinates of its ends at the instant t be x_1 and x_2. Take a system of axes $O'x'y'z'$ which move with the rod, the axis $O'x'$ being in the same line as the axis Ox, and the axes $O'y'$ and $O'z'$ being constantly parallel to the axes Oy and Oz respectively. In this system the length of the rod will be $x'_2-x'_1$, where, of course, x'_2 and x'_1 do not vary with the time. The Lorentz transformation gives

$$x'_2=x_2 \cosh \alpha - ct \sinh \alpha$$
$$x'_1=x_1 \cosh \alpha - ct \sinh \alpha$$

where $\tanh \alpha = (w/c)$. Subtracting, we have

$$x'_2-x'_1=(x_2-x_1) \cosh \alpha = (x_2-x_1)\{1-(w/c)^2\}^{-\frac{1}{2}}$$

or

$$x_2 - x_1 = \sqrt{1-\frac{w^2}{c^2}}\,(x'_2-x'_1).$$

This equation shows that the distance between the ends of the rod, in the system of measurement furnished by the original axes, with reference to which the rod is moving with velocity w, bears the ratio $(1-w^2/c^2)^{\frac{1}{2}} : 1$ to their distance in the system of measurement furnished by the transformed axes, with reference to which the rod is at rest: and this is precisely FitzGerald's hypothesis of contraction. The hypothesis of FitzGerald may evidently be expressed by the statement, that *the equations of the figures of material bodies are covariant with respect to those transformations for which the fundamental equations of the aether are covariant:* that is, for all Lorentz transformations.

Now let us look into Poincaré's remark[2] that the Principle of Relativity requires the creation of a new mechanics in which no velocity can exceed the velocity of light.

Suppose that an inertial system B is being translated relative to an inertial system A with velocity w along the axis of x. Let a point P moving along the axis of x have the co-ordinates $(t, x, 0, 0)$ in system A and $(t', x', 0, 0)$ in system B. Denote the components of velocity dx/dt and dx'/dt' by v_x, v'_x, respectively, and let $w=c \tanh \alpha$. Then the Lorentz transformation gives at once

$$v_x=\frac{dx}{dt}=\frac{c(dx' \cosh \alpha + c dt' \sinh \alpha)}{c\, dt' \cosh \alpha + dx' \sinh \alpha}=\frac{v'_x+w}{1+\frac{v_x w}{c^2}}.$$

[1] *Aether and Matter* (1900), p. 173

[2] cf. p. 31

Now, v_x being the velocity of P relative to A, v'_x the velocity of P relative to B, and w the velocity of B relative to A, in Newtonian kinematics we should have $v_x = v'_x + w$. The denominator $(1 + v_x w/c^2)$ in the relativist formula expresses the difference between Newtonian theory and relativity theory, so far as concerns the composition of velocities. We see that if $v'_x = c$, then $v_x = c$; that is to say, *any velocity compounded with c gives as the resultant c over again*, and therefore that no velocity can exceed the velocity of light.

This result enables us to solve a problem which had perplexed many generations of physicists. It had been supposed that if the correct theory of light is the corpuscular theory, then the corpuscles emitted by a moving star should have a velocity which is compounded of the velocity of the star and the velocity of light relative to a source at rest, just as an object thrown from a carriage window in a moving railway train has a velocity which is obtained by compounding its velocity relative to the carriage with the velocity of the train (the *ballistic* theory); whereas, if the correct theory of light is the wave-theory, the velocity of the light emitted by the star should be unaffected by the velocity of the star, just as the waves created by throwing a stone into a pond move outwards from the point where the stone entered the water, without being affected by the velocity of the stone. The new relativist theory led to the surprising conclusion that the velocity of light would be unaffected by the velocity of its source *even on the corpuscular theory*.

An attempt to explain the Michelson-Morley experiment, and the other evidence which had given rise to relativity theory, without assuming that the velocity of light is independent of the velocity of its source, was made in 1908 by W. Ritz,[1] who postulated that the velocity of light and the velocity of the source are additive, as in the old physics. It is, however, now known certainly that the velocity of light is independent of the motion of the source. The astronomical evidence for this statement has been marshalled by several writers,[2] and further confirmation has been furnished by Majorana by direct experiment.[3] It should be remarked that since in purely terrestrial experiments the light rays always describe closed paths, the results to be expected from 'ballistic' and non-ballistic theories can differ only by quantities of the second order,[4] but the performance of the

[1] *Ann. de chim. et phys.* xiii (1908), p. 145; *Arch. de Génève*, xxvi (1908), p. 232; cf. a careful discussion of it by R. C. Tolman, *Phys. Rev.* xxxv (1912), p. 136

[2] Particularly by R. C. Tolman, *Phys. Rev.* xxxi (1910), p. 26; W. de Sitter, *Amsterdam Proc.* xv (1913), p. 1297; xvi (1913), p. 395; *Phys. ZS.* xiv (1913), pp. 429, 1267; *Bull. of the Astron. Inst. of the Netherlands*, ii (1924), pp. 121, 163; R. S. Capon, *Month. Not. R.A.S.* lxxiv (1914), pp. 507, 658; H. C. Plummer, ibid., p. 660; H. Thirring, *ZS. f. P.* xxxi (1925), p. 133; G. Wataghin, *ZS. f. P.* xl (1926), p. 378

[3] *Comptes Rendus*, clxv (1917), p. 424; clxvii (1918), p. 71; clxix (1919), p. 719; *Phys. Rev.* xi (1918), p. 411; *Phil. Mag.* xxxvii (1919), p. 145; xxxix (1920), p. 488; cf. also Jeans, *Nature*, cvii (1921), pp. 42, 169

[4] cf. P. Ehrenfest, *Phys. ZS.* xiii (1912), p. 317; F. Michaud, *Comptes Rendus*, clxviii (1919), p. 507

Michelson-Morley experiment with light from astronomical sources by R. Tomaschek[1] in 1924 definitely disproved the ballistic hypothesis.

A further result in harmony with the new theory was obtained when Michelson[2] showed experimentally that the velocity of a moving mirror is without influence on the velocity of light reflected at its surface.

It was now recognised that these observational findings, which in the nineteenth century might have been supposed to tell in favour of the wave-theory, were actually without significance one way or the other in the dispute between the wave and corpuscular theories of light. For, according to relativity theory, even on the corpuscular hypothesis, a corpuscle which had a velocity c relative to its source would have the same velocity relative to any observer, whether he shared in the motion of the source or not.

In 1905 Poincaré[3] completed the theorem of Lorentz[4] on the covariance of Maxwell's equations with respect to the Lorentz transformation, by obtaining the formulae of transformation of the electric density ρ and current $\rho\mathbf{v}$. The fundamental equations are

$$\operatorname{div} \mathbf{d} = 4\pi\rho ; \qquad c \operatorname{curl} \mathbf{d} = -\frac{\partial \mathbf{h}}{\partial t}$$

$$\operatorname{div} \mathbf{h} = 0 ; \qquad c \operatorname{curl} \mathbf{h} = \frac{\partial \mathbf{d}}{\partial t} + 4\pi\rho\mathbf{v}$$

and it is desired to find a transformation from the variable $t, x, y, z, \rho, \mathbf{d}, \mathbf{h}, \mathbf{v}$ to new variables $t_1, x_1, y_1, z_1, \rho_1, \mathbf{d}_1, \mathbf{h}_1, \mathbf{v}_1$, such that the equations in terms of these new variables may have the same form as the original equations. The transformations of $t, x, y, z, \mathbf{d}, \mathbf{h}$ have already been found. Poincaré now showed that

$$\begin{aligned} \rho &= \rho_1 \cosh \alpha + (\rho_1 v_{x_1}/c) \sinh \alpha \\ \rho v_x &= \rho_1 v_{x_1} \cosh \alpha + c\rho_1 \sinh \alpha \\ \rho v_y &= \rho_1 v_{y_1} \\ \rho v_z &= \rho_1 v_{z_1}. \end{aligned}$$

When the original variables are by direct substitution replaced by the new variables in the differential equations, the latter take the form

$$\operatorname{div}_1 \mathbf{d}_1 = 4\pi\rho_1, \qquad c \operatorname{curl}_1 \mathbf{d}_1 = -\frac{\partial \mathbf{h}_1}{\partial t_1}$$

$$\operatorname{div}_1 \mathbf{h}_1 = 0, \qquad c \operatorname{curl}_1 \mathbf{h}_1 = \frac{\partial \mathbf{d}_1}{\partial t_1} + 4\pi\rho_1\mathbf{v}_1$$

[1] *Ann. d. Phys.* lxxiii (1924), p. 105
[2] *Astrophys. J.* xxxvii (1913), p. 190
[3] *Comptes Rendus,* cxl (June 1905), p. 1504
[4] cf. p. 33

that is to say, the fundamental equations of aether and electrons retain their form unaltered, when the variables are subjected to the transformation which has been specified.

In the autumn of the same year, in the same volume of the *Annalen der Physik* as his paper on the Brownian motion,[1] Einstein published a paper which set forth the relativity theory of Poincaré and Lorentz with some amplifications, and which attracted much attention. He asserted as a fundamental principle the *constancy of the velocity of light*, i.e. that the velocity of light *in vacuo* is the same in all systems of reference which are moving relatively to each other : an assertion which at the time was widely accepted, but has been severely criticised by later writers.[2] In this paper Einstein gave the modifications which must now be introduced into the formulae for aberration and the Doppler effect.[3]

Consider a star, which is observed from the earth on two occasions. The distance of the star is assumed to be so great that its apparent proper motion in the interval between the observations is negligible. Denote an inertial system of axes at the earth at the time of the first observation by K, and an inertial system of axes at the earth at the time of the second observation by K′ : and choose these axes so that the x-axis has the direction of the velocity w $(=c \tanh \alpha)$ of K′ relative to K. Let ψ be the angle which the ray of light arriving at the earth from the star makes with the x-axis as measured in K, and ψ' the corresponding angle in the system K′. Then the Lorentz transformation gives for the co-ordinates of the star in the two systems

$$\begin{cases} ct' = ct \cosh \alpha - x \sinh \alpha \\ x' = x \cosh \alpha - ct \sinh \alpha \\ y' = y \end{cases}$$

(taking the plane of xy to contain the star) : and since light is propagated with velocity c in both systems, we have $ct = \sqrt{(x^2+y^2)}$, $ct' = \sqrt{(x'^2+y'^2)}$. Thus

$$\cos\psi' = \frac{x'}{\sqrt{(x'^2+y'^2)}} = \frac{x'}{ct'} = \frac{x \cosh \alpha - ct \sinh \alpha}{ct \cosh \alpha - x \sinh \alpha} = \frac{\cos\psi \cosh \alpha - \sinh \alpha}{\cosh \alpha - \cos\psi \sinh \alpha}$$

or

$$\cos\psi' = \frac{c \cos\psi - w}{c - w \cos\psi}.$$

This is the relativist formula for aberration : it may be written

$$\sin\frac{\psi'-\psi}{2} = \tanh\frac{\alpha}{2} \sin\frac{\psi'+\psi}{2}.$$

[1] *Ann. d. Phys.* xvii (Sept. 1905), p. 891

[2] e.g. H. E. Ives, *Proc. Amer. Phil. Soc.* xcv (1951), p. 125; *Sc. Proc. R.D.S.* xxvi (1952), p. 9, at pp. 21–2

[3] *cf.* Vol. I, pp. 368 and 389

When powers of (w/c) above the first are neglected, this gives

$$\psi' - \psi = \frac{w}{c} \sin \psi,$$

which is the aberration-formula of classical physics.

To find the relativist formula for the Doppler effect, we suppose that K′ is an inertial system with respect to which the star is at rest, and K is an inertial system in which the earth is at rest: and choose the axes so that the system K′ is moving with velocity w $(=c \tanh \alpha)$ parallel to the axis of x in the system K. Let ψ be the angle which the line joining the star to the observer makes with the x-axis in the system K, and let ψ' be the corresponding angle in the system K′. Then the phase in the system K is determined by

$$\nu \left(t + \frac{x \cos \psi + y \sin \psi}{c}\right)$$

where ν is the frequency of the light as observed by the terrestrial observer; and as the phase is a physical invariant, we must have

$$\nu \left(t + \frac{x \cos \psi + y \sin \psi}{c}\right) = \nu' \left(t' + \frac{x' \cos \psi' + y' \sin \psi'}{c}\right)$$

when ν' is the frequency of the light as measured by an observer on the star. Thus

$$\nu \left\{t' \cosh \alpha + \frac{x'}{c} \sinh \alpha + \frac{1}{c} \left[(x' \cosh \alpha + ct' \sinh \alpha) \cos \psi + y' \sin \psi\right]\right\}$$

$$= \nu' \left(t' + \frac{x' \cos \psi' + y' \sin \psi'}{c}\right).$$

Equating coefficients of t', we have

$$\nu (\cosh \alpha + \sinh \alpha \cos \psi) = \nu'$$

or

$$\frac{\nu'}{\nu} = \frac{1 + \frac{w}{c} \cos \psi}{\sqrt{\left(1 - \frac{w^2}{c^2}\right)}}.$$

This is the relativist formula for the Doppler effect. When only first-order terms in (w/c) are retained, it gives

$$\nu = \nu' \left(1 - \frac{w_r}{c}\right)$$

where w_r is the radial component of w: which is the older formula for the Doppler effect.

It will be noticed that the relativist formula differs from the older formula by the presence of the factor $\sqrt{(1-w^2/c^2)}$. Now if an observer moving with velocity w relative to an inertial system passes a place P where a clock belonging to the inertial system reads t_1, and if he afterwards passes a place Q where the clock in the inertial system reads t_2, and if t' is the interval of time registered by the observer's clock between the positions P and Q, then it follows at once from the equations of the Lorentz transformation that

$$\frac{t'}{t_2-t_1}=\sqrt{\left(1-\frac{w^2}{c^2}\right)},$$

so that we can (somewhat loosely) speak of the factor $\sqrt{(1-w^2/c^2)}$ as representing the slower rate at which the observer's clock is running as compared with clocks that are at rest on the star. It is obvious that this factor must occur in the relativist formula.

It will be observed that in the relativist formula, the Doppler effect is not zero even when the relative motion of the source and observer is at right angles to the direction of propagation of the light; in this case ($\psi=\frac{1}{2}\pi$) we have

$$\nu=\nu'\left(1-\frac{w^2}{c^2}\right)^{\frac{1}{2}}$$

or in the first approximation

$$\frac{\nu-\nu'}{\nu'}=-\frac{w^2}{2\,c^2}.$$

This is called the *transverse Doppler effect*. In 1907 Einstein suggested [1] that it might be observed by examining the light emitted by canal rays [2] in hydrogen, on which J. Stark [3] had published a paper in 1906. Stark's experimental results, however, did not seem to confirm the theoretical formula: and it was not until more than thirty years later that H. E. Ives and G. R. Stillwell [4] succeeded in carrying out this experiment with any degree of success.

It is clear, from the history set forth in the present chapter, that the theory of relativity had its origin in the theory of aether and electrons. When relativity had become recognised as a doctrine covering the whole operation of physical nature, efforts were made to present it in a form free from any special association with electro-

[1] *Ann. d. Phys.* xxiii (1907), p. 197 [2] cf. Vol. I, p 363

[3] *Ann. d. Phys.* xxi (1906), p. 401

[4] *J. Opt. Soc. Amer.* xxviii (1938), p. 215; xxxi (1941), p. 369

magnetic theory, and deducible logically from a definite set of axioms of greater or less plausibility.[1]

It should be mentioned also that when relativity theory had become generally accepted, the Michelson-Morley experiment was rediscussed with a much more complete understanding and exactitude.[2]

An account may be given here of some experiments performed long after the time with which we are at present mainly concerned, which confirmed in a striking way the predictions of relativity theory. In one of them, due to A. B. Wood, G. A. Tomlinson and L. Essen,[3] a rod in longitudinal vibration was rotated in a horizontal plane, so that its length varied periodically by reason of the FitzGerald contraction. Accurate measurements were made of the vibration frequency, which would have varied with the length, if the length only had been affected. According to relativity theory, however, there should be a complete compensation of the contraction in length, by a modification of the elasticity of the rod according to its orientation with respect to the direction of its motion, so that no change of frequency should be observed. The experiment was carried out with two similar longitudinal piezo-electric quartz oscillators, one rotating and the other stationary, the relative frequency being measured. The experiment yielded a null result within narrow limits of uncertainty of about ± 4 parts in 10^{11}, thus fully confirming the prediction of the Poincaré-Lorentz theory of relativity.

Still later, a prediction of the theory was verified in a striking

[1] Papers on axiomatics are many. Attention may be directed specially to the following: P. Frank and H. Rothe, *Ann. d. Phys.* xxxiv (1911), p. 825; E. V. Huntington, *Phil. Mag.* xxiii (1912), p. 494; L. A. Pars, *Phil. Mag.* xlii (1921), p. 249; C. Carathéodory, *Berlin Sitz.* v (1924), p. 12; V. V. Narliker, *Proc. Camb. Phil. Soc.* xxviii (1932), p. 460; G. J. Whitrow, *Quart. J. Math.* iv (1933), p. 161; L. R. Gomes, *Lincei Rend.* xxi (1935), p. 433; N. R. Sen, *Indian J. of Phys.* x (1936), p. 341; F. Severi, *Proc. Phys.-Math. Soc. Japan,* xviii (1936), p. 257; E. Esclangon, *Comptes Rendus,* ccii (1936), p. 708; *Bull. Astron.* x (1937), p. 1; J. Meurers, *ZS. f. P.* cii (1936), p. 611; V. Lalan, *Comptes Rendus,* ciii (1936), p. 1491; *Bull. Soc. Math. France,* lxv (1937), p. 83; G. Temple, *Quart. J. Math.* ix (1938), p. 283; H. E. Ives, *Proc. Amer. Phil. Soc.* xcv (1951), p. 125. A valuable paper by H. P. Robertson, *Rev. Mod. Phys.* xxi (1949), p. 378, is in a somewhat different category. Robertson discusses the justification of the axioms on the ground of experimental results, and shows that most of the axioms can be based securely on (i) the Michelson-Morley experiment, (ii) the experiment of Ives and Stilwell on the transverse Doppler effect (cf. p. 42), and (iii) an experiment performed in 1932 by R. J. Kennedy and E. M. Thorndike [*Phys. Rev.* xlii (1932), p. 400]; in this, a pencil of homogeneous light was split at a half-reflecting surface into two beams, which, after traversing paths of different lengths, were brought together again and made to interfere; the positions of the fringes in the interference pattern were observed when the velocity of the system was varied owing to the motions of rotation and revolution of the earth. The predictions of relativity theory were verified. An interesting experiment with a rotating interferometer was performed by G. Sagnac in 1913; *Comptes Rendus* clvii (1913), pp. 708, 1410; *J. Phys. Rad.* iv (1914), p. 177; cf. A. Metz, *J. Phys. Rad.* xiii (1952), p. 224.

[2] cf. E. Kohl, *Ann. d. Phys.* xxviii (1909), pp. 259, 662; E. Budde, *Phys. ZS.* xii (1911), p. 979; M. von Laue, *Ann. d. Phys.* xxxiii (1910), p. 186; *Phys. ZS.* xiii (1912), p. 501; A. Right, *Le Radium,* xi (1919), p. 321; *N. Cimento,* xviii (1919), p. 91; J. Villey, *Comptes Rendus,* clxx (1920), p. 1175; clxxi (1920), p. 298; E. H. Kennard and D. E. Richmond, *Phys. Rev.* xix (1922), p. 572; J. L. Synge, *Sci. Proc. Roy. Dub. Soc.* xxvi (1952), p. 45; *Nature* clxx (1952), p. 244

[3] *Proc. R.S.*(A), clviii (1937), p. 606

way. If two events (1) and (2) are considered, and if in an inertial system A these events happen at different points of space, whereas in an inertial system B (moving with velocity w relative to A), the two events happen at the same point of space, then the Lorentz transformation gives

$$t_1^{A} = \frac{t_1^{B} - \dfrac{wx_1^{B}}{c^2}}{\sqrt{\left(1 - \dfrac{w^2}{c^2}\right)}}, \qquad t_2^{A} = \frac{t_2^{B} - \dfrac{wx_2^{B}}{c^2}}{\sqrt{\left(1 - \dfrac{w^2}{c^2}\right)}}.$$

Since $x_1^{B} = x_2^{B}$, these equations give

$$(t_2 - t_1)^{A} = \frac{(t_2 - t_1)^{B}}{\sqrt{\left(1 - \dfrac{w^2}{c^2}\right)}}$$

so the time between the events, measured in system A, is $(1 - w^2/c^2)^{-\frac{1}{2}}$ times greater than the time between the events, measured in system B.

Now certain particles called *cosmic-ray mesons*, discovered observationally in 1937, disintegrate spontaneously; and it may be assumed that the rate of disintegration depends on time as measured by an observer travelling with the meson. Thus to an observer who is stationary with respect to the earth, the rate of disintegration should appear to be slower, the faster the meson is moving. This was found in 1941 to be actually the case.[1]

The study of relativist dynamics was begun in 1906, when Max Planck[2] found the equations which, according to the new theory, should replace the Newtonian equations of motion of a material particle. Considering first the one-dimensional case, let a particle of mass m and charge e be moving along the axis of x with velocity $w(= c \tanh \alpha)$ in the system $Oxyz$, in a field of electric force parallel to Ox. Let $O'x'y'z'$ be axes parallel to these, whose origin O' moves with the particle. The relations between (t, x, y, z) and (t', x', y', z') are

$$\begin{cases} ct' = ct \cosh \alpha - x \sinh \alpha \\ x' = x \cosh \alpha - ct \sinh \alpha \\ y' = y \\ z' = z. \end{cases}$$

The Newtonian equation of motion is assumed to be valid with respect to the axes $O'x'y'z'$, so the equation of motion of the particle is

$$m\frac{d^2x'}{dt'^2} = ed'_{x'} = ed_x$$

[1] B. Rossi and D. B. Hall, *Phys. Rev.* lix (1941), p. 223

[2] *Verh. d. Deutsch, Phys. Ges.* viii (1906), p. 136

where $d'_{x'}$ and d_x denote the electric force in the two systems.[1] Now

$$\frac{dx'}{c\,dt'}=\frac{dx/dt\,\cosh a-c\sinh a}{c\cosh a-dx/dt\,\sinh a}$$

so

$$\frac{d^2x'}{c^2dt'^2}=\frac{\frac{d}{dt}\left\{\frac{(dx/dt)\cosh a-c\sinh a}{c\cosh a-(dx/dt)\sinh a}\right\}}{c\cosh a-\frac{dx}{dt}\sinh a}=\frac{\frac{d^2x}{dt^2}\cosh a}{\left\{c\cosh a-\frac{dx}{dt}\sinh a\right\}^3},$$

remembering that $\frac{dx}{dt}\cosh a-\sinh a=0$. But

$$c\cosh a-\frac{dx}{dt}\sinh a=c\cosh a-\frac{c\sinh^2 a}{\cosh a}=\frac{c}{\cosh a}$$

and therefore

$$\frac{d^2x'}{dt'^2}=\frac{d^2x}{dt^2}\cosh^3 a=\left(1-\frac{w^2}{c^2}\right)^{-\frac{3}{2}}\frac{dw}{dt}=\frac{d}{dt}\left\{\frac{w}{\sqrt{(1-w^2/c^2)}}\right\}.$$

Thus the equation of motion is (writing X for the moving force on the particle, namely ed_x),

$$\frac{d}{dt}\left\{\frac{mw}{\sqrt{(1-w^2/c^2)}}\right\}=\mathrm{X};$$

and extending the investigation to three dimensions, we can show that if the components of velocity are dx/dt, dy/dt, dz/dt, and if their resultant is w, then *the general equations of motion of a particle acted on by a force* (X, Y, Z) are

$$\frac{d}{dt}\left\{\frac{m\,dx/dt}{\sqrt{(1-w^2/c^2)}}\right\}=\mathrm{X},$$

$$\frac{d}{dt}\left\{\frac{m\,dy/dt}{\sqrt{(1-w^2/c^2)}}\right\}=\mathrm{Y}, \qquad (1)$$

$$\frac{d}{dt}\left\{\frac{m\,dz/dt}{\sqrt{(1-w^2/c^2)}}\right\}=\mathrm{Z}.$$

When $c\rightarrow\infty$, these evidently reduce to the Newtonian equations

$$m\frac{d^2x}{dt^2}=\mathrm{X},\qquad m\frac{d^2y}{dt^2}=\mathrm{Y},\qquad m\frac{d^2z}{dt^2}=\mathrm{Z}.$$

[1] cf. p. 33

To obtain the law of conservation of energy, multiply the equations (1) by dx/dt, dy/dt, dz/dt respectively, and add. Thus

$$X\frac{dx}{dt}+Y\frac{dy}{dt}+Z\frac{dz}{dt}$$

$$=\frac{dx}{dt}\frac{d}{dt}\left\{\frac{m\,dx/dt}{\sqrt{(1-w^2/c^2)}}\right\}+\frac{dy}{dt}\frac{d}{dt}\left\{\frac{m\,dy/dt}{\sqrt{(1-w^2/c^2)}}\right\}+\frac{dz}{dt}\frac{d}{dt}\left\{\frac{m\,dz/dt}{\sqrt{(1-w^2/c^2)}}\right\}$$

$$=w^2\frac{d}{dt}\left\{m\left(1-\frac{w^2}{c^2}\right)^{-\frac{1}{2}}\right\}+m\left(1-\frac{w^2}{c^2}\right)^{-\frac{1}{2}}\left(\frac{dx}{dt}\frac{d^2x}{dt^2}+\frac{dy}{dt}\frac{d^2y}{dt^2}+\frac{dz}{dt}\frac{d^2z}{dt^2}\right)$$

$$=\frac{mw^3}{c^2}\left(1-\frac{w^2}{c^2}\right)^{-\frac{3}{2}}\frac{dw}{dt}+m\left(1-\frac{w^2}{c^2}\right)^{-\frac{1}{2}}w\frac{dw}{dt}$$

$$=mw\left(1-\frac{w^2}{c^2}\right)^{-\frac{3}{2}}\frac{dw}{dt}.$$

So that

$$X\frac{dx}{dt}+Y\frac{dy}{dt}+Z\frac{dz}{dt}=\frac{d}{dt}\left\{mc^2\left(1-\frac{w^2}{c^2}\right)^{-\frac{1}{2}}\right\}.$$

The left-hand side of this equation is evidently the rate at which work is being done on the particle, so the right-hand side must represent the rate of increase of the kinetic energy of the particle ; that is, the kinetic energy of the particle is

$$\frac{mc^2}{\sqrt{\left(1-\frac{w^2}{c^2}\right)}}+C$$

where C denotes a constant ; or, expanding the radical by the binomial theorem,

$$mc^2\left(1+\frac{w^2}{2c^2}+\text{higher powers of }\frac{w^2}{c^2}\right)+C.$$

In order that this may agree with the Newtonian value of the kinetic energy, namely, $\frac{1}{2}mw^2$, when the higher powers of w^2/c^2 are neglected, we must have $C=-mc^2$. Thus *the kinetic energy of the particle is*

$$\frac{mc^2}{\sqrt{\left(1-\frac{w^2}{c^2}\right)}}-mc^2. \qquad (2)$$

It is easily seen that the equations (1) may be written

$$\frac{d}{dt}\left\{\frac{\partial L}{\partial\left(\frac{dx}{dt}\right)}\right\}=X, \qquad \frac{d}{dt}\left\{\frac{\partial L}{\partial\left(\frac{dy}{dt}\right)}\right\}=Y, \qquad \frac{d}{dt}\left\{\frac{\partial L}{\partial\left(\frac{dz}{dt}\right)}\right\}=Z$$

where

$$L=-mc^2\sqrt{\left(1-\frac{w^2}{c^2}\right)},$$

so L is the Lagrangean function or kinetic potential. Moreover, if we introduce

$$p_x=\frac{\partial L}{\partial\left(\frac{dx}{dt}\right)}=\frac{m}{\sqrt{\left(1-\frac{w^2}{c^2}\right)}}\frac{dx}{dt} \tag{3}$$

and similar expressions for p_y and p_z, and if we write

$$H=mc^2\sqrt{\left(1+\frac{p_x^2+p_y^2+p_z^2}{m^2c^2}\right)},$$

then the equations of motion may be written

$$\frac{dp_x}{dt}=X, \qquad \frac{dp_y}{dt}=Y, \qquad \frac{dp_z}{dt}=Z,$$

$$\frac{dx}{dt}=\frac{\partial H}{\partial p_x}, \qquad \frac{dy}{dt}=\frac{\partial H}{\partial p_y}, \qquad \frac{dz}{dt}=\frac{\partial H}{\partial p_z},$$

which is the Hamiltonian form.

Remembering that the moving force is the time-rate of the momentum, it is evident from equations (1) that the components of momentum of the particle are

$$\frac{m}{\sqrt{\left(1-\frac{w^2}{c^2}\right)}}\frac{dx}{dt}, \qquad \frac{m}{\sqrt{\left(1-\frac{w^2}{c^2}\right)}}\frac{dy}{dt}, \qquad \frac{m}{\sqrt{\left(1-\frac{w^2}{c^2}\right)}}\frac{dz}{dt} \tag{4}$$

which reduce to the Newtonian expressions $m\,dx/dt$, $m\,dy/dt$, $m\,dz/dt$, when $c \to \infty$. The same result is obtained from equations (3) when we remember that the components of momentum are the derivates of the Lagrangean function with respect to the components of velocity : and it fits in with a remark which Laplace had made more than a century earlier,[1] namely, that if the momentum of a particle, instead of being mw were $m\phi(w)$, then the kinetic energy must be $\int m\phi'(w)w\,dw$.

[1] *Mécanique céleste*, première partie, Livre I (An vii)

For from (4) we have in this case

$$\phi(w) = \frac{w}{\sqrt{\left(1-\frac{w^2}{c^2}\right)}}, \qquad \phi'(w) = \left(1-\frac{w^2}{c^2}\right)^{-\frac{3}{2}},$$

and the kinetic energy $= \int \frac{mwdw}{\left(1-\frac{w^2}{c^2}\right)^{\frac{3}{2}}} = \frac{mc^2}{\sqrt{\left(1-\frac{w^2}{c^2}\right)}} + \text{Constant}$

in agreement with (2).

Equations (2) and (4) fulfil the prediction made by Poincaré in his St Louis lecture of 24 September 1904, that there would be 'a new mechanics, where, the inertia increasing with the velocity, the velocity of light would become a limit that could not be exceeded.'

The arguments by which Planck derived his expressions for the kinetic energy and momentum of a material particle in relativity theory were felt to be perhaps not completely cogent. However, three years afterwards, Gilbert N. Lewis (1875–1946) and Richard C. Tolman (1881–1948)[1] gave a proof of a very different character.

Consider two systems of reference (A) and (B), in relative motion with velocity w parallel to the axes of x and x'. Let a ball P have components of velocity $(0, -u, 0)$ in (A), and let an exactly similar ball Q have components of velocity $(0, u, 0)$ in (B). Let the balls be smooth and perfectly elastic. The experiment is so planned that the balls collide and rebound. From the relativist formulae

$$v_x = \frac{v'_x + w}{1+\frac{v'_x w}{c^2}} \qquad v_y = \frac{v'_y\,(1-w^2/c^2)^{\frac{1}{2}}}{1+\frac{v'_x w}{c^2}}, \qquad v_z = \frac{v'_z\,(1-w^2/c^2)^{\frac{1}{2}}}{1+\frac{v'_x w}{c^2}},$$

we see that the velocity of Q as estimated by (A) before the collision is

$$\left(w,\ u\sqrt{\left(1-\frac{w^2}{c^2}\right)},\ 0\right).$$

The collision is perfectly symmetrical. But as estimated by (A), the y-component of Q's velocity changes from $u\sqrt{(1-w^2/c^2)}$ to $-u\sqrt{(1-w^2/c^2)}$, and the y-component of P's velocity changes from $-u$ to u.

We assume that there exists a vector quantity called the *momentum* depending on the mass and velocity, which is such that the momentum gained by one of the spheres in a collision is equal to the momentum lost by the sphere which collides with it. We assume further that this momentum approximates to the ordinary Newtonian

[1] *Phil. Mag.* xviii (1909), p. 517

momentum when the velocity is very small compared with that of light. So the components of momentum may be written

$$f(v)v_x, \qquad f(v)v_y, \qquad f(v)v_z,$$

where $v = (v^2_x + v^2_y + v^2_z)^{\frac{1}{2}}$, and the function $f(v)$ reduces to the mass m when $v \to 0$. From the law of conservation of momentum, (A) assumes that the ball P experiences the same change of momentum as the ball Q. Therefore

$$f(v_Q)u\left(1 - \frac{w^2}{c^2}\right)^{\frac{1}{2}} = f(v_P)u$$

where v_Q and v_P are the total velocities of Q and P in (A)'s system. Divide by u.

$$f(v_Q)\left(1 - \frac{w^2}{c^2}\right)^{\frac{1}{2}} = f(v_P).$$

Now make u tend to zero. Thus

$$f(w)\left(1 - \frac{w^2}{c^2}\right)^{\frac{1}{2}} = f(0) = m$$

or

$$f(w) = \frac{m}{\sqrt{\left(1 - \frac{w^2}{c^2}\right)}}$$

so the momentum of a particle whose mass is m, and which is moving with velocity (v_x, v_y, v_z) *is*

$$\left\{\frac{mv_x}{\sqrt{\left(1 - \frac{v^2}{c^2}\right)}}, \frac{mv_y}{\sqrt{\left(1 - \frac{v^2}{c^2}\right)}}, \frac{mv_z}{\sqrt{\left(1 - \frac{v^2}{c^2}\right)}}\right\},$$

where $v^2 = v_x^2 + v_y^2 + v_z^2$.

Next consider a collision between two elastic spheres, whose masses are m_1 and m_2 respectively, and which are moving along the axis of x with velocities (u_1, u_2) before the collision, and with velocities (u'_1, u'_2), after the collision. The condition of conservation of momentum gives the equation:

$$\frac{m_1u_1}{\sqrt{\left(1 - \frac{u_1^2}{c^2}\right)}} + \frac{m_2u_2}{\sqrt{\left(1 - \frac{u_2^2}{c^2}\right)}} = \frac{m_1u_1'}{\sqrt{\left(1 - \frac{u'^2_1}{c^2}\right)}} + \frac{m_2u_2'}{\sqrt{\left(1 - \frac{u'^2_2}{c^2}\right)}}. \qquad (1)$$

Now consider another set of axes, which are moving relatively to the first set with velocity $c \tanh \alpha$ parallel to the axis of x. Let the velocities relative to this second set of axes be denoted by grave accents placed over the letters, so that for any one of the u's we have

$$\grave{u} = \frac{u \cosh \alpha - c \sinh \alpha}{\cosh \alpha - (u/c) \sinh \alpha},$$

$$\sqrt{\left(1 - \frac{\grave{u}^2}{c^2}\right)} = \frac{1}{\cosh \alpha - (u/c) \sinh \alpha} \sqrt{\left(1 - \frac{u^2}{c^2}\right)}. \qquad (2)$$

Substituting from (2) in the equation

$$\frac{m_1 \grave{u}_1}{\sqrt{\left(1 - \frac{\grave{u}_1^2}{c^2}\right)}} + \frac{m_2 \grave{u}_2}{\sqrt{\left(1 - \frac{\grave{u}_2^2}{c^2}\right)}} = \frac{m_1 \grave{u}'_1}{\sqrt{\left(1 - \frac{\grave{u}'^2_1}{c^2}\right)}} + \frac{m_2 \grave{u}'_2}{\sqrt{\left(1 - \frac{\grave{u}'^2_2}{c^2}\right)}},$$

we obtain

$$\frac{m_1(u_1 \cosh \alpha - c \sinh \alpha)}{\sqrt{\left(1 - \frac{u_1^2}{c^2}\right)}} + \frac{m_2(u_2 \cosh \alpha - c \sinh \alpha)}{\sqrt{\left(1 - \frac{u_2^2}{c^2}\right)}}$$

$$= \frac{m_1(u'_1 \cosh \alpha - c \sinh \alpha)}{\sqrt{\left(1 - \frac{u'^2_1}{c^2}\right)}} + \frac{m_2(u'_2 \cosh \alpha - c \sinh \alpha)}{\sqrt{\left(1 - \frac{u'^2_2}{c^2}\right)}}.$$

Subtracting this equation from equation (1) multiplied by $\cosh \alpha$, and dividing the resulting equation by $c \sinh \alpha$, we have

$$\frac{m_1}{\sqrt{\left(1 - \frac{u_1^2}{c^2}\right)}} + \frac{m_2}{\sqrt{\left(1 - \frac{u_2^2}{c^2}\right)}} = \frac{m_1}{\sqrt{\left(1 - \frac{u'^2_1}{c^2}\right)}} + \frac{m_2}{\sqrt{\left(1 - \frac{u'^2_2}{c^2}\right)}}. \qquad (3)$$

This equation shows that if the quantity $m\,(1 - u^2/c^2)^{-\frac{1}{2}}$ be calculated for each of the colliding spheres, then the sum of these quantities for the two spheres is unaltered by the impact. We have therefore obtained a new invariant property. Let us see what corresponds to this in Newtonian dynamics. Supposing that (u_1/c) and (u_2/c) are small, and expanding by the binomial theorem, we have

$$m_1 \left(1 + \tfrac{1}{2}\frac{u_1^2}{c^2} + \tfrac{3}{8}\frac{u_1^4}{c^4} + \ldots\right) + m_2 \left(1 + \tfrac{1}{2}\frac{u_2^2}{c^2} + \tfrac{3}{8}\frac{u_2^4}{c^4} + \ldots\right)$$

$$= m_1 \left(1 + \tfrac{1}{2}\frac{u'^2_1}{c^2} + \tfrac{3}{8}\frac{u'^4_1}{c^4} + \ldots\right) + m_2 \left(1 + \tfrac{1}{2}\frac{u'^2_2}{c^2} + \tfrac{3}{8}\frac{u'^4_2}{c^4} + \ldots\right)$$

or

$$\tfrac{1}{2}m_1u_1^2+\tfrac{3}{8}m_1\frac{u_1^4}{c^2}+\ldots+\tfrac{1}{2}m_2u_2^2+\tfrac{3}{8}m_2\frac{u_2^4}{c^2}+\ldots$$
$$=\tfrac{1}{2}m_1u'^2_1+\tfrac{3}{8}m_1\frac{u'^4_1}{c^2}+\ldots+\tfrac{1}{2}m_2u'^2_2+\tfrac{3}{8}m_2\frac{u'^4_2}{c^2}+\ldots$$

When $c\to\infty$, this equation becomes the ordinary equation of conservation of kinetic energy in the collision. We therefore describe (3) as the *equation of conservation of energy* in the relativist theory of the impact, and we call

$$\frac{mc^2}{\sqrt{\left(1-\frac{v^2}{c^2}\right)}}$$

(save for an additive constant) the *kinetic energy* of a particle, whose mass at rest is m, which is moving with velocity v. The c^2 is inserted in the numerator in order to make the expansion in ascending powers of (u/c) begin with the terms [Constant $+\tfrac{1}{2}mu^2$] and thus be assimilated to the Newtonian kinetic energy.

Thus Planck's expressions for the momentum and kinetic energy of a material particle were verified. The quantity m is called the *proper mass.*

We have now to trace the gradual emergence of one of the greatest discoveries of the twentieth century, namely, the connection of mass with energy.

As we have seen,[1] J. J. Thomson in 1881 arrived at the result that a charged spherical conductor moving in a straight line behaves as if it had an additional mass of amount $(4/3c^2)$ times the energy of its electrostatic field.[2] In 1900 Poincaré,[3] referring to the fact that in free aether the electromagnetic momentum is $(1/c^2)$ times the Poynting flux of energy, suggested that electromagnetic energy might possess mass density equal to $(1/c^2)$ times the energy density: that is to say, $E=mc^2$ where E is energy and m is mass: and he remarked that if this were so, then a Hertz oscillator, which sends out electromagnetic energy preponderantly in one direction, should recoil as a gun does when it is fired. In 1904 F. Hasenöhrl[4] (1874–1915) considered a hollow box with perfectly reflecting walls filled with radiation, and found that when it is in motion there is an

[1] Vol. I, pp. 306-310

[2] It was shown long afterwards by E. Fermi, *Lincei Rend.* xxxi$_1$ (1922), pp. 184, 306, that the transport of the stress system set up in the material of the sphere should be taken into account, and that when this is done, Thomson's result becomes

$$\text{Additional mass} = \frac{1}{c^2}\times \text{Energy of field.}$$

The same result was obtained in a different way by W. Wilson, *Proc. Phys. Soc.* xlviii (1936), p. 736.

[3] *Archives Néerland.* v (1900), p. 252

[4] *Ann. d. Phys.* xv (1904), p. 344; *Wien Sitz.* cxiii, 2a (1904), p. 1039

apparent addition to its mass, of amount $(8/3c^2)$ times the energy possessed by the radiation when the box is at rest: in the following year [1] he corrected this to $(4/3c^2)$ times the energy possessed by the radiation when the box is at rest [2]; that is, he agreed with J. J. Thomson's $E = \frac{3}{4}mc^2$ rather than with Poincaré's $E = mc^2$. In 1905 A. Einstein [3] asserted that when a body is losing energy in the form of radiation its mass is diminished approximately (i.e. neglecting quantities of the fourth order) by $(1/c^2)$ times the amount of energy lost. He remarked that it is not essential that the energy lost by the body should consist of radiation, and suggested the general conclusion, in agreement with Poincaré, that the mass of a body is a measure of its energy content: if the energy changes by E ergs, the mass changes in the same sense by (E/c^2) grams. In the following year he claimed [4] that this law is the necessary and sufficient condition that the law of conservation of motion of the centre of gravity should be valid for systems in which electromagnetic as well as mechanical processes are taking place.

In 1908 G. N. Lewis [5] proved, by means of the theory of radiation-pressure, that a body which absorbs radiant energy increases in mass according to the equation

$$dE = c^2 dm$$

and affirmed that the mass of a body is a direct measure of its total energy, according to the equation [6]

$$E = mc^2.$$

As we have seen, Poincaré had suggested this equation but had given practically no proof, while Einstein, who had also suggested it, had given a proof (which, however, was put forward only as approximate) for a particular case: Lewis regarded it as an exact equation, but his proof also was not of a general character. Lewis, however, pointed out that if this principle is accepted, then in Planck's equation of 1906

$$\begin{pmatrix}\text{Kinetic energy of a particle whose}\\ \text{mass when at rest is } m\end{pmatrix} = \frac{mc^2}{\sqrt{\left(1 - \frac{w^2}{c^2}\right)}} - mc^2$$

[1] *Ann. d. Phys.* xvi (1905), p. 589

[2] The moving hollow box filled with radiation was discussed further by K. von Mosengeil (a pupil of Planck), *Ann. d. Phys.* xxii (1907), p. 867, and M. Planck, *Berlin Sitz.* (1907), p. 542, whose formulae essentially involve the general law $E = mc^2$.

[3] *Ann. d. Phys.* xviii (1905), p. 639; his reasoning has, however, been criticised; cf. H. E. Ives, *J. Opt. Soc. Amer.* xlii (1952), p. 540

[4] *Ann. d. Phys.* xx (1906), p. 627; cf. a further paper in *Ann. d. Phys.* xxiii (1907), p. 371

[5] *Phil. Mag.* xvi (1908), p. 705; cf. however the above note on Planck's paper of 1907

[6] A little earlier D. F. Comstock, *Phil. Mag.* xv (1908), p. 1, had obtained $E = \frac{3}{4}mc^2$ in accordance with the formulae of J. J. Thomson and Hasenöhrl, and had remarked that 'assuming the loss of mass accompanying the dissipation of energy, the sun's mass must have decreased steadily through millions of years.'

the last term, mc^2, must be interpreted to mean the energy of the particle when at rest, whereas the difference

$$\frac{mc^2}{\sqrt{\left(1-\frac{w^2}{c^2}\right)}} - mc^2$$

represents the additional energy which it possesses when in motion; and therefore the total energy of the particle when in motion must be simply [1]

$$\frac{mc^2}{\sqrt{\left(1-\frac{w^2}{c^2}\right)}}.$$

For confirmation of this, Lewis referred to experiments by W. Kaufmann [2] and A. H. Bucherer,[3] who studied the magnetic and electric deviations of the β-rays for radio-active substances. The original experiments of Kaufmann [4] showed only that for great velocities the 'mass' of the electron increases with its velocity in general qualitative agreement with the formula $m/\sqrt{(1-w^2/c^2)}$: but Bucherer showed that the formula is accurate to a high degree of precision for values of (w/c) ranging from 0·38 to 0·69.

The mass of a system can therefore be calculated from its total energy by the equation

$$m = \frac{1}{c^2}\mathrm{E}:$$

and the researches that have been described show that in calculating E, we must include energy resident in the aether. In 1911 Lorentz [5] showed that *every* kind of energy must be included—masses, stretched

[1] Lorentz in 1904 (*Amst. Proc.* vi (1904), p. 809] had given the formula

$$m = \frac{m_0}{\sqrt{(1-(w/c)^2)}}$$

for the mass of an electron whose mass when at rest is m_0, and which is moving with velocity w, on the assumption that electrons in their motion experience the FitzGerald contraction.

[2] *Gött. Nach.* (1901), p. 143; (1902), p. 291; (1903), p. 90; *Phys. ZS.* iv (1902), p. 54; *Berlin Sitz.* xlv (1905), p. 949; *Ann. d. Phys.* xix (1906), p. 487; cf. also Planck, *Verh. d. Deutsch, Phys. Ges.* ix (1907), p. 301; Kaufmann, ibid. p. 607; Stark, ibid. x (1908), p. 14

[3] *Berl. Phys. Ges.* vi (1908), p. 688; *Ann. d. Phys.* xxviii (1909), p. 513; *Phys. ZS.* ix (1908), p. 755; cf. also C. Schaefer and G. Neumann, *Phys. ZS.* xiv (1913), p. 1117; G. Neumann, *Ann. d. Phys.* xlv (1914), p. 529; C. Guye and Ch. Lavanchy, *Arch. des Sc.* (Geneva) xlii (1916), p. 286

[4] It may be mentioned that in pre-relativity days the interpretation placed on Kaufmann's experiments was that by means of them it would be possible to find for the electron the proportion of proper mass (which was independent of velocity) to electromagnetic mass (which increased with velocity).

[5] *Amst. Versl.* xx (1911), p. 87

strings, light rays, etc. For example, if a system consisting of two electrically charged spheres, of charges e_1 and e_2, at distance a apart, is considered, then when we calculate the value of E for the system, we do not obtain simply the sum of the values of E for the two spheres separately (as calculated when they are infinitely remote from each other), but we must include also a term representing the electrostatic *mutual* potential energy of the two charges, namely, (e_1e_2/a) : and therefore the mass of the system must include a term[1] (e_1e_2/c^2a).

Similarly, the mass of a system of gravitating bodies is not the sum of their masses taken separately, but includes a term representing $(1/c^2)$ times their mutual potential energy.[2] Thus, if two Newtonian gravitating particles m_1 and m_2 are at rest at a distance a apart, their mass is

$$m_1 + m_2 - \frac{\gamma m_1 m_2}{c^2 a}$$

where γ is the Newtonian constant of gravitation.

The equivalence of mass and energy was expressed by Planck in 1908 [3] in the form of a unified definition of momentum. The flux of energy, he said, is a vector, which when divided by c^2 is the density of momentum. This had long been known in the case of electromagnetic energy, by the relation between the Poynting vector and the momentum density resident in the aether. But Planck now asserted that it was universally true, e.g. in the cases of radiation, or of conduction or convection of heat. In the case of a single particle of proper-mass m and velocity v, the energy is $mc^2/\sqrt{(1-v^2/c^2)}$, the streaming of energy is $mc^2\mathbf{v}/\sqrt{(1-v^2/c^2)}$, and this divided by c^2 is $m\mathbf{v}/\sqrt{(1-v^2/c^2)}$, which is precisely the momentum of the particle. The unified definition of momentum is a more general expression of the equivalence of mass and energy than the equation $E = mc^2$, for the concept of mass becomes more difficult to define when, e.g. momentum and velocity are no longer parallel to each other.

Planck's new conception of momentum was soon found to be capable of explaining some paradoxical consequences which could

[1] A value not agreeing with this was found by L. Silberstein in 1911 [*Phys. ZS.* xii (1911), p. 87], but an error in his method was pointed out by E. Fermi [*Rend. Lincei*, xxxi$_1$ (1922), pp. 184, 306], whose work led to the correct value.

[2] On this problem cf. A. S. Eddington and G. L. Clark, *Proc. R.S.*(A), clxvi (1938), p. 465; Eddington [*Proc. R.S.*(A), clxxiv (1940), p. 16] proposed to define the mass of a system to be that of a point-particle which would produce the same gravitational field as the system at very great distances. This and other definitions were discussed by G. L. Clark, *Proc. R.S. E.* lxii (1949), p. 412. It was shown by Josephine M. Gilloch and W. H. McCrea, *Proc. Camb. Ph. Soc.* xlvii (1951), p. 190, that in the case of a cylinder rotating freely on its axis, the gravitational mass is (to a first approximation) the sum of the proper-mass and $(1/c^2)$ times the kinetic energy; as was to be expected according to the general principle of the equivalence of mass and energy.

[3] *Verh. d. Deutsch. Phys. Ges.* x (1908), p. 728; *Phys. ZS.* ix (1908), p. 828. This statement had been to some extent anticipated (in connection with the moving box containing radiation) by Planck, *Berlin Sitz.* (1907), p. 542, and by F. Hasenöhrl, *Wien Sitz.* cxvi 2a (1907), p. 1391.

apparently be deduced from the theory of relativity. One of these, due to Lewis and Tolman,[1] may be described as follows. Consider a rigid bent lever *abc* at rest, pivoted at *b*, whose arms *ba* and *bc* are equal and perpendicular, and suppose that forces F_x and F_y, each equal to F_0, are applied at *a* and *c* in directions parallel to *bc* and *ba* respectively. The system is thus in equilibrium.

Now let the whole system be referred to axes with respect to which it is moving with velocity *w* in the direction *bc*. Obviously it will still be in equilibrium. But according to the theory of relativity, with reference to the new axes the arm *bc* should experience the FitzGerald contraction, and so should be shortened in the ratio $\sqrt{(1-w^2/c^2)}$ to 1, while *ab* has the same length as at rest. Moreover, if force is defined as the rate of communication of momentum with respect to the time used in the inertial system concerned, we can show that the values of the forces referred to the new axes are

$$F_x = F_0, \qquad F_y = F_0\sqrt{\left(1-\frac{w^2}{c^2}\right)}.$$

Thus the forces produce a moment

$$F_0 \,.\, ba - F_0\sqrt{\left(1-\frac{w^2}{c^2}\right)}\sqrt{\left(1-\frac{w^2}{c^2}\right)} \,.\, bc$$

or

$$\frac{1}{c^2}\, F_0 w^2 \,.\, ba$$

tending to turn the system round *b*; so apparently it would not be in equilibrium.

The paradox is resolved by the following explanation, which is due to Sommerfeld and Laue.[2] At the point *a* the force F_x furnishes the work at the rate wF_0. An energy current of this strength enters the lever at *a*, travels to *b* and then passes into the axis of the lever, since the axis does work at the rate $-wF_0$ on the lever. Corresponding to this flux of energy there is, by Planck's principle, a momentum parallel to *ab*, of amount $(1/c^2)$ times the volume integral of the energy flux, or $(1/c^2) \,.\, ab \,.\, wF_0$. Due to the existence of this momentum there is an angular momentum about a fixed origin *O*, lying in the prolongation of *ab*, of amount $(1/c^2) \,.\, ab \,.\, Ob \,.\, wF_0$, and its rate of increase with respect to the time is

$$\frac{1}{c^2} \,.\, ab \,.\, \frac{d(Ob)}{dt} \,.\, wF_0 \quad \text{or} \quad (1/c^2) \,.\, F_0 w^2 \,.\, ab.$$

Thus we see that the couple $(1/c^2) \,.\, F_0 w^2 \,.\, ba$, produced by the two forces F_x and F_y, is needed in order to account for the rate of increase

[1] *Phil. Mag.* xviii (1909), p. 510. Relativity statics is treated fully by P. S. Epstein. *Ann. d. Phys.* xxxvi (1911), p. 779.

[2] Laue, *Verh. Deutsch, Phys. Gesells.* (1911), p. 513

$(1/c^2) \cdot F_0 w^2 \cdot ab$ of the angular momentum of the lever, and the difficulty is satisfactorily explained.

It may be remarked that if the lever is contained in a case, which supports the axis b of the lever, and also (e.g. by elastic strings attached to points of the case) provides the forces F_x and F_y which act at a and c, then the energy current after leaving the lever at b enters the case there, and after travelling in the case re-enters the lever by the elastic string which is attached to a. The energy current is therefore closed, so the system consisting of case and lever together has not a variable angular momentum. The case and lever in fact exert equal and opposite couples on each other.

This may be regarded as a model of the Trouton-Noble experiment,[1] the electric field being compared to the lever and the material condenser to the case. Neither the elctromagnetic momentum of the field nor the mechanical momentum of the condenser is parallel to the velocity, and both therefore need couples in order to preserve their orientation in translatory motion, but these couples are equal and opposite, and the system condenser plus field requires no couple.

Not long after the publication of Planck's paper of 1906 writers on the theory of relativity began to take advantage of some developments in pure mathematics, of which an account must now be given.

It was Felix Klein (1849–1925) in his famous *Erlanger Programm* of 1872[2] who first clearly indicated the essential nature of a *vector.* Let (p, q, r) be the components of a vector with respect to the rectangular axes $O\,x\,y\,z$. Then $px+qy+rz$ is the product of the lengths of the vectors (p, q, r) and (x, y, z) into the cosine of the angle between them, and is therefore invariant if the axes of reference are changed by a rotation about the origin to any other set of rectangular axes. Klein regarded all geometry as the invariant theory of some definite group, and following him, we can take the property just mentioned as the *definition* of a vector: that is, a set of three numbers (p, q, r) will be called a *vector* if $px+qy+rz$ is invariant under the group of rotations of orthogonal axes. This definition suffices to furnish the laws according to which (p, q, r) are transformed when the axes of reference are changed. Since $\{(x \cdot x)+(y \cdot y)+(z \cdot z)\}$ or $(x^2+y^2+z^2)$ is invariant under a rotation of the axes, we see that (x, y, z) is a particular vector. And since all vectors are transformed in the same way, we may say that (p, q, r) is a vector if its components (p, q, r) are transformed like (x, y, z).

Vectors are not the only physical quantities that are related to direction: another class is represented by *elastic stresses.* If we denote by (X_x, Y_x, Z_x) the components of traction across the yz-plane at a given point P, by (X_y, Y_y, Z_y) the components of traction across the zx-plane at P, and by (X_z, Y_z, Z_z) the components of traction

[1] cf. p. 29

[2] *Programm zum Eintritt in die philosophische Fakultät d. Univ. zu Erlangen,* Erlangen, A. Deichert, 1872. Reprinted in 1893 in *Math. Ann.* xliii, and in Klein's *Ges. Math. Abhandl.* i, p. 460.

across the xy-plane at P, then, as is known, we have $Z_y = Y_z$, $X_z = Z_x$, $Y_x = X_y$, so we can write

$$X_x = a, \quad Y_y = b, \quad Z_z = c, \quad Z_y = Y_y = f, \quad X_z = Z_x = g, \quad Y_x = X_y = h,$$

and the stress can be represented by the six numbers (a, h, c, f, g, h). Now let the axes of reference be changed by any rotation about the origin. Then, as is known, if the components of stress at P with respect to the new axes $Ox'y'z'$ are denoted by (a', b', c', f', g', h'), the expression

$$ax^2 + by^2 + cz^2 + 2fyz + 2gzx + 2hxy$$

is transformed into the expression

$$a'x'^2 + b'y'^2 + a'z'^2 + 2f'y'z' + 2g'z'x' + 2h'x'y'.$$

Any set of six quantities (a, b, c, f, g, h) which, when the axes are changed by a rotation about the origin, changes in this way, that is, in the same way as the coefficients of a quadric surface, is said to constitute a *symmetrical tensor*[1] *of rank* 2. The analogy with the definition of a vector is obvious, and a vector may be called a *tensor of rank* 1. A quantity which is invariant under all rotations of the axes of co-ordinates is called a *scalar* or *tensor of rank zero.*

Since

$$x^2 \,.\, x^2 + y^2 \,.\, y^2 + z^2 \,.\, z^2 + 2yz \,.\, yz + 2zx \,.\, zx + 2xy \,.\, xv = (x^2 + y^2 + z^2)^2$$

is an invariant for rotations of the system of co-ordinate axes, it follows that

$$(x^2, \quad y^2, \quad z^2, \quad yz, \quad zx, \quad xy)$$

is a particular symmetric tensor of rank 2, and since all symmetric tensors of rank 2 are transformed in the same way, we see that *a set of* 6 *quantities* (a, b, c, f, g, h) *constitutes a symmetric tensor of rank* 2, *if* (a, b, c, f, g, h) *are transformed in the same way as* $(x^2, y^2, z^2, yz, zx, xy)$. It is easily shown, for example, that if A, B, C, F, G, H denote the moments and products of inertia of a system of masses with respect to the co-ordinate axes, then (A, B, C, $-$F, $-$G, $-$H) is a symmetric tensor of rank 2.

The definition just given can be generalised, so as to furnish a definition of a tensor of rank 2 which is not necessarily symmetrical. Let (p_1, q_1, r_1) and (p_2, q_2, r_2) be two different vectors. Then a *set of nine numbers*

$$t_{11}, \quad t_{22}, \quad t_{33}, \quad t_{23}, \quad t_{32}, \quad t_{31}, \quad t_{13}, \quad t_{12}, \quad t_{21},$$

[1] Attention was drawn to the properties of sets of quantities obeying these laws of transformation by C. Niven, *Trans. R. S. E.* xxvii (1874), p. 473 ; cf. also W. Thomson (Kelvin), *Phil. Trans.* cxlvi (1856), p. 481 and W. J. M. Rankine, ibid. p. 261. The name *tensor* (with this meaning) is due to J. Willard Gibbs, *Vector Analysis,* New Haven (1881–4), p. 57.

will be called a tensor of rank 2, *if they transform in the same way as*

$$p_1p_2,\ q_1q_2,\ r_1r_2,\ q_1r_2,\ r_1q_2,\ r_1p_2,\ p_1r_2,\ p_1q_2,\ q_1p_2.$$

So far we have considered only tensors which have invariant properties with respect to the rotations of a system of orthogonal co-ordinate axes in three-dimensional space. This theory was generalised into a tensor-calculus applicable to transformations in curved space of any number of dimensions by Gregorio Ricci-Curbastro (1853–1925) of Padua, from 1887 onwards: it first became widely known when a celebrated memoir describing it was published in 1900 by Ricci and Levi-Civita.[1]

Let $x_1, x_2, \ldots x_n$ be any 'generalised co-ordinates' specifying the position of a point in space of n dimensions. Let n new variables $\bar{x}_1, \bar{x}_2, \ldots \bar{x}_n$ be introduced by arbitrary equations

$$\bar{x}_r = f_r(x_1, x_2, \ldots x_n) \qquad (r=1, 2, \ldots n). \qquad (1)$$

Then the differentials of the co-ordinates are transformed according to the equations

$$d\bar{x}_r = \sum_{k=1}^{n} \frac{\partial \bar{x}_r}{\partial x_k} dx_k \qquad (r=1, 2, \ldots n).$$

At a point P of the n-dimensional space we can consider various types of quantities analogous to the scalars, vectors and tensors that we have already considered.

Firstly, there may be a function of position whose value is unchanged when we perform the transformation (1). Such a function is called a *scalar*, or *tensor of rank zero*.

Secondly, we consider a set of n numbers $(V^1, V^2, \ldots V^n)$, which are defined with respect to all co-ordinate systems and which, when we perform the transformation (1), are transformed in the same way as the dx_r, so that

$$\bar{V}^r = \sum_{k=1}^{n} \frac{\partial \bar{x}_r}{\partial x_k} V^k \qquad (r=1, 2, \ldots n)$$

whence

$$V^r = \sum_{k=1}^{n} \frac{\partial x_r}{\partial \bar{x}_k} \bar{V}^k \qquad (r=1, 2, \ldots n).$$

Such a set of n numbers is called a *contravariant tensor of rank* 1, or *contravariant vector*, and the numbers are called its *components*.

Next, consider sets of n numbers $(X_1, X_2, \ldots X_n)$, which are such that if $(V^1, V^2, \ldots V^n)$ is any contravariant tensor of rank 1, the sum $X_1V^1 + X_2V^2 + \ldots + X_nV^n$ is a scalar. Such

[1] *Math. Ann.* liv (1900), p. 125; cf. J. A. Schouten, *Jahresb. d. Deutsch. Math.-Verein.* xxxii (1923), p. 91

a set of n numbers is called a *covariant tensor of rank* 1 or *covariant vector*.

Since

$$\sum_k X_k V^k = \sum_r \overline{X}_r \overline{V}^r = \sum_r \overline{X}_r \sum_k \frac{\partial \overline{x}_r}{\partial x_k} V^k,$$

we have

$$X_k = \sum \frac{\partial \overline{x}_r}{\partial x_k} \overline{X}_r, \quad \text{whence } \overline{X}_k = \sum_r \frac{\partial x_r}{\partial \overline{x}_k} X_r.$$

The covariant or contravariant character is indicated by placing the index in the lower or upper position respectively. In Euclidean space, for rotations of rectangular axes, there is no distinction between contravariant and covariant tensors.

If, at the point P of the n-dimensional space, we have n^2 numbers $(V^{11}, V^{12}, \ldots V^{nn})$ which, when we perform the transformation of co-ordinates, are transformed like $(P^1Q^1, P^1Q^2, \ldots P^nQ^n)$, where $(P^1, \ldots P^n)$ and $(Q^1, \ldots Q^n)$ are two different contravariant tensors of rank 1, then $(V^{11}, V^{12}, \ldots V^{nn})$ are said to be the components of a *contravariant tensor of rank* 2. Similarly n^2 numbers $(X_{11}, X_{12}, \ldots X_{nn})$ which transform like $(X_1Y_1, X_1Y_2, \ldots X_nY_n)$ where $(X_1 \ldots X_n)$ and $(Y_1, \ldots Y_n)$ are two different covariant tensors of rank 1, are said to be the components of a *covariant tensor of rank* 2; while n^2 numbers $(W^1{}_1, W^1{}_2, W^2{}_1, \ldots W^n)$ which transform like $(P^1X_1, P^1X_2, P^2X_1, \ldots P^nX_n)$ where $(P^1, P^2, \ldots P^n)$ is a contravariant tensor of rank 1 and $(X_1, X_2, \ldots X_n)$ is a covariant tensor of rank 1, is called a *mixed tensor of rank* 2. Tensors of rank greater than 2 are defined in a similar way. A tensor whose typical component is, say, X^p_{rs}, is often denoted by (X^p_{rs}).

Consider a tensor such that any two of its components, which may be obtained from each other by a simple interchange of two indices, are equal to each other; thus, $V^{pq} = V^{qp}$. If this property holds for any one system of co-ordinates, it will still hold after any change of the co-ordinate system, as is evident from the equations of transformation. Such a tensor is said to be *symmetric*. If a tensor is such that two components which may be derived from each other by a simple interchange of two indices are equal in magnitude but opposite in sign, thus $V^{pq} = -V^{qp}$, the tensor is said to be *skew*. This property also holds in all systems of co-ordinates, provided it holds in any one system.

Two tensors of the same kind (contravariant, covariant or mixed) and of the same rank, are said to be *equal* if their corresponding components are equal in all co-ordinate systems. This is the case if the corresponding components are equal in any one co-ordinate system.

Consider the transformation of tensors when the co-ordinates are subjected to the particular Lorentz transformation (writing $ct = x_0$, $x = x_1$, $y = x_2$, $z = x_3$)

$$dx_0 = d\bar{x}_0 \cosh \alpha + d\bar{x}_1 \sinh \alpha, \quad dx_1 = d\bar{x}_0 \sinh \alpha + d\bar{x}_1 \cosh \alpha,$$
$$d\bar{x}_2 = dx_2, \quad dx_3 = d\bar{x}_3.$$

It is found at once that :

for any contravariant vector :

$$J^0 = \bar{J}^0 \cosh \alpha + \bar{J}^1 \sinh \alpha,\ J^1 = \bar{J}^0 \sinh \alpha + \bar{J}^1 \cosh \alpha,\ J^2 = \bar{J}^2,\ J^3 = \bar{J}^3.$$

for any covariant vector :

$$J_0 = \bar{J}_0 \cosh \alpha - \bar{J}_1 \sinh \alpha,\ J_1 = -\bar{J}_0 \sinh \alpha + \bar{J}_1 \cosh \alpha,\ J_2 = \bar{J}_2,\ J_3 = \bar{J}_3.$$

for any covariant symmetric tensor of rank 2 *:*

$$\bar{X}_{00} = X_{00} \cosh^2 \alpha + 2X_{01} \cosh \alpha \sinh \alpha + X_{11} \sinh^2 \alpha$$
$$\bar{X}_{11} = X_{00} \sinh^2 \alpha + 2X_{01} \sinh \alpha \cosh \alpha + X_{11} \cosh^2 \alpha$$
$$\bar{X}_{22} = X_{22}, \quad \bar{X}_{33} = X_{33}, \quad \bar{X}_{32} = \bar{X}_{23} = X_{23}$$
$$\bar{X}_{10} = \bar{X}_{01} = X_{00} \cosh \alpha \sinh \alpha + X_{01} (\cosh^2 \alpha + \sinh^2 \alpha) + X_{11} \sinh \alpha \cosh \alpha$$
$$\bar{X}_{20} = \bar{X}_{02} = X_{02} \cosh \alpha + X_{12} \sinh \alpha$$
$$\bar{X}_{30} = \bar{X}_{03} = X_{03} \cosh \alpha + X_{13} \sinh \alpha$$
$$\bar{X}_{12} = \bar{X}_{21} = X_{02} \sinh \alpha + X_{12} \cosh \alpha$$
$$\bar{X}_{13} = \bar{X}_{31} = X_{03} \sinh \alpha + X_{13} \cosh \alpha$$

for any covariant skew tensor of rank 2 *:*

$$\bar{X}_{01} = X_{01}, \quad \bar{X}_{02} = X_{02} \cosh \alpha + X_{12} \sinh \alpha, \quad \bar{X}_{03} = X_{03} \cosh \alpha + X_{13} \sinh \alpha$$
$$\bar{X}_{23} = X_{23}, \quad \bar{X}_{31} = X_{31} \cosh \alpha + X_{30} \sinh \alpha, \quad \bar{X}_{12} = X_{12} \cosh \alpha + X_{02} \sinh \alpha$$

It is evident from these last equations that *a six-vector,*[1] such as is constituted by the electric and magnetic intensities *in vacuo, is a skew tensor of rank* 2. We can write

$$X_{10} = d_x, \quad X_{20} = d_y, \quad X_{30} = d_z, \quad X_{23} = h_x, \quad X_{31} = h_y, \quad X_{12} = h_z$$

From the definition of a tensor, it is evident that if two tensors of the same type are taken, say (X^p_{rs}), and (Y^p_{rs}), then the quantities formed by adding corresponding components of these tensors

$$Z^p_{rs} = X^p_{rs} + Y^p_{rs}$$

are the components of a tensor of the same type, which is called the *sum* of the tensors (X^p_{rs}) and (Y^p_{rs}).

[1] cf. p. 35

Moreover it is evident from the definitions that if two tensors, say of rank λ and rank μ, are given in n-dimensional space, and if we multiply each of the n^λ components of one by each of the n^μ components of the other, then the $n^{\lambda+\mu}$ products so formed are the components of a new tensor of rank $(\lambda+\mu)$, thus :

$$X_{ij}\ Y^p_{rs} = U^p_{ijrs}.$$

The tensor (U^p_{ijrs}) is called the *outer product* of the tensors (X_{ij}) and (Y^p_{rs}). It may properly be called a product, since the distributive law

$$X(Y+Z) = XY + XZ$$

holds. We can form in this way the outer product of any number of tensors.

An arbitrary tensor cannot in general be expressed as an outer product of tensors of rank 1, since there would not be enough quantities at our disposal to satisfy all the conditions. Thus, if a tensor (X^p_{rs}) is given, we cannot in general find tensors of rank 1 (Y_r), (Z_s), and (V^p), such that

$$X^p_{rs} = Y_r\, Z_s\, V^p \qquad (p, r, s = 1, 2 \ldots n):$$

but the sum of any number of outer products of this type will be a tensor of the type (X^p_{rs}) ; and by taking the number of such products sufficiently great, we shall have enough quantities at our disposal to represent any tensor (X^p_{rs}) in the form

$$X^p_{rs} = Y_r\, Z_s\, V^p + H_r\, K_s\, L^p + E_r\, F_s\, G^p + \ldots$$

Next consider a tensor which has both contravariant and covariant indices, e.g. (X^{lk}_{pqr}). Make one of the upper or contravariant indices identical with one of the lower or covariant indices, and sum with respect to this index, thus :

$$\sum_{p=1}^{n} X^{pk}_{pqr}.$$

Then we can show that *the numbers thus obtained,* when $k, q, r = 1, 2 \ldots n$, *are the components of a new tensor* (Y^k_{qr}), which is two units lower in rank than (X^{lk}_{pqr}). To prove this, we remark that (X^{lk}_{pqr}) can be expressed as a sum of outer products of tensors of rank 1, and the theorem will therefore evidently be true in general if it is true for the case when (X^{lk}_{pqr}) is a *single* outer product of tensors of rank 1, say

$$X^{lk}_{pqr} = Y_p\, Z_q\, T_r\, U^l\, V^k.$$

Then we have

$$\sum_{p=1}^{n} X^{pk}_{pqr} = \sum_{p=1}^{n} \left(Y_p U^p\right) Z_q T_r V^k :$$

and since $\sum_{p=1}^{n} T_p U^p$ is a scalar, these quantities are the components of a tensor of type (Y^k_{qr}) ; which establishes the theorem. This process is called *contraction*.

By forming the outer product of any number of tensors, and then contracting (once or oftener) the tensor thereby obtained, we obtain results such as

$$\sum_{abc, \ldots \alpha\beta\gamma \ldots} \left(X^{abc \ldots rst \ldots}_{\alpha\beta\gamma \ldots \rho\sigma\tau \ldots} Y^{\alpha\beta\gamma \ldots lmn \ldots}_{abc \ldots jhk \ldots}\right) = Z^{rst \ldots lmn \ldots}_{\rho\sigma\tau \ldots jhk \ldots}$$

This process is called *transvection*.

The spaces we consider will generally be supposed each to possess a *metric*, that is to say, there will be an equation expressing an element ds of arc-length at any point of the space in terms of the infinitesimal differences of the co-ordinates between the ends of the arc-element : thus in ordinary Euclidean three-dimensional space with rectangular co-ordinates (x, y, z), we have

$$(ds)^2 = (dx)^2 + (dy)^2 + (dz)^2,$$

and with spherical-polar co-ordinates (r, θ, ϕ), we have

$$(ds)^2 = (dr)^2 + r^2(d\theta)^2 + r^2 \sin^2\theta (d\phi)^2.$$

We assume generally that the square of the line-element ds is a homogeneous quadratic form in the differentials of the co-ordinates. These differentials will be written $(dx^1, dx^2, \ldots dx^n)$, the index being placed above since $(dx^1, \ldots dx^n)$ is a contravariant vector : thus

$$(ds)^2 = \sum_{p,q} g_{pq} dx^p dx^q.$$

Since $(ds)^2$ is a scalar, it is obvious from this equation that the numbers g_{pq} $(p, q = 1, 2, \ldots n)$ must be the components of a co-variant symmetric tensor of rank 2, (g_{pq}) ; this is called the *covariant fundamental tensor*.

Let g denote the determinant $\|g_{pq}\|$ of the coefficients g_{pq}, and let g^{pq} denote $(1/g)$ times the co-factor of g_{pq} in g, so that $\sum_{p=1} g_{pr}g^{pq} = \delta_r^{\ q}$, where $\delta_r^{\ q}$ is equal to 1 or 0 according as q is equal to, or different from, r. Then

$$\sum_{p,\,q} g_{qs}g_{pr}g^{pq} = \sum_{q} g_{qs}\delta_r^{\ q} = g_{rs}.$$

Now if X_p and Y_p are two arbitrary covariant vectors, and if

$$X_p = \sum_r g_{pr}X^r$$

so that X^r is a contravariant vector, we have

$$\sum_{pq} g^{pq}X_pY_q = \sum_{pqrs} g_{pr}g_{qs}g^{pq}X^rY^s$$

$$= \sum_{rs} g_{rs}X^rY^s$$

which is a scalar: and therefore the g^{pq} are the components of a contravariant tensor of rank 2. It is called the *contravariant fundamental tensor.*

Moreover, if U^p is a contravariant vector, and X_q is any covariant vector, we have

$$\sum_{pq} \delta_q^{\ p}U^pX_q = \sum_p U^pX_p = \text{a scalar}$$

and therefore $(\delta_p^{\ q})$ is a tensor of rank 2, covariant with respect to the index p and contravariant with respect to the index q. It is called the *mixed fundamental tensor.*

By aid of the fundamental tensor (g^{pq}) we can derive from any covariant tensor $(X_{p_1 p_2 \ldots p_m})$ a contravariant tensor of the same rank by writing

$$X^{q_1 q_2 \ldots q_m} = \sum_{p_1, p_2, \ldots p_m} g^{p_1 q_1} g^{p_2 q_2} \cdots g^{p_m q_m} X_{p_1 p_2 \ldots p_m}.$$

It is easily shown that this equation is equivalent to

$$X_{p_1 p_2 \ldots p_m} = \sum_{q_1, q_2, \ldots q_m} g_{p_1 q_1} g_{p_2 q_2} \cdots g_{p_m q_m} X^{q_1 q_2 \ldots q_m}$$

Thus to every contravariant tensor we can correlate a definite covariant tensor; and we may say that the distinction between covariant and contravariant tensors loses most of its importance when the fundamental tensor is given, i.e. in a *metrical* space, since it is not the tensors that are essentially different, but only their mode of expression, i.e. their components. For example, we regard (g^{pq}), (g_{pq}), and $(\delta_p{}^q)$, as essentially the *same* tensor.

If two vectors (X) and (Y) are such that when (X) is expressed in covariant form (X_p) and the other in contravariant form (Y^q), we have

$$\sum_p X_p Y^p = 0,$$

then the two vectors are said to be *orthogonal*.

After this rather long excursus on Ricci's tensor calculus, we can return to physics. A contribution of great importance to relativity theory was made in 1908 by Hermann Minkowski (1864–1909).[1] Its ostensible purpose, as indicated in its title, was to show that the differential equations of the electromagnetic field in moving ponderable bodies under the most general conditions (e.g. of magnetisation) can be derived from the differential equations for the same system of bodies at rest, by the principle of relativity : and to criticise some of the formulae that had been given by Lorentz. But these were not actually the most important elements in the paper ; the great advances made by Minkowski [2] were connected with his formulation of physics in terms of a four-dimensional manifold, the use of tensors in this manifold, and the discovery of some of the more important of these tensors.[3]

The phenomena studied in natural philosophy take place each at a definite location at a definite moment, the whole constituting a four-dimensional world of space and time. The theory of relativity had now made it clear that the separation of this four-dimensional world into a three-dimensional world of space and an independent one-dimensional world of time may be effected in an infinite number of ways, each of which is distinguished from the others only by characteristics that are merely arbitrary and accidental. In order to represent natural phenomena without introducing this contingent element, it is necessary to abandon the customary three-dimensional system of co-ordinates, and to operate in four dimensions.

If (t_1, x_1, y_1, z_1) and (t_2, x_2, y_2, z_2) are the time-and-space co-

[1] *Gött. Nach.* (1908), p. 53 ; cf. also *Math. Ann.* lxviii (1910), p. 472

[2] Minkowski had been to some extent anticipated by Poincaré, who had substantially introduced the metric

$$ds^2 = c^2dt^2 - dx^2 - dy^2 - dz^2 = -\sum_{r=1}^{4} dx_r{}^2$$

(where $x_1 = x$, $x_2 = y$, $x_3 = z$, $x_4 = ct\sqrt{-1}$) in *Rend. circ. Palermo*, xxi (1906), p. 129.

[3] The principle of treating the time co-ordinate on the same level as the other co-ordinates was introduced and developed simultaneously with Minkowski's paper by R. Hargreaves, [*Camb. Phil. Trans.* xxi (1908), p. 107] : his work suggests the use of space-time vectors just as Minkowski's does. For comments on this point, cf. H. Bateman, *Phys. Rev.* xii (1918), p. 459.

ordinates of two point-events referred to an inertial system, then, as we have seen, the expression

$$(t_2 - t_1)^2 - \frac{1}{c^2}\left\{(x_2 - x_1)^2 + (y_2 - y_1)^2 + (z_2 - z_1)^2\right\}$$

is invariant under all Lorentz transformations, and therefore has the same value *whatever be the inertial framework of reference.* This quantity is therefore an invariant of the two point-events, which is the same for all observers : and we can make our four-dimensional space-time suited to describe nature when we impose a metric on it, which we do by taking the *interval* (the four-dimensional analogue of length) between the two events (t_1, x_1, y_1, z_1) and (t_2, x_2, y_2, z_2) to be [1]

$$\left[(t_2 - t_1)^2 - \frac{1}{c^2}\left\{(x_2 - x_1)^2 + (y_2 - y_1)^2 + (z_2 - z_1)^2\right\}\right]^{\frac{1}{2}}.$$

Taking any point in the four-dimensional manifold as origin, the cone

$$x^2 + y^2 + z^2 - c^2t^2 = 0$$

which is called the *null cone*, partitions space-time into two regions, of which one is defined by the inequality

$$c^2t^2 < x^2 + y^2 + z^2$$

and includes the hyperplane $t = 0$: the directions at the origin satisfying this inequality are said to be *spatial* : directions at the origin in the other region are said to be *temporal.* Lorentz transformations are simply the rotations and translations in this manifold.

Now consider tensors in the manifold.

Minkowski had not properly assimilated the Ricci tensor-calculus as applied to non-Euclidean manifolds, and in order to be able to work with a space of Euclidean type, he used the device of writing x_4 for $ct\sqrt{-1}$ (the space-co-ordinates being denoted by x_1, x_2, x_3), so that the expression

$$(dx)^2 + (dy)^2 + (dz)^2 - c^2(dt)^2$$

which is invariant under all Lorentz transformations, became

$$(dx_1)^2 + (dx_2)^2 + (dx_3)^2 + (dx_4)^2 \ :$$

[1] The metric of space-time thus introduced is that of a four-dimensional Cayley-Klein manifold which has for absolute (in homogeneous co-ordinates)

$$\left.\begin{aligned} x^2 + y^2 + z^2 - c^2t^2 &= 0 \\ w^2 &= 0 \end{aligned}\right\}$$

a double hyperplane at infinity containing a quadric hypersurface, which is real but with imaginary generators, like an ordinary sphere.

this enabled him to take as his metric

$$(ds)^2 = (dx_1)^2 + (dx_2)^2 + (dx_3)^2 + (dx_4)^2$$

which defines a four-dimensional Euclidean manifold.[1]

It is, however, simpler to work with the real value of the time, and to express Minkowski's results in terms of tensors which exist in the non-Euclidean four-dimensional manifold we have introduced, whose metric is specified by

$$(ds)^2 = c^2(dt)^2 - (dx)^2 - (dy)^2 - (dz)^2$$

which we may write

$$(ds)^2 = (dx^0)^2 - (dx^1)^2 - (dx^2)^2 - (dx^3)^2$$

so

$$g_{00} = 1, \quad g_{11} = g_{22} = g_{33} = -1,$$
$$g^{00} = 1, \quad g^{11} = g^{22} = g^{33} = -1.$$

His greatest discovery[2] was that at any point in the electromagnetic field *in vacuo* there exists a tensor of rank 2 of outstanding physical importance, which in its mixed form ($E_p{}^q$) may be defined by the equation

$$E_p{}^q = \frac{1}{16\pi}\delta_p{}^q \sum_{\alpha\beta} X_{\alpha\beta}X^{\alpha\beta} - \frac{1}{4\pi}\sum_t X_{pt}X^{qt}$$

where X_{pq} is the electromagnetic six-vector, that is to say, if (d_x, d_y, d_z) and (h_x, h_y, h_z) are the electric and magnetic intensities respectively, then[3]

$$d_x = X^{01} = -X_{01}, \quad d_y = X^{02} = -X_{02}, \quad d_z = X^{03} = -X_{03}$$

$$h_x = X^{23} = X_{23}, \quad h_y = X^{31} = X_{31}, \quad h_z = X^{12} = X_{12}.$$

Substituting in the equation which defines $E_p{}^q$, we find the values of the components of this tensor, namely,

$$E_0{}^0 = \frac{1}{8\pi}(d_x^2 + d_y^2 + d_z^2 + h_x^2 + h_y^2 + h_z^2):$$

this represents the density of electromagnetic energy, discovered by W. Thomson (Kelvin) in 1853[4];

$$E_0{}^1 = \frac{1}{4\pi}(d_yh_z - d_zh_y), \quad E_0{}^2 = \frac{1}{4\pi}(d_zh_x - d_xh_z), \quad E_0{}^3 = \frac{1}{4\pi}(d_xh_y - d_yh_x):$$

[1] Minkowski's use of $x_4 = ct\sqrt{-1}$ led some philosophers to an outpouring of metaphysical nonsense about time being an imaginary fourth dimension of space.

[2] loc. cit., equation (74)

[3] X_{pq} is immediately derived from X^{pq} by the formula

$$X_{pq} = \sum_{ts} g_{ps}\, g_{qt}\, X^{st}.$$

[4] cf. Vol. I, pp. 222, 224

(E_0^1, E_0^2, E_0^3) represents $(1/c)$ times the flux of electromagnetic energy, discovered by Poynting and Heaviside in 1884[1];

$$E_1^0 = -\frac{1}{4\pi}(d_y h_z - d_z h_y),\ E_2^0 = -\frac{1}{4\pi}(d_z h_x - d_x h_z),\ E_3^0 = -\frac{1}{4\pi}(d_x h_y - d_y h_x):$$

$(-E_1^0, \quad -E_2^0, \quad -E_3^0)$ represents c times the density of electromagnetic momentum, discovered by J. J. Thomson in 1893[2];

$$E_1^1 = \frac{1}{8\pi}(d_x^2 - d_y^2 - d_z^2 + h_x^2 - h_y^2 - h_z^2),$$

and similarly for E_2^2 and E_3^3;

$$E_2^3 = E_3^2 = \frac{1}{4\pi}(d_y d_z + h_y h_z),$$

and similarly for E_3^1, E_1^3, E_1^2, E_2^1.

The nine quantities

$$\begin{matrix} E_1^1 & E_2^1 & E_3^1 \\ E_1^2 & E_2^2 & E_3^2 \\ E_1^3 & E_2^3 & E_3^3 \end{matrix}$$

represent the components of stress in the aether, discovered by Maxwell in 1873.[3] Thus, *each component of the tensor* E_p^q *has a physical interpretation*, which in every case had been discovered many years before Minkowski showed that these 16 components constitute a tensor of rank 2. The tensor E_p^q is called the *energy tensor* of the electromagnetic field.

Since $E_{qp} = \sum_r g_{pr} E_q^r = g_{pp} E_q^p$ for this metric, we have

$$E_{0p} = E_0^p,\ E_{1p} = -E_1^p,\ E_{2p} = -E_2^p,\ E_{3p} = -E_3^p,$$

and hence we find $E_{01} = E_{10}$ and generally $E_{pq} = E_{qp}$, that is, E_{pq} *is a symmetric tensor.*

Moreover we can show that if ρ is the density of electricity and $\mathbf{v}$ its velocity, then

$$\left.\begin{aligned} \frac{\partial E_0^0}{\partial x^0} + \frac{\partial E_0^1}{\partial x^1} + \frac{\partial E_0^2}{\partial x^2} + \frac{\partial E_0^3}{\partial x^3} &= -\frac{\rho}{c}(v_x d_x + v_y d_y + v_z d_z) \\ \frac{\partial E_1^0}{\partial x^0} + \frac{\partial E_1^1}{\partial x^1} + \frac{\partial E_1^2}{\partial x^2} + \frac{\partial E_1^3}{\partial x^3} &= \frac{\rho}{c}(d_x + v_y h_z - v_z h_y) \\ \frac{\partial E_2^0}{\partial x^0} + \frac{\partial E_2^1}{\partial x^1} + \frac{\partial E_2^2}{\partial x^2} + \frac{\partial E_2^3}{\partial x^3} &= \frac{\rho}{c}(d_y + v_z h_x - v_x h_z) \\ \frac{\partial E_3^0}{\partial x^0} + \frac{\partial E_3^1}{\partial x^1} + \frac{\partial E_3^2}{\partial x^2} + \frac{\partial E_3^3}{\partial x^3} &= \frac{\rho}{c}(d_z + v_x h_y - v_y h_x). \end{aligned}\right\} \quad \text{(A)}$$

[1] cf. Vol. I, pp. 313-4 [2] cf. Vol. I, p. 317 [3] cf. Vol. I, pp. 271-2

The first of these equations is

$$\frac{\partial}{c\partial t}\frac{1}{8\pi}(d_x^2+d_y^2+d_z^2+h_x^2+h_y^2+h_z^2)+\frac{\partial}{\partial x}\frac{1}{4\pi}(d_yh_z-d_zh_y)$$
$$+\frac{\partial}{\partial y}\frac{1}{4\pi}(d_zh_x-d_xh_z)+\frac{\partial}{\partial z}\frac{1}{4\pi}(d_xh_y-d_yh_x)$$

$$=-\frac{\rho}{c}(v_xd_x+v_yd_y+v_zd_z)$$

or

$\partial/\partial t$ (density of electromagnetic energy) $+\,\partial/\partial x$ (x-component of flux of electromagnetic energy $+\,\partial/\partial y$ (y-component of flux of electromagnetic energy) $+\,\partial/\partial z$ (z-component of flux of electromagnetic energy)

$$=-\rho(v_xd_x+v_yd_y+v_zd_z)$$

or

rate of increase of electromagnetic energy in unit volume + rate at which energy is leaving unit volume
= – (work done by the electromagnetic forces on electric charges within the unit volume)
and this is clearly nothing but the *equation of conservation of energy*. Similarly the other three of the equations (A) are the equations of conservation of x-momentum, y-momentum and z-momentum respectively.[1]

In an appendix to his paper,[2] Minkowski threw a new light on the equations of the relativistic dynamics of a material particle, which had been discovered by Planck two years earlier.[3] Denoting by (x, y, z) the co-ordinates of the particle at the instant t, he introduced the notion of the *proper-time* τ of the particle, whose differential is defined by the equation

$$(d\tau)^2=(dt)^2-\frac{1}{c^2}\left\{(dx)^2+(dy)^2+(dz)^2\right\}.$$

It is evident from this equation that $d\tau$ is invariant under all Lorentz transformations of (t, x, y, z), i.e. it is, in the language of the tensor-calculus, a scalar. Now writing $x^0=ct$, $x^1=x$, $x^2=y$, $x^3=z$, we know that

$$(dx^0,\quad dx^1,\quad dx^2,\quad dx^3)$$

[1] On the energy tensor cf. also A. Sommerfeld, *Ann. d. Phys.* xxxii (1910), p. 749; xxxiii (1910), p. 649; and M. Abraham, *Palermo Rend.* xxx (1910), p. 33

[2] loc. cit. [3] cf. p. 44

is a contravariant vector, and therefore

$$\left(\frac{dx^0}{d\tau}, \frac{dx^1}{d\tau}, \frac{dx^2}{d\tau}, \frac{dx^3}{d\tau}\right)$$

is a contravariant vector. But

$$\frac{d\tau}{dx^0} = \frac{1}{c}\left(1 - \frac{v^2}{c^2}\right)^{\frac{1}{2}},$$

where v denotes the velocity of the particle. Therefore

$$\frac{c}{\sqrt{\left(1 - \frac{v^2}{c^2}\right)}}, \quad \frac{v_x}{\sqrt{\left(1 - \frac{v^2}{c^2}\right)}}, \quad \frac{v_y}{\sqrt{\left(1 - \frac{v^2}{c^2}\right)}}, \quad \frac{v_z}{\sqrt{\left(1 - \frac{v^2}{c^2}\right)}}$$

are the components of a contravariant vector.

Now Planck had shown that if m is the mass of the particle, its energy E is $mc^2(1 - v^2/c^2)^{-\frac{1}{2}}$, and its components of momentum are

$$p_x = \frac{mv_x}{\sqrt{\left(1 - \frac{v^2}{c^2}\right)}}, \quad p_y = \frac{mv_y}{\sqrt{\left(1 - \frac{v^2}{c^2}\right)}}, \quad p_z = \frac{mv_z}{\sqrt{\left(1 - \frac{v^2}{c^2}\right)}}.$$

Thus

$$(E/c, \quad p_x, \quad p_y, \quad p_z)$$

is a contravariant vector. This is called the *energy-momentum vector.*

The Newtonian and relativist definitions of *force* may be compared as follows. In Newtonian physics the momentum (p_x, p_y, p_z) is a vector and the time t is a scalar, so dp_x/dt, dp_y/dt, dp_z/dt is a vector, namely, the Newtonian *force*. In relativist physics, as we have seen, instead of a momentum vector (p_x, p_y, p_z), we have the contravariant energy-momentum vector $(E/c, p_x, p_y, p_z)$, or in its covariant form $(E/c, -p_x, -p_y, -p_z)$, and the scalar which takes the place of the time is the interval of proper-time,

$$d\tau = \left[(dt^2) - \frac{1}{c^2}\left\{(dx)^2 + (dy)^2 + (dz)^2\right\}\right]^{\frac{1}{2}}.$$

Thus it is natural to represent a *force* in relativity by the covariant vector

$$(F_k) = \left(-\frac{1}{c}\frac{dE}{d\tau}, \frac{dp_x}{d\tau}, \frac{dp_y}{d\tau}, \frac{dp_z}{d\tau}\right).$$

Now the equations of motion of a particle, as found by Planck, were

$$m\frac{d}{dt}\left\{\left(1-\frac{v^2}{c^2}\right)^{-\frac{1}{2}}\frac{dx}{dt}\right\}=X, \qquad m\frac{d}{dt}\left\{\left(1-\frac{v^2}{c^2}\right)^{-\frac{1}{2}}\frac{dy}{dt}\right\}=Y,$$

$$m\frac{d}{dt}\left\{\left(1-\frac{v^2}{c^2}\right)^{-\frac{1}{2}}\frac{dz}{dt}\right\}=Z,$$

or

$$\frac{dp_x}{dt}=X, \qquad \frac{dp_y}{dt}=Y, \qquad \frac{dp_z}{dt}=Z.$$

Comparing these results, we see that we must take the last three components of the relativist force to be

$$F_1=\left(1-\frac{v^2}{c^2}\right)^{-\frac{1}{2}}X, \quad F_2=\left(1-\frac{v^2}{c^2}\right)^{-\frac{1}{2}}Y, \quad F_3=\left(1-\frac{v^2}{c^2}\right)^{-\frac{1}{2}}Z,$$

and then the last three relativist equations of motion will be

$$m\frac{d^2x}{d\tau^2}=F_1, \qquad m\frac{d^2y}{d\tau^2}=F_2, \qquad m\frac{d^2z}{d\tau^2}=F_3.$$

Since

$$\left(c\frac{d^2t}{d\tau^2}, -\frac{d^2x}{d\tau^2}, -\frac{d^2y}{d\tau^2}, -\frac{d^2z}{d\tau^2}\right)$$

is a covariant vector, the first relativist equation of motion must evidently be

$$-mc\frac{d^2t}{d\tau^2}=F_0.$$

This completes Minkowski's set of equations of motion. The last equation may be written

$$mc^2\frac{d^2t}{d\tau^2}=\frac{dE}{dt},$$

where E is the energy ; which is evidently true, since $E=mc^2\,dt/d\tau$.

Since

$$dE=X\,dx+Y\,dy+Z\,dz,$$

we have

$$\left(1-\frac{v^2}{c^2}\right)^{-\frac{1}{2}}dE=F_1dx+F_2dy+F_3dz$$

or

$$-c\,d\tau\left(1-\frac{v^2}{c^2}\right)^{-\frac{1}{2}}F_0=F_1dx+F_2dy+F_3dz$$

or

$$F_0 c\,dt + F_1 dx + F_2 dy + F_3 dz = 0$$

an equation which may be expressed geometrically by the statement that *the vector* (F_0, F_1, F_2, F_3) is *orthogonal to the vector which represents the velocity of the particle, namely,*

$$\left(c\frac{dt}{d\tau}, \frac{dx}{d\tau}, \frac{dy}{d\tau}, \frac{dz}{d\tau}\right).$$

We can now obtain a simple expression for the ponderomotive force on a particle of charge e and velocity $\mathbf{v}$ in electromagnetic theory. In Newtonian physics the three components are

$$e\left(d_x + \frac{h_z}{c}\frac{dy}{dt} - \frac{h_y}{c}\frac{dz}{dt}\right), \quad e\left(d_y + \frac{h_x}{c}\frac{dz}{dt} - \frac{h_z}{c}\frac{dx}{dt}\right), \quad e\left(d_z + \frac{h_y}{c}\frac{dx}{dt} - \frac{h_x}{c}\frac{dy}{dt}\right).$$

The corresponding force in relativity theory will have for its last three components these quantities multiplied by $dt/d\tau$. So if the relativist force is (F_k), we have

$$F_1 = \frac{e}{c}\left(d_x \frac{c\,dt}{d\tau} + h_z\frac{dy}{d\tau} - h_y\frac{dz}{d\tau}\right) = \frac{e}{c}\left(X_{10}V^0 + X_{12}V^2 + X_{13}V^3\right)$$

where
$$(V^q) = \left(c\frac{dt}{d\tau}, \frac{dx}{d\tau}, \frac{dy}{d\tau}, \frac{dz}{d\tau}\right)$$

is the contravariant vector representing the relativist velocity of the particle. This may be written

$$F_1 = \frac{e}{c}\sum_q X_{1q}V^q$$

and similarly we have

$$F_2 = \frac{e}{c}\sum_q X_{2q}V^q, \qquad F_3 = \frac{e}{c}\sum_q X_{3q}V^q.$$

These are the last three components of the covariant vector which is obtained by transvecting the electromagnetic six-vector X_{pq} with the particle's velocity V_q. Clearly the first component of the force must be the first component of this transvectant: and therefore *the relativist force on a particle of charge e and velocity* V_q *is*

$$F_k = \frac{e}{c}\sum_q X_{kp}V^q.$$

The fact that energy-density occurs as the component E_0^0 of Minkowski's energy-tensor, while energy occurs as the first component of his energy-momentum vector, leads naturally to an inquiry into the connection between these two tensors. This can be investigated as follows.

Consider a system occupying a finite volume and involving energy of any kind (e.g. electromagnetic energy, or stress-energy, or gravitational energy) for which we can define an energy tensor T_p^q such that T_0^0 is the energy-density, (T_0^1, T_0^2, T_0^3) is $(1/c)$ times the flux of the energy, (T_1^0, T_2^0, T_3^0) is $(-c)$ times the density of momentum, and $(T_1^1, T_2^2, T_3^3, T_2^3, T_3^2, T_1^3, T_3^1, T_2^1, T_1^2)$ are the components of flux of momentum, just as in the case of Minkowski's energy-tensor of the electromagnetic field *in vacuo*: and suppose that the following conditions are satisfied:

(i) the system is rigidly-connected, and is considered in the first place as being at rest:

(ii) its state does no vary with the time:

(iii) there is conservation of momentum, so

$$\frac{\partial T_r^1}{\partial x} + \frac{\partial T_r^2}{\partial y} + \frac{\partial T_r^3}{\partial z} = 0 \quad (r = 1, 2, 3) \tag{1}$$

(iv) there is no flux of energy in the state of rest, so

$$T_r^0 = 0, \qquad T_0^r = 0 \quad (r = 1, 2, 3). \tag{2}$$

From (1) we have

$$\iiint T_1^1 dx\, dy\, dz = \iiint \left\{ \frac{\partial}{\partial x}(xT_1^1) + \frac{\partial}{\partial y}(xT_1^2) + \frac{\partial}{\partial z}(xT_1^3) \right\} dx\, dy\, dz$$

where the integration is taken over the whole volume occupied by the system and therefore

$$\iiint T_1^1 dx\, dy\, dz = \iint x(lT_1^1 + mT_1^2 + nT_1^3)\, dS$$

where the last integral is taken over a surface S enclosing the whole system, and (l, m, n) are the direction-cosines of the outward-drawn normal to S.

If we suppose the surface S so large that it includes the whole of the space in which there are any sensible effects due to the system, then T_p^q is zero on S, and therefore the last integral vanishes: so we have

$$\iiint T_1^1 dx\, dy\, dz = 0. \tag{3}$$

Now suppose that (t, x, y, z) is the frame of reference relative to which the system is at rest, and let $(\bar{t}, \bar{x}, \bar{y}, \bar{z})$ be a frame of reference such that relative to it, the system is in motion parallel to the axis of $\bar{x}$ with velocity $w=c \tanh a$. The axes of $\bar{x}, \bar{y}, \bar{z}$, are taken to be parallel to the axes of x, y, z respectively. Then the two sets of co-ordinates are connected by the equations

$$\begin{aligned} t &= \bar{t} \cosh a - (\bar{x}/c) \sinh a \\ x &= \bar{x} \cosh a - c\bar{t} \sinh a \\ y &= \bar{y}, \qquad z = \bar{z}, \end{aligned}$$

and the equations of transformation of the mixed tensor $T_p{}^q$ give

$$\begin{aligned} \bar{T}_0{}^0 &= \cosh^2 a\, T_0{}^0 - c \sinh a \cosh a\, T_1{}^0 + \frac{1}{c} \cosh a \sinh a\, T_0{}^1 - \sinh^2 a\, T_1{}^1 \\ &= \cosh^2 a\, T_0{}^0 - \sinh^2 a\, T_1{}^1 \end{aligned}$$

by (2). Thus

$$\begin{aligned} \iiint \bar{T}_0{}^0\, d\bar{x}\, d\bar{y}\, d\bar{z} &= \iiint (\cosh^2 a\, T_0{}^0 - \sinh^2 a\, T_1{}^1) d\bar{x}\, d\bar{y}\, d\bar{z} \\ &= \iiint (\cosh^2 a\, T_0{}^0 - \sinh^2 a\, T_1{}^1) \operatorname{sech} a\, dx\, dy\, dz, \end{aligned}$$

since $\partial(x, y, z)/\partial(\bar{x}, \bar{y}, \bar{z}) = \cosh a$, it being understood that $\bar{x}, \bar{y}, \bar{z}$ are measured over the field at a constant value of $\bar{t}$. So by (3),

$$\iiint \bar{T}_0{}^0\, d\bar{x}\, d\bar{y}\, d\bar{z} = \cosh a \iiint T_0{}^0\, dx\, dy\, dz.$$

Now let $U = \iiint T_0{}^0\, dx\, dy\, dz$, so U represents the total energy of the system when at rest. Then since $\cosh a = (1 - w^2/c^2)^{-\frac{1}{2}}$, the result now becomes :

Total energy associated with the moving system $= U\,(1 - w^2/c^2)^{-\frac{1}{2}}$.

This may be regarded as an extension, to systems of finite size, of the formula that in relativity theory the energy of a particle of proper-mass m, moving with velocity w, is

$$\frac{mc^2}{\sqrt{\left(1 - \frac{w^2}{c^2}\right)}}.$$

Now consider the momentum. The equation of transformation of a mixed tensor of rank 2 gives

$$\bar{T}_1{}^0 = -\frac{1}{c} \sinh a \cosh a\, T_0{}^0 + \frac{1}{c} \cosh a \sinh a\, T_1{}^1,$$

since T_0^1 and T_1^0 are zero. Therefore

$$\iiint \overline{T}_1^0 \, d\bar{x} \, d\bar{y} \, d\bar{z} = -\frac{1}{c} \sinh \alpha \cosh \alpha \iiint (T_0^0 - T_1^1) \, d\bar{x} \, d\bar{y} \, d\bar{z}$$

$$= -\frac{1}{c} \sinh \alpha \iiint (T_0^0 - T_1^1) \, dx \, dy \, dz$$

$$= -\frac{1}{c} \sinh \alpha \iiint T_0^0 \, dx \, dy \, dz, \quad \text{by (3)}$$

$$= - \frac{wU}{c^2 \sqrt{\left(1 - \frac{w^2}{c^2}\right)}}.$$

Now $-\overline{T}_1^0$ represents the density of $\bar{x}$-momentum. Therefore the total momentum of the moving system parallel to the x-axis is

$$\frac{w}{\sqrt{\left(1 - \frac{w^2}{c^2}\right)}} \frac{U}{c^2}.$$

This may be regarded as an extension, to systems of finite size, of the formula for the x-component of momentum of a particle.

The above analysis shows how the components of the energy-momentum vector (now no longer restricted by the condition that it is to apply only to a single particle) can be derived from those of the energy-tensor. It is evident that whereas the energy-tensor is *localised* (i.e. each of its components is a function of position in space), the *energy-momentum vector is not localised.*[1]

Before the discovery of relativity theory, physicists were accustomed to think of energy not as a component of a tensor, but as a scalar: and indeed even in relativity theory, energy *as observed by a particular observer* is a scalar. For let an observer be moving along the axis of x with velocity v, so that the covariant vector representing his velocity, namely,

$$\left(c \frac{dt}{d\tau}, -\frac{dx}{d\tau}, -\frac{dy}{d\tau}, -\frac{dz}{d\tau}\right) \quad \text{or} \quad (\xi_0, \xi_1, \xi_2, \xi_3)$$

is given by

$$\xi_0 = c\left(1 - \frac{v^2}{c^2}\right)^{-\frac{1}{2}}, \quad \xi_1 = -v\left(1 - \frac{v^2}{c^2}\right)^{-\frac{1}{2}}, \quad \xi_2 = 0, \quad \xi_3 = 0.$$

[1] On the relation of the energy-momentum vector of a particle to the energy-tensor of a continuous field, see further H. P. Robertson, *Proc. Edin. Math. Soc.*(2) v (1937), p. 63, and M. Mathisson, *Proc. Camb. Phil. Soc.* xxxvi (1940), p. 331.

Let a particle of proper-mass m be moving in the same straight line with velocity w, so that its contravariant energy-momentum vector is

$$\eta^0 = mc\left(1-\frac{w^2}{c^2}\right)^{-\frac{1}{2}}, \quad \eta^1 = mw\left(1-\frac{w^2}{c^2}\right)^{-\frac{1}{2}}, \quad \eta^2 = 0, \quad \eta^3 = 0.$$

The transvectant of these vectors, namely, $\xi_0\eta^0+\xi_1\eta^1+\xi_2\eta^2+\xi_3\eta^3$, is a scalar : its value is

$$m(c^2 - vw)\left(1-\frac{v^2}{c^2}\right)^{-\frac{1}{2}}\left(1-\frac{w^2}{c^2}\right)^{-\frac{1}{2}}. \tag{A}$$

Now the relative velocity of the particle and the observer is, by the relativist formula,

$$\frac{v-w}{1-\dfrac{vw}{c^2}}$$

and the energy of a particle moving with this velocity relative to the axes of reference is

$$mc^2\left\{1-\frac{(v-w)^2}{c^2(1-vw/c^2)^2}\right\}^{-\frac{1}{2}}$$

which reduces at once to the expression (A). Thus we see that *the energy of an observed particle may properly be regarded as a scalar, being the transvectant of the particle's energy-momentum vector and the observer's velocity.*

A vector which is of importance in electromagnetic theory may be introduced in the following way. We have seen that the electric intensity (d_x, d_y, d_z) and the magnetic intensity (h_x, h_y, h_z), at a point in free aether, are parts of a six-vector

$$d_x = X^{01}, \quad d_y = X^{02}, \quad d_z = X^{03}, \quad h_x = X^{23}, \quad h_y = X^{31}, \quad h_z = X^{12}.$$

Now if ϕ is the electric potential, and (a_x, a_y, a_z) the vector-potential, we have

$$d_x = -\frac{\partial\phi}{\partial x} - \frac{\partial a_x}{c\partial t}, \qquad h_x = \frac{\partial a_z}{\partial y} - \frac{\partial a_y}{\partial z}$$

and four similar equations. The question therefore suggests itself, what is the character of the potential ϕ, a_x, a_y, a_z, from the point of view of the tensor-calculus ? The answer, which is easily verified by examining the effects of Lorentz transformations, is that *if*

$$\phi_0 = \phi, \qquad \phi_1 = -a_x, \qquad \phi_2 = -a_y, \qquad \phi_3 = -a_z,$$

then $(\phi_0, \phi_1, \phi_2, \phi_3)$ *is a covariant vector*: its connection with the six-vector is given by the equations

$$X_{pq} = \frac{\partial \phi_p}{\partial x^q} - \frac{\partial \phi_q}{\partial x^p}.$$

It was discovered in 1915 by D. Hilbert[1] that the energy-tensor of a system can be expressed in terms of the Lagrangean function of the system. This theorem was developed further by E. Schrödinger[2] and H. Bateman[3] in 1927: the rule was given by Schrödinger as follows:

Let (a_0, a_1, a_2, a_3) *be one of the four-vectors on which the Lagrangean* L *depends* (as e.g. the Lagrangean in electromagnetic theory depends on the electromagnetic potential-vector), *and let* a_{pq} *denote the derivative of* a_p *with respect to the co-ordinate* x_q: *then the components of the energy-tensor are given by*

$$E_p{}^q = \sum \left(\sum_{t=0}^{3} a_{tp} \frac{\partial L}{\partial a_{tq}} + \sum_{t=0}^{3} a_{pt} \frac{\partial L}{\partial a_{qt}} + a_p \frac{\partial L}{\partial a_q} \right) - \delta_p{}^q L,$$

where $\delta_p{}^q = 0$ *or* 1 *according as q is, or is not, different from p, and the summation is taken over all the four-vectors a.*

For example, consider the electromagnetic field in free aether, for which the Lagrangean function is

$$L = \frac{1}{8\pi} \left(d_x^2 + d_y^2 + d_z^2 - h_x^2 - h_y^2 - h_z^2 \right),$$

or, if (a_0, a_1, a_2, a_3) denotes the covariant electromagnetic potential vector,

$$L = \frac{1}{8\pi} \left[\begin{array}{l} \left(-\frac{\partial a_0}{\partial x_1} + \frac{\partial a_1}{\partial x_0}\right)^2 + \left(-\frac{\partial a_0}{\partial x_2} + \frac{\partial a_2}{\partial x_0}\right)^2 + \left(-\frac{\partial a_0}{\partial x_3} + \frac{\partial a_3}{\partial x_0}\right)^2 \\ -\left(-\frac{\partial a_3}{\partial x_2} + \frac{\partial a_2}{\partial x_3}\right)^2 - \left(-\frac{\partial a_1}{\partial x_3} + \frac{\partial a_3}{\partial x_1}\right)^2 - \left(-\frac{\partial a_2}{\partial x_1} + \frac{\partial a_1}{\partial x_2}\right)^2 \end{array} \right];$$

from this we have

$$\frac{\partial L}{\partial a_{kn}} = \frac{1}{4\pi} X^{nk}$$

[1] *Gött. Nach.* (1915), p. 395; cf. also F. Klein, *Gött. Nach.* (1917), p. 469 and Hilbert, *Math. Ann.* xcii (1924), p. 1

[2] *Ann. d. Phys.* lxxxii (1927), p. 265

[3] *Proc. Nat. Acad. Sci.* xiii (1927), p. 326

Thus Schrödinger's formula gives

$$E_p{}^q = \frac{1}{4\pi}\sum_{t=0}^{3}\frac{\partial a_t}{\partial x_p}X^{qt} + \frac{1}{4\pi}\sum_{t=0}^{3}\frac{\partial a_p}{\partial x_t}X^{tq} - \delta_p{}^q L$$

$$= -\frac{1}{4\pi}\sum_t\left(\frac{\partial a_p}{\partial x_t} - \frac{\partial a_t}{\partial x_p}\right)X^{qt} - \delta_p{}^q L$$

$$= -\frac{1}{4\pi}\sum_t X_{pt}X^{qt} + \frac{1}{16\pi}\delta_p{}^q\sum_{\alpha,\beta}X_{\alpha\beta}X^{\alpha\beta}$$

which is the usual formula for Minkowski's energy tensor.

Chapter III

THE BEGINNINGS OF QUANTUM THEORY

At the end of the nineteenth century the theory of radiation was in a most unsatisfactory state. For the energy per cm.^{3} of pure-temperature or black-body radiation, in the range of wave-lengths from λ to $\lambda+d\lambda$, two different formulae had been proposed. Firstly, that of Wien,[1]

$$E=C\lambda^{-5}e^{-b/\lambda T}d\lambda$$

where λ is wave-length, T is absolute temperature, and b and C are constants. This formula is asymptotically correct in the region of short waves (more precisely, when λT is small); but, as O. Lummer and E. Pringsheim showed,[2] is irreconcilable with the observational results for long waves. Secondly, that of Rayleigh and Jeans [3]

$$E=8\pi kT\lambda^{-4}d\lambda$$

where k is Boltzmann's constant; which, as shown by the experiments of Rubens and Kurlbaum,[4] is asymptotically correct for the long waves, but is inapplicable at the other end of the spectrum. What was wanted was a formula which for the extreme limits $\lambda\to 0$ and $\lambda\to\infty$ would tend asymptotically to Wien's and Rayleigh's formulae respectively, and which would agree with the experimental values over the whole range of wave-lengths.

In the spring and summer of 1900 attempts were made to construct such a formula empirically by M. Thiesen,[5] by O. Lummer and E. Jahnke,[6] and by O. Lummer and E. Pringsheim.[7] These formulae were of the type

$$E=CT^{5-\mu}\lambda^{-\mu}e^{-b/(\lambda T)^{\nu}}$$

which for $\mu=5$, $\nu=1$, gives Wien's law, and for $\mu=4$, $\nu=1$, $b=0$, gives Rayleigh's.

The correct law was first given by Max Karl Ernst Ludwig Planck (1858–1947) in a communication which was read on 19 October 1900 before the German Physical Society.[8] Planck was the son of

[1] Vol. I, p. 381
[2] *Verh. d. deutsch phys. Ges.* i (1899), p. 215; ii (1900), p. 163
[3] Vol. I, p. 384
[4] loc. cit.
[5] *Verh. d. deutsch. phys. Ges.* ii (1900), p. 65
[6] *Ann. d. Phys.* iii (1900), p. 283
[7] *Verh. d. deutsch. phys. Ges.* ii (1900), p. 163
[8] *Verh. d. deutsch. phys. Ges.* ii (1900), p. 202

a professor of law at Kiel, later translated to Munich; he was educated at the University of Munich, but for one year attended the lectures of Helmholtz and Kirchhoff at Berlin. After four years as *professor extraordinarius* at Kiel, he was called in 1889 to succeed Kirchhoff at Berlin, where the rest of his academic life was spent.

In the study of pure-temperature radiation, his starting-point was the known fact that in a hollow chamber at a given temperature, the distribution of radiant energy among wave-lengths is altogether independent of the material of which the chamber is composed; and he was therefore free to suppose the walls of the chamber to have any constitution which was convenient for the calculations, so long as they were capable of absorbing and emitting radiation, and thereby making possible the exchange of energy between matter and aether. He chose them to be of the simplest type imaginable, namely an aggregate of Hertzian vibrators,[1] each with one proper frequency. Each vibrator absorbs energy from any surrounding radiation which is nearly of its own proper frequency, and acts as a resonator, emitting radiant energy.

He first calculated (by classical electrodynamics) the average absorption and emission of a vibrator of frequency[2] ν which is immersed in, and statistically in equilibrium with, a field of radiation, and found that if the average energy-density of the radiation, in the interval of frequency ν to $\nu + d\nu$, is E, then [3]

$$E = \frac{8\pi\nu^2}{c^3} U d\nu, \tag{1}$$

where U is the average energy of the vibrator.

While most of the other workers on radiation were attempting to find the relations between energy, wave-length and temperature, by direct methods, Planck, who was a master of thermodynamics, felt that the concept of entropy must play a fundamental part: and he examined the relation between the energy of a vibrator and its entropy S, showing that if S is known as a function of U, then the law of distribution of energy in the spectrum of pure-temperature radiation can be determined.

We have, from thermodynamics, for a system of constant volume,

$$dS = \frac{dU}{T} \quad \text{or} \quad \frac{dS}{dU} = \frac{1}{T}, \tag{2}$$

while Wien's law of radiation, namely

$$E = a\nu^3 e^{-\beta\nu/T} d\nu$$

[1] This, of course, does not mean (as it has sometimes been wrongly interpreted to mean) that actual matter necessarily has this character.

[2] It will be remembered that ν is the number of oscillations in one second, that is, c multiplied by the wave-number $1/\lambda$.

[3] *Ann. d. Phys.* i (1900), p. 69, equation (34): *Phys. ZS.* ii (1901), p. 530

requires by (1) that we should have

$$U = \gamma \nu e^{-\beta\nu/T}$$

where γ is a constant: so by (2)

$$\frac{dS}{dU} = -\frac{1}{\beta\nu}\log\frac{U}{\gamma\nu}$$

whence

$$\frac{d^2S}{dU^2} = \frac{\text{Constant}}{U}. \tag{3}$$

Planck had earlier attempted [1] to give a proof of Wien's law of radiation based on this equation (3), which he obtained independently by thermodynamical reasoning: but when confronted by Lummer and Pringsheim's experimental results he realised that Wien's could not be the true law of radiation; and he now proposed to modify (3), which he did by writing

$$\frac{d^2S}{dU^2} = \frac{\alpha}{U(\beta + U)} \tag{4}$$

where α and β are constants. This is the simplest of all the expressions which give dS/dU as a logarithmic function of U (as suggested by the probability theory of entropy), and which for small values of U agrees with equation (3). Moreover, if Rayleigh's law of radiation had been taken instead of Wien's, we should have obtained

$$\frac{d^2S}{dU^2} = \frac{\text{Constant}}{U^2},$$

which again is a case of (4). From (4) we have, by (2),

$$\frac{1}{T} = \frac{dS}{dU} = \text{Const. } \log\left(\frac{\text{Const.} + U}{U}\right)$$

or

$$U = \frac{\text{Const.}}{e^{\text{Const.}/T} - 1}. \tag{5}$$

This equation does not give the way in which the frequency ν enters into the formula for U. But as Wien had shown in 1893,[2] E must be of the form $T^5\phi(T\lambda)d\lambda$, or $\nu^3\psi(\nu/T)d\nu$, so by (1), U must be of the form

$$U = \nu\psi\left(\frac{\nu}{T}\right). \tag{6}$$

[1] *Berlin Sitz.* xxv (1899), p. 440

[2] cf. Vol. I, p. 380

Thus equation (5) must have the form

$$U = \text{Const.} \frac{\nu}{e^{l\nu/T} - 1}$$

and therefore by (1) the average energy-density of the radiation in the frequency-range ν to $\nu + d\nu$ is

$$E = \frac{g\nu^3 d\nu}{e^{l\nu/T} - 1} \tag{7}$$

where g and l are constants. *This is Planck's formula* which agreed with the experimental determinations of Lummer and Pringsheim, and also of H. Rubens and F. Kurlbaum,[1] and F. Paschen,[2] so well that it soon displaced all other suggested laws of radiation.

It was, however, as yet hardly more than an empirical formula, since equation (4) had no complete theoretical justification. This defect was remedied on 14 December of the same year (1900), when Planck read to the German Physical Society a paper [3] which placed his new law on a sound foundation, and in so doing created a new branch of physics, the quantum theory.

He considered a system consisting of a large number of simple Hertzian vibrators, in a hollow chamber enclosed by reflecting walls: let N of the vibrators have the frequency ν, N′ the frequency ν' and so on. Suppose that an amount A of energy is in the vibrators of frequency ν. Planck assumed that this energy is constituted of equal discrete elements, each of amount ϵ, and that there are altogether P such elements in the N vibrators, so that

$$A = P\epsilon.$$

Thus he assumed that the emission and absorption of radiation by these vibrators takes place not continuously, but by jumps of amount ϵ.

Any distribution of these P elements among the N vibrators may be called a *complexion.* The number of possible complexions is the number of possible ways of distributing P objects among N containers, when we do not take account of which particular objects lie in particular containers, but only of the number contained in each. This number is, by the ordinary theory of permutations and combinations,

$$\frac{(N + P - 1)!}{(N - 1)!\, P!}.$$

[1] *Berlin Sitz.*, 25 Oct. 1900, p. 929

[2] *Ann. d. Phys.* iv (1901), p. 277

[3] *Verh. deutsch. phys. Ges*, ii (1900), p. 237. This and the paper of 19 October were re-edited and printed in the new form in *Ann. d. Phys.* iv (1901), p. 553.

As N and P are very large numbers, we can use Stirling's approximate value for the factorials, namely,

$$\log \{(z-1)!\} = (z-\tfrac{1}{2}) \log z - z + \tfrac{1}{2} \log (2\pi)$$

so the number of complexions is approximately

$$\left\{\frac{N}{2\pi P(N+P)}\right\}^{\frac{1}{2}} \frac{(N+P)^{N+P}}{N^N\, P^P}.$$

We assume that all complexions have equal probability, so the probability W of any state of the system of N vibrators is proportional to the number of complexions corresponding to it; that is, with sufficient approximation for our present purpose, we have

$$\log W = (N+P) \log (N+P) - N \log N - P \log P.$$

Now the entropy in any state of a system depends on the inequality of the distribution of the total energy among the individual members of the system: and Boltzmann had shown by his work on the kinetic theory of gases[1] that the entropy S_N in any state of a system such as these vibrators is closely connected with the probability W of the state. Planck developed this discovery into the equation

$$S_N = k \log W \qquad (8)$$

where the *thermodynamic probability* W is always an integer, and k denotes the gas-constant for one molecule, or Boltzmann constant.[2] Thus

$$S_N = k\{(N+P) \log (N+P) - N \log N - P \log P\}.$$

Now $P = NU/\epsilon$, where U is the average, taken over the N oscillators, of the energy of one of them. Thus, retaining only the most important terms, and ignoring terms which do not involve U, we have

$$S_N = kN\left\{\left(1+\frac{U}{\epsilon}\right) \log \left(1+\frac{U}{\epsilon}\right) - \frac{U}{\epsilon} \log \frac{U}{\epsilon}\right\}$$

[1] cf. L. Boltzmann, *Vorlesungen über Gastheorie*, i (1896), § 6. This is essentially Boltzmann's 'H-theorem.'

[2] cf. Vol. I, p. 382. With Boltzmann the factor k did not occur, since his calculations referred not to individual molecules but to gramme-molecules, and with him the entropy was undetermined as regards an additive constant (i.e. there was an undetermined factor of proportionality in the probability W), whereas with Planck the entropy had a definite absolute value. This was a step of fundamental importance, and, as we shall see, led directly to the hypothesis of 'quanta.' The occurrence of the logarithm in the formula is explained by the circumstance that in compound systems a multiplication of probabilities corresponds to an addition of entropies.

so the entropy of a single oscillator of the set is

$$S = \frac{S_N}{N} = k\left\{\left(1+\frac{U}{\epsilon}\right)\log\left(1+\frac{U}{\epsilon}\right) - \frac{U}{\epsilon}\log\frac{U}{\epsilon}\right\}.$$

Thus from equation (2) above,

$$\frac{1}{T} = \frac{dS}{dU} = \frac{k}{\epsilon}\log\frac{\epsilon+U}{U}$$

or

$$U = \frac{\epsilon}{e^{\epsilon/kT}-1}. \qquad (9)$$

But by equation (5) U must be of the form $\nu\psi(\nu/T)$. This condition can be satisfied only if

$$\epsilon = h\nu \qquad (10)$$

where h is a constant independent of ν. Thus *the average energy of any simple-harmonic Hertzian vibrator of frequency* ν *must be an integral multiple of* $h\nu$, *and the smallest amount of energy that can be emitted or absorbed by it is* $h\nu$.

From (9) and (10) we have

$$U = \frac{h\nu}{e^{h\nu/kT}-1}$$

and therefore by (1) *the average energy-density of black-body radiation in the interval of frequency between* ν *and* $\nu + d\nu$ *is*

$$E = \frac{8\pi h}{c^3}\frac{\nu^3 d\nu}{e^{h\nu/kT}-1} \quad \text{or} \quad E = \frac{8\pi ch\lambda^{-5}d\lambda}{e^{ch/k\lambda T}-1}, \qquad (11)$$

which is *Planck's formula.* This agrees with his earlier result (7), but the constants which were unknown in (7) are now replaced by h and k, which are important constants of nature and appear in many other connections.

When $\nu \to 0$, the formula gives

$$E = \frac{8\pi\nu^2}{c^3}kTd\nu, \quad \text{or} \quad E = 8\pi k\lambda^{-4}Td\lambda$$

which is Rayleigh's law ; and when $\nu \to \infty$ it gives

$$E = \frac{8\pi h}{c^3}\nu^3 e^{-h\nu/kT}d\nu, \quad \text{or} \quad E = 8\pi ch\lambda^{-5}\, e^{-hc/k\lambda T}d\lambda$$

which is Wien's law, now expressed in terms of the constants h and k.

To obtain Wien's displacement law,[1] we proceed as follows: Let λ_m denote the wave-length corresponding to the maximum ordinate of a graph in which energy-density of radiation is plotted against wave-length. Then by Planck's formula (11), λ_m is the value of λ given by

$$0 = \frac{\partial}{\partial \lambda} \frac{\lambda^{-5}}{e^{ch/k\lambda T} - 1},$$

or

$$0 = -5 + \frac{ch/k\lambda T}{1 - e^{ch/k\lambda T}}.$$

Let q be the root of the equation

$$\frac{x}{1 - e^{-x}} = 5$$

so

$$q = 4{\cdot}965114 \ldots.$$

Then

$$\frac{hc}{k\lambda_m T} = q,$$

or

$$\lambda_m T = \frac{(hc/k)}{q} = \frac{1{\cdot}4384}{4{\cdot}965114} \text{cm. degree} = 0.28971 \text{ cm. degree},$$

which is Wien's displacement law.

Planck determined the values of the constants h and k by comparing his formula (11) with the measurements of F. Kurlbaum[2] and O. Lummer and E. Pringsheim,[3] the results obtained being

$$h = 6{\cdot}55 \times 10^{-27} \text{ erg. sec.,} \qquad k = 1{\cdot}346 \times 10^{-16} \text{ ergs per degree.}$$

He used this determination of k in order to calculate the number of molecules in a gramme-molecule (Avogadro's number)[4]: from the equation

$$S = k \log W$$

we can calculate the entropy of one gramme-molecule of an ideal gas, and from this can derive thermodynamically the relation

$$p = \frac{kNT}{V}$$

where p denotes the pressure of the gas, V its volume, and N denotes Avogadro's number: this shows that if R is the absolute gas-constant, then

$$R = kN.$$

[1] cf. Vol. I, p. 380
[2] *Verh. d. deutsch phys. Ges.* ii (1900), p. 176
[3] *Ann. d. Phys.* lxv (1898), p. 759
[4] cf. pp. 8, 18

From the known values of R and k, Planck found

$$N = 6{\cdot}175 \times 10^{23};$$

this agreed satisfactorily with the value $6{\cdot}40 \times 10^{23}$ which had been given by O. E. Meyer.[1]

Moreover, the knowledge of N so obtained leads to a new method of finding the charge of an electron. For the charge which is carried in electrolysis by one gramme-ion, that is by N ions, was known, being at that time believed to be 9,658 electromagnetic units. Thus if e is the charge of an electron in electrostatic units, we have

$$Ne = 9{,}658 \times 3 \times 10^{10}$$

which gives

$$e = 4{\cdot}69 \times 10^{-10} \text{ e.s.u.}$$

J. J. Thomson had found $e = 6 \times 10^{-10}$ e.s.u. two years earlier[2]; Planck's value was actually much nearer to the later determinations, which gave approximately $4{\cdot}77 \times 10^{-10}$ e.s.u.

Planck's law made it possible to give a more accurate formulation of the Stefan-Boltzmann law[3] for the total radiation per second from unit surface of a black body at temperature T. For[4] the element of this radiation in the range of wave-lengths λ to $\lambda + d\lambda$ is, by Planck's law

$$\frac{2\pi hc^2\lambda^{-5}d\lambda}{e^{hc/k\lambda T} - 1}$$

so the total radiation for all wave-lengths is

$$2\pi hc^2\int_0^\infty \frac{\lambda^{-5}d\lambda}{e^{hc/k\lambda T} - 1}$$

or

$$\frac{2\pi h}{c^2}\int_0^\infty \frac{\nu^3 d\nu}{e^{h\nu/kT} - 1}.$$

Now if B_n is the nth Bernoullian number, we have

$$B_n = 4n\int_0^\infty \frac{t^{2n-1}dt}{e^{2\pi t} - 1},$$

[1] *Die Kinetische Theorie der Gase*, 2 Aufl. (1899), p. 337; Planck's actual result, *Ann. d. Phys.* IV (1901), p. 564, is that the number of oxygen molecules in 1 cm^3 at 760 mm. pressure and 15°C., is $2{\cdot}76.\ 10^{19}$.

[2] cf. Vol. I, pp. 364–5

[3] cf. Vol. I, p. 374

[4] Vol. I, p. 373

whence, remembering that $B_2 = 1/30$, we have

$$\int_0^\infty \frac{s^3 ds}{e^{2\pi ps} - 1} = \frac{1}{240p^4}.$$

Putting $p = h/2\pi kT$, we see that *the total radiation per second from unit surface of a black body at temperature* T *is*

$$\frac{2\pi^5 k^4}{15c^2h^3} T^4;$$

this is the precise expression of the Stefan-Boltzmann formula in terms of the universal constants c, h and k.

A deeper insight into the physical conceptions underlying Planck's law of radiation was furnished by a later proof of it.[1] As is well known, in the kinetic theory of gases, it is shown that the probability that for a particular molecule the x-component of velocity will lie between u and $u + du$, its y-component of velocity between v and $v + dv$ and its z-component between w and $w + dw$, is

$$\left(\frac{m}{2k\pi T}\right)^{\frac{3}{2}} e^{-U/kT} du\, dv\, dw$$

where m is the mass of the molecule, U its kinetic energy, k is Boltzmann's constant, and T the absolute temperature. This result was generalised by Josiah Willard Gibbs[2] (1839–1903) into the following theorem: if we consider a large number of similar dynamical systems (which for simplicity we shall suppose to be linear oscillators), which are in statistical equilibrium with a large reservoir of heat at temperature T, and if q is the co-ordinate in an oscillator (e.g. the elongation of a vibrating electron) and p the momentum (defined as $\partial L/\partial(\partial q/\partial t)$ where L is the kinetic potential), then the probability that for any particular oscillator the co-ordinate lies between q and $q + dq$ and the momentum lies between p and $p + dp$ is[3]

$$\frac{e^{-U/kT} dq\, dp}{\int e^{-U/kT} dq\, dp}$$

where U is the energy of the oscillator, and the integration is to be taken over all possible values of q and p.

The theorem corresponding to this in the quantum theory is that

[1] cf. Lorentz, *Phys. ZS.* xi (1910), p. 1234; F. Reiche, *Die Quantentheorie* (Berlin 1921), Note 48

[2] *Elementary Principles in Statistical Mechanics* (New York, 1902)

[3] This is Gibbs's *canonical distribution*

if the energy of an oscillator can take only the discrete set of values U_0, U_1, U_2, U_3, . . ., then the probability that the energy of a particular oscillator is U_s is [1]

$$e^{-U_s/kT} \Big/ \sum_{s=0}^{\infty} e^{-U_s/kT}.$$

Thus if $U_s = sh\nu$ for $s = 0, 1, 2, \ldots$, the probability is

$$e^{-sh\nu/kT}(1 - e^{-h\nu/kT}).$$

The *mean* energy of an oscillator is therefore

$$h\nu(1 - e^{-h\nu/kT}) \sum_{s=0}^{\infty} s e^{-sh\nu/kT},$$

which has the value

$$\frac{h\nu}{e^{h\nu/kT} - 1}.$$

This leads at once, as before, to Planck's formula that the energy-density of black-body radiation in the frequency-interval from ν to $\nu + d\nu$ is

$$E = \frac{8\pi h}{c^3} \frac{\nu^3 d\nu}{e^{h\nu/kT} - 1}.$$

Other derivations of the law, based on many different assumptions, were given by various writers.[2] Some of them will be discussed later.

The next important advance in quantum theory was made by Einstein,[3] in the same volume of the *Annalen der Physik* as his papers

[1] If to an energy-level U_s there belongs a number g_s of permissible states, then the level U_s is said to be *degenerate*, and g_s is called the *weight* of the state. Taking the possibility of degenerate states into account, the above formula should be written

$$\frac{g_s e^{-U_s/kT}}{\sum_{s=0}^{\infty} g_s e^{-U_s/kT}}.$$

[2] Special reference may be made to the following: J. Larmor, *Proc. R.S.*(A), lxxxiii (1909), p. 82; P. Debye, *Ann. d. Phys.* xxxiii (1910), p. 1427, completed by A. Rubinowicz, *Phys. ZS.* xviii (1917), p. 96; P. Franck, *Phys. ZS.* xiii (1912), p. 506; A. Einstein and O. Stern, *Ann. d. Phys.* xl (1913), p. 551; M. Wolfke, *Verh. d. deutsch. phys. Ges.* xv (1913), pp. 1123, 1215: *Phys. ZS.* xv (1914), pp. 308, 463; A. Einstein, *Phys. ZS.* xviii (1917), p. 121; C. G. Darwin and R. H. Fowler, *Phil. Mag.* xliv (1922), pp. 450, 823: *Proc. Camb. Phil. Soc* xxi (1922), p. 262; S. N. Bose, *ZS. f. P.* xxvi (1924), p. 178, xxvii (1924), p. 384; A. S. Eddington, *Phil. Mag.* l (1925), p. 803

[3] *Ann. d. Phys.* xvii (1905), p. 132: cf. also *Ann. d. Phys.* xx (1906), p. 199

on the Brownian motion[1] and Relativity.[2] Einstein supposed monochromatic radiation of frequency ν and of small density (within the range of values of ν/T for which Wien's formula of radiation is applicable) to be contained in a hollow chamber of volume v_0 with perfectly-reflecting walls, its total energy being E : and, investigating by use of Wien's formula the dependence of the entropy on the volume, he found for the difference of the entropies when the radiation occupies the volume v_0 and when it occupies a smaller volume v, the equation

$$S - S_0 = \frac{Ek}{h\nu} \log \frac{v}{v_0}.$$

Now by inverting the Boltzmann-Planck relation

$$\text{entropy} = k \times \text{logarithm of probability}$$

he calculated the relative probability from the difference of entropies, and found that the probability that at an arbitrarily-chosen instant of time, the whole of the energy of the radiation should be contained within a part v of the volume v_0, is

$$\left(\frac{v}{v_0}\right)^{\frac{E}{h\nu}}.$$

This formula he studied in the light of a known result in the kinetic theory of gases, namely that if a gas contained in a volume v_0 consists of n molecules, the probability that at an arbitrarily-chosen instant of time, all the n molecules should be collected together within a a part v of the volume, is

$$\left(\frac{v}{v_0}\right)^n.$$

Comparing these formulae, he inferred that the radiation behaves as if it consisted of $E/h\nu$ quanta of energy or *photons*,[3] each of amount $h\nu$. The probability that *all* the photons are found at an arbitrary instant in the part v of the volume v_0 is the product of the probabilities (v/v_0) that a single one of them is in the part v : which shows that they are completely independent of each other.

Now it will be remembered that according to Planck's theory, a vibrator of frequency ν can emit or absorb energy only in multiples of $h\nu$. Planck regarded the quantum property as belonging essentially to the interaction between radiation and matter : free radiation he supposed to consist of electromagnetic waves, in accordance with

[1] cf. p. 9 [2] cf. p. 40

[3] The word *photon* was actually introduced much later, namely, by G. N. Lewis, *Nature*, 18 Dec., 1926 ; but it is so convenient that we shall adopt it now.

Maxwell's theory. Einstein in this paper put forward the hypothesis that parcels of radiant energy of frequency ν and amount $h\nu$ occur not only in emission and absorption, but that they have an independent existence in the aether.

It was shown by P. Ehrenfest[1] of Leiden, by A. Joffé[2] of St Petersburg, by L. Natanson[3] of Cracow and by G. Krutkow[4] of Leiden that Einstein's hypothesis leads not to Planck's law of radiation but to Wien's, at any rate if we assume that each of the light-quanta or photons of frequency ν has energy $h\nu$ and that they are completely independent of each other. In order to obtain Planck's formula it is necessary to assume that the elementary photons of energy $h\nu$ form aggregates, or photo-molecules as we may call them, of energies $2h\nu$, $3h\nu$, . . ., respectively, and that the total energy of radiation is distributed, on the average, in a regular manner between the photons and the different kinds of photo-molecules. This will be discussed more fully later.

Einstein applied his ideas in order to construct a theory of photo-electricity.[5] As we have seen,[6] in 1899 J. J. Thomson and P. Lenard showed independently that the emission from a metal irradiated by ultra-violet light consists of negative electrons : and in 1902 Lenard,[7] continuing his researches, showed that the number of electrons liberated is proportional to the intensity of the incident light, so long as its frequency remains the same : and that the initial velocity of the electrons is altogether independent of the intensity of the light, but depends on its frequency.

Knowledge regarding photo-electricity had reached this stage when in 1905 Einstein's paper appeared. Considering a metal surface illumined by radiation of frequency ν, he asserted that the radiation consists of parcels of energy ; when one such parcel or photon falls on the metal, it may be absorbed and liberate a photo-electron : and that the maximum kinetic energy of the photo-electron at emission is $(h\nu - e\phi)$, where $e\phi$ is the energy lost by the electron in escaping from its original location to outside the surface. This of course implies that no photo-electrons will be generated unless the frequency of the light exceeds a certain 'threshold' value $e\phi/h$.

Einstein's equation was verified in 1912 by O. W. Richardson and K. T. Compton[8] and by A. L. Hughes,[9] and with great care in 1916 by R. A. Millikan.[10] For many metals, the threshold frequency is in the ultra-violet : but for the electro-positive metals, such as the alkali metals, it is in the visible spectrum : for sodium, it is in the green.

[1] *Ann. d. Phys.* xxxvi (1911), p. 91

[2] ibid, p. 534

[3] *Phys. ZS.* xii (1911), p. 659

[4] *Phys. ZS.* xv (1914), p. 133

[5] cf. Vol. I, pp. 356–7

[6] cf. Vol. I, p. 365

[7] *Ann. d. Phys.* viii (1902), p. 149; also E. R. Ladenburg (1878–1908), *Ann. d. Phys.* xii (1903), p. 558

[8] *Phil. Mag.* xxiv (1912), p. 575

[9] *Phil. Trans.* ccxii (1912), p. 205

[10] *Phys. Rev.* vii (1916), p. 355. cf. also M. de Broglie, *J. de Phys.* ii (1921), p. 265, and J. Thibaud, *Comptes Rendus*, clxxix (1924), pp. 165, 1053, 1322

The function ϕ is closely connected with the thermionic work-function measured at the same temperature:[1] in fact, the thermionic work-function is equal to h times the least frequency which will eject an electron from the metal:[2] and ϕ is therefore connected with the contact potential-differences between two metallic surfaces:[3] the difference of the functions ϕ for the two metals is equal to the contact difference of potential (reversing the order of the metals) together with the (small) coefficient of the Peltier effect at the junction between them.

Gases and vapours also exhibit the photo-electric effect, if the frequency of the incident radiation is sufficiently great: and the phenomenon can be observed for individual atoms by use of X-rays with the Wilson chamber. This effect is simply ionisation: and the law regarding the threshold frequency becomes the assertion that for ionisation to take place, the energy of the incident photon must be not less than the ionisation energy of the atom or molecule concerned. The electrons chiefly affected photo-electrically are the strongly-bound ones in the K-shell: the electrons in the outer shells, being more feebly-bound, do not absorb radiation to the same degree. The function ϕ in the equation

$$\text{maximum kinetic energy of electron at emission} = h\nu - e\phi$$

is now no longer connected with thermo-electric phenomena or contact differences of potential, but has different values depending on the shell in the atom from which the electron has come.

If a photo-electron is liberated from the K-shell, it may happen that the vacant place is filled by an electron from an outer shell, creating a photon whose energy is equal to the difference of the energies of the electron in the two shells: and this photon may in its turn be absorbed in another shell, giving rise to a second photo-electron, so that two electrons are ejected together. This effect, which was discovered by P. Auger,[4] is called the *compound photo-electric effect.*

The photo-electric effect cannot be explained classically, because the time-lag required by the classical theory, due to the necessity for accumulating sufficient energy from the radiation, is found not to occur.[5]

A hypothesis closely allied to Einstein's light-quantum explanation of the photo-electric effect was put forward in 1908 by J. Stark[6]: namely, that the frequency of the violet edge of the band-spectrum

[1] cf. Vol. I, pp. 426–8
[2] O. W. Richardson and K. T. Compton, *Phil. Mag.* xxiv (1912), p. 595
[3] O. W. Richardson, *Phil. Mag.* xxiii (1912), pp. 263, 594
[4] *Comptes Rendus*, clxxxii (1926), p. 1215
[5] On the time-lag, cf. E. Meyer and W. Gerlach, *Archives des sc. phys. et nat.* xxxvi (1914), p. 253.
[6] *Phys. ZS.* ix (1908), p. 85

of a gas is connected with the ionisation-potential of the gas (measured by the potential-fall necessary to give sufficient kinetic energy to the ionising electron) by the formula

$$I = h\nu.$$

Experiments in agreement with this relation were published in the following year by W. Steubing.[1]

Another hypothesis of the same type, also proposed by Stark[2] in 1908, and elaborated by Einstein[3] in 1912, related to photochemical decomposition: it asserted that when a molecule is dissociated as a result of absorbing radiation of frequency ν, the amount of energy absorbed by the molecule is $h\nu$. There must, therefore, be a lower limit to the frequency of light capable of producing a given chemical reaction, and a relation between the amount of reaction and the amount of light absorbed. The law is applicable only within the range of validity of Wien's law and when the decomposition is purely a thermal effect. Experiments designed to test this hypothesis were made by E. Warburg,[4] with results on the whole favourable.

From Einstein's doctrine that the energy of a photon of frequency ν is $h\nu$, combined with Planck's principle[5] that flux of energy is momentum, it follows at once that in free aether, where the velocity of the photon is c, its momentum[6] must be $h\nu/c$, in the direction of propagation of the light. Long afterwards it was shown experimentally by R. Frisch[7] that when an atom absorbs or emits a photon, the atom experiences a change of momentum of the magnitude and direction attributed to the photon by Einstein: but there had never been any doubt about the matter, since Einstein's value was assumed in the theory of many phenomena, and predictions based on the theory were experimentally verified.

It may be remarked that the above relation between the energy and the momentum of a photon is in agreement with the classical electromagnetic theory of light: for if a beam of light is propagated in free aether in a certain direction, the electric vector **E** and the magnetic vector **H** are equal, and at right angles to each other and to the direction of propagation; and therefore Kelvin's energy-density[8] $1/8\pi\ (E^2+H^2)$ is $E^2/4\pi$, while J. J. Thomson's momentum-density[9] $1/4\pi c\ [\mathbf{E}.\mathbf{H}]$ is $E^2/4\pi c$: and the latter is equal to the former divided by c.

[1] *Phys. ZS.* x (1909), p. 787
[2] *Phys. ZS.* ix (1908), p. 889: *Ann. d. Phys.* xxxviii (1912), p. 467; cf. E. Warburg, *Verh. d. deutsch. phys. Ges.* ix (1907) p. 753
[3] *Ann. d. Phys.* xxxvii (1912), p. 832: xxxviii (1912), p. 881
[4] *Berlin Sitz.*, 1911, p. 746: 1912, p. 216: 1913, p. 644: 1914, p. 872: 1915, p. 230: 1916, p. 314: 1918, pp. 300, 1228
[5] cf. p. 54
[6] cf. A. Einstein, *Phys. ZS.* x (1909), pp. 185, 817: J. Stark, ibid. pp. 579, 902
[7] *ZS. f. P.* lxxxvi (1933), p. 42
[8] cf. Vol. I, p. 222
[9] cf. Vol. I, p. 317

The formula for the momentum of a photon is also in agreement with the equation

$$\text{momentum} = \text{mass} \times \text{velocity}:$$

for since the energy is $h\nu$, the mass is $h\nu/c^2$; and the velocity is c; so the momentum is $h\nu/c$.

The corpuscular theory of light thus formulated by Einstein leads at once to the relativist formulation of the Doppler effect.[1] For suppose that a star is moving with velocity w relative to an observer P, and that a quantum of light emitted towards the observer P has a frequency ν' as measured by an observer P′ on the star, and a frequency ν as measured by the observer P.

Let the direction-cosines of the line PP′, referred to rectangular axes in P's system of measurement, of which the x-axis is in the direction of the velocity w, be (l, m, n). Then the energy and momentum of the light-quant as observed by P, that is, in a system of reference in which P is at rest, are $(h\nu, -h\nu l/c, -h\nu m/c, -h\nu n/c)$: and the energy of the light-quant as observed by P′, that is, in a system of reference in which P′ is at rest, is $h\nu'$. But the (energy)/c and the three components of momentum form a contravariant four-vector which transforms according to the Lorentz transformation, in which

$$t' = \frac{t - \dfrac{w}{c^2}x}{\left(1 - \dfrac{w^2}{c^2}\right)^{\frac{1}{2}}}$$

and therefore

$$\frac{h\nu'}{c^2} = \frac{\dfrac{h\nu}{c^2} + \dfrac{wh\nu l}{c^3}}{\left(1 - \dfrac{w^2}{c^2}\right)^{\frac{1}{2}}}$$

or

$$\frac{\nu'}{\nu} = \frac{1 + \dfrac{w}{c}l}{\left(1 - \dfrac{w^2}{c^2}\right)^{\frac{1}{2}}}$$

which is the relativist formula for the Doppler effect. Thus *the Doppler effect is simply the Lorentz transformation of the energy-momentum four-vector of the light-quant.*

In the earlier years of the development of quantum-theory, much attention was given to the relation of light-propagation to space,

[1] cf. p. 41

the aim being to find a conception of the mode of propagation which would account both for those experiments which were most naturally explained by the wave-theory, and also for those which seemed to require a corpuscular theory. J. Stark [1] and A. Einstein [2] discussed a fact which seemed very difficult to reconcile with the wave-theory, namely, that when cathode rays fall on a metal plate, and the X-rays there generated fall on a second metal plate, they generate cathode rays whose velocity is of the same order of magnitude as that of the primary cathode rays.

More precisely, let the X-rays be excited by a stream of electrons striking an anticathode (this is, of course, the process inverse to the photo-electric effect). Suppose that the energy of the electrons is what would be obtained by a fall through a potential-difference V, so that the kinetic energy of the electrons is eV. When the electrons are stopped, they give rise to the X-rays, whose frequencies form a continuous [3] spectrum with a limit ν_{max} on the side of high frequencies given by the equation [4]

$$h\nu_{max} = eV.$$

That is, X-rays of frequency ν are not produced unless energy $h\nu$ is available. It is reasonable to suppose that the maximum value of the frequency is obtained when the whole of the energy eV of the electron is converted into energy of radiation (X-radiation of lower frequency is also obtained, because the incident electron may spend part of its energy in causing changes in the atoms of the anticathode). Since the energy of an electron ejected by the X-ray in the photo-electric effect is (save for differences due to other circumstances which need not be considered at the moment) equal to $h\nu_{max}$, we see that it is equal to the energy of the electrons in the cathode rays which had originally excited the X-ray: so that no energy is lost in the changes from electron to X-ray and back to electron again. The X-ray must therefore carry its energy over its whole track in a compact bundle, without any diminution due to spreading: as had been asserted in 1910 by W. H. Bragg (cf. p. 17).

On the other hand, X-rays are certainly of the nature of ordinary light, and can be diffracted: so one would expect them to show the spreading characteristic of waves. The apparent contradiction between the wave-properties of radiation and some of its other properties had been considered by J. J. Thomson in his Silliman lectures of 1903 [5]; ' Röntgen rays,' he said, ' are able to pass very

[1] *Phys. ZS.* x (1909), pp. 579, 902 : xi (1910), pp. 24, 179

[2] *Phys. ZS.* x (1909), p. 817 ; cf. H. A. Lorentz, *Phys. ZS.* xi (1910), p. 1234

[3] Regarding the discontinuous spectrum of characteristic X-rays, cf. D. L. Webster, *Phys. Rev.* vii (1916), p. 599

[4] cf. W. Duane and F. L. Hunt, *Phys. Rev.* vi (1915), p. 166. The value of h was calculated on the basis of this property by F. C. Blake and W. Duane, *Phys. Rev.* ix (1917), p. 568 : x (1917), pp. 93, 624.

[5] J. J. Thomson, *Electricity and Matter* (1904), pp. 63–5, cf. his *Conduction of Electricity through Gases* (1903), p. 258

long distances through gases, and as they pass through the gas they ionise it: the number of molecules so split up is, however, an exceedingly small fraction, less than one-billionth, even for strong rays, of the number of molecules in the gas. Now, if the conditions in the front of the wave are uniform, all the molecules of the gas are exposed to the same conditions: how is it, then, that so small a proportion of them are split up?' His answer was: 'The difficulty in explaining the small ionisation is removed if, instead of supposing the front of the Röntgen ray to be uniform, we suppose that it consists of specks of great intensity separated by considerable intervals where the intensity is very small.'

In this passage Thomson originated the conception of *needle* radiation,[1] i.e. that *in the elementary process of light-emission, the radiations from a source are not distributed equally in all azimuths, but are concentrated in certain directions.* This hypothesis was now adopted by Einstein,[2] who, as we shall see, developed it further in 1916.

When, however, the phenomena of interference were taken into account, the conception of needle radiation, and indeed the whole quantum principle of radiation, met with difficulties which were not resolved for many years. It was shown experimentally[3] that when a classical interference-experiment was performed with light so faint that only a single photon was travelling through the apparatus at any one time, the interference-effects were still produced. This was interpreted at first to mean that a single photon obeys the laws of partial transmission and reflexion at a half-silvered mirror and of subsequent re-combination with the phase-difference required by the wave-theory of light. It is evident, however, that such an explanation would be irreconcilable with the fundamental principle of the quantum theory, according to which interaction between the light and matter at a particular point on the screen can take place only by the absorption or emission of whole quanta of light.

A further objection to the view that coherent beams of light, which are capable of yielding interference-phenomena, could be identical with single photons, appeared when it was found that the volume of a beam of light over which coherence can extend, was much greater than the nineteenth-century physicists had supposed. In 1902 O. Lummer and E. Gehrcke,[4] using green rays from a mercury lamp, obtained interference-phenomena with a phase-difference of 2,600,000 wave-lengths—a distance of the order of one

[1] cf. Thomson's further papers in *Proc. Camb. Phil. Soc.* xiv (1907), p. 417; *Phil. Mag.*(6) xix (1910), p. 301; and N. R. Campbell, *Proc. Camb. Phil. Soc.* xv (1910), p. 310. For an interesting application of the relativity equations to needle radiation, cf. H. Bateman, *Proc. Lond. Math. Soc.*(2) viii (1910), p. 469.

[2] loc. cit.

[3] G. I. Taylor, *Proc. Camb. Phil. Soc.* xv (1909), p. 114; most completely by A. J. Dempster and H. F. Batho, *Phys. Rev.* xxx (1927), p. 644; cf. E. H. Kennard, *J. Franklin Inst.*, ccvii (1929), p. 47

[4] *Verh. deutsch. phys. Ges.* iv (1902), p. 337; cf. M. von Laue, *Ann. d. Phys.*(4) xiii (1904), p. 163, §6

metre. Regarding the lateral extension of a coherent beam, since all the light from a star that enters a telescope-objective takes part in the formation of the image, it is evident that this light must be coherent : and a still greater estimate was obtained in 1920, when interference methods were used at Mount Wilson Observatory to determine the angular diameter of Betelgeuse, and interference was obtained between beams which arrived from the star twenty feet apart. It seemed impossible that these very large coherent beams could be single photons.

The alternative hypothesis seemed to be, that the photons in a coherent beam form a regular aggregate, possessing a quality equivalent to the coherence.[1] One supposition was that the motion of the photons is subject to a system of probability corresponding to the wave-theory explanation of interference, so that a large number of them is directed to the bright places of the interference-pattern, and few or none are directed to the dark places. This explanation was, however, unsatisfactory : for it is not until the two interfering beams of light have actually met that the interference-pattern is determined, and therefore the guiding of the motion of the photons cannot take place during the propagation of the interfering beams, but must happen later, perhaps at the screen itself—a process difficult to imagine. It was therefore suggested that while the photons are being propagated in the beams which are later destined to interfere, they are characterised by a quality corresponding to what in the wave-theory is called *phase*. In order to construct a definite theory based on this idea, however, it would be necessary first to consider whether the photons are to be regarded as points (in which case any particular photon would always retain the same phase, since it travels with the velocity of light, but different photons would have different phases) or whether the photons are to be regarded as extending over finite regions (in which case the phase would presumably vary from one point of a photon to another). In order to account for interference, it would be necessary to suppose that photons, or parts of photons, in opposite phases, neutralise each other. But if the photons are regarded as points, the mutual annulment of two complete photons would be incompatible with the principle of conservation of energy, while if the photons are regarded as extending over finite regions, the annulment of part of one by part of another would seem to be incompatible with the integral character of photons. There is, moreover, a difficulty created by the observation that interference can take place when only a single photon is travelling through the apparatus at any one instant, for this seems to require that the effect of a quantum on an atom persists for some time.

These various attempts—none of them entirely satisfactory—to combine the new and the old conceptions of light, created a doubt

[1] This was first put forward by J. Stark, loc. cit.

as to whether it was possible to construct, within the framework of space and time, a picture or model which would be capable of representing every known phenomenon in optics.[1]

In December 1906 Einstein[2] initiated a new development of quantum theory, by carrying its principles outside the domain of radiation, to which they had hitherto been confined, and applying them to the study of the specific heats of solids. We have seen[3] that according to the classical law of equipartition of energy, in a state of statistical equilibrium at absolute temperature T, with every degree of freedom of a dynamical system there is associated on the average a kinetic energy $\frac{1}{2}kT$, where k is Boltzmann's constant. Now the thermal motions of a crystal are constituted by the elastic vibrations of its atoms about their positions of equilibrium: one of these atoms, since it has three kinetic degrees of freedom and also three potential-energy degrees of freedom, will have a mean kinetic energy $\frac{3}{2}kT$ and a mean potential energy $\frac{3}{2}kT$, or a total mean energy $3kT$. Thus a gramme-atom of the crystal will have a mean energy $3kNT$, where N is Avogadro's number[4]; or 3RT, where R is the gas-constant per gramme-atom. The atomic heat of the crystal (i.e. the amount of heat required to raise the temperature of a gramme-atom by one degree) is therefore 3R. Now

$$R = 8{\cdot}3136 \times 10^7 \text{ ergs} = 1{\cdot}986 \text{ cal.}$$

so the atomic heat (at constant volume) is 5·958 cal. This law had been discovered empirically by P. L. Dulong and A. T. Petit[5] in 1819.

While Dulong and Petit's law is approximately true for a great many elements at ordinary temperatures, exceptions to it had long been known, particularly in the case of elements of low atomic weight, such as C, Bo, Si, for which at ordinary temperatures the atomic heats are much smaller than 5·958 cal.: and shortly before this time it had been shown, particularly by W. Nernst and his pupils, that at very low temperatures *all* bodies have small atomic heats, while at sufficiently high temperatures even the elements of low atomic weight obey the normal Dulong-Petit rule, as was shown e.g. by experiments with graphite at high temperatures.

[1] The state of the coherence problem twenty years after Einstein's paper of 1905 may be gathered from G. P. Thomson, *Proc. R.S.*(A), civ (1923), p. 115; A. Landé, *ZS. f. P.* xxxiii (1925), p. 571; E. C. Stoner, *Proc. Camb. Phil. Soc.* xxii (1925), p. 577; W. Gerlach and A. Landé, *ZS. f. P.* xxxvi (1926), p. 169; E. O. Lawrence and J. W. Beams, *Proc. N.A.S.* xiii (1927), p. 207.

[2] *Ann. d. Phys.* xxii (1907), pp. 180, 800

[3] Vol. I, p. 382. The argument given here is substantially due to Boltzmann, *Wien Sitz.* lxiii (Abth. 2) (1871), p. 712.

[4] cf. pp. 8, 18

[5] *Ann. de chim. et de. phys.* x (1819), p. 395; *Phil. Mag.* liv (1819), p. 267

Einstein now pointed out that if we write Planck's formula in the form :

Energy-density of radiation in the frequency-range ν to $\nu+d\nu$

$$=8\pi\lambda^{-4}d\lambda\frac{h\nu}{e^{h\nu/k\mathrm{T}}-1},$$

then on comparing this with Rayleigh's derivation of his law of radiation,[1] we see that in order to obtain Planck's formula, a mode of vibration of frequency ν must be counted as possessing the average energy

$$\tfrac{1}{2}k\mathrm{T}\frac{x}{e^x-1},$$

where $x=h\nu/k\mathrm{T}$, instead of (as Rayleigh assumed) $\frac{1}{2}k\mathrm{T}$. If then in the above proof of Dulong and Petit's law we replace $\frac{1}{2}k\mathrm{T}$ by $\frac{1}{2}k\mathrm{T}\ x/(e^x-1)$, we find that a gramme-atom of the crystal will have a mean energy

$$3\mathrm{RT}\frac{x}{e^x-1}$$

(if for simplicity we assume that all the atomic vibrators have the same frequency ν) and therefore the atomic heat is

$$\frac{d}{d\mathrm{T}}\cdot\left\{3\mathrm{RT}\frac{x}{e^x-1}\right\},\quad\text{or}\quad 3\mathrm{R}\frac{x^2e^x}{(e^x-1)^2},\quad\text{or}\quad 5{\cdot}958\,\frac{x^2e^x}{(e^x-1)^2}.$$

This is Einstein's formula. As the temperature falls, x increases and $x^2e^x/(e^x-1)^2$ decreases, so the decrease of atomic heat with temperature is accounted for. As the absolute zero of temperature is approached, the atomic heats of all solid bodies tend to zero.[2]

The determination of the quantity x, i.e. the determination of ν, was studied by E. Madelung,[3] W. Sutherland,[4] F. A. Lindemann,[5] A. Einstein,[6] W. Nernst,[7] E. Grüneisen,[8] C. E. Blom[9] and H. S. Allen.[10]: relations were found connecting ν approximately with the cubical compressibility of the crystal, with its melting-point and with the 'residual rays' which are strongly reflected from it.

[1] Vol. I, pp. 383–4

[2] This is a special case of a more general theorem discovered and developed by W. Nernst in 1910 and the following years, namely, that all the properties of solids which depend on the average behaviour of the atoms (including the thermodynamic functions) become independent of the temperature at very low temperatures. Thus, at the absolute zero of temperature, the entropy of every chemically homogeneous body is zero; cf. Nernst, *Die theoretischen und experimentellen Grundlagen des neuen Wärmesatzes* (Halle, 1918).

[3] *Gött. Nach.* (1909), p. 100; *Phys. ZS.* xi (1910), p. 898

[4] *Phil. Mag.*(6) xx (1910), p. 657

[5] *Phys. ZS.* xi (1910), p. 609

[6] *Ann. d. Phys.* xxxiv (1911), p. 170; xxxv (1911), p. 679

[7] *Ann. d. Phys.* xxxvi (1911), p. 395

[8] *Ann. d. Phys.* xxxix (1912), p. 257

[9] *Ann. d. Phys.* xlii (1913), p. 1397

[10] *Proc. R.S.*(A), xciv (1917), p. 100; *Phil. Mag.* xxxiv (1917), pp. 478, 488

There was, however, one obvious imperfection in all this work, which was pointed out by Einstein in the second of the papers just referred to; namely, that the vibrations of the atoms in a crystal do not all have the same frequency ν. The mean energy of a gramme-atom of a crystal will therefore not be

$$3\mathrm{RT}\frac{x}{e^x - 1}$$

where x has a single definite value, but

$$k\mathrm{T}\sum_r \frac{x_r}{e^{x_r} - 1},$$

when the summation is taken over all the frequencies (three for each atom), and k is Boltzmann's constant. The atomic heat, obtained from this by differentiating with respect to T (remembering that $x = h\nu/k\mathrm{T}$), is therefore

$$\sum_r \frac{k x_r^2 e^{x_r}}{(e^{x_r} - 1)^2}.$$

In order to determine the frequencies of the natural vibrations of the atoms of a body, and so to be able to evaluate these expressions, P. Debye[1] (*b.* 1884) took, as an approximation to the actual body, an elastic solid, and considered the elastic waves in it. He showed that for a fixed isotropic body of volume V, the number of natural periods or modes of vibration in the frequency-range ν to $\nu + d\nu$ is

$$4\pi \mathrm{V}\nu^2 d\nu\left(\frac{2}{c_t^3} + \frac{1}{c_l^3}\right)$$

where c_t is the velocity of transverse waves in the solid and c_l the velocity of longitudinal waves; and he assumed that the energies of these different sound waves vary in the same way as the energies of the light waves in Planck's formula, so that each of these modes has the energy

$$\frac{h\nu}{e^{h\nu/k\mathrm{T}} - 1};$$

thus the energy per unit volume of waves within the frequency-range ν to $\nu + d\nu$ is

$$4\pi h\nu^3\left(\frac{2}{c_t^3} + \frac{1}{c_t^3}\right)\frac{d\nu}{e^{h\nu/k\mathrm{T}} - 1}.$$

[1] *Ann. d. Phys.* xxxix (1912), p. 789

The total energy is to be obtained by integrating this with respect to ν. But here a difficulty presents itself: for the number of natural frequencies of a continuous body is infinitely great: ν extends from zero to infinity. Debye (somewhat arbitrarily) dealt with this situation by taking, in the integration with respect to ν, an upper limit ν_m, such that the total number of frequencies less than ν_m is equal to 3N, where N is the number of atoms in the body. Thus ν_m is to be determined from the equation

$$3N = \int_0^{\nu_m} 4\pi V\left(\frac{2}{c_t^3}+\frac{1}{c_l^3}\right)\nu^2 d\nu = \frac{4\pi V}{3}\left(\frac{2}{c_t^3}+\frac{1}{c_l^3}\right)\nu_m^3,$$

and the atomic heat of the body is

$$4\pi k V\left(\frac{2}{c_t^3}+\frac{1}{c_l^3}\right)\int_0^{\nu_m}\frac{x^2e^x}{(e^x-1)^2}\nu^2 d\nu \quad \text{where} \quad x=\frac{h\nu}{kT}$$

or

$$\frac{9R}{x_m^3}\int_0^{x_m}\frac{x^4e^x dx}{(e^x-1)^2}, \quad \text{where} \quad x_m=\frac{h\nu_m}{kT}.$$

If we write

$$\frac{h\nu_m}{k}=\Theta,$$

we have $x_m=\Theta/T$, and *the atomic heat is a universal function of the ratio* T/Θ, *that is, the temperature* T *divided by a temperature* Θ *which is characteristic of the body.*

Debye's theory is in good general accord with the experimental results for many elements.[1]

When the temperature T is very great, x_m is very small, and the above formula for the atomic heat becomes

$$\frac{9R}{x_m^3}\int_0^{x_m} x^2 dx \quad \text{or} \quad 3R,$$

which is Dulong and Petit's law, as would be expected.

When on the other hand T is very small, we escape the difficulties which arise from the fact that the body, as contrasted with the continuous elastic solid, has only a finite number of degrees of freedom. The above formula shows that the energy of the body is proportional at low temperatures to the integral

$$\int_0^{\infty}\frac{\nu^3 d\nu}{e^{h\nu/kT}-1} \quad \text{or} \quad \int_0^{\infty}\nu^3 e^{-h\nu/kT}d\nu,$$

that is, it is proportional to T^4; *so the atomic heat of a body at low*

[1] cf. the exhaustive report by E. Schrödinger, *Phys. ZS.* xx (1919), pp. 420, 450, 474, 497, 523

temperatures is proportional to the cube of the absolute temperature. This law has been carefully verified.[1]

In the same year in which Debye's theory of atomic heats was published, Max Born (*b.* 1882) and Theodor v. Kármán (*b.* 1881)[2] attacked the problem from another angle; instead of replacing the body by a continuous elastic medium, as Debye had done, they made a dynamical study of crystals, regarded as Bravais space-lattices of atoms, in order to determine their natural periods. The formulae obtained were of considerable complexity: Debye's method is simpler, though Born and Kármán's is more general and stringent. In the case of low temperatures, Born and Kármán confirmed Deybe's result that the atomic heat is proportional to T^3 when T is small.[3]

The behaviour, as found by experiment, of the molecular heat of gases (i.e. the amount of heat required to raise by one degree the temperature of one gramme-molecule of the gas, at constant volume) can be explained by quantum theory in much the same way as that of solid bodies. According to classical theory,[4] a monatomic gas (such as helium, argon or mercury vapour) has three degrees of freedom (namely, the three required for translatory motion), and to each of them should correspond a mean kinetic energy $\frac{1}{2}kT$, so a gramme-molecule should have an energy $\frac{3}{2}kNT$ when N is Avogadro's number, or $\frac{3}{2}RT$ where R is the gas-constant per gramme-molecule: thus the molecular heat should be $\frac{3}{2}R$ or approximately three calories, a result verified empirically.[5]

For the chief diatomic gases—H, N, O etc.—the molecular heat is 5 calories, which is explained classically by supposing that they have 3 translatory and 2 rotational degrees of freedom. The molecule may be pictured as a rigid dumbbell, having no oscillations along the line joining the atoms, and no rotations about this line as axis. It was, however, found experimentally by A. Eucken[6] that the molecular heat of hydrogen at temperatures below 60° abs. falls to 3 calories, the same value as for monatomic gases. This evidently implies that the part of the molecular heat which is due to the two rotational degrees of freedom of the molecule falls to zero at low temperatures. The quantum theory supplies an obvious explanation

[1] A. Eucken and F. Schwers, *Verh. deutsch phys. Ges.* xv (1913), p. 578; W. Nernst and F. Schwers, *Berlin Ber.* (1914), p. 355; W. H. Keesom and H. Kamerlingh Onnes, *Amsterdam Proc.* xvii (1915), p. 894; xviii (1915), p. 484

[2] *Phys. ZS.* xiii (1912), p. 297; xiv (1913), pp. 15, 65

[3] The diamond was studied specially by Born, *Ann. d. Phys.* xliv (1914), p. 605. cf. Born, *Dynamik der Kristallgitter* (Leipzig, 1915), and many later papers. The detailed study of crystal-theory is beyond the scope of the present work.

[4] cf. Vol. I, p. 383

[5] The quantity which is usually determined directly by experiment (from the velocity of sound in the gas) is the ratio of the molecular heat at constant pressure to the molecular heat at constant volume, or $(2+x)/x$, where x is the molecular heat at constant volume.

[6] *Berlin Sitz.* (1912), p. 141. cf. also K. Scheel and W. Heuse, *Ann. d. Phys.* xl (1913), p. 473, who examined the specific heats of helium, and of nitrogen, oxygen and other diatomic gases, between +20° and −180°; and F. Reiche, *Ann. d. Phys.* lviii (1919), p. 657.

of this behaviour: it is, that no vibration can be excited except by the absorption of quanta of energy that are whole multiples of $h\nu$, where ν is the proper frequency of the vibration: and at low temperatures, the energy communicated by molecular impacts is insufficient to do this, so far as the rotational degrees of freedom are concerned. For the translational degrees of freedom, on the other hand, ν is effectively zero, so no limitation is imposed.

At very high temperatures the molecular heats of the permanent gases rise above 5 calories—to 6 or nearly 7—which evidently signifies that some additional degrees of freedom have come into action, e.g. vibrations along the line joining the two atoms in the molecule. For chlorine and bromine, this phenomenon is observed even at ordinary temperatures, a fact which may be explained by reference to the looser connection of the atoms in the molecules of these elements.

A new prospect opened in 1909, when Einstein [1] discussed the fluctuations in the energy of radiation in an enclosure which is at a given temperature T. From general thermodynamics it can be shown that at any place in the enclosure the mean square of the fluctuations of energy per unit volume in the frequency-range from ν to $\nu + d\nu$, which we may denote by $\overline{\epsilon^2}$, is $kT^2\, dE/dT$, where k is Boltzmann's constant and E is the mean energy per unit volume. Now by Planck's law we have

$$E = \frac{8\pi h\nu^3}{c^3}\,\frac{d\nu}{e^{h\nu/kT} - 1}$$

whence for $\overline{\epsilon^2}$ we obtain the value

$$\frac{8\pi h^2\nu^4 d\nu}{c^3}\left\{\frac{1}{e^{h\nu/kT} - 1} + \frac{1}{(e^{h\nu/kT} - 1)^2}\right\}$$

or

$$h\nu E + \frac{c^3 E^2}{8\pi\nu^2 d\nu}.$$

If instead of Planck's law of radiation we had taken Wien's law,[2] we should have obtained

$$\overline{\epsilon^2} = h\nu E$$

while if we had taken Rayleigh's law,[3] we should have obtained

$$\overline{\epsilon^2} = \frac{c^3 E^2}{8\pi\nu^2 d\nu}.$$

Thus *the mean-square of the fluctuations according to Planck's law is the sum of the mean-squares of the fluctuations according to Wien's law and Rayleigh's law*, a result which, seen in the light of the principle that fluctuations due to independent causes are additive, suggests that

[1] *Phys. ZS.* x (1909), p. 185 [2] cf. Vol. I, p. 381 [3] cf. Vol. I, p. 384

the causes operative in the case of high frequencies (for which Wien's law holds) are independent of those operative in the case of low frequencies (for which the law is Rayleigh's). Now Rayleigh's law is based on the wave-theory of light, and in fact the value $c^3E^2/(8\pi\nu^2 d\nu)$ for the mean-square fluctuation was shown by Lorentz [1] to be a consequence of the interferences of the wave-trains which, according to the classical picture, are crossing the enclosure in every direction: whereas the value $h\nu E$ for the mean-square fluctuation is what would be obtained if we were to take the formula for the fluctuation of the number of molecules in unit volume of an ideal gas, and suppose that each molecule has energy $h\nu$: that is, the expression is what would be obtained by a corpuscular-quantum theory. Moreover, the ratio of the particle-term to the wave-term in the complete expression for the fluctuation is $e^{h\nu/kT}-1$: so when $h\nu/kT$ is small, i.e. at low frequencies and high temperatures, the wave-term is predominant, and when $h\nu/kT$ is large, i.e. when the energy-density is small, the particle-term is predominant. The formula therefore suggests that light cannot be represented completely either by waves or by particles, although for certain classes of phenomena the wave-representation is practically sufficient, and for other classes of phenomena the particle-representation. The undulatory and corpuscular theories are in some sense both true.

Some illuminating remarks on Einstein's formula were made by Prince Louis Victor de Broglie [2] (*b.* 1892). Planck's formula

$$E = \frac{8\pi h\nu^3}{c^3}\frac{d\nu}{e^{h\nu/kT}-1}$$

may be written

$$E = \frac{8\pi h\nu^3}{c^3}(e^{-h\nu/kT}+e^{-2h\nu/kT}+e^{-3h\nu/kT}+\ldots)d\nu$$

$$= E_1 + E_2 + E_3 + \ldots$$

where

$$E_s = \frac{8\pi h\nu^3}{c^3}e^{-sh\nu/kT}d\nu.$$

Now Einstein's formula is

$$\overline{\epsilon^2} = \frac{8\pi h^2\nu^4 d\nu}{c^3}\left\{\frac{1}{e^{h\nu/kT}-1}+\frac{1}{(e^{h\nu/kT}-1)^2}\right\}$$

$$= \frac{8\pi h^2\nu^4 d\nu}{c^3}\{e^{-h\nu/kT}+2e^{-2h\nu/kT}+3e^{-3h\nu/kT}+\ldots\}$$

$$= \sum_{s=1}^{\infty} sh\nu E_s.$$

[1] cf. Lorentz, *Les théories statistiques en thermodynamique* (Leipzig, 1916), p. 114

[2] *Comptes Rendus*, clxxv (1922), p. 811; *J. de phys.* iii (1922), p. 422. cf. W. Bothe, *ZS. f. P.* xx (1923), p. 145; *Naturwiss.* xi (1923), p. 965

This resembles the first term $h\nu E$ in Einstein's formula, but it is now summed for all values of s. So it is precisely the result we should expect if the energy E_s were made up of light-quanta each of energy $sh\nu$. Thus de Broglie suggested that the term E_1 should be regarded as corresponding to energy existing in the form of quanta of amount $h\nu$, that the second term E_2 should be regarded as corresponding to energy existing in the form of quanta of amount $2h\nu$, and so on. So *Einstein's formula for the fluctuations may be obtained on the basis of a purely corpuscular theory of light, provided the total energy of the radiation is suitably allocated among corpuscles of different energies $h\nu$, $2h\nu$, $3h\nu$, ...*[1]

The theory of the fluctuations of the energy of radiation was developed further in many subsequent papers.[2]

In spite of the many triumphs of the quantum theory, its discoverer Planck was in 1911 still dissatisfied with it, chiefly because it could not be reconciled with Maxwell's electromagnetic theory of light. In that year he proposed[3] a new hypothesis, namely, that although emission of radiation always takes place discontinuously in quanta, absorption on the other hand is a continuous process, which takes place according to the laws of the classical theory. Radiation while in transit might therefore be represented by Maxwell's theory, and the energy of an oscillator at any instant might have any value whatever. When an oscillator has absorbed an amount $h\nu$ of energy, it has a chance of emitting this exact amount : but it does not necessarily take the opportunity, so that emission is a matter of probability. In the new theory, therefore, there were no discontinuities in *space*, although the act of emission involved a discontinuity in *time*.

The system based on these principles is generally called *Planck's Second Theory*. He showed that it can lead to the same formula for black-body radiation as the original theory of 1900 ; but there is a notable difference, in that the mean-energy of a linear oscillator of frequency ν is now

$$\tfrac{1}{2}h\nu\,\frac{e^{h\nu/kT}+1}{e^{h\nu/kT}-1},$$

which is greater by $\frac{1}{2}h\nu$ than the value given by the earlier theory : so that *at the absolute zero of temperature, the mean energy of the oscillator is $\frac{1}{2}h\nu$*. This was the first appearance in theoretical physics of the

[1] This corpuscular theory had been proposed in the previous year by M. Wolfke, *Phys. ZS.* xxii (1921), p. 375.

[2] M. von Laue, *Verh. d. deutsch phys. Ges.* xvii (1915), p. 198 ; W. Bothe, *ZS. f. P.* xx (1923), p. 145 ; M. Planck, *Berlin Sitz.* xxxiii (1923), p. 355 ; *Ann. d. Phys.* lxxiii (1924), p. 272 ; P. Ehrenfest, *ZS. f. P.* xxxiv (1925), p. 362 ; M. Born, W. Heisenberg, u. P. Jordan, *ZS. f. P.* xxxv (1926), p. 557 ; S. Jacobsohn, *Phys. Rev.* xxx (1927), pp. 936, 944 ; J. Solomon, *Ann. de phys.* xvi (1931), p. 411 ; W. Heisenberg, *Leipzig Ber.* lxxxiii (1931), Math.-Phys. Klasse, p. 3 ; M. Born and K. Fuchs, *Proc. R.S.* clxxii (1939), p. 465

[3] *Verh. d. deutsch. phys. Ges.* xiii (1911), p. 138

doctrine of *zero-point energy*, which later assumed great importance. In 1913 A. Einstein and O. Stern [1] made it the basis of a new proof of Planck's radiation-formula, and in 1916 W. Nernst [2] suggested that the aether everywhere might be occupied by zero-point energy.

Planck's Second Theory was criticised in 1912 by Poincaré,[3] and in 1914 Planck [4] came to the conclusion that emission by quanta could scarcely be reconciled with classical doctrines, so he now made a new proposal (known as his *Third Theory*), namely, that the emission as well as the absorption of radiation by oscillators is continuous, and is ruled by classical electrodynamics, and that quantum discontinuities take place only in exchanges of energy by collisions between the oscillators and free particles (molecules, ions and electrons). A year later, however, he [5] abandoned the Third Theory, having become convinced by a paper of A. D. Fokker [6] that the calculation of the stationary state of a system of rotating rigid electric dipoles in a given field of radiation, when the calculation was performed according to the rules of classical electrodynamics, led to results that were in direct contradiction with experiment. The Second Theory fell from favour with most physicists about the same time, when the experiments of Franck and Hertz [7] showed the strong analogy between optical absorption and the undoubtedly quantistic phenomena which take place when slow electrons collide with molecules.

Meanwhile, in a Report presented to the Physical Section of the 83rd Congress of German men of science at Karlsruhe on 25 September 1911, Sommerfeld [8] made a suggestion which was the first groping towards a new method. Referring to the name *Quantum of Action* which had been given to the quantity h by Planck, on account of the fact that its dimensions were those of (Energy $\times$ Time) or *Action*, he remarked that there should be some connection between h and the integral which appears in Hamilton's Principle, namely,

$$\int (\mathrm{T}-\mathrm{V})\,dt$$

where T denotes the kinetic and V the potential energy of the mechanical system considered. He proposed to achieve this by making the following hypothesis: *In every purely molecular process, a certain definite amount of Action is absorbed or emitted, namely, the amount*

$$\int_0^\tau \mathrm{L}\,dt = \frac{1}{2\pi}h$$

[1] *Ann. d. Phys.* xl (1913), p. 551

[2] *Verh. d. deutsch. phys. Ges.* xviii (1916), p. 83

[3] *J. de phys.*(5) ii (1912), p. 5, at p. 30

[4] *Berlin Sitz.* 30 July 1914, p. 918

[5] *Berlin Sitz.*, 8 July 1915, p. 512

[6] *Ann. d. Phys.* xliii (1914), p. 810

[7] Described below in Chapter IV

[8] *Verh. deutsch. phys. Ges.* xiii (1911), p. 1074

where L = T − V *is the kinetic potential or Lagrangean function, and where* τ *is the duration of the process*,. A discussion of the photo-electric effect in the light of this principle was given in 1913 by Sommerfeld and Debye.[1] The principle itself, however, was superseded in the later development of the subject.

[1] *Ann. d. Phys.* xli (1913), p. 873

Chapter IV

SPECTROSCOPY IN THE OLDER QUANTUM THEORY

In the nineteenth century, it was generally supposed that the luminous vibrations represented by line spectra were produced in the same way as sounds are produced by the free vibrations of a material body. That is to say, the atom was regarded as an electrical system of some kind, which had a large number of natural periods of oscillation, corresponding to the aggregate of its spectral lines. The first physicist to break with this conception was Arthur William Conway (1875–1950), professor of mathematical physics in University College, Dublin, who in 1907, in a paper of only two and a half pages,[1] enunciated the principles on which the true explanation was to be based : namely, that the spectrum of an atom does not represent the free vibrations of the atom as a whole, but that an atom produces spectral lines one at a time, so that the production of the complete spectrum depends on the presence of a vast number of atoms. In Conway's view, an atom, in order to be able to generate a spectral line, must be in an abnormal or disturbed state : and in this abnormal state, a single electron, situated within the atom, is stimulated to produce vibrations of the frequency corresponding to the spectral line in question. The abnormal state of the atom does not endure permanently, but lasts for a time sufficient to enable the active electron to emit a fairly long train of vibrations.

Conway had not at his disposal in 1907 certain facts about spectra and atoms which were indispensable for the construction of a satisfactory theory of atomic spectra : for until after the publication of Ritz's paper of 1908 [2] physicists did not realise that the frequencies of the lines in the spectrum of an element are the differences, taken in pairs, of certain numbers called ' terms ' ; and it was not until 1911 that Rutherford [3] introduced his model atom, constituted of a central positively-charged nucleus with negative electrons circulating round it. But the revolutionary general principles that Conway introduced were perfectly sound, and showed a remarkable physical insight.

These principles were reaffirmed in 1910 by Penry Vaughan Bevan (1875–1913), in a paper[4] recording experiments on the anomalous dispersion, by potassium vapour, of light in the region of the red lines of the potassium spectrum. He first attempted to explain his results. theoretically in accordance with Lorentz's modernised version [5] of the Maxwell-Sellmeier theory, and found

[1] *Sci Proc. R. Dubl. Soc.* xi (March 1907), p. 181
[2] cf. Vol. I, p. 378
[3] cf. p. 22
[4] *Proc. R.S.*(A), lxxxiv (1910), p. 209
[5] cf. Vol. I, p. 401

that it required him to postulate an impossibly great number of electrons per molecule : deriving from this contradiction the correct conclusion, that spectroscopic phenomena are to be explained by the presence of a very great number of atoms, which at any one instant are in different states, and each of which at that instant is concerned not with the whole spectrum, but at most with only one line of it.

The next advances were made by John William Nicholson[1] at that time of Trinity College, Cambridge. He introduced into spectroscopic theory the model atom which had been proposed a few months before by Rutherford, namely a very small nucleus carrying practically all the mass of the atom, surrounded by negative electrons circling round it like planets round the sun. This was so much more precisely defined than earlier model atoms that it might now be possible to calculate exact numerical values for the wave-lengths of lines in atomic spectra. Nicholson's second advance was to recognise the fundamental fact, that the production of atomic spectra is essentially a quantum phenomenon. 'The fundamental physical laws,' he said, 'must lie in the quantum or unit theory of radiation, recently developed by Planck and others, according to which, interchanges of energy between systems of a periodic kind can only take place in certain definite amounts determined by the frequencies of the systems'[2]; and he discovered the form which the quantum principle should take in its application to the Rutherford atom : '*the angular momentum of an atom can only rise or fall by discrete amounts*.'[3] Moreover, following Conway and Bevan, he asserted that the different lines of a spectrum are produced by different atoms : 'the lines of a series may not emanate from the same atom, but from atoms whose internal angular momenta have, by radiation or otherwise, run down by various discrete amounts from a standard value. For example, in this view there are various kinds of hydrogen atom, identical in chemical properties and even in weight, but different in their internal motions.'[4] In other words, an atom of a given chemical element may exist in many different *states*, resembling in many ways the energy-levels of Planck's oscillators. And 'the incapacity' of an atom 'for radiating in a continuous way will secure sharpness of the lines.'[5] Nicholson did not, however, fully appropriate Conway's idea that a single electron (among the many present in the atom) is alone concerned in the production of a spectral line : on the contrary, he studied the vibrations of a number of electrons circulating round a nucleus by methods recalling the work of Maxwell on Saturn's rings, retaining classical ideas for the actual computation of the motion, and identifying the frequency of spectral lines with the frequency of vibration of a dynamical system.

[1] *Mon. Not. R.A.S.* lxxii, pp. 49 (November 1911), 139 (December 1911), 677, 693 (June 1912), 729 (August 1912)
[2] loc. cit. p. 729 [3] loc. cit. p. 679 [4] loc. cit. p. 730 [5] ibid.

The first successful application of quantum principles to spectroscopy was in the domain not of atomic but of molecular spectra. In 1912 Niels Bjerrum [1] applied quantum ideas in order to explain certain characteristics of the absorption spectra that had been observed with hydrochloric and hydrobromic acids in their gaseous forms. For these and similar compounds, two widely-separated regions of absorption had been found in the infra-red, of which one, that of longest wave-length, was assigned by Bjerrum, following Drude,[2] to rotations of the molecules. For the absorption in the short-wave infra-red he gave a new explanation. He assumed that the two atoms constituting a molecule are positively and negatively charged respectively, and that they oscillate relatively to each other along the line joining them, say with frequency ν_0; incident radiation of this frequency is absorbed. Moreover (following a suggestion of Lorentz), he assumed that the line joining the atoms rotates in a plane, and that the rotational energy must be a multiple of $h\nu$, where ν is the number of revolutions per second.[3] Denoting by J the moment of inertia of the rotating system, the rotational energy is $\frac{1}{2}\mathrm{J}\,.\,(2\pi\nu)^2$, and we have

$$\tfrac{1}{2}\mathrm{J}\,.\,(2\pi\nu)^2 = nh\nu \qquad (n=0, 1, 2, 3, \ldots):$$

or, denoting this value of ν by ν_n, we have

$$\nu_n = \frac{nh}{2\pi^2\mathrm{J}}.$$

By comparing the linear and rotational motions,[4] oscillations are obtained of frequencies ν_n, $\nu_0+\nu_n$, and $\nu_0-\nu_n$: so the absorption-spectrum should contain in the short-wave infra-red the equidistant frequencies

$$\nu_0, \quad \nu_0 \pm \frac{h}{2\pi^2\mathrm{J}}, \quad \nu_0 \pm \frac{2h}{2\pi^2\mathrm{J}}, \quad \nu_0 \pm \frac{3h}{2\pi^2\mathrm{J}}, \quad \text{etc.}$$

Bjerrum's theory stimulated more careful experimental measurements by W. Burmeister,[5] Eva von Bahr,[6] J. B. Brinsmade and E. C. Kemble,[7] and E. S. Imes.[8] Burmeister, and afterwards Imes, found that the central frequency ν_0 was not observed, which would seem to indicate that rotation is always present.[9]

[1] *Nernst Festschrift*, 1912, p. 90

[2] *Ann. d. Phys.*(4) xiv (1904), p. 677

[3] The quantification of molecular rotations had been suggested by Nernst, *ZS. f. Elektrochem.* xvii (1911), p. 265.

[4] The principle of this composition is due to Rayleigh, *Phil. Mag.*(5) xxiv (1892), p. 410.

[5] *Verh. d. deutsch. phys. Ges.* xv (1913), p. 389

[6] *Verh. d deutsch. phys. Ges.* xv (1913), pp. 710, 731, 1150

[7] *Proc. Nat. Acad. Sci.* iii (1917), p. 420

[8] *Astrophys. J.* l (1919), p. 251

[9] It was later found necessary to modify Bjerrum's theory by associating the lines not with the actual rotations, but with *transitions* from one state of rotation to another.

In the year following Bjerrum's paper, P. Ehrenfest [1] improved the quantum theory of rotation by assuming that if ν is the number of revolutions per second, then the rotational energy must be a multiple of $\frac{1}{2}h\nu$ (not $h\nu$, as Bjerrum had supposed); the factor $\frac{1}{2}$ being inserted because the rotational energy is purely kinetic, and not, as in the case of an oscillator, half potential.[2] If J is the moment of inertia, so that the rotational energy is $\frac{1}{2}J \cdot (2\pi\nu)^2$, we have

$$\tfrac{1}{2}J \cdot (2\pi\nu)^2 = n \cdot \tfrac{1}{2}h\nu \qquad (n=0, 1, 2, 3, \ldots)$$

or

$$\nu = \frac{nh}{4\pi^2 J}.$$

The angular momentum $2\pi\nu J$ therefore has the value $nh/2\pi$. The quantity $h/2\pi$ is now generally denoted[3] by $\hbar$. Thus *the law of quantification of angular momentum is that it must be a whole multiple of* $\hbar$.

The culmination of the efforts to explain atomic spectra came in July 1913, when Niels Bohr (*b.* 1885), a Danish research student of Rutherford's at Manchester, found[4] the true solution of the problem. With unerring instinct Bohr seized upon whatever was right in the ideas of his predecessors, and rejected what was wrong, adjoining to them precisely what was needed in order to make them fruitful, and eventually producing a theory which has been the starting-point of all subsequent work in spectroscopy.

Bohr accepted Conway's principles that (1) atoms produce spectral lines one at a time, and (2) that a single electron is the agent in the process, together with Nicholson's principles that (3) the Rutherford atom provides a satisfactory basis for exact calculations of wave-lengths of spectral lines, (4) the production of atomic spectra is a quantum phenomenon, (5) an atom of a given chemical element may exist in different *states*, characterised by certain discrete values of its angular momentum and also discrete values of its energy. He discovered independently[5] Ehrenfest's principle (6) that in quantum-theory, angular momenta must be whole multiples of $\hbar$. He further adopted a principle which was suggested by Ritz's law of spectral 'terms,' and had been in some degree adumbrated but perhaps not clearly grasped by Nicholson, namely (7) that *two* distinct states of the atom are concerned in the production of a spectral line: and he recognised an exact correspondence between the 'terms' into which the spectra were analysed

[1] *Verh. d. deutsch. phys. Ges.* xv (1913), p. 451

[2] Ehrenfest's assumption was confirmed by E. C. Kemble, *Phys. Rev.*(2) viii (1916), p. 689.

[3] Read 'crossed *h*'

[4] *Phil. Mag.* xxvi (1913), pp. 1, 476, 875; xxvii (1914), p. 506; xxix (1915), p. 332; xxx (1915), p. 394

[5] Ehrenfest's paper was published on 15 June 1913, and Bohr's in the July number of the *Phil. Mag.*

by Ritz and the states or energy-levels of the atom described by Nicholson. He also assumed (8) that the Planck-Einstein equation $E = h\nu$ connecting energy and radiation-frequency holds for *emission* as well as absorption: and finally he introduced (9) the principle that *we must renounce all attempts to visualise or to explain classically the behaviour of the active electron during a transition of the atom from one stationary state to another.* This last principle, which had not been dreamt of by any of his predecessors, was the decisive new element that was required for the creation of a science of theoretical spectroscopy.

Let us now see how the Balmer series of hydrogen is explained by Bohr's theory. The atom of hydrogen consists of one proton with one negative electron circulating round it, the charges being e and $-e$ respectively. We suppose that by some event such as a collision, the atom has been thrown into an 'excited' state in which the electron describes an orbit more remote from the proton than its normal orbit, and that a spectral line is emitted when the electron falls back into an orbit closer to the proton. Considering any particular orbit, supposed circular for simplicity, let m denote the mass and v the velocity of the electron, and r the radius of the orbit. Then the electrostatic force e^2/r^2 between the proton and electron must be equal to the centripetal force mv^2/r required to hold the electron in its orbit, so

$$mv^2r = e^2.$$

The quantum condition is that the angular momentum of the motion must be a whole multiple of $\hbar$, say

$$mvr = n\hbar.$$

These equations give

$$v = \frac{e^2}{n\hbar}, \qquad r = n^2\frac{\hbar^2}{me^2}.$$

The kinetic energy of the electron, $\frac{1}{2}mv^2$, is $me^4/2n^2\hbar^2$. If the electron makes a transition to an orbit nearer the nucleus for which the angular momentum is $p\hbar$, there is a gain of kinetic energy

$$\frac{me^4}{2\hbar^2}\left(\frac{1}{p^2} - \frac{1}{n^2}\right)$$

but a loss of potential energy

$$e^2\left(\frac{1}{r_p} - \frac{1}{r_n}\right) \quad \text{or} \quad \frac{me^4}{\hbar^2}\left(\frac{1}{p^2} - \frac{1}{n^2}\right),$$

so altogether (remembering that $h = 2\pi\hbar$) the loss of energy of the atom is

$$\frac{2\pi^2 me^4}{h^2}\left(\frac{1}{p^2}-\frac{1}{n^2}\right).$$

The equation $E = h\nu$ shows that the frequency of the homogeneous radiation emitted is

$$\nu = \frac{2\pi^2 e^4 m}{h^3}\left(\frac{1}{p^2}-\frac{1}{n^2}\right).$$

This can be identified with the formula for Balmer's series [1]

$$\nu = R\left(\frac{1}{4}-\frac{1}{n^2}\right)$$

provided we take $p = 2$ and

$$R = \frac{2\pi^2 e^4 m}{h^3}$$

so we have obtained an expression for Rydberg's constant R in terms of e, m and h.

The value

$$e = 4{\cdot}78 \times 10^{-10}$$

had been obtained by R. A. Millikan,[2] the value

$$\frac{e}{m} = 5{\cdot}31 \times 10^{17}$$

by P. Gmelin [3] and A. H. Bucherer.[4] On the basis of Planck's theory, Bohr obtained [5]

$$\frac{e}{h} = 7{\cdot}27 \times 10^{16}.$$

[1] cf. Vol. I, pp. 376–8. In theoretical researches frequencies are often employed instead of wave-numbers, so e.g. Balmer's formula would be written

$$\nu = R\left(\frac{1}{4}-\frac{1}{m^2}\right)$$

where R stands for c times the R of the wave-number formula.

It may be mentioned that three years previously A. E. Haas, *Wien. Sitz.* cxix (1910), Abth. IIa, p. 119, had conceived the idea that Rydberg's constant should be expressible in terms of the constants e, m, h, which were already known. By an argument which was highly speculative, he obtained the result that this constant has the value $16\pi^2 e^4 m/h^3$, which differs from the correct value only by a simple numerical factor. However, this is not surprising in view of the fact that $e^4 mh^{-3}$ is the only product of powers e, m and h, which has the same dimensions as the Rydberg constant.

[2] *B.A. Rep.* 1912, p. 410

[3] *Ann. d. Phys.* xxviii (1909), p. 1086

[4] *Ann. d. Phys.* xxxvii (1912), p. 597

[5] Calculated from the experiments of E. Warburg, G. Leithaüser, E. Hapka and C. Müller, *Ann. d. Phys.* xl (1913), p. 611

Using these values, he got

$$R = 3{\cdot}26 \times 10^{15}\ (\text{sec})^{-1}$$

in close agreement with observation.[1] This successful prediction had an effect which may be compared with the effect of Maxwell's calculation of the velocity of light from his electromagnetic theory.

Bohr pointed out that while by taking $p=2$ we obtain Balmer's series, by taking $p=3$, $n=4, 5, 6, \ldots$, we obtain a series in the infra-red whose first two members had been discovered by F. Paschen[2] in 1908. At the time of Bohr's paper, the series obtained by taking $p=1$ was not known observationally, but it was discovered in 1914 by Th. Lyman,[3] and the series obtained by taking $p=4$, $n=5, 6, 7, \ldots$, was observed in 1922 by F. Brackett.[4]

Bohr remarked that the frequency of revolution of the electron in the nth state of the hydrogen atom is $v/2\pi r$, or

$$\frac{4\pi^2 me^4}{n^3 h^3}.$$

But the frequency of the radiation emitted in the transition from the $(n+1)^{th}$ state to the nth is

$$\frac{2\pi^2 me^4}{h^3}\left\{\frac{1}{n^2}-\frac{1}{(n+1)^2}\right\}$$

which when n is great has approximately the value

$$\frac{4\pi^2 me^4}{n^3 h^3}.$$

Thus *for great values of n, the frequency of the radiation emitted in the transition from the nth orbit to the next orbit is equal to the frequency of revolution in the nth orbit.* So we have an asymptotic connection between the classical and quantum theories.

Evidence in favour of Bohr's theory was obtained from certain facts regarding absorption-spectra. It was known that the lines of the Balmer series do not appear in the absorption-spectrum of hydrogen under ordinary terrestial conditions, and this was at once explained by the circumstance that the hydrogen atoms have normally no electrons in the orbits for which $p=2$, and therefore no electrons can be raised from these orbits to higher orbits, as would be necessary for the production of a Balmer absorption line. In the

[1] For a comparison with the observational values of 1950, cf. R. T. Birge, *Phys. Rev.* lxxix (1950), p. 193.

[2] *Ann. d. Phys.* xxvii (1908), p. 565. This is the *Bergmann series* for hydrogen; cf. Vol. I, p. 379.

[3] *Phys. Rev.* iii (1914), p. 504; *Phil. Mag.* xxix (1915), p. 284

[4] *Nature* cix (1922), p. 209

atmospheres of the stars, on the other hand, there are excited hydrogen atoms having electrons in the orbits for which $p=2$, and these are capable of giving the Balmer lines in absorption. Considerations of the same kind can be applied to explain why under ordinary terrestial conditions the principal series of the alkali metals can be obtained in absorption-spectra, but the two subordinate series can not.

In the early days of quantum theory no-one troubled about the fact that most of the equations used were not invariant with respect to the transformations of relativity theory. This defect was evident in the case of Bohr's frequency-equation $\delta E = h\nu$, which connects the loss of energy in the transition with the frequency of the emitted radiation. It was, however, shown in 1924 by P. A. M. Dirac [1] that, *provided the radiation is emitted in a definite direction*, the frequency equation can be expressed in a form which is independent of the frame of reference; it can in fact be written as a vector equation in four-dimensional space-time

$$\delta E^i = h\nu^i$$

where the direction of this vector in space-time is the same as that of the radiation.[2]

Some other problems which had been raised by the observational spectroscopists were solved in Bohr's first papers. In 1896 the American astronomer Edward Charles Pickering (1846–1919) discovered [3] in the spectrum of the star ζ Puppis, together with the Balmer series, a series of lines which had the same convergence-number as the Balmer series. Now it is a property [4] of the 'diffuse' and 'sharp' subordinate series of the alkali metals, that the more refrangible members of the doublets converge to the same limit in the two series; and this fact suggested to Rydberg [5] that the Pickering and Balmer series were actually the 'sharp' and 'diffuse' subordinate series respectively of hydrogen. If this were true, then the wave-lengths of the lines of the principal series would be immediately calculable, the first of them being at $\lambda 4687{\cdot}88$: and in fact a line was observed in the spectrum of ζ Puppis at $\lambda 4686$, very near this position: the higher members of the series could not be expected to be seen, since they were beyond the limit for which our atmosphere is transparent.

The line $\lambda 4686$ was observed by the English spectroscopist Alfred Fowler [6] (1868–1940) at the Indian eclipse of 22 January 1898, in the spectrum of the sun's chromosphere. In 1912, in an ordinary

[1] *Proc. Camb. Phil. Soc.* xxii (1924), p. 432

[2] On the relation of Bohr's theory to relativity theory, cf. K. Försterling, *ZS. f. P.* iii (1920), p. 404, and E. Schrödinger, *Phys. ZS.* xxiii (1922), p. 301.

[3] *Astroph. J.* iv (1896), p. 369; v (1897), p. 92

[4] cf. Vol. I, p. 377

[5] *Astroph. J.* vi (1897), p. 233; cf. also H. Kayser, ibid. v (1897), pp. 95, 243

[6] *Phil. Trans.* cxcvii (1901), p. 202

discharge tube containing a mixture of hydrogen and helium, he found [1] other lines very near the positions calculated by Rydberg for the supposed 'principal series of hydrogen,' and, moreover, found a new series in the ultra-violet which had the same convergence-limit as this supposed principal series, and which he provisionally named the 'second principal series of hydrogen.' The small discrepancies in wave-length with Rydberg's calculations were, however, unexplained, and also the fact that the presence of helium appeared to be necessary.

Bohr accounted in the most natural way for Pickering's ζ Puppis series, and the two series found by Fowler, by suggesting that they were not due to hydrogen at all, but to ionised helium. The helium nucleus has a charge $2e$, and in ionised helium this is accompanied by one negative electron circulating round it. A calculation similar to that carried out for the hydrogen atom shows that in this case the frequency of the radiation emitted is

$$4R\left\{\frac{1}{p^2}-\frac{1}{n^2}\right\}, \quad \text{where as before } R=\frac{2\pi^2e^4m}{h^3}.$$

If in this we take $p=3$ and $n=4, 5, 6, \ldots$, we obtain a series which includes the two series found by Fowler : while if we take $p=4$ and $n=5, 6, 7, \ldots$, we obtain the series observed by Pickering in the spectrum of ζ Puppis. Every alternate line in the series thus calculated would be identical with a line in the Balmer series of hydrogen.

There still, however, remained the difficulty arising from the slight discrepancies in wave-length. This was accounted for by Bohr, who remarked that the geometrical centre of the circular orbits of the electron is, strictly speaking, not the nucleus but the centre of gravity of the nucleus and the electron. This makes it necessary to multiply the value of the Rydberg constant by the ratio of the mass of the nucleus to the combined mass of the nucleus and electron. Remembering that the nucleus of the helium atom has a mass four times as great as the mass of the hydrogen nucleus, the slight difference of the Rydberg constants for hydrogen and for helium can be calculated. As Bohr [2] pointed out, ionised helium must be expected to emit a series of lines closely but not exactly coinciding with lines of the ordinary hydrogen spectrum : the alternate members of the ζ Puppis series cannot be superposed on the Balmer hydrogen lines, but should be slightly displaced with respect to them. Thus near the hydrogen lines $H_\alpha(\lambda 6563)$, $H_\beta(\lambda 4861)$, $H_\gamma(\lambda 4340{\cdot}5)$, $H_\delta(\lambda 4102)$, there are helium lines at $\lambda 6560{\cdot}37$, $\lambda 4859{\cdot}53$, $\lambda 4338{\cdot}86$, $\lambda 4100{\cdot}22$ respectively, and the observed discrepancies are completely explained.

[1] *Mon. Not. R.A.S.* lxxiii (1912), p. 62 ; *Phil. Trans.* ccxiv (1914), p. 225

[2] *Nature* xcii (1913), p. 231. The attribution of the line $\lambda 4686$ to helium was confirmed experimentally by E. J. Evans, ibid. p. 5.

A. Fowler [1] noticed that some lines which he had observed in the spectrum of magnesium could be arranged in a series with a Rydberg constant which, like the Rydberg constant for ionised helium, was approximately four times the normal Rydberg constant. The explanation obviously was that they were produced by ionised atoms. The nucleus of magnesium has a positive charge $12e$; and when the metal is ionised, the outermost electron, which describes the orbits concerned, is under the influence of the nucleus together with 10 negative electrons in the inner orbits, that is, the effective central charge is $2e$; and so, as in the case of ionised helium, the Rydberg constant must be multiplied approximately by 4.

In 1923 this principle was carried further by F. Paschen and A. Fowler. Paschen [2] found a series in the spark spectrum of aluminium capable of being represented by a series of the Rydberg type in which the Rydberg constant had nine times its normal value. Since the nucleus of aluminium has a positive charge $13e$, this series obviously belonged to doubly-ionised atoms, for which there would be 10 inner negative electrons, and the effective central charge on the outermost or active electron would be $3e$.

Fowler [3] went further still and discovered in the case of silicon not only series with a Rydberg constant 9 R, due to double ionisation, but also series with a Rydberg constant 16R, due to trebly-ionised atoms of silicon.

Another type of experimental investigation which led to results confirming Bohr's theory must now be mentioned. In the course of an investigation on the ionisation of gases by collisions of electrons with the atoms, James Franck (*b.* 1882) and Gustav Hertz (*b.* 1887) (a nephew of Heinrich Hertz) found [4] that the collision of slow electrons with the atoms of mercury vapour led in some cases to emission of the mercury line $\lambda 2536$. So long as the kinetic energy of the electrons is smaller than $h\nu$, where ν is the frequency of the line, they are reflected elastically by the mercury atoms; but when the kinetic energy is greater than this, light of frequency ν is emitted. Evidently the collision brings about an excited state of the atom, and the radiation is emitted when the atom falls back into its normal state. The subject was pursued in many papers by these and other authors,[5] with the following general results: *Every transition of an electron from one orbit to another, corresponding to a line of the atom's spectrum, can be brought about by the collision of a free electron with the atom, the electron losing an amount $h\nu$ of kinetic energy, where ν is the frequency of the line. Which transitions occur, depends on the state of excitement of the atom. With a normal or unexcited atom, the transitions are those that correspond to the lines of the absorption-spectrum of the unexcited atom.*

[1] *Nature*, xcii (1913), p. 232; *Phil. Trans.* ccxiv (1914), p. 225

[2] *Ann. d. Phys.* lxxi (May 1923), p. 142

[3] *Proc. R.S.* ciii (June 1923), p. 413

[4] *Verh. deutsch. phys. Ges.* xvi (May and June 1914), pp. 457, 517

[5] An extensive bibliography of the experimental work is given at the end of a paper by Franck and Hertz, *Phys. ZS.* xx (1919), p. 132.

Moreover, there is a direct connection between the ionisation of an atom and its spectrum. *If* eV *denotes the energy that must be given to an electron in order that it may ionise the atom, so that* V *is the 'ionisation-potential,' then*

$$eV = h\nu$$

where ν *denotes the frequency of the ultra-violet limit of a series in the spectrum: for the unexcited atom, the ionisation-potential is given by this equation, where* ν *is the limiting frequency of the absorption series of the unexcited atom.* Thus in the case of mercury vapour, the ionisation-potential [1] is 10·27 volts, corresponding to the ultra-violet limit at λ1188 of the series of which λ2536 is the first term.[2]

In one notable case, namely that of helium, the value derived spectroscopically proved more reliable than that obtained in the first place by direct observation of ionisation brought about by electronic bombardment. The latter value was believed until 1922 to be 25·3 volts, whereas Lyman found in that year [3] that the limit of the spectral series concerned was at λ504, which would correspond to an ionisation-potential of 24·5 volts. The discrepancy led Franck [4] to re-examine his experimental data, with the result that he found a source of error which led to reduction of the value 25·3 volts by 0·8 volts, bringing it into coincidence with the spectroscopic value.

Bohr's theory proved adequate also to account for some typical phenomena of fluorescence observed by R. J. Strutt.[5] It was well-known that if sodium vapour is illumined by the D-light emitted by a sodium flame, it emits D-light as a resonance radiation. Now the D-lines constitute the first doublet in the principal series of sodium, the second doublet being in the ultra-violet, at λ3303. Strutt asked the question, whether stimulation of sodium vapour by this second doublet would give rise to D-light? He found that it would. Moreover, he noticed that there is a line in the spectrum of zinc which practically coincides with the less refrangible member of the sodium doublet at λ3303, but that there is no zinc line coinciding with the more refrangible member. So by making use of a zinc spark he was able to stimulate the sodium vapour with light of the wave-length of the less refrangible member only of the sodium doublet. It was found that *both* the D-lines were emitted. The explanation depends on the fact that both lines of both doublets correspond to transitions down to a certain orbit which is the same in all four cases, and which may be called orbit I. By the absorption of the ultra-violet light of the second doublet, an electron is raised from orbit I to a higher orbit, and collisions with other atoms may shake it into somewhat lower orbits, from which it falls into orbit I

[1] F. N. Bishop, *Phys. Rev.* x (1917), p. 244

[2] In making the calculation it is useful to note that according to the equation $eV = h\nu$, one electron volt corresponds to λ12336. This would make λ1188 correspond to 10·4 volts.

[3] *Nature*, cx (1922), p. 278; *Astroph. J.* lx (1924), p. 1

[4] *ZS f. P.* xi (1922), p. 155

[5] *Proc. R.S.*(A), xci (1915), p. 511

with emission of the D-lines. Bohr's theory is thus seen to be of fundamental importance in regard to fluorescence.

In the first year of the century W. Voigt[1] had predicted the existence of an electric analogue to the Zeeman effect—a splitting of spectral lines by an intense electric field. However, discussing the matter by classical physics, he concluded that with a potential-fall of 300 volts per cm. in the field, the effect would be only about the 20,000th part of the separation between the D-lines of sodium, and hence would be unobservable. Notwithstanding this unfavourable opinion, in 1913 Johannes Stark[2] (*b.* 1874), when investigating the light emitted by the particles which constitute the canal rays, examined the influence of an electric field on this light, and observed a measurable effect. By spectroscopic observation in a direction perpendicular to the field, it was found that the hydrogen lines H_β, H_γ were each split into five components, the oscillations of the three inner components (which were of feeble intensity) being parallel to the electric field and the oscillations of the two outer components (which were stronger) being at right angles to the field. The distance between the components was proportional to the electric force. For helium, it was found that the effect of the electric field on the lines of the principal series and the sharp series was very small and hardly distinguishable, but the effect on the lines of the diffuse series[3] was of the same order of magnitude as for the hydrogen lines, though of a different type.

The *Stark-effect*, as it has since been called, was extensively studied from the experimental side in the years immediately following its discovery, chiefly by Stark and his disciples.[4] Some curious properties were noticed, such as that in some cases, where no splitting could be observed with the fields employed, terms were displaced towards the ultra-violet; that terms belonging to the same series were not as a rule affected in the same way, but that there was often a similarity in behaviour between terms which belonged to different but homologous series, and which had the same term-number: and that when the spectroscopic observation was in a direction parallel to the field, only those components of an electrically-split line appeared which, when the observation was at right angles to the field, oscillated linearly perpendicularly to the field; but that these components were now unpolarised.

[1] *Ann. d. Phys.* iv (1901), p. 197

[2] *Berlin Sitz.* 20 November 1913, p. 932; reprinted *Ann. d. Phys.* xliii (1914), p. 965. The effect was discovered independently at the same time by A. Lo Surdo, *Rend. Lincei* xxii (1913), p. 664; xxiii (1914), p. 82.

[3] It may be noted that lines of the diffuse series are broadened when the gas-pressure is increased, while those of the other series are not much affected.

[4] J. Stark, *Verh. deutsch. phys. Ges.* xvi (1914), p. 327; J. Stark and G. Wendt, *Ann. d. Phys.* xliii (1914), p. 983; J. Stark and H. Kirschbaum, ibid. pp. 991, 1017; J. Stark, *Ann. d. Phys.* xlviii (1915), pp. 193, 210; H. Nyquist, *Phys. Rev.* ii (1917), p. 226; J. Stark, O. Hardtke and G. Liebert, *Ann. d. Phys.* lvi (1918), p. 569; J. Stark, ibid. p. 577; G. Liebert, ibid. pp. 589, 610; J. Stark and O. Hardtke, *Ann. d. Phys.* lviii (1919), p. 712; J. Stark, ibid. p. 723

A study of the Stark effect, from the standpoint of Bohr's theory, was undertaken by E. Warburg [1] and by Bohr himself.[2] It was assumed that the field influences the stationary states of the emitting system, and thereby the energy possessed by the system in these states, splitting a term into two or more components. Warburg showed that the effect to be expected, according to quantum theory, of an electric field on the spectral lines of hydrogen, would be of the same order of magnitude as that observed experimentally by Stark ; but the investigations were imperfect as compared with those published two years afterwards by Schwarzschild and Epstein, which will be described later.

The next great advance in theoretical spectroscopy was the removal of a limitation characteristic of the original Bohr theory, namely that it took into consideration only a single set of circular orbits round each atom : it was obviously desirable to extend Bohr's principles by taking into account more than one degree of freedom. This was achieved independently and almost simultaneously in 1915 by William Wilson [3] (*b.* 1875) and Sommerfeld.[4] Their idea recalls that which had inspired Sommerfeld's paper of 1911.[5]

In the circular orbits of the steady states in Bohr's theory of the hydrogen atom, let q denote the angle which the line joining the electron to the proton makes with a fixed line ; then the momentum p corresponding to the co-ordinate q is the angular momentum of the electron round the proton, and for steady states this must be a multiple of $\hbar$. Thus, since $h=2\pi\hbar$, we have

$$p\int dq = \text{a multiple of } h$$

where the integration is taken once round the circle ; or, remembering that $\int pdq$ is the definition of the Action, we have

Increase of Action in going once round the orbit = a multiple of h.

Wilson and Sommerfeld generalised this into the statement that under certain circumstances, in a system with several degrees of freedom, if $q_1, q_2, \ldots$ are the co-ordinates, and $p_1, p_2, \ldots$ are the corresponding momenta, then *the steady states of the system are such that* $\int p_1 dq_1, \int p_2 dq_2, \ldots$, *are multiples of* h, when the integrations are extended over periods corresponding to the co-ordinates.[6]

[1] *Verh. d. deutsch. phys. Ges.* xv (December 1913), p. 1259

[2] *Phil. Mag.* xxvii (March 1914), p. 506 ; xxx (1915), p. 404

[3] *Phil. Mag.* xxix (1915), p. 795 ; xxxi (1916), p. 156

[4] *München Sitz.* 1915, pp. 425, 459 ; *Ann. d. Phys.* li (1916), p. 1. At almost the same time Jun Ishiwara, *Tôkyô Sûgaki-But. Kizi* (2) viii, No. 4, p. 106, *Proc. Math. Phys. Soc. Tôkyô*, viii (1915), p. 318 published proposals which in some respects resembled those of Wilson and Sommerfeld ; cf. also the work of Planck on systems with several degrees of freedom, *Verh. d. deutsch. phys. Ges.* xvii (1915), pp. 407, 438 ; *Ann. d. Phys.* l (1916), p. 385.

[5] cf. p. 104

[6] The rule as thus broadly stated is evidently not independent of the choice of co-ordinates, a point on which see Einstein, *Verh. d. deutsch. phys. Ges.* xix (1917), p. 82, and other papers discussed later in the present chapter.

Sommerfeld now considered electrons moving round the nucleus in non-circular orbits, in fact Keplerian ellipses. Denoting by (r,θ) the radius vector and vectorial angle of the electron, by m its mass, and by Ze the positive charge on the nucleus, the Lagrangean function of the motion is

$$L=\tfrac{1}{2}mv^2-V$$

where

$$v^2=\left(\frac{dr}{dt}\right)^2+r^2\left(\frac{d\theta}{dt}\right)^2, \qquad V=-\frac{Ze^2}{r},$$

so the momenta corresponding to θ and r are $mr^2\, d\theta/dt$ and $m\, dr/dt$ respectively. The quantum conditions specifying a steady state were therefore taken by Sommerfeld to be

$$\int mr^2\frac{d\theta}{dt}\,d\theta=nh \tag{1}$$

and

$$\int m\frac{dr}{dt}\,dr=n'h \tag{2}$$

where n and n' are whole numbers, and the integrations are to be taken once round the orbit.

From (1) and (2) we have

$$(n+n')h=\int mv^2 dt.$$

But from the known properties of Keplerian elliptic motion, we have

$$\int mv^2 dt=2\pi e(Zma)^{\frac{1}{2}},$$

denoting by a the major semi-axis of the orbit. Therefore

$$(n+n')\,h=2\pi e(Zma)^{\frac{1}{2}}. \tag{3}$$

Also from (1)

$$nh=mr^2\frac{d\theta}{dt}\int d\theta=2\pi mr^2\frac{d\theta}{dt}$$

so

$$\frac{n}{n+n'}=\frac{1}{e(Zma)^{\frac{1}{2}}}mr^2\frac{d\theta}{dt}$$

$$=\frac{b}{a} \tag{4}$$

from the known properties of elliptic motion, where b denotes the minor semi-axis. Equations (3) and (4) connect the quantum numbers n and n' with the semi-axes of the orbit.

$$\text{Also, total energy of electron} = \tfrac{1}{2}mv^2 - \frac{Ze^2}{r}$$

$$= -\frac{Ze^2}{2a}$$

$$= -\frac{2\pi^2 e^4 Z^2 m}{h^2(n+n')^2}$$

$$= -\frac{RZ^2 h}{(n+n')^2} \tag{5}$$

where R is Rydberg's number.

The total number of possible steady orbits has been greatly increased by Sommerfeld's introduction of ellipses in addition to Bohr's circular orbits: but from equation (5) it seems as if the total number of possible values of the energy has *not* been increased, since the energy depends only on the single number $(n+n')$. Thus apparently in the case of the hydrogen atom (for which $Z=1$) we should get exactly the same spectra as before. However, as Sommerfeld pointed out, the theory has so far supposed the orbit to be a Keplerian ellipse, the electron being regarded as a particle of constant mass; whereas, since the velocity in the orbit is great, the relativistic increase of mass with velocity ought to be taken into account.[1] Sommerfeld showed that the orbit is an ellipse with a moving perihelion, the motion of the perihelion being great or small according as (for the quantified ellipses) the quantity $e^2/c\hbar$ is great or small. Actually $e^2/c\hbar$, which is generally denoted by α and called the *fine-structure constant*, has a value[2] which at that time was believed to be about $7 \cdot 10^{-3}$, and has since been found to be 1/137. It represents the ratio of the velocity of the electron in the first Bohr orbit to the velocity of light.

The formula for an energy-level or 'term' of the hydrogen atom now becomes, in the first approximation,

$$T = T_0 + T_1$$

where T_0 is the uncorrected value, so $T_0 = R/n^2$, n being the 'principal quantum number,' and T_1 is a correction term given by

$$T_1 = \frac{R\alpha^2}{n^4}\left(\frac{n}{k} - \frac{3}{4}\right)$$

[1] The necessity for a relativity correction had been pointed out already by Bohr, *Phil. Mag.* xxix (1915), p. 332.

[2] The fact that $e^2/c\hbar$ is a pure number had been pointed out by Jeans, *Brit. Ass. Rep.* 1913, p. 376, who suspected that it might have the value $1/(4\pi)^2$.

where k is a second quantum number. The frequency of a radiation corresponding to the fall of an electron from one energy-level to another will depend on the values of n and k for both levels. It is therefore evident that to each spectral line of the original theory (which depended only on the principal quantum numbers) there will now correspond a group of lines very close together; these predicted lines were found to agree remarkably well with lines observed [1] in what was called the *fine-structure* of the spectrum.

As we shall see later, Sommerfeld's theory is not the complete explanation of the fine-structure, which depends also on a property discovered later, that of the spin of the electron; but it was rightly acclaimed at the time as a great achievement. The quantum number k now introduced in connection with hydrogen was found later to account for the distinction between the principal and subordinate series of e.g. the alkali metals.

Sommerfeld's extension of Bohr's theory led Karl Schwarzschild [2] (1873–1916), director of the Astrophysical Observatory at Potsdam, and Paul Sophus Epstein [3] (*b.* 1883), a former student of Sommerfeld's at Munich, to investigate theoretically the Stark effect. In a stationary state of an atom the active electron has three degrees of freedom, and we may therefore expect that the state will be specified by three quantum numbers. Success in the mathematical treatment of the problem depends on the possibility of finding three pairs of variables q_i, p_i (where q_i is a co-ordinate and p_i the corresponding momentum), such that the Hamilton's partial differential equation belonging to the classical problem [4] can be solved by separation of variables. The integrals $\int p_i dq_i$ corresponding to these separate pairs of variables are then equated to multiples of h. By carrying out this programme, Epstein found in the case of the Balmer series of hydrogen the following simple formula for the displacements of the Stark components from the original position of the spectral line:

$$\Delta\nu = \frac{3h}{8\pi^2 ce\mu} \mathrm{EZ}$$

where

$$\mathrm{Z} = (m_1 + m_2 + m_3)(m_1 - m_2) - (n_1 + n_2 + n_3)(n_1 - n_2).$$

Here E is the applied electric force (in the electrostatic system of units), (m_1, m_2, m_3) are the quantum numbers of the outer orbit from which the electron falls into an inner orbit of quantum numbers (n_1, n_2, n_3), μ is the mass of the electron, and ν is the reciprocal

[1] Especially by the measurements, by F. Paschen, *Ann. d. Phys.* l (1916), p. 901, of the fine structure of Fowler's spectra of ionised helium; cf. E. J. Evans and C. Croxson, *Nature*, xcvii (1916), p. 56

[2] *Berlin Sitz.* April 1916, p. 548

[3] *Phys. ZS.* xvii (1916), p. 148; *Ann. d. Phys.* l (1916), p. 489

[4] cf. Whittaker, *Analytical Dynamics*, § 142

of the wave-length. The sums $(m_1+m_2+m_3)$ and $(n_1+n_2+n_3)$ correspond to the ordinal number of the term in the Balmer series, so $n_1+n_2+n_3=2$, while $m_1+m_2+m_3=3$ for the line H_α, 4 for H_β, 5 for H_γ, etc. In the case of the H_α-line, Z could take the values 0, 1, 2, 3, 4, 5; and the wave-numbers calculated by the formula agreed well with Stark's observations, not only as regards the magnitude of the displacements but also as regards polarisation, which depends on whether (m_3-n_3) is odd or even. Similar satisfactory results were found for the other Balmer lines.

The theory of the Stark effect given by Schwarzschild and Epstein is valid so long as the electric field is strong enough to make the Stark separation of the components much greater than the separation due to the fine-structure. The case when the field is so weak that this condition is not satisfied was treated by H. A. Kramers,[1] who found that the fine-structure lines were split into several components, whose displacements were proportional to the *square* of the electric field-strength. Kramers also investigated the Stark effect on series lines in the spectra of elements of higher atomic number.

The methods of quantification employed by Sommerfeld, Schwarzschild and Epstein may be justified by a theory known as the *theory of adiabatic invariants*, which was originally created by the founders of thermodynamics,[2] who studied quantities that are invariant during adiabatic changes. It was now developed by Paul Ehrenfest[3] (1880–1933) of Leiden, whose starting-point was a theorem proved by Wien in the establishment of his displacement law,[4] namely that if radiation is contained in a perfectly-reflecting hollow sphere which is slowly contracting, then the frequency ν_p and the energy ϵ_p of each of the (infinitely-many) principal modes of vibration of the cavity increase together during the compression in such a way that their ratio remains constant. Ehrenfest showed that this property is really the basis of the law that the ratio of energy to frequency ϵ/ν can take only the values 0, h, $2h$, . . . : for a different assumption, such as that ϵ/ν^2 is proportional to 0, h, $2h$, . . ., could be shown to lead to a conflict with the second law of thermodynamics. What impressed Ehrenfest was the circumstance that although Wien's theorem was deduced by purely classical methods, it was valid in quantum theory, and actually indicated a fundamental quantum law. He was thus led to suspect that if in

[1] *ZS. f. Phys.* iii (1920), p. 199

[2] cf. R. Clausius, *Ann. d. Phys.* cxlii (1871), p. 433; cxlvi (1872), p. 585; cl (1873), p. 106; English translations, *Phil. Mag.* xlii (1871), p. 161; xliv (1872), p. 365; xlvi (1873), pp. 236, 266; C. Szily, *Ann. d. Phys.* cxlv (1872), p. 295 = *Phil. Mag.* xliii (1872), p. 339; L. Boltzmann, *Vorlesungen über die Principe der Mechanik*, ii Teil (Leipzig, 1904), ch. iv

[3] *Ann. d. Phys.* xxxvi (1911), p. 91; li (1916), p. 327; *Verh. d. deutsch. phys. Ges.* xv (1913), p. 451; *Proc. Amst. Acad*, xvi (1914), p. 591; xix (1917), p. 576; *Phys. ZS.* xv (1914), p. 657; *Phil. Mag.* xxxiii (1917), p. 500

[4] cf. Vol. I, p. 379

a dynamical system some parameter is slowly changed (as was the radius of the sphere in Wien's theorem), and if some quantity J is found to be invariant according to classical physics during this change (as was ϵ/ν in Wien's case), then the system may be quantified by putting J equal to a multiple of h. An example of 'adiabatic change' is furnished by the motion of a simple pendulum when the length of the suspending cord is very slowly altered.

The general problem he formulated as follows. Let there be a system of differential equations (for simplicity taking two independent variables)

$$\frac{dx_1}{dt} = X_1(x_1, x_2, t, m), \qquad \frac{dx_2}{dt} = X_2(x_1, x_2, t, m)$$

where the letter m stands for one or more constant parameters. These equations will possess integrals

$$F_1(x_1, x_2, t, m) = c_1, \qquad F_2(x_1, x_2, t, m) = c_2.$$

Then it is possible that *some functions of the constants of integration* c_1, c_2, *and of the parameters m, exist, which maintain their values unaltered when the parameters m are varied in an arbitrary manner, provided the variation is very slow.* Such functions are called *adiabatic invariants*; and Ehrenfest now showed that *the quantities which had been put equal to whole-number multiples of h in earlier papers on quantum theory were always adiabatic invariants*; and, moreover, that *the rule thus indicated was valid generally.* In other words, *stationary orbits are adiabatically invariant.*

Now it had been shown by J. Willard Gibbs [1] and Paul Hertz,[2] that in the case of a dynamical system with one degree of freedom whose solutions are periodic and whose equations of motion are

$$\frac{dq}{dt} = \frac{\partial H}{\partial p}, \qquad \frac{dp}{dt} = -\frac{\partial H}{\partial q},$$

so that the trajectory is represented in the (q, p) plane by a curve of constant energy

$$H(q, p) = W,$$

then the area $\iint dq\, dp$ enclosed by this curve is an adiabatic invariant; or, $\int p dq$ taken round the curve is an adiabatic invariant: that is, *the increment of the Action, in a complete period, is an adiabatic invariant.*

This theorem is true even when the system has more than one degree of freedom, provided the solution is periodic. As an example,[3]

[1] *Principles of statistical mechanics*, 1902, p. 157 [2] *Ann. d. Phys.* xxxiii (1910), p. 537

[3] Other examples of adiabatic invariants are given by W. B. Morton, *Phil. Mag.* viii (1929), p. 186 and by P. L. Bhatnagar and D. S. Kothari, *Indian J. Phys.* xvi (1942), p. 271.

consider the case of a planet describing an elliptic orbit under the Newtonian law of force to the focus; and suppose that the mass m of the planet is slowly increased by moving through a dusty atmosphere, while gradual changes also take place in the strength μ of the centre of force. It is well known that the velocity v, the radius vector r and the mean distance a, are connected by the equation

$$v^2 = \mu\left(\frac{2}{r} - \frac{1}{a}\right).$$

Varying this, we have

$$\frac{2\delta v}{v} = \frac{\delta\mu}{\mu} + \frac{\mu\delta a}{a^2 v^2}$$

or

$$\frac{\delta a}{a^2} = \frac{v^2}{\mu}\left(\frac{2\delta v}{v} - \frac{\delta\mu}{\mu}\right).$$

When the planet picks up a small particle δm previously at rest, by the theorem of conservation of linear momentum we have

$$\delta(mv) = 0, \quad \text{or} \quad \frac{\delta v}{v} = -\frac{\delta m}{m},$$

so the previous equation becomes

$$\frac{\delta a}{a^2} = -\frac{v^2}{\mu}\left(\frac{2\delta m}{m} + \frac{\delta\mu}{\mu}\right).$$

Suppose the changes in m and μ to be brought about so gradually that the increments δm and $\delta\mu$ are spread over a large number of orbital revolutions; we may then, before integrating the last equation, replace v^2 by its time-average. Now since the average value of $1/r$ is $1/a$, we see from the equation

$$v^2 = \mu\left(\frac{2}{r} - \frac{1}{a}\right)$$

that the average value of v^2 is μ/a. When this is inserted, the variational equation becomes

$$\frac{\delta a}{a} + \frac{2\delta m}{m} + \frac{\delta\mu}{\mu} = 0.$$

Integrating, we have

$$m^2\mu a = \text{Constant}:$$

and since the increment of the Action in a complete period is known to be $2\pi m\mu^{\frac{1}{2}} a^{\frac{1}{2}}$, this equation verifies for Keplerian motion the theorem that the increment of the Action in a complete period is an adiabatic invariant.

In Keplerian motion, the Action

$$\int m\left\{\left(\frac{dr}{dt}\right)^2+r^2\left(\frac{d\theta}{dt}\right)^2\right\}dt$$

is the sum of two integrals, namely $\int m\,(dr/dt)\,dr$ and $\int mr^2\,(d\theta/dt)\,d\theta$: and since the alterations of μ and m do not affect the angular momentum round the centre of force, we see that the second of these integrals, taken singly, must be an adiabatic invariant. Whence it follows that the first must be also. Thus *a justification is provided for Sommerfeld's choice of integrals to be equated to integral multiples of h.*

We shall now establish a connection (not restricted to Keplerian motion) between the adiabatic invariant representing the increment of Action in a complete period (which we shall denote by J), the total energy of the motion (which we shall denote by W) and the frequency ν of the motion (i.e. the reciprocal of the periodic time).

The solution of the differential equations of motion will consist in representing the original Hamiltonian variables in terms of $Q\equiv\nu t+\epsilon$ (where ϵ is an arbitrary constant) and a variable P which is conjugate to Q and is actually constant: the transformation from (q, p) to (Q, P) being a contact-transformation, so that

$$p\,dq-P\,dQ=d\Omega$$

where $d\Omega$ is the differential of a function which resumes its value after describing the periodic orbit.

Integrating this equation once round the orbit, we have

$$\int p\,dq-P\int dQ=\int d\Omega=0.$$

But

$$\int dQ=\int\frac{dQ}{dt}dt=\nu\int dt=1.$$

Therefore $P=\int p\,dq$ integrated round the orbit: that is, P is the adiabatic invariant which we have called J. The Hamiltonian equations in terms of the variables P and Q are

$$\frac{dQ}{dt}=\frac{\partial H}{\partial P},\qquad \frac{dP}{dt}=\frac{\partial H}{\partial Q}=0.$$

The former equation is

$$\nu=\frac{dH}{dP}$$

so we have

$$\nu=\frac{dW}{dJ}$$

where W *denotes the total energy.*

The methods employed by Schwarzschild and Epstein in the mathematical treatment of the Stark effect led other workers in quantum theory [1] to study dynamical systems which can be solved by separation of variables, a subject which had been studied extensively in 1891 and the following years by Paul Stäckel of Kiel.[2] In such systems the integral of Action,

$$\int 2T\,dt$$

where T denotes the kinetic energy, can be separated into a sum of integrals each of which depends on one only of the co-ordinates,

$$\sum_k \int \sqrt{\left\{F_k(q_k)\right\}}\,dq_k.$$

In general, the motion of each co-ordinate is a *libration*, i.e. it oscillates between two fixed limits, the values of which are determined by the integrated equations of motion.[3] For such systems it was proved by J. M. Burgers [4] that *the single integrals*

$$J_k = \int \sqrt{\left\{F_k(q_k)\right\}}\,dq_k,$$

where the integration is taken over a range in which q_k oscillates once up and down between its limits, are adiabatic invariants [5]: and therefore the rule that must be followed in order to quantify is to write

$$J_k = n_k h$$

where n_k is a whole number, and is called a *quantum number*.

Denoting by T the average value of the kinetic energy, we have

$$2\overline{T} = \frac{1}{A+B}\int_{-A}^{B} 2T\,dt$$

where A and B are large numbers. Hence

$$2\overline{T} = \frac{1}{A+B}\int_{-A}^{B} \sum_k \sqrt{\left\{F_k(q_k)\right\}}\,dq_k.$$

Now

$$J_k = \int \sqrt{\left\{F_k(q_k)\right\}}\,dq_k$$

[1] P. Debye, *Gött. Nach.* 1916, p. 142; *Phys. ZS.* xvii (1916), pp. 507, 512; A. Sommerfeld, *Phys. ZS.* xvii (1916), p. 491

[2] *Habilitationsschrift*, Halle, 1891; *Comptes Rendus*, cxvi (1893), p. 485; cxxi (1895), p. 489; *Math. Ann.* xlii (1893), p. 545

[3] There is, however, often among the co-ordinates an azimuthal angle which can increase indefinitely but with respect to which the configuration of the system is periodic: an increase of 2π in it takes the place of the libration of the other co-ordinates.

[4] *Phil. Mag.* xxxiii (1917), p. 514; *Ann. d. Phys.* lii (1917), p. 195

[5] Leaving aside some special cases of degeneration

taken over a range of integration described in the periodic time of the oscillation of q_k. So

$$2T = \frac{1}{A+B}\sum_k \frac{A+B}{T_k} J_k$$

or

$$2T = \sum_k \nu_k J_k$$

where ν_k is the frequency associated with the co-ordinate q_k.

Now it was shown in 1887 by Otto Staude [1] of Dorpat (for the case of two degrees of freedom) and by Stäckel [2] in 1901 (for any number of degrees of freedom) that a dynamical system, for which Hamilton's partial differential equation can be integrated by separation of variables, is *multiply periodic*,[3] that is to say, the co-ordinates ($q_1, q_2, \ldots, q_n$) can be expressed by generalised Fourier series of the type

$$q_r = \sum Q_r \cos\{2\pi(\tau_1\nu_1 + \tau_2\nu_2 + \ldots + \tau_n\nu_n)t + \gamma_r\}$$

where Q_r and γ_r are functions of $\tau_1, \tau_2, \ldots \tau_n$, where the ν_k have the same meanings as before, and where summation is over integral values of the parameters $\tau_1, \tau_2, \ldots \tau_n$.

It can be shown, as in the case of a system with one degree of freedom, that

$$\nu_k = \frac{\partial W}{\partial J_k} \qquad (k = 1, 2, \ldots n)$$

where W denotes the total energy, so that the frequency of the radiation emitted in the transition from a state W_r to a state W_s is given by the equation

$$h\nu_{rs} = W_r - W_s.$$

Now suppose that the quantum numbers belonging to the states W_r and W_s are large, and that the quantum numbers of the state W_r differ from those of the state W_s by $\tau_1, \tau_2, \ldots \tau_n$, respectively, so that

$$\Delta J_1 = \tau_1 h, \quad \Delta J_2 = \tau_2 h, \ \ldots, \ \Delta J_n = \tau_n h.$$

Since consecutive orbits with large quantum numbers do not differ greatly from each other, we can represent the increment $W_r - W_s$ of the energy approximately by

$$\Delta W = \frac{\partial W}{\partial J_1}\Delta J_1 + \frac{\partial W}{\partial J_2}\Delta J_2 + \ \ldots \ + \frac{\partial W}{\partial J_n}\Delta J_n$$

[1] *Math. Ann.* xxix (1887), p. 468 ; *Sitz. d. Dorpater Naturfor.*, April 1887
[2] *Math. Ann.* liv (1901), p. 86
[3] The German term, introduced by Staude, loc. cit., is *bedingt-periodisch.*

so we have

$$h\nu_{rs} \sim \frac{\partial W}{\partial J_1}\tau_1 h + \frac{\partial W}{\partial J_2}\tau_2 h + \ . \ . \ . \ + \frac{\partial W}{\partial J_n}\tau_n h$$

or

$$\nu_{rs} \sim \nu_1\tau_1 + \nu_2\tau_2 + \ . \ . \ . \ + \nu_n\tau_n :$$

that is to say, *the frequency of the spectroscopic line emitted in the transition* $r \rightarrow s$ *is approximately* $\nu_1\tau_1 + \nu_2\tau_2 + \ldots + \nu_n\tau_n$, *which is the frequency of one term in the multiple Fourier series representing the classical solution of the problem, namely that for which*

$$\tau_1 = {}^r n_1 - {}^s n_1, \quad \tau_2 = {}^r n_2 - {}^s n_2, \ . \ . \ . \ \tau_n = {}^r n_n - {}^s n_n$$

where $({}^r n_1, {}^r n_2, \ldots)$ *are the quantum numbers in the state* r, *and* $({}^s n_1, {}^s n_2, \ldots)$ *are the quantum numbers in the state* s. This was called by Ehrenfest the *correspondence theorem for frequencies.*

In the same year (1916) in which Schwarzschild and Epstein published their explanations of the Stark effect, Sommerfeld [1] and Debye [2] showed that the Zeeman effect also could be brought within the compass of the quantum theory. If we consider the motion of an electron under the influence of a fixed electric charge at the origin, and a magnetic field H parallel to the axis of z, the classical equations of motion may be written in the Hamiltonian form

$$\frac{dq_r}{dt} = \frac{\partial K}{\partial p_r}, \qquad \frac{dp_r}{dt} = -\frac{\partial K}{\partial q_r} \qquad (r = 1, 2, 3)$$

where q_1 is the radius vector from the origin to the electron, q_2 is the angle between q_1 and the intersection of the plane of xy with the plane of instantaneous motion of the electron (which we may call the plane of the orbit), q_3 is the angle between the fixed axis Ox and the intersection of the plane of xy with the plane of the orbit, p_1 is the component of linear momentum along the radius vector, p_2 is the angular momentum of the electron round the origin, and p_3 is the angular momentum of the electron about the axis of z. These variables are separable. During the motion, the plane of the orbit precesses [3] uniformly round the axis of z with angular velocity eH/2mc, so the dynamical equations involving q_3 and p_3 must be

$$\frac{dq_3}{dt} = \frac{eH}{2mc}, \qquad \frac{dp_3}{dt} = 0$$

[1] *Phys. ZS.* xvii (1916), p. 491

[2] *Phys. ZS.* xvii (1916), p. 507; cf. also A. W. Conway, *Nature*, cxvi (1925), p. 97 and T. van Lohuizen, *Amst. Proc.* xxii (1919), p. 190

[3] This is the *Larmor precession,* described in Vol. I, pp. 415–6

and therefore the term involving q_3 and p_3 in the Hamiltonian function K must be

$$\frac{eHp_3}{2mc}.$$

Proceeding now to the quantification, there will be three quantum conditions of the kind usual in problems for which the variables can be separated, and the third of them will be

$$p_3 = m_j\hbar$$

where m_j is a whole number which will be called the *magnetic quantum number*: so the existence of the magnetic quantum number is an assertion that the component of angular momentum in the direction of the magnetic field can take only values which are whole-number multiples of $\hbar$. Since this component attains its greatest value when it is equal or opposite to the total angular momentum, we see that m_j can take only the values $-j, -j+1, \ldots, j-1, j$, where j is the quantum number which specifies the total angular momentum.

Thus the Hamiltonian function K, which represents the energy of the motion, is increased (as compared with the case when the magnetic field is absent) by

$$\frac{m_j e\hbar H}{2mc}.$$

Supposing that we are dealing with the hydrogen atom, the part of the energy corresponding to the unperturbed motion is (as in Bohr's original theory, neglecting the fine structure)

$$-\frac{2\pi^2 me^4}{n^2h^2}$$

and adding to this the part we have just found, due to the magnetic field, we have for the total energy in the stationary state specified by the quantum numbers n, m_j,

$$-\frac{2\pi^2 me^4}{n^2h^2} + \frac{m_j ehH}{4\pi mc},$$

and therefore the frequency of the spectral line emitted in the transition from the state (n, m_j) to the state (n', m_j') is

$$\nu = \frac{2\pi^2 me^4}{h^3}\left(\frac{1}{n'^2} - \frac{1}{n^2}\right) + \frac{eH}{4\pi cm}\left(m_j - m_j'\right).$$

Thus *when the magnetic field is applied, in place of the single spectral line specified by* (n, n'), *we have a number of lines depending on* m_j *and* m_j'.

These are the Zeeman components. Their number, intensity and state of polarisation are furnished by a principle which was not discovered until 1918, and which will be described presently. It will appear that so far as the number, position and state of polarisation of the components are concerned, the quantum theory gives (for lines such as those we are now considering, namely single lines which are not members of doublets or triplets) exactly the same results as Lorentz's original theory. It will be noted that this became possible because Planck's constant h cancelled out in the magnetic part of the above expression for the frequency.

We may note that since in the classical problem the angle a which the plane of the orbit makes with Oz is given by

$$\cos a = \frac{p_3}{p_2},$$

therefore in the quantified problem, when both the total angular momentum and its component in the direction of Oz are whole-number multiples of $\hbar$, *this angle* a *can take only certain discrete values.* This is an example of what is called *space quantification* or *direction quantification*; the plane of the orbit is permitted to be inclined at only certain definite angles to the direction of the field.

The principle of space quantification was strikingly confirmed by an experiment performed in 1921 by O. Stern [1] and W. Gerlach,[2] working in the department of Max Born at Frankfort-on-Main. Let a ray of atoms of silver produced by boiling silver in a furnace and passing the vapour through two fine slits, be travelling in the x-direction, and suppose that the ray encounters a non-uniform magnetic field parallel to the axis of z. In the non-uniform magnetic field H_z, a particle, whose magnetic moment in the z-direction is M, experiences a mechanical force $M\partial H_z/\partial z$ in the z-direction. Space-quantification ensures, however, that atoms orient themselves in the magnetic field in certain ways, in fact that M can take only values which correspond to the directions parallel and antiparallel to H_z, and so are equal and opposite. Hence the original ray of silver atoms is split into two rays in the plane of xz, corresponding to these two opposite values of the z-component of the magnetic moment: in the experiment, these rays strike a plane to which the atoms adhere and so produce an image which is observed.

It may be noted that the Stern-Gerlach effect concerns only a single state of the atom, not (like the Zeeman effect) a transition from one state to another.

The effect of crossed electric and magnetic fields on the radiation from a hydrogen atom was discussed by O. Klein,[3] W. Lenz [4] and N. Bohr.[5]

[1] *ZS. f. P.* vii (1921), p. 249
[2] *ZS. f. P.* viii (1921), p. 110; ix (1922), pp. 349, 352
[3] *ZS. f. P.* xxii (1924), p. 109
[4] *ZS. f. P.* xxiv (1924), p. 197
[5] *Proc. Phys. Soc.* xxxv (1923), p. 275

Before 1918 the Bohr theory had been applied to determine only the *frequencies* of lines in spectra, and had not yielded any results regarding their *intensity*. Some definite questions concerning intensity were, however, suggested by spectroscopic observations ; in particular, certain lines, whose existence might be expected according to the Bohr theory, were found to be absent in the spectrum as observed : which suggested that transitions of the active electron from certain orbits to certain other orbits never took place, so that the corresponding lines could not appear.

The explanation of this phenomenon was given by Adalbert Rubinowicz[1] (*b.* 1889), a Pole then working at Munich, who, considering atoms of the hydrogen type, and defining a stationary orbit by the numbers n, n', introduced by Sommerfeld's quantum conditions

$$\int mr^2 \frac{d\theta}{dt} d\theta = nh, \qquad \int m \frac{dr}{dt} dr = n'h,$$

remarked that the angular momentum of the atom was $n\hbar$, and that in a transition between an orbit of quantum numbers (m, m') and an orbit of quantum numbers (n, n'), this angular momentum would change by $|m-n|\hbar$. But by the principle of conservation of angular momentum, any change in the angular momentum of the atom must be balanced by the angular momentum carried off by the radiation associated with the transition. By an argument based partly on classical electrodynamics and partly on quantum theory (and therefore perhaps not very secure), Rubinowicz found that the angular momentum radiated, when the radiation is circularly polarised, is $\hbar$; and when the radiation is linearly polarised, the angular momentum radiated is zero. Thus we have

$$|m-n|\hbar = 0 \text{ or } 1,$$

and we obtain a *selection-principle*, namely that *the azimuthal quantum number n can only change by* 1, 0 or −1.

In a footnote appended to his paper, Rubinowicz explained that when his paper was ready for press there had appeared the first Part of a memoir by Bohr,[2] in which the same problem was approached from quite a different standpoint, depending on the close relations which exist between the quantum theory and the classical theory for very great quantum numbers. We have seen an example of these relations in the correspondence theorem for frequencies ; Bohr now extended this theorem by assuming that there is a relation between the *intensity* of the spectral line radiated and the amplitude of the corresponding term in the classical multiple-Fourier expansion: in fact, that the *transition-probability* associated with the genesis of

[1] *Phys. ZS.* xix, p. 441 (15 Oct. 1918), and p. 465 (1 November 1918)

[2] *D. Kgl. Danske Vid. Selsk. Skr., Nat. og Math. Afd.*, 8 Raekke, iv, 1 (1918) ; cf. N. Bohr, *Ann. d. Phys.* lxxi (1923), p. 228

the spectral line contains a factor proportional to the square of the corresponding coefficient in the Fourier series. Moreover, he extended this correspondence principle for intensities by assuming its validity not only in the region of high-quantum numbers but over the whole range of quantum numbers: so that *if any term in the classical multiple-Fourier expansion is absent, the spectral line, which corresponds to it according to the correspondence-theorem for frequencies, will also be absent.* This is a *selection-principle* of wide application.

He postulated also that the *polarisation* of the emitted spectral line may be inferred from the nature of the conjugated classical vibration. Thus [1] considering in particular the Zeeman splitting of a spectral line of hydrogen, when the transition is such that the magnetic quantum number is unchanged, the Zeeman component will occupy the same position as the original line, and the radiation will correspond to that emitted in classical electrodynamics by an electron performing linear oscillations parallel to the magnetic field; while in the case when the magnetic quantum number changes by ± 1, (which is the only other possibility permitted by the correspondence-principle), we shall obtain Zeeman components symmetrically situated with respect to the original line, and the radiation will correspond to that emitted by a classical electron describing a circular orbit in a plane at right angles to the magnetic field, in one or the other direction of circulation. The polarisation of the emitted line will therefore in all three cases be the same as that predicted by Lorentz [2] on the classical theory.

An extensive memoir by H. A. Kramers [3] supplied convincing evidence of the validity of Bohr's correspondence-principle for the calculation of the intensities of spectral lines: while W. Kossel and A. Sommerfeld [4] showed that the deductions from the selection principle were confirmed by experiment in the case of many different kinds of atoms.

The correspondence principle was extended to absorption by J. H. van Vleck.[5]

We must now consider developments in the theory of quantum numbers. We have seen that Sommerfeld specified an energy-level or 'term' of an atom by two quantum numbers (leaving aside for the moment the magnetic quantum number). The first of these is the *principal quantum number n*, which had been introduced by Bohr in 1913, and which increases by unity when we pass from a term of a spectral series to the next higher term: and the other is the *azimuthal quantum number k*, which had been introduced by Sommerfeld himself in 1915, and which distinguishes the different series from each other: thus in the sodium spectrum [6] the terms of the 'sharp' series have the frequencies $R/(n+s)^2$, where R is Rydberg's constant

[1] N. Bohr, *Proc. Phys. Soc.* xxxv (1923), p. 275
[2] cf. Vol. I, p. 412
[3] *D. Kgl. Dansk. Vid. Selsk. Skr., Nat. og Math. Afd.*, 8 Raekke, iii, 3 (1919)
[4] *Verh. d. deutsch phys. Ges.* xxi (1919), p. 240
[5] *Phys. Rev.* xxiv (1924), p. 330
[6] cf. Vol. I, p. 378

and n is a positive whole number: those of the 'principal' series have the frequencies $R/(n+p_1)^2$ and $R/(n+p_2)^2$; those of the 'diffuse' series have the frequencies $R/(n+d)^2$, and those of the 'fundamental' series have the frequencies $R/(n+f)^2$: and these series correspond respectively to $k=1, 2, 3, 4$. It has become customary to use in place of k the letter l, where $l=k-1$: the reason being that when there is one active electron, its orbital angular momentum is $l\hbar$. The series of energy-levels of the atom for which $l=0, 1, 2, 3, \ldots$, were denoted by $s, p, d, f, \ldots$, these being the initial letters of the words *sharp*, *principal*, *diffuse*, *fundamental*, etc. l has the selection rule that in a transition it can change only to $l+1$ or $l-1$.

Evidently, however, the two quantum numbers n and l did not suffice for the description of the terms of the alkali spectra, for the principal series was a series of doublets. To meet this situation, Sommerfeld [1] in 1920 introduced a third number j, which he called the *inner quantum number* and which is different for the two terms of a doublet. This number must arise from the quantification of a motion in some third degree of freedom, and it was natural to suppose that besides the orbital angular momentum of the atom, which was accounted for by l, there was yet another independent angular momentum. This was at first conjectured to be the angular momentum of the atom's core, was denoted by $s\hbar$ and was supposed to have (for the alkalis) the value $\frac{1}{2}\hbar$; so that when it was compounded with the angular momentum $l\hbar$, with which space-quantification compels it to be either parallel or anti-parallel, the resultant total angular momentum of the atom, $j\hbar$, could have either of the two values [2] $(l-\frac{1}{2})\hbar$ and $(l+\frac{1}{2})\hbar$. There is a selection-rule that j may pass only to $(j+1)$, j, or $(j-1)$, and moreover transitions in which j remains zero are forbidden.

In the spectra of the alkaline earths there are series of triplets which were accounted for in the same way by supposing that the angular momentum $s\hbar$ can take the values 0 and $\hbar$, giving for the total angular momentum the three possibilities $(l-1)\hbar$, $l\hbar$, $(l+1)\hbar$, and so for j the three possibilities $j=l-1$, $j=l$, $j=l+1$. In other atoms, every value [3] of l was supposed to yield a set of energy-levels or terms corresponding to the values

$$j=l+s, \quad l+s-1, \quad . \quad . \quad . \quad |l-s|+1, \quad |l-s|.$$

We have said that at first the independent angular momentum $s\hbar$, which is compounded with the orbital angular momentum $l\hbar$ in order to produce the resultant total angular momentum $j\hbar$, was supposed to be the angular momentum of the 'core' of the atom, i.e. possibly the nucleus together with the innermost closed shells

[1] *Ann. d. Phys.* lxiii (1920), p. 221; lxx (1923), p. 32
[2] Except when $l=0$, in which case there is only one value of j, namely $\frac{1}{2}$.
[3] $l=0$ yields only a single term

of electrons. This hypothesis was overthrown by Wolfgang Pauli (*b.* 1900), a Viennese who, after studying with Sommerfeld at Munich and with Bohr at Copenhagen, had become a privat-dozent in the University of Hamburg. Pauli showed [1] that if the angular momentum $s\hbar$ belonged to the atomic core, there would follow a certain dependence of the Zeeman effect on the atomic number, and this effect was not observed: he inferred that the angular momentum $s\hbar$ must be due to a new quantum-theoretic property of the electron, which he called 'a two-valuedness not describable classically.' This remark suggested later in the same year to two pupils of Ehrenfest, G. E. Uhlenbeck and S. Goudsmit of Leiden [2] the adoption of a proposal which had been made in 1921 by Arthur H. Compton,[3] an American who was at that time working with Rutherford at Cambridge, namely that *the electron itself possesses an angular momentum or spin, and a magnetic moment.* Uhlenbeck and Goudsmit proposed as the amount of angular momentum $\frac{1}{2}\hbar$; and they suggested that the values $(l+\frac{1}{2})\hbar$ and $(l-\frac{1}{2})\hbar$ which are possible for the total angular momentum $j\hbar$ in e.g. the alkali spectra, are obtained by compounding the angular momentum $l\hbar$ with the electron-spin, which (since it exists in the magnetic field created by the orbital revolution of the electron) is compelled by space-quantification to take orientations either parallel or anti-parallel to $l\hbar$. Associated with the spin there is a magnetic moment whose value they asserted (for reasons to be discussed presently) to be $e\hbar/2mc$.

The discovery of electron-spin raised a question as to the validity of Sommerfeld's explanation of the fine-structure of the hydrogen lines. For if the electron has a spin with a magnetic moment, then two different orientations of the spin must be allowed, and to these must correspond two different energy-levels for the atom, causing a further resolution of each fine-structure component into a doublet. The measurements of Paschen had shown, however, that Sommerfeld's formula expressed the experimental data for the hydrogen spectrum satisfactorily: and it was found that the two corrections which in a more complete theory should be made to Sommerfeld's analysis, namely (1) replacing the particle-dynamics of Sommerfeld by quantum-mechanics, and (2) taking account of the spin magnetic moment, more or less neutralised each other, producing only a replacement of the quantum number k by $(j+\frac{1}{2})$.

The theory of spectra was much advanced by investigations arising out of the new experimental work on the Zeeman effect.

In the first decade of the twentieth century no great progress was made in the observational field, though in 1907 C. Runge [4]

[1] *ZS. f. P.* xxxi (1925), p. 373

[2] *Naturwiss*, xiii (1925), p. 953; *Nature*, cxvii (1926), p. 264. It is said that R. de L. Kronig had the same idea somewhat earlier, but finding it received unsympathetically by a colleague, did not publish it.

[3] *Phil. Mag.* xli (1921), p. 279; *J. Frankl. Inst.* cxcii (1921), p. 144

[4] *Phys. ZS.* viii (1907), p. 232

studied a number of spectral lines which show complex types of resolution, and found that the distances (measured as differences of frequency) of the components from the centre of the undisturbed line were connected by simple numerical relations with the frequency-difference which was given by Lorentz's theory of the Zeeman triplet, namely $eH/4\pi cm$, where H is the external magnetic force in electromagnetic units. In 1912, however, F. Paschen and E. Back [1] studied the Zeeman effect in lines which are members of doublets or triplets in series spectra (e.g. of the alkali atoms), and found that so long as the magnetic field is not strong the Zeeman splitting is very complicated. The different lines of a doublet or triplet behave differently, though the separations between the components always increase proportionally to the field-strength. When, however, the field has increased to such a strength that the separations between the Zeeman components are of the same order of magnitude as the separations between the components of the original doublet or triplet, the individual Zeeman components become diffuse and tend to amalgamate : and ultimately, at very great field-strengths, the whole system reduces to three components constituting a normal Zeeman triplet,[2] and having its centre at the centre of the original doublet or triplet. The difference in character between the Zeeman effect in weak and strong external magnetic fields was seen at once to be connected with the fact that the space-quantification of the electron spin is governed by the magnetic field of the orbital motions when the external magnetic field is weak, but is governed by the external field when that is sufficiently strong.

The experimental knowledge regarding the splitting that is found with comparatively weak fields in doublets and triplets, or the *anomalous Zeeman effect* as it was called, was reduced to a mathematical formula by Alfred Landé [3] of Tübingen in 1923. He showed that the frequency of any one of the Zeeman components can, like the frequencies of the lines of the original spectrum, be represented by the difference of two ' terms ' or energy-levels. In a magnetic field, each term W of the original spectrum is split into several terms W+Z. If (supposing for simplicity that there is only one active or valence electron) the term W has the quantum numbers n, l, j, s, then for its Zeeman components we have

$$Z = \frac{m_j e g \hbar H}{2mc} \text{ ergs,}$$

where e, $\hbar$, m, c have their usual meanings ; g, which is called the *Landé splitting factor*, is given by the equation

$$g = 1 + \frac{j(j+1) + s(s+1) - l(l+1)}{2j(j+1)}$$

[1] *Ann. d. Phys.* xxxix (1912), p. 897 ; xl (1913), p. 960

[2] For a discussion of the phenomena taking place during the passage from weak to strong fields, cf. W. Pauli, *ZS. f. P.* xx (1923), p. 371.

[3] *ZS. f. P.* xv (1923), p. 189

and m_j is the magnetic quantum number,[1] which, as we have already seen, can take the values

$$m_j = j, j-1, \quad . \; . \; ., \; -j$$

so one spectral term splits into $(2j+1)$ equidistant terms. As we have seen, in transitions m_j can change only by 0 or ± 1, and the lines so arising are polarised in the same way as the lines of a Lorentz Zeeman triplet.

If the original term is a singlet level, we have $s=0$ and $l=j$, so $g=1$; and lines resulting from combinations of levels in singlet series have the appearance of Lorentz Zeeman triplets.

Observation of the Zeeman effect is of great assistance in determining the character of any particular spectral line, since the observation furnishes the value of the Landé splitting factor, and this knowledge generally leads to the determination of the quantum number l.

One naturally inquires why the quantum theory of the Zeeman effect given earlier (which in fact is valid only for lines which belong to series of singlets) does not apply in general. The reason must obviously be, that the Larmor procession, whose value was assumed to be $eH/2mc$ in the earlier proof, has not always this value: and this again can only mean that the ratio of magnetic moment to mechanical angular momentum is somehow different in the case of the anomalous Zeeman effect from what is asserted in Larmor's theory, where the ratio is that corresponding to the revolution of an electron in an orbit large compared with its own size. We conclude therefore that the existence of a g-factor different from unity indicates that the ratio of magnetic moment to angular momentum is not the same for the intrinsic spin of the electron as it is for the orbital motion. Now a state of the atom in which the angular momentum is due solely to electron-spin is specified by $l=0$, $s=\frac{1}{2}$, $j=\frac{1}{2}$: and the g-factor then has the value 2: so we are led to suspect that for the electron-spin, the ratio of magnetic moment to angular momentum is twice as great as it is in the case of an electron circulating in an orbit: that is, it is e/mc instead of $e/2mc$. *The magnetic moment of the spinning electron is therefore conjectured to be* $e\hbar/2mc$; and it is found that with this assumption the Landé g-factor can be satisfactorily explained in all cases. The cause of the anomalous Zeeman effect was therefore now revealed.

Some striking confirmations of the validity of the correspondence-principle were obtained in a series of papers which followed a discovery made in 1922 by Miguel A. Catalán,[2] a Spanish research student of Alfred Fowler's at the Imperial College in London. When investigating the spectrum of manganese, Catalán noticed that there was a marked tendency for lines of similar character to

[1] cf. A. Sommerfeld and W. Heisenberg, *ZS. f. P.* xxxi (1922), p.131
[2] *Phil. Trans.* ccxxiii (1922), p. 127 (at p. 146)

appear in groups, and that these groups included some of the most intense lines in the spectrum: for instance, a group of nine lines between λ4455 and λ4462. For this kind of regularity he suggested the name *multiplet*. The lines arise from the combination of multiplet energy-levels. Spectral lines, particularly in multiplets, their relative intensities and their Zeeman components, were studied by many writers [1] in 1924–5, with results satisfactorily in accord with the predictions of the correspondence principle.

The rapid development of spectroscopy from 1913 onwards led to a much fuller understanding of the system of electrons which surrounds the nucleus of an atom. Investigators of this subject naturally based their work on the Newlands-Mendeléev periodic table,[2] and were stimulated by the attempts of A. M. Mayer and J. J. Thomson [3] to explain it in terms of stable configurations of electrons. It was obvious from chemical evidence that the two electrons possessed by the helium atom form a very stable configuration, which may be regarded as constituting a complete 'shell' of electrons surrounding the nucleus. The next atom in order of atomic number, lithium, must have this shell together with one loosely-attached electron outside it, and for the succeeding elements further electrons are added to this second shell until it contains eight electrons, when the tenth element neon is formed, thereby completing the shell and arriving again at a very stable configuration. The eleventh element sodium has these two complete shells together with one loosely-attached electron outside them, and so on.

The chemical evidence on atomic structure was marshalled in 1916–19 by two Americans, G. N. Lewis [4] and Irving Langmuir [5] (*b.* 1881). Lewis began by considering the different kinds of bonds that unite atoms into molecules, interpreting one kind of chemical bond as a couple of electrons held in common by two atoms; many facts, such as the tetrahedral carbon atom which is necessary for

[1] F. M. Walters, *J. Opt. Soc. Amer.* viii (1924), p. 245; O. Laporte, *ZS. f. P.* xxiii (1924), p. 135; xxvi (1924), p. 1. These two writers analysed the iron spectrum. H. C. Burger and H. B. Dorgels, *ZS. f. P.* xxiii (1924), p. 258; L. S. Ornstein and H. C. Burger, *ZS. f. P.* xxiv (1924), p. 41; xxviii (1924), p. 135; xxix (1924), p. 241; xxxi (1925), p. 355; W. Heisenberg, *ZS. f. P.* xxxi (1925), p. 617; xxxii (1925), p. 841; S. Goudsmit and R. de L. Kronig, *Proc. Amst. Ac.* xxviii (1925), p. 418; H. Hönl, *ZS. f. P.* xxxi (1925), p. 340; R. de L. Kronig, *ZS. f. P.* xxxi (1925), p. 885; xxxiii (1925), p. 261; A. Sommerfeld and H. Hönl, *Berlin Sitz.* (1925), p. 141; H. N. Russell, *Nature*, cxv (1925), p. 835; *Proc. N.A.S.* xi (1925), pp. 314, 322; H. N. Russell and F. A. Saunders, *Astroph. J.* lxi (1925), p. 38. This paper, arising out of an investigation of groups of lines in the arc spectrum of calcium, was of great importance for the study of complex spectra. It took into account the simultaneous action of two displaced electrons; F. Hund, *ZS. f. P.* xxxiii (1925), p. 345; xxxiv (1925), p. 296.

[2] cf. p. 11

[3] cf. p. 21

[4] *Journ. Amer. Chem. Soc.* xxxviii (1916), p. 762; Lewis, *Valence and the Structure of Atoms and Molecules*, New York, 1923

[5] *Journ. Amer. Chem. Soc.* xli (1919), p. 868. Somewhat similar ideas were published by W. Kossel, *Ann. d. Phys.* (1916), p. 229, who studied the transfer of electrons from electropositive to electronegative atoms, resulting in the formation of ions.

the understanding of chemical processes in organic substances, indicate that the atom must have a structure in three-dimensional space (in contradistinction to a flat ring system) [1]; and Lewis favoured a cubical form. Langmuir, continuing Lewis's work, concluded that the electrons in any atom are arranged in a series of nearly spherical shells. The outermost occupied shell consists of those electrons that do not belong to a closed configuration. These play the principal part in spectroscopic phenomena, and are known as the *active electrons*: they also play the principal part in chemical phenomena, in which connection they are known as *valence electrons*. The properties of the atoms depend much on the ease with which they are able to revert to more stable forms by giving or taking up electrons. The shells that are completed in helium and neon respectively have already been mentioned: argon (atomic number 18) in addition to the innermost shell of 2 and the next shell of 8, has a third shell of 8: while krypton has four shells of 2, 8, 8 and 18 electrons: and so on. In the light of this model atom, Langmuir explained the chemical properties of the elements, and also their physical properties such as boiling-points, electric conductivity and magnetic behaviour.

It was early realised, however, that the formation of the shells cannot be quite regular: it was suggested in 1920 by R. Ladenburg [2] that in the case of the elements of atomic numbers 21 to 28 inclusive (scandium to nickel) the electrons newly added are not placed in the outermost shell but are used in building up a shell interior to this. This implies that a shell begins to be formed with potassium (atomic number 19) and calcium (atomic number 20) before the third shell is really complete.

It was, however, from the study of X-ray spectra (i.e. the characteristic X-rays which are emitted by solid chemical elements, or compounds of them, when bombarded by a beam of high-energy electrons) [3] that the greatest help was obtained in relating the shell-structure to the chemical elements. It was known that the characteristic X-rays constituted an additive atomic property and therefore the X-ray spectra must belong to the *atoms* of the anti-cathode. Moreover, it was known that they consisted of lines which could be arranged in series, like those of optical spectra: and Moseley's law connecting the progression of X-ray spectra from element to element with the amount of nuclear charge suggested that their origin should be sought in the innermost layers of the atom.

Almost immediately after the publication of Bohr's theory, W. Kossel [4] explained them as being due, like the radiations of optical spectra, to transition-processes: they arise when the atom

[1] On this see also Born, *Verh. d. deutsch. phys. Ges.* xx (1918), p. 230; Landé, *Verh. d. deutsch. phys. Ges.* xxi (1919), p. 2; E. Madelung, *Phys. ZS.* xix (1918), p. 524

[2] *Naturwiss*, viii (1920), p. 5; *ZS. f. Elektrochem*, xxvi (1920), p. 262

[3] cf. p. 16

[4] *Verh. d. deutsch. phys. Ges.* xvi (November 1914), p. 953; xviii (1916), p. 339

is restored to its original state after a disturbance which consists primarily in dislodging an electron from one of the innermost shells: the place thus vacated is filled by an electron which falls from a shell at a greater distance from the nucleus, and the place of this is again filled by an electron from a shell still more remote, and so on. The lowest level may be called, in harmony with Barkla's nomenclature of three years earlier,[1] the K-shell, and the fall of an electron into a vacant place in this shell yields an X-ray of the K-series. The next lowest levels are three known as L_{I}, L_{II} and L_{III}, and so for the others.[2]

Now if Z denotes the atomic number of the atom considered, by taking account of Z in the calculation on page 111, we see that according to Bohr's original theory, the frequency of the radiation emitted when an electron passes from a circular orbit of angular momentum $n\hbar$ to one of angular momentum $p\hbar$ is

$$\frac{2\pi^2 me^4 Z^2}{h^3}\left(\frac{1}{p^2}-\frac{1}{n^2}\right).$$

But if we put $p=1$, this formula represents precisely the frequencies of the K-lines. If $p=2$, it represents the L-lines; and so on: from which fact we infer that the electrons in the K, L, M, . . . shells move in orbits which have respectively the principal quantum numbers 1, 2, 3, . . .[3] *The principal quantum number increases by unity from each shell to the next.* That the characteristic X-rays are of high frequency as compared with the radiations in optical spectra is explained by the presence of the factor Z^2 in the formula: a factor whose presence accounts at once for Moseley's law that the square root of the frequency of any particular line, such as the K_α-line, is proportional to the atomic number.

In the early days of X-ray spectroscopy, lines of the K-series could not be observed for the lighter elements (atomic numbers below 11), because their wave-lengths were longer than those of X-rays and yet shorter than that of ultra-violet light. The gap between X-rays and the ultra-violet was filled about 1928 by Jean Thibaud[4] of Paris, Erik Bäcklin[5] of Upsala, and A. P. R. Wadlund[6] of Chicago, and it was then found possible to trace the K-lines continuously down to the lightest elements. As might be expected, since the K-lines represent transitions down to the level of principal number 1, they pass into the Lyman series of hydrogen, while the L-lines, which represent transitions down to the principal number 2,

[1] cf. p. 16

[2] The γ-rays emitted by radio-active bodies come from the nucleus of the atom, and depend on nuclear levels, so they constitute a phenomenon altogether different from the characteristic X-rays of the K. L, M, . . . series.

[3] Sommerfeld, *Ann. d. Phys.*(4) li (1916), pp. 1 and 125

[4] *Phys. ZS.* xxix (1928), p. 241; *Journ. Opt. Soc. Amer.* xvii (1928), p. 145

[5] *Inaug. Diss. Uppsala Universitets Årsskrift*, 1928

[6] *Proc. Nat. Ac. Sc.* xiv (1928), p. 588

pass into the Balmer series, and the M-lines, which represent transitions down to the principal number 3, pass into the Paschen series of hydrogen.

The Bohr theory, as applied by Kossel, yields at once the laws of absorption of X-rays. Remembering that the first line of the K-series is emitted when an electron falls from the L-shell to the K-shell, it is obvious that this line could not be expected to appear in the absorption-spectrum : for if in absorption an electron were dislodged from the K-shell, there would normally be no vacant place in the L-shell to receive it : indeed the K-absorption only sets in suddenly, when the incident energy is sufficient to separate a K-electron completely from the atom : that is, *the absorption-edge coincides with the series-limit.* This explains why in X-ray spectra there are no absorption-lines,[1] but only absorption-edges : and *the frequency of every line in the X-ray emission-spectrum is the difference of the frequencies of two absorption-edges.*

What was known in 1921 regarding the structure of the atom in relation to the physical and chemical properties of the elements was set forth in an extensive survey by Bohr,[2] whose principles were vindicated in the following year in a somewhat dramatic way. One of the missing elements in the Newlands-Mendeléev table was that of atomic number 72. Now the elements immediately preceding (of atomic numbers 57 to 71 inclusive) belong to the group of the 'rare earths,' and it was expected by many chemists that number 72 would also belong to this group. Indeed in 1911 G. Urbain,[3] by fractionation of the earths of gadolinite, believed that he had discovered a new rare earth to which he gave the name of *Celtium* ; and this was later identified by Urbain and Dauvillier with the missing element 72. Bohr, however, gave a rational interpretation of the occurrence of the rare earths in the periodic system, asserting that they represent a gradual completion of the shell of electrons for which the principal quantum number is 4, while the number of electrons in the shells of principal quantum numbers 5 and 6 remains unchanged. With the rare-earth lutecium (71) the shell $n=4$ attains its full complement of 32 electrons, and it follows that the element 72 cannot be a rare earth, but must have an additional electron in the shells $n=5$ or $n=6$: it must in fact be a homologue of zirconium. In 1922 D. Coster and G. Hevesy of Copenhagen verified this prediction[4] : examining the X-ray spectrum of a Norwegian zirconium mineral, they found lines which, by Moseley's rule, must certainly belong to an element of atomic number 72 : and for this they proposed the name of *Hafnium* (Hafniae = Copenhagen). Its chemical properties showed that it was undoubtedly analogous to titanium (22) and zirconium (40).

[1] For a refinement of this general statement cf. Kossel, *ZS. f. P.* i (1920), p. 119.
[2] *Fysisk Tideskrift*, xix (1921), p. 153 = *Theory of Spectra and Atomic Constitution*, Cambridge, 1922 [3] *Comptes Rendus*, clii (1911), p. 141 ; *Chem. News*, ciii (1911), p. 73
[4] *Nature*, cxi (1923), p. 79

In 1922 and the following years great additions were made to the accurate knowledge of X-ray spectra and their relation to atomic number.[1] Guided by this work, in 1923 N. Bohr and D. Coster [2] introduced new symbols for the layers of electrons in the atoms, based on their spectroscopic behaviour with respect to X-rays. The innermost group was now denoted by $1(1,1)\mathrm{K}$, the next groups by $2(1,1)\mathrm{L_{I}}$, $2(2,1)\mathrm{L_{II}}$, $2(2,2)\mathrm{L_{III}}$, and so on outward. Here the letters with the Roman subscript numbers indicate the previously-recognised shells with their subsidiary levels, while the symbols of the form $n(k, j)$ define the subsidiary levels more closely, n being Bohr's principal quantum number, while k and j are whole numbers which were later to be identified with Sommerfeld's azimuthal quantum number and a function of Sommerfeld's inner quantum number respectively. Then in January 1924 A. Dauvillier [3] showed experimentally (by examining the absorption relative to the level) that the 8 electrons in the L-level must be partitioned into sub-groups of 2, 2 and 4. These results quickly led to the understanding of the orbits and energies of the various groups of electrons which was finally accepted, and which was proposed in 1924–5 by Edmund C. Stoner of Cambridge [4] (whose arguments were based on physical reasoning) and J. D. Main Smith of Birmingham [5] (who approached the matter from the chemical side). According to this system, to each complete shell corresponds a definite value of the principal quantum number n. Within this shell the subsidiary quantum number l can take the values $0, 1, \ldots (n-1)$: and the inner quantum number j of an electron can then take the values $(l+\frac{1}{2})$ or $(l-\frac{1}{2})$, (unless $l=0$, in which case j can take only the value $\frac{1}{2}$). The number of electrons in the sub-group (n, l, j) is simply $(2j+1)$, and therefore the total number of electrons with the quantum numbers n and l is $2(2l+1)$, and the total number of electrons in the n-shell is

$$2\{1+3+5+7+\ldots+(2n-1)\}$$

or

$$2n^2.$$

Helium has altogether 2 electrons, forming a complete K-shell. For neon there are 10 electrons, namely 2 forming a complete K-shell and 8 forming a complete L-shell. For argon there are 18 electrons, of which 2 form a complete K-shell, 8 form a complete L-shell and 8 form an incomplete M-shell : the M-shell is, however, complete

[1] Dirk Coster, *Phil. Mag.* xliii (1922), p. 1070 ; xliv (1922), p. 546 ; A. Landé, *ZS. f. P.* xvi (1923), p. 391 ; Manne Siegbahn and A. Žáček, *Ann. d. Phys.* lxxi (1923), p. 187 ; M. Siegbahn and B. B. Ray, *Ark. f. Mat, Ast. och Fys.* xviii (1924), No. 19 ; M. Siegbahn and R. Thoraeus, *Phil. Mag.* xlix (1925), p. 513 ; cf. L. de Broglie and A. Dauvillier, *Phil. Mag.* xlix (1925), p. 752

[2] *ZS. f. P.* xii (1923), p. 342

[3] *Comptes Rendus*, clxxviii (1924), p. 476

[4] *Phil. Mag.* xlviii (1924), p. 719

[5] *J. Chem. Ind.* xliii (1924), p. 323 ; xliv (1925), p. 944 ; *Chemistry and Atomic Structure*, London, 1924 ; cf. A. Sommerfeld, *Ann. d. Phys.* lxxvi (1925), p. 284 ; *Phys. ZS.* xxvi (1925), p. 70

in the next noble gas, krypton, which has 2 K-, 8 L-, 18 M-, and 8 N-electrons. For any complete shell the orbital, spin and total angular momentum are all zero.

We have seen that the number of electrons in the sub-group (n, l, j) is $(2j+1)$: but for a given value of j, the magnetic quantum number m_j can take precisely the $(2j+1)$ values $j, j-1, \ldots, -j$. Thus there is one and only one electron corresponding to each distinct state or energy-level, i.e. each distinct set of the four quantum numbers (n, l, j, m_j). In 1924 Pauli[1] based on this fact a general principle, that *two electrons in a central field can never be in states of binding which have the same four quantum numbers.* This assertion can be extended to systems in which there is not a single central field : e.g. it applies to electrons which are in the field of two nuclei at the same time : the states of these electrons can be described by quantum numbers, and it is still true that no two electrons can have the same state, i.e. be described by the same set of quantum numbers. The statement in this general form is called *Pauli's exclusion principle.* It is valid for protons as well as for electrons, and indeed for all elementary particles whose spin is $\frac{1}{2}\hbar$.

Another discovery of Pauli's, made in the same year,[2] related to a structure much finer than the ordinary multiplet structure, which is observed in some spectra, and which is called the *hyperfine structure* of spectral lines. Pauli showed that this is to be ascribed not to the electron-shells, but to the influence of the atomic nucleus, which may itself have an angular momentum and a magnetic moment : and in the case when the spectrum is that of a mixture of isotopes, the differences in nuclear mass of the isotopes will also cause small differences of position of lines in their spectra, and so contribute to the hyperfine structure.

In 1923 the domain of quantum theory was enlarged, when the diffraction of a parallel beam of radiation by a grating was explained on quantum principles by William Duane.[3] Consider an infinite grating with the spacing d between its rulings. If the grating moves with constant velocity in a direction in its own plane perpendicular to the rulings, it will return to its original aspect when it has moved through a distance d : so we can regard it as a periodic system to which the Wilson-Sommerfeld quantum rule

$$\int p\,dq = nh$$

can be applied, where p denotes momentum in the direction of the surface of the grating perpendicular to its rulings, the spacing d being the domain over which the integration must be extended : and therefore

$$pd = nh,$$

[1] *ZS. f. P.* xxxi (1925), p. 765 [2] *Naturwiss,* xii (1924), p. 741

[3] *Proc. N. A. S.* ix (1923), p. 158. Duane's treatment was considerably improved as regards its justification by A. H. Compton, *ibid.* p. 359.

so *the grating can pick up momentum* p *only in multiples of* h/d: the momentum of radiation is transferred to and from matter in quanta. If a photon of energy $h\nu$, and therefore of momentum $h\nu/c$, falls on the grating in a direction making an angle i with the normal, and is diffracted in a direction making an angle r with the normal, then taking the components of momentum in a direction of the surface of the grating at right angles to its rulings, we have the equation of conservation of momentum

$$(h\nu/c)\,(\sin i - \sin r) = nh/d$$

which in the language of the wave-theory would be

$$n\lambda = d\,(\sin i - \sin r).$$

This is the ordinary equation giving the directions of the diffracted radiation, now obtained from the corpuscular (photon) theory of light.

It may be noted that Duane's equation $pd = nh$ can be applied to a photon, if d be interpreted as wave-length, so we obtain $p\lambda = h$: since $\lambda = c/\nu$, this shows that the momentum of the photon is $h\nu/c$.

Duane's principle is closely related to a principle introduced later in the same year (1923) by L. de Broglie, which will be considered in Chapter VI.

The Duane method was extended to finite gratings, including even the case of only two reflecting points, by P. S. Epstein and P. Ehrenfest in 1924.[1]

[1] *Proc. N. A. S.* x (1924), p. 133; xiii (1927), p. 400

Chapter V

GRAVITATION

We have seen [1] that for many years after its first publication, the Newtonian doctrine of gravitation was not well received. Even in Newton's own University of Cambridge, the textbook of physics in general use during the first quarter of the eighteenth century was still Cartesian: while all the great mathematicians of the Continent—Huygens in Holland, Leibnitz in Germany, Johann Bernoulli in Switzerland, Cassini in France—rejected the Newtonian theory altogether.

This must not be set down entirely to prejudice: many well-informed astronomers believed, apparently with good reason, that the Newtonian law was not reconcilable with the observed motions of the heavenly bodies. They admitted that it explained satisfactorily the first approximation to the planetary orbits, namely that they are ellipses with the sun in one focus: but by the end of the seventeenth century much was known observationally about the departures from elliptic motion, or *inequalities* as they were called, which were presumably due to mutual gravitational interaction: and some of these seemed to resist every attempt to explain them as consequences of the Newtonian law.

The inequalities were of two kinds: first, there were disturbances which righted themselves after a time, so as to have no cumulative effect: these were called *periodic* inequalities. Much more serious were those derangements which proceeded continually in the same sense, always increasing the departure from the original type of motion: these were called *secular* inequalities. The best known of them was what was called the *great inequality of Jupiter and Saturn*, of which an account must now be given.

A comparison of the ancient observations cited by Ptolemy in the *Almagest* with those of the earlier astronomers of Western Europe and their more recent successors, showed that for centuries past the mean motion, or average angular velocity round the sun, of Jupiter, had been continually increasing, while the mean motion of Saturn had been continually decreasing. This indicated some striking consequences in the remote future. Since by Kepler's third law the square of the mean motion is proportional to the inverse cube of the mean distance, the decrease in the mean motion of Saturn implied that the radius of his orbit must be increasing, so that this planet, the most distant of those then known, would be always becoming more remote, and would ultimately, with his attendant ring and

[1] cf. Vol. I, pp. 29–31

satellites, be altogether lost to the solar system. The orbit of Jupiter, on the other hand, must be constantly shrinking, so that he must at some time or other either collide with one of the interior planets, or must be precipitated on the incandescent surface of the sun.

No explanation of the secular inequality of Jupiter and Saturn could be obtained by any simple and straightforward application of Newton's gravitational law, and the French Academy of Sciences offered a prize in 1748, and again in 1752, for a memoir relating to these two planets. On each occasion Euler made considerable advances [1] in the general treatment of planetary perturbations, and received the award : but the result of his investigations was to make the observed secular accelerations of Jupiter and Saturn more mysterious than ever, for they appeared to be quite inconsistent with the tolerably complete theory which he created. Lagrange, who wrote on the problem [2] in 1763, and gave a still more complete discussion, likewise failed to obtain a satisfactory agreement with the observations.

In 1773 the matter was taken up by Laplace.[3] He began by carrying the approximation to a higher order than his predecessors, and was surprised to find that in the final expression for the effect of Jupiter's disturbing action upon the mean motion of Saturn, the terms cancelled each other out. The same result, as he showed, held for the effect of any planet upon the mean motion of any other: thus *the mean motions of the planets cannot have any secular accelerations whatever as a result of their mutual attractions.* Laplace accordingly concluded that the accelerations observed in the case of Jupiter and Saturn could not be genuinely secular : they must really be periodic, though the period might be immensely long.

With this key to the mystery, he completely solved it, in a great memoir of 1784.[4] He realised that an inequality of long period could be produced only by a term of long period in the perturbing function : denoting this term by $p \sin qt$, then in order that it may be of long period, q must be extremely small. By a double integration with respect to the time, such as happens in the course of solving the differential equations, this term would become $(p/q^2) \sin qt$, and (p/q^2) might be quite large even though p were very small. Thus a great inequality of long period might be produced by a term in the perturbing function which was so small that it had been neglected altogether by preceding investigators. This explained why Euler and Lagrange had failed to solve the problem, and nothing remained to be done except to inquire more closely into the identity of the term in the perturbing function. Now five times the mean motion of Saturn is very nearly equal to twice the mean motion of Jupiter : so if n, n' are the mean motions, then $5n-2n'$

[1] *Recueil des pièces qui ont remporté les prix de l'Acad.*, tome vii (1769)
[2] *Mélanges de phil. et de math. de la Soc. Roy. de Turin pour l'année* 1763, p. 179 (1766)
[3] *Mém. des Savans étrang.* vii (1776). Read 10 Feb. 1773
[4] *Mém. de l'Acad.*, 1784, p. 1

is very small: and the term in the perturbing function whose argument is $(5n - 2n')t$, which would have an extremely small coefficient, would satisfy all the conditions required. Thus Laplace was able to assert that *the great inequality of Jupiter and Saturn is not a secular inequality, but is an inequality of long period, in fact* 929 *years: and it is due to the fact that the mean motions of the two planets are nearly commensurable.* When the results of his calculations were compared with the observations, the agreement was found to be perfect.[1]

The story of the great inequality of Jupiter and Saturn illustrates a distinctive feature of the situation, namely that the truth of the Newtonian or any other law of gravitation cannot be tested by means of controlled experiments in a laboratory, and its verification must depend on the comparison of astronomical observations, extended over centuries, with mathematical theories of extreme complexity.

After the triumphant conclusion of Laplace's researches on the great inequality of Jupiter and Saturn, there was still outstanding one unsolved problem which formed a serious challenge to the Newtonian theory, namely the secular acceleration of the mean motion of the moon. From a study of ancient eclipses recorded by Ptolemy and the Arab astronomers, Halley[2] had concluded in 1693 that the mean motion of the moon has been becoming continually more rapid ever since the epoch of the earliest recorded observations. The mean distance of our satellite must therefore have been continually decreasing, and it seemed that at some time in the remote future the moon must be precipitated on the earth. The Academy of Sciences of Paris proposed the subject for the prize in 1770, and again in 1772 and 1774, and prizes were awarded to Euler[3] and Lagrange,[4] who made valuable contributions to general dynamical astronomy: on the question proposed, however, they found only the negative results that no secular inequality could be produced by the action of Newtonian gravitation when the heavenly bodies were regarded as spherical, and, moreover, that the observed phenomena could not be explained by taking into account the departures of the figures of the earth and moon from sphericity. Laplace now took up the matter, and showed, first, that the effect was not due to any retardation of the earth's diurnal rotation due to the resistance of the aether: he then investigated the consequences of another supposition, namely that gravitational effects are propagated with a velocity which is finite[5]: but this also led to no satisfactory issue, and at last he found the true solution,[6] which

[1] As Laplace's discovery showed, the existence of 'small divisors' makes it a matter of great difficulty to investigate the convergence of the series that occur in Celestial Mechanics; cf E. T. Whittaker, *Proc. R.S. Ed.* xxxvii (1917), p. 95.

[2] Phil. Trans. xvii (1693), p. 913

[3] *Recueil des pièces qui ont remporté les prix de l'Acad.* ix (1777)

[4] *Recueil des pièces qui ont remporté les prix de l'Acad.* ix (1777), for the competition of 1772; *Mém. des Savans Étrang.* vii, for that of 1774

[5] cf. Vol. I, p. 207

[6] It was presented to the Academy on 19 March 1787; *Mém. de l'Acad.* 1786, p. 235 (published 1788).

may be described as follows. The mean motion of the moon round the earth depends mainly on the moon's gravity to the earth, but is slightly diminished by the action of the sun upon the moon. This solar action, however, depends to a certain extent on the eccentricity of the terrestial orbit, which is slowly diminishing, as a result of the action of the planets on the earth. Consequently the sun's mean action on the moon's mean motion must also be diminishing, and hence the moon's mean motion must be continually increasing, which is precisely the phenomenon that is observed. The acceleration of the moon's mean motion will continue as long as the earth's orbit is approaching a circular form: but as soon as this process ceases, and the orbit again becomes more elliptic, the sun's mean action will increase and the acceleration of the moon's motion will be converted into a retardation. The inequality is therefore not truly secular, but periodic, though the period is immensely long, in fact millions of years. This striking vindication of the Newtonian theory came exactly a century after the publication of the *Principia*.

'The moon, in the present day,' wrote Robert Grant,[1] 'is about two hours later in coming to the meridian than she would have been if she had retained the same mean motion as in the time of the earliest Chaldean observations. It is a wonderful fact in the history of science that those rude notes of the priests of Babylon should escape the ruin of successive empires, and, finally, after the lapse of nearly 3,000 years, should become subservient in establishing a phenomenon of so refined and complicated a character as the inequality we have just been considering.'

In the nineteenth and twentieth centuries, however, many astronomers formed the opinion that Laplace's great memoir had not completely cleared up the situation. The fact to be explained is that the moon has relative to the sun an apparent acceleration of its mean motion of about 22″ per century per century.[2] Laplace's theoretical value was of about this amount, but J. C. Adams [3] found, by including terms of higher order in the calculation, that Laplace's value was much too great, the amount explicable by purely gravitational causes being only 12·2″ per century per century: and his conclusion was substantiated by later workers in lunar theory. In 1905 P. H. Cowell [4] redetermined (from ancient eclipses) the observed secular acceleration, and found that it was almost twice the theoretical value, and that there was also deducible from observation a secular acceleration of the sun (i.e. of the earth's orbital motion). His own tentative explanation [5] depended on the

[1] *History of Physical Astronomy* (London, 1852), p. 63

[2] There is some confusion of language on this subject. The coefficient of t^2 in the expression for the longitude is about 11″, and the true secular acceleration is therefore about 22″; but many writers speak of the acceleration as '11″ in a century.'

[3] *Phil. Trans.* cxliii (1853), p. 397; *Mon. Not. R.A.S.* xl (1880), p. 472

[4] *Mon. Not. R.A.S.* lxv (1905), p. 861

[5] *Mon. Not. R.A.S.* lxvi (1906), p. 352

notion of tidal friction, while J. H. Jeans [1] suggested a modification of the Newtonian law, and E. A. Milne [2] proposed a dependence of the Newtonian constant of gravitation on the age of the universe. These matters were discussed by J. K. Fotheringham [3] in 1920 and in 1939 by H. Spencer Jones,[4] who showed that the secular accelerations of the sun, Mercury, and Venus are proportional to their mean motions, and can be accounted for by the retardation of the earth's axial rotation by tidal friction.[5] This retardation must also produce, in addition to its direct apparent effect, a real secular acceleration of the moon's mean motion, in order that the total angular momentum of the earth-moon system may be conserved: but the amount of this acceleration cannot be predicted theoretically.

We may now refer to other phenomena in the solar system which were not explained with complete satisfaction by Newton's formula. The anomalous motion of the perihelion of the planet Mercury has already been referred to [6]: it might possibly be accounted for if the inverse square law is modified by adding a term involving the velocities of the bodies [7]: or it might, as H. Seeliger [8] showed, be explained by the attraction of the masses forming the zodiacal light. The node of the orbit of Venus was also found by Newcomb [9] to have a secular acceleration which was five times the probable error, and for which no explanation could be offered: and a secular increase in the mean motion of the inner satellite of Mars, discovered in 1945 by B. P. Sharpless,[10] is so far not accounted for.

Certain comets also present problems. Among them the best known is an object which was discovered by Jean-Louis Pons in 1818, but is generally called *Encke's comet* from a long series of memoirs [11] devoted to it by J. F. Encke, who showed in 1819 that it was periodic, with a period of 1,207 days, and later that its motion showed an acceleration which was not explicable by the Newtonian theory. Encke himself proposed to explain this by postulating a resisting medium whose density was inversely proportional to the

[1] *Mon. Not. R.A.S.* lxxxiv (1923), p. 60 (*at* p. 75)

[2] *Proc. R.S.*(A), clvi (1936), p. 62 (*at* p. 81)

[3] *Mon. Not. R.A.S.* lxxx (1920), p. 578

[4] *Mon. Not. R.A.S.* xcix (1939), p. 541

[5] The retardation of the earth's rotation due to tidal friction increases the length of the day by about 1/1000 of a second per century, so each century is 36½ seconds longer than the one preceding. There is, moreover, a variability in the rate of rotation, which is slower in February than in August. This fluctuation, which was discovered by means of clocks formed of vibrating quartz crystals, is probably of meteorological origin.

[6] cf. Vol. I, p. 208

[7] For work more recent than that referred to in Vol. I, cf. Paul Gerber, *ZS. f. Math. u. Phys.* xliii (1898), p. 93; *Ann. d. Phys.* lii (1917), p. 415, and the comments on the latter paper by H. Seeliger, *Ann. d. Phys.* liii (1917), p. 31; liv (1917), p. 38, and S. Oppenheim, ibid. liii (1917), p. 163

[8] *München Ber.* 1906, p. 595

[9] S. Newcomb, *The elements of the four inner planets*; Supplement to the American Ephemeris and Nautical Almanac for 1897; Washington, 1895

[10] *Ast. J.* li (1945), p. 185

[11] Mostly in the *Berlin Abhandlungen* and the *Astronomische Nachrichten*; cf. specially *Comptes Rendus*, xlviii (1858), p. 763

square of the distance from the sun : but O. Backlund, who devoted many years to the study of the comet, showed that Encke's assumption is impossible.[1] An alternative hypothesis is that the comet has encounters with a swarm of meteors.

In 1910 P. H. Cowell and A. C. D. Crommelin [2] computed with great care the motion of another periodic comet, that of Halley, between 1759 and 1910, and predicted the time of perihelion for its return in 1910 : the time deduced later from actual observations was about 2·7 days later than this. The discrepancy could not be accounted for by any defect in the calculations, and it would seem therefore that there is some small disturbing cause or causes at work, other than the gravitational attraction expressed by Newton's law.[3]

At the meeting of the Amsterdam Academy of Sciences on 31 March 1900, Lorentz communicated a paper entitled *Considerations on Gravitation*,[4] in which he reviewed the problem as it appeared at that time—a problem which, as the above recital shows, was still far from a completely satisfactory solution. So many phenomena had been successfully accounted for by applications of electromagnetic theory that it seemed natural to seek in the first place an explanation in terms of electric and magnetic actions. As we have seen,[5] the assumptions of Laplace's investigation, which led him to conclude that the velocity of propagation of gravitation must be vastly greater than that of light, do not to twentieth-century minds seem very plausible [6] : and Lorentz felt free to put forward a theory depending on electromagnetic actions propagated with the speed of light. The first possibility he considered was suggested by Le Sage's concept of ultra-mundane corpuscles.[7] Since it had been found that a pressure against a body could be produced as well by trains of electric waves as by moving projectiles, and that the X-rays with their remarkable penetrating power were essentially electric waves, it was natural to replace Le Sage's corpuscles by vibratory motions. Why should there not exist radiations far more penetrating than even the X-rays, which might account for a force which, so far as is known, is independent of all intervening matter ?

Lorentz therefore calculated the interaction between two ions on the assumption that space is traversed in all directions by trains of electric waves of very high frequency. If an ion P is alone in a

[1] Backlund's conclusions are summarised in *Bull. astronomique*, xi (1894), p. 473; cf. also A. Wilkens, *Astr. Nach.* cxcvi (1914), p. 57. On the perturbations of Encke's comet cf. D. Brouwer, *Ast. J.* lii (1947), p. 190.

[2] *Investigation of the motion of Halley's Comet from 1759 to 1910*; Appendix to the 1909 volume of *Greenwich Observations* (1910). *Essay on the return of Halley's comet, Publ. Astr. Ges., Lpz.*, No. 23 (1910)

[3] On the effect of loss of mass by evaporation when a comet is near the sun; cf. F. Whipple, *Astroph. J.* cxi (1950), p. 375

[4] *Proc. Amst. Acad.* ii (1900), p. 559; French translation in *Arch. Néerl.* vii (1902), p. 325

[5] Vol. I, pp. 207–8

[6] The same remark applies to the ideas of R. Lehmann-Filhes, *München. Ber.* xxv (1895), p. 71.

[7] cf. Vol. I, p. 31

field in which the propagation of waves takes place equally in all directions, the mean force on it will vanish. But the situation will be different as soon as a second ion Q has been placed in the neighbourhood of P : for then, in consequence of the vibrations emitted by Q after it has been exposed to the rays, there might be a force on P, of course in the direction of the line QP. It was found, however, that this force could exist only if in some way or other electromagnetic energy were continually disappearing : and after full consideration, Lorentz concluded that the assumptions he had made could not provide a satisfactory explanation of gravitation.

He then considered a second hypothesis, which may be regarded as having been foreshadowed in the one-fluid electrical theory of Watson, Franklin and Aepinus.[1] According to this theory, as developed in 1836 by O. F. Mossotti[2] (1791–1863), electricity is conceived as a continuous fluid, whose atoms repel each other. Material molecules are also supposed to repel each other, but to have with the aether-atoms a mutual attraction, which is somewhat greater than the mutual repulsion of the particles which repel. The composition of these forces accounts for gravitation, except at very small distances, where the same mechanism accounts for cohesion.

Wilhelm Weber (1804–91) of Göttingen and Friedrich Zöllner[3] (1834–82) of Leipzig developed this conception into the idea that all ponderable molecules are associations of positively and negatively charged electrical corpuscles, with the condition that the force of attraction between corpuscles of unlike sign is somewhat greater than the force of repulsion between corpuscles of like sign. If the force between two electric units of like charge at a certain distance is α dynes, and the force between a positive and a negative unit charge at the same distance is γ dynes, then, taking account of the fact that a neutral atom contains as much positive as negative electric charge, it was found that $(\gamma-\alpha)/\alpha$ need only be a quantity of the order 10^{-35} in order to account for gravitation as due to the difference between α and γ.

At the time of Lorentz's paper, no strong physical reason for an assumption of this kind could be given. Many years afterwards Eddington[4] suggested one. He had taken to heart a warning uttered by Mach.[5] 'Even in the simplest case, in which apparently we deal with the mutual action between only two particles, it is impossible to disregard the rest of the universe. Nature does not begin with elements, as we are forced to do. Certainly it is fortunate for us

[1] cf. Vol. I, p. 50

[2] *Sur les forces qui régissent la constitution intérieure des corps, apperçu pour servir à la détermination de la cause et des lois de l'action moléculaire* (Turin, 1836).

[3] *Erklärung der universellen Gravitation*, Leipzig, 1882 ; pp. 67–82 deal with Weber's contributions ; cf. also J. J. Thomson, *Proc. Camb. Phil. Soc.* xv (1910), p. 65

[4] *Fundamental Theory* (Cambridge, 1946), p. 102

[5] E. Mach, *Die Mechanik in ihrer Entwickelung* (5th edn., 1904), p. 249 ; English translation (London, 1893), p. 235

that we can sometimes turn away from the overwhelming All, and allow ourselves to study isolated facts. But we must not forget ultimately to amend and complete our views by taking into account what had been omitted.' Eddington applied Mach's general principle to the interaction between two electric charges. If they are of opposite sign, all their lines of force run from one to the other, and the two together may be regarded as a self-contained system which is independent of the rest of the universe: but if the two charges are of the same sign, then the lines of force from each of them must terminate on other bodies in the universe, and it is natural to expect that these other bodies will have some influence on the nature of the interaction between the charges. Following up this idea by a calculation, Eddington arrived at the conclusion that when two protons are at a distance r apart, which is of the same order of magnitude as the radius of an atomic nucleus, their mutual energy contains, in addition to the ordinary electrostatic energy corresponding to the inverse-square law, a term of the form

$$Ae^{-\frac{r^2}{k^2}}$$

where A and k are constants: if it could be supposed that this is correct, and is an asymptotic approximation, valid for values of r of nuclear dimensions, to a function whose asymptotic approximation, valid for values of r large compared with nuclear dimensions, is inversely proportional to r, then there would obviously be a possibility of accounting on these lines for gravitation.

From 1904 onwards the Newtonian law of gravitation was examined in the light of the relativity theory of Poincaré and Lorentz. This was done first by Poincaré,[1] who pointed out that if relativity theory were true, gravity must be propagated with the speed of light, and who showed that this supposition was not contradicted by the results of observation, as Laplace had supposed it to be. He suggested modifications of the Newtonian formula, which were afterwards discussed and further developed by H. Minkowski[2] and by W. de Sitter.[3] It was found that relativity theory would require secular motions of the perihelia of the planets, which however would be of appreciable amount only in the case of Mercury, and even in that case not great enough to account for the observed anomalous motion.

In 1907 Planck[4] broke new ground. It had been established by the careful experiments of R. v. Eötvös[5] that *inertial mass* (which determines the acceleration of a body under the action of a given

[1] *Comptes Rendus*, cxl (1905), p. 1504; *Palermo Rend.* xxi (1906), p. 129

[2] *Gött. Nach.* 1908, p. 53

[3] *Mon. Not. R.A.S.* lxxi (1911), p. 388; cf. also F. Wacker, *Inaug. Diss.*, Tübingen, 1909

[4] *Berl. Sitz.* 13 June 1907, p. 542, specially at p. 544

[5] *Math. u. nat. Ber. aus Ungarn*, viii (1891), p. 65

force) and *gravitational mass* (which determines the gravitational forces between the body and other bodies) are always exactly equal: which indicates that *the gravitational properties of a body are essentially of the same nature as its inertial properties.* Now, said Planck, all energy has inertial properties, and therefore *all energy must gravitate.* Six months later Einstein[1] published a memoir in which he introduced[2] what he later called the *Principle of Equivalence*, which may be thus described:

Consider an observer who is enclosed in a chamber without windows, so that he is unable to find out by direct observation whether the chamber is in motion relative to an outside world or not. Suppose the observer finds that any object in the chamber, whatever be its chemical or physical nature, when left unsupported, falls towards one particular side of the chamber with an acceleration f which is constant relative to the chamber. The observer would be justified in putting forward either of two alternative explanations to account for this phenomenon:

(i) he might suppose that the chamber is 'at rest,' and that there is a *field of force*, like the earth's gravitational field, acting on all bodies in the chamber, and causing them if free to fall with acceleration f: or

(ii) he might explain the observed effects by supposing that *the chamber is in motion*: if he postulates that in the outside world there are co-ordinate axes (C) relative to which there is no field of force, and if he moreover supposes the chamber to be in motion relative to these axes (C) with an acceleration equal in magnitude but opposite in direction to f, then it is obvious that free bodies inside the chamber would have an acceleration f relative to the chamber.

The observer has no criterion enabling him to tell which of these two explanations is the true one. If we could say definitely that the chamber is at rest, then explanation (i) would be true, while if we could say definitely that the axes (C) are at rest, then explanation (ii) would be true. But by the Principle of Relativity, we cannot give a preference to one of these sets of axes over the other: we cannot say that one of them is moving and the other at rest: and we must therefore regard the two explanations as equally valid, or, in other words, must assert that a homogeneous field of force is equivalent to an apparent field which is due to the accelerated motion of one set of axes relative to another: *a uniform gravitational field is physically equivalent to a field which is due to a change in the co-ordinate system.*

In this paper Einstein also showed[3] by combining Doppler's principle with the principle of equivalence, that a spectral line generated by an atom situated at a place of very high gravitational potential, e.g. at the sun's surface, has, when observed at a place

[1] *Jahrb. d. Radioakt.* iv (4 Dec. 1907), p. 411; cf. Einstein, *Ann. d. Phys.* xxxv (1911), p. 898

[2] At p. 454 [3] At pp. 458–9

of lower potential, e.g. on the earth, a greater wave-length than the corresponding line generated by an identical atom on the earth. This may also be shown very simply as follows. Denoting by Ω the gravitational potential at the sun's surface, the energy lost by a photon of frequency ν in escaping from the sun's gravitational field is $\Omega \times$ the mass of the photon, or $\Omega h\nu/c^2$. Remembering that the energy $h\nu$ is hc/λ, we see that the wave-length of the solar radiation as measured by the terrestial observer is $1+(\Omega/c^2)$ times the wave-length of the same radiation when produced on earth.[1]

In 1911 Einstein followed up this work by an important memoir,[2] in which he argued that since light is a form of (electromagnetic) energy, therefore light must gravitate, that is, *a ray of light passing near a powerfully gravitating body such as the sun, must be curved*: and *the velocity of light must depend on the gravitational field.*

Einstein's paper was the starting-point of a theory published shortly afterwards by Max Abraham.[3] Accepting the principles that the velocity of light c depends on the gravitational potential, and that the law of gravitation might be expressed by a differential equation satisfied by c, he postulated that the negative gradient of c indicates the direction of the gravitational force, and that the energy-density in a statical gravitational field is proportional to $c^{-1}(\text{grad } c)^2$. Einstein himself at almost the same time published[4] a somewhat different theory, in which the equations of motion of a particle in a statical gravitational field, when gravity only is acting, are

$$\frac{d}{dt}\left(\frac{1}{c^2}\frac{dx}{dt}\right) = -\frac{1}{c}\frac{\partial c}{\partial x},$$

and similar equations in y and z.

To the same period belong the theories of G. Nordström[5] and Gustav Mie.[6] Though Mie's theory has not survived as the permanent basis of mathematical physics, it had a marked influence on thought, and some of its ideas appeared later in the researches of other workers. It aimed at being a complete theory of physics,

[1] cf. also J. M. Whittaker, *Proc. Camb. Phil. Soc.* xxiv (1928), p. 414. The red-displacement due to a gravitational field with arbitrary motion of the source and of the observer was calculated by H. Weyl in the fifth edition (1923) of his *Raum Zeit Materie*, Anhang III.

[2] *Ann. d. Phys.*(4) xxxv (21 June 1911), p. 898

[3] *Lincei Atti*, xx (Dec. 1911), p. 678; *Phys. ZS.* xiii (1912), pp. 1, 4, 176, 310, 311, 793; *N. Cimento*(4) iv (Dec. 1912), p. 459

[4] *Ann. d. Phys.* xxxviii (Feb. 1912), pp. 355, 443. A controversy followed, for which see Abraham, *Ann. d. Phys.* xxxviii (1912), p. 1056; xxxix (1912), p. 444; and Einstein, *Ann. d. Phys.* xxxviii (1912), p. 1059; xxxix (1912), p. 704.

[5] *Phys. ZS.* xiii (Nov. 1912), p. 1126; *Ann. d. Phys.*(4) xl (April 1913), p. 856; ibid. xlii (Oct. 1913), p. 533; *Phys. ZS.* xv (1914), p. 375; cf. A. Einstein and A. D. Fokker, *Ann. d. Phys.*(4) xliv (1914), p. 321; M. v. Laue, *Jahrb. d. Rad. u. El.* xiv (1917), p. 263

[6] *Ann. d. Phys.* xxxvii (1912), p. 511; xxxix (1913), p. 1; xl (1913), p. 1; *Phys. ZS.* xv (1914), pp. 115, 169, 263; Festschrift für J. Elster u. H. Geitel (Braunschweig, 1915), pp. 251–68; cf. A. Einstein, *Phys. ZS.* xv (1914), p. 176. There is a good short account of Mie's theory in H. Weyl, *Raum Zeit Materie*, 4th Aufl., (Berlin, 1921), § 26.

based on the principle that electric and magnetic fields and electric charges and currents suffice completely to describe all that happens in the material world, so that matter can be constructed from these elements. Moreover, he originated the notion of a single *world-function* from which, by the aid of the Calculus of Variations, all the laws of physical processes could be derived: as we shall see presently, this conception was developed afterwards by Hilbert.

In Mie's theory all happenings, both in the field and in matter, are described by twenty functions, constituting two six-vectors which describe the field and two four-vectors which describe matter: namely

(i) a six-vector formed of the components of the electric displacement **D** and the magnetic force **H**

(ii) a six-vector formed of the components of the magnetic induction **B** and the electric force **E**

(iii) a four-vector formed of the electric charge and current

(iv) a four-vector formed of the electric scalar and vector potentials.

In the Maxwell-Lorentz theory this last four-vector plays merely a mathematical part: but in Mie's theory its components are physical realities. In Chapter IV of his series of memoirs, Mie discussed quantum theory, and in Chapter V, gravitation.

The next advance owed much to a paper that had been written in 1909 by Harry Bateman[1] (1882–1946). At any place in the earth's gravitational field, take moving rectangular axes (x^1, x^2, x^3) and a measure of time (x^0), such that these axes constitute an inertial system (A), so that the path of a free particle relative to them is (at any rate near the origin) a straight line, and the vanishing of the differential form

$$c^2(dx^0)^2 - (dx^1)^2 - (dx^2)^2 - (dx^3)^2$$

is the condition that a luminous disturbance originating at the point (x^1, x^2, x^3) at the instant x^0, should arrive at the point $(x^1 + dx^1, x^2 + dx^2, x^3 + dx^3)$, at the instant $(x^0 + dx^0)$.

In free aether, where there is no field of force, two different inertial systems either can be derived from each other by simple translation and rotation in ordinary three-dimensional space, or else they have a uniform motion of translation relative to each other (or, of course, a combination of these methods of derivation). But when we move to a distant place in a field of force, e.g. if we move to the antipodes in the earth's gravitational field, although we can here again find axes (say (B)) which are inertial (that is, free particles in their vicinity move relatively to them with uniform velocity in straight lines), a framework (B) does not move with uniform velocity

[1] *Proc. L.M.S.*(2) viii (1910), p. 223; cf. also *Amer. J. Math.* xxxiv (1912), p. 325

relative to a framework (A) (in fact the two frameworks are in accelerated motion relative to each other), so the relation between two inertial frameworks which holds in the relativity theory of Poinacré and Lorentz does not hold when a gravitational field is present. We cannot therefore find co-ordinates (x^0, x^1, x^2, x^3) describing position and time over *the whole* field such that the interval ds at any place in the field is given by the equation

$$(ds)^2 = (dx^0)^2 - \frac{1}{c^2}\left\{(dx^1)^2 + (dx^2)^2 + (dx^3)^2\right\}.$$

Instead of this, we can now find at every place in the field a *local* framework of inertial axes (X^0, X^1, X^2, X^3), such that the interval will be given approximately for points in the neighbourhood of the origin by

$$(ds)^2 = (dX^0)^2 - \frac{1}{c^2}\left\{(dX^1)^2 + (dX^2)^2 + (dX^3)^2\right\}.$$

Let (x^0, x^1, x^2, x^3) now be any co-ordinates specifying position and time over the whole field. Then at each place, the differentials dX^p will be expressible in terms of the x^p and the dx^p by equations of the form

$$dX^p = \sum_{r=0}^{3} a_{pr}\, dx^r.$$

Substituting this in the expression for $(ds)^2$, we have

$$(ds)^2 = \sum_{p,q=0}^{3} g_{pq}\, dx^p dx^q \tag{1}$$

where

$$g_0 = a_{00}^2 - \frac{1}{c^2}(a_{10}^2 + a_{20}^2 + a_{30}^2) \text{ etc.}$$

The vanishing of this form (1) is now the condition that a luminous disturbance originating at the space-time point (x^0, x^1, x^2, x^3) should arrive at the space-time point $(x^0 + dx^0, x^1 + dx^1, x^2 + dx^2, x^3 + dx^3)$. The form (1) must be invariant for *all* transformations of the co-ordinates (x^0, x^1, x^2, x^3); and its coefficients g_{pq}, which are functions of (x^0, x^1, x^2, x^3), are characteristic of the field.

Bateman realised the connection of his work with the tensor-calculus of Ricci and Levi-Civita[1]: in fact, since (dx^0, dx^1, dx^2, dx^3) is a contravariant vector, it follows from the invariance of the quadratic differential form that the set of the g_{pq} is a symmetric covariant tensor of rank 2.[2]

[1] cf. p. 58

[2] These ideas were applied by Bateman in order to investigate a scheme of fundamental electromagnetic equations which are not altered by very general transformations.

Bateman's ideas were carried over into a more profound treatment of the problem of gravitation in the second half of the year 1913 by Einstein,[1] who in the years 1912–14 worked in partnership (as regarded the mathematics) with a Zürich geometer, Marcel Grossmann. In these papers the theory was put forward, that just as the rectilinear motion of a particle in free aether when there is no field is determined by the equation

$$\delta\left(\int ds\right)=0$$

where δ is the symbol of the Calculus of Variations, and

$$(ds)^2=c^2(dt)^2-(dx)^2-(dy)^2-(dz)^2,$$

so now (making a step analogous to that in Bateman's paper), the motion of a free material particle in a gravitational field is determined by the equation

$$\delta\left(\int ds\right)=0$$

where

$$(ds)^2=\sum_{p,q=0}^{3} g_{pq}\,dx^p\,dx^q,$$

the coefficients g_{pq} being characteristic of the state at the point (x^0, x^1, x^2, x^3) in space-time, and ds being invariant with respect to arbitrary transformations of (x^0, x^1, x^2, x^3). As with Bateman, $ds=0$ is the condition that a luminous disturbance originally at the world-point (x^0, x^1, x^2, x^3) should arrive at the world-point $(x^0+dx^0,\ x^1+dx^1,\ x^2+dx^2,\ x^3+dx^3)$. In geometrical language, *the path of a free material particle in a gravitational field is a geodesic in the four-dimensional curved space whose metric is specified by the equation* [2]

$$(ds)^2=\sum_{p,q=0}^{3} g_{pq}\,dx^p\,dx^q.$$

This was a tremendous innovation, because it implied the abandonment of the time-honoured belief that a gravitational field can be specified by a single scalar potential-function: instead, it proposed

[1] A. Einstein and M. Grossmann, *ZS. f. M. u. P.* lxii (1913), p. 225; lxiii (1914), p. 215; A. Einstein, *Vierteljahr. d. Nat. Ges. Zürich*, lviii (1913), p. 284; *Archives des sc. phys. et nat.*(4) xxxvii (1914), p. 5; *Phys. ZS.* xiv (15 Dec. 1913), p. 1249; *Berlin Sitz.* 1914, p. 1030

[2] A theory that matter consists in 'crinkles' of space had been published by W. K. Clifford in 1870; cf. *Proc. Camb. Phil. Soc.* ii (1876), p. 157 = Clifford's *Math. Papers*, p. 21.

to specify the gravitational field by the ten functions g_{pq} which could now be spoken of as the *gravitational potentials.*

Einstein justified[1] this new departure by showing that the theory of a single scalar gravitational potential led to inacceptable inferences. He compared, for instance, two systems, in the first of which a moveable hollow box with perfectly-reflecting walls is filled with pure-temperature radiation, while in the second the same radiation is contained inside a fixed vertical pit which is closed at the top and bottom by moveable pistons connected by a rod so as to be always at a fixed distance apart, the pit walls and pistons all being perfectly-reflecting: and he showed that on the single-scalar-potential theory the work necessary to raise the radiation upwards against the force of gravity would in the second system be only one-third of the work required in the first system: a conclusion which was obviously wrong. He admitted, however, that in his own mind the strongest reason for rejecting the single-scalar-potential theory was his conviction that relativity in physics exists not only with respect to the Lorentz group of linear orthogonal transformations but with respect to a much wider group.

The ten coefficients g_{pq} not only specify the force of gravitation, but they determine also the scale of distance in every direction, and the rate of clocks. The metric defined by

$$(ds)^2 = \sum_{p,\,q=0}^{3} g_{pq}dx^p dx^q$$

is not, in general, Euclidean: and since its non-Euclidean qualities determine the gravitational field, we may say that *gravitational theory is reduced to geometry*, in accordance with an idea expressed by FitzGerald[2] in 1894 in the words 'Gravity is probably due to a change of structure of the aether, produced by the presence of matter.' The 'aether' of FitzGerald was called by Einstein simply 'space' or 'space-time': and FitzGerald's somewhat vague term 'structure' became with Einstein the more precise 'curvature.' Thus we obtain the central proposition of the Einsteinian theory: 'Gravity is due to a change in the curvature of space-time, produced by the presence of matter.'

In comparing FitzGerald's statement with Einstein's, it may be remarked that if we consider a gravitational field which is *statical*, i.e. such as would be produced by gravitating masses that are permanently at rest relative to each other, then feeble[3] electromagnetic phenomena taking place in it can be shown to happen exactly in accordance with the ordinary Maxwellian theory of electromagnetic phenomena taking place in a medium whose specific inductive capacity and magnetic permeability are aelotropic and vary from

[1] § 7 of the paper in the *ZS. f. M. u. P.*

[2] FitzGerald's *Works*, p. 313

[3] i.e. so feeble that they do not appreciably change the curvature of the field.

point to point. In particular,[1] if we consider an electric point-charge at rest in the field of a single gravitating point-mass, the electric field is the same as would be obtained, in ordinary electrostatics, by supposing that the specific inductive capacity and magnetic permeability of the medium vary with the distance from the gravitating mass according to the law $(r+1)^3/r^3(r-1)$.

It is possible that when FitzGerald said 'Gravity is probably due to a change of structure of the aether,' he was actually thinking of a change which would show itself in alterations of the dielectric constant and magnetic permeability, and that he had in mind an electrical constitution of matter, on account of which matter would be subject to forces depending on the values of the dielectric constant and magnetic permeability: by analogy with the fact that in a liquid whose dielectric constant varies from point to point, an electrified body moves from places of lower to places of higher dielectric constant.[2]

What differentiates the Einsteinian theory from all previous conceptions is that the older physicists had regarded gravity as merely one among many types of natural force—electric, magnetic, etc.—each of which influenced in its own way the motion of material particles. Space, whose properties were set forth in Euclidean geometry, was, so to speak, the stage on which the forces played their parts. But in the new theory gravity was no longer one of the players, but part of the structure of the stage. A gravitational field consisted essentially in a replacement of the Euclidean properties by a much more complicated kind of geometry: space was no longer homogeneous or isotropic. An analogy may be drawn from the game of bowls. Bowling-greens, in the north of England, are not flat, but rise to a slight elevation in the centre. An observer who failed to notice the central elevation would find that a bowl (supposed without bias) always described a path convex toward the centre of the green, and he might account for this by postulating a centre of repellent force there. A better-informed observer would attribute the phenomenon to a geometrical feature—the slope. The two explanations correspond respectively to the Newtonian and the Einsteinian conceptions of gravity: for Newton it is a force, for Einstein it is a modification of the geometry of space.

When the metric of space-time is specified by an equation

$$(ds)^2 = \sum_{p\ q=0}^{3} g_{pq} dx^p dx^q,$$

an observer moving in any manner will have a *world-line* consisting of the points of space-time which he successively occupies : and at

[1] E. T. Copson, *Proc. R.S.*(A), cxviii (1928), p. 184

[2] This idea was later developed by E. Wiechert, *Ann. d. Phys.* lxiii (1920), p. 301.

any point of his world-line he will have in his immediate neighbourhood an *instantaneous three-dimensional space*, formed by the aggregate of all the elements of length which are orthogonal to his world-line at the point: orthogonality being defined, as already explained,[1] by the statement that two vectors (X) and (Y) are said to be orthogonal if

$$\sum_{p=0}^{3} X_p Y^p = 0,$$

where (X_p) is the covariant form of one vector and (Y^p) is the contravariant form of the other.

Einstein laid down the principle that the equations which describe any physical process must satisfy the condition that their covariance with respect to arbitrary substitutions can be deduced from the invariance of ds. In other words, the laws of nature must be represented by equations which are covariantive for the form $\sum_{p,q} g_{pq} dx^p dx^q$ with respect to all point-transformations of co-ordinates. Laws of nature are assertions of *coincidences* in space-time, and therefore must be expressible by covariant equations.

It might be thought that by following up the consequences of this principle we should obtain important positive results. However, Ricci and Levi-Civita[2] had shown long before that from practically *any* assumed law we can derive another law which does not differ from it in any way that can be tested by observation, but which is covariant. The fact that a formula has the covariant property does not, therefore, tell us anything as to whether it is correct or not. We are, however, perhaps justified in believing that a conjectural law which can be expressed readily and simply in covariant form is more worthy of attention (as being more likely to be true) than one whose covariant form is awkward and complicated.

Not only must the general laws of physics be covariant, it is also necessary that every single assertion which has a physical meaning must be covariant with respect to arbitrary transformations of the co-ordinate system. Thus the assertion that an electron is at rest for an interval of time of duration unity cannot have a physical meaning, since this assertion is not covariant.[3]

In Einstein's general theory, the velocity of light at any place has always the value c *with respect to any inertial frame of reference* for this neighbourhood, and the velocity of any material body is less than c. Thus there is no difficulty in the fact that the fixed stars have velocities greater than c with respect to axes fixed in the rotating earth: for such axes are not inertial.

[1] See p. 64

[2] *Math. Ann.* liv (1901), p. 125; cf. E. Kretschmann, *Ann. d. Phys.* liii (1917), p. 575, and A. Einstein, *Ann. d. Phys.* lv (1918), p. 241

[3] D. Hilbert, *Math. Ann.* xcii (1924), p. 1

Some physicists called attention to the fact that when light is propagated in a medium where there is anomalous dispersion, the index of refraction may be less than unity, whence it seemed as if the velocity of light in the dispersive medium might be greater than the velocity of light *in vacuo*. The difficulty was removed when it was pointed out by L. Brillouin [1] and A. Sommerfeld [2] that the velocity of light with which the index of refraction is concerned is the *phase* velocity, whereas the velocity of a signal is the *group* velocity, which is never greater than c.

It has sometimes been supposed, by a misunderstanding, that the general Einsteinian theory requires us to regard the Copernican conception of the universe as no more true than the Ptolemaic, and that it is indifferent whether we regard the earth as rotating on her axis or regard the stellar universe as performing a complete revolution about the earth every twenty-four hours. The root of the matter, by which everything is explained, is that the Copernican axes are inertial, while the Ptolemaic are not. The earth rotates *with respect to the local inertial axes.*[3]

In his first paper in the *Zeitschrift für Math. u. Phys.*,[4] Einstein gave the form which Maxwell's equations of the electromagnetic field must take when the metric of space-time is given by a quadratic differential form

$$(ds)^2 = \sum_{p, q=0}^{3} g_{pq}\, dx^p\, dx^q.$$

To obtain these, it will be necessary to introduce some other concepts of Ricci's absolute differential calculus, or *tensor-calculus* [5] as Einstein henceforth called it. Suppose that we are given a quadratic differential form in any number of variables

$$(ds)^2 = \sum_{p, q=1}^{n} g_{pq}\, dx^p\, dx^q$$

then, following Elwin Bruno Christoffel [6] (1829–1900), we introduce what are called *Christoffel symbols of the first kind*, defined as

$$\begin{bmatrix} p\,q \\ l \end{bmatrix} = \tfrac{1}{2}\left(\frac{\partial g_{pl}}{\partial x^q} + \frac{\partial g_{ql}}{\partial x^p} - \frac{\partial g_{pq}}{\partial x^l}\right) \qquad (p, q, l = 1, 2, \ldots n),$$

[1] *Comptes Rendus*, clvii (1913), p. 914

[2] *Ann. d. Phys.* xliv (1914), p. 177

[3] cf. G. Giorgi and A. Cabras, *Rend. Lincei*, ix (1929), p. 513

[4] At page 241

[5] The word *tensor* had been used by W. Voigt in 1898, in connection with the elasticity of crystals.

[6] *J. für Math.* lxx (1869), pp. 46, 241

and *Christoffel symbols of the second kind*, defined as

$$\left\{ {p\,q \atop l} \right\} = \sum_{r=1}^{n} g^{rl} \left[{p\,q \atop r} \right] \qquad (p, q, l = 1, 2, \ldots n).$$

As Christoffel showed, the Christoffel symbols enable us to form new tensors from known tensors by a process of generalised differentiation. If $(X_1, X_2, \ldots X_n)$ is any covariant vector, then *the quantities* $(X_p)_q$ *defined by*

$$(X_p)_q = \frac{\partial X_p}{\partial x^q} - \sum_{r=1}^{n} \left\{ {p\,q \atop r} \right\} X_r$$

constitute a covariant tensor of rank 2. This process was called *covariant differentiation* by Ricci. Similarly if (X_{pq}) is any covariant tensor of rank 2, then *the quantities* $(X_{pq})_s$ *defined by the equations*

$$(X_{pq})_s = \frac{\partial X_{pq}}{\partial x^s} - \sum_{r=1}^{n} \left\{ {p\,s \atop r} \right\} X_{rq} - \sum_{r=1}^{n} \left\{ {q\,s \atop r} \right\} X_{pr}$$

constitute a covariant tensor of rank 3 *which is called the covariant derivative of* X_{pq}. The covariant differentiation of sums and products is effected by rules similar to those that apply to ordinary differentiation. Moreover if (X^l) is any contravariant vector, then the *quantities* $(X^l)_s$ *defined by*

$$(X^l)_s = \frac{\partial X^l}{\partial x^s} + \sum_{q=1}^{n} \left\{ {q\,s \atop l} \right\} X^q$$

define a mixed tensor of rank 2 which is called the covariant derivative of (X^l).

We know that Maxwell's equations in Euclidean space consist of the Ampère-Maxwell tetrad

$$\frac{\partial d_x}{\partial x} + \frac{\partial d_y}{\partial y} + \frac{\partial d_z}{\partial z} = 4\pi\rho$$

$$\frac{\partial h_z}{\partial y} - \frac{\partial h_y}{\partial z} = \frac{1}{c}\frac{\partial d_x}{\partial t} + 4\pi\rho v_x$$

$$\frac{\partial h_x}{\partial z} - \frac{\partial h_z}{\partial x} = \frac{1}{c}\frac{\partial d_y}{\partial t} + 4\pi\rho v_y$$

$$\frac{\partial h_y}{\partial x} - \frac{\partial h_x}{\partial y} = \frac{1}{c}\frac{\partial d_z}{\partial t} + 4\pi\rho v_z$$

and the Faraday tetrad

$$\frac{\partial h_x}{\partial x}+\frac{\partial h_y}{\partial y}+\frac{\partial h_z}{\partial z}=0$$

$$\frac{\partial d_z}{\partial y}-\frac{\partial d_y}{\partial z}=-\frac{1}{c}\frac{\partial h_x}{\partial t}$$

$$\frac{\partial d_x}{\partial z}-\frac{\partial d_z}{\partial x}=-\frac{1}{c}\frac{\partial h_y}{\partial t}$$

$$\frac{\partial d_y}{\partial x}-\frac{\partial d_x}{\partial y}=-\frac{1}{c}\frac{\partial h_z}{\partial t}.$$

Now write $x^0=ct$, $x^1=x$, $x^2=y$, $x^3=z$. We have seen in Chapter II that the electric and magnetic vectors together constitute a six-vector

$$d_x=X^{01},\quad d_y=X^{02},\quad d_z=X^{03},\quad h_x=X^{23},\quad h_y=X^{31},\quad h_z=X^{12}.$$

Thus the Ampère-Maxwell tetrad becomes

$$\frac{\partial X^{01}}{\partial x^1}+\frac{\partial X^{02}}{\partial x^2}+\frac{\partial X^{03}}{\partial x^3}=4\pi\rho$$

$$\frac{\partial X^{10}}{\partial x^0}+\frac{\partial X^{12}}{\partial x^2}+\frac{\partial X^{13}}{\partial x^3}=4\pi\rho v_x$$

$$\frac{\partial X^{20}}{\partial x^0}+\frac{\partial X^{21}}{\partial x^1}+\frac{\partial X^{23}}{\partial x^3}=4\pi\rho v_y$$

$$\frac{\partial X^{30}}{\partial x^0}+\frac{\partial X^{31}}{\partial x^1}+\frac{\partial X^{32}}{\partial x^2}=4\pi\rho v_z.$$

Now if (T^p) is a contravariant vector, the quantity $\sum (T^p)_p$ (where the suffix outside the bracket denotes covariant differentiation) is a scalar which is called the *divergence* of (T^p), and denoted by div (T^p): we can easily show that

$$\text{div}\,(T^p)=\frac{1}{\sqrt{g}}\sum_p \frac{\partial}{\partial x^p}\left(\sqrt{g}\,T^p\right).$$

If (T^{pq}) is a contravariant tensor of rank 2, then the quantity

$$\sum_q (T^{pq})_q$$

which is a vector, is called the *vectorial divergence* of (T^{pq}), and is denoted by $\Delta\text{iv}\,(T^{pq})$. We can show that if (T^{pq}) is skew, i.e. is a six-veetor, then

$$\Delta\text{iv}\,(T^{pq}) = \frac{1}{\sqrt{g}}\sum_q \frac{\partial}{\partial x^q}\left(\sqrt{g}\,T^{pq}\right).$$

Remembering that when the space is Euclidean, $\sqrt{g}$ is a constant, we see that for a six-vector (T^{pq}) in Euclidean space we have

$$\Delta\text{iv}\,(T^{pq}) = \sum_q \frac{\partial T^{pq}}{\partial x^q}.$$

Comparing this with the above form of the Ampère-Maxwell tetrad, we see that the tensorial (i.e. covariant) form of this tetrad must be

$$\Delta\text{iv}\,(X^{pq}) = 4\pi J^p \qquad \text{(A)}$$

where J^p denotes the four-vector which represents the electric charge and current, namely $\rho_0\, dx^p/ds$, where ρ_0 is the proper density of the charge, i.e. the charge divided by the volume it occupies, as measured by an observer moving with it.

Next consider the Faraday tetrad of equations, which may clearly be written

$$\frac{\partial X_{23}}{\partial x^1} + \frac{\partial X_{31}}{\partial x^2} + \frac{\partial X_{12}}{\partial x^3} = 0$$

$$\frac{\partial X_{23}}{\partial x^0} + \frac{\partial X_{30}}{\partial x^2} + \frac{\partial X_{02}}{\partial x^3} = 0$$

$$\frac{\partial X_{31}}{\partial x^0} + \frac{\partial X_{10}}{\partial x^3} + \frac{\partial X_{03}}{\partial x^1} = 0$$

$$\frac{\partial X_{12}}{\partial x^0} + \frac{\partial X_{20}}{\partial x^1} + \frac{\partial X_{01}}{\partial x^2} = 0.$$

Now it can be shown that if X^{rs} is a six-vector, and if (p, q, r, s) is an even permutation of the numbers (0, 1, 2, 3), then a six-vector Y_{pq} can be defined by the equations

$$Y_{pq} = \sqrt{(-g)}\,X^{rs}$$

and that we then have

$$X_{pq} = -\sqrt{(-g)}\,Y^{rs}.$$

The two six-vectors X_{pq} and Y_{pq} are said to be *dual* to each other. If Y_{pq} is the six-vector dual to the electromagnetic six-vector X_{pq}, so that in Euclidean space

$$Y^{10}=h_x, \quad Y^{20}=h_y, \quad Y^{30}=h_z, \quad Y^{23}=d_x, \quad Y^{31}=d_y, \quad Y^{12}=d_z,$$

then the Faraday tetrad may evidently be written

$$\frac{\partial Y^{01}}{\partial x^1}+\frac{\partial Y^{02}}{\partial x^2}+\frac{\partial Y^{03}}{\partial x^3}=0$$

$$\frac{\partial Y^{10}}{\partial x^0}+\frac{\partial Y^{12}}{\partial x^2}+\frac{\partial Y^{13}}{\partial x^3}=0$$

$$\frac{\partial Y^{20}}{\partial x^0}+\frac{\partial Y^{21}}{\partial x^1}+\frac{\partial Y^{23}}{\partial x^3}=0$$

$$\frac{\partial Y^{30}}{\partial x^0}+\frac{\partial Y^{31}}{\partial x^1}+\frac{\partial Y^{32}}{\partial x^2}=0$$

or

$$\Delta\text{iv}\,(Y^{pq})=0 \qquad \text{(B)}.$$

Since the equations (A) and (B) are tensor equations, and are therefore covariant with respect to all transformations of the coordinates (x^0, x^1, x^2, x^3), we may assume with Einstein that *they represent the equations of the electromagnetic field in space-time of any metric whatever.*

The six-vector of the electromagnetic field may be expressed in terms of potentials, in the same way as in Euclidean space it is expressed by the equations

$$d_x=-\frac{\partial\phi}{\partial x}-\frac{\partial a_x}{c\partial t} \quad \text{etc.,} \qquad h_x=\frac{\partial a_z}{\partial y}-\frac{\partial a_y}{\partial z} \quad \text{etc.}$$

For if we write $(\phi_0, \phi_1, \phi_2, \phi_3)$ for $(\phi, -a_x, -a_y, -a_z)$, these equations become, as we have seen on page 76,

$$X_{pq}=\frac{\partial\phi_p}{\partial x^q}-\frac{\partial\phi_q}{\partial x^p} \qquad (p, q=0, 1, 2, 3).$$

Writing this

$$X_{pq}=(\phi_p)_q-(\phi_q)_p$$

where the suffixes outside the brackets represent covariant differentiation, we see that *the potential* $(\phi_0, \phi_1, \phi_2, \phi_3)$ *is a covariant vector.*

Electromagnetic theory leads naturally to physical optics, and this again to geometrical optics. Now in the relativity theory of Poincaré and Lorentz, for which the line-element $d\tau$ in the world of space-time is given by

$$(d\tau)^2 = (dt)^2 - \frac{1}{c^2}\left\{(dx)^2 + (dy)^2 + (dz)^2\right\},$$

the geodesics of the world are straight lines, and the null geodesics (i.e. the geodesics for which $d\tau$ vanishes) are the straight lines for which

$$\frac{(dx)^2 + (dy)^2 + (dz)^2}{(dt)^2} = c^2,$$

so the null geodesics are the tracks of rays of light. When Einstein created his new general theory of relativity, in which gravitation was taken into account, he carried over this principle by analogy, and asserted its truth for gravitational fields. The principle was, however, not proved at that time : and indeed there was the obvious difficulty in proving it, that strictly speaking there are no 'rays' of light—that is to say, electromagnetic disturbances which are filiform, or drawn out like a thread—except in the limit when the frequency of the light is infinitely great : in all other cases diffraction causes the 'ray' to spread out.

The matter was investigated in 1920 by M. von Laue,[1] who, starting from the partial differential equations of electromagnetic phenomena in a gravitational field, obtained a particular solution which corresponded to light of infinitely high frequency, and showed that the path of this disturbance satisfied the differential equations of the null geodesics : thus for the first time proving the truth of Einstein's assertion. It was afterwards shown by E. T. Whittaker[2] that the law is really an immediate deduction from the theory of the characteristics of partial differential equations, and that it is not necessary to introduce the notion of frequency at all : in fact, that *in a gravitational field, any electromagnetic disturbance which is filiform must necessarily have the form of a null geodesic of space-time.*

At any point of space-time, the directions which issue from it may be classified into those that have a spatial character and those that have a temporal character : the two classes are separated from each other by a cone whose generators are the paths of rays of light : this is called the *null-cone.*

It can be shown mathematically that if we know the geodesics of space-time (i.e. the paths of free material particles) and also know which of them are null geodesics (i.e. paths of rays of light), then the coefficients g_{pq} of the equation defining the metric

$$(ds)^2 = \sum_{p,\,q=0}^{3} g_{pq}\,dx^p dx^q$$

[1] *Phys. ZS.* xxi (1920), p. 659

[2] *Proc. Camb. Phil. Soc.* xxiv (1928), p. 32

are completely determinate. In fact, when the null geodesics are given, we can infer the metric save for a factor, say

$$(ds)^2 = \lambda(x^0, x^1, x^2, x^3) \times (\text{a determinate quadratic form in the } dx^p)$$

where λ is unknown: and when we are also given the non-null geodesics, the factor $\lambda(x^0, x^1, x^2, x^3)$ can be determined.

In the papers we have referred to, which are of date earlier than November 1915, Einstein gave, as we have seen, a satisfactory account of the behaviour of mechanical and electrical systems in a field of gravitation which is supposed given: his formulae were derived fundamentally from the principle of equivalence, i.e. the principle that the systems behave just as if there were no gravitational field, but they were referred to a co-ordinate-system with an acceleration equal and opposite to the acceleration of gravity. But he had not as yet succeeded in obtaining an entirely satisfactory set of fundamental equations for the gravitational field itself, i.e. equations which would play the same part in his theory that Poisson's equation [1]

$$\frac{\partial^2 V}{\partial x^2} + \frac{\partial^2 V}{\partial y^2} + \frac{\partial^2 V}{\partial z^2} = -4\pi\rho$$

played in the Newtonian theory. This defect was repaired, and the theory (now known as *General Relativity*) substantially completed, in a series of short papers published in November-December 1915, in the *Berlin Sitzungsberichte*.[2]

Let us first inquire what covariants of the form $\sum_{p,q} g_{pq} dx^p dx^q$ can be formed from the g_{pq}'s and their derivatives alone. It can be shown that these can all be derived from a certain tensor of rank 4, known as the *Riemann tensor*, which must now be introduced.

Let D_s denote the operation which when applied to the Christoffel symbol of the first kind is

$$D_s \begin{bmatrix} rt \\ u \end{bmatrix} = \frac{\partial}{\partial x^s} \begin{bmatrix} rt \\ u \end{bmatrix} - \sum_l \begin{Bmatrix} su \\ l \end{Bmatrix} \begin{bmatrix} rt \\ l \end{bmatrix};$$

then we define the Riemann tensor [3] by the equation

$$K_{pqrs} = D_q \begin{bmatrix} pr \\ s \end{bmatrix} - D_r \begin{bmatrix} pq \\ s \end{bmatrix}.$$

[1] cf. Vol. I, p. 61

[2] *Berlin Sitz.* 1915, pp. 778, 799, 831, 844. An eight-line abstract, dated 25 March 1915, is given at p. 315.

[3] It was discovered by Riemann, in a memoir *Commentata mathematica qua respondere tentatur* . . ., which was sent to the Paris Academy in 1861 and published posthumously, *Werke* (1892), p. 401; it was afterwards used by Christoffel, loc. cit.

From the Riemann tensor we can obtain a tensor of rank 2 which is defined by the equation

$$K_{pq} = \sum_{r,s} g^{rs} K_{pqrs};$$

it is called the *Ricci-tensor* or *contracted curvature-tensor*: and from the Ricci tensor we obtain what is called the *scalar curvature* of the space, defined by the equation

$$K = \sum_{p,q} g^{pq} K_{pq}.$$

For a two-dimensional space, e.g. a surface in Euclidean three-dimensional space, $-\frac{1}{2}K$ is the ordinary Gaussian measure of curvature, $1/\rho_1\rho_2$.

In applications to the theory of gravitation, we are dealing with the four-dimensional world of space-time. For such a world, K may be defined geometrically in the following way. At any world-point, take any four directions that are mutually orthogonal with respect to the metric of the world. These four directions, taken in pairs, determine six surface-elements or orientations, in each of which there is an aggregate of geodesics issuing from the point, forming a 'geodesic surface': then K is minus twice the sum of the Gaussian measures of curvature of these six surfaces at the given world-point. It is independent of the choice of the four orthogonal directions.

Now Mach had introduced long before a principle, that inertia must be reducible to the interaction of bodies: and Einstein generalised this into what he called *Mach's principle*, namely that the field represented by the ten potentials g_{pq} is determined *solely* by the masses of bodies. The word 'mass' is here to be understood in the sense given to it by the theory of relativity, that is, as equivalent to energy: and as energy is expressed covariantively by Minkowski's energy-tensor T_{pq}, it follows that in the fundamental equations of gravitation, corresponding to the equation $\nabla^2 V = -4\pi\rho$ of the Newtonian theory, we may expect the tensor T_{pq} or some linear function of it to take the place of Poisson's ρ. We expect to find on the other side of the equation, corresponding to $\nabla^2 V$, a tensor of the same rank as T_{pq}, that is, the second rank, containing second derivatives of the potentials but no higher derivatives. The only covariant tensors of this character are the Ricci-tensor K_{pq}, with Kg_{pq} and g_{pq}. Einstein first supposed that K_{pq} might be a simple constant multiple of T_{pq}: but this is not satisfactory for reasons that will be more evident later[1]: and he finally proposed the equations

$$K_{pq} = -\kappa(T_{pq} - \tfrac{1}{2} g_{pq} T) \qquad (p, q = 0, 1, 2, 3)$$

[1] The divergence of T_{pq} is zero, and of K_{pq} is not in general zero.

where $T = \sum_{p,q} g^{pq} T_{pq}$, and κ is a constant depending on the Newtonian constant of gravitation. *These are the general field-equations of gravitation.*

Multiplying them by g^{pq}, summing with respect to p and q, and remembering that $\sum_{p,q} g^{pq} g_{pq} = 4$, we have $K = \kappa T$, so the equations may be written

$$K_{pq} - \tfrac{1}{2} g_{pq} K = -\kappa T_{pq} \qquad (p, q = 0, 1, 2, 3).$$

These are ten equations for the ten unknowns g_{pq} : there are four identities between them, as might be expected : for four of the g_{pq}'s can be assigned arbitrarily as functions of the x^p, corresponding to the fact that the equations are invariant under the most general transformation of co-ordinates.

According to Mach's principle as adopted by Einstein, the curvature of space is governed by physical phenomena, and we have to ask .whether the metric of space-time may not be determined *wholly* by the masses and energy present in the universe, so that space-time cannot exist at all except in so far as it is due to the existence of matter. The point at issue may be illustrated by the following concrete problem : if all matter were annihilated except one particle which is to be used as a test-body, would this particle have inertia or not? The view of Mach and Einstein is that it would not : and in support of this view it may be urged that, according to the deductions of general relativity, the inertia of a body is increased when it is in the neighbourhood of other large masses : it seems needless, therefore, to postulate other sources of inertia, and simplest to suppose that *all* inertia is due to the presence of other masses. When we confront this hypothesis with the facts of observation, however, it seems that the masses of whose existence we know—the solar system, stars and nebulae—are insufficient to confer on terrestial bodies the inertia that they actually possess : and therefore if Mach's principle were adopted, it would be necessary to postulate the existence of enormous quantities of matter in the universe which have not been detected by astronomical observation, and which are called into being simply in order to account for inertia in other bodies. This is, after all, no better than regarding some part of inertia as intrinsic.

The relation of Einstein's to Newton's laws of motion in the general case was discussed by L. Silberstein,[1] who showed that the differential equations of a geodesic in General Relativity are rigorously identical with the Newtonian equations of motion of a particle,

$$\frac{d^2\xi_p}{dt^2} = \frac{\partial\Omega}{\partial\xi_p} \qquad (p = 1, 2, 3)$$

[1] *Nature*, cxii (1923), p. 788

so long as the frame of reference for the differential equations of the geodesic is a system which is momentarily at rest relative to the particle. For simplicity consider a 'statical' world, specified by a metric

$$(ds)^2 = V^2(dt)^2 - \frac{1}{c^2} \sum_{p, q=1}^{3} a_{pq}\, dx^p dx^q$$

where V and the a_{pq} are functions of (x^1, x^2, x^3) only; and consider an observer who is stationary, i.e. whose co-ordinates (x^1, x^2, x^3) do not vary. Then the components of the gravitational force on the particle can be shown to be

$$g^0 = 0, \qquad g^p = -c^2 \sum_q \frac{a^{pq}}{V} \frac{\partial V}{\partial x^q} \qquad (p = 1, 2, 3),$$

and it can be shown from Einstein's fundamental equation

$$K_{pq} - \tfrac{1}{2} g_{pq} K = -\kappa T_{pq}$$

that

$$\frac{\Delta_2 V}{V} = \tfrac{1}{2}\kappa \left(\frac{T_{00}}{V^2} + \sum_{p, q=1}^{3} a^{pq} T_{pq} \right),$$

where $\Delta_2 V$ denotes the Second Differential Parameter of V in the three-dimensional space.

Now suppose that the field is of the kind considered in Poisson's equation, namely that the space is approximately Euclidean and that there is a volume-density ρ of matter at rest, and no radiation; then the only sensible element of the energy-tensor is the energy-density due to the equivalence of mass and energy, which has the value $c^2\rho$. Moreover we can take $V = 1 + \gamma$, where γ is small; so the above equation becomes

$$\Delta_2 \gamma = \tfrac{1}{2} \kappa c^2 \rho.$$

But writing x, y, z for the co-ordinates, since

$$(ds)^2 = \frac{1}{c^2} \left\{ (dx)^2 + (dy)^2 + (dz)^2 \right\}$$

approximately, we have

$$\Delta_2 \gamma = c^2 \left(\frac{\partial^2 \gamma}{\partial x^2} + \frac{\partial^2 \gamma}{\partial y^2} + \frac{\partial^2 \gamma}{\partial z^2} \right).$$

Thus

$$\frac{\partial^2 \gamma}{\partial x^2} + \frac{\partial^2 \gamma}{\partial y^2} + \frac{\partial^2 \gamma}{\partial z^2} = \tfrac{1}{2} \kappa \rho.$$

Now in the light of the above equations for the gravitational force, we have

$$-c^2\gamma = \text{the gravitational potential } \Omega,$$

so

$$\frac{\partial^2\Omega}{\partial x^2} + \frac{\partial^2\Omega}{\partial y^2} + \frac{\partial^2\Omega}{\partial z^2} = -\tfrac{1}{2}\kappa c^2 \rho.$$

Comparing this with Poisson's equation

$$\frac{\partial^2\Omega}{\partial x^2} + \frac{\partial^2\Omega}{\partial y^2} + \frac{\partial^2\Omega}{\partial z^2} = -4\pi\beta\rho$$

where β is the Newtonian constant of gravitation, we have

$$\kappa = \frac{8\pi\beta}{c^2}$$

an equation connecting the Einsteinian and Newtonian constants of gravitation. Since

$$\beta = 6{\cdot}67 \times 10^{-8} \text{ gr}^{-1} \text{ cm}^3 \text{ sec}^{-2}$$

this gives

$$\kappa = 1{\cdot}87 \times 10^{-27} \text{ cm.gr}^{-1}.$$

Almost simultaneously with Einstein's discovery of General Relativity, David Hilbert[1] (1862–1943) gave a derivation of the whole theory from a unified principle. Defining a point in space-time by generalised co-ordinates (x^0, x^1, x^2, x^3), he adopted Einstein's ten gravitational potentials g_{pq} and a four-vector $(\phi_0, \phi_1, \phi_2, \phi_3)$ representing the electrodynamic potential: and assumed the following axioms:

Axiom I (Mie's axiom of the world-function). *All physical happenings* (gravitational, electrical, etc.) *in the universe are determined by a scalar function* H (called the *world-function*) *which involves the arguments* g_{pq} *and their first and second derivatives with respect to the x's, and involves also the* ϕ*'s and their first derivatives with respect to the x's: and the laws of physical processes are obtained by annulling the variation of the integral*

$$\iiiint \mathrm{H}\sqrt{g}\, dx^0 dx^1 dx^2 dx^3$$

(where g denotes the determinant of the g_{pq}) *for each of the fourteen potentials* g_{pq}, ϕ_s. The reason for the occurrence of the factor $\sqrt{g}$

[1] *Gött. Nach.*, 1915, p. 395; read 20 Nov. 1915. The investigation was carried further by: H. A. Lorentz, *Proc. Amst. Ac.* xix (1916), p. 751; J. Tresling, *Proc. Amst. Ac.* xix (1916), p. 892; A. Einstein, *Berlin Sitz*, 1916, p. 1111; F. Klein, *Gött. Nach.* 1917, p. 469; H. Weyl, *Ann. d. Phys.* liv (1917), p. 117; A. D. Fokker, *Proc. Amst. Ac.* xix (1917), p. 968; A. Palatini, *Palermo Rend.* xliii (1919), p. 203; D. Hilbert, *Math. Ann.* xcii (1924), p. 1; E. T. Whittaker, *Proc. R.S.*(A), cxiii (1927), p. 496.

is that $\sqrt{g}\, dx^0 dx^1 dx^2 dx^3$, which is called the *invariant hypervolume*, is invariant under all transformations of co-ordinates.

Axiom II (axiom of general invariance). *The world-function* H *is invariant with respect to arbitrary transformations of the* x^r.

These axioms represented a distinct advance on Einstein's methods, in which Hamilton's Principle had played only a very subordinate part.

We suppose that the world-function H is a sum of two terms, of which the first represents gravitation, i.e. whatever is inherent in the intrinsic structure of space-time, while the second term represents all other physical effects. The first term we take to be proportional to K, the scalar curvature of space-time at the point (x^0, x^1, x^2, x^3); this amounts to supposing that the mutual energy of all the gravitating masses in the world can be expressed analytically as an integral taken over the whole of space, namely a numerical multiple of $\iiiint K \sqrt{g}\, dx^0 dx^1 dx^2 dx^3$. With regard to the second term, we shall suppose for simplicity that the only other physical effect to be considered is an electromagnetic field in free aether. Now in Euclidean space there is, in the field, electric energy $(d_x^2 + d_y^2 + d_z^2)/8\pi$ per unit volume, and magnetic energy $(h_x^2 + h_y^2 + h_z^2)/8\pi$ per unit volume; and since these are of opposite type, in the sense in which kinetic and potential energy are of opposite type, we should expect their difference to occur in the world-function. In general co-ordinates this difference is represented by

$$L = -\frac{1}{16\pi} \sum_{b\,p} X^{pq} X_{pq}$$

where X_{pq} is the electromagnetic six-vector. We shall take the second part of the world-function to be a numerical multiple of L. We must therefore have

$$\delta \iiiint (K + 2\kappa L) \sqrt{g}\, dx^0 dx^1 dx^2 dx^3 = 0$$

where κ is a constant: or

$$\iiiint \left\{ \delta K \,.\, \sqrt{g} + K \,.\, \delta \sqrt{g} + 2\kappa \delta (L \sqrt{g}) \right\} dx^0 dx^1 dx^2 dx^3 = 0.$$

A calculation shows that

$$\iiiint \delta K \,.\, \sqrt{g}\, dx^0 dx^1 dx^2 dx^3 = \iiiint \sum_{p,\,q} K_{pq} \sqrt{g}\, \delta g^{pq}\, dx^0 dx^1 dx^2 dx^3$$

where K_{pq} is the Ricci tensor: and

$$\delta \sqrt{g} = -\tfrac{1}{2} \sqrt{g} \sum_{p,\,q} g_{pq} \delta g^{pq}.$$

Let us now find the value of

$$\delta \iiiint L \sqrt{g}\, dx^0 dx^1 dx^2 dx^3.$$

The part that depends on the variations of the g^{pq} is

$$\iiiint \sum_{p,q} \zeta \frac{\partial(L\sqrt{g})}{\partial g^{pq}} \delta g^{pq}\, dx^0 dx^1 dx^2 dx^3$$

where ζ is 1 or $\frac{1}{2}$ according as p is equal or unequal to q: or

$$\tfrac{1}{2} \sum_{p,q} \iiiint T_{pq}\, \delta g^{pq} \,.\, \sqrt{g}\, dx^0 dx^1 dx^2 dx^3$$

where

$$T_{pq} = \frac{2}{\sqrt{g}} \zeta \frac{\partial(L\sqrt{g})}{\partial g^{pq}} = 2\zeta \frac{\partial L}{\partial g^{pq}} + \frac{2}{\sqrt{g}} \zeta \frac{\partial(\sqrt{g})}{\partial g^{pq}} L.$$

Now it may easily be shown that

$$\zeta \frac{\partial g}{\partial g^{pq}} = -g\, g_{pq}$$

so

$$T_{pq} = 2\zeta \frac{\partial L}{\partial g^{pq}} - g_{pq}\, L.$$

Moreover, when we regard L as a function of the g^{pq} and the ϕ_s, remembering that X_{pq} is a function of the ϕ_s only, not involving the g^{pq}, we must write

$$L = -\frac{1}{16\pi} \sum_{k,l,p,q} g^{pk}\, g^{ql}\, X_{pq}\, X_{kl}$$

so we have

$$\zeta \frac{\partial L}{\partial g^{pq}} = -\frac{1}{8\pi} \sum_s X_{qs}\, X_p{}^s$$

and therefore

$$T_{pq} = -\frac{1}{4\pi} \sum_s X_{qs}\, X_p{}^s + \frac{g_{pq}}{16\pi} \sum_{r,s} X^{rs}\, X_{rs}$$

or

$$T^p_q = -\frac{1}{4\pi} \sum_s X_{qs}\, X^{ps} + \frac{\delta^p_q}{16\pi} \sum_{r,s} X^{rs}\, X_{rs}$$

which is the expression previously found for Minkowski's electromagnetic energy-tensor.

Thus the variational integral, in so far as the variations of the g^{pq} are concerned, is

$$\sum_{p,q=0}^{3} \iiiint \left(K_{pq} - \tfrac{1}{2} g_{pq} K + \kappa T_{pq}\right) \delta g^{pq} . \sqrt{g}\, dx^0 dx^1 dx^2 dx^3 = 0$$

and therefore *the variational equations derived from the terms in* g^{pq} *are*

$$K_{pq} - \tfrac{1}{2} g_{pq} K = -\kappa T_{pq} \qquad (p, q = 0, 1, 2, 3)$$

which are identical with Einstein's gravitational equations, previously given.

Hilbert showed moreover that *the Ricci tensor satisfies the four identical relations expressed by the equation*

$$\Delta\mathrm{iv}\, (K_p^q - \tfrac{1}{2} \delta_p^q K) = 0.$$

We must now find the part of the variational integral that involves the variations of the electromagnetic potentials. It is (disregarding a constant multiplier)

$$\iiiint \sum_{p,q} X^{pq} \sqrt{g}\, \delta X_{pq}\, dx^0 dx^1 dx^2 dx^3.$$

But

$$\delta X_{pq} = \delta \left(\frac{\partial \phi_p}{\partial x^q}\right) - \delta \left(\frac{\partial \phi_q}{\partial x^p}\right)$$

so the part of the variational integral with which we are concerned is

$$\iiiint \sum_{p,q} X^{pq} \sqrt{g} \left\{ \delta\left(\frac{\partial \phi_p}{\partial x^q}\right) - \delta\left(\frac{\partial \phi_q}{\partial x^p}\right) \right\} dx^0 dx^1 dx^2 dx^3$$

which after integration by parts yields

$$-\iiiint \sum_{p,q} \left\{ \frac{\partial (X^{pq} \sqrt{g})}{\partial x^q} \delta \phi_p - \frac{\partial (X^{pq} \sqrt{g})}{\partial x^p} \delta \phi_q \right\} dx^0 dx^1 dx^2 dx^3.$$

Interchanging p and q in the summation of the second term, this becomes

$$-2 \iiiint \sum_{p,q} \frac{\partial (X^{pq} \sqrt{g})}{\partial x^q} \delta \phi_p\, dx^0 dx^1 dx^2 dx^3,$$

and therefore the variational equations obtained from the variations of the potentials are

$$\sum_q \frac{\partial (X^{pq} \sqrt{g})}{\partial x^q} = 0 \qquad (p = 0, 1, 2, 3)$$

or

$$\Delta\text{iv}\,(X^{pq}) = 0$$

and these are precisely the Ampère-Maxwell tetrad of electromagnetic equations in free aether. The Faraday tetrad of course follows from the expression of the X_{pq} in terms of the potentials. Thus *both Einstein's gravitational equations and the electromagnetic equations can be obtained by the variation of the integral of Hilbert's world-function.*

It was proposed by Cornel Lanczos[1] that the world-function H occurring in the integral $\iiiint H\sqrt{g}\,dx^0dx^1dx^2dx^3$ should be formed in a way different from that adopted by Hilbert, namely that it should be a *quadratic* function of the Ricci tensor K_{pq}, and in fact should be of the form

$$H = \sum_{p,q} K_{pq}K^{pq} + CK^2$$

where C is a numerical constant. The advantage of this form is that in the course of the analysis, a four-vector is found to occur naturally, and this vector can be identified with the electromagnetic potential-vector: thus electromagnetic theory can be unified with the theory of General Relativity.

In 1920 a criticism of General Relativity was published in the *Times Educational Supplement*[2] by the mathematician and philosopher Alfred North Whitehead (1861–1947). 'I doubt,' he said, 'the possibility of measurement in space which is heterogeneous as to its properties in different parts. I do not understand how the fixed conditions for measurement are to be obtained.' He followed up this idea by devising an alternative theory which he set forth in a book, *The Principle of Relativity*,[3] in 1922. 'I maintain,' he said, 'the old-fashioned belief in the fundamental character of simultaneity. But I adapt it to the novel outlook by the qualification that the meaning of simultaneity may be different in different individual experiences.' This statement of course admits the relativity theory of Poincaré and Lorentz, but it is not compatible with the General Theory of Relativity, in which an observer's domain of simultaneity is usually confined to a small region of his immediate neighbourhood. Whitehead therefore postulated two fields of natural relations, one of them (namely space and time relations) being isotropic, universally uniform and not conditioned by physical circumstances: the other comprising the physical relations expressed by laws of nature, which are contingent.

A profound study of Whitehead's theory was published in 1952 by J. L. Synge.[4] As he remarked, the theory of Whitehead offers something between the two extremes of Newtonian theory on the

[1] *ZS. f. P.* lxxiii (1931), p. 147; lxxv (1932), p. 63; *Phys. Rev.* xxxix (1932), p. 716
[2] *Times Educ. Suppl.*, 12 Feb. 1920, p. 83
[3] *The Principle of Relativity, with applications to physical science*, Camb. Univ. Press, 1922
[4] *Proc. R.S.*(A), ccxi (1952), p. 303

one hand and the General Theory of Relativity on the other. It conforms to the requirement of Lorentz invariance (thus overcoming the major criticism against the Newtonian theory), but it does not reinstate the concept of force, with the equality of action and reaction, so that its range of applicability remains much lower than that of Newtonian mechanics. However, it does free the theorist from the nearly impossible task of solving a set of non-linear partial differential equations whenever he seeks a gravitational field. It is not a field theory, in the sense commonly understood, but a theory involving action at a distance, propagated with the fundamental velocity *c*.

Whitehead's doctrine, though completely different from Einsteins' in its formulation, may be described very loosely as fitting the Einsteinian laws into a flat space-time ; and no practicable observational test has hitherto been suggested for discriminating between the two theories.[1]

The idea of mapping the curved space of General Relativity on a flat space, and making the latter fundamental, was revived many years after Whitehead by N. Rosen.[2] He and others[3] who developed it claimed that in this way it was possible to explain more directly the conservation of energy, momentum, and angular momentum, and also possibly to account for certain unexplained residuals in repetitions of the Michelson-Morley experiment.[4]

In 1916 K. Schwarzschild made an important advance in the Einsteinian theory of gravitation, by discovering the analytical solution of Einstein's equations for space-time when it is occupied by a single massive particle.[5]

The field being a statical one, we can take the quadratic form which specifies the metric of space-time to be

$$(ds)^2 = V^2(dt)^2 - \frac{1}{c^2}(dl)^2$$

where *dl* is the line-element in the three-dimensional space, S, so

$$(dl)^2 = \sum_{p,\,q=1}^{3} a_{pq}dx^p dx^q$$

and the functions V, a_{pq}, (p, $q=1, 2, 3$), do not involve t. There will be a one-to-one correspondence between the points of the space S, which contains the massive particle, and the points of a

[1] cf. G. Temple, *Proc. Phys. Soc.* xxxvi (1924), p. 176 ; W. Band, *Phys. Rev.* lxi (1942), p. 698

[2] *Phys. Rev.* lvii (1940), pp. 147, 150, 154

[3] M. Schoenberg, *Phys. Rev.* lix (1941), p. 616 ; G. D. Birkhoff, *Proc. Nat. Ac. Sci.* xxix (1943), p. 231 ; A. Papapetrou, *Proc. R.I.A.*, lii (A) (1948), p. 11

[4] D. C. Miller, *Proc. Nat. Ac. Sci.* xi (1925), p. 306

[5] K. Schwarzschild, *Berl. Sitz.* 1916, p. 189 ; cf. also : A. Einstein, *Berl. Sitz.* 1915, p. 831 ; D. Hilbert, *Gött. Nach.* 1917, p. 53 ; J. Droste, *Proc. Amst. Ac.* xix (1917), p. 197 ; A. Palatini, *N. Cimento*, xiv (1917), p. 12 ; C. W. Oseen, *Ark. f. Mat. Ast. och. Fys.* xv (1921), No. 9 ; J. L. Synge, *Proc. R.I.A.* liii (1950), p. 83.

space S′ which is completely empty. Now the latter space can be specified by co-ordinates (r, θ, ϕ) in such a way that its line-element is given by

$$(dl')^2 = (dr)^2 + r^2\{(d\theta)^2 + \sin^2\theta(d\phi)^2\},$$

the origin of S′ corresponding to the location of the massive particle in S. The effect of the presence of the particle will be to distort S as compared with S′, but the distortion will be symmetrical with respect to the particle, so the line-element in S will be expressible in the form

$$(dl)^2 = f(r)(dr)^2 + g(r)r^2\{(d\theta)^2 + \sin^2\theta(d\phi)^2\}$$

and if we take a new variable R such that $\sqrt{g(r)}\,.\,r = \mathrm{R}$, this may be written

$$(dl)^2 = \mathrm{A}^2(d\mathrm{R})^2 + \mathrm{R}^2\{(d\theta)^2 + \sin^2\theta(d\phi)^2\}$$

where A is some function of R. By symmetry, V is also a function of R, and the problem is to determine A and V.

Thus, now writing r for R, the quadratic form which defines the metric in the space around the particle is

$$(ds)^2 = \sum_{p,q} g_{pq}dx^p dx^q$$

where

$$x^0 = t, \qquad x^1 = r, \qquad x^2 = \theta, \qquad x^3 = \phi$$

$$g_{00} = \mathrm{V}^2, \qquad g_{11} = -\frac{\mathrm{A}^2}{c^2}, \qquad g_{22} = -\frac{r^2}{c^2}, \qquad g_{33} = -\frac{r^2\sin^2\theta}{c^2}$$

and g_{pq} is zero when p is different from q.

Since the energy-tensor is zero everywhere except at the origin, and since $\mathrm{K} = 0$ when $\mathrm{K}_{pq} = 0$, Einstein's gravitational equations reduce to

$$\mathrm{K}_{pq} = 0.$$

Calculating the Ricci tensor, and denoting differentiations by dashes, we find

$$\mathrm{K}_{00} = c^2\left(-\frac{\mathrm{VV}''}{\mathrm{A}^2} + \frac{\mathrm{A}'\mathrm{VV}'}{\mathrm{A}^3} - \frac{2\mathrm{VV}'}{\mathrm{A}^2 r}\right)$$

$$\mathrm{K}_{11} = \frac{\mathrm{V}''}{\mathrm{V}} - \frac{\mathrm{A}'\mathrm{V}'}{\mathrm{AV}} - \frac{2\mathrm{A}'}{\mathrm{A}r}$$

$$\mathrm{K}_{22} = c^2\left(\frac{1}{\mathrm{A}^2} + \frac{r\mathrm{V}'}{\mathrm{A}^2\mathrm{V}} - \frac{r\mathrm{A}'}{\mathrm{A}^3} - 1\right).$$

Forming the combination $(A^2/c^2V^2)K_{00}+K_{11}$, and equating it to zero, we have

$$\frac{V'}{V}+\frac{A'}{A}=0$$

so AV = a constant; and since when $r\to\infty$ the space tends to that defined by the quadratic form

$$(ds)^2=(dt)^2-\frac{1}{c^2}\{(dr)^2+r^2(d\theta)^2+r^2\sin^2\theta(d\phi)^2\},$$

that is, $A\to 1$ and $V\to 1$, we must have

$$AV=1.$$

The equation $K_{22}=0$ now becomes

$$\frac{1}{A^2}-\frac{2rA'}{A^3}-1=0.$$

so denoting $(1/A^2)-1$ by u, we have

$$u+r\frac{du}{dr}=0, \qquad \text{or} \qquad ru=\text{Constant},$$

so

$$\frac{1}{A^2}=\left(1-\frac{a}{r}\right).$$

Thus *the metric in the space around the particle is specified by*

$$(ds)^2=\left(1-\frac{a}{r}\right)(dt)^2-\frac{1}{c^2}\left\{\frac{(dr)^2}{1-\frac{a}{r}}+r^2(d\theta)^2+r^2\sin^2\theta(d\phi)^2\right\}.$$

This is Schwarzschild's solution.

If at any instant, in the plane $\theta=\frac{1}{2}\pi$, we consider a circle on which r is constant, we see that the length of an element of its arc is given by the equation

$$(dl)^2=r^2(d\phi)^2$$

so $dl=rd\phi$, and the circumference of the whole circle is $2\pi r$. This determines the physical meaning of the co-ordinate r. But if we consider a radius vector from the origin, the element of length along this radius is given by

$$(dl)^2=\frac{(dr)^2}{1-\frac{a}{r}},$$

so when r, in decreasing, tends to a, then $dl \to \infty$: that is to say, *the region inside the sphere $r = a$ is impenetrable.*

The differential equations of motion of a particle (supposed to be so small that it does not disturb the field), under the influence of a single gravitating centre, are the differential equations of the geodesics in space-time with Schwarzschild's matric. These equations can be written down and integrated in terms of elliptic functions.[1]

A particular case is that of *rectilinear orbits along a radius vector from the central mass.* We then have $d\phi = 0$, and can readily derive the integral

$$\left(1 - \frac{a}{r}\right)\frac{dt}{ds} = \text{Constant} = \frac{1}{\mu} \quad \text{say,}$$

and we have

$$(ds)^2 = \left(1 - \frac{a}{r}\right)(dt)^2 - \frac{(dr)^2}{c^2\left(1 - \frac{a}{r}\right)}.$$

Eliminating ds between these equations, we have

$$\frac{1}{c^2\left(1 - \frac{a}{r}\right)^3}\left(\frac{dr}{dt}\right)^2 = \frac{1}{1 - \frac{a}{r}} - \mu^2.$$

Let dl denote the element of length along the radius, and let $d\tau$ denote the element of time, so

$$(dl)^2 = \frac{(dr)^2}{1 - \frac{a}{r}}, \qquad (d\tau)^2 = \left(1 - \frac{a}{r}\right)(dt)^2;$$

then the preceding equation becomes

$$\frac{1}{c^2}\left(\frac{dl}{d\tau}\right)^2 = 1 - \mu^2 + \frac{a\mu^2}{r}$$

which obviously corresponds to the Newtonian equation of the conservation of energy. Differentiating it, we have

$$\frac{d^2l}{d\tau^2} = -\frac{a\mu^2c^2}{2r^2}\left(1 - \frac{a}{r}\right)^{\frac{1}{2}}.$$

[1] cf. W. de Sitter, *Proc. Amst. Ac.* xix (1916), p. 367; T. Levi-Civita, *Rend. Lincei*, xxvi (i) (1917), pp. 381, 458; xxvi (ii) (1917), p. 307; xxvii (i) (1918), p. 3; xxvii (ii) (1918), pp. 183, 220, 240, 283, 344; xxviii (i) (1919), pp. 3, 101; A. R. Forsyth, *Proc. R.S.*(A), xcvii (1920), p. 145 (Integration by Jacobian elliptic functions); F. Morley, *Am. J. Math.* xliii (1921), p. 29 (by Weierstrassian elliptic functions); C. de Jans, *Mém. de l'Ac. de Belgique*, vii (1923) (by Weierstrassian functions); K. Ogura, *Jap. J. of Phys.* iii (1924), pp. 75, 85; a very complete study of the trajectories of a small particle in the Schwarzschild field was made by Y. Hagihara, *Jap. J. of Ast. and Geoph.* viii (1931), p. 67.

When r is large compared with a, this becomes approximately

$$\frac{d^2r}{dt^2} = -\frac{ac^2}{2r^2}$$

which is the Newtonian law of attraction.

Comparing it with the Newtonian law of attraction to the sun,

$$\frac{d^2r}{dt^2} = -\frac{\gamma M}{r^2}$$

where γ is the Newtonian constant of attraction and M is the sun's mass, we have

$$a = \frac{2\gamma M}{c^2}$$

which gives

$$a = 2{\cdot}95 \text{ km.}$$

This is the value of the constant a for the sun.

By approximating from the elliptic-function solution for the general case of motion round a gravitating centre, we can study the orbits when the distance of the small particle from the centre of force is large compared with a, in which case we can have orbits differing little from the ellipses described by the planets under the Newtonian attraction of the sun. It is found that the line of apsides of such an orbit is not fixed, but slowly rotates: if l is the semi-latus-rectum, then the advance of the perihelion in one complete revolution is $3\pi a/l$, which for the case of the planet Mercury revolving round the sun amounts to about $0''{\cdot}1$: since Mercury makes about 420 revolutions in a century, the secular advance of the perihelion is 42″, which agrees well with the observed value of 43″.[1]

The paths of rays of light in the field of a single gravitating centre are, of course, the null geodesics of the Schwarzschild field. Many of them have remarkable forms. It is found that a ray of light can be propagated perpetually in a circle of radius $\frac{3}{2}a$ about the gravitating centre: and there are trajectories spirally asymptotic to this circle both externally and internally, the other terminus of the trajectory being on the circle $r = a$ in the former case and at infinity in the latter. There are also trajectories of which one

[1] Einstein first showed that the new gravitational theory would explain the anomalous motion of the perihelion of Mercury in the third of his papers in the *Berlin Sitz.* of 1915, p. 831. On the whole subject cf. G. M. Clemence, *Proc. Amer. Phil. Soc.* xciii (1949), p. 532. The secular motion of the earth's perihelion, as revealed by observation, was found by H. R. Morgan, *Ast. J.* li (1945), p. 127, to agree with that calculated by Einstein's theory, which is $3''{\cdot}84$.

terminus is at infinity and the other on the circle $r = \alpha$, trajectories which at both termini meet the circle $r = \alpha$, and quasi-hyperbolic trajectories both of whose extremities are at infinity. For these last-named trajectories it is found that if a is the apsidal distance, the angle between the asymptotes is approximately equal to $2\alpha/a$. A ray coming from a star and passing close to the sun's gravitational field, when observed by a terrestrial observer, will therefore have been deflected through an angle $2\alpha/a$.[1] If we take the radius of the solar corona to be 7×10^5 km., so that $a = 7 \times 10^5$ km. and $\alpha = 3$ km., *then* Einstein found that *the displacement of the star is* 2.3/7.10[5] *in circular measure, or* 1″·75.

The notion that light possesses gravitating mass, and that therefore a ray of light from a star will be deflected when it passes near the sun, was far from being a new one, for it had been put forward in 1801 by J. Soldner,[2] who calculated that a star viewed near the sun would be displaced by 0″·85. Einstein's prediction was tested by two British expeditions to the solar eclipse of May 1919, who found for the deflection the values 1″·98 ± 0″·12 and 1″·61 ± 0″·30 respectively, so the prediction was regarded as confirmed observationally: and this opinion was strengthened by the first reports regarding the Australian eclipse of 1922 September 21. Three different expeditions found for the shift at the sun's limb the values 1″·72 ± 0″·11, 1″·90 ± 0″·2 and 1″·77 ± 0″·3, all three results differing from Einstein's predicted value by less than their estimated probable errors. However, a re-examination of the 1922 measures gave about 2″·2: at the Sumatra eclipse of 1929 the deflection was found[3] to be 2″·0 to 2″·24: and at the Brazilian eclipse of May 1947 a value of 2″·01 ± 0″·27 was obtained[4]: while it must not be regarded as impossible that the consequences of Einstein's theory may ultimately be reconciled with the results of observation, it must be said that at the present time (1952) there is a discordance.

A second observational test proposed was the anomalous motion of the perihelion of Mercury. This is quantitatively in agreement with Einstein's theory,[5] but as we have seen,[6] there are alternative explanations of it.

A third proposed test was the displacement to the red of spectral lines emitted in a strong gravitational field. This, however, was, as we have seen,[7] explained before General Relativity was discovered, and does not, properly speaking, constitute a test of it in contra-

[1] cf. F. D. Murnaghan, *Phil. Mag.* xliii (1922), p. 580; T. Shimizu, *Jap. J. Phys.* iii (1924), p. 187; R. J. Trumpler, *J.R.A.S. Canada*, xxiii (1929), p. 208

[2] *Berliner Astr. Jahrb.* 1804, p. 161; reprinted *Ann. d. Phys.* lxv (1921), p. 593

[3] E. Freundlich, H. v. Klüber and A. v. Brunn, *ZS. f. Astroph.* iii (1931), p. 171; vi (1933), p. 218; xiv (1937), p. 242; cf. J. Jackson, *Observatory*, liv (1931), p. 292

[4] G. van Biesbroek, *Ast. J.* lv (1950), p. 49; cf. E. Finlay-Freundlich and W. Ledermann, *Mon. Not. R.A.S.* civ (1944), p. 40

[5] cf. G. M. Clemence, *Ast. J.* l (1944)

[6] cf. p. 148

[7] p. 152–3; cf. G. Y. Rainich, *Phys. Rev.* xxxi (1928), p. 448

distinction to other theories.[1] In any case, the observed effect so far as solar lines[2] are concerned is complicated by other factors.[3] With regard to stellar lines, the greatest effect might be expected from stars of great density. Now Eddington showed that in a certain class of stars, the 'white dwarfs,' the atoms have lost all their electrons, so that only the nuclei remain : and under the influence of the gravitational field, the nuclei are packed together so tightly that the density is enormous. The companion of Sirius is a star of this class : its radius is not accurately known, but the star might conceivably have a density about 53,000 times that of water, in which case the red-displacement of its lines would be about 30 times that predicted for the sun. The comparison of theory with observation of the red-displacement can, however, hardly be said to furnish a quantitative test of the theory.[4]

We may remark that it is easy to construct models which behave so as to account for the observed facts of astronomy. For example, it has been shown by A. G. Walker[5] that if a particle is supposed to be moving in ordinary Euclidean space in which there is a Newtonian gravitational potential ϕ (so that the contribution to ϕ from a single attracting mass M is $-\gamma M/r$, where γ is the Newtonian constant of attraction) and if the kinetic energy of the particle is

$$T = \tfrac{1}{2}e^{-2\phi/c^2}v^2 \quad \text{where} \quad v^2 = \left(\frac{dx}{dt}\right)^2 + \left(\frac{dy}{dt}\right)^2 + \left(\frac{dz}{dt}\right)^2$$

(which reduces to $T = \tfrac{1}{2}v^2$ when $c \to \infty$), and if the potential energy of the particle is

$$V = \tfrac{1}{2}c^2\left\{1 - e^{-2\phi/c^2}\right\}$$

[1] The derivation of the effect from Einstein's equations may readily be constructed from the following indications. For an atom at rest in a gravitational field (e.g. on the sun) the proper-time is given by

$$d\tau^2 = V^2 dt^2,$$

where, as we have seen, $V = 1 - (1/c^2)\,\Omega$ if Ω is the gravitational potential ; while for an atom at rest at a great distance from the gravitational field (e.g. on the earth) we have

$$V = 1 \quad \text{and} \quad d\tau^2 = dt^2.$$

The period of the radiation, as measured at its place of emission, is to its period as measured by the terrestrial observer, in the ratio that $d\tau/dt$ at the place of emission bears to $d\tau/dt$ at the place of the observer, that is, the ratio V to unity ; so, as before, we find that the wave-length of the radiation produced in the strong gravitational field and observed by the terrestrial observer is $(1 + \Omega/c^2)$ times the wave-length of the same radiation produced on earth.

[2] C. E. St. John, *Astroph. J.* lxvii (1928), p. 195

[3] cf. Miss M. G. Adam of Oxford, *Mon. Not. R.A.S.* cviii (1948), p. 446

[4] For a discussion as to whether the perihelion motion of Mercury, the deflection of light rays that pass near a gravitating body, and the displacement of spectral lines in the gravitational field, are to be regarded as establishing General Relativity, cf. E. Wiechert, *Phys. ZS.* xvii (1916), p. 442, and *Ann. d. Phys.* lxiii (1920), p. 301.

[5] *Nature*, clxviii (1951), p. 961

(which reduces to ϕ when $c\to\infty$), then the Lagrangean equations of motion of the particle are

$$\frac{d}{dt}\left\{\frac{1}{1-v^2/c^2}\frac{dx}{dt}\right\} = -\frac{1+v^2/c^2}{1-v^2/c^2}\frac{\partial\phi}{\partial x}$$

and two similar equations in y and z (which reduce to $d^2x/dt^2 = -\partial\phi/\partial x$ and two similar equations when $c\to\infty$), and that the trajectories have the following properties:

(i) the velocity of the particle can never exceed c;

(ii) a particle moving initially with velocity c continues to move with this velocity;

(iii) the perihelion of an orbit advances according to the same formula as is derived from General Relativity;

(iv) if the particles moving with speed c are identified with photons, they are deflected towards a massive gravitating body according to the same formulae as is derived from General Relativity.

It is unwise to accept a theory hastily on the ground of agreement between its predictions and the results of observation in a limited number of instances: a remark which perhaps is specially appropriate to the investigations of the present chapter.

An astronomical consequence of General Relativity which has not yet been mentioned was discovered in 1921 by A. D. Fokker,[1] namely that, as a result of the curvature of space produced by the sun's gravitation, the earth's axis has a precession, additional to that deducible from the Newtonian theory, of amount 0″·019 per annum; unlike the ordinary precession, it would be present even if the earth were a perfect sphere. Its existence cannot however be tested observationally, since the shape and internal constitution of the earth are not known with sufficient accuracy to give a reliable theoretical value for the ordinary precession, and the relativistic precession is only about $\frac{1}{2500}$ of this.

Besides the Schwarzschild solution, a number of other particular solutions of the equations of General Relativity were obtained in the years following 1916, notably those corresponding to a particle which has both mass and electric charge,[2] and to fields possessing axial symmetry,[3] especially an infinite rod,[4] and two particles.[5] In 1938 A. Einstein, L. Infeld and B. Hoffman published a method[6] for finding, by successive approximation, the field due to n bodies.

[1] *Proc. Amst. Ac.* xxiii (1921), p. 729; cf. H. A. Kramers, ibid. p. 1052, and G. Thomsen, *Rend. Lincei*, vi (1927), p. 37. The existence of an effect of this kind had been predicted in 1919 by J. A. Schouten, *Proc. Amst. Ac.* xxi (1919), p. 533.

[2] H. Reissner, *Ann. d. Phys.* l (1916), p. 106; H. Weyl, *Ann d. Phys.* liv (1917), p. 117; G. Nordström, *Proc. Amst. Ac.* xx (1918), p. 1236; C. Longo, *N. Cimento*, xv (1918), p. 191; G. B. Jeffery, *Proc. R.S.*(A), xcix (1921), p. 123.

[3] T. Levi-Civita, *Rend. Lincei*(5) xxviii (i) (1919), pp. 4, 101

[4] W. Wilson, *Phil. Mag.*(6) xl (1920), p. 703

[5] H. E. J. Curzon, *Proc. L.M.S.*(2) xxiii (1924–5), pp. xxix and 477

[6] *Ann. of Math.* xxxix (1938), p. 65; cf. H. P. Robertson, ibid., p. 101 and T. Levi-Civita, *Mém. des Sc. Math.*, fasc. cxvi (1950)

Static isotropic solutions of the field equations, and symmetric distributions of matter, have been discussed by M. Wyman.[1]

In 1916 and the following years attention was given [2] to the propagation of disturbances in a gravitational field. If the distribution of matter in space is changed, e.g. by the circular motion of a plate in its own plane, gravitational waves are generated, which are propagated outwards with the speed of light.

If such waves impinge on an electron which is at rest, the principle of equivalence shows that the physical situation is the same as if the electron were moving with a certain acceleration, and therefore *an electron exposed to gravitational waves must radiate.*[3]

In 1917 Einstein [4] pointed out that the field-equations of gravitation, as he had given them in 1915, do not satisfy Mach's Principle, according to which, no space-time could exist except in so far as it is due to the existence of matter (or energy). Einstein's equations of 1915, however, admit the particular solution

$$g_{pq} = \text{Constant}, \qquad T_{pq} = 0 \qquad (p, q = 0, 1, 2, 3)$$

so that a field is thinkable without any energy to generate it. He therefore proposed now [5] to modify the equations by writing them

$$K_{pq} - \tfrac{1}{2} g_{pq} K - \lambda g_{pq} = -\kappa T_{pq} \qquad (p, q = 0, 1, 2, 3).$$

The effect of the λ-term is to add to the ordinary gravitational attraction between particles a small repulsion from the origin varying directly as the distance: at very great distances this repulsion will no longer be small, but will be sufficient to balance the attraction: and in fact, as Einstein showed, it is possible to have a statical universe, spherical in the spatial co-ordinates, with a uniform distribution of matter in exact equilibrium.[6] This is generally called the *Einstein universe.* The departure from Euclidean metric is measured by the radius of curvature R_0 of the spherical space, and this is connected with the total mass M of the particles constituting the universe by the equation

$$\frac{\gamma M}{c^2} = \tfrac{1}{4}\pi R_0$$

[1] *Phys. Rev.* lxvi (1944), p. 267; lxxv (1949), p. 1930

[2] A. Einstein, *Berlin Sitz.* 1916, p. 688; 1918, p. 154; H. Weyl, *Raum Zeit Materie*, 4th edn. (Berlin, 1921) p. 228 (English edn., p. 252); A. S. Eddington, *Proc. R.S.*(A), cii (1922), p. 268; H. Mineur, *Bull. S.M. Fr.* lvi (1928), p. 50; A. Einstein and N. Rosen, *J. Frankl. Inst.* ccxxiii (1937), p. 43; M. Brdička, *Proc. R.I.A.* liv (1951), p. 137.

[3] This problem was investigated by W. Alexandrow, *Ann. d. Phys.* lxv (1921), p. 675.

[4] *Berlin Sitz.*, 1917, p. 142; *Ann. d. Phys.* lv (1918), p. 241

[5] The new equations can be derived variationally by adding a constant to Hilbert's World-Function.

[6] The suggestion that our universe might be an Einsteinian space-time of constant *spatial* curvature seems to have been made first by Ehrenfest in a conversation with de Sitter about the end of 1916.

where γ denotes the Newtonian constant of gravitation. The total volume of this universe is $2\pi^2 R_0^3$.

It was shown by J. Chazy [1] and by E. Trefftz [2] that when the λ-term is included in the gravitational equations, Schwarzschild's metric for the space-time about a single gravitating centre must be modified to

$$(ds)^2 = \left(1 - \frac{a}{r} - \frac{\lambda}{3} r^2\right)(dt)^2 - \frac{1}{c^2}\left\{\frac{(dr)^2}{1 - \frac{a}{r} - \frac{\lambda}{3} r^2} + r^2(d\theta)^2 + r^2 \sin^2\theta (d\phi)^2\right\}.$$

If in this we put $a = 0$, which amounts to supposing that there is no mass at the origin, so the world is completely empty,[3] we obtain

$$(ds)^2 = \left(1 - \frac{\lambda}{3} r^2\right)(dt)^2 - \frac{1}{c^2}\left\{\frac{(dr)^2}{1 - \frac{\lambda}{3} r^2} + r^2(d\theta)^2 + r^2 \sin^2\theta (d\phi)^2\right\}.$$

Now this metric had been discovered by W. de Sitter[4] in 1917. *It is the metric of a four-dimensional space-time of constant curvature.*

The *de Sitter world,* as it was called, was the subject of many papers, partly on account of its intrinsic geometrical interest to the pure mathematician, and partly because of the possibility that some or all of its features might be similar to those of our actual universe as revealed by astronomical observation.[5]

Let us consider the universe as it would be if all minor irregularities were smoothed out: just as when we say that the earth is a spheroid, we mean that the earth would be a spheroid if all mountains were levelled and all valleys filled up. In the case of the universe the levelling is a more formidable operation, since we have to smooth out the earth, the sun and all the heavenly bodies, and to reduce the world to a complete uniformity. But after all, only a very small fraction of the cosmos is occupied by material bodies: and it is

[1] *Comptes Rendus,* clxxiv (1922), p. 1157

[2] *Math. Ann.* lxxxvi (1922), p. 317; cf. M. von Laue, *Berlin Sitz.,* 1923, p. 27

[3] This, of course, shows the invalidity of the reason Einstein had originally given for introducing the λ-term.

[4] *Proc. Amst. Acad.* xix (31 March 1917), p. 1217; xx (1917), pp. 229, 1309; *Mon. Not. R.A.S.* lxxviii (Nov. 1917), p. 3

[5] cf. F. Klein, *Gött. Nach.* 6 Dec. 1918, = *Ges. Math. Abh.* i, p. 604; K. Lanczos, *Phys. ZS.* xxiii (1922), p. 539; H. Weyl, *Phys. ZS.* xxiv (1923), p. 230; *Phil. Mag.*(6) xlviii (1924), p. 348; *Phil. Mag.*(7) ix (1930), p. 936; P. du Val, *Phil. Mag.*(6) xlvii (1924), p. 930; M. von Laue and N. Sen, *Ann. d. Phys.* lxxiv (1924), p. 252; L. Silberstein, *Phil. Mag.*(6) xlvii (1924), p. 907; H. P. Robertson, *Phil. Mag.*(7) v (1928), p. 835; R. C. Tolman, *Astroph. J.* lxix (1929), p. 245; G. Castelnuovo, *Lincei Rend.* xii (1930), p. 263; M. von Laue, *Berlin Sitz.* 1931, p. 123; E. T. Whittaker, *Proc. R.S.*(A), cxxxiii (1931), p. 93; H.S. Coxeter, *Amer. Math. Monthly,* l (1943), p. 217

interesting to inquire what space-time as a whole is like when we simply ignore them.[1]

The answer must evidently be, that it is a manifold of constant curvature. This means that it is isotropic (i.e. the curvature is the same for all orientations at the same point) and is also homogeneous. As a matter of fact, there is a well-known theorem that any manifold which is isotropic in this sense is necessarily also homogeneous, so that the two properties are connected. A manifold of constant curvature is a projective manifold, i.e. ordinary projective geometry is valid in it when we regard geodesics as straight lines: and it is possible to move about in it any system of points, discrete or continuous, rigidly, i.e. so that the mutual distances are unaltered.

The simplest example of a manifold of constant curvature is the surface of a sphere in ordinary three-dimensional Euclidean space: and the easiest way of constructing a model of the de Sitter world is to take a pseudo-Euclidean manifold of five dimensions in which the line-element is specified by the equation

$$-(ds)^2 = (dx)^2 + (dy)^2 + (dz)^2 - (du)^2 + (dv)^2,$$

and in this manifold to consider the four-dimensional hyper-pseudo-sphere [2] whose equation is

$$x^2 + y^2 + z^2 - u^2 + v^2 = R^2.$$

The pseudospherical world thus defined has a constant Riemannian measure of curvature $-1/R^2$.[3]

The de Sitter world may be regarded from a slightly different mathematical standpoint as having a Cayley-Klein metric, governed by an Absolute whose equation in four-dimensional homogeneous co-ordinates is

$$x^2 + y^2 + z^2 - u^2 + v^2 = 0$$

where u is time. Hyperplanes which do not intersect the Absolute are spatial, so spatial measurements are elliptic, i.e. the three-dimensional world of space has the same kind of geometry as the surface of a sphere, differing from it only in being three-dimensional instead of two-dimensional. In such a geometry there is a natural unit of length, namely the length of the complete straight line,

[1] The curvature of space at any particular place due to the general curvature of the universe is quite small compared to the curvature that may be imposed on it locally by the presence of energy. By a strong magnetic field we can produce a curvature with a radius of less than 100 light-years, and of course in the presence of matter the curvature is stronger still. So the universe is like the earth, on which the local curvature of hills and valleys is far greater than the general curvature of the terrestrial globe.

[2] The prefix *hyper-* indicates that we are dealing with geometry of more than the usual three dimensions, and the prefix *pseudo-* refers to the occurrence of negative signs in the equation.

[3] The world of the Poincaré-Lorentz theory of relativity can be regarded as a four-dimensional hyperplane in the five-dimensional hyperspace.

just as on the surface of a sphere there is a natural unit of length, namely the length of a complete great circle.

We are thus brought to the question of the dimensions of the universe: what is the length of the complete straight line, the circuit of all space? Since 1917 there has seemed to be a possibility that, by the combination of theory with astrophysical observation, this question might be answered.

Different investigations of the de Sitter world, however, reached conclusions which apparently were not concordant. The origin of some of the discordances could be traced to the ambiguities which were involved in the use of the terms 'time,' 'spatial distance' and 'velocity,' when applied by an observer to an object which is remote from him in curved space-time. The 'interval' which is defined by $(ds)^2 = \sum_{p,q} g_{pq}dx^p dx^q$ involves space and time blended together: and although any particular observer at any instant perceives in his immediate neighbourhood an 'instantaneous three-dimensional space' consisting of world-points which he regards as simultaneous, and within which the formulae of the Poincaré-Lorentz relativity theory are valid, yet this space cannot be defined beyond his immediate neighbourhood: for with a general metric defined by a quadratic differential form, it is not in general possible to define simultaneity (with respect to a particular observer) over any *finite* extent of space-time.

The concept of 'spatial distance between two material particles' is, however, not really dependent on the concept of 'simultaneity.'[1] When the astronomer asserts that 'the distance of the Andromeda nebula is a million light-years,' he is stating a relation between the world-point occupied by ourselves at the present instant and the world-point occupied by the Andromeda nebula at the instant when the light left it which arrives here now, that is, he is asserting a relation between two world-points such that a light-pulse, emitted at one, arrives at the other; or, in geometrical language, between two world-points which lie on the same null geodesic. The spatial distance of two material particles in a general space-time may, then, be thought of as *a relation between two world-points which are on the same null geodesic.* It is obviously right that 'spatial distance' should exist only between two world-points which are on the same null geodesic, for it is only then that the particles at these points are in direct physical relation with each other. This statement brings out into sharp relief the contrast between 'spatial distance' and the 'interval' defined by $(ds)^2 = \sum_{p,q} g_{pq}dx^p dx^q$: for between two points on the same null geodesic the 'interval' is always zero. Thus '*spatial distance' exists when, and only when, the 'interval' is zero.*

In order to define 'spatial distance' conformably to these ideas, with a general metric for space-time, it is necessary to translate

[1] E. T. Whittaker, *Proc. R.S.*(A), cxxxiii (1931), p. 93

into the language of differential geometry the principle by which astronomers actually calculate the ' distance ' of very remote objects such as the spiral nebulae. The principle is this : first, the absolute brightness of the object (the ' star ' as we may call it) is determined[1] : then this is compared with the apparent brightness (i.e. the brightness as actually seen by the observer). The *distance* of the star is then defined to be proportional to the square root of the ratio of the absolute brightness to the apparent brightness.

In adopting this principle into differential geometry, we take a ' star ' A and an observer B, which are on the same null geodesic, and we consider a thin pencil of null geodesics (rays of light) which issue from A and pass near B. This pencil intersects the observer B's ' instantaneous three-dimensional space,' giving a two-dimensional cross-section : the ' spatial distance AB ' is then defined to be proportional to the square root of this cross-section. Distance, as thus defined, is an invariant, i.e. it is independent of the choice of the co-ordinate system. This invariant, however, involves not only the position of the star and the position of the observer, but also the motion of the observer, since his ' instantaneous three-dimensional space ' is determinate only when his motion is known. Thus the ' spatial distance ' of a star from an observer depends on the motion of the observer : but this is quite as it should be, and indeed had always been recognised in the relativity theory of Poincaré and Lorentz : for in that theory the spatial distance of a star from an observer is $(X^2+Y^2+Z^2)^{\frac{1}{2}}$, where (T, X, Y, Z) are co-ordinates referred to *that particular inertial system with respect to which the observer is at rest* : the necessity for the words in italics shows that the distance depends on the observer's motion.

When the de Sitter world is studied in the light of this definition of distance (the mass of any material particles concerned being supposed to be so small that they do not sensibly affect the geometrical character of the universe), some remarkable results are found. Thus a freely moving star and a freely moving observer cannot remain at a constant spatial distance apart in the de Sitter world. When the observer first sees the star, its spatial distance is equal to the radius of curvature of the universe. After this the star is continuously visible, the distance passing through a minimum value, after which it increases again indefinitely : that is, the star's apparent brightness ultimately decreases to zero, it becomes too faint to be seen. When this happens the star is at a point which is not the terminus of its own world-line, so the star continues to exist after it has ceased to have any relations with this particular observer.

The Einstein world, which as we have seen is a statical solution of the gravitational equations, spherical in the spatial co-ordinates,

[1] For this purpose, astronomers in practice use the known relation between the period and absolute magnitude of Cepheid variables, or (in the method of spectroscopic parallaxes) the known relations between absolute magnitude and spectral behaviour.

was generalised by A. Friedman [1] into a solution in which the curvature depends on the time—what in fact came to be known later as an *expanding universe.* Not much notice was taken of this paper until, five years afterwards, the Abbé Georges Lemaitre,[2] a Belgian priest who had been a research student of Eddington's at Cambridge, proved that the Einstein universe is unstable, and that when its equilibrium is disturbed, the world progresses through a continuous series of intermediate states, towards a limit which is no other than the de Sitter universe.

Eddington, who at the beginning of 1930 was investigating, in conjunction with his research student G. C. McVittie, the stability of the Einstein universe, found in Lemaitre's paper the solution of the problem, and at once [3] saw that it provided an explanation of the observed scattering apart of the spiral nebulae. Since 1930 the theory of the expanding universe has been of central importance in cosmology.

The scheme of general relativity, as put forward by Einstein in 1915, met with some criticism as regards the unsatisfactory position occupied in it by electrical phenomena. While gravitation was completely fused with metric, so that the notion of a mechanical force on ponderable bodies due to gravitational attraction was entirely abolished, the notion of a mechanical force acting on electrified or magnetised bodies placed in an electric or magnetic field still persisted as in the old physics. This seemed, at any rate from the aesthetic point of view, to be an imperfection, and it was felt that sooner or later everything, including electromagnetism, would be re-interpreted and represented in some way as consequences of the pure geometry of space and time. In 1918 Weyl [4] proposed to effect this by rebuilding geometry once more on a new foundation, which we must now examine.

Weyl fixed attention in the first place on the 'light-cone,' or aggregate of directions issuing from a world-point P, in which light-signals can go out from it. The light-cone separates those world-points which can be affected by happenings at P, from those points whose happenings can affect P: it, so to speak, separates past from future. Now the light-cone is represented by the equation $(ds)^2=0$, where ds is the element of interval, and Weyl argued that this equation, rather than the quantity $(ds)^2$ itself, must be taken as the starting-point of the subject: in other words, it is the *ratios* of the ten coefficients g_{pq} in $(ds)^2$, and not the *actual values* of these coefficients,

[1] *ZS. f. P.* x (1922), p. 377. Aberration and parallax in the universes of Einstein, de Sitter and Friedman, were calculated by V. Fréedericksz and A. Schechter, *ZS. f. P.* li (1928), p. 584.

[2] *Ann. de la Soc. sc. de Bruxelles*, xlviiA (April 1927), p. 49

[3] *Mon Not. R.A.S.* xc (1930), p. 668

[4] *Berl. Sitz.* 1918, Part I, p. 465; *Math. ZS.* ii (1918), p. 384; *Ann. d. Phys.* lix (1919), p. 101; *Phys. ZS.* xxi (1920), p. 649, xxii (1921), p. 473; *Nature*, cvi (1921), p. 781; cf. A. Einstein, *Berl. Sitz.* 1918, p. 478; W. Pauli. *Phys. ZS.* xx (1919), p. 457; *Verh. d. phys. Ges.* xxi (1919), p. 742; A. Einstein, *Berl. Sitz.* 1921, p. 261; L. P. Eisenhart, *Proc. N.A.S.* ix (1923), p. 175

which are to be taken as determined by our most fundamental physical experiences. This leads to the conclusion that comparisons of length at different times and places may yield discordant results according to the route followed in making the comparison.

Following up this principle, Weyl devised a geometry more general than the Riemannian geometry that had been adopted by Einstein : instead of being specified, like the Riemannian geometry, by a single quadratic differential form

$$\sum_{p,q} g_{pq}dx^p dx^q$$

it is specified by a quadratic differential form $\sum_{p,q} g_{pq}dx^p dx^q$ and a linear differential form $\sum_p \phi_p dx^p$ together. The coefficients g_{pq} of the quadratic form can be interpreted, as in Einstein's theory, as the potentials of gravitation, while the four coefficients ϕ_p of the linear form can be interpreted as the four components of the electromagnetic potential-vector. Thus Weyl succeeded in exhibiting both gravitation and electricity as effects of the metric of the world.[1]

The enlargement of geometrical ideas thus achieved was soon followed by a still wider extension due to Eddington.[2] This, and most subsequent constructions in the same field, are based on an analysis of the notion of *parallelism*, which must now be considered.

The question of parallelism in curved spaces was raised in an acute form by the discovery of General Relativity theory : for it now became necessary for the purposes of physics to create a theory of vectors in curved space, and of their variation from point to point of the space. Now in Euclidean space if U and V are two vectors at the same point, P, we can find the vector which is their difference by using the triangle of vectors. But if U is a vector at a point P and V is a vector at a different point Q, and if we want to find the vector which is the difference of U and V, it is necessary (in principle) first to transfer U parallel to itself from the point P to the point Q, and then to find the difference of the two vectors at Q. Thus a process of *parallel transport* is necessary for finding the difference of vectors at different points, and hence for the spatial differentiation of vectors : and this is true whether the space is Euclidean or non-Euclidean.

The spatial differentiation of vectors in curved space has already been discussed[3] analytically under the name of *covariant differentiation.* Evidently this covariant differentiation must really be based on a parallel transport of vectors in the curved space : and in 1917 this particular form of parallel transport was definitely formulated

[1] It does not seem necessary to describe this geometry in detail, since Weyl himself later expressed the opinion (*Amer. J. Math.* lxvi (1944), p. 591) that it does not (at least in its original form) provide a satisfactory unification of electromagnetism and gravitation.

[2] *Proc. R.S.*(A), xcix (1921), p. 104

[3] cf. p. 161

by T. Levi-Civita.[1] It may be regarded as providing a geometrical interpretation for the Christoffel symbols.

After this the idea[2] was developed that a space may be regarded as formed of a great number of small pieces cemented together, so to speak, by a parallel transport, which states the conditions under which a vector in one small piece is to be regarded as parallel to a vector in a neighbouring small piece. Thus every type of differential geometry must have at its basis a definite parallel-transport or ' connection.'

To illustrate these points, let us consider geometry on the surface of the earth, regarded as a sphere. It is obvious that directions U and V at two different points cannot (unless one of the points happens to be the antipodes of the other) be parallel in the ordinary Euclidean sense, i.e. parallel in the three-dimensional space in which the earth is immersed : but a new kind of parallelism can be defined, and that in many different ways. We might, for instance, say that V is derived by parallel transport from U if V and U have the same compass-bearing, so that, e.g. the north-east direction at one place on the earth's surface would be said to be parallel to the north-east direction at any other place.

Having defined parallel-transport on the earth in this way, there are now two different kinds of curve on the earth's surface which may be regarded as analogous to the straight line in the Euclidean plane. If we define a straight line by the property that it is the shortest distance between two points, then its analogue on the earth is a great circle, since this gives the shortest distance between two points on the spherical surface. But if we define a straight line by the property that it preserves the same direction along its whole length, or, more precisely, that its successive elements are derived from each other by parallel transport, then its analogue on the earth is the track of a ship whose compass-bearing is constant throughout her voyage : this is the curve called a *loxodrome*. The existence of the two families of curves, the great circles and the loxodromes, may be assimilated to the fact that if an electrostatic and a gravitational field coexist, the lines of electric force and the lines of gravitational force are two families of curves in space ; and this rough analogy may serve to suggest how different physical phenomena may be represented simultaneously in terms of geometrical conceptions.[3]

Weyl's proposal for a unified theory of gravitation and electromagnetism was followed up in another direction by Th. Kaluza,[4]

[1] *Palermo Rend.* xlii (1917), p. 173

[2] This idea is due essentially to G. Hessenberg, *Math. Ann.* lxxviii (1917), p. 187, though he did not explicitly introduce Levi-Civita's notion of parallel-transport.

[3] Eddington's generalisation of Weyl's theory was the starting-point of important papers by A. Einstein, *Berl. Sitz.* (1923), pp. 32, 76, 137, and by Jan Arnoldus Schouten of Delft (*b.* 1883), *Amst. Proc.* xxvi (1923), p. 850. The latter represented the electromagnetic field as a vortex connected with the *torsion* (in the sense of Cartan) of four-dimensional space ; on this cf. H. Eyraud, *Comptes Rendus*, clxxx (1925), p. 127.

[4] *Berlin Sitz.* (1921), p. 966

in whose theory the ten gravitational potentials g_{pq} of Einstein and the four components ϕ_p of the electromagnetic potential were expressed in terms of the line-element in a space of five dimensions, in such a way that the equations of motion of electrified particles in an electromagnetic field became the equations of geodesics. This *five-dimensional theory of relativity* was afterwards developed by Oskar Klein [1] and others.[2] To ordinary space-time a fifth dimension is adjoined, but the curves in this dimension are very small, of the order of 10^{-30} cm., so that the universe is cylindrical, and indeed filiform, with respect to the fifth dimension, and the variations of the fifth variable are not perceptible to us, its non-appearance in ordinary experiments being due to a kind of averaging over it. A connection with quantum theory was made by assuming that Planck's quantum of Action h is due to periodicity in the fifth dimension: and the atomicity of electricity was presented as a quantum law, the momentum conjugate to the fifth co-ordinate having the two values e and $-e$.

In 1930 Oswald Veblen and Banesh Hoffman of Princeton [3] showed that the Kaluza-Klein theory may be interpreted as a four-dimensional theory based on projective instead of affine geometry.

Weyl's geometry, and especially his type of parallel-transport, suggested to W. Wirtinger [4] philosophical considerations which led him to a very original form of infinitesimal geometry: this, however, perhaps on account of its extreme generality, has not as yet found applications in theoretical physics.

Besides the investigations that have been described, the work of Weyl and Eddington has led, during the last thirty years, to a vast number of other investigations aimed at expressing gravitational and electromagnetic theory together in terms of a system of differential geometry. Some of them, such as those of the Princeton school [5] in America, led by O. Veblen and L. P. Eisenhart, the

[1] *ZS. f. P.* xxxvi (1926), p. 835; xxxvii (1926), p. 895; xlv (1927), p. 285; xlvi (1927), p. 188; *J. de Phys. et le Rad.* viii (1927), p. 242; *Ark. f. Mat. Ast. Fys* xxxiv (1946), No. 1

[2] A. Einstein, *Berlin Sitz.* (1927), pp. 23, 26; L. de Broglie, *J. de Phys. et le Rad.* viii (1927), p. 65; G. Darrieus, *J. de Phys. et le Rad.* viii (1927), p. 444; F. Gonseth et G. Juvet, *Comptes Rendus*, clxxxv (1927), p. 341; H. Mandel, *ZS. f. P.* xxxix (1926), p. 136; xlv (1927), p. 285; xlix (1928), p. 697; liv (1929), p. 564; lx (1930), p. 782; W. Wilson, *Proc. R.S.*(A), cxviii (1928), p. 441; J. W. Fisher, *Proc. R.S.*(A), cxxiii (1929), p. 489; H. T. Flint, *Proc. R.S.*(A), cxxiv (1929), p. 143; J. W. Fisher and H. T. Flint, *Proc. R.S.*(A), cxxvi (1930), p. 644

[3] *Phys. Rev.* xxxvi (1930), p. 810; cf. also J. A. Schouten and D. van Dantzig, *Proc. Amst. Ac.* xxv (1932), pp. 642, 843; *ZS. f. P.* lxxviii (1932), p. 639; J. A. Schouten, *ZS. f. P.* lxxxi (1933), pp. 129, 405; lxxxiv (1933), p. 92; W. Pauli, *Ann. d. Phys.* xviii (1933), p. 305.

[4] *Trans. Camb. Phil. Soc.* xxii (1922), p. 439; *Abh. aus dem math. Sem. der Hamb.* iv (1926), p. 178.

[5] cf. L. P. Eisenhart and O. Veblen, *Proc. N.A.S.* viii (1922), p. 19; O. Veblen, *Proc. N.A.S.* viii (1922), p. 192; ix (1923), p. 3; L. P. Eisenhart, *Proc. N.A.S.* viii (1922), p. 207; ix (1923), p. 4; xi (1925), p. 246; *Annals of Math.* xxiv (1923), p. 367; O. Veblen and J. M. Thomas, *Proc. N.A.S.* xi (1925), p. 204; T. Y. Thomas, *Proc. N.A.S.* xi (1925), p. 199; xiv (1928), p. 728; *Math. ZS.* xxv (1926), p. 723, and many papers in the succeeding years, in the same journals, by these authors and their associates

Dutch school at Delft, founded by J. A. Schouten,[1] and the French school whose most distinguished representative was E. Cartan [2] (1869–1951), have made discoveries which have great potentialities, but the significance of which at present appears to lie in pure mathematics rather than in physics, and which are therefore not described in detail here. The work of E. Bortolotti [3] of Cagliari should also be referred to. The outlook of the Germans, and of some of the Americans, has been, broadly speaking, more physical.[4]

It must be said, however, that, elegant though the mathematical developments have undoubtedly been, their relevance to fundamental physical theory must for the present be regarded as hypothetical.

This chapter has been concerned, for the most part, with General Relativity, which is essentially a geometrisation of physics. It may be closed with some account of a movement in the opposite direction, seeking to abolish the privileged position of geometry in physics, and indeed inquiring how far it may be possible to construct a physics independent of geometry. Since the notion of metric is a complicated one, which requires measurements with clocks and scales, generally with *rigid* bodies, which themselves are systems of great complexity, it seems undesirable to take metric as fundamental, particularly for phenomena which are simpler and actually independent of it.

The movement was initiated by Friedrich Kottler of Vienna, who in 1922 published two papers *Newton'sches Gesetz und Metrik* [5] and *Maxwell'sche Gleichungen und Metrik.*[6] Kottler first remarked on

[1] cf. J. A. Schouten, *Proc. Amst. Ac.* xxvii (1924), p. 407; xxix (1926), p. 334; *Palermo Rend.* l (1926), p. 142; J. A. Schouten and D. van Dantzig, *Proc. Amst. Ac.* xxxiv (1932), p. 1398; xxxv (1932), p. 642; *ZS. f. P.* lxxviii (1932), p. 639; lxxxi (1933), pp. 129, 405

[2] cf. E. Cartan, *Ann. Éc. Norm.* xl (1923), p. 325; xli (1923), p. 1; *Bull. Soc. M. France*, lii (1924), p. 205

[3] *Atti Inst. Veneto*, lxxxvi (1926–7), p. 459; *Rend. Lincei*, ix (1929), p. 530

[4] cf. R. Weitzenbock, *Wien. Ber.* cxxix 2a (1920), pp. 683, 697; cxxx 2a (1921), p. 15; H. Weyl, *Gött. Nach.* (1921), p. 99; F. Jüttner, *Math. Ann.* lxxxvii (1922), p. 270; E. Reichenbächer, *ZS. f. P.* xiii (1923), p. 221; A. Einstein, *Berlin Sitz.* (1925), p. 414; D. J. Struik and O. Wiener, *J. Math. Phys.* vii (1927), p. 1; E. Wigner, *ZS. f. P.* liii (1929), p. 592; A. Einstein, *Berlin Sitz.* (1928), pp. 217, 224; (1929), pp. 2, 156; N. Wiener and M. S. Vallarta, *Proc. N.A.S.* xv (1929), pp. 353, 802; M. S. Vallarta, *Proc. N.A.S.* xv (1929), p. 784; H. Weyl, *Bull. Am. M.S.* xxxv (1929), p. 716; N. Rosen and M. S. Vallarta, *Phys. Rev.* xxxvi (1930), p. 110; A. Einstein, *Berlin Sitz.* (1930), p. 18; *Math. Ann.* cii (1930), p. 685; A. Einstein and W. Mayer, *Berlin Sitz.* (1930), p. 110; (1931), pp. 257, 541; (1932), p. 130; W. Pauli, *Ann. d. Phys.*(5) xviii (1933), p. 305; E. Schrödinger, *Proc. R.I.A.* xlix (1943), pp. 43, 135; G. D. Birkhoff, *Proc. N.A.S.* xxix (1943), p. 231; xxx (1944), p. 324; A. Barajas, G. D. Birkhoff, C. Graef and M. S. Vallarta, *Phys. Rev.*(2) lxvi (1944), p. 138; H. Weyl, *Am. J. Math.* lxvi (1944), p. 591; E. Schrödinger, *Nature*, cliii (1944), p. 572; *Proc. R.I.A.* xlix (A) (1944), pp. 225, 237, 275; A. Einstein and V. Bargmann, *Annals of M.* xlv (1944), p. 1; A Einstein, *Annals of M.* xlv (1944), p. 15; E. Schrödinger, *Proc. R.I.A.* l (1945), pp. 143, 223; li (1946), p. 41; li (1947), pp. 147, 163; li (1948), p. 205; lii (1948), p. 1; A. Einstein, Appendix II in the third edition of his *The Meaning of Relativity* (Princeton, 1949); on the further development of this, cf. A. Papapetrou and E. Schrödinger, *Nature*, clxvii (1951), p. 40, and W. B. Bonnor, *Proc. R.S.*(A), ccix (1951), p. 353; ccx (1952), p. 427

[5] *Wien. Sitz.*, Abt. IIa, cxxxi (1922), p. 1

[6] ibid., p. 119

the independence of metric which characterises the analysis of skew tensors (i.e. tensors for which an interchange of any two indices produces a reversal of sign) : thus the divergence of a six-vector (X^{pq}) (which is a skew tensor of rank 2) is

$$\frac{1}{\sqrt{g}}\sum_q \frac{\partial}{\partial x^q}\left(\sqrt{g}\, X^{pq}\right)$$

and this is clearly unchanged if the metric

$$(ds)^2 = \sum_{p,q} g_{pq}\, dx^p\, dx^q$$

is replaced by

$$(ds)^2 = \lambda^2 \sum_{p,q} g_{pq}\, dx^p\, dx^q$$

where λ is any constant. While he recognised that it is impossible to abolish metric from physics entirely, he aimed at expressing the laws of nature as far as possible in terms of skew tensors, and ascertaining in each case the precise point where they cease to be adequate and the introduction of a metric becomes inevitable. In the first paper he considered the Newtonian theory of gravitation : Poisson's equation for the gravitational potential is

$$\frac{\partial^2\phi}{\partial x_1^2}+\frac{\partial^2\phi}{\partial x_2^2}+\frac{\partial^2\phi}{\partial x_3^2} = -4\pi\rho$$

or, if the metric is given by

$$(ds)^2 = \sum_{p,q} a_{pq} dx^p dx^q \qquad (p,\ q = 1,\ 2,\ 3),$$

it is

$$\frac{1}{\sqrt{a}}\sum_p \frac{\partial}{\partial x^p}\left(\sqrt{a}\sum_q a^{pq}\frac{\partial\phi}{\partial x^q}\right) = -4\pi\rho.$$

But $\sum_q a^{pq}\frac{\partial\phi}{\partial x^q}$ is the contravariant vector corresponding to the covariant vector $\partial\phi/\partial x^p$: call it (L^p). Then Poisson's equation is

$$\sum_p \frac{\partial}{\partial x^p}\left(\sqrt{a}\,.\,L^p\right) = -4\pi\rho\sqrt{a}.$$

In order to transform this into an equation depending on a skew tensor, Kottler made use of a theorem of tensor-calculus which may be stated thus : in space of three dimensions, where the metric is given by

$$(ds)^2 = \sum_{p,q} a_{pq} dx^p dx^q \qquad (p,\ q = 1,\ 2,\ 3),$$

let δ_{pqr} have the value 1 or -1 according as we obtain (pqr) from (123) by an even or an odd substitution: and when p, q, r are not all different, let δ_{pqr} be defined to be zero. *Then* $\sqrt{a} \,.\, \delta_{pqr}$ *is a skew tensor of rank* 3.[1] We shall denote it by G_{pqr}.

Now write

$$F_{pq} = \sum_r G_{pqr} L^r$$

so F_{pq} is a covariant skew tensor of rank 2: its three numerically distinct components are

$$F_{12} = \sqrt{a} \,.\, L^3, \qquad F_{31} = \sqrt{a} \,.\, L^2, \qquad F_{23} = \sqrt{a} \,.\, L^1.$$

Thus Poisson's equation may be written

$$\frac{\partial F_{23}}{\partial x^1} + \frac{\partial F_{31}}{\partial x^2} + \frac{\partial F_{12}}{\partial x^3} = \mu_{123}$$

where $\mu_{123} = -4\pi\rho G_{123}$.

Now if S be any simple surface enclosing a volume V, then by Green's theorem (which is independent of any system of metric) we have

$$\iiint_V \left(\frac{\partial F_{23}}{\partial x^1} + \frac{\partial F_{31}}{\partial x^2} + \frac{\partial F_{12}}{\partial x^3}\right) dx^1 dx^2 dx^3$$

$$= \iint_S \left\{ F_{23} \frac{\partial(x_2, x_3)}{\partial(u, v)} + F_{31} \frac{\partial(x_3, x_1)}{\partial(u, v)} + F_{12} \frac{\partial(x_1, x_2)}{\partial(u, v)} \right\} du dv$$

where (u, v) are any parameters fixing position on the surface. Thus we obtain the equation

$$\iint_S \left\{ F_{23} \frac{\partial(x_2, x_3)}{\partial(u, v)} + F_{31} \frac{\partial(x_3, x_1)}{\partial(u, v)} + F_{12} \frac{\partial(x_1, x_2)}{\partial(u, v)} \right\} du\, dv = \iiint_V \mu_{123} dx^1 dx^2 dx^3.$$

Kottler interpreted the left-hand side of this equation as the total *flux*, through the surface S, of the skew tensor F_{pq}: and he regarded the equation as representing the relation between the occurrences inside S and the field (expressed by a skew tensor) acting on S, *in a form independent of metric.* It is of course essentially a form of Gauss's theorem.[2] Kottler regarded this equation as the starting-point of

[1] It is generally called the *skew fundamental tensor of Ricci and Levi-Civita.*

[2] On Gauss's theorem and the concept of mass in General Relativity, cf: E. T. Whittaker, *Proc. R.S.*(A), cxlix (1935), p. 384; H. S. Ruse, *Proc. Edin. Math. Soc.*(2) iv (1935), p. 144; J. T. Combridge, *Phil. Mag.*(7) xx (1935), p. 971; G. Temple, *Proc. R.S.*(A), cliv (1936), p. 354; J. L. Synge, *Proc. R.S.*(A), clvii (1936), p. 434; *Proc. Edin. Math. Soc.*(2) v (1937), p. 93; A. Lichnerowicz, *Comptes Rendus*, ccv (1937), p. 25

a fundamental non-metrical formulation of gravitational theory: metric is introduced, at a later stage, only when the concept of *work*, as distinguished from that of *flux*, is introduced: that is, when the force acting on a particle comes to be considered.

His second paper, on Maxwell's equations, presented an easier problem, since the electric and magnetic forces together constitute a six-vector, which is the kind of tensor required for a non-metrical presentation of the subject. Moreover there were available the results of papers written in 1910 by E. Cunningham[1] and H. Bateman,[2] who, starting from the proposition that the fundamental equations of electrodynamics are invariant with respect to Lorentz transformations, i.e. to rotations in the four-dimensional world of space-time, had remarked that these electrodynamic equations are invariant with respect to a much wider group, namely all the transformations for which the equation

$$(dx^2+(dy)^2+(dz)^2-c^2(dt)^2=0$$

is invariant. By writing $\sqrt{-1}\,.\,cdt=du$, this last equation becomes

$$(dx)^2+(dy)^2+(dz)^2+(du)^2=0.$$

The transformations which leave this latter equation invariant are what are called the *conformal* group of transformations in space of four dimensions: they had been studied long before by Sophus Lie, who had shown that they are composed of reflections, translations, rotations, magnifications and inversions with respect to the hyperspheres of the four-dimensional space. Types of motion of an electromagnetic system may be derived from one another by such transformations: they are in general more complicated than those considered in the relativity theory of Poincaré and Lorentz: for in the latter case a fixed three-dimensional configuration is transformed into one, every point of which has the same velocity of translation: but in the conformal case, under the simplest operation of the group, a fixed system becomes one in which the whole is expanding or contracting in a certain way.[3]

Cunningham and Bateman's work was now developed by Kottler,[4] H. Weyl,[5] J. A. Schouten and J. Haantjes,[6] and particularly by D. van Dantzig,[7] into a *complete theory of electromagnetism, independent*

[1] *Proc. L.M.S.*(2) viii, p. 77

[2] ibid., p. 223; cf. also Bateman's book, *Electrical and Optical Wave-motions* (Camb., 1915); H. Bateman, *Phys. Rev.* xii (1918), p. 459; *Proc. L.M.S.*(2) xxi (1920), p. 256; E. Bessel-Hagen, *Math. Ann.* lxxxiv (1921), p. 258; F. D. Murnaghan, *Phys. Rev.* xvii (1921), p. 73; G. Kowalewski, *J. für Math.* clvii (1927), p. 193

[3] It is characteristic of these transformations that a sphere which is expanding with the velocity of light transforms into a sphere expanding with the same velocity.

[4] loc. cit. [5] *Raum, Zeit, Materie*, 4th Aufl., p. 260 [6] *Physica* I (1934), p. 869

[7] *Proc. Camb. Phil. Soc.* xxx (1934), p. 421; *Proc. Amst. Ac.* xxxvii (1934), pp. 521, 526, 644, 825; xxxix (1936), pp. 126, 785; *Cong. Int. des Math.* Oslo (1936), II, p. 225; cf. also J. A. Schouten, *Tensor Calculus for Physicists*, Oxford, 1951

of metrical geometry, and in fact needing the ideas neither of metric nor of parallelism. It is characteristic of theories such as this that *differential* relations are generally replaced by integral relations: thus for the statement that under certain circumstances 'the divergence of the flux of energy vanishes' is substituted the statement that 'the integral of the flux over a closed surface vanishes,' which is [1] a mathematical form of the physical statement that 'the algebraic sum of the energies of all the particles crossing through a closed surface vanishes.'

[1] cf. D. van Dantzig, *Erkenntnis*, vii (1938), p. 142

Chapter VI

RADIATION AND ATOMS IN THE OLDER QUANTUM THEORY

In 1916 Einstein[1] published a new and extremely simple proof of Planck's law of radiation, and at the same time obtained some important results regarding the emission and absorption of light by molecules. The train of thought followed was more or less similar to that adopted by Wien in the derivation[2] of his law of radiation, but Einstein now adapted it to the new situation created by Bohr's theory of spectra.

Consider a molecule of a definite kind, disregarding its orientation and translational motion: according to quantum theory, it can take only a discrete set of states $Z_1, Z_2, \ldots Z_n, \ldots$ whose internal energies may be denoted by $\epsilon_1, \epsilon_2, \ldots \epsilon_n, \ldots$ If molecules of this kind belong to a gas at temperature T, then the relative frequency W_n of the state Z_n is given by the formula of Gibbs's canonical distribution as modified for discrete states,[3] namely,[4]

$$W_n = e^{-\frac{\epsilon_n}{kT}}.$$

Now suppose that the probability of a single molecule in the state Z_m passing in time dt spontaneously, i.e., without excitation by external agencies (as in the emission of γ rays by radio-active bodies) to the state of lower energy Z_n with emission of radiant energy $\epsilon_m - \epsilon_n$ is

$$A_m^n \, dt. \tag{A}$$

Suppose also that the probability of a molecule under the influence of radiation of frequency ν and energy-density ρ passing in time dt from the state of lower energy Z_n to the state of higher energy Z_m, by absorbing the radiant energy $\epsilon_m - \epsilon_n$, is

$$B_n^m \rho \, dt \tag{B}$$

and suppose that the probability of a molecule under the influence of this radiation-field passing in time dt from the state of higher

[1] *Mitt. d. phys. Ges. Zürich*, No. 18 (1916); *Phys. ZS*. xviii (1917), p. 121; cf. A. S. Eddington, *Phil. Mag.* l (1925), p. 803

[2] cf. Vol. I, p. 382

[3] cf. *supra*, p. 87

[4] For simplicity we omit consideration of the statistical 'weight' of the state.

energy Z_m to the state of lower energy Z_n, with emission of the radiant energy $\epsilon_m - \epsilon_n$, is

$$B^n_m \rho \, dt. \tag{B'}$$

This is called *stimulated emission*; its existence was recognised here for the first time.

Now the exchanges of energy between radiation and molecules must not disturb the canonical distribution of states as given above. So in unit time, on the average, as many elementary processes of type (B) must take place as of types (A) and (B′) together. We must therefore have

$$e^{-\frac{\epsilon_n}{kT}} B^m_n \rho = e^{-\frac{\epsilon_m}{kT}} (B^n_m \rho + A^n_m).$$

We assume that ρ increases to infinity with T, so this equation gives

$$B^m_n = B^n_m, \tag{1}$$

and the preceding equation may therefore be written

$$\rho = \frac{(A^n_m / B^n_m)}{e^{(\epsilon_m - \epsilon_n)/kT} - 1}.$$

This is evidently Planck's law of radiation: in order that it may pass asymptotically into Rayleigh's law for long wave-lengths, and Wien's law for short wave-lengths, we must have

$$\epsilon_m - \epsilon_n = h\nu$$

and

$$A^n_m = \frac{8\pi h\nu^3}{c^3} B^n_m. \tag{2}$$

The two equations (1) *and* (2), first given in this paper of Einstein's, *are fundamental in the theory of the exchanges of energy between matter and radiation*, and have been extensively used in the later development of quantum theory.[1] It may be remarked that as there is spontaneous emission, but not spontaneous absorption, there is asymmetry as between past and future: but so far as transitions stimulated by radiation are concerned, there is a symmetrical probability, $B^n_m = B^m_n$.[2]

[1] If the weights of the energy-levels are g_m, g_n, the relation (1) must be replaced by $g_n B^m_n = g_m B^n_m$. Relation (2) is unaffected.

[2] Einstein's formulae were extended to the case of non-sharp energy-levels by R. Becker, *ZS. f. P.* xxvii (1924), p. 173, and to the laws of interaction between radiation and free electrons by A. Einstein and P. Ehrenfest, *ZS. f. P.* xix (1923), p. 301.

One of the chief problems of quantum theory is to compute coefficients, such as these Einstein coefficients, from data regarding atoms and molecules. The relation (2) has been verified experimentally by a comparison of the strengths of absorption and emission lines. The B's have been found from measures of the intensities of the components of multiplets in spectra by L. S. Ornstein and H. C. Burger.[1]

Another important result established in this paper related to exchanges of momentum between molecules and radiation. Einstein showed that *when a molecule, in making a transition from the state* Z_n *to* Z_m, *receives the energy* $\epsilon_m - \epsilon_n$, *it receives also momentum of amount* $(\epsilon_m - \epsilon_n)/c$ *in a definite direction;* and, moreover, that *when a molecule, in the transition from* Z_m *to the state of lower energy* Z_n, *emits radiant energy of amount* $(\epsilon_m - \epsilon_n)$, *it acquires momentum of amount* $(\epsilon_m - \epsilon_n)/c$ *in the opposite direction.* Thus the processes of emission and absorption are *directed* processes: there seems to be no emission or absorption of spherical waves.

Einstein's theory of the coefficients of emission and absorption enabled W. Bothe[2] to give an instructive fresh calculation of the numbers of quanta $h\nu$ in black-body radiation which are associated as 'photo-molecules' in pairs $2h\nu$, trios $3h\nu$, etc. He considered a cavity filled with black-body radiation at temperature T, in which there were a number of gas-molecules, each of which was either in the state of energy Z_1 or the state of energy Z_2, where $Z_2 - Z_1 = h\nu$, their average relative numbers being given by the law of canonical distribution at temperature T. He assumed that when a single quant $h\nu$ of the radiation causes a stimulated emission, the emitted quant moves with the same velocity and in the same direction as the quant that causes it, so they become a quant-pair $2h\nu$; if the exciting quant itself already belongs to a pair, then there arises a trio $3h\nu$, and so on. The absorption of a quant from a quant-pair leaves a single quant, and a spontaneous emission produces a single quant. Writing down the conditions that the average numbers of single quants, of quant-pairs, etc. do not change in time, and using Einstein's relations between the coefficients of spontaneous emission, stimulated emission and absorption, we obtain a set of equations from which it follows that in unit volume and in the frequency-range $d\nu$ the average number of single quants which are united into s-fold quant-molecules $sh\nu$ is

$$\frac{8\pi\nu^2}{c^3} e^{-\frac{sh\nu}{kT}} d\nu,$$

in agreement with the result previously obtained.[3] The average total energy per unit volume in the range $d\nu$ is therefore

$$\frac{8\pi h\nu^3}{c^3}\left(\sum_{s=1}^{\infty} e^{-\frac{sh\nu}{kT}}\right) d\nu$$

[1] *ZS. f. P.* xxiv (1924), p. 41 [2] *ZS. f. P.* xx (1923), p. 145 [3] cf. p. 103 *supra*

or

$$\frac{8\pi h\nu^3}{c^3}\frac{d\nu}{e^{h\nu/kT}-1}$$

in agreement with Planck's law of radiation.

Einstein's theory of the coefficients of emission and absorption also enabled theoretical physicists within a few years to create a satisfactory quantum theory of the scattering, refraction and dispersion of light.[1] In 1921 Rudolf Ladenburg[2] (*b.* 1882), a former pupil of Röntgen's at Munich, who afterwards settled in America at Princeton, introduced quantum concepts into the theory.

It was necessary first for him to find a quantum-theoretic interpretation for the number that in classical theory represented the number of electrons bound to atoms by forces of restitution, for it was these electrons which were responsible for light-scattering, refraction, and dispersion. Let us call them *dispersion-electrons*. This he achieved by calculating the emitted and absorbed energy of a set of molecules in temperature-equilibrium with radiation, on the basis of classical theory on the one hand and of quantum theory on the other.

Suppose then that there are $\mathfrak{N}$ dispersion-electrons per cm.^{3}, capable of oscillating freely with frequency ν_1. Now for the harmonic oscillator of frequency ν_1 if the displacement x at the instant t is $x_0 \cos 2\pi\nu_1 t$, the mean value of the energy $\frac{1}{2}m\,\{(dx/dt)^2 + 4\pi^2\nu_1^2x^2\}$ is $\overline{U} = 2\pi^2 m\nu_1^2 x_0^2$, and therefore[3] the average energy radiated per second is

$$\frac{16\pi^4e^2}{3c^3}\nu_1^4x_0^2 \quad \text{or} \quad \frac{8\pi^2e^2\nu_1^2}{3mc^3}\overline{U},$$

and therefore the energy radiated in one second by the $\mathfrak{N}$ dispersion-electrons is

$$J_{el} = \frac{8\pi^2e^2\nu_1^2}{3mc^3}\mathfrak{N}\overline{U}.$$

If the molecules are in equilibrium with radiation at temperature T, and if we regard the electrons as spatial oscillators with three degrees

[1] The classical theory of the scattering of light by small particles had been given by Lord Rayleigh in 1871 (*Phil. Mag.*(4) xli, pp. 107, 274, 447) on the basis of the elastic-solid theory of light, and in 1881 (*Phil. Mag.*(5) xii, p. 81) and 1899 (*Nature*, lx, p. 64; *Phil. Mag.*(5) xlvii, p. 375) on the basis of Maxwell's electromagnetic theory; cf. also J. J. Thomson, *Conduction of Electricity through Gases*, 2nd edn. (1906), p. 321.

[2] *ZS. f. P.* iv (1921), p. 451; cf. also R. Ladenburg and F. Reiche, *Naturwiss*, xi (1923), p. 584; R. Ladenburg, *ZS. f. P.* xlviii (1928), p. 15

[3] cf. Vol. I, p. 326

of freedom,[1] then [2] we have between $\overline{U}$ and the radiation-density ρ the relation

$$\overline{U}=\frac{3c^3}{8\pi\nu_1{}^2}\rho.$$

Thus

$$J_{el}=\frac{\pi e^2}{m}\mathfrak{N}\rho.$$

The energy absorbed in equilibrium is of course equal to the energy radiated.

Now in quantum-theory the emission from a molecule is due to transitions from a state of higher energy (2) to a state of lower energy (1). The number of spontaneous transitions per second is, in the notation we have already used,

$$N_2A_2^1$$

where N_2 is the number of molecules in the state (2). The number of transitions from the lower to the higher state per second is

$$N_1B_1^2\rho,$$

where N_1 is the number of molecules in the state (1): and therefore the absorbed energy is

$$J_Q=h\nu_1N_1B_1^2\rho.$$

By Einstein's relations we have [3]

$$B_1^2=\frac{c^3}{8\pi h\nu_1{}^3}A_2^1.$$

Thus we have

$$J_Q=N_1\frac{c^3}{8\pi\nu_1{}^2}A_2^1\rho.$$

Equating J_Q to J_{el}, we have

$$\mathfrak{N}=N_1\frac{mc^3}{8\pi^2e^2\nu_1{}^2}A_2^1.$$

[1] For a three-dimensional oscillator in an isotropic radiation-field we obtain three times the value for a linear oscillator in the same field; on this point, cf. F. Reiche, *Ann. d. Phys.* lviii (1919), p. 693.

[2] cf. Planck, p. 79 *supra*

[3] Supposing for simplicity that the statistical weights of the energy-levels (1) and (2) are equal.

This formula expresses the constant $\mathfrak{N}$ (which may be derived experimentally from measurements of emission, absorption, anomalous dispersion and magnetic rotation, and which in classical theory is interpreted as the number of dispersion-electrons) *in terms of the quantum-theoretic quantities* N_1 (the number of molecules in the lower of the two states) *and* A_2^1 (the probability of the spontaneous transitions which give rise to radiation of frequency ν_1). Thus from measurements of e.g. anomalous dispersion at different lines of a series in a spectrum, we can make inferences regarding the probability of the various transitions.

Ladenburg's result enables us to replace the dispersion-formula found in Vol. I [1] by

$$\mu^2 = 1 + \frac{c^3}{8\pi^3}\sum_i \frac{N_i A_k^i}{\nu_i^2(\nu_i^2 - \nu^2)}$$

where (i) denotes the lower level and (k) the upper level of energy in the transition corresponding to the frequency ν_i.

Now consider the scattering of light by an atom. We suppose the wave-length of the light to be much greater than the dimensions of the atom, so that at any instant the field of force is practically uniform over these dimensions. We suppose also that the atom contains an electron of charge $-e$ and mass m, which is bound to the atom so that when the electron is displaced a distance ξ parallel to the x-axis from its equilibrium position, a force of restitution $4\pi^2 m\nu_1{}^2\xi$ is called into play. Then when the atom is irradiated by a light-wave whose electric field is

$$Ee^{2\pi i\nu t}$$

it can easily be shown that according to classical theory it radiates secondary wavelets such as would be produced by an electric doublet of moment

$$\frac{e^2 Ee^{2\pi i\nu t}}{4\pi^2 m(\nu_1{}^2 - \nu^2)}.$$

These secondary spherical wavelets, which are coherent with the incident wave, constitute the scattered radiation.

In quantum theory we are concerned with scattering not by a single bound electron but by an atom, and therefore in order to transfer this expression to quantum theory we must first multiply it by $\mathfrak{N}/N_1$, the number of classical dispersion-electrons per atom (the atom being supposed to be in the state (1) when scattering).

[1] At p. 401 : there is a change of notation, the N, k, n of Vol. I being here represented by $\mathfrak{N}_i$, $2\pi\nu_i\sqrt{m}$, and $2\pi\nu$ respectively :] and we suppose now that in the classical case there are several kinds of dispersion-electrons with different natural periods. The formula had been confirmed by the experiments of R. W. Wood, *Phil. Mag.*(6) viii (1904), p. 293 and P. V. Bevan, *Proc. R.S.* lxxxiv (1910), p. 209 ; lxxxv (1911), p. 58, on the dispersion of light in the vapours of the alkali metals.

Thus we obtain for the amplitude of the doublet-moment

$$\frac{\mathfrak{N}}{N_1}\frac{Ee^2}{4\pi^2 m(\nu_1{}^2-\nu^2)}.$$

Applying Ladenburg's result, this becomes

$$\frac{c^3 A_2^1 E}{32\pi^4\nu_1{}^2(\nu_1{}^2-\nu^2)}.$$

We must sum over all higher states (2) : and thus we have the result that *an atom in state* (1), *when irradiated by a light-wave whose electric field is* $Ee^{2\pi i\nu t}$, *emits secondary wavelets like an electric oscillator of frequency* ν, *whose electric moment has the amplitude*

$$P=\frac{c^3 E}{32\pi^4}\sum_i \frac{A_i^1}{\nu_i{}^2(\nu_i{}^2-\nu^2)},$$

where A_i^1 *denotes the probability of the atom performing spontaneously in unit time the transition to the state* (1) *of energy* E_1 *from a higher state* (i) *of energy* E_i, *and where*

$$\nu_i=\frac{E_i-E_1}{h}.$$

This result was further modified by Henrik Antony Kramers[1] (1894–1952), who, taking the above Ladenburg formula as correct when the atom was in the normal state, took into consideration the case when the atom is excited, and proposed to deal with it by taking the summation not only over the stationary states i which have higher energy-levels than the state (1), but also over the states j which have lower energy-levels than the state (1), so that the formula becomes

$$P=\frac{c^3 E}{32\pi^4}\left\{\sum_i \frac{A_i^1}{\nu_i{}^2(\nu_i{}^2-\nu^2)}-\sum \frac{A_1^j}{\nu_j{}^2(\nu_j{}^2-\nu^2)}\right\}$$

where[2]

$$\nu_j=\frac{E_1-E_j}{h}.$$

This formula of course relates to a single atom, and a factor must be adjoined representing the number of atoms in this state.

Now as we have seen, in the classical theory of scattering, the

[1] *Nature*, cxiii (1924), p. 673 ; cxiv (1924), p. 310

[2] The formula as given by Kramers contained an additional factor 3 ; this was a consequence of his assumption that the free oscillations are parallel to the incident field, whereas the above formula assumes all orientations of the atom to be equally probable.

atom behaves like an electric doublet, the amplitude of whose moment is

$$\frac{e^2 E}{4\pi^2 m(\nu_1^2 - \nu^2)}.$$

Comparing the above Ladenburg-Kramers formula with this, we see that according to quantum theory the atom behaves with respect to the incident radiation as if it contained a number of bound electric charges constituting harmonic oscillators as in the classical theory, one of these oscillators corresponding to each possible transition between the state of the atom and another stationary state. Thus we can describe the behaviour of an atom in dispersion by means of a doubly-infinite (i.e. depending on two quantum numbers m and n) set of virtual harmonic oscillators, the displacement in the oscillator (m, n) being represented by

$$q(m, n) = Q(m, n)e^{2\pi i\nu(m, n)t}$$

where $\nu(m, n)$ denotes the frequency of this oscillator. The aggregate of these virtual harmonic oscillators was called by A. Landé [1] the *virtual orchestra.* The virtual orchestra is thus a classical substitution-formalism for the radiation, and so indirectly becomes a representative of the quantum radiator itself.

In place of the classical e^2/m we have the value $c^3 A_i^1/8\pi^2\nu_i^2$ for one of the 'absorption oscillators,' i.e. those corresponding to transitions between the state (1) and higher states, but we have the value $-c^3 A_1^j/8\pi^2\nu_j^2$ for one of the 'emission oscillators,' i.e. those corresponding to transitions between the state (1) and lower states: so that there is a kind of 'negative dispersion' arising from the emission oscillators, which may be regarded as analogous to the 'negative absorption' represented by Einstein's coefficient B_2^1. Another way of putting the matter is to say that a quantum-oscillator which is in the higher state, when irradiated by a light-wave which is not markedly absorbed by it, emits a spherical wave, which differs from that emitted in the lower state only by being displaced 180° in phase.

Almost immediately after the appearance in May 1924 of Kramers's paper, Max Born [2] gave a general method, to which he gave the name *quantum mechanics,*[3] for translating the classical theory of the perturbations of a vibrating system into the corresponding quantum formulae. In particular he studied the problem of an oscillator exposed to an external field of radiation: his method was based on the correspondence-principle, the frequency belonging to a transition between states characterised by the quantum numbers

[1] *Naturwiss.* xiv (1926), p. 455

[2] *ZS. f. P.* xxvi (1924), p. 379, communicated 13 June 1924

[3] This was the first occurrence of the term in the literature of theoretical physics.

n and n' corresponding to the overtone $|n-n'|\nu$ of the classical solution: but the results he obtained were afterwards shown to be correct in the light of the later form of quantum theory.

A theory of scattering is also essentially a theory of refraction. For when (in classical theory) light is scattered by atoms, the scattered light has the same frequency as the incident light and interferes with it: the total effect produced is the same as if the primary wave alone existed, but was propagated with a different velocity: and this change of velocity is the essential feature in refraction. In the Ladenburg-Kramers formula, the terms

$$-\sum_j \frac{A_1^j}{\nu_j^2(\nu_j^2-\nu^2)}$$

affect the refraction in the opposite sense to the other terms. This was verified in 1928 by R. Ladenburg[1] and by H. Kopfermann and R. Ladenburg,[2] who studied the refractive index, in the neighbourhood of a certain spectral line of neon. When a current was passed through the gas, many of the atoms were thereby raised to an excited level, and it was found that the refractive index dropped. the refraction due to the ordinary fall in energy-level being partly counterbalanced by that due to the rise.

Even before the appearance of Kramers's paper, new possibilities in regard to scattering had been indicated by A. Smekal.[3] It may happen that the emission of scattered radiation is associated with a quantum transition in the scattering structure, in which case there will be a difference of frequency between the scattered and primary rays of the same order of magnitude as the spectral frequencies of the scattering atomic system; if ν denotes the frequency of the primary radiation, and $h\nu_k$ or $h\nu_l$ denotes the change of energy of the atom in the transition, according as this change happens in the positive or negative direction, then there may be scattered radiation (of low intensity) of frequency $\nu+\nu_k$ or $\nu-\nu_l$. At the time of Smekal's paper, his conjecture of *anomalous scattering*, as he called it, had not been verified experimentally: but in 1928 Sir Chandrasakara V. Raman[4] showed that light scattered in water and other transparent substances contains radiations of frequency quite different from that of the incident light, and it was at once seen that this was the effect predicted by Smekal: almost simultaneously G. Landsberg and L. Mandelstam[5] found the same effect experimentally, working with quartz.

In general, radiation of a definite frequency ν generates scattered

[1] *ZS. f. P.* xlviii (1928), p. 15
[2] ibid., pp. 26, 51; *ZS. f. phys. Chem.* cxxxix (1928), p. 375
[3] *Naturwiss.* xi (1923), p. 873
[4] *Ind. Journ. of Phys.* ii (1928), p. 387; C. V. Raman and K. S. Krishnan, ibid., p. 399
[5] *Naturwiss.* xvi (1928), p. 557; *ZS. f. P.* l (1928), p. 769

radiation of several modified frequencies, all of the form $\nu \pm \nu'$, where ν' is either an infra-red frequency in the absorption spectrum of the scattering material, or a difference of such frequencies.[1]

In the year following the publication of Smekal's paper, H. A. Kramers and W. Heisenberg[2] made an exhaustive study of the radiation which an atom emits when irradiated by incident light. Their method, as in Born's paper published a few months earlier, was first to study by classical theory the perturbation by incident radiation of an atom regarded as a multiply-periodic dynamical system, and then to employ the correspondence-principle in order to translate the results into their quantum-theoretic form.

According to classical theory, under the irradiation the system emits a scattered radiation, whose intensity is proportional to the intensity of the incident light: when it is analysed into harmonic components, a component involves the sum or difference of the frequency of the incident light and frequencies occurring in the undisturbed motion (as in the Smekal-Raman effect). The quantum-theoretic formulae must satisfy the condition that in the region of great quantum numbers, where successive stationary states differ comparatively little from each other, the quantum scattering must tend asymptotically to coincide with the classical scattering. This condition is satisfied by interpreting certain derivatives which occur in the classical formulae as differences of two quantities: in this way, Kramers and Heisenberg obtained equations which involved only the frequencies characteristic of transitions, while all symbols relating to the mathematical theory of multiply-periodic dynamical systems disappeared.

It was shown that when irradiated with monochromatic light, an atom emits not only spherical waves of the same frequency as the incident light, and coherent with it, but also systems of non-coherent spherical waves, whose frequencies are combinations of that frequency with other frequencies, which correspond to thinkable transitions to other stationary states. These additional systems of spherical waves occur as scattered light, but they do not contribute to the phenomena of dispersion and absorption of the incident radiation.

An interesting comment on the Kramers-Heisenberg formula was made by P. Jordan,[3] who remarked that it remained valid even when the incident radiation consisted of very long electromagnetic waves, which in the limit of zero frequency tend to a field-strength constant in time: and that in this limiting case, the formula actually yields the changes of frequency in spectral lines which are observed in the Stark effect. He drew the moral that discontinuous jumps must

[1] A review of literature on the Smekal-Raman effect to Feb. 1931 was given by K. W. F. Kohlrausch, *Phys. ZS.* xxxii (1931), p. 385. On the Smekal-Raman effect in molecules and crystals, cf. E. Fermi, *Mem. Acc. Ital. Fis.* Nr. 3 (1932), p. 1.

[2] *ZS. f. P.* xxxi (1925), p. 681

[3] P. Jordan, *Anschauliche Quantentheorie* (Berlin, 1936), p. 85

not be regarded as the essential characteristic of quantum theory: for phenomena in which they occur can be connected continuously by intermediate types with phenomena in which there is no discontinuity.

About this time much attention was given to scattering of a different type. In 1912 C. A. Sadler and P. Mesham,[1] working in L. R. Wilberforce's laboratory at Liverpool University, showed that when a homogeneous beam of X-rays is scattered by a substance of low atomic weight, the scattered rays are of a softer type (i.e. of longer wave-length).[2] Moreover, in the case of the γ-rays produced by a radium salt, it was shown by J. A. Gray[3] that the secondary or scattered γ-rays were less penetrating than the primary, that this 'softening' was due to a real change in the character of the primary rays when the secondary rays were formed, and that it increased with the angle between the primary and secondary rays (generally called the *angle of scattering*).[4] An explanation of these phenomena favoured by physicists at the time was that the primary beam was not truly homegeneous, and that its softer components were more strongly scattered than the harder ones: this, however, was explicitly denied by Sadler and Mesham. In 1917 C. G. Barkla[5] propounded the hypothesis, that there existed a new series of characteristic radiations which were of shorter wave-length than the K- and L-series which he had discovered previously, and which he named the *J-series*: and that the softening observed in the scattered radiation from light elements was due to the admixture of this J-radiation with radiation of the same hardness as the primary.[6] This explanation, however, lost credit when it was found[7] that the J-series had no critical absorption limit similar to the absorption limit of the K- and L-series, and that it showed under spectroscopic observation[8] no spectral lines such as might have been expected. When the J-series explanation was dismissed,[9] there still seemed to be a

[1] *Phil. Mag.* xxiv (1912), p. 138; cf. J. Laub, *Ann. d. Phys.* xlvi (1915), p. 785, and J. A. Gray, *J. Frank. Inst.* cxc (1920), p. 633

[2] It will be remembered that the secondary X-rays in general consist of a mixture of the characteristic rays discovered by Barkla (K, L, etc.) and of truly scattered rays. The wave-length of the former depends solely on the chemical nature of the scattering substance, but the wave-length of the latter is the same (or nearly the same) as that of the primary rays. For the heavier elements the characteristic rays predominate, but for elements of low atomic weight at that time only the truly scattered rays were normally observable.

[3] *Phil. Mag.* xxvi (1913), p. 611

[4] These results were confirmed as regards γ-rays by D. C. H. Florance (working at Manchester under Rutherford) in *Phil. Mag.* xxvii (1914), p. 225, and Arthur H. Compton, *Phil. Mag.* xli (1921), p. 749.

[5] *Phil. Trans.* ccxvii (1917), p. 315

[6] cf. C. G. Barkla and Margaret P. White, *Phil. Mag.* xxxiv (1917), p. 270. They found an abnormally great mass absorption coefficient for aluminium at 0·37Å, and regarded this as additional proof of the reality of J-radiation.

[7] cf. F. K. Richtmyer and Kerr Grant, *Phys. Rev.* xv (1920), p. 547; C. W. Hewlett, *Phys. Rev.* xvii (1921), p. 284

[8] W. Duane and T. Shimizu, *Phys. Rev.* xiii (1919), p. 289; xiv (1919), p. 389

[9] Further proof of this, depending on the polarisation of the scattered rays, was given by A. H. Compton and C. F. Hagenow, *J. Opt. Soc. Amer.* viii (1924), p. 487.

possibility of explaining the increase of wave-length in the scattered light without departing from classical theory: for[1] radiation-pressure (which had not been taken into account in the previous classical theories of scattering) might set the electrons in motion in the direction of the primary radiation, and the wave-length of the scattered light would then be increased by the Doppler effect. On performing the calculations, it was found that the increase in wave-length should follow the law actually verified by experiment, namely that it would be proportional to $\sin^2 \frac{1}{2}\theta$ where θ denotes the angle of scattering: but its magnitude as calculated was not in agreement with observation.

There seemed to be a likelihood therefore that some new type of physical explanation was required. In October 1922 a Bulletin[2] of the National Research Council of the U.S.A. appeared, written by A. H. Compton and containing a full discussion of secondary radiations produced by X-rays: in this the author suggested that when an X-ray quantum is scattered, it spends all its energy and momentum upon some particular electron. This electron in turn re-radiates the whole quantum (degraded in frequency) in some definite direction; the change in momentum of the X-ray quantum due to the change in its frequency and direction of propagation is associated with a recoil of the scattering electron. The ordinary conservation laws of energy and momentum are obeyed,[3] so that the energy of the recoil electron accounts for the difference between the energy of the incident photon and the energy of the scattered photon.

Compton's theory was communicated to a meeting of the American Physical Society on 1–2 December 1922,[4] and published more fully in May of the following year.[5] On examining the scattered rays[6] from light elements spectroscopically, he found that their spectra showed lines corresponding to those in the primary rays, and also other lines corresponding to these but displaced slightly towards the longer wave-lengths; and that the difference in wave-length increased rapidly at large angles of scattering. It is these displaced lines which represent the *Compton effect*.

In the Compton effect the electron is effectively free, i.e. it is so feebly bound to the nucleus of the atom that the binding-energy can be neglected in comparison with the energy $h\nu$ of the incident quant; this condition is realised in the scattering of hard X-rays by elements of low atomic number. When the frequency is decreased, or the atomic number is increased, the binding forces can no longer be neglected in comparison with the energy of the incident photon,

[1] cf. O. Halpern, *ZS. f. P.* xxx (1924), p. 153; G. Wentzel, *Phys. ZS.* xxvi (1925), p. 436

[2] Vol. IV, No. 20

[3] It may be noted that while the photo-electric effect shows that the *energy* of radiation is transferred in quanta, the Compton effect shows that *momentum* also is transferred in quanta.

[4] *Phys. Rev.* xxi (1923), p. 207

[5] *Phys. Rev.* xxi (1923), p. 483

[6] *Phys. Rev.* xxii (Nov. 1923), p. 409

and the phenomenon passes over into the photo-electric effect (in the case when the energy of the incident photon is transferred to the electron) or ordinary scattering (in the case when the photon retains its energy and changes only its direction).

The Compton effect may be discussed in an elementary way by light-quantum methods as follows : [1]

Let the incident light-quantum, of frequency ν, propagated in the positive direction of the axis of x, encounter at O the electron, which recoils in a direction making an angle θ with the axis Ox, with velocity v, while the light-quantum, degraded to frequency ν', is scattered in a direction making an angle $-\phi$ with Ox. Then (using the relativist formulae for energy and momentum), the equation of conservation of energy is

$$h\nu = h\nu' + mc^2\left\{\left(1-\frac{v^2}{c^2}\right)^{-\frac{1}{2}} - 1\right\},$$

while the equations of conservation of momentum are

$$\frac{h\nu}{c} = \frac{h\nu'}{c}\cos\phi + \frac{mv}{\sqrt{\left(1-\frac{v^2}{c^2}\right)}}\cos\theta,$$

$$\frac{h\nu'}{c}\sin\phi = \frac{mv}{\sqrt{\left(1-\frac{v^2}{c^2}\right)}}\sin\theta.$$

From these equations we have to calculate ν' in terms of ν and ϕ, which are supposed given. We readily find

$$\nu' = \frac{\nu}{1+\frac{h\nu}{mc^2}(1-\cos\phi)}.$$

The increment of wave-length $\Delta\lambda$ or $c/\nu' - c/\nu$ is

$$\Delta\lambda = \frac{2h}{mc}\sin^2\frac{\phi}{2},$$

a formula which was definitely confirmed by observation.

[1] In addition to the papers quoted above and below, cf. the following papers of 1923 and 1924 : P. Debye, *Phys. ZS.* xxiv (April 1923), p. 161, who discovered the theory independently. A. H. Compton, *Phil. Mag.* xlvi (1923), p. 897 ; *J. Frank. Inst.* cxcviii (1924), p 57. G. E. M. Jauncey, *Phys. Rev.* xxii (1923), p 233. P. A. Ross, *Proc. N.A.S.* ix (1923), p. 246 ; x (1924), p. 304 ; *Phys. Rev.* xxii (1923), p. 524. M. de Broglie, *Proc. Phys. Soc.* xxxvi (1924), p. 423. D. Skobelzyn, *ZS. f. P.* xxiv (1924), p. 393 ; xxviii (1924), p. 278. A. H. Compton and A. W. Simon, *Phys. Rev.* xxv (1925), p. 306

Eliminating ν' and ν, we have

$$\left(1+\frac{h\nu}{mc^2}\right)\tan\frac{\phi}{2}=\cot\theta.$$

The kinetic energy of the recoiling electron is

$$\frac{h\nu\,\dfrac{2h\nu}{mc^2}\cos^2\theta}{\left(1+\dfrac{h\nu}{mc^2}\right)^2-\dfrac{h^2\nu^2}{m^2c^4}\cos^2\theta}.$$

The quantity (h/mc), which has the dimensions of a length, is called the *Compton wave-length.* Its value is $0{\cdot}0242\times10^{-8}$ cm. Since the mass of a quantum is $h\nu/c^2$ or $h/\lambda c$, it is seen at once that *a quantum of radiation, whose wave-length is the Compton wave-length, has a mass equal to the mass of the electron.*

The recoil electrons of the Compton effect were studied by C. T. R. Wilson,[1] using his cloud-expansion method. He found that X-radiation of wave-length less than about 0·5Å in air produced two classes of β-ray tracks, namely (*a*), those of electrons ejected with initial kinetic energy comparable to a quantum of the incident radiation : these were photo-electrons : and (*b*), tracks of very short range, which were the recoil electrons. A. H. Compton and J. C. Hubbard,[2] discussing Wilson's results, showed that the motion of the recoil electrons corresponds precisely to Compton's theory.

We have seen that in the spectrum of scattered X-rays there are, in addition to the lines corresponding to the Compton effect, lines corresponding exactly to the primary X-rays. These unshifted lines (which are the only lines to appear when the primary rays are those of visible light) may be explained by supposing that some electrons are closely attached to the nucleus and must scatter while nearly at rest. The theory of this state of affairs was investigated by Compton,[3] who also explained certain results which had been obtained by Duane and his collaborators,[4] and which appeared to be inconsistent with the original Compton theory.

A more accurate treatment of the Compton effect, making use of later developments in general theory, was given in 1929 by

[1] *Proc. R.S.*(A), civ (1923), p. 1 ; cf. also W. Bothe of Charlottenburg, *ZS. f. P.* xvi (1923), p. 319 ; xx (1923), p. 237

[2] *Phys. Rev.* xxiii (1924), p. 439

[3] *Phys. Rev.* xxiv (1924), p. 168 ; *Nature,* cxiv (1924), p. 627

[4] G. L. Clark and W. Duane, *Proc. N.A.S.* ix (1923), pp. 413, 419 ; x (1924), pp. 41, 92 ; xi (1925), p. 173. G. L. Clark, W. W. Stifler and W. Duane, *Phys. Rev.* xxiii (1924), p. 551. A. H. Armstrong, W. Duane and W. W. Stifler, *Proc. N.A.S.* x (1924), p. 374. S. K. Allison, G. L. Clark and W. Duane, *Proc. N.A.S.* x (1924), p. 379. J. A. Becker, *Proc. N.A.S.* x (1924), p. 342. S. K. Allison and W. Duane, *Proc. N.A.S.* xi (1925), p. 25. cf. P. A. Ross and D. L. Webster, *Proc. N.A.S.* xi (1925), pp. 56, 61. A. H. Compton and J. A. Bearden, *Proc. N.A.S.* xi (1925), p. 117

O. Klein and Y. Nishina.[1] The formulae they obtained have been found experimentally to be accurate even with the hardest type of radiation.[2]

The Compton effect raised in an acute form the controversy regarding the reality of light-darts as contrasted with spherical waves of light: for in Compton's explanation, the incident and diffracted X-ray quanta were supposed to have definite directions of propagation. Opinion among theoretical physicists was divided: 'In a recent letter to me,' wrote A. H. Compton in 1924[3] 'Sommerfeld has expressed the opinion that this discovery of the change of wave-length of radiation, due to scattering, sounds the death knell of the wave theory. On the other hand, the truth of the spherical wave hypothesis indicated by interference experiments has led Darwin and Bohr, in conversation with me, to choose rather the abandonment of the principles of conservation of energy and momentum.' The latter policy was embodied in a hypothesis put forward in 1924 by N. Bohr, H. A. Kramers and J. C. Slater,[4] in which it was accepted, that in atomic processes, energy and momentum are only statistically conserved.

They abandoned the principle, common to all previous physical theories, that an atom which is emitting or absorbing radiation must be losing or gaining energy: in its place they introduced the notion of *virtual radiation*, which is propagated in spreading waves as in the electromagnetic theory of light, but which does not transmit energy or momentum: it has indeed no connection with physical reality except the capacity to generate in atoms a probability for the occurrence of transitions: and transitions of the atoms are the only phenomena actually observable. A transition of an atom from one state to another is accompanied by changes of energy and momentum, but is *not* accompanied by radiation: thus the part played by the atom in its relations with radiation reduces to interaction with the field of virtual radiation, while the atom remains in a stationary state. An atom in a stationary state is continually emitting virtual radiation, compounded of all the frequencies corresponding to possible transitions between this state and lower states: this radiation is emitted both spontaneously and by stimulation (in accordance with Einstein's principles of 1917). While in this state, the atom is also capable of *absorbing* radiation corresponding to transitions to states of higher energy. The absorption is performed by *virtual oscillators* situated in the atoms, the frequencies of these oscillators corresponding to the energy-differences between the state of the atom and all

[1] *ZS. f. P.* lii (1929), p. 853

[2] cf. H. C. Trueblood and D. H. Loughridge, *Phys. Rev.* liv (1938), p. 545; Z. Bay and Z. Szepesi, *ZS. f. P.* cxii (1939), p. 20

[3] *J. Frank. Inst.* cxcviii (1924), p. 57

[4] J. C. Slater, *Nature*, cxiii (1924), p. 307. N. Bohr, H. A. Kramers, and J. C. Slater, *Phil. Mag.* xlvii (1924), p. 785; *ZS. f. P.* xxiv (1924), p. 69. J. H. Van Vleck, *Phys. Rev.* xxiv (1924), p. 330. R. Becker, *ZS. f. P.* xxvii (1924), p. 173. J. C. Slater, *Phys. Rev.* xxv (1925), p. 395

higher states. When a virtual oscillator is absorbing virtual radiation, the atom to which it belongs has a certain *probability* of making a transition to the higher state corresponding to the frequency of this virtual oscillator. A transition marks the change from the continuous radiation appropriate to the old state, to the continuous radiation appropriate to the new state: simultaneously with the transition, some virtual oscillators disappear and others come into being: the transition has no other influence on the radiation. The occurrence of a transition in a given atom depends on the initial state of this atom and on the states of the atoms with which it is in communication through the field of virtual radiation, but not on transition processes in the latter atoms: so there is no *direct connection* between the transition of one atom from a higher to a lower state, and the transition of another atom from a lower to a higher state: the principles of energy and momentum are retained in a *statistical* sense, though not in individual interactions of atoms with radiation. The atoms scatter radiation which is incident on them, acting as secondary sources of virtual radiation which interferes with the incident radiation. In any transition, say between states (p) and (q), the energy of the atom changes by $h\nu_{pq}$ and its momentum by $h\nu_{pq}/c$. If the transition is a spontaneous one, the direction of this momentum is random: but if it is stimulated, i.e. induced by the surrounding virtual radiation, the direction of the momentum is the same as that of the wave propagation in this virtual field.

The Bohr-Kramers-Slater theory was wrecked when it was shown to be inconsistent with the results of more refined experiments relating to the Compton effect. One of these was performed by W. Bothe and Hans Geiger.[1] According to Compton's theory, a recoil electron is emitted simultaneously with the scattering of every quantum; while according to the Bohr-Kramers-Slater theory, the connection was much less close, the recoil electrons being emitted only occasionally, while the scattering of virtual radiation is continuous. In Bothe and Geiger's experiment two different Geiger counters counted respectively the recoil electrons, and the photo-electrons produced by the scattered photons. A great many coincidences in time were observed, so many that the probability of their occurrence on the Bohr-Kramers-Slater theory was only 1/400,000. It was therefore concluded that the conservation of energy and momentum holds in individual encounters, and the Bohr-Kramers-Slater theory could not be true.

Another experiment was performed by A. H. Compton and A. W. Simon,[2] who remarked that if in a Wilson cloud-experiment the quantum of scattered X-rays produces a photo-electron in the chamber, then a line drawn from the beginning of the recoil track to the beginning of the track of the photo-electron gives the direction

[1] *ZS. f. P.* xxvi (1924), p. 44; xxxii (1925), p. 639
[2] *Phys. Rev.* xxvi (1925), p. 289

of the quantum after scattering. It was therefore possible to test the truth of the equation

$$\left(1+\frac{h\nu}{mc^2}\right)\tan\frac{\phi}{2}=\cot\theta$$

which connects the directions of the scattered quantum and the recoil electron. If the energy of the scattered X-rays were propagated in spreading waves of the classical type, there would be no correlation whatever between the directions in which the recoil electrons proceed and the directions of the points at which the photo-electrons are ejected by the scattered photons. The results of Compton and Simon's experiment showed that the scattered photons proceed in definite directions and that the above equation connecting ϕ and θ is true.[1] And this result, like that of Bothe and Geiger, is fatal to the Bohr-Kramers-Slater hypothesis.

The discovery of the Compton effect opened a new prospect of solving a problem which had for some years baffled theoretical physicists. The thermal equilibrium between radiation and electrons in a reflecting enclosure had been investigated by H. A. Lorentz[2] and A. D. Fokker[3] on the basis of classical electrodynamics : and they had shown that Planck's law of spectral distribution of radiation could not persist in such a reflecting enclosure, if an electron were present : and, moreover, that if the Planck distribution were artificially maintained, the electron could not maintain the amount $\frac{3}{2}k\text{T}$ of mean translational kinetic energy required by the statistical theory of heat : so the classical theory failed to account for the interaction of pure-temperature radiation with free electrons. W. Pauli now,[4] basing his investigation on the work of Compton and Debye, attacked the problem of finding a quantum-theoretic mechanism for the interaction of radiation with free electrons, which should satisfy the thermodynamic requirement that electrons with the Maxwellian distribution of velocities can be in equilibrium with radiation whose spectral distribution is determined by Planck's radiation-formula. He found that the probability of a Compton interaction between a photon $h\nu$ and an electron could be represented as the sum of two expressions, one of which was proportional to the radiation-density of the primary frequency ν, while the other was proportional to the product of this radiation-density and the radiation-density of the frequency ν' which arises through the Compton process. The latter term was puzzling from the philosophic point of view, since it seemed to imply that the probability of something happening depended on something that had not yet happened. However, it was shown by W. Bothe[5] that Pauli's second term was a mistake, arising from

[1] cf. R. S. Shankland, *Phys. Rev.* lii (1937), p. 414
[2] At the Solvay conference in Brussels in 1911
[3] *Diss.*, Leiden, 1913 ; *Arch. Néerl*, iv (1918), p. 379
[4] *ZS. f. P.* xviii (1923), p. 272
[5] *ZS. f. P.* xxiii (1924), p. 214

Pauli's assumption that the photons scattered all have the energy $h\nu$: whereas, as we have seen earlier,[1] in pure-temperature radiation there are also definite proportions of photons having the energies $2h\nu$, $3h\nu$, etc.: if it is assumed that each of these is scattered as a whole, in exactly the same way as the photons of energy $h\nu$, then Pauli's second term falls out, and the theory becomes much simpler.[2]

In the latter half of 1923 Louis de Broglie introduced a new conception which proved to be of great importance in quantum theory. The analogy of Fermat's Principle in Optics with the Principle of Least Action in Dynamics suggested to him the desirability of studying more profoundly the parallelism between corpuscular dynamics and the propagation of waves, and attaching to it a physical meaning. He developed this idea first in a series of notes in the *Comptes Rendus*,[3] then in a doctorate thesis sustained in 1924,[4] and in other papers.[5]

In de Broglie's theory, with the motion of any electron or material particle there is associated a system of plane waves, such that the velocity of the electron is equal to the group-velocity[6] of the waves. Let m be the mass and v the velocity of the particle. It is assumed that the frequency ν of the waves is given by Planck's relation

$$E = h\nu,$$

when

$$E = \frac{mc^2}{\sqrt{\left(1 - \frac{v^2}{c^2}\right)}}$$

is the kinetic energy of the particle, so

$$h\nu = \frac{mc^2}{\sqrt{\left(1 - \frac{v^2}{c^2}\right)}}. \tag{1}$$

Since v is equal to the group-velocity of the waves, we have

$$v = \frac{d\nu}{d(1/\lambda)} \tag{2}$$

where λ is the wave-length of the waves, so that $\lambda = V/\nu$ where V is the phase-velocity of the waves.

[1] cf. p. 103

[2] A simple treatment of the equilibrium between a Maxwellian distribution of atoms, and radiation obeying Planck's law, is given by P. Jordan, *ZS. f. P.* xxx (1924), p. 297.

[3] *Comptes Rendus*, clxxvii (10 Sept. 1923), p. 507; ibid. (24 Sept. 1923), p. 548; ibid. (8 Oct. 1923), p. 630; clxxix (7 July 1924), p. 39; ibid. (13 Oct. 1924), p. 676; ibid. (17 Nov. 1924), p. 1039

[4] *Thèse*, Paris, Edit. Musson et Cie., 1924

[5] *Phil. Mag.* xlvii (Feb. 1924), p. 446; *Annales de phys.*(10) iii (1925), p. 22.

[6] cf. Vol. I, p. 253, *note* 4. For the purpose of calculating group-velocity, ν is regarded as a function of λ with c and m/h as fixed constants.

From (1) and (2) we have

$$d\left(\frac{1}{\lambda}\right)=\frac{mc^2}{hv}\,d\left(1-\frac{v^2}{c^2}\right)^{-\frac{1}{2}}=\frac{m}{h}\left(1-\frac{v^2}{c^2}\right)^{-\frac{3}{2}}dv=\frac{1}{h}dp$$

where

$$p=\frac{mv}{\sqrt{\left(1-\frac{v^2}{c^2}\right)}}$$

is the momentum of the particle. Integrating, we have $1/\lambda=p/h$, or $\lambda=h/p$. *This equation gives the wave-length of the de Broglie wave associated with a particle of momentum p.*

The phase-velocity V of the de Broglie wave is

$$V=\lambda\nu=\frac{h}{p}\cdot\frac{E}{h}=\frac{E}{p}$$

or

$$V=\frac{c^2}{v}$$

an equation which gives the phase-velocity of the de Broglie wave.[1]

Now a wave-motion of frequency $E/2\pi\hbar$ and of wave-length $2\pi\hbar/p$, where p has the components (p_x, p_y, p_z), is represented by a wave-function

$$\psi=\exp.\left\{\frac{i}{\hbar}(Et-p_xx-p_yy-p_zz)\right\}.$$

[1] The following derivation of de Broglie's result was given by Einstein, *Berlin Sitz.* (1925), p. 3.

A material particle of mass m is first correlated to a frequency ν_0 conformably to the equation

$$mc^2=h\nu_0$$

The particle now rests with respect to a Galilean system K′, in which we imagine an oscillation of frequency ν_0 everywhere synchronous. Relative to a system K′, with respect to which K′ with the mass m is moved with velocity v along the positive X-axis, there exists a wave-like process of the kind

$$\sin\left(2\pi\nu_0\frac{t-\frac{v}{c^2}x}{\sqrt{\left(1-\frac{v^2}{c^2}\right)}}\right).$$

The frequency ν and phase-velocity V of this process are thus given by

$$\nu=\frac{\nu_0}{\sqrt{\left(1-\frac{v^2}{c^2}\right)}},\qquad V=\frac{c^2}{v}.$$

The relation

$$(\text{phase velocity})\times(\text{group velocity})=c^2$$

was shown by R. W. Ditchburn, *Revue optique* xxvii (1948), p. 4, and J. L. Synge, *Rev. Opt.* xxxi (1952), p. 121, to be a necessary consequence of relativity theory.

This is the analytical expression of the de Broglie wave associated with a particle whose kinetic energy is E *and whose momentum is* (p_x, p_y, p_z).

But a wave of ordinary light of frequency ν in the direction (l, m, n) is represented by a wave-function

$$\psi = \exp.\left[2\pi i\nu\left\{t - \left(\frac{1}{c}\right)(lx + my + nz)\right\}\right]$$

or

$$\psi = \exp.\left\{\frac{i}{\hbar}(Et - p_x x - p_y y - p_z z)\right\}$$

when E denotes the energy $h\nu$ and (p_x, p_y, p_z) the momentum $h\nu/c$ of the corresponding photon. Comparing this with the above expression for the de Broglie wave, we see that *an ordinary wave of light is simply the de Broglie wave belonging to the associated photon.* It follows that if a de Broglie wave is regarded as a quantum effect, then *the interference and diffraction of light must be regarded as essentially quantum effects.* It is, in fact, a mistake to speak of the wave-theory of light as the 'classical' theory: that it is usually so called is due to the historical accident that the wave-theory of light happened to be discovered before the photon theory, which is the corpuscular theory. When interference is treated by the corpuscular theory (Duane's method, p. 142), then its quantum character is shown by the fact that quantum jumps of momentum make their appearance. Moreover, the interference and diffraction of light are evidently of the same nature as the interference and diffraction of electron beams and of molecular rays, and the latter phenomena are undoubtedly quantum effects.

The principle of Fermat applied to the wave may be shown to be identical with the principle of Least Action applied to the particle; in fact, $\delta\int ds/\lambda = 0$ is equivalent to $\delta\int pds = 0$ if p is a constant multiple of $1/\lambda$.

De Broglie now showed that his theory provided a very simple interpretation of Bohr's quantum condition for stationary states of the hydrogen atom. That condition was, that the angular momentum of the atom should be a whole-number multiple of $\hbar$ or $h/2\pi$, say $nh/2\pi$. But if r denotes the radius of the orbit and p the linear momentum of the electron, the angular momentum is rp: so the condition becomes

$$2\pi rp = nh$$

or

$$2\pi r = n\lambda$$

where λ is the wave-length of the de Broglie wave associated with the electron. This equation, however, means simply that *the circumference of the orbit of the electron must be a whole-number multiple of the wave-length of the de Broglie wave.*

More generally, the Wilson-Sommerfeld quantum condition is that the Action must be a multiple of h. But the Action is $\int p\,ds$ where p denotes the momentum and ds the element of the path: so $h\int ds/\lambda$ must be a multiple of h, where λ is the de Broglie wave-length: or, $\int ds/\lambda$ must be a whole number. We can express this by saying that *the de Broglie wave must return to the same phase when the electron completes one revolution of its orbit.*

The connection between the Wilson-Sommerfeld condition and the wave-theory can be seen also in the case of the diffraction of light or electron beams or corpuscular rays by an infinite reflecting plane grating. The solution of this problem by the corpuscular theory (Duane's method) depends on the Wilson-Sommerfeld condition, which yields (cf. p. 143)

$$(pd/h)\ (\sin i - \sin r) = \text{a whole number},$$

where p now denotes the total momentum of the quantum of light. The solution by the wave-theory of Young and Fresnel, on the other hand, depends on the principle that $(1/2\pi)$ times the difference in phase between the rays reflected from adjacent spacing-intervals must be a whole number, and this at once gives

$$(d/\lambda)\ (\sin i - \sin r) = \text{a whole number}.$$

These two equations are identical if $(p/h) = (1/\lambda)$, which is assured by de Broglie's relation.

In July 1925 a research student named Walter Elsasser, working under James Franck at Göttingen, made an important contribution[1] to the theory. Franck had been told by his colleague Max Born of an investigation made in America by Clinton J. Davisson and C. H. Kunsman,[2] who had studied the angular distribution of electrons reflected at a platinum plate, and had found at certain angles strong maxima of the intensity of the electronic beam. Born, who knew of de Broglie's theory, mentioned it in this connection, and Franck proposed to Elsasser that he should examine the question whether Davisson's maxima could be explained in some way by de Broglie's waves. Elsasser showed that they could in fact be interpreted as an effect due to interference of the waves. There were strong maxima, which with increasing electron-velocity approximated to the positions of the maxima which would be observed if light of the wave-length given by de Broglie's law, namely, $\lambda = h/mv$ were diffracted at an optical plane grating, the constants of the grating being those of platinum crystals. This was the first confirmation of de Broglie's theory which was based on comparison

[1] *Naturwiss*, xiii (1925), p. 711

[2] *Phys. Rev.* xxii (1923), p. 242

with experiment. Elsasser remarked further that de Broglie's theory provided solutions for some other puzzles of current physics. He discussed an effect discovered by C. Ramsauer[1] in Germany, and independently but later by J. S. Townsend and V. A. Bailey[2] at Oxford, and by R. N. Chaudhuri[3] working in O. W. Richardson's laboratory in London, namely, that the mean free path of an electron in the inert gases becomes exceedingly long when the velocity of the electron is reduced: thus when an electron moving with a velocity of about 10^8 cm./sec. collides with a molecule of a non-inert gas, in general it loses more than one per cent of its energy, but when it collides with a molecule of argon, it loses only about one ten-thousandth part of its energy: and the mean free path of such an electron in argon is about ten times as long as that calculated from the kinetic theory. Elsasser explained this by showing that when slow electrons are scattered by atoms of the inert gases, the effect follows the same laws as the classical scattering of radiation, of the associated de Broglie wave-length, by small spheres whose radius is the same as that of the atom.

Elsasser further suggested that electrons reflected at a single large crystal of some substance might show the diffraction-effect of the de Broglie waves decisively.[4] The phenomenon thus predicted was found experimentally in 1927 by Clinton J. Davisson and Lester H. Germer[5] of the Bell Telephone Co., who found that a beam of slow electrons, reflected from the face of a target cut from a single crystal of nickel, gave well-defined beams of scattered electrons in various directions in front of the target: in fact, diffraction phenomena were observed precisely similar to those obtained with X-rays, of a wave-length connected with the momentum of the electrons by de Broglie's formula.

In Nov.–Dec. 1927, George Paget Thomson[6] examined the scattering of cathode rays by a very thin metallic film, which could be regarded as a microcrystallic aggregate, and confirmed the fact that a beam of electrons behaves like a wave: from the size of the rings in the diffraction-pattern it was possible to deduce the wave-length of the waves causing them, and in all cases he obtained the value $\lambda = h/p$. If a stream of electrons is directed at a screen in which

[1] *Ann. d. Phys.* lxiv (1921), p. 451; lxvi (1921), p. 546; lxxii (1923), p. 345

[2] *Phil. Mag.* xliii (1922), p. 593

[3] *Phil. Mag.* xlvi (1923), p. 461

[4] Another experimental investigation which could be explained by de Broglie's theory was that of E. G. Dymond, *Nature*, cxviii (1926), p. 336, on the scattering of electrons in helium; the moving electrons could be associated with plane de Broglie waves, whose interference governed the scattering. I. Langmuir, *Phys. Rev.* xxvii (1906), p. 806 had shown that inelastic collisions in several gases lead to very small angles of scattering.

[5] *Phys. Rev.* xxx (1927), p. 707; *Nature*, cxix (16 April 1927), p. 558; *Proc. N.A.S.* xiv (1928), p. 317; cf. K. Schaposchnikow, *ZS. f. P.* lii (1928), p. 451; E. Rupp, *Ann. d. Phys.* i (1929), p. 801; C. J. Davisson. *J. Frank. Inst.*, ccv (1928), p. 597

[6] G. P. Thomson and A. Reid, *Nature*, cxix (1927), p. 890. G. P. Thomson, *Proc. R.S.*(A), cxvii (1928), p. 600; cxxviii (1930), p. 641. A. Reid, *Proc. R.S.*(A), cxix (1928), p. 663. R. Ironside, ibid., p. 668. S. Kikuchi, *Proc. Tokyo Ac.* iv (1928), pp. 271, 354, 471

there are two small holes close together, an interference-pattern is produced just as with light.

The experimental verification of de Broglie's formula was further extended by E. Rupp,[1] who succeeded in obtaining electronic diffraction by a ruled grating.

The electron-energies for which the formula has been verified range from 50 to 10^6 electron-volts.[2]

These effects have been observed not only with streams of electrons (cathode rays and the β-rays from radio-active sources), but also with streams of material particles: I. Estermann and O. Stern[3] in 1930 found that molecular rays of hydrogen and helium, impinging on a crystal face of lithium fluoride, were diffracted, giving a distribution of intensity corresponding to the spectra formed by a crossed grating. The wave-length, calculated from the known constants of the crystal, was in agreement with de Broglie's formula.

Theoretical papers on the diffraction of electrons at crystals were published not long after the Davisson-Germer experiment by Hans Bethe,[4] a pupil of Sommerfeld, and C. G. Darwin.[5]

The next notable advance in physical theory was made by Satyandra Nath Bose[6] of Dacca University, in a short paper giving a new derivation of Planck's formula of radiation: Einstein, who seems to have translated it into German from an English manuscript sent to him by Bose, recognised at once its importance and its connection with de Broglie's theory.

Bose regarded the radiation as composed of photons, which for statistical purposes could be treated like the particles of a gas, but with the important difference that photons are indistinguishable from each other, so that instead of considering the allocation of individual distinguishable photons among a set of states, he fixes attention on the number of states that contain a given number of photons. He assumes that the total energy E of the photons is given, and that they are contained in an enclosure of unit volume.

A photon $h\nu$ may be specified by its co-ordinates (x, y, z) and the three components of its momentum (p_x, p_y, p_z). Since the total momentum is $h\nu/c$, we have

$$p_x^2 + p_y^2 + p_z^2 = r^2 \quad \text{where} \quad r = h\nu/c.$$

Let volume in the six-dimensional space of (x, y, z, p_x, p_y, p_z) be

[1] *ZS. f. P.* lii (1928), p. 8; *Phys. ZS.* xxix (1928), p. 837

[2] J. V. Hughes, *Phil. Mag.* xix (1935), p. 129

[3] *ZS. f. P.* lxi (1930), p. 95: cf. T. H. Johnson, *Phys. Rev.* xxxi (1928), p. 103: F. Knauer and O. Stern, *ZS. f. P.* ciii (1929), p. 779

[4] *Ann. d. Phys.* lxxxvii (1928), p. 55

[5] *Proc. R.S.*(A), cxx (1928), p. 631. This paper is concerned with the polarisation of electron-waves, on which, see also C. Davisson and L. H. Germer, *Phys. Rev.* xxxiii (1929), p. 760, and E. Rupp, *ZS. f. P.* liii (1929), p. 548.

[6] *ZS. f. P.* xxvi (1924), p. 178

called *phase-space.* Then to the frequency-range from ν to $\nu+d\nu$ there corresponds the phase-space

$$\int dx\,dy\,dz\,dp_x\,dp_y\,dp_z = 4\pi r^2 dr = 4\pi\,\frac{h^3\nu^2}{c^3}\,d\nu.$$

Bose now assumes that this phase-space is partitioned into cells of volume h^3, so there are $4\pi\nu^2 d\nu/c^3$ cells. In order to take account of polarisation we must double the number, so we obtain $8\pi\nu^2 d\nu/c^3$ cells.

Now let N_s be the number of photons in the frequency-range $d\nu_s$, and consider the number of ways in which these can be allocated among the cells belonging to $d\nu_s$. Let p_0^s be the number of vacant cells, p_1^s the number of cells that contain one photon, p_2^s the number of cells that contain two photons etc. Then the number of possible ways of choosing a set of p_0^s cells, a set of p_1^s cells, etc. out of a total of $8\pi\nu^2 d\nu/c^3$ cells is

$$\frac{A_s!}{p_0^s!\,p_1^s!\,p_2^s!\,\ldots}, \quad \text{where} \quad A_s = \frac{8\pi\nu_s^2}{c^3}\,d\nu_s,$$

and we have

$$N_s = \sum_r r p_r^s.$$

As the fundamental assumption of his statistics, Bose assumes that if a particular quantum state is considered, then all values for the number of particles in that state are equally likely, so the probability of any distribution specified by the p_r^s is measured by the number of different ways in which it can be realised. Hence the probability of the state specified by the p_r^s (now taking into account the whole range of frequencies) is

$$W = \prod_s \frac{A_s!}{p_0^s!\,p_1^s!\,p_2^s!\,\ldots}.$$

Since the p_r^s are large, we can use Stirling's approximation

$$\log n! = n \log n - n$$

so $$\log W = \sum_s A_s \log A_s - \sum_s \sum_r p_r^s \log p_r^s, \quad \text{since} \quad A_s = \sum_r p_r^s.$$

This expression is to be a maximum subject to the condition

$$E = \sum_s N_s h\nu_s, \quad \text{when} \quad N_s = \sum_r r p_r^s.$$

The usual conditions become in this case

$$\sum_s \sum_r \delta p_r^s (1 + \log p_r^s) = 0, \qquad \sum_s \delta N_s . h\nu_s = 0,$$

where

$$\delta N_s = \sum_r r\delta p_r^s, \qquad \sum_r \delta p_r^s = 0,$$

which give

$$\sum_s \sum_r \delta p_r^s \left\{(1 + \log p_r^s + \lambda^s) + \frac{rh\nu_s}{\beta}\right\} = 0$$

where β and the λ^s are constants : so

$$p_r^s = B_s e^{-\frac{rh\nu_s}{\beta}}$$

where the B_s are constants.

Therefore

$$A_s = \sum_r p_r^s = \sum_r B_s e^{-\frac{rh\nu_s}{\beta}} = B_s\left(1 - e^{-\frac{h\nu_s}{\beta}}\right)^{-1}$$

or

$$B_s = A_s\left(1 - e^{-\frac{h\nu_s}{\beta}}\right)$$

while $N_s = \sum_r rp_r^s = A_s \sum_r re^{-\frac{rh\nu_s}{\beta}}\left(1 - e^{-\frac{h\nu_s}{\beta}}\right) = A_s e^{-\frac{h\nu_s}{\beta}}\left(1 - e^{-\frac{h\nu_s}{\beta}}\right)^{-1}.$

Thus

$$E = \sum_s N_s h\nu_s = \sum_s \frac{8\pi h\nu_s^3}{c^3} e^{-\frac{h\nu_s}{\beta}}\left(1 - e^{-\frac{h\nu_s}{\beta}}\right)^{-1} d\nu_s.$$

The entropy is

$$S = k \log W$$

which gives

$$S = k\left\{\frac{E}{\beta} - \sum_s A_s \log\left(1 - e^{\frac{h\nu_s}{\beta}}\right)\right\}.$$

Since $\partial S/\partial E = 1/T$ where T denotes the absolute temperature, we have

$$\beta = kT$$

so

$$E = \sum_s \frac{8\pi h\nu_s^3}{c^3} \frac{1}{e^{\frac{h\nu_s}{kT}} - 1} d\nu_s$$

which is equivalent to Planck's formula.

Bose's paper therefore showed that in order to obtain Planck's law of radiation, we must assume that photons obey a particular kind of statistics. This point may be illustrated by an analogy, as follows. Consider an empty railway train standing at a platform, with passengers on the platform who get into the train ; and suppose that when they have all taken their places, p_0 compartments are vacant, p_1 compartments have one passenger apiece, p_2 compartments have two passengers apiece, and so on : so that if A is the number of compartments in the train, we have

$$A = p_0 + p_1 + p_2 + - \ldots,$$

and if N is the number of passengers, we have

$$N = 0 \,.\, p_0 + 1 \,.\, p_1 + 2 \,.\, p_2 + \ldots.$$

For simplicity, we shall assume that comparatively few people are travelling, so that the number of compartments is greater than the number of passengers. Let us inquire, what is the probability of this particular distribution specified by the numbers $(p_0, p_1, p_2, \ldots)$. Evidently the probability depends on the assumption that we make regarding the motives which influence passengers in their choice of a compartment. Three such assumptions are as follows :

(i) We might assume that each passenger chooses his compartment at random, without regard to whether there are already any other passengers in it, or not.

(ii) We might assume that each passenger likes to have a compartment to himself, so he refuses to enter any compartment which already has an occupant.

(iii) We might assume that among the passengers there are small family parties whose members wish to be together in the same compartment, so that if we know that at least one place in a compartment is occupied, there is a certain probability (arising from this fact) that other seats in it will also be occupied. We may regard each family party or unattached traveller as a unit, and assume that each unit chooses its compartment at random, without regard to whether there are already other passengers in it, or not.

It is evident that these three different assumptions will give quite different values for the probability of any particular distribution $(p_0, p_1, p_2, \ldots)$: this we express by saying that they give rise to *different statistics*. The difference between *classical statistics* and the different kinds of *quantum statistics* may be illustrated by this analogy.

Bose's discovery was immediately extended by Einstein,[1] to the study of a monatomic ideal gas. The difference between Bose's

[1] *Berlin Sitz.* (1924), p. 261 ; (1925), pp. 3, 18

photons and Einstein's gas-particles is that for photons the energy is c times the momentum, whereas for particles a different equation holds: and, moreover, in Bose's problem the total energy is fixed but the number of photons is not fixed, whereas in Einstein's problem the total number of particles is definite. These differences, however, do not affect the general plan of the investigation. The analysis leads to the following conclusions: the average number of particles of mass m in unit volume with energies in the range ϵ to $\epsilon + d\epsilon$ is

$$\frac{2\pi}{h^3}\frac{(2m)^{\frac{3}{2}}\epsilon^{\frac{1}{2}}d\epsilon}{e^{\epsilon/kT+\mu}-1} \tag{1}$$

where μ is a constant: whence *the total number of particles in unit volume is*

$$n=\frac{2\pi}{h^3}(2m)^{\frac{3}{2}}\int_0^\infty\frac{\epsilon^{\frac{1}{2}}d\epsilon}{e^{\epsilon/kT+\mu}-1}=\frac{2\pi}{h^3}(2mkT)^{\frac{3}{2}}\int_0^\infty\frac{x^{\frac{1}{2}}dx}{e^{x+\mu}-1} \tag{2}$$

and *the total energy in unit volume is*

$$E=\frac{2\pi}{h^3}(2m)^{\frac{3}{2}}\int_0^\infty\frac{\epsilon^{\frac{3}{2}}d\epsilon}{e^{\epsilon/kT+\mu}-1}=\frac{2\pi}{h^3}(2m)^{\frac{3}{2}}(kT)^{\frac{5}{2}}\int_0^\infty\frac{x^{\frac{3}{2}}dx}{e^{x+\mu}-1}. \tag{3}$$

These are the fundamental formulae of what is generally called *Bose-Einstein* statistics.

Now consider the relation between these formulae and the formulae of the classical (Maxwellian) kinetic theory of gases. We should expect the classical formulae to be obtained as the limiting case when $h \to 0$, in which case it is evident from (1) that μ must tend to infinity in such a way that h^3e^μ has a finite value, say λ. From (2) we then have, in this limiting case,

$$n=\frac{2\pi}{\lambda}(2mkT)^{\frac{3}{2}}\int_0^\infty x^{\frac{1}{2}}e^{-x}\,dx=\frac{\pi^{\frac{3}{2}}}{\lambda}(2mkT)^{\frac{3}{2}} \tag{4}$$

while (1) states in the limiting case that the total number of particles in unit volume with energies between ϵ and $\epsilon + d\epsilon$ is

$$\frac{2\pi}{\lambda}(2m)^{\frac{3}{2}}e^{-\epsilon/kT}\epsilon^{\frac{1}{2}}\,d\epsilon$$

or, by (4), out of a total of n particles, the number with energies between ϵ and $\epsilon + d\epsilon$ is

$$\frac{2n}{\pi^{\frac{1}{2}}(kT)^{\frac{3}{2}}}e^{-\epsilon/kT}\epsilon^{\frac{1}{2}}\,d\epsilon$$

which is precisely the Maxwellian formula.

Moreover (3) becomes in this limiting case

$$E = \frac{2\pi}{\lambda}.(2m)^{\frac{3}{2}}.(kT)^{\frac{5}{2}}.\,T(\tfrac{5}{2}) = \tfrac{3}{2}nkT$$

which is the Maxwellian formula for the total energy.

We see that in Bose statistics, the slower molecules are more numerous, as compared with the faster ones, than is the case in Maxwell's theory.

The similarity in statistical behaviour between Bose's photons and the particles of a gas, revealed by this investigation, was examined further by Einstein. He pointed out [1] that de Broglie's discovery made it possible to correlate to any system of material particles a (scalar) wave-field: and he showed the close connection between the fluctuations of energy in systems of waves and in systems of particles. We have seen [2] that the mean-square of the fluctuations of energy per unit volume in the frequency-range from ν to $\nu + d\nu$ in the radiation in an enclosure at temperature T is

$$\overline{\epsilon^2} = h\nu E + \frac{c^3 E^2}{8\pi\nu^2 d\nu},$$

the first term representing the fluctuation in the number of molecules in unit volume of an ideal gas on the classical theory, when each molecule has energy $h\nu$. Einstein now showed that when the gas-particles are assumed to satisfy the Bose-Einstein statistics, *both* terms appear. In other words, a Bose-Einstein gas differs from a Maxwellian gas in precisely the same way as radiation obeying Planck's formula differs from radiation obeying the law of Wien.

The next advance in the theory of quantum statistics was made by Enrico Fermi [3] (*b.* 1901). He remarked that in the Maxwellian kinetic theory of gases, the average kinetic energy per molecule is $\frac{3}{2}kT$, and hence the molecular heat at constant volume (i.e. the heat that must be communicated to one gramme-molecule in order that its temperature may be raised one degree, the volume remaining unchanged), calculated from this theory, is $c_v = \frac{3}{2}R$, where R is the gas-constant. If, however, Nernst's thermodynamical law, which requires $(dE/dT) \to 0$ as $T \to 0$, is applicable to an ideal gas, then c_v must vanish in the limit when $T \to 0$, and therefore (as Einstein had remarked in 1906) the Maxwellian theory cannot be true at very low temperatures. The reason for this must, he argued, be sought in the quantification of molecular motions, and this quantification be made to depend on Pauli's exclusion principle, that one system can

[1] At page 9 of the first paper in *Berlin Sitz.* (1925). [2] cf. p. 101 *supra*

[3] *Lincei Rend.* iii (7 Feb. 1926), p. 145; *ZS. f. P.* xxxvi (1926), p. 902. A contribution of great importance was made by P.A.M. Dirac somewhat later in the year, *Proc. R.S.*(A), cxii (1926), p. 661, on account of which the type of statistics introduced by Fermi is often called the Fermi-Dirac statistics; but as Dirac's paper involves the ideas of wave-mechanics, its description is postponed for the present.

never contain two elements of equal value with exactly the same set of quantum numbers. Thus he asserted that at most one molecule with specified quantum numbers could be present in an ideal gas: where by quantum numbers he understood not only those which relate to the internal motions of the molecules, but also those which specify the molecule's motion of translation.

The quantification may be performed as follows: suppose the gas is contained in a cubical vessel whose edge is of length l, so that the possible quantum values of the components of momentum of the particle are $p_x=(s_1h/l)$, $p_y=(s_2h/l)$, $p_z=(s_3h/l)$, where s_1, s_2, s_3 are whole numbers or zero; then Pauli's principle asserts that in the whole gas there can be at most one particle with specified quantum numbers s_1, s_2, s_3.

The energy of the particle is

$$\epsilon=\frac{h^2}{2l^2m}\ (s_1^2+s_2^2+s_3^2)$$

and as in the case of Bose statistics, the number of quantum states of the particle which correspond to kinetic energy between ϵ_s and $\epsilon_s+d\epsilon_s$ is

$$R_s=\frac{2\pi V}{h^3}\ (2m)^{\frac{3}{2}}\ \epsilon_s^{\frac{1}{2}}\ d\epsilon_s.$$

Now suppose that N_s particles have energies between ϵ_s and $\epsilon_s+d\epsilon_s$, so of the R_s states, N_s are occupied (by one particle each) and the rest unoccupied. The number of ways in which this can occur is

$$\frac{R_s\,!}{N_s\,!\ (R_s-N_s)\,!}$$

so the total number of ways in which the allocation specified by the N_s can be realised is

$$W=\prod_s\frac{R_s\,!}{N_s\,!\ (R_s-N_s)\,!}$$

and we assume that this is proportional to the probability of this allocation. Thus

$$\log W=\sum_s\left\{\log R_s\,!-\log N_s\,!-\log\,(R_s-N_s)\,!\right\}.$$

Using Stirling's approximation to the log. of a factorial, this gives

$$\log W=\sum_s\left\{R_s\log R_s-N_s\log N_s-(R_s-N_s)\log\,(R_s-N_s)\right\}.$$

This is to be made a maximum, subject to

$\sum_s N_s=n$, where n is the total number of particles

and

$$\sum_s \epsilon_s N_s = E, \text{ where E is the total energy.}$$

The usual procedure yields

$$N_s = \frac{R_s}{e^{\beta\epsilon_s + \mu} + 1}$$

where β and μ are independent of s. Now the entropy is

$$S = k \log W$$

and

$$\frac{dS}{dE} = \frac{1}{T},$$

whence we find

$$\frac{1}{T} = k\beta.$$

Hence *the number of particles per unit volume with kinetic energy between ϵ and $\epsilon + d\epsilon$* is

$$n(\epsilon)d\epsilon = \frac{2\pi(2m)^{\frac{3}{2}}}{h^3} \cdot \frac{\epsilon^{\frac{1}{2}} d\epsilon}{e^{\epsilon/kT + \mu} + 1}$$

This is the fundamental equation of the Fermi statistics.[1]

The *total density of particles* is therefore

$$n = \int_0^\infty n(\epsilon)d\epsilon = \frac{(2mkT)^{\frac{3}{2}} 2\pi}{h^3} \int_0^\infty \frac{x^{\frac{1}{2}} dx}{e^{x+\mu} + 1}$$

an equation which determines μ as a function of the density and temperature.

The parameter μ has a thermodynamical significance. If for a finite mass of gas U is the total energy, S the entropy, T the temperature, p the pressure and v the volume then

$$G = U - TS + pv$$

is called the *Gibbs's thermodynamical potential*; and the *Gibbs's thermodynamical potential per molecule* is defined as $\psi = \partial G/\partial n$, where n denotes the number of molecules, and the temperature and pressure are kept constant in the differentiation. The *Fermi constant* μ is now given [2] by

$$\mu = -\frac{\psi}{kT}.$$

[1] This may be modified, e.g. when the particles concerned have different possibilities of spin.

[2] W. Pauli, *ZS. f. P.* xli (1927), p. 81, *at* p. 91

The *total kinetic energy of the particles in unit volume* is

$$E=\int_0^\infty \epsilon n(\epsilon)d\epsilon=\frac{2\pi}{h^3}(2m)^{\frac{3}{2}}(kT)^{\frac{5}{2}}\int_0^\infty \frac{x^{\frac{3}{2}}dx}{e^{x+\mu}+1}.$$

The close similarity of these equations to the corresponding equations in the Bose statistics is evident, the only difference being in the occurrence of $+1$ instead of -1 in the denominator. In exactly the same way as in the case of Bose statistics, we see that in the limit when $h\to 0$, $\mu\to+\infty$, Fermi statistics pass into Maxwellian statistics: speaking physically, a Fermi gas approximates to a classical gas at sufficiently high temperatures and low pressures. The deviation of Fermi statistics from classical behaviour is in the opposite direction to the deviation in the case of Bose statistics.[1]

The pressure of a gas in Fermi statistics is related to the energy-density by the same equation as in classical theory,

$$p=\tfrac{2}{3}\frac{E}{V}.$$

Classical statistics is not the only limiting case of Fermi statistics: there is another limiting case at the opposite extreme, namely when μ is very large and *negative*. If we write $-\alpha$ for μ, the integrals occurring in the theory are of the form

$$\int_0^\infty \frac{x^p dx}{e^{x-\alpha}+1}$$

and it can be shown that

$$\lim_{\alpha\to\infty}\left\{\alpha^{-p-1}\int_0^\infty \frac{x^p dx}{e^{x-\alpha}+1}\right\}=\frac{1}{p+1}$$

so the total density of particles becomes

$$n=\frac{4\pi}{3h^3}(2mkT\alpha)^{\frac{3}{2}}$$

and the total kinetic energy of the particles in unit volume is

$$E=\frac{4\pi}{5h^3}(2m)^{\frac{3}{2}}(kT\alpha)^{\frac{5}{2}}.$$

Eliminating α, we have

$$E=\frac{3}{40}\left(\frac{6}{\pi}\right)^{\frac{2}{3}}\frac{h^2}{m}n^{\frac{5}{3}}\text{ approximately,}$$

the neglected terms involving T^2 and higher powers[2] of T.

[1] The thermodynamical functions for a gas with Fermi statistics were studied by E. C. Stoner, *Phil. Mag.* xxviii (1939), p. 257.

[2] For an electron-gas, e.g. in metals, there is an extra factor 2 arising from the two possible values of the spin of the electron.

It is evident from this equation that at the zero of absolute temperature the particles of a gas with Fermi statistics are not at rest, but have a definite zero-point energy.

In this limiting case we see from the above value of n that

$$nh^3(mkT)^{-\frac{3}{2}} \gg 1,$$

which is realised if the temperature is low, the density n is large and the mass m of the individual particles is small. The deviation from classical behaviour is due fundamentally to the circumstance that the quantum states of very small momentum are all occupied by particles. In fact, since (in accordance with Nernst's heat theorem) the entropy S vanishes at the absolute zero of temperature, and since $S = k \log W$, it follows that $W = 1$ when $T = 0$, i.e. there is only one way of distributing the particles: they occupy every quantum state in the neighbourhood of the state of zero momentum.

This fact has an interesting application in astrophysics. In 1844 Bessel concluded from irregularities in the motion of Sirius that it was one component of a double star, the other member of the pair being invisible with the telescopes then available. Some years later the companion was observed telescopically, and found to be a star between the 8th and 9th magnitude. Its mass is not much less than that of the sun, and it is a 'white' star, so that its surface-brightness must be greater than the sun's; but its total radiation is only about $\frac{1}{360}$th of that of the sun: hence the area of its surfaces must be very much smaller than the sun's, and in fact not much larger than the earth's. Its density must therefore be very great—about 60,000 times that of water. This surprising inference was confirmed in 1925, when Walter S. Adams[1] of Mount Wilson found in the spectrum of the companion of Sirius a displacement of the lines which might well be the decrease of frequency that is to be expected in lines that have been emitted in an intense gravitational field.

The explanation[2] of this abnormal density is that matter can exist in such a dense state if it has so much energy that the electrons escape from the nuclei, so they are no longer bound in ordinary atomic orbits but are, in the main, free. The density of the matter is then limited no longer by the size of atoms, but only by the sizes of the electrons and atomic nuclei: and as the volumes of these are perhaps 10^{-14} of the volumes of atoms, we may conclude that densities of as much as 10^{14} times that of terrestrial matter may be possible: this is much greater than that of any of the 'white dwarfs,' as stars like the companion of Sirius are called.

According to the classical theory of the relation between energy and temperature, such a star, having excessively great energy, would have a very high temperature, and would therefore radiate intensely. It was, however, shown by R. H. Fowler[3] that this is

[1] *Proc. N.A.S.* xi (1925), p. 382

[2] Due chiefly to Eddington

[3] *Mon. Not. R.A.S.* lxxxvii (1926), p. 114

not the case. The radiation depends on the temperature, and depends on the energy only in so far as the energy determines the temperature: the apparent difficulty was due to the use of the classical correlation between energy and temperature: the correct relation, for such dense stellar matter, is that of the above-mentioned limiting case of Fermi's statistics, when μ is very large and negative: the energy is still very great, but the temperature is not correspondingly great, and indeed ultimately approaches zero: so radiation, which depends on temperature, stops when the dense matter still has ample energy.

The absolutely final state (the 'black dwarf') is one in which there is only one possible configuration left: the star is then analogous to one gigantic molecule in its lowest quantum state, and the temperature (which ceases to have any meaning) may be said to be zero.

It was shown in 1927 by L. S. Ornstein and H. A. Kramers[1] of Utrecht that the formulae of Fermi statistics may be derived in a totally different way by considering the kinetics of reactions.

Let the possible values of the energy of a gas-molecule in an enclosure be $\epsilon_1, \epsilon_2, \epsilon_3, \ldots$; for simplicity we suppose these values all different. Consider two molecules in the states k' and l' respectively, which by their interaction are changed to the states k'' and l'' respectively. Let the *a priori* probability that this transition should take place in unit time be denoted by $a_{\prime\prime}{}'$. In order that the transition may actually take place, it is necessary that there should exist a molecule in the k' state and one in the l' state, and, moreover (by the Pauli principle), that neither a molecule in the k'' state nor one in the l'' state is present. Denoting by $\bar{n}_k$ the probability that a molecule is in the k state, the mean frequency for the process in which k', l' tend to k'', l'', is

$$\mathrm{A}_{\prime\prime}{}' = a_{\prime\prime}{}' \, \bar{n}_{k'} \, \bar{n}_{l'} \, (1-\bar{n}_{k''}) \, (1-\bar{n}_{l''})$$

while the frequency of the reverse process is

$$\mathrm{A}_{\prime}{}'' = a_{\prime}{}'' \, \bar{n}_{k''} \, \bar{n}_{l''} \, (1-\bar{n}_{k'}) \, (1-\bar{n}_{l'}).$$

In thermodynamic equilibrium we must have

$$\mathrm{A}_{\prime\prime}{}' = \mathrm{A}_{\prime}{}''$$

and if we postulate further, that the *a priori* probability for reverse processes is equal, i.e.

$$a_{\prime\prime}{}' = a_{\prime}{}'',$$

[1] *ZS. f. P.* xlii (1927), p. 481. On the kinetic relations of quantum statistics, cf. also: P. Jordan, *ZS. f. P.* xxxiii (1925), p. 649; xli (1927), p. 711. W. Bothe, *ZS. f. P.* xlvi (1928), p. 327. L. Nordheim, *Proc. R.S.*(A), cxix (1928), p. 689. S. Kikuchi and L. Nordheim, *ZS. f. P.* lx (1930), p. 652. S. Flügge, *ZS. f. P.* xciii (1935), p. 804

then if we write

$$\frac{\bar{n}_k}{1-\bar{n}_k} = m_k,$$

the above equations give

$$m_{k'}\, m_{l'} = m_{k''}\, m_{l''}.$$

This equation must hold for all the values of k', l', k'', l'', for which the equation of conservation of energy

$$\epsilon_{k'} + \epsilon_{l'} = \epsilon_{k''} + \epsilon_{l''}$$

holds : whence it follows that the only possible reasonable distribution is given by

$$m_k = e^{-a-\beta\epsilon_k}$$

where a and β are constants : whence we have

$$\bar{n}_k = \frac{e^{-a-\beta\epsilon_k}}{1+e^{-a-\beta\epsilon_k}} = \frac{1}{e^{a+\beta\epsilon_k}+1}$$

which is Fermi's law of distribution.

Another application of the Fermi statistics was to the theory of electrons in metals.

In 1927 Pauli [1] succeeded in accounting for the observed character of the paramagnetism of the alkali metals, which is feeble and nearly independent of the temperature, by assuming that the conducting electrons in a metal may be regarded as an electron-gas which is degenerate in the sense of Fermi's statistics (i.e. μ is large and negative). In the following year Sommerfeld [2] discussed the main problems of the electron-theory of metals on the same assumption : the conducting electrons may be supposed to have free paths of the order of 100 times the atomic distances.

To see that the assumption is justified, we have to show that

$$nh^3(mkT)^{-\frac{3}{2}} \gg 1.$$

Now we may suppose that the number n of free electrons is of the same order as the number of atoms, say 10^{22} per cm^3. The mass of

[1] *ZS. f. P.* xli (1927), p. 81 ; communicated 16 Dec. 1926 ; published 10 Feb. 1927

[2] Preliminary note in *Naturwiss*, xv (14 Oct. 1927), p. 825 ; xvi (1928), p. 374 ; *ZS. f. P.* xlvii (1928), p. 1. His associate C. Eckart, ibid. p. 38, discussed the Volta-effect, and his associate W. V. Houston, *ZS. f. P.* xlviii (1928), p. 449, discussed electric conduction. Sommerfeld's theory has been developed and improved by : E. Kretschmann, *ZS. f. P.* xlviii (1928), p. 739. F. Block, *ZS. f. P.* lii (1928), p. 555 ; lix (1930), p. 208. L. W. Nordheim, *Proc. R.S.*(A), cxix (1928), p. 689. R. Peierls, *ZS. f. P.* liii (1929), p. 255 ; *Ann. d. Phys.* iv (1930), p. 121 ; v (1930), p. 244. L. Brillouin, *Les Statistiques quantiques* (Paris, 1930), Tome II. L. Nordheim, *Ann. d. Phys.* ix (1931), p. 607. A. H. Wilson, *Proc. R.S.*(A), cxxxviii (1932), p. 594. On semi-conductors, cf. A. H. Wilson, *Proc. R.S.*(A), cxxxiii (1931), p. 458 ; cxxxiv (1931), p. 277.

the electron m is 9×10^{-28} *gr.*, Planck's constant h is $6{\cdot}6\times10^{-27}$ erg. sec., Boltzmann's constant k is $1{\cdot}4\times10^{-16}$ erg./deg. Thus the condition is roughly

$$10^{-57}\,(10^{-43}\,\mathrm{T})^{-\frac{3}{2}}\gg1$$

or

$$\mathrm{T}\ll10^{5}$$

which is satisfied at all ordinary temperatures. Thus *the electron-gas inside a metal has a degenerate Fermi distribution.* At the zero of absolute temperature, there is a finite energy given by the equation

$$\mathrm{E}=\frac{3}{40}\left(\frac{3}{\pi}\right)^{\frac{2}{3}}\frac{h^2}{m}\,n^{\frac{5}{3}},$$

and the derivative $d\mathrm{E}/d\mathrm{T}$ is zero in accordance with Nernst's theorem: the specific heat of the electron-gas is in fact proportional to T for low temperatures, whereas the classical statistics made it a constant. Thus it is understood why the specific heat of the electrons in metals is exceedingly small: the specific heat of a metal is usually what would be expected if it were due to the metallic atoms alone.

The replacement of classical by Fermi statistics does not appreciably change the theoretical ratio of the thermal and electrical conductivities.

We have seen that in Fermi statistics, the probability that a quantum state whose energy is ϵ is occupied

is $$(e^{\epsilon/k\mathrm{T}+\mu}+1)^{-1},$$

and that $\mu=-\psi/k\mathrm{T}$ where ψ is Gibbs's thermodynamical potential per molecule. For the electron-gas inside a metal, this quantity ψ is called the *electrochemical potential.* It has the property that the electrochemical potentials of electrons in any two regions which are in thermal equilibrium (e.g. electrons in different metals which are in contact with each other at the same temperature) must be equal.

In a second paper [1] in the *Zeitschrift für Physik*, Sommerfeld discussed thermo-electric phenomena, obtaining new formulae for the Peltier and Thomson effects, and showing that the new expression for the Thomson heat agreed with the experimental values much better than the old. Shortly before this a new thermo-electric effect had been discovered by P. W. Bridgman,[2] namely, that when an electric current passes across an interface where the crystal orientation changes, an ' internal Peltier heat ' is developed. A theoretical discussion of this and other thermoelectric phenomena was given in 1929 by P. Ehrenfest and A. J. Rutgens.[3]

[1] *ZS. f. P.* xlvii (1928), p. 43
[2] *Proc. N.A.S.* xi (1925), p. 608 ; *Phys. Rev.* xxxi (1928), p. 221
[3] *Proc. Amst. Ac.* xxxii (1929), p. 698

The twentieth century brought considerable developments in the subject of thermionics. Richardson had deduced his original formula[1] from the assumption that the free electrons in a metal have the same energy as is attributed in the kinetic theory of gases to gas-molecules at the same temperature as the metal. H. A. Wilson[2] showed that the phenomena could not be explained completely by this theory, and proposed to replace Richardson's treatment by one analogous to the theory of evaporation. Now the latent heat of vaporisation of a liquid may be shown by thermodynamical methods to be given by what is called the *equation of Clapeyron and Clausius*, which may be written in the form

$$\chi = \mathrm{RT}^2 . \frac{1}{p} \frac{dp}{d\mathrm{T}}$$

where χ is the latent heat of evaporation per gramme-molecule at the absolute temperature T, R is the gas-constant per gramme-molecule (= 1·987 cal./deg.), and p is the vapour-pressure at temperature T: and Wilson proposed to apply this equation. Under the influence of these ideas, Richardson[3] suggested as an alternative to his earlier formula that the saturation emission per unit area, in ampères per cm.^{2}, should be represented by an equation

$$\mathrm{I} = \mathrm{AT}^2 e^{-\frac{\chi}{k\mathrm{T}}}$$

where A is a constant, k is Boltzmann's constant, and χ represents the energy required to get an electron through the surface: χ, which is called the *work-function*, is analogous to the latent heat of evaporation of a monatomic gas: it is usually reckoned in electron-volts and is often written $e\phi$ where ϕ is expressed in volts, and so is in fact the potential in volts necessary to impart to an electron the kinetic energy required for evaporation.

It may be remarked that although the conducting electrons inside a metal have a degenerate Fermi distribution, the external electrons produced by thermionic emission have a distribution which is practically Maxwellian: this is explained by their much smaller concentration.

In 1923 Saul Dushman,[4] by use of the Clapeyron-Clausius formula, and on the assumption that the electrons within the metal obey Maxwellian statistics, obtained Richardson's second formula, showing, however, that it is not rigorously true if χ is a function of temperature (though for clean metals the deviations are small);

[1] cf. Vol. I, pp. 425–8

[2] *Phil. Trans.*(A), ccii (1903), p. 243

[3] *Phil. Mag.* xxviii (1914), p. 633

[4] *Phys. Rev.* xxi (1923), p. 623; cf. earlier papers by: W. Schottky, *Phys. ZS.* xv (1914), p. 872; xx (1919), pp. 49, 220. M. von Laue, *Ann. d. Phys.* lviii (1919), p. 695. R. C. Tolman, *J. Am. Chem. Soc.* xlii (1920), p. 1185; xliii (1921), p. 866. And cf. later papers by: P. W. Bridgman, *Phys. Rev.* xxvii (1926), p. 173; xxxi (1928), p. 90. L. Tonks and I. Langmuir, *Phys. Rev.* xxix (1927), p. 524. L. Tonks, *Phys. Rev.* xxxii (1928), p. 284. K. F. Herzfeld, *Phys. Rev.* xxxv (1930), p. 248

in fact χ should be understood as relating to the work-function for the zero of absolute temperature: and he showed that A is a universal constant, having the value

$$A = 2\frac{\pi mek^2}{h^3} = 60{\cdot}2 \text{ amp./cm.}^2 \text{ deg}^2.$$

Sommerfeld's discovery that the electrons in a metal have degenerate Fermi statistics naturally led[1] to a new treatment of thermionics, but the formula finally obtained had the same general character as Richardson's formula of 1914. L. W. Nordheim[2] found

$$i = A(1-r)T^2 e^{-\frac{\chi}{kT}}$$

the quantities in this equation being defined as follows: A has twice the value of Dushman's universal constant A, so

$$A = \frac{4\pi mek^2}{h^3} = 120 \text{ amp./cm.}^2 \text{ deg.}^2:$$

the factor 2 arises when we take account of the two possible values of the electron-spin. It might be thought easy to discriminate experimentally between Dushman's and Nordheim's formulae for A: but small uncertainties in the values of χ and T, which occur in the exponential, affect the value of i so much that the experimental determination of A is very uncertain.[3] r is the reflection coefficient, i.e. the ratio of the number of electrons reflected internally at the surface to the total number reaching it, only those electrons being considered that have velocity components normal to the surface sufficient to allow them to escape. r is small and need not usually be considered, since the experimental value of A is so doubtful. χ corresponds to Richardson's work-function: but χ now has the form

$$\chi = W_a - W_i$$

where W_a is e times the difference in electrostatic potential between the inside and outside of the metal, and W_i is an energy which depends on the pressure of the electron-gas, and which assists the escape of the electrons from the metal: we may take W_i to be the kinetic energy of the highest filled level at the absolute zero of temperature.

Now we saw that in a Fermi distribution, the number of quantum

[1] A. Sommerfeld, *ZS. f. P.* xlvii (1928), p. 1

[2] *ZS. f. P.* xlvi (1928), p. 833; *Phys. ZS.* xxx (1929), p. 177

[3] L. A. du Bridge, *Phys. Rev.* xxxi (1928), pp. 236, 912, found that for clean platinum the thermionic constant A has the value 14,000 amp./cm.^{2} deg^2, which is 230 times Dushman's theoretical value 60·2; cf. L. A. du Bridge, *Proc. N.A.S.* xiv (1928), p. 788, and R. H. Fowler, *Proc. R.S.*(A), cxxii (1929), p. 36.

states of a particle (in unit volume) which correspond to kinetic energy between ϵ and $\epsilon + d\epsilon$ is

$$\frac{2\pi}{h^3}(2m)^{\frac{3}{2}}\epsilon^{\frac{1}{2}}d\epsilon$$

or inserting a factor 2 on account of the two values of the electron-spin, it is

$$\frac{4\pi}{h^3}(2m)^{\frac{3}{2}}\epsilon^{\frac{1}{2}}d\epsilon.$$

At zero temperature, all the quantum states are occupied up to a certain level of kinetic energy, say W_i : so if n be the number of free electrons per unit volume, we have

$$n = \frac{4\pi}{h^3}(2m)^{\frac{3}{2}}\int_0^{W_i}\epsilon^{\frac{1}{2}}d\epsilon = \frac{8\pi}{3h^3}(2mW_i)^{\frac{3}{2}}$$

so

$$W_i = \frac{h^2}{2m}\left(\frac{3}{8\pi}\right)^{\frac{2}{3}}n^{\frac{2}{3}}.$$

It can be shown that

$$W_i = \mu + e\Phi_i$$

where μ is the electrochemical potential of the electrons just inside the metal, and Φ_i is the electrostatic potential just inside the metal.[1]

The function W_a, the true total energy necessary to liberate an electron from the metal, may be determined independently by a method which depends on the diffraction of beams of electrons by metallic crystals.[2] For although the experiments of Davisson and Germer, already referred to, were explained qualitatively by de Broglie's theory of the wave-behaviour of electrons, there was a quantitative discrepancy, which was resolved by assuming that the wave-length of the electron-waves in the metal was different from the wave-length *in vacuo*, or in other words, that the metal had a refractive index for the electron-waves: this refractive index, which could be determined from the diffraction-experiments, determined the grating-potential and thence the function W_a : W_a was found to be considerably greater than the work-function χ.

In a complete treatment of thermionic and photoelectric phenomena, it is necessary to take account of factors that cannot be fully discussed here, e.g. the fact that an electron just outside a metal is

[1] The use of the electrochemical potential in thermionics is due to W. Schottky and H. Rothe, *Handbuch d. Experimentalphysik* (Leipzig, 1928), XII/2.

[2] cf. C. Eckart, *Proc. N.A.S.* xii (1927), p. 460; H. Bethe, *Naturwiss*, xv (1927), p. 787; L. Rosenfeld and E. E. Witmer, *ZS. f. P.* xlix (1928), p. 534

influenced by a force due to its own image in the metal :[1] in the case of a flat perfectly-conducting surface this force is of amount $e^2/4z^2$, where z is the distance of the electron from the surface.

The constants and functions that occur in thermionics occur also in related branches of physics. If two metals A and B at the same temperature are in contact, electrons flow from one to the other until the electrochemical potentials of the electrons in the two metals are equal, and there is a difference of potential (the *Volta effect* or *contact potential-difference*[2]) between a point just outside A and a point just outside B. Since the work-function χ is

$$W_a - W_i \quad \text{or} \quad -e\Phi_0 + e\Phi_i - (\mu + e\Phi_i)$$

where Φ_0 is the electrostatic potential at a point just outside the metal, we see that *the Volta effect* $(\Phi_0)_A - (\Phi_0)_B$ *is equal to* $(\chi_B - \chi_A)/e$, *where* χ_A *and* χ_B *are the thermionic work-functions of the two metals*: so if i_A and i_B are the thermionic electric saturation currents per unit area at the same temperature, the contact potential-difference is[3]

$$\frac{kT}{e} \log \frac{i_B}{i_A}.$$

Thermionics is closely connected also with photoelectricity, as we have seen in Chapter III.[4] The frequency of light which will just eject electrons photoelectrically, but with zero velocity, is called the *threshold frequency*, and the corresponding wave-length is called the *photoelectric long wave-length limit.* R. A. Millikan[5] verified, by a series of careful experiments, that if ν_0 is the threshold frequency, the photoelectric quantity $h\nu_0$ is equal to the thermionic work-function $e\phi$ measured at the same temperature. L. A. Du Bridge[6] found that for clean platinum the photoelectric long wave-length limit is 1962Å, which by the above equation corresponds to 6·30 volts, while the thermionic work-function is 6·35 volts: the two are thus in agreement within the limits of error.

The theory of the connection between photoelectric threshold frequency and thermionic work-function was studied in the light of Sommerfeld's electron theory of metals by R. H. Fowler.[7]

[1] The theory of electric images is due to William Thomson (Kelvin). This effect had been considered in P. Lenard's early work on the photoelectric effect, *Ann. d. Phys.* viii (1902), p. 149; cf. P. Debye, *Ann. d. Phys.* xxxiii (1910), p. 441, and Walter Schottky, *Phys. ZS.* xv (1914), p. 872.

[2] cf. Vol. I, p. 71

[3] O. W. Richardson, *Phil. Mag.* xxiii (1912), p. 265; verified by O. W. Richardson and F. S. Robertson, *Phil. Mag.* xliii (1922), p. 557

[4] cf. p. 90 *supra*

[5] *Phys. Rev.* xviii (1921), p. 236

[6] *Phys. Rev.* xxxi (1928), pp. 236, 912; cf. also A. H. Warner, *Proc. N.A.S.* xiii (1927), p. 56, who worked with tungsten.

[7] *Proc. R.S.*(A), cxviii (March, 1928), p. 229; cf. G. Wentzel, *Sommerfeld Festschrift* (1928), p. 79

The effect of temperature on the photoelectric sensibility of a clean metal near the threshold was studied by J. A. Becker and D. W. Mueller[1] in 1928, by E. O. Lawrence and L. B. Linford[2] in 1930 and by R. H. Fowler in 1931.[3] It was shown that the photoelectric long wave-length limit is shifted towards the red.

Some interesting experiments on the photoelectric properties of thin films were carried out in 1929 by H. E. Ives and A. R. Olpin.[4] They found that the long-wave limit of photoelectric action in the case of films of the alkali metals on platinum varies with the thickness of the film. As the film accumulates, the long-wave limit moves towards the red end of the spectrum, reaches an extreme position and then recedes again to the final position characteristic of the metal in bulk. The wave-length of the maximum excursion of the long-wave limit was found in every case to coincide with the first line of the principal series of the metal in the form of vapour, i.e. the resonance potential. This seemed to suggest that photoelectric emission is caused when sufficient energy is given to the atom to produce its first stage of excitation.

Richardson[5] in 1912 studied the photoelectric emission from a surface exposed to black-body radiation corresponding to some high temperature T: this has been called the *complete photoelectric effect.* It was shown by A. Becker[6] that the relative distribution of the velocities of the electrons released from platinum photoelectrically under these conditions is identical with that of electrons released thermionically. If there is equilibrium, the surface must itself be at temperature T, and then the total number of electrons leaving the surface, whether as a result of thermionic or photoelectric action, must be given by the Richardson equation.

Sommerfeld's discovery that the conducting electrons in metals obey Fermi statistics made possible a satisfactory theory of a phenomenon which had long been known[7] but which, in some of its features, had not been explained : namely, the extraction of electrons from cold metals by intense electric fields.[8] Experimental work by R. A. Millikan and his coadjutors[9] had shown that the currents obtained at a given voltage are independent of the temperature, provided the latter is not so high as to approach the temperature at which thermionic emission becomes appreciable : whence Millikan

[1] *Phys. Rev.* xxxi (1928), p. 431

[2] *Phys. Rev.* xxxvi (1920), p. 482

[3] *Phys. Rev.* xxxviii (1931), p. 45 ; cf. J. A. Becker and W. H. Brittain, *Phys. Rev.* xlv. (1934), p. 694

[4] *Phys. Rev.* xxxiv (1929), p. 117 ; cf. Hughes and du Bridge, *Photoelectric Phenomena* (New York, 1932), p. 178

[5] *Phil. Mag.* xxiii (1912), p. 594

[6] *Ann. d. Phys.* lx (1919), p. 30

[7] Knowledge of it seems to have evolved gradually from R. F. Earhart's experiments on very short sparks. *Phil. Mag.* i (1901), p. 147.

[8] A first approximate theory was given by W. Schottky, *ZS. f. P.* xiv (1923), p. 63, following on J. E. Lilienfeld, *Phys. ZS.* xxiii (1922), p. 306.

[9] R. A. Millikan and C. F. Eyring, *Phys. Rev.* xxvii (1926), p. 51 ; R. A. Millikan and C. C. Lauritsen, *Proc. N.A.S.* xiv (1928), p. 45 ; cf. also N. A. de Bruyne, *Proc. Camb. Phil. Soc.* xxiv (1928), p. 518

concluded that the conduction electrons extracted in this way do not share in the thermal agitation, whereas the thermions, which are expelled at high temperatures, do.

Millikan and Lauritsen showed that if I is the current obtained by an applied electric force F, then log I when plotted against I/F yields approximately a straight line. In 1928 the subject was attacked in several theoretical papers,[1] most completely by R. H. Fowler and L. Nordheim.[2] They calculated the emission coefficient (i.e. the ratio of the number of electrons going through the surface, to the number of electrons incident from inside the metal) and the reflection coefficient, at the surface, and integrated over all incident electrons according to Sommerfeld's electron theory of metals; and they showed that it is the electrons with small energies that are pulled out by strong fields. Now these electrons with small energies have Fermi statistics, and that is why the intensity of the emitted current is, at ordinary temperatures, independent of the temperature. The formula obtained theoretically by Fowler and Nordheim for the current I was

$$I = \frac{\epsilon}{2\pi h} \frac{\psi^{\frac{1}{2}}}{(\chi+\psi)\chi^{\frac{1}{2}}} F^2 e^{-4\kappa\chi^{\frac{3}{2}}/3F}$$

where ϵ is the electron-charge, $\kappa^2 = 8\pi^2 m/h^2$, ψ is Gibbs's thermodynamical potential per electron, χ is the thermionic work-function and F is the applied electric force. This formula agrees with the experimental results.[3]

[1] R. H. Fowler, *Proc. R.S.*(A), cxvii (1928), p. 549; L. Nordheim, *ZS. f. P.* xlvi (1928), p. 833; W. V. Houston, *ZS. f. P.* xlvii (1928), p. 33 (working with Sommerfeld); O. W. Richardson, *Proc. R.S.*(A), cxvii (1928), p. 719; W. S. Pforte, *ZS. f. P.* xlix (1928), p. 46

[2] *Proc. R.S.*(A), cxix (1928), p. 173; cf. also J. R. Oppenheimer, *Phys. Rev.* xxxi (1928), p. 66

[3] cf. also: O. W. Richardson, *Proc. R.S.*(A), cxix (1928), p. 531; N. A. de Bruyne, *Proc. R.S.*(A), cxx (1928), p. 423; A. T. Waterman, *Proc. R.S.*(A), cxxi (1928), p. 28; L. W. Nordheim, *Proc. R.S.*(A), cxxi (1928), p. 626; T. E. Stern, B. S. Gossling and R. H. Fowler, *Proc. R.S.*(A), cxxiv (1929), p. 699; W. V. Houston, *Phys. Rev.* xxxiii (1929), p. 361

Chapter VII

MAGNETISM AND ELECTROMAGNETISM, 1900–26

At the end of the nineteenth century the classical theory of electrons was well established, and one magnetic phenomenon, namely, the Zeeman effect, had been explained in terms of it by Lorentz and Larmor. The time was evidently ripe for the consideration of diamagnetic, paramagnetic and ferromagnetic phenomena in the light of electron-theory.

It will be remembered[1] that Weber had explained diamagnetism (the magnetic polarisation of bodies induced in a direction opposite to that of the magnetising field) by postulating the existence in molecules of circuits whose electric resistance is zero (but whose self-induction is not zero), so that the creation of an external field causes induced currents in them: the total magnetic flux through the circuits remains zero, and therefore by Lenz's law the direction of the induced currents corresponds to diamagnetism. Paramagnetism was explained by postulating Ampèrean electric currents in the molecules, whose planes were orientated by the magnetising field: and Ewing had developed on this basis an explanation of ferromagnetism.

The re-statement of these ideas in terms of the theory of electrons was undertaken in 1901–3 by W. Voigt[2] and J. J. Thomson.[3] They studied the effect of an external magnetic field on the motion of a number of electrons, which are situated at equal intervals round the circumference of a circle, and are rotating in its plane with uniform velocity round its centre; and they found that if a substance contained a uniform distribution of such systems, the coefficient of magnetisation of the substance would be zero; so that it would be impossible to explain the magnetic or diamagnetic properties of bodies by supposing that the atoms contain charged particles circulating in closed periodic orbits under the action of central forces.

The origin of the difference between the effects produced by charged particles freely describing orbits, and those produced by constant electric currents flowing in circular circuits, as in Ampère's theory of magnetism, is that in the case of the particles describing their orbits we get, in addition to the effects due to the constant electric currents, effects of the same character as those due to the induction of currents in conductors by the variation of the magnetic field: these induced currents tend to make the body diamagnetic,

[1] cf. Vol. I, pp. 208–11 [2] *Gött. Nach.* (1901), p. 169; *Ann. d. Phys.* ix (1902), p. 115
[3] *Phil. Mag.* (6) vi (1903), p. 673

while the Ampèrean currents tend to make it magnetic ; and in the case of the particles describing free orbits, these tendencies balance each other.

Voigt showed that spinning electrons, obstructed in their motion by continual impacts, would lead to paramagnetism or diamagnetism according as they possessed, immediately after the impact, an average excess of potential or kinetic energy. However, besides the complexity of this representation, there is the objection that it attributes to the same cause phenomena so different as paramagnetism and diamagnetism, and that it offers no interpretation of certain laws, established experimentally by P. Curie,[1] namely, that the paramagnetic susceptibility varies inversely as the absolute temperature, whereas diamagnetism is in all observed cases except bismuth, rigorously independent of T.

The first successful application of electron-theory to the general problem of magnetism was made in 1905 by Paul Langevin.[2] He accepted Weber's view that diamagnetism is really a property possessed by all bodies, and that the so-called paramagnetic and ferromagnetic substances are those in which the diamagnetism is masked by vastly greater paramagnetic and ferromagnetic effects. The condition for the absence of paramagnetism and ferromagnetism is that the molecules of the substance should have no magnetic moment, so that they have no tendency to orient themselves in an external magnetic field.

In order to explain diamagnetism, we observe that the external applied magnetic field H creates a Larmor precession,[3] each electron acquiring an additional angular velocity $eH/2mc$ in its orbit[4]; whence it follows, as Langevin showed, that the increase of the magnetic moment of a molecule due to one particular electron circulating in it is

$$\Delta M = -\frac{He^2}{4mc^2}\overline{r^2}$$

where r is the distance of the electron from the atomic nucleus, projected in a plane perpendicular to H, and $\overline{r^2}$ is an average extended over the duration of several revolutions. The negative sign shows that the effect is diamagnetic. In order to obtain the total effect on a molecule we must sum over all the electrons in it. Thus if N denotes Avogadro's number, i.e. the number of molecules in a gramme-molecule (which is the same for all elements and compounds), then *the diamagnetic susceptibility per gramme-molecule is*

$$-\frac{e^2N}{4mc^2}\sum\overline{r^2}$$

[1] *Ann. chim. phys.* v (1895), p. 289

[2] *Comptes Rendus*, cxxxix (26 Dec. 1904), p. 1204 ; *Soc. Française de phys., Résumés* (1905), p. 13 * ; *Ann. chim. phys.*(8) v (1905), p. 70

[3] cf. Vol. I, pp. 415–6

[4] As usual, H and M are in electromagnetic units, and e in electrostatic units.

where the summation is taken over all the electrons in the atom or molecule, and (r, z, ϕ) are the cylindrical co-ordinates of the electron when the z-axis is taken in the direction of the magnetic field H.[1] Since atoms which exhibit diamagnetism without paramagnetism have spherical symmetry, $\sum \overline{r^2}$ can be replaced by $\frac{2}{3}\sum \overline{R^2}$, where R denotes the distance of the electron from the centre of the atom.

The smallness of the diamagnetic effects is accounted for by the smallness of the radius r, which is necessarily less than the molecular dimensions.

The diamagnetic property is acquired instantaneously, at the moment of the creation of the external field.

The intramolecular motions of the electrons depend very little on the temperature, as is shown by the fixity of spectral lines: so the diamagnetic susceptibility should vary very little with the temperature, in agreement with Curie's experimental result. It varies also very little with the physical or chemical state: *diamagnetism is an atomic property*.

The exception presented by solid bismuth, whose diamagnetic susceptibility diminishes approximately linearly when the temperature rises, was attributed by Langevin (following J. J. Thomson) to the presence in the metal of free conduction-electrons.

The fact that diamagnetism and the Zeeman effect both depend on the Larmor precession shows that the two phenomena (at any rate when considered classically) are very closely connected, and indeed may be regarded as different aspects of the same phenomenon.

If the intrinsic magnetic moment of a molecule is not null, there is superposed on the diamagnetic effect another phenomenon, due to the orientation of the molecular magnets by the external field; this *paramagnetic* effect, when it exists, is large compared with the diamagnetic, and completely masks it. In discussing it we shall assume that the molecular magnets are not associated in aggregates within which their mutual actions are of importance (this is the case with *ferromagnetic* substances, which will be considered later): and in fact we shall suppose that they can be treated in the same way as the molecules in the kinetic theory of gases. In an external magnetic field H, a molecule whose magnetic moment is M has a potential energy $-\mathrm{MH}\cos\alpha$, where α is the angle between M and H; the potential energy being thus least when the molecular magnet is parallel to the external field. The situation may now be compared with that of a gas composed of heavy molecules and acted on by gravity, where the ascent of a molecule causes an increase in its gravitational potential energy and a decrease in its kinetic energy; similarly the magnetic molecule experiences a change of kinetic energy when α changes, and the partition of kinetic energy between the molecules becomes incompatible with thermal

[1] W. Pauli, *ZS. f. P.* ii (1920), p. 201

equilibrium. Through the mediation of collisions, a rearrangement is brought about, by which the mean kinetic energy of a molecule is made independent of its orientation: the temperature becomes uniform, and the molecular magnets are directed preferentially in the direction of H; though this orientation is not universal, on account of the thermal agitation and the collisions. Unlike diamagnetism, paramagnetism does not appear instantaneously, since collisions are required to create it.

Langevin found that if μ denotes the intrinsic magnetic moment of a molecule, T the absolute temperature, k Boltzmann's constant, N Avogadro's number, M the magnetisation in one gramme-molecule, then

$$M = N\mu\left(\coth X - \frac{1}{X}\right) \quad \text{where} \quad X = \frac{\mu H}{kT}.$$

When X is small, so that we need retain only the first term in $(\coth X - 1/X)$, this gives for the molecular susceptibility χ,

$$\chi = \frac{M}{H} = \frac{N\mu^2}{3kT};$$

this formula involves Curie's law, that the paramagnetic susceptibility varies inversely as the absolute temperature.[1]

From the point of view of the Rutherford-Bohr theory of atomic structure, paramagnetism, being due to the possession of intrinsic magnetic moment, is associated with *incomplete shells* of electrons[2]; for every closed shell has spherical symmetry. This explains why the alkaline and alkaline-earth elements in the metallic state are paramagnetic, while their salts are diamagnetic: for the electrons which do not belong to closed shells in the metals are taken into gaps in the shells of the elements with which they are combined. In the case of the incomplete inner shells of the rare earths, however, this kind of compensation does not take place, so the salts are paramagnetic.[3]

In 1907 Langevin's theory was extended so as to give an account of ferromagnetism, by Pierre Weiss[4] (*b.* 1865). Ferromagnetism is a property of *molecular aggregates* (crystals), so Weiss took into consideration the *internal* magnetic field, assuming that each molecule is acted on by the surrounding molecules with a force equal to that which it would experience in a uniform field proportional to the intensity of the magnetisation and in the same direction. On this

[1] On the magnetic susceptibility of oxygen, hydrogen and helium, cf. A. P. Wills and L. G. Hector, *Phys. Rev.*, xxiii (1924), p. 209; for helium, neon, argon and nitrogen, cf. L. G. Hector, *Phys. Rev.*, xxiv (1924), p. 418. For theoretical predictions regarding these gases, cf. G. Joos, *ZS. f. P.* xix (1923), p. 347.

[2] cf. N. W. Taylor and G. N. Lewis, *Proc. N.A.S.*, xi (1925), p. 456

[3] F. Hund, *ZS. f. P.* xxxiii (1925), p. 855

[4] *Bull. des séances de la soc. fr. de phys., Année* 1907, p. 95; *Journ. de phys.* vi (1907), p. 661

assumption Weiss showed that a ferromagnetic body when heated to a certain critical temperature, the *Curie point*,[1] ceases to be ferromagnetic, and that for higher temperatures the inverse of the susceptibility is a linear function of the excess of the temperature above the Curie point.

It follows from Weiss's work that in a ferromagnetic substance there must be *domains* (i.e. regions larger than a single atom or molecule) which are inherently magnetic, although if the magnetic moments of the domains are not oriented in any preferential direction, a finite block of the substance may show no magnetisation.

Weiss found [2] that the magnetic moments of all known molecules were multiples of a certain greatest common divisor, which he called a *magneton*. This is called *Weiss's magneton* in order to distinguish it from the natural unit of magnetic moment to which W. Pauli [3] in 1920 gave the name *Bohr magneton*, and which has the value

$$\frac{he}{4\pi mc}.$$

The Bohr magneton is nearly five times the Weiss magneton. For the vapours of the elements in the first column of the Newlands-Mendeléev table (i.e. the alkalis and Cu, Ag, Au), it was shown by W. Gerlach and O. Stern,[4] by W. Gerlach and A. C. Cilliers,[5] and by J. B. Taylor,[6] that the intrinsic magnetic moment is one Bohr magneton. As we have seen,[7] the Bohr magneton was found in 1925 to be the magnetic moment of an electron; and from this time the Weiss magneton ceased to figure in physical theory.[8]

The researches of Langevin and Weiss represented notable advances in the theory of magnetism, and the agreement of their results with experimental data was striking: yet it was shown in Niels Bohr's inaugural-dissertation [9] in 1911 and by Miss H. J. van Leeuwen,[10] a pupil of Lorentz's, in 1919, that the validity of these results could be explained only by supposing that Langevin and Weiss had not consistently applied classical statistics to all the degrees of freedom concerned: in other words, they had made assumptions of a quantistic character. The difficulty was eventually removed only by the development of quantum mechanics.

An account must now be given of some questions which had been

[1] The name was introduced by P. Weiss and H. Kamerlingh Onnes, *Comm. phys. Labor. Leiden*, No. 114 (1910), p. 3.

[2] *Journ. de phys.* i (1911), pp. 900, 965; *Phys. ZS.* xii (1911), p. 935; cf. B. Cabrera, *Journ. de phys.* iii (1922), p. 443

[3] *Phys. ZS.* xxi (1920), p. 615

[4] *Ann. d. Phys.* lxxiv (1924), p. 673

[5] *ZS. f. P.* xxvi (1924), p. 106

[6] *Phys. Rev.* xxviii (1926), p. 576

[7] cf. p. 136 *supra*

[8] cf. W. Gerlach, *Phys. ZS.* xxiv (1923), p. 275

[9] N. Bohr, *Studier over Metallernes Elektronteori* (120 pp.), Copenhagen, 1911

[10] *Inaugural dissertation*, Leiden, 1919: her argument, which was based on Boltzmann's H-theorem, was substantially reproduced in her paper *J. de phys. et le radium*, ii (1921), p. 361; cf. also J. N. Kroo, *Ann. d. Phys.* xlii (1913), p. 1354, who argued that the electron theory could explain only diamagnetism; E. Holm, *Ann. d. Phys.* xliv (1914), p. 241; R. Gans, *Ann. d. Phys.* xlix (1916), p. 149.

left unsettled in the nineteenth century and were definitely answered in the twentieth.

Maxwell had suggested[1] the possibility that an electromotive force might be produced by simply altering the velocity of a conductor. This effect was not known to exist until 1916, when it was observed by Richard C. Tolman and T. Dale Stewart.[2] They rotated a coil of copper wire about its axis at a high speed and then suddenly brought it to rest, so that a pulse of current was produced at the instant of stopping by the tendency of the free electrons to continue in motion. The ends of the coil were connected to a sensitive ballistic galvanometer, which enabled the experimenters to measure the pulse of current thus produced. Denoting by R the total resistance in the circuit, by l the length of the rotating coil, by v the rim speed of the coil, by Q the pulse of electricity which passes through the galvanometer at the instant of stopping, by F the charge carried in electrolysis by one gramme-ion, that is Ne/c electromagnetic units, where N is Avogadro's number, and by M the effective mass of the carrier of the current (the electron), they found the equation

$$Q = \frac{Mvl}{FR},$$

by use of which they inferred from the experiments that the carriers of electric current in metals have approximately the same ratio of mass to charge as an electron.

Maxwell[3] mentioned another possible effect which might be caused by the carriers of electricity in conductors. He proposed to take a circular coil of a great many windings and suspend it by a fine vertical wire, so that the windings are horizontal and the coil is capable of rotating about a vertical axis. A current is supposed to be conveyed into the coil by means of the suspending wire, and, after passing round the windings, to complete its circuit by passing downwards through a wire which is in the same line with the suspending wire and dips into a cup of mercury. If a current is sent through the coil, then at the moment of starting it, a force would require to be supplied in order to produce the angular momentum of the carriers of electricity passing round the coil; and as this must be supplied by the elasticity of the suspending wire, the coil must rotate in the opposite direction.

Maxwell failed to detect this phenomenon experimentally, but it was successfully observed in 1931 by S. J. Barnett[4] (*b.* 1873). Like the preceding effects of electron-inertia, it provides a measure of m/e.

[1] In § 577 of his *Treatise*

[2] *Phys. Rev.*(2) viii (1916), p. 97; ix (1917), p. 164; cf. R. C. Tolman, S. Karrer and E. W. Guernsey, *Phys. Rev.* xxi (1923), p. 525; R. C. Tolman and L. M. Mott-Smith, *Phys. Rev.* xxviii (1926), p. 794

[3] *Treatise*, § 574

[4] *Phil. Mag.* xii (1931), p. 349

In 1908 O. W. Richardson[1] suggested the existence of a mechanical effect accompanying the act of magnetisation. He imagined a long thin cylindrical bar of iron suspended by a fibre, so that it is capable of small rotations about a vertical axis. When the bar is not magnetised, the electrons which are moving in closed orbits in the molecules (Ampère's molecular currents) will not possess any resultant angular momentum, since one azimuth is as probable as another for the orbits. Now consider the effect of suddenly applying a vertical magnetic field: the orbits will orient themselves so as to leave a balance in favour of the plane perpendicular to the direction of the field, and thus an angular momentum of electrons will be created about the axis of suspension. This must be balanced by an equal reaction elsewhere, and therefore a twisting of the suspended system as a whole is to be expected.

Richardson himself did not succeed in obtaining the effect, which was first observed by A. Einstein and W. J. de Haas[2] in 1915. They were followed by many other experimenters, particularly J. Q. Stewart[3] of Princeton and W. Sucksmith and L. F. Bates,[4] working with Professor A. P. Chattock in Bristol. Sucksmith and Bates concluded that the *gyromagnetic ratio*, i.e. the ratio of the angular momentum of an elementary magnet to its magnetic moment, has a value only slightly greater than mc/e. This is only half the value of the ratio for an electron circulating in an orbit, and has been taken to imply that the magnetic elements responsible for the phenomenon are chiefly not orbital electrons but are electrons spinning on their own diameters.[5]

The converse of the Richardson effect has also been observed, namely, it has been found possible to magnetise iron rods by spinning them about their axes. John Perry[6] said in 1890 'Rotating a large mass of iron rapidly in one direction and then in the other in the neighbourhood of a delicately-suspended magnetic needle ought, I think, to give rise to magnetic phenomena. I have hitherto failed to obtain any trace of magnetic action, but I attribute my failure to the comparatively slow speed of rotation which I have employed, and to the want of delicacy of my magnetometer.'

The effect predicted by Perry was anticipated also by Schuster[7] in 1912, but was first observed in 1914–15 by S. J. Barnett[8]: the Ampèrean molecular currents, since they possess angular momentum,

[1] *Phys. Rev.* xxvi (1908), p. 248

[2] *Verh. d. deutsch. phys. Ges.* xvii (1915), p. 152; cf. A. Einstein, ibid. xviii (1916), p. 173 and W. J. de Haas, ibid. xviii (1916), p. 423

[3] *Phys. Rev.* xi (1918), p. 100

[4] *Proc. R.S.*(A), civ (1923), p. 499; cf. W. Sucksmith, *Proc. R.S.*(A), cviii (1925), p. 638; also Emil Beck, *Ann. d. Phys.* lx (1919), p. 109; and G. Arvidsson, *Phys. ZS.* xxi (1920), p. 88

[5] On the gyromagnetic effect for paramagnetic substances, cf. W. Sucksmith, *Proc. R.S.*(A), cxxviii (1930), p. 276; cxxxv (1932), p. 276

[6] J. Perry, *Spinning Tops*, p. 112

[7] *Proc. Phys. Soc.* xxiv (1911–12), p. 121

[8] *Phys. Rev.*(2) vi (1915), p. 239; *Bull. Nat. Res. Council*, iii (1922), p. 235; cf. S. J. Barnett and L. J. H. Barnett, *Phys. Rev.* xx (1922), p. 90; *Nature*, cxii (1923), p. 186

behave like the wheels of gyroscopes, changing their orientation, with a tendency to make their rotations become parallel to the impressed rotation. Thus a preponderance of magnetic moment is caused along the axis of the impressed rotation.

From time to time papers appeared on questions belonging to the same class as the problem of *unipolar induction*. This problem, which had first been considered by Faraday,[1] may be stated as follows. A magnet, symmetrical with regard to an axis about which it rotates, has sliding contact with the ends A and B of a stationary wire ACB, at two points A and B not in the same equatorial plane of the magnet. As Faraday found, a steady current flows through the wire. The question is, do the lines of magnetic induction rotate with the magnet, so that the electromotive force is produced when they cut the stationary wire ACB : or do the lines of magnetic induction remain fixed, so that the electromotive force is produced when the moving part of the circuit (the magnet) rotates through them? Faraday believed the latter explanation to be the true one: but Weber, who introduced the name *unipolar induction*, took the contrary view.[2] It was known in the nineteenth century that the experimental results could be explained equally well on either hypothesis.

More generally, we can consider a magnet which is capable of being rotated about its axis of symmetry (whether it is actually so rotated or not) and a conducting circuit composed of two parts, one of which is rotated about the axis of the magnet (in the case considered above it is the magnet itself) while the other, ACB, remains fixed. It is found that if the moving part of the circuit is distinct from the magnet, the electromotive force is independent of whether the magnet is rotating or not. The electromotive force round the circuit is in all cases, if ω denotes the angular velocity of the moving part of the circuit, $(\omega/2\pi c)$ times the flux of magnetic induction through any cylindrical surface having as boundaries the circles described round the axis of revolution by the two contact points A and B of the fixed with the rotating part of the circuit.

In the twentieth century some new types of experiment were devised. In 1913 Marjorie Wilson and H. A. Wilson[3] constructed a non-conducting magnet by embedding a large number of small steel spheres in a matrix of wax, and rotated this in a magnetic field whose direction was parallel to the axis of rotation. Their measures of the induced electromotive force were in satisfactory agreement with the predictions of the electromagnetic theory of moving bodies published by A. Einstein and J. Laub[4] in 1908.

Further contributions to problems of this class were made by

[1] cf. Vol. I, pp. 173–4
[2] loc. cit., Vol. I
[3] *Proc. R.S.*(A), lxxxix (1913), p. 99
[4] *Ann. d. Phys.* xxvi (1908), p. 532

S. J. Barnett,[1] E. H. Kennard,[2] G. B. Pegram,[3] and W. F. G. Swann,[4] and in 1922 a comprehensive review of the whole subject was published by J. T. Tate.[5]

The problems of classical electromagnetic theory which were studied in the early years of the twentieth century related chiefly to the expression of the field due to a moving electron, and the rate at which energy is radiated from it outwards. Formulae for the electric and magnetic vectors of the field were given by many writers.[6] The most interesting terms are those which involve the acceleration of the electron. Denote by (v_x, v_y, v_z) the velocity and by $(w_x, w_y\ w_z)$ the acceleration of the electron at the instant $\bar{t}$ when it emits the radiation which reaches the point (x, y, z) at the instant t: let the co-ordinates of the electron at time t' be (x', y', z'), and let $x'(\bar{t}), y'(\bar{t}), z'(\bar{t})$, be denoted by $(\bar{x}', \bar{y}', \bar{z}')$; and let

$$\bar{r}^2 = (\bar{x}' - x)^2 + (\bar{y}' - y)^2 + (\bar{z}' - z)^2,$$

so
$$\bar{t} = t - \bar{r}/c.$$

Then the terms in the x-component of the electric force which involve the acceleration are

$$D_x = -\frac{ew_x}{\bar{r}s^2} + \frac{e\{(\bar{x}' - x)w_x + (\bar{y}' - y)w_y + (\bar{z}' - z)w_z\}}{\bar{r}^2\, s^3}\left\{\frac{\bar{x}' - x}{\bar{r}} + \frac{v_x}{c}\right\}$$

where
$$s = 1 + \frac{(\bar{x}' - x)v_x + (\bar{y}' - y)v_y + (\bar{z}' - z)v_z}{c\bar{r}}.$$

From this we have at once

$$D_x\,(\bar{x}' - x) + D_y\,(\bar{y}' - y) + D_z\,(\bar{z}' - z) = 0$$

so *the vector* (D_x, D_y, D_z) *is perpendicular to* $\bar{\mathbf{r}}$. Moreover, the part of the magnetic vector which depends on the acceleration of the electron (call it **H**) *is perpendicular to both* **D** *and* $\bar{\mathbf{r}}$, *and equal in magnitude to* **D**. So the *wave of acceleration*, as Langevin [7] called the field specified by **D** and **H**, *has all the characters of a wave of light.*

At great distances from the electron, the intensity of **D** and **H** decreases like $1/r$, whereas the intensity of the terms in the electric

[1] *Phys. ZS.* xiii (1912), p. 803 ; *Phys. Rev.* xxxv (1912), p. 323
[2] *Phys. ZS.* xiii (1912), p. 1155 ; *Phys. Rev.*(2) i (1913), p. 355
[3] *Phys. Rev.* x (1917), p. 591
[4] *Phys. Rev.* xv (1920), p. 365
[5] *Bull. Nat. Res. Council*, iv, Part 6 (1922), p. 75
[6] K. Schwarzschild, *Gött. Nach.* (1903), p. 132. G. Herglotz, *Gött. Nach.* (1903), p. 357. H. A. Lorentz, *Proc. Amst. Acad.* v (1903), p. 608. A. W. Conway, *Proc. Lond. M.S.*(2) i (1903), p. 154. A. Sommerfeld, *Gött. Nach.* (1904), p. 99. P. Langevin, *J. de phys.* iv (1905), p. 165. H. Poincaré, *Palermo Rend.* xxi (1906), p. 129. G. A. Schott, *Ann. d. Phys.* xxiv (1907), p. 637 ; xxv (1908), p. 63. F. R. Sharpe, *Bull. Amer. Math. Soc.* xiv (1908), p. 330. A. Sommerfeld, *Munich Sitz.* (1911), p. 51. A. W. Conway, *Proc. R.I.A.* xxix, A (1911), p. 1 ; *Proc. R.S.*(A), xciv (1918), p. 436
[7] loc. cit.

and magnetic vectors which do not involve w decrease like $1/r^2$: so *at great distances the field consists solely of the wave of acceleration*, which represents the *radiation* emitted by the electron.

The rate of loss of energy by radiation from a charge e moving with an acceleration w and a velocity small compared with c is, as we have seen.[1]

$$\frac{2}{3}\frac{e^2w^2}{c^3}.$$

When v/c is no longer neglected, it was shown by Heaviside[2] that *the energy radiated per second is*

$$\frac{2}{3}\frac{e^2w^2}{c^3}\cdot\frac{1-(v^2/c^2)\sin^2\widehat{vw}}{(1-v^2/c^2)^3}.$$

The question as to a reaction or 'back pressure' experienced by a moving mass on account of its own emission of radiant energy was discussed by O. Heaviside,[3] M. Abraham,[4] J. H. Poynting,[5] A. W. Conway,[6] J. Larmor[7] and Leigh Page.[8]

A striking unification of electromagnetic theory was published in 1912 by Leigh Page[9] (*b.* 1884). It had been realised long before by Priestley[10] that from the experimental fact that there is no electric force in the space inside a charged closed hollow conductor, it is possible to deduce the law of the inverse square between electric charges, and so the whole science of electrostatics. It was now shown by Page that if a knowledge of the relativity theory of Poincaré and Lorentz is assumed, the effect of electric charges in motion can be deduced from a knowledge of their behaviour when at rest, and thus the existence of magnetic force may be inferred from electrostatics: magnetic force is in fact merely a name introduced in order to describe those terms in the ponderomotive force on an electron which depend on its velocity. In this way Page showed that Ampère's law for the force between current-elements, Faraday's law of the induction of currents and the whole of the Maxwellian electromagnetic theory, can be derived from the simple assertion of the absence of electric effects within a charged closed hollow conductor.

In 1914 two new representations of electromagnetic actions were introduced, both of which were evidently inspired by Maxwell's *Encyclopaedia Britannica* article on the aether, in which it was regarded as composed of corpuscles, moving in all directions with the velocity

[1] cf. Vol. I, p. 396
[2] *Nature*, lxvii (1902), p. 6
[3] *Nature*, lxvii (1902), p. 6
[4] *Ann. d. Phys.* x (1903), p. 156; *Brit. Ass. Rep., Cambridge* 1904, p. 436
[5] *Phil. Trans.* ccii (1904), p. 525
[6] *Proc. R.I.A.* xxvii (1908), p. 169
[7] *Proc. Int. Cong. Math., Cambridge* 1912, Vol. I, p. 213; *Nature*, xcix (1917), p. 404
[8] *Phys. Rev.* xi (1918), p. 376
[9] *Amer. J. Sci.* xxxiv (1912), p. 57; *Phys. ZS.* xiii (1912), p. 609
[10] cf. Vol. I, p. 53

of light, never colliding with each other, and possessing some vector quality such as rotation.

The first of these representations was due to Ebenezer Cunningham of Cambridge[1] (*b.* 1881). Cunningham remarked that if the aether is supposed to be at rest, as it is in Lorentz's theory, then the transfer of energy represented by the Poynting vector cannot be identified with the rate of work of the stress in the aether. This, indeed, can be done only if the aether is supposed to be in motion; for a stress on a stationary element of area does not transmit any energy across that element. He therefore proposed to assign a velocity to the aether at every point, such that a state of stress in the medium would account both for the transference of momentum and for the flow of energy.

He found that the component of the aether-velocity which is in the direction of the Poynting vector must be the smaller root of the quadratic equation $(c^2+x^2)g=2\mathrm{W}x$ where g is the value of the momentum $(1/c)$ $[\mathbf{E}\,.\,\mathbf{H}]$, and W is the density of energy: that the other component lies in a definite direction in the plane of **E** and **H**: and that the *total* velocity of the aether at every point must be equal to c: so the direction and magnitude of the aethereal velocity are completely determined. The scheme is relativistically invariant: so that an objective aether is not necessarily foreign to the point of view of the principle of relativity: and the mechanical categories of momentum, energy, and stress can also be maintained in their entirety. The only essential modification of the ordinary theory of material media is that (as in relativity generally) we no longer take momentum to be in the direction of, or proportional to, the velocity.

The *true stress* in the aether may be defined as differing from the Faraday-Maxwell stress by an amount representing the rate at which momentum is *convected* by the aether across stationary elements of area. This 'true stress' consists of a tension P, defined by the equation

$$\mathrm{P}^2=\mathrm{W}^2-c^2g^2$$

in the direction of the component velocity of the aether in the plane of **E** and **H**, together with an equal pressure P in all directions at right angles to this. It is this 'true stress' which does work by acting on moving elements, and so transfers energy.

Cunningham gave examples of the determination of the aether-velocity in some simple cases. For a train of plane waves of light the stress P vanishes, and the system is one of pure convection: the aether moves as a whole in the direction of propagation of the waves. For a moving point charge, the aether moves as if continually emitted from the charge with velocity c, every element travelling uniformly in a straight line after emission.

[1] E. Cunningham, *The Principle of Relativity*, Cambridge, 1914; *Relativity and the Electron Theory*, London, 1915

The continual emission from an electron of corpuscles moving with velocity c in all directions is the chief feature also of an *emission theory of electromagnetism* published in the same year (1914) by Leigh Page.[1] Page's work is in some ways reminiscent of ideas which had been put forward by J. J. Thomson in an Adamson lecture[2] delivered in Manchester University in 1907, when the concept of aether in motion with velocity c was brought into relation with Thomson's favourite concept of moving lines of electric force.

Page proposed to regard an electron, when viewed in an inertial system in which it is at rest, as being like a sphere whose surface is studded with *emittors* distributed uniformly over it: each emittor is continually projecting into the surrounding space a stream of *corpuscles*, each corpuscle moving radially in a straight line with the velocity of light: it is assumed that the emittors have no rotation relative to the inertial system. When the electron is in motion in any way, the stream of corpuscles that have been ejected from any one emittor at successive instants form a curve in space which, as Page showed, is a *line of electric force* in the field due to the moving electron.

As in Thomson's theory,[3] each electron is supposed to possess its own system of lines of force, independently of other electrons, so that in general there will be as many lines of force crossing at a point as there are electrons in the field.

Owing to the motion of the electron, the direction of the line of force at a point of space does not generally coincide with the direction of motion of the corpuscles at the point. The component of the electric vector **d** in any direction is measured by the number of lines of force which cross unit area at right angles to this direction. The magnetic vector **h** is defined as in Thomson's papers, by the equation

$$\mathbf{h} = \frac{1}{c}\left[\mathbf{c}.\,\mathbf{d}\right],$$

where **c** is the (vector) velocity of the corpuscles at the point. Page worked out the consequences of these assumptions for an electron of given velocity and acceleration, and found expressions for **d** and **h** at any point of the field at any time, which agreed exactly with the values deduced by previous investigators from the Maxwell-Lorentz equations.

In 1913 the existence of a hitherto unknown phenomenon was deduced by John Gaston Leathem[4] (1871–1923) of Cambridge,

[1] *Amer. J. Sci.* xxxviii (1914), p. 169; L. Page, *An Introduction to Electrodynamics*, Boston, 1922; L. Page and N. I. Adams, *Electrodynamics*, London, 1941

[2] *Manchester Univ. Lectures*, No. 8 (Manchester Univ. Press, 1908): reprinted in the *Smithsonian Report* for 1908, p. 233; cf. also J. J. Thomson, *Phil. Mag.* xxxix (1920), p. 679, and *Mem. and Proc. Manchester Lit. and Phil. Soc.* lxxv (1930–1), p. 77

[3] cf. J. J. Thomson, *Proc. Camb. Phil. Soc.* xv (1909), p. 65

[4] *Proc. R.S.*(A), lxxxix (1913), p. 31; cf. Larmor's *Collected Papers*, Vol. II, p. 72

from classical electrodynamics, namely, a minute mechanical force exerted by a varying electric field on a magnetic dipole. On the assumption that the magnetism is due to molecular electric currents, he found that the ponderomotive force[1] on the dipole is

$$\left(m_x\frac{\partial}{\partial x}+m_y\frac{\partial}{\partial y}+m_z\frac{\partial}{\partial z}\right)\mathbf{H}+4\pi\left[\mathbf{m},\frac{\partial\mathbf{D}}{\partial t}\right]$$

where **m** is the magnetic moment of the dipole, **H** is the magnetic force, and **D** is the electric displacement. The first term is the ordinary formula for the force exerted on a magnetic dipole, regarded as a polarised combination of positive and negative magnetism, by a field of magnetic force. The second term was unexpected: it represents a mechanical force exerted on a magnet by a displacement-current, at right angles to the displacement-current and to the magnetic moment, and proportional to the product of the two and the sine of the angle between them.

It might seem possible to test the matter by hanging a small magnet horizontally between the horizontal plates of a charged condenser and then effecting a non-oscillatory discharge of the condenser. If the upper plate were originally charged with positive electricity, the displacement-current on discharge would be upwards, and an eastwards impulse on the magnet might be looked for: the effect would, however, be too small to be observed.

In the first quarter of the twentieth century several interesting results in classical electrodynamics were discovered by Richard Hargreaves (1853–1939). In 1908 he proved[2] that Lorentz's fundamental equations (with $c=1$) can be replaced by two integral-equations

$$\iint(\mathrm{H}_x dydz+\mathrm{H}_y dzdx+\mathrm{H}_z dxdy+\mathrm{E}_x dxdt+\mathrm{E}_y dydt+\mathrm{E}_z dzdt)=0 \qquad \text{(I)}$$

and

$$\iint(\mathrm{E}_x dydz+\mathrm{E}_y dzdx+\mathrm{E}_z dxdy-\mathrm{H}_x dxdt-\mathrm{H}_y dydt-\mathrm{H}_z dzdt)$$

$$=-\iiint(\rho w_x dydzdt+\rho w_y dzdxdt+\rho w_z dxdydt-\rho dxdydz). \qquad \text{(II)}$$

Here (t, x, y, z) denotes the co-ordinates of a point in space-time, **E** is the electric and **H** the magnetic vector, and **w** is the velocity of the charge-density ρ. Let any closed two-dimensional manifold S_2 in the four-dimensional space-time be assigned, and let S_2 be the boundary of a three-dimensional manifold S_3 in which the co-ordinates t, x, y, z are functions of three parameters α, β, γ, of which $\gamma=0$ on S_2 and $\gamma<0$ in S_3: so that on S_2 the co-ordinates

[1] In electromagnetic units

[2] *Trans. Camb. Phil. Soc.* xxi (1908), p. 107

are functions of α, β only. Then any term such as $\iint H_x dydz$ may be interpreted to mean $\iint H_x\, \partial(y, z)/\partial(\alpha, \beta)\, d\alpha d\beta$ taken over S_2, and any term such as $\iiint \rho dxdydz$ may be interpreted to mean

$$\iiint \rho \frac{\partial(x, y, z)}{\partial(\alpha, \beta, \gamma)} d\alpha d\beta d\gamma$$

taken over S_3. In the general case the quantities occurring in these equations are evaluated at different points of space *at different times* : the integrals are thus more general than the usual surface and volume integrals.

Let S be an arbitrary closed surface in the (x, y, z) space, and let t be expressed in terms of (x, y, z) by an arbitrary law $t = t(x, y, z)$. Each particle is supposed to be within S at the moment when its charge is evaluated, but since the charges on the particles are evaluated at different times, the particles need not all be within the closed surface at a given time. This explains why the *total charge on the particles* is not $\iiint \rho dxdydz$, but is the triple integral on the right-hand side of equation (II).

Hargreaves showed further that the equations

$$E_x = -\frac{\partial \Phi}{\partial x} - \frac{\partial A_x}{\partial t}, \qquad H_x = \frac{\partial A_z}{\partial y} - \frac{\partial A_y}{\partial z} \quad \text{etc.,}$$

which express the electric and magnetic vectors in terms of the scalar and vector potentials, are equivalent to the single integral equation

$$\int (A_x dx + A_y dy + A_z dz - \Phi dt)$$
$$= \iint (H_x dydz + H_y dzdx + H_z dxdy + E_x dxdt + E_y dydt + E_z dzdt).$$

Here we are considering a closed curve, and a surface bounded by this curve, and we can suppose that t is expressed in terms of (x, y, z) by an arbitrary known law: then the line-integral may be understood to mean

$$\int\left[\left(A_x - \Phi\frac{\partial t}{\partial x}\right)dx + \left(A_y - \Phi\frac{\partial t}{\partial y}\right)dy + \left(A_z - \Phi\frac{\partial t}{\partial z}\right)dz\right].$$

In 1920 Hargreaves[1] generalised Maxwell's theory of light-pressure[2] by showing that when a general electromagnetic field is present near the surface of any perfectly-reflecting (i.e. perfectly-conducting) body, which is in motion in any way, the pressure is normal, and is measured by the difference between the magnetic and electric energies (per unit volume) at the surface.

[1] *Phil. Mag.* xxxix (1920), p. 662 [2] cf. Vol. I, pp. 274–5

In 1922 Hargreaves[1] discovered remarkable expressions for the scalar and vector potentials of a moving electron. Let $x'(t)$, $y'(t)$, $z'(t)$ specify the position of the electron at time $\bar{t}$: let (t, x, y, z) be the world-point for which the potentials are to be calculated: let $\bar{t}$ be the value of t for the electron at the instant when it emits actions which reach the point (x, y, z) at the instant t: let $\bar{x}'$ denote $x'(\bar{t})$, let $\bar{y}'$ denote $y'(\bar{t})$ and let $\bar{z}'$ denote $z'(\bar{t})$. Then Hargreaves showed that *the scalar potential has the value*

$$\Phi = -\tfrac{1}{2}\, ec \left(\frac{\partial^2}{\partial x^2} + \frac{\partial^2}{\partial y^2} + \frac{\partial^2}{\partial z^2} - \frac{1}{c^2}\frac{\partial^2}{\partial t^2}\right)(\bar{t} - t),$$

and *the x-component of the vector-potential* is

$$A_x = -\tfrac{1}{2}e\left(\frac{\partial^2}{\partial x^2} + \frac{\partial^2}{\partial y^2} + \frac{\partial^2}{\partial z^2} - \frac{1}{c^2}\frac{\partial^2}{\partial t^2}\right)(\bar{x}' - x).$$

Other properties found by him were :

$$\left(\frac{\partial \bar{t}}{\partial x}\right)^2 + \left(\frac{\partial \bar{t}}{\partial y}\right)^2 + \left(\frac{\partial \bar{t}}{\partial z}\right)^2 - \frac{1}{c^2}\left(\frac{\partial \bar{t}}{\partial t}\right)^2 = 0$$

and

$$\left(\frac{\partial^2}{\partial x^2} + \frac{\partial^2}{\partial y^2} + \frac{\partial^2}{\partial z^2} - \frac{1}{c^2}\frac{\partial^2}{\partial t^2}\right)^2(\bar{t} - t) = 0 :$$

in fact, the harmonic operator applied twice to any function of $\bar{t}$ yields a zero result, except at the source.

[1] *Mess. of Math.* lii (1922), p. 34

Chapter VIII

THE DISCOVERY OF MATRIX-MECHANICS

A young German named Werner Heisenberg (*b.* 1901), shortly after taking his doctor's degree in 1923 under Sommerfeld at Munich, moved to Niels Bohr's research school at Copenhagen. Here he became closely associated with H. A. Kramers, who in 1924 made important contributions to the theory of dispersion,[1] in the development of which Heisenberg took part.

The concepts employed in this theory suggested to Heisenberg a new approach to the general problems of atomic theory. It will be remembered that the quantum theory of dispersion had originated in Ladenburg's successful translation into quantum language of the analysis that was used in the classical theory. In place of classical electrons in motion within the atom, Ladenburg introduced into the formulae transitions between stationary states: so that instead of the atom being regarded as a Rutherford planetary system of nucleus and electrons obeying the laws of classical dynamics, its behaviour with respect to incident radiation was predicted by means of calculations based on the 'virtual orchestra.'[2] The great advantage thereby gained depended on the circumstance that the motions of the electrons in the Rutherford planetary system were completely unobservable, and did not yield directly the frequencies of the spectral lines emitted by the atom: whereas the virtual orchestra emitted radiations of the frequencies that were actually observed, and thus was much more closely related to physical experiments.

Heisenberg saw that this idea of replacing the classical dynamics of the Rutherford atom by formulae based on the virtual orchestra could be applied in a far wider connection. He took as his primary aim to lay the foundations of a quantum-theoretic mechanics which should be based exclusively on relations between quantities that are actually observable. Previous investigators had found integrals of the classical equations of motion of the atomic system, and so had obtained formulae for the co-ordinates and velocities of the electrons as functions of the time. These formulae Heisenberg now abandoned, on the ground that they do not represent anything that is accessible to direct observation: and in their place he proposed to make the virtual orchestra the central feature of atomic theory.

By taking this step, he made it no longer necessary to find first the classical solution of a problem and then translate it into quantum language: he proved, in fact, that it is possible to translate the

[1] cf. p. 203 *supra* [2] cf. p. 204 *supra*

classical problem into a quantum problem at the very beginning, before solving it: that is to say, he translated the fundamental laws of classical dynamics into a system of fundamental quantum laws constituting what Born[1] had adumbrated under the name *quantum mechanics.*

He considered problems defined by differential equations, such as, for instance, the problem of the anharmonic oscillator, which is defined by the differential equation

$$\frac{d^2x}{dt^2}+\omega_0{}^2 x+\lambda x^2=0$$

and inquired how a solution of this equation can be obtained which, like a virtual orchestra, refers to a doubly-infinite aggregate of transitions between stationary states. Let us suppose that x can be *represented* by an aggregate of terms x_{mn}, where x_{mn} is associated with the transition between the stationary states m and n: and let these terms be arranged in a double array thus:

$$\begin{array}{ccccc} x_{11} & x_{12} & x_{13} & x_{14} & \cdot\ \cdot \\ x_{21} & x_{22} & x_{23} & x_{24} & \cdot\ \cdot \\ x_{31} & x_{32} & x_{33} & x_{34} & \cdot\ \cdot \\ \cdot & \cdot & \cdot & \cdot & \cdot\ \cdot \\ \cdot & \cdot & \cdot & \cdot & \cdot\ \cdot \end{array}$$

The differential equation involves x^2, and therefore Heisenberg suggested that x^2 should be capable of being represented by a similar array

$$\begin{array}{cccc} (x^2)_{11} & (x^2)_{12} & (x^2)_{13} & \cdot\ \cdot\ \cdot\ \cdot \\ (x^2)_{21} & (x^2)_{22} & (x^2)_{23} & \cdot\ \cdot\ \cdot\ \cdot \\ (x^2)_{31} & (x^2)_{32} & (x^2)_{33} & \cdot\ \cdot\ \cdot\ \cdot \\ \cdot & \cdot & \cdot & \cdot\ \cdot\ \cdot\ \cdot \\ \cdot & \cdot & \cdot & \cdot\ \cdot\ \cdot\ \cdot \end{array}$$

The question is, what is the relation between the elements $(x^2)_{mn}$ and the elements x_{mn}? Now since x_{mn} refers to the transition between the stationary states m and n, we may expect it to have the time-factor $e^{2\pi i\nu(m,\,n)t}$, where $\nu(m, n)=(W_m-W_n)/h$ is the frequency of

[1] cf. p. 204 *supra*

the spectral line associated with this transition. Guided by the principle that the time-factor of $(x^2)_{mn}$ must be the same as the time-factor of x_{mn}, Heisenberg suggested that the element $(x^2)_{mn}$ could be expressed in terms of the elements of the x-array by the equation

$$(x^2)_{mn} = \sum_r x_{mr}\, x_{rn}.$$

He went further and proposed that if x and y are two different co-ordinates, then

$$(xy)_{mn} = \sum_r x_{mr}\, y_{rn}.$$

Since this involves the equation

$$(yx)_{mn} = \sum_r y_{mr}\, x_{rn},$$

it is evident that in general the products xy and yx are represented by different arrays: *multiplication of the quantum-theoretic x and y is not commutative.*

During the month of June 1925 Heisenberg, then on holiday in the island of Heligoland, worked out the solution of the anharmonic oscillator equation according to these ideas, and was delighted to find that principles such as the conservation of energy could be fitted into his system. In the first week of July 1925,[1] he wrote a paper embodying his new theory. Being then on Professor Max Born's staff at Göttingen, he took the MS to Born and asked him to read it, at the same time asking for leave of absence for the rest of the term (which ended about 1 August), as he had been invited to lecture at the Cavendish Laboratory in Cambridge. Born, who granted the leave of absence and then read the MS., at once recognised and identified the law of multiplication which Heisenberg had introduced for the virtual-orchestra arrays: for having attended lectures on non-commutative algebras by Rosanes in Breslau and then discussed the subject with Otto Toeplitz in Göttingen, Born saw that Heisenberg's law was simply the law of multiplication of *matrices*. For the benefit of readers who are not acquainted with matrix-theory, some simple explanations may be given here. Consider any square array

$$\begin{pmatrix} a_{11} & a_{12} & a_{13} & \cdots & a_{1n} \\ a_{21} & a_{22} & a_{23} & \cdots & a_{2n} \\ \cdot & \cdot & \cdot & \cdots & \cdot \\ \cdot & \cdot & \cdot & \cdots & \cdot \\ a_{n1} & a_{n2} & a_{n3} & \cdots & a_{nn} \end{pmatrix}$$

[1] *ZS. f. P.* xxxiii (1925), p. 879 (received 29 July 1925)

formed of ordinary (real or complex) numbers $a_{11}, a_{12}, \ldots,$ which we shall call the *elements*. This array we shall call a *matrix*,[1] and we shall regard it as capable of undergoing operations such as addition and multiplication—in fact, as a kind of generalised *number*, which can be represented by a single letter. The number n of rows or columns is called the *order*. Two matrices of the same order

$$A \equiv \begin{pmatrix} a_{11} & a_{12} & \ldots & a_{1n} \\ a_{21} & a_{22} & \ldots & a_{2n} \\ \cdot & \cdot & \cdot & \cdot \\ \cdot & \cdot & \cdot & \cdot \\ a_{n1} & a_{n2} & \ldots & a_{nn} \end{pmatrix} \qquad B \equiv \begin{pmatrix} b_{11} & b_{12} & \ldots & b_{1n} \\ b_{21} & b_{22} & \ldots & b_{2n} \\ \cdot & \cdot & \cdot & \cdot \\ \cdot & \cdot & \cdot & \cdot \\ b_{n1} & b_{n2} & \ldots & b_{nn} \end{pmatrix}$$

are said to be *equal* when their elements are equal, each to each: that is

$$a_{pq} = b_{pq} \qquad (p, q = 1, 2, \ldots n).$$

The *sum* of A and B is defined to be the matrix

$$\begin{pmatrix} a_{11}+b_{11} & a_{12}+b_{12} & \ldots & a_{1n}+b_{1n} \\ a_{21}+b_{21} & a_{22}+b_{22} & \ldots & a_{2n}+b_{2n} \\ \cdot & \cdot & \cdot & \cdot \\ \cdot & \cdot & \cdot & \cdot \\ a_{n1}+b_{n1} & a_{n2}+b_{n2} & \ldots & a_{nn}+b_{nn} \end{pmatrix}$$

It is denoted by $A+B$. Evidently addition so defined is *commutative* (that is, $A+B=B+A$) and *associative* (that is, $[A+B]+C=A+[B+C]$). The *null matrix* is defined to be a matrix all of whose elements are zero: and the *unit matrix* is defined to be the matrix

$$\begin{pmatrix} 1 & 0 & 0 & \ldots & 0 \\ 0 & 1 & 0 & \ldots & 0 \\ 0 & 0 & 1 & \ldots & 0 \\ \cdot & \cdot & \cdot & \cdot & \cdot \\ \cdot & \cdot & \cdot & \cdot & \cdot \\ 0 & 0 & 0 & \ldots & 1 \end{pmatrix}$$

If k denotes an ordinary real or complex number, the matrix

$$\begin{pmatrix} ka_{11} & ka_{12} & \ldots & ka_{1n} \\ ka_{21} & ka_{22} & \ldots & ka_{2n} \\ \cdot & \cdot & \cdot & \cdot \\ \cdot & \cdot & \cdot & \cdot \\ ka_{n1} & ka_{n2} & \ldots & ka_{nn} \end{pmatrix}$$

[1] The term *matrix* was introduced in 1850 by J. J. Sylvester (Papers I, p. 145); but the theory was really founded by A. Cayley, *J. für Math.* 1 (1855), p. 282 and *Phil. Trans.* cxlviii (1858), p. 17. Under the name *linear vector operator* the same idea had been developed previously by Sir W. R. Hamilton: some of the more important theorems are given in his *Lectures on Quaternions* (1852).

which is obtained by multiplying every element of the matrix A by k, is called the *product* of A by k, and denoted by kA.

We shall now define the multiplication of two matrices. With the above matrices A and B we can associate two linear substitutions

$$\begin{cases} y_1 = a_{11}x_1 + a_{12}x_2 + \quad . \; . \; . \quad + a_{1n}x_n \\ y_2 = a_{21}x_1 + a_{22}x_2 + \quad . \; . \; . \quad + a_{2n}x_n \\ . \; . \; . \; . \; . \; . \; . \; . \; . \; . \\ . \; . \; . \; . \; . \; . \; . \; . \; . \; . \\ y_n = a_{n1}x_1 + a_{n2}x_2 + \; . \; . \; . \quad + a_{nn}x_n \end{cases}$$

and

$$\begin{cases} u_1 = b_{11}y_1 + b_{12}y_2 + \quad . \; . \; . \quad + b_{1n}y_n \\ u_2 = b_{21}y_1 + b_{22}y_2 + \quad . \; . \; . \quad + b_{2n}y_n \\ . \; . \; . \; . \; . \; . \; . \; . \; . \; . \\ . \; . \; . \; . \; . \; . \; . \; . \; . \; . \\ u_n = b_{n1}y_1 + b_{n2}y_2 + \; . \; . \; . \quad + b_{nn}y_n. \end{cases}$$

A matrix may in fact be regarded as a symbol of linear operation, representing the associated substitution.

The effect of performing these substitutions in succession is represented by the equations

$$\begin{cases} u_1 = (b_{11}a_{11} + b_{12}a_{21} + \; . \; . \; .)x_1 + (b_{11}a_{12} + b_{12}a_{22} + \; . \; . \; .)x_2 + \; . \; . \; . \; . \\ . \; . \; . \; . \; . \; . \; . \\ u_n = (b_{n1}a_{11} + b_{n2}a_{21} + \; . \; . \; .)x_1 + (b_{n1}a_{12} + b_{n2}a_{22} + \; . \; . \; .)x_2 + \; . \; . \; . \; . \end{cases}$$

This suggests that the *product* BA of the two matrices should be defined to be the matrix which has for its element in the pth row and q^{th} column the quantity

$$b_{p1}a_{1q} + b_{p2}a_{2q} + \; . \; . \; . \; + b_{pn}a_{nq},$$

so in multiplying matrices we multiply the *rows* of the first factor, element by element, into the *columns* of the second factor. Multiplication so defined is always possible and unique, and satisfies the associative law

$$\mathrm{A(BC) = (AB)C}.$$

It does not, however, satisfy the commutative law : that is, AB and BA are in general two different matrices. The distributive law connecting multiplication with addition, namely,

$$\mathrm{A(B+C) = AB + AC}$$
$$\mathrm{(B+C)A = BA + CA}$$

is always satisfied.

The derivation, with respect to the time, of a matrix whose elements are A_{mn} is the matrix whose elements are $\partial \mathrm{A}_{mn}/\partial t$. The derivative of a function of a matrix with respect to the matrix of which it is a function may be defined by the equation

$$\frac{df(\mathrm{A})}{d\mathrm{A}} = \lim_{k \to 0} \frac{1}{k}\{f(\mathrm{A}+k1) - f(\mathrm{A})\}$$

where 1 denotes the unit matrix: whence we readily see that matrix differentiation obeys the same formal laws as ordinary differentiation.

The above elementary account refers to matrices of *finite* order. The matrices which represent the co-ordinates in quantum theory are of *infinite* order, and differ in some of their properties from finite matrices: in particular, the multiplication of infinite matrices is not necessarily associative (though the associative law holds for row-finite matrices, i.e. those in which each row has only a finite number of non-zero elements).

Let us now return to Born's train of thought. He considered dynamical systems with one degree of freedom, represented classically by the differential equations

$$\frac{dq}{dt} = \frac{\partial \mathrm{H}}{\partial p}, \qquad \frac{dp}{dt} = -\frac{\partial \mathrm{H}}{\partial q},$$

where q denotes the co-ordinate, p the momentum, and H the Hamiltonian function: and following Heisenberg, he retained the form of these equations, but assumed that q and p can be represented by matrices, (representing a matrix by the element in its nth row and m^{th} column)

$$\mathbf{q} = \{q(n, m)e^{2\pi i\nu(n, m)t}\}$$

$$\mathbf{p} = \{p(n, m)e^{2\pi i\nu(n, m)t}\}$$

where $\nu(n, m)$ denotes the frequency belonging to the transition between the stationary states with the quantum numbers n and m, so

$$h\nu(n, m) = \mathrm{W}_n - \mathrm{W}_m$$

where W_n and W_m denote the values of the energy in the states.[1]

Since $\nu(n, n) = 0$, the elements in the leading diagonals of the matrices do not involve the time. This suggests the question, what is the physical meaning of these diagonal-elements? Now in the classical Fourier expansion of a variable, the constant term is equal to the average value (with respect to the time) of that variable in

[1] Since $\nu(m, n) = -\nu(n, m)$, it is natural to assume that the matrices are *Hermitean*, i.e. the elements obtained by interchanging rows and columns are conjugate complex quantities.

It is obvious that if two matrices which have equal frequencies in corresponding elements (m, n) are multiplied together, the frequency of the corresponding element (m, n) in the resulting product-matrix will again be the same as in the factor matrices.

the type of motion considered: so by the correspondence-principle we infer that *a non-temporal element x_{nn} of the matrix representing a variable x is to be interpreted as the average value of the variable x in the stationary state corresponding to the quantum number n, when all phases are equally probable.* The non-diagonal elements have not in general an equally direct physical interpretation: but it is obvious that a knowledge of the non-diagonal elements of the matrix representing a variable x would be necessary in order to calculate the diagonal elements of (say) x^2: and, moreover, we must not lose sight of the connection of the non-diagonal element x_{mn} with the transition between the stationary states m and n: this will be referred to later.

Conversely, if the derivative of a matrix with respect to the time vanishes, the matrix must be a diagonal matrix. Hence the equation of conservation of energy $(d\mathrm{H}/dt)=0$, has in general[1] the consequence, that *the matrix which represents the energy* H *is a diagonal matrix.* The diagonal element H_{nn} of this matrix represents the average value of H in the stationary state of the system for which the energy is H_n, i.e. it is H_n itself: that is, $\mathrm{H}_{nn}=\mathrm{H}_n$. So Bohr's frequency-condition may be written

$$\nu(n, m)=\frac{\mathrm{H}(n, n)-\mathrm{H}(m, m)}{h}.$$

As might be expected in matrix-calculus, the products **pq** and **qp** are not equal. Now Heisenberg had given a formula, derived originally by W. Kuhn[2] of Copenhagen and W. Thomas[3] of Breslau, which constituted a translation, into the new quantum theory, of the Wilson-Sommerfeld relation

$$\int p dq=nh:$$

and from this formula Born deduced that the terms in the leading diagonal of the matrix

$$\mathbf{pq}-\mathbf{qp}$$

must all be equal and must each have the value $h/2\pi i$. He could not, however, obtain the values of the non-diagonal elements in the matrix, though he suspected that they might all be zero.

At this stage (about the middle of July 1925) he called in the help of his other assistant, Pascual Jordan (*b.* 1902), who in a few days succeeded in showing from the canonical equations of motion

$$\frac{d\mathbf{q}}{dt}=\frac{\partial \mathbf{H}}{\partial \mathbf{p}}, \qquad \frac{d\mathbf{p}}{dt}=-\frac{\partial \mathbf{H}}{\partial \mathbf{q}}$$

[1] The case of degenerate systems requires further consideration.
[2] *ZS. f. P.* xxxiii (1925), p. 408
[3] *Naturwiss,* xiii (1925), p. 627

that the derivative of $\mathbf{pq}-\mathbf{qp}$ with respect to the time must vanish: $\mathbf{pq}-\mathbf{qp}$ must therefore be a diagonal matrix, and Born's guess was correct. *The equation thus arrived at, namely,*

$$\mathbf{pq}-\mathbf{qp}=\frac{h}{2\pi i}.\mathbf{1}$$

where **1** *denotes the unit matrix, corresponds in the matrix-mechanics to the Wilson-Sommerfeld quantum condition*

$$\int p\,dq=nh$$

of the older quantum theory. It is known as the *commutation rule.* A proof of it, much simpler than the method by which Born and Jordan established it, is as follows: from the equation

$$\mathbf{q}=\{q(n,\,m)e^{2\pi i\nu(n,\,m)t}\}$$

we have

$$\frac{\partial\mathbf{q}}{\partial t}=\frac{i}{\hbar}\{(\mathrm{W}_n-\mathrm{W}_m)q(n,\,m)e^{2\pi i\nu(n,\,m)t}\}$$

$$=\frac{i}{\hbar}(\mathbf{Hq}-\mathbf{qH})$$

or

$$\mathbf{Hq}-\mathbf{qH}=\frac{\hbar}{i}\frac{\partial\mathbf{H}}{\partial\mathbf{p}}$$

and therefore the equation

$$\mathbf{fq}-\mathbf{qf}=\frac{\hbar}{i}\frac{\partial\mathbf{f}}{\partial\mathbf{p}} \qquad (1)$$

is valid for $\mathbf{f}=\mathbf{H}$ and $\mathbf{f}=\mathbf{q}$. Now suppose that (1) is valid for any two values of $\mathbf{f}$, say $\mathbf{f}=\mathbf{a}$ and $\mathbf{f}=\mathbf{b}$: then we can show easily that it must be valid for $\mathbf{f}=\mathbf{a}+\mathbf{b}$ and for $\mathbf{f}=\mathbf{ab}$. Since all matrix functions depend on repeated additions and multiplications, we conclude that equation (1) is valid when $\mathbf{f}$ is any function of $\mathbf{H}$ and $\mathbf{q}$. Now the equation $\mathbf{H}=\mathbf{H}(\mathbf{q},\,\mathbf{p})$ determines $\mathbf{p}$ as a function of $\mathbf{H}$ and $\mathbf{q}$: and therefore equation (1) is valid for $\mathbf{f}=\mathbf{p}$: that is, we have [1]

$$\mathbf{pq}-\mathbf{qp}=\frac{\hbar}{i}\mathbf{1}$$

It may be remarked that this relation could not hold if $\mathbf{p}$ and $\mathbf{q}$ were finite matrices, since in that case the sum of the diagonal elements of $\mathbf{pq}-\mathbf{qp}$ would be zero.

Born and Jordan published their discoveries in a paper which was received by the editor of the *Zeitschrift für Physik* on 27 September.[2]

[1] On the physical meaning of the commutation-rule, cf. H. A. Kramers, *Physika*, v (1925), p. 369

[2] *ZS. f. P.* xxxiv (1925), p. 858

In it they applied the theory to the case of the harmonic oscillator. For the harmonic oscillator the Hamiltonian function is

$$H = \frac{1}{2m}(p^2 + m^2\omega^2 q^2) \tag{1}$$

so the matrix equations of motion are

$$\frac{d\mathbf{q}}{dt} = \frac{\partial \mathbf{H}}{\partial \mathbf{p}} = \frac{1}{m}\mathbf{p}, \qquad \frac{d\mathbf{p}}{dt} = -\frac{\partial \mathbf{H}}{\partial \mathbf{q}} = -m\omega^2\mathbf{q}. \tag{2}$$

Hence

$$\frac{d^2\mathbf{q}}{dt^2} + \omega^2\mathbf{q} = 0$$

of which the solution is

$$\mathbf{q} = \mathbf{A}e^{i\omega t} + \mathbf{B}e^{-i\omega t} \tag{3}$$

where **A** and **B** are matrices not involving the time.

Now in a matrix which has the time-factor $e^{i\omega t}$, in the system of reference in which **H** is represented by a diagonal-matrix, the only possible non-zero elements are those to which the time-factor $e^{i\omega t}$ belongs, i.e. those in the sub-diagonal immediately below the principal diagonal. Thus

$$\mathbf{A} = \begin{pmatrix} 0 & 0 & 0 & 0 & \cdots \\ \alpha_1 & 0 & 0 & 0 & \cdots \\ 0 & \alpha_2 & 0 & 0 & \cdots \\ 0 & 0 & \alpha_3 & 0 & \cdots \\ \cdot & \cdot & \cdot & \cdot & \cdot \\ \cdot & \cdot & \cdot & \cdot & \cdot \end{pmatrix} \text{ and } \mathbf{B} = \begin{pmatrix} 0 & \beta_1 & 0 & 0 & \cdots \\ 0 & 0 & \beta_2 & 0 & \cdots \\ 0 & 0 & 0 & \beta_3 & \cdots \\ \cdot & \cdot & \cdot & \cdot & \cdot \\ \cdot & \cdot & \cdot & \cdot & \cdot \end{pmatrix} \tag{4}$$

From (2) and (3)

$$\mathbf{p} = m\frac{d\mathbf{q}}{dt} = im\omega(\mathbf{A}e^{i\omega t} - \mathbf{B}e^{-i\omega t}). \tag{5}$$

From (3) and (5),

$$\mathbf{pq} - \mathbf{qp} = 2im\omega(\mathbf{AB} - \mathbf{BA})$$

or

$$\frac{\hbar}{i} = 2im\omega \begin{pmatrix} -\alpha_1\beta_1 & 0 & 0 & \cdots \\ 0 & \alpha_1\beta_1 - \alpha_2\beta_2 & 0 & \cdots \\ 0 & 0 & \alpha_2\beta_2 - \alpha_3\beta_3 & \cdots \\ \cdot & \cdot & \cdot & \cdot \\ \cdot & \cdot & \cdot & \cdot \end{pmatrix}$$

whence

$$\alpha_1\beta_1 = \frac{\hbar}{2m\omega}, \quad \alpha_2\beta_2 = \frac{2\hbar}{2m\omega}, \quad \alpha_3\beta_3 = \frac{3\hbar}{2m\omega}, \quad \text{etc.} \tag{6}$$

Also

$$\mathbf{H} = \frac{1}{2m}(\mathbf{p}^2 + m^2\omega^2\mathbf{q}^2) = m\omega^2(\mathbf{A}\,\mathbf{B} + \mathbf{B}\,\mathbf{A})$$

$$= m\omega^2 \begin{pmatrix} \alpha_1\beta_1 & 0 & 0 & 0 \ldots \\ 0 & \alpha_1\beta_1 + \alpha_2\beta_2 & 0 & 0 \ldots \\ 0 & 0 & \alpha_2\beta_2 + \alpha_3\beta_3 & 0 \ldots \\ \ldots & \ldots & \ldots & \ldots \\ \ldots & \ldots & \ldots & \ldots \end{pmatrix}$$

$$= \frac{\omega\hbar}{2} \begin{pmatrix} 1 & 0 & 0 & 0 \ldots \\ 0 & 3 & 0 & 0 \ldots \\ 0 & 0 & 5 & 0 \ldots \\ \ldots & \ldots & \ldots & \ldots \\ \ldots & \ldots & \ldots & \ldots \end{pmatrix} \tag{7}$$

so **H** is a diagonal matrix, as it should be.

Moreover, **q** must be a Hermitean matrix, so α_r and β_r must be complex-conjugates : and therefore from (6)

$$\alpha_1 = \left(\frac{\hbar}{2m\omega}\right)^{\frac{1}{2}} e^{i\delta_1}, \quad \alpha_2 = \left(\frac{2\hbar}{2m\omega}\right)^{\frac{1}{2}} e^{i\delta_2}, \quad \alpha_3 = \left(\frac{3\hbar}{2m\omega}\right)^{\frac{1}{2}} e^{i\delta_3}, \ldots$$

$$\beta_1 = \left(\frac{\hbar}{2m\omega}\right)^{\frac{1}{2}} e^{-i\delta_1}, \quad \beta_2 = \left(\frac{2\hbar}{2m\omega}\right)^{\frac{1}{2}} e^{-i\delta_2}, \quad \beta_3 = \left(\frac{3\hbar}{2m\omega}\right)^{\frac{1}{2}} e^{-i\delta_3}, \ldots$$

where δ_1 δ_2, δ_3, . . . are arbitrary real numbers. If we write $\delta_r = \gamma_r - \pi/2$, from (3), (4) and (8) we have

$$\mathbf{q} = i\left(\frac{\hbar}{2m\omega}\right)^{\frac{1}{2}} \begin{pmatrix} 0 & e^{-i(\omega t+\gamma_1)} & 0 & 0 & \ldots \\ e^{-i(\omega t+\gamma_1)} & 0 & \sqrt{2}.e^{-i(\omega t+\gamma_2)} & 0 & \ldots \\ 0 & -\sqrt{2}e^{i(\omega t+\gamma_2)} & 0 & \sqrt{3}e^{-i(\omega t+\gamma_3)} & \ldots \\ 0 & 0 & -\sqrt{3}e^{i(\omega t+\gamma_3)} & 0 & \ldots \\ \ldots & \ldots & \ldots & \ldots & \ldots \\ \ldots & \ldots & \ldots & \ldots & \ldots \end{pmatrix}$$

and then

$$\mathbf{p} = \left(\frac{m\omega\hbar}{2}\right)^{\frac{1}{2}} \begin{pmatrix} 0 & e^{-i(\omega t+\gamma_1)} & 0 & 0 & \ldots \\ e^{i(\omega t+\gamma_1)} & 0 & \sqrt{2}e^{-i(\omega t+\gamma_2)} & 0 & \ldots \\ 0 & \sqrt{2}e^{i(\omega t+\gamma_2)} & 0 & \sqrt{3}e^{-i(\omega t+\gamma_3)} & \ldots \\ 0 & 0 & \sqrt{3}e^{i(\omega t+\gamma_3)} & 0 & \ldots \\ \ldots & \ldots & \ldots & \ldots & \ldots \\ \ldots & \ldots & \ldots & \ldots & \ldots \end{pmatrix}$$

Thus *the matrices which represent the co-ordinate and the momentum in the problem of the harmonic oscillator are determined.* Equation (7) shows that *the values of the energy, corresponding to the stationary states of the oscillator*, are

$$\frac{\omega\hbar}{2}, \quad \frac{3\omega\hbar}{2}, \quad \frac{5\omega\hbar}{2} \ldots$$

or in general

$$(n+\tfrac{1}{2})\omega\hbar.$$

The squares of the moduli of the elements of the matrix which represents the electric moment of an atom are in general the measure of its *transition-probabilities*. A connection is thus set up with Einstein's coefficients A^n_m and with Planck's theory of radiation.

Born and Jordan's paper represented a great advance: it contained the formulation of matrix mechanics, the discovery of the commutation law, some simple applications to the harmonic and anharmonic oscillator and (in its last section) a discussion of the quantification of the electromagnetic field.

When the paper had been sent off to the *Zeitschrift für Physik*, Born went for a holiday with his family to the Engadine. On his return in September to Göttingen, there began a hectic time of collaboration with Jordan, and also by correspondence with Heisenberg, who was now in Copenhagen. Born had received and accepted an invitation to lecture during the winter at the Massachusetts Institute of Technology, and was bound to leave at the end of October; but the joint paper was finished in time before his departure, and was received by the editor of the *Zeitschrift für Physik* on 16 November.[1]

The first important result in it was a general method for the solution of quantum-mechanical problems, analogous to the general Hamiltonian theory of classical dynamics. A *canonical transformation* of the variables $\mathbf{p}, \mathbf{q}$ to new variables $\mathbf{P}, \mathbf{Q}$ was defined to be a transformation for which

$$\mathbf{pq}-\mathbf{qp}=\mathbf{PQ}-\mathbf{QP}=-i\hbar\,;$$

when this equation is satisfied then the canonical equations

$$\frac{d\mathbf{q}}{dt}=\frac{\partial\mathbf{H}}{\partial\mathbf{p}}, \qquad \frac{d\mathbf{p}}{dt}=-\frac{\partial\mathbf{H}}{\partial\mathbf{q}}$$

transform into

$$\frac{d\mathbf{Q}}{dt}=\frac{\partial\mathbf{H}}{\partial\mathbf{P}}, \qquad \frac{d\mathbf{P}}{dt}=-\frac{\partial\mathbf{H}}{\partial\mathbf{Q}}.$$

[1] M. Born, W. Heisenberg and P. Jordan, *ZS. f. P.* xxxv (1926), p. 557

A general transformation which satisfies this condition is

$$\mathbf{P} = \mathbf{SpS}^{-1}$$
$$\mathbf{Q} = \mathbf{SqS}^{-1}$$

where **S** is an arbitrary quantum-theoretic quantity. We then have for any function $f(\mathbf{P}, \mathbf{Q})$.

$$f(\mathbf{P}, \mathbf{Q}) = \mathbf{S}f(\mathbf{p}, \mathbf{q})\mathbf{S}^{-1}.$$

The importance of canonical transformations depends on the following theorem : *matrices* **p** *and* **q** *which satisfy the commutation-rule*

$$\mathbf{pq} - \mathbf{pq} = -i\hbar$$

and which, when substituted in $\mathbf{H}(\mathbf{p}, \mathbf{q})$ *make it a diagonal matrix, represent solutions of the equations*

$$\frac{d\mathbf{q}}{dt} = \frac{\partial \mathbf{H}}{\partial \mathbf{p}}, \qquad \frac{d\mathbf{p}}{dt} = -\frac{\partial \mathbf{H}}{\partial \mathbf{q}}.$$

Therefore if we take any pair of matrices $\mathbf{p}_0$, $\mathbf{q}_0$, which satisfy the commutation-rule (for instance, we might take $\mathbf{p}_0$, $\mathbf{q}_0$, to be the solution of the problem of the harmonic oscillator), then we can reduce the problem of the integration of the canonical equations for a Hamiltonian $\mathbf{H}(\mathbf{p}, \mathbf{q})$ to the following problem : *To determine a matrix* **S** *such that when*

$$\mathbf{p} = \mathbf{S}\mathbf{p}_0\mathbf{S}^{-1}. \qquad \mathbf{q} = \mathbf{S}\mathbf{q}_0\mathbf{S}^{-1},$$

the function

$$\mathbf{H}(\mathbf{p}, \mathbf{q}) \equiv \mathbf{SH}(\mathbf{p}_0, \mathbf{q}_0)\mathbf{S}^{-1}$$

is a diagonal matrix. This last equation is analogous to the Hamilton's partial differential equation of classical dynamics, and **S** corresponds in some measure to the Action-function.

The next question taken up in the memoir of the three authors was the theory of perturbations. Let a problem be defined by a Hamiltonian function

$$\mathbf{H} = \mathbf{H}_0(\mathbf{p}, \mathbf{q}) + \lambda\mathbf{H}_1(\mathbf{p}, \mathbf{q}) + \lambda^2\mathbf{H}_2(\mathbf{p}, \mathbf{q}) + \ldots$$

and suppose that the solution is known of the problem defined by the Hamiltonian function $\mathbf{H}_0(\mathbf{p}, \mathbf{q})$, so that matrices $\mathbf{p}_0$, $\mathbf{q}_0$ are known which satisfy the commutation-rule and make $\mathbf{H}_0(\mathbf{p}_0, \mathbf{q}_0)$ a diagonal matrix. Then it is required to find a transformation-matrix **S** such that if

$$\mathbf{p} = \mathbf{S}\mathbf{p}_0\mathbf{S}^{-1} \quad \text{and} \quad \mathbf{q} = \mathbf{S}\mathbf{q}_0\mathbf{S}^{-1}$$

then

$$\mathbf{H}(\mathbf{p}, \mathbf{q}) = \mathbf{SH}(\mathbf{p}_0, \mathbf{q}_0)\mathbf{S}^{-1}$$

will be a diagonal-matrix. It was shown how to solve this problem by successive approximations, and one of the formulae of Kramers' theory of dispersion was derived.

The theory was then extended to systems with more than one degree of freedom. If the Hamiltonian equations are

$$\frac{d\mathbf{q}_k}{dt}=\frac{\partial \mathbf{H}}{\partial \mathbf{p}_k}, \qquad \frac{d\mathbf{p}_k}{dt}=-\frac{\partial \mathbf{H}}{\partial \mathbf{q}_k},$$

then the commutation-rules are

$$\begin{cases} \mathbf{p}_k\mathbf{q}_l-\mathbf{q}_l\mathbf{p}_k=-i\hbar\delta_{kl}, \text{ where } \delta_{kl}=1 \text{ or } 0 \text{ according as } k=l \text{ or } k\neq l \\ \mathbf{p}_k\mathbf{p}_l-\mathbf{p}_l\mathbf{p}_k=0 \\ \mathbf{q}_k\mathbf{q}_l-\mathbf{q}_l\mathbf{q}_k=0. \end{cases}$$

Among those who listened to Heisenberg's lectures at Cambridge in the summer of 1925 was a young research student named Paul Adrien Maurice Dirac (*b.* 1902), who by a different approach arrived at a theory essentially equivalent to that devised by Born and Jordan. On 7 November he sent to the Royal Society a paper[1] in which Heisenberg's ideas were developed in an original way.

Dirac investigated the form of a quantum operation (denote it by d/dv) that satisfies the laws

$$\frac{d}{dv}\cdot(x+y)=\frac{d}{dv}\cdot x+\frac{d}{dv}\cdot y$$

and

$$\frac{d}{dv}(xy)=\left(\frac{d}{dv}\cdot x\right)y+x\cdot\left(\frac{d}{dv}\cdot y\right).^{2}$$

It was found that the most general operation satisfying these laws is

$$\frac{d}{dv}\cdot x=xy-yx$$

where y is some other quantum variable. By considering the limit, when for large quantum numbers the quantum theory passes into the classical theory, Dirac showed that the corresponding expression in classical physics is

$$i\hbar\sum_r\left(\frac{\partial x}{\partial q_r}\frac{\partial y}{\partial p_r}-\frac{\partial y}{\partial q_r}\frac{\partial x}{\partial p_r}\right)$$

where the p's and q's are a set of canonical variables of the system. This is $i\hbar$ multiplied by the well-known *Poisson-bracket expression*[3]

[1] *Proc. R.S.*(A), cix (1925), p. 642

[2] Note that the order of x and y is preserved in the second equation.

[3] cf. Whittaker, *Analytical Dynamics*, § 130

of the functions x and y. Thus, if in quantum theory we define a quantity $[x, y]$ by the equation

$$xy - yx = i\hbar[x, y],$$

then $[x, y]$ is analogous to the Poisson-bracket expressions of classical theory.

Now the Hamiltonian equations of classical theory,

$$\frac{dq_r}{dt} = \frac{\partial H}{\partial p_r}, \qquad \frac{dp_r}{dt} = -\frac{\partial H}{\partial q_r} \qquad (r = 1, 2, \ldots n)$$

may be written in terms of the classical Poisson-brackets (x, y)

$$\frac{dq_r}{dt} = (q_r, H), \qquad \frac{dp_r}{dt} = (p_r, H).$$

The fundamental postulate on which Dirac built his theory was, that *the whole of classical dynamics, so far as it can be expressed in terms of Poisson-brackets instead of derivatives, may be taken over immediately into quantum theory*. Thus, for any quantum-theoretic quantity $\mathbf{x}$ the equation of motion is

$$\frac{d\mathbf{x}}{dt} = [\mathbf{x}, \mathbf{H}] = \frac{1}{i\hbar}(\mathbf{xH} - \mathbf{Hx}).$$

Moreover, if

$$\mathbf{p}_r, \mathbf{q}_r \qquad (r = 1, 2, \ldots n)$$

are any set of canonical variables of the classical system, then we have for the classical Poisson-brackets the values

$$(q_r, q_s) = 0, \qquad (p_r, p_s) = 0, \qquad (q_r, p_s) = \delta_{rs}$$

and therefore in quantum theory we must have

$$\mathbf{q}_r\mathbf{q}_s - \mathbf{q}_s\mathbf{q}_r = 0, \qquad \mathbf{p}_r\mathbf{p}_s - \mathbf{p}_s\mathbf{p}_r = 0, \qquad \mathbf{q}_r\mathbf{p}_s - \mathbf{p}_s\mathbf{q}_r = i\hbar\delta_{rs}\mathbf{1}.$$

Thus Dirac arrived at all the fundamental equations of Heisenberg, Born and Jordan, without explicitly introducing matrices. He introduced the name *q-numbers* for the quantum-mechanical quantities whose multiplication is not in general commutative, and *c-numbers* for ordinary numbers.

In a second paper[1] he applied the theory to the hydrogen atom, and obtained the formula for the Balmer spectrum. This was done at about the same time by Pauli,[2] who observed, moreover, the automatic disappearance of certain difficulties, which had been

[1] *Proc. R.S.*(A), cx (1926), p. 561
[2] *ZS. f. P.* xxxvi (1926), p. 336

found [1] to occur in examining the simultaneous action on the hydrogen atom of crossed electric and magnetic fields. Pauli also gave a matrix-mechanical derivation of the Stark effect.

In the same year Heisenberg and Jordan [2] applied the theory (using the hypothesis of the rotating electron) to the problem of the anomalous Zeeman effect, and obtained Landé's *g*-formula.[3] The fine-structure of spectral doublets in the absence of an external field was also completely explained. Dirac [4] treated the Compton effect by quantum mechanics, and obtained formulae for the angular distribution of the recoil electrons and the scattered radiation which agreed completely with experiment.

Meanwhile Born had been in America since November 1925, and had there met Norbert Wiener (*b.* 1894), who worked with him at the problems of continuous spectra and aperiodic phenomena. In February 1926 they wrote a joint paper [5] on the formulation of the laws of quantification. As they pointed out, the representation of the quantum laws by matrices incurs serious difficulties in the case of aperiodic phenomena. For example, in uniform rectilinear motion, since no periods are present, the matrix which represents the co-ordinate can have no element outside the principal diagonal. They therefore tried to generalise the quantum rules in such a way as to cover these cases, and this they effected by developing a theory of *operators*, representing a co-ordinate by a linear operator instead of by a matrix ; and they enunciated the general principle that *to every physical quantity there corresponds an operator*. The case of unaccelerated motion in one dimension, which has no periodic components, was shown to be as amenable to their methods as a periodic motion.

The development of matrix mechanics in the year following Heisenberg's first paper was amazingly rapid : and, only eight months from the date of his discovery, it was supplemented by a parallel theory, which will be described in the next chapter.[6]

[1] O. Klein, *ZS. f. P.* xxii (1924), p. 109 ; W. Lenz, *ZS. f. P.* xxiv (1924), p. 197
[2] *ZS. f. P.* xxxvii (1926), p. 263
[3] On the Zeeman effect, cf. also S. Goudsmit and G. E. Uhlenbeck, *ZS. f. P.* xxxv (1926), p. 618 ; L. H. Thomas, *Phil. Mag.* iii (1927), p. 13.
[4] *Proc. R.S.*(A), cxi (1926), p. 405
[5] *J. Math. Phys. Mass. Inst. Tech.* v (1926), p. 84 ; *ZS. f. P.* xxxvi (1926), p. 174
[6] On the state of matrix-mechanics at the end of 1926, cf. Dirac, *Physical interpretation of quantum dynamics, Proc. R.S.*(A), cxiii (1 Jan. 1927), p. 621

Chapter IX

THE DISCOVERY OF WAVE-MECHANICS

IN 1926 a new movement began, which developed from de Broglie's principle,[1] that with any particle of (relativistic) energy E and momentum (p_x, p_y, p_z) there is associated a wave represented by a wave-function

$$\psi = e^{(i/\hbar)(Et - p_x x - p_y y - p_z z)}.$$

De Broglie had not hitherto extended his theory to the point of introducing a medium—a kind of aether—whose vibrations might be regarded as constituting the wave, and whose behaviour could be specified by partial differential equations. A step equivalent to this was now taken.

With the above value of ψ, we have

$$\frac{\partial^2\psi}{\partial x^2} = -\frac{p_x^2}{\hbar^2}\psi,$$

so

$$\frac{\partial^2\psi}{\partial x^2} + \frac{\partial^2\psi}{\partial y^2} + \frac{\partial^2\psi}{\partial z^2} = -\frac{p^2}{\hbar^2}\psi$$

and

$$\frac{1}{c^2}\frac{\partial^2\psi}{\partial t^2} = -\frac{E^2}{c^2\hbar^2}\psi,$$

so

$$\frac{1}{c^2}\frac{\partial^2\psi}{\partial t^2} - \frac{\partial^2\psi}{\partial x^2} - \frac{\partial^2\psi}{\partial y^2} - \frac{\partial^2\psi}{\partial z^2} = -\left(\frac{E^2}{c^2} - p^2\right)\frac{\psi}{\hbar^2} = -\frac{m^2c^2}{\hbar^2}\psi,$$

where m denotes the mass of the particle. Thus *the de Broglie wave-function satisfies the partial differential equation*

$$\frac{1}{c^2}\frac{\partial^2\psi}{\partial t^2} = \frac{\partial^2\psi}{\partial x^2} + \frac{\partial^2\psi}{\partial y^2} + \frac{\partial^2\psi}{\partial z^2} - \frac{m^2c^2}{\hbar^2}\psi.$$

This equation was discovered by L. de Broglie[2] and others.[3] It is satisfied by all possible de Broglie waves belonging to the particle, but it does not specify their passage into each other as the particle moves—that is, it yields no information as to the variation from moment to moment of the energy and momentum of

[1] cf. p. 214

[2] *Comptes Rendus*, clxxxiii (26 July 1926), p. 272

[3] E. Schrödinger, *Ann. d. Phys.* lxxxi (1926), p. 109; O. Klein, *ZS. f. P.* xxxvii (1926), p. 895, *at* p. 904; cf. also W. Gordon, *ZS. f. P.* xl (1926), p. 117

the particle. What was really wanted was a partial differential equation for the de Broglie waves associated with a particle, which would yield the equations of motion of the particle by a limiting process similar to that by which geometrical optics is derived from physical optics : a partial differential equation, in fact, whose *filiform solutions* (i.e. solutions which are zero everywhere except at points very close to some *curve* in space-time) are the trajectories of the particle. Such an equation can be obtained, at any rate, in the non-relativistic approximation. For from the above equations we have at once

$$\frac{\partial^2\psi}{\partial x^2}+\frac{\partial^2\psi}{\partial y^2}+\frac{\partial^2\psi}{\partial z^2}=\frac{p^2}{E^2}\,\frac{\partial^2\psi}{\partial t^2}.$$

Substituting the values for p and E in terms of the velocity v of the particle, this becomes

$$\frac{\partial^2\psi}{\partial x^2}+\frac{\partial^2\psi}{\partial y^2}+\frac{\partial^2\psi}{\partial z^2}=\frac{v^2}{c^4}\,\frac{\partial^2\psi}{\partial t^2}. \qquad \text{(A)}$$

This equation is not relativistically invariant, so we shall restrict ourselves to the non-relativistic case of a particle m moving in a field of potential V, whose equation of energy is

$$\tfrac{1}{2}mv^2+V=\epsilon$$

where ϵ is its total non-relativistic energy. The equation (A) has filiform solutions, each of which is[1] a null geodesic of the metric for which the square of the element of interval is

$$(c^4/v^2)\,(dt)^2-(dx)^2-(dy)^2-(dz)^2.$$

It can be shown that the null geodesics of this metric are the curves which satisfy the differential equations

$$m\frac{d^2x}{dt^2}=-\frac{\partial V}{\partial x} \qquad \text{and similar equations in } y \text{ and } z\,;$$

but these are the ordinary Newtonian equations of motion of the particle. Thus *the filiform solutions of equation* (A) *are precisely the trajectories of the particle in the given potential field.* The solution ψ of equation (A) is therefore the wave-function we are seeking.

The time factor in ψ is $\exp\{(i/\hbar)Et\}$, so if ψ now denotes the wave-function deprived of its time-factor, we have from (A)

$$\frac{\partial^2\psi}{\partial x^2}+\frac{\partial^2\psi}{\partial y^2}+\frac{\partial^2\psi}{\partial z^2}=-\frac{v^2}{c^4}\,\frac{E^2}{\hbar^2}\psi$$
$$=-\frac{m^2v^2}{\hbar^2(1-v^2/c^2)}\psi.$$

[1] E. T. Whittaker, *Proc. Camb. Phil. Soc.* xxiv (1928), p. 32

Since we are considering only the non-relativistic approximation, we can replace the factor $(1-v^2/c^2)$ in the denominator by unity, and thus obtain

$$\frac{\partial^2\psi}{\partial x^2}+\frac{\partial^2\psi}{\partial y^2}+\frac{\partial^2\psi}{\partial z^2}=-\frac{m^2v^2}{\hbar^2}\psi$$

or

$$\frac{\partial^2\psi}{\partial x^2}+\frac{\partial^2\psi}{\partial y^2}+\frac{\partial^2\psi}{\partial z^2}+\frac{2m}{\hbar^2}(\epsilon-V)\psi=0.$$

This equation, which was published by Erwin Schrödinger[1] in March 1926, gave the first impetus to the study of *wave-mechanics*. Schrödinger's own approach, which was different from that given above, laid stress on a connection with the theory of Hamilton's Principal Function in dynamics. This will now be considered.

He considered a particle of mass m with momentum p and total energy ϵ in a field of force of potential $V(x, y, z)$, so that the (non-relativist) equation of energy is

$$\frac{1}{2m}p^2+V=\epsilon, \quad \text{or} \quad p=\sqrt{\{2m(\epsilon-V)\}}.$$

If we associate with this moving particle a frequency ν given by $\epsilon=h\nu$, and a de Broglie wave-length λ given by $\lambda=h/p$, then the phase-velocity of the de Broglie wave is

$$\varpi=\nu\lambda=\frac{\epsilon}{h}\cdot\frac{h}{p}=\frac{\epsilon}{\sqrt{\{2m(\epsilon-V)\}}}.$$

Take any *phase-surface*, or surface of constant phase, at the instant $t=0$; and suppose that the equation of the phase-surface at the instant t, which has been derived (as in Huygens' Principle) by wave-propagation with the phase-velocity ϖ from this original phase-surface, has the equation

$$\tau(x, y, z)=t.$$

Then by elementary analytical geometry we have

$$\left(\frac{\partial\tau}{\partial x}\right)^2+\left(\frac{\partial\tau}{\partial y}\right)^2+\left(\frac{\partial\tau}{\partial z}\right)^2=\frac{1}{\varpi^2}$$

and therefore

$$\left(\frac{\partial\tau}{\partial x}\right)^2+\left(\frac{\partial\tau}{\partial y}\right)^2+\left(\frac{\partial\tau}{\partial z}\right)^2=\frac{2m(\epsilon-V)}{\epsilon^2}. \tag{1}$$

This equation can, however, be obtained in a very different way.

[1] *Ann. d. Phys.*(4) lxxix (1926), pp. 361, 489

The Hamilton's partial differential equation associated with the particle[1] is

$$\frac{\partial W}{\partial t}+H\left(x, y, z, \frac{\partial W}{\partial x}, \frac{\partial W}{\partial y}, \frac{\partial W}{\partial z}\right)=0,$$

where H is the Hamiltonian function for the particle, namely,

$$H=\frac{1}{2m}(p_x^2+p_y^2+p_z^2)+V(x, y, z)$$

and W is *Hamilton's Principal Function.* The Hamilton's equation is therefore in this case

$$\frac{\partial W}{\partial t}+\frac{1}{2m}\left\{\left(\frac{\partial W}{\partial x}\right)^2+\left(\frac{\partial W}{\partial y}\right)^2+\left(\frac{\partial W}{\partial z}\right)^2\right\}+V(x, y, z)=0.$$

The Principal Function may be written

$$W=-\epsilon t+S(x, y, z)$$

where ϵ denotes the total energy and S is *Hamilton's Characteristic Function.* The equation for S now becomes

$$-\epsilon+\frac{1}{2m}\left\{\left(\frac{\partial S}{\partial x}\right)^2+\left(\frac{\partial S}{\partial y}\right)^2+\left(\frac{\partial S}{\partial z}\right)^2\right\}+V(x, y, z)=0$$

or

$$\left(\frac{\partial S}{\partial x}\right)^2+\left(\frac{\partial S}{\partial y}\right)^2+\left(\frac{\partial S}{\partial z}\right)^2=2m(\epsilon-V).$$

Comparing this with equation (1), we see that *the equation of the phase-surfaces of the de Broglie waves associated with a particle, namely,*

$$\tau(x, y, z)=t$$

is obtained by equating to zero the Hamilton's Principal Function of the particle. Thus *the theory of Hamilton's Principal Function in Dynamics corresponds to Huygens' Principle in Optics.*

This investigation so far belongs to what may be called the 'geometrical optics' of the de Broglie waves; a similar equation exists in ordinary optics, the surfaces $\tau(x, y, z)=$ constant being the wave-fronts, whose normals are the 'rays.' Schrödinger now put forward the idea that the failure of classical physics to account for quantum phenomena is analogous to the failure of geometrical optics to account for interference and diffraction: and he proposed to create in connection with de Broglie waves a theory analogous to Physical Optics.

Considering the stationary states of an atom, we have seen[2] that the de Broglie wave associated with an electron in an atom

[1] cf. Whittaker, *Analytical Dynamics*, § 142

[2] cf. p. 216 *supra*

returns to the same phase when the electron completes one revolution of an orbit belonging to a stationary state. Therefore at any one point of space the de Broglie disturbance in a stationary state is purely periodic, with the frequency $\nu=\epsilon/h$, where ϵ is the energy of the stationary state; and it can be represented by a *wave-function* of the form

$$\psi = e^{(i\epsilon/\hbar)[-t+\tau(x,\,y,\,z)]}.$$

Differentiating this twice, we have

$$\frac{\partial^2\psi}{\partial x^2} = -\frac{\epsilon^2}{\hbar^2}\left(\frac{\partial\tau}{\partial x}\right)^2\psi + \frac{i\epsilon}{\hbar}\frac{\partial^2\tau}{\partial x^2}\psi,$$

which gives, using (1),

$$\frac{\partial^2\psi}{\partial x^2}+\frac{\partial^2\psi}{\partial y^2}+\frac{\partial^2\psi}{\partial z^2} = -\frac{2m(\epsilon-V)}{\hbar^2}\psi + \frac{i\epsilon}{\hbar}\psi\left(\frac{\partial^2\tau}{\partial x^2}+\frac{\partial^2\tau}{\partial y^2}+\frac{\partial^2\tau}{\partial z^2}\right).$$

The ratio of the second term to the first term on the right-hand side of this equation is excessively small if $\hbar$ is excessively small, as it is: so we may neglect the second term, and write

$$\frac{\partial^2\psi}{\partial x^2}+\frac{\partial^2\psi}{\partial y^2}+\frac{\partial^2\psi}{\partial z^2} = -\frac{2m(\epsilon-V)}{\hbar^2}\psi,$$

which is again *Schrödinger's equation for the wave-function* ψ *of the particle.*

The potential function V which occurs in this equation will possess singularities at certain points or at infinity, and these points (with others) will in general be singular points of the solution ψ of the differential equation. But if ψ is to represent the de Broglie wave of a stationary state, it must be free from singularities; and therefore Schrödinger laid down the condition that *the solution of the wave-equation corresponding to a stationary state must be one-valued, finite and free from singularities even at the singularities of* $V(x, y, z)$.[1]

Now it is known that *the partial differential equation possesses a solution of this character only for certain special values of the constant* ϵ, *which are known as the proper-values.*[2] These proper-values will be the only values of the energy ϵ which the atom can have when it is in a stationary state; and thus was justified the title which Schrödinger gave to his first paper, *Quantification as a Problem of Proper-values.*[3]

[1] The conditions laid down in Schrödinger's earlier papers were unnecessarily stringent: on this, cf. G. Jaffé, *ZS. f. P.* lxvi (1930), p. 770; R. H. Langer and N. Rosen, *Phys. Rev.* xxxvii (1931), p. 658; E. H. Kennard, *Nature*, cxxvii (1931), p. 892.

[2] The undesirable hybrid word *eigenvalues* has often been used, but *proper-values*, which is in every way preferable, is used in the English translation of Schrödinger's *Collected Papers on Wave Mechanics* (Blackie and Son, 1928).

[3] 'In the winter of 1926, Born and Jordan having just announced a new development in quantum mechanics, I found more than twenty Americans in Göttingen at this fount of quantum wisdom. A year later they were at Zürich, with Schrödinger. A couple of years later, Heisenberg at Leipzig, and then Dirac at Cambridge, held the Elijah mantle of quantum theory.' (K. T. Compton, *Nature*, cxxxix (1937), p. 222.)

The first problem that Schrödinger investigated by his new method was that of the hydrogen atom, consisting of a proton and an electron ; denoting their distance apart by r, so that the potential energy is $-e^2/r$, and neglecting the motion of the nucleus, the wave-equation is (now writing E for the total energy)

$$\frac{\partial^2\psi}{\partial x^2}+\frac{\partial^2\psi}{\partial y^2}+\frac{\partial^2\psi}{\partial z^2}+\frac{2m}{\hbar^2}\left(\mathrm{E}+\frac{e^2}{r}\right)\psi=0.$$

We want the values of E for which solutions of this equation exist that are one-valued and everywhere finite.[1] To find these, we introduce spherical-polar co-ordinates defined by

$$x=r\sin\theta\cos\phi,\qquad y=r\sin\theta\sin\phi,\qquad z=r\cos\theta,$$

and try to obtain a solution of the form

$$\psi=\mathrm{RY}$$

where R is a function of r alone, and Y is a function of θ and ϕ only. The wave-equation now becomes

$$\frac{1}{r^2\mathrm{R}}\frac{d}{dr}\left(r^2\frac{d\mathrm{R}}{dr}\right)+\frac{1}{r^2\mathrm{Y}}\left[\frac{1}{\sin\theta}\frac{\partial}{\partial\theta}\left(\sin\theta\frac{\partial\mathrm{Y}}{\partial\theta}\right)+\frac{1}{\sin^2\theta}\frac{\partial^2\mathrm{Y}}{\partial\phi^2}\right]+\frac{2m}{\hbar^2}\left(\mathrm{E}+\frac{e^2}{r}\right)=0,$$

and this may be broken up into the two equations :

$$\frac{1}{\mathrm{R}}\frac{d}{dr}\left(r^2\frac{d\mathrm{R}}{dr}\right)+\frac{2m}{\hbar^2}(\mathrm{E}r^2+e^2r)=\mathrm{C}$$

$$\frac{1}{\sin\theta}\frac{\partial}{\partial\theta}\left(\sin\theta\frac{\partial\mathrm{Y}}{\partial\theta}\right)+\frac{1}{\sin^2\theta}\frac{\partial^2\mathrm{Y}}{\partial\phi^2}+\mathrm{CY}=0.$$

The latter is the well-known equation of surface harmonics,[2] and the requirements of one-valuedness, finiteness and continuity, make it necessary that C should have the value $n(n+1)$, where n is zero or a positive whole number. The surface-harmonic $\mathrm{Y}(\theta, \phi)$ is then a sum of terms of the form

$$\mathrm{P}_n{}^{\mu}(\cos\theta)e^{\pm i\mu\phi}$$

where μ is one of the numbers 0, 1, 2, . . . n, and where $\mathrm{P}_n{}^{\mu}(\cos\theta)$

[1] The problem was first solved by Schrödinger, *Ann. d. Phys.* lxxix (1926), p. 361 ; lxxx (1926), p. 437. For a rigorous proof of the completeness of the set of proper functions, cf. T. H. Gronwall, *Annals of Math.* xxxii (1931), p. 47. It will of course be understood that the solution given above was improved later when relativity, electron-spin, etc. were taken into account.

[2] cf. Whittaker and Watson, *Modern Analysis*, § 18·31

is the Associated Legendre Function.[1] The differential equation for R now becomes

$$\frac{d}{dr}\left(r^2\frac{dR}{dr}\right)+\frac{2m}{\hbar^2}(Er^2+e^2r)R-n(n+1)R=0.$$

From the elementary Bohr theory of the hydrogen atom, we know that for the energy-levels whose differences give rise to the Balmer lines, the total energy E is negative. Suppose this to be the case: writing k for $(e^2/2\hbar)\sqrt{(-2m/E)}$ and z for $(2/\hbar)\sqrt{(-2mE)}r$, the equation becomes

$$\frac{d^2R}{dz^2}+\frac{2}{z}\frac{dR}{dz}+\left\{-\frac{1}{4}+\frac{k}{z}-\frac{n(n+1)}{z^2}\right\}R=0$$

of which the solution that remains finite as $r\to\infty$ is

$$\frac{1}{z}W_{k,\,n+\frac{1}{2}}(z)$$

where $W_{k,\,n+\frac{1}{2}}(z)$ is the confluent hypergeometric function.[2] Thus ψ must be a constant multiple of

$$\frac{1}{r}W_{k,\,n+\frac{1}{2}}(z)P_n^{\mu}(\cos\theta)e^{\pm i\mu\phi}.$$

It is, however, necessary also that ψ should be finite at $r=0$. This condition requires that the asymptotic expansion of $W_{k,\,n+\frac{1}{2}}(z)$,[3] namely,

$$W_{k,\,n+\frac{1}{2}}(z)\sim e^{-\frac{1}{2}z}z^k\left[1+\frac{(n+\frac{1}{2})^2-(k-\frac{1}{2})^2}{1\,!\,z}\right.$$
$$\left.+\frac{\{(n+\frac{1}{2})^2-(k-\frac{1}{2})^2\}\{(n+\frac{1}{2})^2-(k-\frac{3}{2})^2\}}{2\,!\,z^2}+\ldots\right]$$

should terminate; which evidently can happen only if

$$\pm(n+\tfrac{1}{2})=k-\tfrac{1}{2},\quad\text{or}\quad k-\tfrac{3}{2},\quad\text{or}\quad k-\tfrac{5}{2},\quad\ldots,$$

that is, k *must be a whole number*: and on account of the factor z^k, it must be a *positive* whole number: thus

$$\frac{e^2}{2\hbar}\sqrt{\left(-\frac{2m}{E}\right)}$$

must be a positive whole number: so

$$E=-\frac{me^4}{2\hbar^2k^2}\qquad\text{where } k=1,\,2,\,3,\,4,\,\ldots.$$

[1] cf. Whittaker and Watson, *Modern Analysis*, § 15·5
[2] cf. Whittaker and Watson, *Modern Analysis*, § 16·1
[3] cf. Whittaker and Watson, *Modern Analysis*, § 16·3

These values of E *are precisely the energy levels of the stationary orbits in Bohr's theory ; to each value of* E *there corresponds a finite number of particular solutions ;* and thus we obtain *the line spectrum of the hydrogen atom.* It is evident from this equation for E that *k must be identified with the total quantum number.*

The question may now be raised as to the physical significance of the wave function ψ. Schrödinger at first, in a paper[1] received 18 March 1926, supposed that if ψ^* denotes the complex quantity conjugate to ψ, then the space-density of electricity is given by the real part of $\psi\, \partial\psi^*/\partial t$. In a paper[2] received on 10 May, however, he corrected this to $\psi\psi^*$, basing his new result on the fact that the integral of $\psi\psi^*$ taken over all space is, like the charge, constant in time. This interpretation of ψ was, however, soon again modified. The notion of waves which do not transmit energy or momentum, but which determine probability, had become familiar to theoretical physicists from the Bohr-Kramers-Slater theory of 1924 : and in a paper[3] received 25 June, and one received 21 July,[4] both dealing with the treatment of collisions by wave-mechanics, Max Born adopted this conception, and proposed that $\psi\psi^*$ should be interpreted in terms of probability ; to be precise, that $\psi\psi^*\, dx\, dy\, dz$ should be taken to be the *probability* that an electron is in the infinitesimal volume-element $dx\, dy\, dz$. This interpretation was soon universally accepted.

It is convenient to *normalise* the wave-functions by the condition

$$\int \psi_n \psi_n^* d\tau = 1$$

where $d\tau$ denotes the element of volume and the integration is taken over all space. As an example of normalisation, consider the fundamental state of the hydrogen atom, for which $k=1$, $n=0$, $\mu=0$. The wave-function is now some multiple of $(1/r)W_{1,\frac{1}{2}}(z)$, and since $W_{1,\frac{1}{2}}(z)=e^{-\frac{1}{2}z}z$, this gives for the wave-function a multiple of $e^{-\frac{1}{2}z}$, or $e^{-(r/a)}$ where a is the radius of the first circular orbit in Bohr's theory of the hydrogen atom. The wave-function may therefore be written

$$\psi = Ce^{-\frac{r}{a}} \quad \text{where C is a constant.}$$

The normalisation condition is

$$C^2 \int_0^\infty \int_0^\pi \int_0^{2\pi} e^{-\frac{2r}{a}} r^2 \sin\theta\, dr d\theta d\phi = 1$$

which gives

$$\pi a^3 C^2 = 1$$

[1] *Ann. d. Phys.*(4) lxxix (1926), p. 734, equation (36)
[2] *Ann. d. Phys.*(4) lxxx, p. 437, note on p. 476
[3] *ZS. f. P.* xxxvii (1926), p. 863
[4] *ZS. f. P.* xxxviii (1926), p. 803

and therefore *the normalised wave-function of the fundamental state of the hydrogen atom is*

$$\psi = \pi^{-\frac{1}{2}} a^{-\frac{3}{2}} e^{-\frac{r}{a}}.$$

Of course this is to be multiplied by the time-factor $e^{-(i/\hbar)E_1 t}$, where E_1 is the energy of the state.

Similarly the normalised wave-function of the hydrogen atom in the state $k=2$, $n=0$, $\mu=0$, is

$$\psi = 2^{-\frac{1}{2}} \pi^{-\frac{1}{2}} a^{-\frac{3}{2}} e^{-\frac{r}{2a}} \left(\frac{r}{2a} - 1\right) e^{-\frac{i}{\hbar}E_2 t}$$

where E_2 is the energy of the state; and the normalised wave-function of the hydrogen atom in the state $k=2$, $n=1$, $\mu=0$, is

$$\psi = (32\pi a^5)^{-\frac{1}{2}} e^{-\frac{r}{2a}} r \cos\theta \, e^{-\frac{i}{\hbar}E_2 t}$$

where E_2 is the energy of the state.

Now suppose that to two different proper-values of the total energy, say E_n and E_m, there correspond respectively solutions ψ_n and ψ_m of the wave equation, so that

$$\nabla^2\psi_n + \frac{2m}{\hbar^2}(E_n - V)\psi_n = 0$$

$$\nabla^2\psi_m + \frac{2m}{\hbar^2}(E_m - V)\psi_m = 0.$$

Multiplying these equations by ψ_m and ψ_n respectively, subtracting and integrating over all space, we have

$$\int(\psi_m\nabla^2\psi_n - \psi_n\nabla^2\psi_m)d\tau + \frac{2m}{\hbar^2}(E_n - E_m)\int\psi_n\psi_m d\tau = 0.$$

The first integral may by Green's theorem be transformed into a surface-integral taken over the surface at infinity, which vanishes; and thus we have

$$\int\psi_n\psi_m d\tau = 0,$$

that is, *two wave-functions* ψ_m *and* ψ_n, *corresponding to different values of* E, *are orthogonal to each other.*

Since in this equation we can replace ψ_m by its complex-conjugate $\psi_m{}^*$, we can combine it with the normalisation-equation into the equation

$$\int\psi_n\psi_m{}^* d\tau = \delta_m{}^n$$

where $\delta_m{}^n = 1$ or 0 according as m and n are the same or different numbers.[1]

In the case of the hydrogen atom, a single proper-value of the energy, say

$$E = -\frac{me^4}{2\hbar^2 k^2}$$

where k is a definite whole number, corresponds to many wave-functions

$$\frac{1}{r} W_{k,\, n+\frac{1}{2}}(z) P_n{}^{\mu}(\cos\theta) e^{\pm i\mu\phi},$$

n and μ being able to take different whole-number values. This phenomenon—the correspondence of several different stationary states to the same proper-value of the energy—is called *degeneracy*, a term already introduced in Chapter III. It is known from the general theory of partial differential equations that we can always choose linear combinations of the wave-functions which belong to the same value of the energy, in such a way that these combinations are orthogonal among themselves [2] (and, of course, orthogonal to the wave-functions which belong to all other proper-values of the energy). The degeneracy of the hydrogen atom, so far as it is due to the fact that many different values of n correspond to the same value of k, can be removed, exactly as in Sommerfeld's theory of 1915, by taking account of the relativist increase of mass with velocity : while the degeneracy, so far as it is due to the choice of different values of μ, can be removed by applying a magnetic field, as in Sommerfeld and Debye's theory of 1916. It is evident that n is essentially the azimuthal quantum number, and that μ is the magnetic quantum number.

Now consider the case when the total energy E of the electron in the hydrogen atom is positive. $\sqrt{(-E)}$ is now imaginary, and all the confluent hypergeometric functions which are solutions of the differential equation for rR remain finite as $r \to \infty$, so R tends to zero as $r \to \infty$. Moreover, at least one solution of the differential equation exists which is finite at $r = 0$. Thus *every positive value of* E *is a proper-value*, to which correspond wave-functions possessing azimuthal and magnetic quantum numbers, in the same way as the discrete wave-functions. Physically, this case corresponds to the complete ionisation of the hydrogen atom, or to the reverse process, i.e. the capture of free electrons.

[1] If the proper-values of Schrödinger's wave-equation have also a continuous spectrum, the above equations persist in a modified form, for which, see H. Weyl, *Math. Ann.* lxviii (1910), p. 220, and *Gött. Nach.* (1910), p. 442 ; E. Fues, *Ann. d. Phys.* lxxxi (1926), p. 281.

[2] There is a certain degree of arbitrariness in the choice, the arbitrariness being represented by an orthogonal transformation which may be performed on the wave-functions of the set.

In a further paper,[1] Schrödinger extended his theory by finding the wave-equation for problems more general than those considered hitherto. Still restricting ourselves for simplicity to the motion of a single particle, suppose that its Hamiltonian Function $H(q_1, q_2, q_3, p_1, p_2, p_3)$ is the sum of (1) a kinetic energy represented by a general quadratic form T in the momenta (p_1, p_2, p_3), with coefficients which depend on the co-ordinates (q_1, q_2, q_3), and (2) a potential energy V which depends only on (q_1, q_2, q_3). Then the wave-equation may be derived by a process similar to that followed already in the case of the particle referred to Cartesian co-ordinates. It will be evident from what was there proved that the wave-function must be of the form

$$\psi = e^{\frac{i}{\hbar}\{-Et+S(x,y,z)\}}$$

where $S(x, y, z)$ is Hamilton's Characteristic Function, satisfying the equation

$$H\left(q_1, q_2, q_3, \frac{\partial S}{\partial q_1}, \frac{\partial S}{\partial q_2}, \frac{\partial S}{\partial q_3}\right) = E.$$

Differentiating the expression for ψ, we have

$$\frac{\partial^2 \psi}{\partial q_r^2} = -\frac{1}{\hbar^2}\left(\frac{\partial S}{\partial q_r}\right)^2 \psi + \frac{i}{\hbar}\frac{\partial^2 S}{\partial q_r^2}\psi.$$

As before, the second term on the right is very small compared with the first, and may be neglected. Thus

$$T\left(q_1, q_2, q_3, \frac{\hbar}{i}\frac{\partial}{\partial q_1}, \frac{\hbar}{i}\frac{\partial}{\partial q_2}, \frac{\hbar}{i}\frac{\partial}{\partial q_3}\right)\psi = T\left(q_1, q_2, q_3, \frac{\partial S}{\partial q_1}, \frac{\partial S}{\partial q_2}, \frac{\partial S}{\partial q_3}\right)\psi,$$

and therefore

$$H\left(q_1, q_2, q_3, \frac{\hbar}{i}\frac{\partial}{\partial q_1}, \frac{\hbar}{i}\frac{\partial}{\partial q_2}, \frac{\hbar}{i}\frac{\partial}{\partial q_3}\right)\psi = H\left(q_1, q_2, q_3, \frac{\partial S}{\partial q_1}, \frac{\partial S}{\partial q_2}, \frac{\partial S}{\partial q_3}\right)\psi$$

or as we may write it

$$H\left(q, \frac{\hbar}{i}\frac{\partial}{\partial q}\right)\psi = E\psi.$$

This is the extension of the wave-equation, applicable to the more general form of the Hamiltonian Function.

Since

$$\frac{\partial \psi}{\partial t} = -\frac{i}{\hbar}E\psi,$$

[1] *Ann. d. Phys.*(4) lxxix (1926), p. 734

we have

$$H\left(q, \frac{\hbar}{i}\frac{\partial}{\partial q}\right)\psi = i\hbar\frac{\partial\psi}{\partial t}.$$

This equation, which does not involve E explicitly, may be called the *general wave-equation.*[1] It is clearly the equation that would be obtained by replacing p by $\hbar/i\ \partial/\partial q$ and E by $-\hbar/i\ \partial/\partial t$ in the equation of energy, and then operating on ψ.

It would seem from the foregoing derivations that Schrödinger's wave-equation is connected with Hamilton's Principal and Characteristic Functions by equations which are not exact, but only approximate, involving the neglect of powers of $\hbar$. It was shown[2] many years after the discovery of the wave-equation that this conclusion is incorrect: Schrödinger's equation is *rigorously equivalent* to Hamilton's partial differential equation for the Principal Function, provided the symbols are understood in a certain way.

Considering for simplicity a conservative system with one degree of freedom, let the co-ordinate at the instant t be q, and let Q be the value of the co-ordinate at a previous instant T. Then, as Hamilton showed, there exists in classical dynamics a *Principal Function* $W(q, Q, t-T)$, which has the properties

$$\frac{\partial W(q, Q, t-T)}{\partial q} = p$$

$$\frac{\partial W(q, Q, t-T)}{\partial Q} = -P$$

$$\frac{\partial W(q, Q, t-T)}{\partial t} = -H$$

where p and P are the values of the momentum at the instants t and T respectively, and H is the Hamiltonian Function.

Now consider the quantum-mechanical problem which is specified by the Hamiltonian $H(q, p)$. As explained in Chapter VIII, the variables q and p are no longer ordinary algebraic quantities, but are non-commuting variables satisfying the commutation-rule

$$pq - qp = \frac{\hbar}{i}.$$

The quantity Q, which represents the co-ordinate at the instant T, also does not commute with q or p. A function of q and Q is said to be *well-ordered*, when it is arranged (as of course it can be, by

[1] With regard to the general character of Schrödinger's theory, Heisenberg [*ZS. f. P.* xxxviii (1926), p. 411, *at* p. 412] said 'So far as I can see, Schrödinger's procedure does not represent a consistent wave-theory of matter in de Broglie's sense. The necessity for waves in space of f dimensions (for a system with f degrees of freedom), and the dependence of the wave-velocity on the mutual potential energy of particles, indicates a loan from the conceptions of the corpuscular theory.'

[2] E. T. Whittaker, *Proc. R.S. Edin.* lxi (1940), p. 1

use of the commutation-rules) as a sum of terms, each of the form $f(q)g(Q)$, the factors being in this order.[1] Then it can be shown that there exists a well-ordered function $U(q, Q, t-T)$, which formally satisfies exactly the same equations as Hamilton's Principal Function, namely,[2]

$$\frac{\partial U}{\partial q} = p, \qquad \frac{\partial U}{\partial Q} = -P, \qquad \frac{\partial U}{\partial t} = -H.$$

U may be called the *quantum-mechanical Principal Function.* Although it satisfies the same equations as Hamilton's Principal Function, its expression in terms of the variables q, Q, $t-T$, is quite different from the expression of the classical function W : the reason being that the above equations for U are true only on the understanding that all the quantities occurring in them are well-ordered in q and Q: but on substituting the well-ordered expression for p in the Hamiltonian $H(q, p)$, we shall need to invert the order of factors in many terms, by use of the commutation-rules, in order to reduce H to a well-ordered function of q and Q; and this introduces new terms which do not occur in the classical equations. This explains why, although the equations of quantum mechanics are formally identical with those of classical mechanics, the solutions in the two cases are altogether different.

Now introduce a function $R(q, Q, t-T)$, which is obtained by taking the well-ordered function U and replacing the non-commuting variables q and Q by ordinary algebraic quantities q and Q. It may be called the *Third Principal Function.* By what has been said, it is quite different from the classical Principal Function W belonging to the same Hamiltonian. Then if we write

$$\psi(q, Q, t-T) = e^{\frac{i}{\hbar}R(q, Q, t-T)}$$

it may be shown that $\psi\ (q, Q, t-T)$ *satisfies Schrödinger's differential equation for the wave-function belonging to the Hamiltonian* $H(q, p)$. Thus *the relation between Principal Function and Schrödinger's wave-function is rigorous, not requiring the neglect of any powers of* $\hbar$, *provided the Principal Function is understood to be the Third Principal Function* R, *and not Hamilton's classical Principal Function* W.

We must now consider how the proper-values and wave-functions of Schrödinger's equation are to be determined. The following method for determining them, at least approximately, was given in 1926 by G. Wentzel,[3] H. M. Kramers[4] and L. Brillouin.[5]

[1] This conception is due to Jordan, *ZS. f. P.* xxxviii (1926), p. 513

[2] The first two of these equations were given substantially by Dirac, *Phys. Zeits. Sowjetunion*, iii (1933), p. 64.

[3] *ZS. f. P.* xxxviii (1926), p. 518

[4] *ZS. f. P.* xxxix (1926), p. 828

[5] *Comptes Rendus*, clxxxiii (1926), p. 24; *J. de phys. et le rad.* vii (1926), p. 353. The method was to a great extent anticipated by H. Jeffreys, *Proc. L.M.S.*(2) xxiii (1925), p. 428; cf. also R. E. Langer, *Bull. Amer. M.S.* xl (1934), p. 545, who indicated a correction in Jeffreys's paper.

Consider a one-dimensional oscillatory motion defined by a Hamiltonian

$$H = \frac{p^2}{2m} + V(q).$$

The Schrödinger equation for the wave-function ψ is

$$\frac{\hbar^2}{2m}\frac{d^2\psi}{dq^2} + (E - V)\psi = 0$$

where E is the proper-value of the energy. Suppose that the range in which the particle would oscillate according to classical dynamics, namely, that defined by $E - V = 0$, is $q_1 \leqq q \leqq q_2$, so that q_1 and q_2 are roots of the equation $V(q) = E$; within this range put

$$\left\{\frac{2m}{\hbar^2}(E - V)\right\}^{\frac{1}{2}} = \zeta(q),$$

so the wave-equation becomes

$$\frac{d^2\psi}{dq^2} + \left\{\zeta(q)\right\}^2 \psi = 0.$$

If V, and therefore ζ, were constant, the solution would be of the form

$$\psi = \text{Constant} \times \sin(\zeta q + \text{Constant}):$$

this suggests that even when ζ is variable, we should try within the range $q_1 \leqq q \leqq q_2$ to represent ψ by a sine-curve of slowly varying amplitude and wave-length, say

$$\psi = A(q) \sin S(q).$$

Substituting in the wave-equation, we have

$$(A'' - AS'^2 + A\zeta^2) \sin S + (2A'S' + AS'') \cos S = 0.$$

Let us try to satisfy this equation by imposing on A and S the two conditions

$$A'' - AS'^2 + A\zeta^2 = 0, \qquad 2A'S' + AS'' = 0.$$

Now suppose that A' varies so slowly that $|A''|$ is very small compared to $\zeta^2|A|$ (it is at this stage that the approximation enters): then we may neglect the first term in the former of these equations, which now becomes

$$S'^2 = \zeta^2$$

giving

$$S = \int_{q_1}^{q} \zeta dq + \alpha, \quad \text{where } \alpha \text{ is a constant.}$$

The second condition

$$2A'S' + AS'' = 0$$

gives at once

$$A = c(S')^{-\frac{1}{2}} \quad \text{where } c \text{ is a constant}$$

or

$$A = c(\zeta)^{-\frac{1}{2}}.$$

The approximate expression for ψ is therefore

$$\psi = c(\zeta)^{-\frac{1}{2}} \sin\left(\int_{q_1}^{q} \zeta dq + \alpha\right).$$

Attention must be given, however, to the behaviour of ψ at the points for which $q = q_1$ and $q = q_2$, say P and Q respectively, since at these points ζ vanishes, and A″ cannot be neglected in comparison with $A\zeta^2$. A closer consideration of this difficulty[1] shows that the function ψ must have at P the phase $\pi/4$, so the approximate expression of ψ in the range PQ is

$$\psi = c(\zeta)^{-\frac{1}{2}} \sin\left(\int_{q}^{q} \zeta dq + \frac{\pi}{4}\right).$$

This is the Wentzel-Kramers-Brillouin approximate solution of Schrödinger's equation.

Similarly the function ψ must have at Q the phase $3\pi/4$: so when $q = q_2$,

$$\int_{q_1}^{q} \zeta dq + \frac{\pi}{4} = n\pi + \frac{3\pi}{4}$$

where n is a whole number, representing the number of nodes comprised between P and Q; that is to say,

$$\int_{q_1}^{q_2} \zeta dq = (n + \tfrac{1}{2})\pi.$$

This equation determines the proper-values E *of the energy, in the Wentzel-Kramers-Brillouin approximation.*

Between P and Q the graph of ψ oscillates like a sine-curve, whereas to the left of P and to the right of Q it decreases exponentially.

The Wentzel-Kramers-Brillouin approximation throws light on the connection between wave-mechanics and the formula of quantification enunciated by Wilson and Sommerfeld in the older quantum

[1] For which cf. E. Persico, *N. Cimento*, xv (1938), p. 133

theory, namely that $\int pdq$, where the integration is taken round an orbit, is a multiple of Planck's constant h.

For the momentum, calculated according to classical dynamics from the equation

$$\frac{p^2}{2m} + V(q) = E$$

is

$$p = \pm\{2m(E - V)\}^{\frac{1}{2}} = \pm\hbar\zeta(q)$$

where the upper sign must be taken for the semi-oscillation from q_1 to q_2, and the lower sign from q_2 to q_1. Thus

$$\int pdq \text{ round the orbit} = 2\int_{q_1}^{q_2} pdq = 2\hbar\int_{q_1}^{q_2}\zeta(q)dq,$$

and thus the Wentzel-Kramers-Brillouin condition becomes

$$\frac{1}{2\hbar}\int pdq = (n + \tfrac{1}{2})\pi$$

or

$$\int pdq = (n + \tfrac{1}{2})h,$$

which is the Wilson-Sommerfeld condition, completed by the term $\frac{1}{2}$ which appears in the more accurate theory.

So far we have connected Schrödinger's wave-equation only with the stationary states of the atom, to which correspond proper-values of the total energy. We shall now consider more general states.

In Bohr's theory a stationary state meant a particular kind of orbital motion, so that an atom could be in only one stationary state at one time. In Schrödinger's theory, on the other hand, the stationary states correspond to different solutions of a linear partial differential equation, and therefore the various stationary states can be *superposed* just as overtones can be superposed on the fundamental tone of a violin string. We have to consider what is the physical interpretation of this superposition.

Suppose that plane-polarised light whose vibrations are in a direction α passes through a Nicol prism, which resolves it into vibrations in directions β and γ respectively parallel and perpendicular to the plane of polarisation of the Nicol, and permits only the former to pass through. Fixing our attention on a single photon, which is initially polarised in the direction α, we can regard this state as a *superposition* of two states, namely, that of polarisation in the direction β and polarisation in the direction γ.[1] We can speak of the

[1] The superposition here spoken of must not be confused with superposition in classical mechanics; in classical mechanics, the superposition of a certain state of vibration on itself gives a vibration of twice the amplitude, but in quantum-mechanics it gives merely the same state of vibration.

probability of the photon being in either of the states β and γ, these probabilities being such as to account for the observed intensity of light polarised in the direction β, which emerges from the Nicol.

This connection between superposition and probability will now be extended to wave-functions. Consider two stationary states of an atom, for which the energy has the proper-values E_1 and E_2 respectively. Let ψ_1 and ψ_2 be the corresponding normalised solutions of the wave-equation, with their appropriate time-factors $e^{-(iE_1/\hbar)t}$ and $e^{-(iE_2/\hbar)t}$, and let $\psi_3 = c_1\psi_1 + c_2\psi_2$ where c_1 and c_2 are arbitrary complex numbers. Then the state or physical situation represented by ψ_3 is said to be formed by the *superposition* of the states represented by ψ_1 and ψ_2. ψ_3 of course satisfies the general wave-equation

$$H\left(q, \frac{\hbar}{i}\frac{\partial}{\partial q}\right)\psi = i\hbar\frac{\partial\psi}{\partial t}.$$

Let us find the condition that must be satisfied by c_1 and c_2 if ψ_3 is normalised. We have

$$1 = \int\psi_3\psi_3^* d\tau = \int(c_1\psi_1 + c_2\psi_2)\,(c_1^*\psi_1^* + c_2^*\psi_2^*)d\tau$$

$$= c_1c_1^*\int\psi_1\psi_1^* d\tau + c_2c_2^*\int\psi_2\psi_2^* d\tau$$

$$= |c_1|^2 + |c_2|^2.$$

This equation can be interpreted to mean that *there is an uncertainty as to the value which would be found by a measurement of the energy of the atom, either of the values* E_1 *or* E_2 *being possible: and their respective probabilities are* $|c_1|^2$ *and* $|c_2|^2$.

More generally, if $\psi_0, \psi_1, \psi_2, \ldots$ are the normalised wave-functions (with their appropriate time-factors) belonging respectively to the proper-values $E_0, E_1, E_2, \ldots$ (supposed for simplicity to be all different) of the energy of the atom, and if a normalised solution ψ of the general wave-equation is expanded in the form

$$\psi = c_0\psi_0 + c_1\psi_1 + c_2\psi_2 + \ldots,$$

then *in the physical situation defined by the wave-function* ψ, *the probability that a measurement of the energy will yield the value* E_n is $|c_n|^2$. On account of the relation $\int\psi_n\psi_m^* d\tau = \delta_m^n$, it is seen at once that the value of the coefficient c_n is $\int\psi\psi_n^* d\tau$.

The equations which we have found lead to a certain connection [1] with the classical electromagnetic theory of the emission of radiation. Suppose that we consider an atom in a stationary state, so that the wave-function ψ involves the time through a factor (say) $e^{2\pi i\nu t}$.

[1] E. Fermi, *Rend. Lincei*, v (May 1927), p. 795

Then ψ^* will have the time-factor $e^{-2\pi i\nu t}$, so in the product $\psi\psi^*$ these time-factors will destroy each other : that is to say, the distribution of electric charge in the atom, and therefore its electric moment, will not vary with the time : and therefore *according to the classical theory*, the atom, when it is in a stationary state, will not emit radiation.

Suppose next, however, that the atom is not in a pure stationary state, but is in a state which is represented by the superposition of two stationary states, for which the wave-function has the time-factors $e^{2\pi i\nu_1 t}$ and $e^{2\pi i\nu_2 t}$ respectively, so that

$$\psi = Ae^{2\pi i\nu_1 t} + Be^{2\pi i\nu_2 t}$$

where A and B do not involve the time. Then we have

$$\psi\psi^* = (Ae^{2\pi i\nu_1 t} + Be^{2\pi i\nu_2 t})(A^*e^{-2\pi i\nu_1 t} + B^*e^{-2\pi i\nu_2 t})$$

$$= AA^* + BB^* + AB^*e^{2\pi i(\nu_1-\nu_2)t} + A^*Be^{-2\pi i(\nu_1-\nu_2)t},$$

and hence the electric moment of the atom, which depends on an integral involving t only in the combination $\psi\psi^*$, will be periodic, with frequency $(\nu_1 - \nu_2)$, and consequently the atom, *according to the classical theory*, will emit radiation of frequency $(\nu_1 - \nu_2)$. This radiation will continue until the consequent exhaustion of energy has again reduced the atom to a single pure stationary state.

If E_1 and E_2 are the energies associated with the two stationary states, then $E_1 = h\nu_1$ and $E_2 = h\nu_2$, so the radiation is of frequency $(1/h)\ (E_1 - E_2)$, just as in Bohr's theory, and this result has now been obtained by what are essentially classical methods—the classical theory of the solutions of partial differential equations—without doing violence to the electromagnetic theory of light.

Schrödinger did not, at the outset of his researches, suspect any connection between his theory and the theory of matrix-mechanics.[1] He now, however,[2] showed that the two theories are actually equivalent. In the first place, the commutation-rules of matrix-mechanics, which in the case of systems with one degree of freedom reduce to

$$qp - pq = i\hbar,$$

become obvious identities if we write $p = (\hbar/i)\ \partial/\partial q$, since

$$q\left(\frac{\hbar}{i}\frac{\partial}{\partial q}\right)\psi - \frac{\hbar}{i}\frac{\partial}{\partial q}(q\psi) = i\hbar\psi.$$

To any physical quantity $\zeta(q, p)$ we can correlate a differential

[1] He says so in *Ann. d. Phys.*(4) lxxix (1926), p. 734 ; 'I naturally knew about his [Heisenberg's] theory, but was discouraged by what appeared to me as very difficult methods of transcendental algebra.'

[2] loc. cit. : cf. also C. Eckart, *Phys. Rev.* xxviii (1926), p. 711

operator $\zeta(q, \hbar/i.d/dq)$. Let us consider the operation of this quantity on a wave-function $\psi(q)$. As we have seen, it is possible to expand $\psi(q)$ as a series of normalised orthogonal wave-functions, in the form

$$\psi(q) = c_1\psi_1 + c_2\psi_2 + \ldots$$

where

$$c_n = \int \psi(q)\psi_n^*(q)dq.$$

If $\zeta(q, \hbar/i\ d/dq)\psi(q)$ has a corresponding expression

$$\zeta\left(q, \frac{\hbar}{i}\frac{d}{dq}\right)\psi(q) = c_1'\psi_1 + c_2'\psi_2 + \ldots,$$

then evidently we must have

$$c_n' = \int \psi_n^*(q)\zeta\left(q, \frac{\hbar}{i}\frac{d}{dq}\right)\psi(q)dq$$

$$= \int \psi_n^*(q)\zeta\left(q, \frac{\hbar}{i}\frac{d}{dq}\right)(c_1\psi_1 + c_2\psi_2 + \ldots)dq$$

$$= e_{n1}c_1 + e_{n2}c_2 + e_{n3}c_3 + \ldots$$

where

$$e_{nr} = \int \psi_n^*(q)\zeta\left(q, \frac{\hbar}{i}\frac{d}{dq}\right)\psi_r(q)dq.$$

But this equation shows that the column-vector $(c_1', c_2', c_3', \ldots)$ is derived from the column-vector $(c_1, c_2, c_3, \ldots)$ by operating on it with a matrix whose element in the nth row and r^{th} column is e_{nr}. Thus *the physical quantity* $\zeta(q, p)$, *or the operator* $\zeta(q, \hbar/i\ d/dq)$, *may be correlated to the matrix whose element in the* n^{th} *row and* r^{th} *column* is

$$e_{nr} = \int \psi_n^*(q)\zeta\left(q, \frac{\hbar}{i}\frac{d}{dq}\right)\psi_r(q)dq,$$

in the sense that the performance of the operator on any wave-function $\psi(q)$ *is equivalent to the operation of the matrix* (e_{nr}) *on the column-vector of the coefficients* c_r, *which express* $\psi(q)$ *in terms of the normalised orthogonal wave-functions* $\psi_1(q)$, $\psi_2(q)$, ...

We observe that the matrix thus found for the physical quantity $\zeta(q, p)$ is specially associated with the set of normalised orthogonal wave-functions $\psi_n(q)$ which belong to a particular Schrödinger's equation

$$H\left(q, \frac{\hbar}{i}\frac{\partial}{\partial q}\right)\psi = i\hbar\frac{\partial\psi}{\partial t}.$$

We may call this set of wave-functions the *basis* of the matrix.

In order to establish completely the identity of this correlation (between physical quantities and matrices) with matrix-mechanics, it is necessary to prove some other mathematical theorems, e.g. that the result of operating with the product of two operators $\zeta(q, \hbar/i\ d/dq)$ and $\eta(q, \hbar/i\ d/dq)$ on a wave-function is the same as the result of operating on the column of coefficients of the wave-function with the matrix-product of the matrices corresponding to ζ and η. The proofs will be omitted here.

Suppose now that the basis of the matrix representing $\zeta(q, p)$ is the set of wave-functions belonging to the Hamiltonian function $\zeta(q, p)$, so it is the set of wave-functions of the Schrödinger equation

$$\zeta\left(q, \frac{\hbar}{i}\frac{\partial}{\partial q}\right)\psi = i\hbar\frac{\partial\psi}{\partial t}.$$

Then the matrix-elements are given by

$$e_{nr} = \int\psi_n^*(q)\zeta\left(q, \frac{\hbar}{i}\frac{d}{dq}\right)\psi_r(q)dq$$

$$= \int\psi_n^*(q)E_r\psi_r(q)dq$$

where E_r is the proper-value of the energy corresponding to the wave-function $\psi_r(q)$. Thus

$$e_{nr} = E_r\delta_r^n$$

so *the matrix is now a diagonal matrix whose elements are the proper-values of the energy for the Schrödinger equation*

$$\zeta\left(q, \frac{\hbar}{i}\frac{\partial}{\partial q}\right)\psi = i\hbar\frac{\partial\psi}{\partial t}.$$

Thus *the physical quantity $\zeta(q, p)$, when expressed as a matrix in terms of this basis, is a diagonal matrix whose elements are its proper-values.*

Therefore the problem as formulated in matrix-mechanics, namely, *to reduce the matrix for $\zeta(q, p)$ to a diagonal matrix*, is solved when we have found a solution of the problem as formulated in wave-mechanics, namely, *to find the proper values of* E *and the corresponding wave-functions for the Schrödinger equation*

$$\zeta\left(q, \frac{\hbar}{i}\frac{\partial}{\partial q}\right)\psi = E\psi.$$

The basis constituted by the wave-functions of this equation enables us, by the above formulae, to calculate the matrices that represent q and p. Thus *matrix-mechanics and wave-mechanics are equivalent.*

Let us now investigate more closely the physical meaning of the matrix-elements. Consider the electric moment $\mathbf{M}_{PQ}$ of the classical oscillator which would emit the radiation associated with a transition from a stationary state P to a stationary state Q of the atom: let $\mathbf{M}$ denote $\sum e\mathbf{r}$, the electric moment of the atom, expressed in terms of the co-ordinates of the electrons: and let ψ_P and ψ_Q be the wave-functions of the states P and Q. We may expect that $\mathbf{M}_{PQ}$ will depend in some way on $\mathbf{M}$, ψ_P and ψ_Q. Now we know that the time-factor in $\mathbf{M}_{PQ}$ is $e^{-2\pi i\nu_{PQ}t}$, where $\nu_{PQ} = (1/h)(E_P - E_Q)$, the time-factor in ψ_P is $e^{-iE_Pt/\hbar}$, the time-factor in ψ_Q is $e^{-iE_Qt/\hbar}$, and $\mathbf{M}$ has no time-factor. We therefore expect that the expression for $\mathbf{M}_{PQ}$ will involve the product $\psi_P\mathbf{M}\psi_Q{}^*$, where $\psi_Q{}^*$ is the complex-conjugate of ψ_Q. The explicit expression of the co-ordinates has to be removed from this, which can evidently be done by integrating over space. Thus we obtain the expression

$$\int \psi_P \mathbf{M} \psi_Q{}^* d\tau,$$

and *this matrix-element we shall identify with* $\mathbf{M}_{PQ}$, *the electric moment of the classical oscillator which would emit the radiation associated with the transition* P→Q. This identification is justified by comparison with the results of experiments.

Let us now illustrate the equivalence of matrix-mechanics and wave-mechanics by considering the harmonic oscillator in one dimension, for which the Hamiltonian function is

$$H = \frac{1}{2m}(p^2 + m^2\omega^2 q^2)$$

where m and ω are constants. The Schrödinger wave-equation is

$$\hbar^2 \frac{d^2\psi}{dq^2} + (2mE - m^2\omega^2 q^2)\psi = 0$$

where E is the total energy of the motion.

Writing $q = (\hbar/2m\omega)^{\frac{1}{2}}z$, this may be written

$$\frac{d^2\psi}{dz^2} + \left(\frac{E}{\hbar\omega} - \tfrac{1}{4}z^2\right)\psi = 0$$

which is the well-known differential equation of the parabolic-cylinder functions.[1] It has a solution which is finite for all real values of z only when

$$\frac{E}{\hbar\omega} = n + \tfrac{1}{2}$$

[1] cf. Whittaker and Watson, *Modern Analysis*, § 16·5

where n is a whole number,[1] and the solution is then a constant multiple of the parabolic-cylinder function $D_n(z)$, which is a polynomial multiplied by $e^{-\frac{1}{4}z^2}$. Thus *the proper-values of* H *are*

$$E=(n+\tfrac{1}{2})\hbar\omega \quad \textit{where} \quad n=0, 1, 2, 3, \ldots$$

and the corresponding wave-functions are

$$\psi_n(q)=\lambda D_n\left(q\sqrt{\frac{2m\omega}{\hbar}}\right)$$

where λ is a constant to be determined by the normalising condition

$$\int_{-\infty}^{\infty}|\psi_n(q)|^2dq=1.$$

Since

$$\int_{-\infty}^{\infty}\{D_n(z)\}^2dz=(2\pi)^{\frac{1}{2}}n!,$$

this gives

$$\lambda=\left(\frac{m\omega}{\pi\hbar}\right)^{\frac{1}{4}}(n\,!)^{-\frac{1}{2}}e^{i\alpha_n}$$

where α_n is real ; so the normalised wave-functions are

$$\psi_n(q)=\left(\frac{m\omega}{\pi\hbar}\right)^{\frac{1}{4}}(n\,!)^{-\frac{1}{2}}D_n\left(q\sqrt{\frac{2m\omega}{\hbar}}\right)e^{i\alpha_n}.$$

The definition of the element in the lth row and n^{th} column of the matrix representing the co-ordinate q, in terms of the wave-functions, is

$$\int_{-\infty}^{\infty}q\psi_l^*(q)\psi_n(q)dq.$$

Using the known properties of the parabolic-cylinder functions

$$zD_n(z)=D_{n+1}(z)+nD_{n-1}(z)$$

and

$$\int_{-\infty}^{\infty}D_m(z)D_n(z)dz=(2\pi)^{\frac{1}{2}}n\,!\,\delta_n^m,$$

[1] The Wilson-Sommerfeld quantum condition would lead us to expect the Action to be a whole multiple of h, that is, energy $\times$ period to be a multiple of h, or E equal to a multiple of $\hbar\omega$. The occurrence of $(n+\frac{1}{2})$ instead of n is characteristic of the quantum-mechanical solution.

the value of the matrix-element can readily be calculated, and we obtain (inserting the time-factor)

$$q=\left(\frac{\hbar}{2m\omega}\right)^{\frac{1}{2}}\begin{pmatrix} \cdot & e^{-i(\omega t+\beta_1)} & \cdot & \vdots \\ e^{i(\omega t+\beta_1)} & \cdot & 2^{\frac{1}{2}}e^{-i(\omega t+\beta_2)} & \vdots \\ \cdot & 2^{\frac{1}{2}}e^{i(\omega t+\beta_2)} & \cdot & \vdots \\ \cdot & \cdot & \cdot & \cdot \end{pmatrix}$$

in agreement with the value found in chapter VIII.

Similarly the element in l^{th} row and n^{th} column of the matrix representing the momentum p is expressed in terms of the wave-functions by

$$\frac{\hbar}{i}\int_{-\infty}^{\infty}\psi_l^*(q)\frac{d}{dq}\psi_n(q)dq$$

which when evaluated gives the same value as was found in Chapter VIII.

It was shown by C. Eckart [1] by use of the wave-equation that in general problems, an element of the matrix representing a co-ordinate can be expressed as a series of the form

$$\chi(m, n)=\chi_0(m, n)+h\chi_1(m, n)+h^2\chi_2(m, n)+\ldots$$

whereas $n\to\infty$, $\chi_0(m, n)$ tends to the coefficient of the $(m-n)^{th}$ harmonic in the Fourier expansion of the co-ordinate as determined by the classical theory of the motion. This is in accord with Bohr's correspondence principle.

In a later paper [2] he illustrated this theorem by calculating the matrix-elements of the radius vector for the hydrogen atom, and showing that their limiting values, as $h\to 0$ and the quantum numbers tend to infinity, coincide with the Fourier coefficients of the classical motion. In particular, the diagonal terms of the matrix tend to the constant term of the Fourier expansion.

In a paper entitled 'The Continuous Transition from Micro- to Macro-mechanics,' [3] Schrödinger showed how to construct for the harmonic oscillator a *wave-packet*, i.e. a group of wave-functions of high-quantum number and small quantum-number-differences such that the electric density is very nearly concentrated at a single point. He proved that the differential equation of the wave-functions of the harmonic oscillator, namely,

$$\frac{1}{2m}\left(-\hbar^2\frac{\partial^2\psi}{\partial q^2}+m^2\omega^2q^2\right)=i\hbar\frac{\partial\psi}{\partial t}$$

is satisfied by

$$\psi(t)=\exp\left(-\tfrac{1}{2}i\omega t+ke^{-i\omega t}q\sqrt{\left(\frac{2m\omega}{\hbar}\right)}-\frac{m\omega q^2}{2\hbar}-\frac{k^2}{2}-\frac{k^2}{2}e^{-2i\omega t}\right)$$

[1] *Proc. Nat. Ac. Sci.* xii (1926), p. 684 ; cf. P. Debye, *Phys. ZS.* xxviii (1927), p. 170
[2] *ZS. f. P.* xlviii (1928), p. 295
[3] *Naturwiss*, xxviii (1926), p. 664

when k is any constant. This gives

$$\psi\psi^* = \exp\left[-\frac{\{q-(2\hbar/m\omega)^{\frac{1}{2}}k\cos\omega t\}^2}{\hbar/m\omega}\right].$$

The probability that at the instant t the co-ordinate q lies in the range from q to $q+dq$ is $\psi\psi^* dq$: and clearly if $\hbar$ is very small (as it is) and k is very large, while $\hbar k^2\omega$ is finite, say equal to E, so $(2\hbar/m\omega)^{\frac{1}{2}}k$ becomes $(2E/m\omega^2)^{\frac{1}{2}}$, then $\psi\psi^*$ is negligibly small except when the numerator of the argument of the exponential is approximately zero; that is, the wave-function represents a wave-packet whose position at time t is given by

$$q=\sqrt{\left(\frac{2E}{m\omega^2}\right)}\cos\omega t,$$

which is precisely the equation determining the position of the particle in the classical problem when the energy is E.

The *quantum-mechanical theory of collisions* was founded in two papers of 1926 by Max Born.[1] A classical treatment of the problem had been given in 1911 by Rutherford,[2] whose investigation of the scattering of particles by a Coulomb field will first be described.

Consider the scattering of a narrow beam of α-rays by a sheet of metal foil on which the beam impinges at right angles. The scattered particles afterwards strike a screen of zinc sulphide, and the number of scintillations on each square millimetre of the screen is observed.

Let the number of atoms per unit volume in the foil be n, and let q be the thickness of the foil, so there are nq atoms per unit cross-section of the beam. The area of that part of this unit cross-section which is within a distance B of the centre of an atom is therefore πnqB^2, and out of σ incident particles the number that are scattered at distances between B and $B+dB$ is $2\pi\sigma nqBdB$. Now according to classical dynamics the path of a particle is a hyperbola having the centre of the atom as its external focus: and since the perpendicular from the focus on an asymptote of the hyperbola is equal to the minor semi-axis, it follows that B is the minor semi-axis of the hyperbola. The angle θ through which the particle is scattered is equal to the external angle between the asymptotes of the hyperbola, so

$$\cot\tfrac{1}{2}\theta=\frac{B}{A}$$

where A is the major semi-axis. The number of particles scattered in the annulus between angles θ and $\theta+d\theta$ is therefore

$$\pi\sigma nqA^2\cot\tfrac{1}{2}\theta\,\operatorname{cosec}^2\tfrac{1}{2}\theta\,d\theta.$$

[1] *ZS. f. P.* xxxvii (1926), p. 863; xxxviii (1926), p. 803
[2] *Phil. Mag.* xxi (1911), p. 669

These fall on the screen, which is supposed to be at a distance l beyond the foil, on an annulus of area

$$d(\pi l^2 \tan^2\theta) \quad \text{or} \quad 2\pi l^2 \tan\theta \sec^2\theta \, d\theta.$$

The number of scintillations observed per unit area is therefore

$$\frac{\sigma nq A^2 \cos^3\theta \cos\frac{1}{2}\theta}{2l^2 \sin\theta \sin^3\frac{1}{2}\theta} \quad \text{or} \quad \frac{\sigma nq A^2 \cos^3\theta}{4l^2 \sin^4\frac{1}{2}\theta}.$$

Writing r = distance of the scintillation from the atom = $l/\cos\theta$, we see that the number of scintillations per unit area perpendicular to r is

$$\frac{\sigma nq A^2}{4r^2 \sin^4\frac{1}{2}\theta}.$$

Now if E be the charge at the centre, e the charge and m the mass of the incident particle, and v_0 the velocity at infinity, the usual formulae of hyperbolic motion give $A = eE/mv_0^2$: so the number of scintillations per unit area perpendicular to v is

$$\frac{\sigma nq e^2 E^2}{4m^2 v_0^4 \, r^2 \sin^4\frac{1}{2}\theta}.$$

Thus if the beam of incident particles is of such intensity that one particle crosses unit area in unit time, and the beam falls on a single scattering centre, then the probability that in unit time a particle should be scattered into the element of solid angle $\sin\theta \, d\theta d\phi$ (whose θ, ϕ are spherical-polar co-ordinates, the polar axis being the direction of the incident particle) is

$$\frac{e^2 E^2}{4m^2 v_0^4 \sin^4\frac{1}{2}\theta} \sin\theta \, d\theta d\phi.$$

This formula, given in Rutherford's paper, was verified experimentally. It afforded a means of determining the charge E on the nucleus of an atom, and led to the conclusion that the nuclear charge is equal to the electronic charge multiplied by the atomic number.

The quantum-mechanical treatment of collision problems originated in the two papers of Max Born already referred to. Consider the scattering of a beam of particles by a fixed centre of force, the potential energy of one particle at a distance r from the scattering centre being $V(r)$. Schrödinger's wave-equation for a particle in presence of the scattering centre is

$$\nabla^2\psi + \left\{k^2 - \frac{2m}{\hbar^2}V(r)\right\}\psi = 0$$

where $k = mv_0/\hbar$, m being the mass and v_0 the initial velocity of the incident particles. The general solution of the equation

$$\nabla^2\psi + k^2\psi = f(x, y, z)$$

where f is a given function, is known to be

$$\psi = C(x, y, z) - \frac{1}{4\pi}\int \frac{e^{ik\rho}}{\rho} f(x', y', z')d\tau'$$

where $C(x, y, z)$ is the general solution of the equation $\nabla^2\psi + k^2\psi = 0$, and where $\rho^2 = (x' - x)^2 + (y' - y)^2 + (z' - z)^2$, and $d\tau'$ is the volume-element $dx'dy'dz'$, the integration being extended over all space. So a solution of the wave-equation for the particle is a solution of the integral-equation

$$\psi(x, y, z) = C(x, y, z) - \frac{m}{2\pi\hbar^2}\int \frac{e^{ik\rho}}{\rho} V(r')\psi(x', y', z')d\tau'. \qquad (1)$$

Now if the incident beam is directed parallel to the z-axis, it may be represented by a wave-function, which involves z only: since this wave-function must satisfy the Schrödinger equation

$$\nabla^2\psi + k^2\psi = 0,$$

it must satisfy $(d^2\psi/dz^2) + k^2\psi = 0$, so it must be a linear combination of e^{ikz} and e^{-ikz}: a stream moving in the positive direction along the axis of z, with electron-density unity, will be represented by the term e^{ikz} alone. The incident beam is thus represented as a plane monochromatic ψ-wave, whose wave-length is inversely proportional to the momentum of the particles.

The complete wave-function ψ of equation (1) must represent this incident beam together with the scattered wave, which is a wave diverging from the scattering centre, of the form

$$\frac{e^{ikr}}{r} g(\theta),$$

where (r, θ, ϕ) are spherical-polar co-ordinates, with the scattering centre as origin, so that θ is the angle of scattering. Evidently the second term in (1) represents the scattered wave, and therefore $C(x, y, z)$ must represent the incident wave. Thus (1) becomes

$$\psi(x, y, z) = e^{ikz} - \frac{m}{2\pi\hbar^2}\int \frac{e^{ik\rho}}{\rho} V(r')\psi(x', y', z')d\tau'. \qquad (2)$$

Let us find the asymptotic form of the second term on the right at great distances from the origin. When r is very large we have approximately

$$\rho = r - r' \cos(\widehat{rr'}),$$

so (2) becomes

$$\psi(x, y, z) = e^{ikz} - \frac{m}{2\pi\hbar^2}\frac{e^{ikr}}{r}\int e^{-ikr'\cos(\widehat{rr'})}V(r')\psi(x', y', z')d\tau'.$$

When the particles of the incident beam are travelling so fast that the scattering is not very great, the function $\psi(x', y', z')$ in the integral may be replaced by the wave-function of the incident wave alone, so the last equation becomes

$$\psi(x, y, z) = e^{ikz} + \frac{e^{ikr}}{r}g(\theta)$$

where

$$g(\theta) = -\frac{m}{2\pi\hbar^2}\int e^{ikr'(\cos\widehat{r'z} - \cos\widehat{r'r})}V(r')d\tau'.$$

Making use of Gegenbauer's formula,[1] this integral becomes

$$g(\theta) = -\frac{2m}{\hbar^2}\int_0^\infty \frac{\sin p}{p}V(r)r^2dr, \qquad (3)$$

where $p = 2kr\sin\frac{1}{2}\theta$.

Now denoting the electric density $e\psi\psi^*$ by ρ, and denoting the quantity

$$\frac{e\hbar}{2mi}(\psi^*\,\mathrm{grad}\,\psi - \psi\,\mathrm{grad}\,\psi^*)$$

by $\mathbf{s}$, it can be shown from the wave-equation that

$$\frac{\partial\rho}{\partial t} + \mathrm{div}\,\mathbf{s} = 0$$

and this equation suggests that $\mathbf{s}$ may be interpreted as the *electric current* [2]; and therefore the number of particles which in unit time pass through a unit cross-section is the component, normal to the cross-section, of the vector

$$\mathbf{S} = \frac{\hbar}{2mi}(\psi^*\,\mathrm{grad}\,\psi - \psi\,\mathrm{grad}\,\psi^*)$$

so

$$S_x = \frac{\hbar}{2mi}\left(\psi^*\frac{\partial\psi}{\partial x} - \psi\frac{\partial\psi^*}{\partial x}\right) \text{ etc.}$$

For the incident plane wave, $\psi = e^{ikz}$, so $S_z = k\hbar/m = v_0$, so the number

[1] G. N. Watson, *Bessel functions*, p. 378

[2] Schrödinger, *Ann. d. Phys.* lxxxi (1926), p. 109. Born, *ZS. f. P.* lxxxviii (1926), p. 803; xl (1926), p. 167. Gordon, *ZS. f. P.* xl (1926), p. 117

of particles which in unit time pass through a unit cross-section at right angles to the beam is v_0, the velocity of the particles, as it must be. For the scattered wave we have

$$\psi = \frac{e^{ikr}}{r} g(\theta)$$

so

$$S_r = \frac{k\hbar}{m} \frac{|g(\theta)|^2}{r^2} = v_0 \frac{|g(\theta)|^2}{r^2}.$$

Thus, *with an incident beam in which one particle crosses unit area transverse to the path in unit time, the number of particles scattered into the solid angle* $\sin\theta\, d\theta d\phi$ *in unit time is* $|g(\theta)|^2 \sin\theta\, d\theta d\phi$, *where* $g(\theta)$ *has the value given by* (3).

Of course in this treatment, which is called the *Born approximation*, the interaction $V(r)$ has been treated as a small perturbation, and only the first approximation has been retained.[1] Born's formula was applied by Wentzel[2] to the scattering of a beam of electrically-charged particles by an electrically-charged centre. Suppose the charges of the centre and of a particle are Ze and $Z'e$, where e is the electronic charge ; we suppose the Coulomb force of the centre to be modified by 'shielding,' so we can write

$$V(r) = \frac{e^2 ZZ'}{r} e^{-\frac{r}{a}}$$

where the factor $e^{-\frac{r}{a}}$ represents the shielding, a being the *effective radius* of the atom. Evaluating (3) on this assumption, we find

$$g(\theta) = -\frac{2me^2ZZ'}{\hbar^2\left\{(2k\sin\tfrac{1}{2}\theta)^2 + \frac{1}{a^2}\right\}}.$$

In the usual experiments, $1/a^2$ is small compared with $(2k\sin\frac{1}{2}\theta)^2$, so may be omitted. Thus

$$g(\theta) = -\frac{me^2ZZ'}{2\hbar^2k^2\sin^2\frac{1}{2}\theta}$$

$$= -\frac{e^2ZZ'}{2mv_0^2\sin^2\frac{1}{2}\theta},$$

and this is precisely the formula obtained classically by Rutherford. Agreement with experiment confirmed the interpretation of $\psi\psi^*$ that had been used in the derivation.

[1] The wave-function ψ of the quantum-mechanical treatment of the problem was derived analytically from the classical solution (by considering the Action) by W. Gordon, *ZS. f. P.* xlviii (1928), p. 180.
[2] *ZS. f. P.* xl (1926), p. 590

Born's method was applied in 1927 and following years to investigate the collision of an incident beam of electrons with the atoms of a gas, the atoms being possibly raised to excited states, and the electrons scattered : this work, however, falls outside the limits of the present volume.[1]

On 10 May 1926 Schrödinger contributed another paper to the *Annalen der Physik*,[2] in which he set forth a general method for treating perturbations (i.e. solving problems which are very closely related to problems that have already been solved), and applied his theory to investigate the Stark effect in the Balmer lines of hydrogen. The method was essentially the same as that given by Born, Heisenberg and Jordan, a few months earlier, or indeed as that used long before by Lord Rayleigh[3] in discussing the vibrations of a string whose density has small inhomogeneities. The wave-equation for the unperturbed system is

$$\left\{\mathrm{H}\left(\frac{\hbar}{i}\frac{\partial}{\partial q}, q\right) - \mathrm{E}\right\}\psi = 0.$$

Let the proper-values of E, and the corresponding normalised wave-functions ψ (which are supposed to be known) be E_1, E_2, E_3, . . . ; $\psi_1(q)$, $\psi_2(q)$, $\psi_3(q)$, ... Let the perturbation be represented by the adjunction of a small term $\lambda r\psi$ to the left-hand side of the wave-equation, where λ denotes a small constant and r is a known function of the q's, so that the wave-equation for the perturbed system is

$$\left\{\mathrm{H}\left(\frac{\hbar}{i}\frac{\partial}{\partial q}, q\right) + \lambda r - \mathrm{E}\right\}\psi = 0.$$

Let the new proper-values of the energy be E'_1, E'_2, . . ., where

$$E'_s = E_s + \lambda\epsilon_s + \text{terms involving higher powers of } \lambda,$$

and let the new wave-functions be ψ'_1, ψ'_2, . . ., where

$$\psi'_s = \psi_s + \lambda v_s + \text{terms involving higher powers of } \lambda.$$

Substituting in the wave-equation, and retaining only the first power of λ, we have

$$\{\mathrm{H} - E_s + \lambda(r - \epsilon_s)\}(\psi_s + \lambda v_s) = 0.$$

Since $(\mathrm{H} - E_s)\psi_s = 0$, this gives

$$(\mathrm{H} - E_s)v_s = -(r - \epsilon_s)\psi_s.$$

[1] Though Born himself obtained important results in *Gött. Nach.* 1926, p. 146.
[2] *Ann. d. Phys.*(4) lxxx (1926), p. 437
[3] *Theory of Sound*, 2nd edn. (London, 1894), i, p. 115

Now in Rayleigh's problem of the vibrating string, if there is resonance between the applied force and a proper vibration of the unperturbed system, the oscillation increases without limit, and no finite solution exists : it can be shown that corresponding to this in the present case, in order that there may be a finite solution, the right-hand side of the last equation must be orthogonal to the wave-function which is the solution ψ_s of the equation obtained by equating the left-hand side to zero ; so we must have

$$\int (r-\epsilon_s)\psi_s\psi_s^* dq = 0$$

or, since $\int \psi_s\psi_s^* dq = 1$,

$$\epsilon_s = \int r\psi_s\psi_s^* dq.$$

This equation gives the new proper-value of the energy as

$$E'_s = E_s + \lambda \int r\psi_s\psi_s^* dq.$$

We now proceed to find the corresponding wave-function, the new part of which is given by the equation

$$(H - E_s)v_s = -(r-\epsilon_s)\psi_s.$$

The integration of this presents no difficulty, and gives for the perturbed system the wave-function

$$\psi'_s = \psi_s + \lambda \sum_k{}' \frac{\int r\psi_k^*\psi_s dq}{E_s - E_k}\psi_k$$

where the prime above the $\sum$ means that the term corresponding to $k=s$ is to be omitted.

Schrödinger then considered the case of degeneracy, when to a single proper-value E_s of the energy there correspond several wave-functions : in the perturbed system, the degeneracy will in general be wholly or partly removed, so that a single original energy-level gives rise to a number of energy-levels close together, and spectral lines are split, as in the Stark and Zeeman effects.

In applying this theory to the investigation of the Stark effect in the Balmer lines of hydrogen, we suppose that there is an electric field of intensity F in the positive z-direction, so the wave-equation is

$$\nabla^2\psi + \frac{2m}{\hbar^2}\left(E + \frac{e^2}{r} - eFz\right)\psi = 0.$$

Schrödinger followed Epstein [1] in introducing space-parabolic co-ordinates λ, μ, ϕ by the equations

$$x=(\lambda\mu)^{\frac{1}{2}}\cos\phi, \qquad y=(\lambda\mu)^{\frac{1}{2}}\sin\phi, \qquad z=\tfrac{1}{2}(\lambda-\mu)$$

for which the volume-element $dxdydz$ is $\frac{1}{4}(\lambda+\mu)d\lambda d\mu d\phi$. The wave-equation now becomes

$$\frac{\partial}{\partial\lambda}\left(\lambda\frac{\partial\psi}{\partial\lambda}\right)+\frac{\partial}{\partial\mu}\left(\mu\frac{\partial\psi}{\partial\mu}\right)+\frac{1}{4}\left(\frac{1}{\lambda}+\frac{1}{\mu}\right)\frac{\partial^2\psi}{\partial\phi^2}$$

$$+\frac{m}{2\hbar^2}\{\mathrm{E}(\lambda+\mu)+2e^2-\tfrac{1}{2}e\mathrm{F}(\lambda^2-\mu^2)\}\psi=0.$$

To solve this equation we write

$$\psi=\Lambda\mathrm{M}\Phi$$

where Λ is a function of λ only, M is a function of μ only and Φ is a function of ϕ only. The equation for Φ is

$$\frac{d^2\Phi}{d\phi^2}=-n^2\Phi$$

where n is a constant which (in order that Φ and its derivatives should be single-valued and continuous functions of ϕ) must be a whole number, so $n=0, 1, 2, 3, \ldots$. The equations for Λ and M are both of the form

$$\frac{d}{d\xi}\left(\xi\frac{d\Lambda}{d\xi}\right)+\left(\mathrm{D}\xi^2+\mathrm{A}\xi+2\mathrm{B}-\frac{n^2}{4\xi}\right)\Lambda=0.$$

The term $\mathrm{D}\xi^2$ represents the Stark-effect perturbation. So for the unperturbed system

$$\frac{d}{d\xi}\left(\xi\frac{d\Lambda}{d\xi}\right)+\left(\mathrm{A}\xi+2\mathrm{B}-\frac{n^2}{4\xi}\right)\Lambda=0.$$

Put $\Lambda=\xi^{-\frac{1}{2}}u$, $2\xi\sqrt{(-\mathrm{A})}=\eta$, and $\mathrm{B}(-\mathrm{A})^{-\frac{1}{2}}=p$. The equation becomes

$$\frac{d^2u}{d\eta^2}+\left(-\frac{1}{4}+\frac{p}{\eta}+\frac{1-n^2}{4\eta^2}\right)u=0$$

the solution of which is the confluent hypergeometric function

$$u=\mathrm{W}_{p,\frac{1}{2}n}(\eta),$$

[1] cf. p. 121 *supra*

so in order that the solution may be finite everywhere we must have

$$p=\tfrac{1}{2}(n+1)+k \qquad (k=0, 1, 2, \ldots),$$

and the corresponding solutions are

$$u_k(\eta)=W_{k+(n+1)/2,\, n/2}(\eta).$$

There are two values of k, corresponding to the equations for Λ and M respectively: call them k_1 and k_2. Then n, k_1 and k_2 are the three quantum numbers of the unperturbed motion. Denoting the values of B in the Λ- and M-equations respectively by B_1 and B_2, we have

$$B_1=\left(\frac{n+1}{2}+k_1\right)\sqrt{(-A)}$$

$$B_2=\left(\frac{n+1}{2}+k_2\right)\sqrt{(-A)}$$

so

$$B_1+B_2=(n+1+k_1+k_2)\sqrt{(-A)}.$$

But

$$B_1+B_2=\frac{me^2}{2\hbar^2} \quad \text{and} \quad A=\frac{mE}{2\hbar^2}$$

so the energy in the unperturbed motion is

$$E=-\frac{me^4}{2\hbar^2(n+1+k_1+k_2)^2}$$

an equation from which it is evident that $(n+1+k_1+k_2)$ is the principal quantum number.

If we now carry out the process described above for finding the energy-levels in the perturbed motion, we obtain

$$E=-\frac{me^4}{2\hbar^2(n+1+k_1+k_2)^2}-\frac{3\hbar^2F(k_2-k_1)(n+1+k_1+k_2)}{2me}$$

which is precisely the formula found by Epstein[1] for the energy-levels in the Stark effect of the Balmer lines of hydrogen.

Schrödinger proceeded to discuss the intensity of the Stark components. In order to find the intensity of a line, it is necessary to calculate the matrix-element relating to the corresponding transition. This matrix-element can be determined, as we have seen, from the wave-functions: and it was found to vindicate the selection and polarisation rules given by Epstein. Some of the theoretical results on intensities were confirmed experimentally in the following year by J. S. Foster.[2]

[1] cf. p. 121 *supra* [2] *Proc. R.S.*(A), cxiv (1927), p. 47; cxvii (1927), p. 137

The terms proportional to the square of the applied electric force in the Stark effect were calculated in the following year by G. Wentzel[1] and I. Waller.[2] The Stark effect was investigated in 1926 also by Pauli[3] from the standpoint of Heisenberg's quantum-mechanics.

Schrödinger's treatment, like that of Schwarzschild and Epstein, ignored the fine-structure of the hydrogen lines: the theory was completed in this respect not long afterwards by R. Schlapp,[4] whose work was based on the new explanation of the fine-structure which had been provided by the theory of electron-spin.

Another phenomenon which is to be treated by the quantum-mechanical theory of perturbations is the Zeeman effect.[5] If in a first treatment of the subject we ignore electron-spin, and consider simply the motion of an electron attracted by a fixed nucleus of charge e and under the influence of a magnetic field of intensity H parallel to the axis of z, then the Lagrangean function is

$$\tfrac{1}{2}m\left\{\left(\frac{dx}{dt}\right)^2+\left(\frac{dy}{dt}\right)^2+\left(\frac{dz}{dt}\right)^2\right\}-\frac{e\mathrm{H}}{2c}\left(x\frac{dy}{dt}-y\frac{dx}{dt}\right)+\frac{e^2}{r}$$

and therefore the Hamiltonian function is[6]

$$\frac{1}{2m}\left\{\left(p_x-\frac{e\mathrm{H}}{2c}y\right)^2+\left(p_y+\frac{e\mathrm{H}}{2c}x\right)^2+p_z^2\right\}-\frac{e^2}{r}$$

(measuring e in electrostatic and H in electromagnetic units, as usual) or

$$\frac{1}{2m}\left\{p_x^2+p_y^2+p_z^2+\frac{e\mathrm{H}}{c}(xp_y-yp_x)+\frac{e^2\mathrm{H}^2}{4c^2}(x^2+y^2)\right\}-\frac{e^2}{r}.$$

The corresponding wave-equation, obtained by replacing p_x by $(\hbar/i)\ (\partial/\partial x)$ etc., is

$$\nabla^2\psi+\frac{2m}{\hbar^2}\left(\mathrm{E}+\frac{e^2}{r}\right)\psi+\frac{ie\mathrm{H}}{c\hbar}\left(x\frac{\partial\psi}{\partial y}-y\frac{\partial\psi}{\partial x}\right)-\frac{e^2\mathrm{H}^2}{4c^2\hbar^2}(x^2+y^2)\psi=0.$$

Introducing spherical-polar co-ordinates (r, θ, ϕ), with the z-axis as polar axis, this becomes

$$\frac{1}{r^2}\frac{\partial}{\partial r}\left(r^2\frac{\partial\psi}{\partial r}\right)+\frac{1}{r^2\sin\theta}\frac{\partial}{\partial\theta}\left(\sin\theta\frac{\partial\psi}{\partial\theta}\right)+\frac{1}{r^2\sin^2\theta}\frac{\partial^2\psi}{\partial\phi^2}+\frac{2m}{\hbar^2}\left(\mathrm{E}+\frac{e}{r^2}\right)\psi$$
$$+\frac{ie\mathrm{H}}{c\hbar}\frac{\partial\psi}{\partial\phi}-\frac{e^2\mathrm{H}^2}{4c^2\hbar^2}r^2\sin^2\theta\,\psi=0.$$

[1] *ZS. f. P.* xxxviii (1927), p. 518
[2] *ZS. f. P.* xxxviii (1927), p. 635
[3] *ZS. f. P.* xxxvi (1926), p. 336; cf. also C. Lanczos, *ZS. f. P.* lxii (1930), p. 518; lxv (1930), p. 431; lxviii (1931), p. 204
[4] *Proc. R.S.*(A), cxix (1928), p. 313
[5] cf. Schrödinger, *Phys. Rev.* xxviii (1926), p. 1049
[6] cf. V. Fock, *ZS. f. P.* xxxviii (1926), p. 242

We shall ignore the term[1] quadratic in the magnetic intensity H: and we shall regard the linear term in H as a perturbation: so the wave-equation of the unperturbed system is the ordinary wave-equation of the hydrogen atom, which has already been considered: the wave-function is a constant multiple of

$$\frac{1}{r}W_{k,n+\frac{1}{2}}(z)P_n^{\mu}(\cos\theta)e^{\pm i\mu\phi}$$

and the energy-levels are

$$E_k = -\frac{me^4}{2\hbar^2 k^2} \quad \text{where } k = 1, 2, 3, 4, \ldots.$$

Now in the above wave-equation the perturbation-term implies the replacement of E by $E - \lambda r$ where

$$\lambda r = -\frac{ie\hbar H}{2mc}\frac{\partial}{\partial\phi}:$$

so the term $\lambda\epsilon_k$ to be added to E_k on account of the perturbation is (by the general theory)

$$\lambda\epsilon_k = \int\left\{-\frac{i\hbar eH}{2mc}(\pm i\mu)\right\}\psi_k\psi_k^* dq$$

when ψ_k is supposed to be normalised, so $\int\psi_k\psi_k^* dq = 1$. Thus *the displacement of the energy-level in the Zeeman effect is*

$$\lambda\epsilon_k = \pm\frac{e\mu\hbar H}{2mc}$$

where μ *is the magnetic quantum number*: which is precisely the value found in Lorentz's classical theory.[2]

The above derivation accounts only for the normal Zeeman effect; as might be expected, since we know[3] that the anomalous Zeeman effect requires for its explanation the assumption of electron-spin. The problem thus presented was solved by W. Heisenberg and P. Jordan[4] by matrix-mechanical methods in 1926, and by C. G. Darwin[5] by wave-mechanics in 1927.

Darwin's model consists of a charged spinning spherical body moving in a central field of force. Denoting the charge by $-e$,

[1] The effect of the quadratic term was considered by O. Halpern and Th. Sexl, *Ann. d. Phys.*(5) iii (1929), p. 565.

[2] Vol. I, p. 412; cf. P. S. Epstein, *Proc. N.A.S.* xii (1926), p. 634; A. E. Ruark, *Phys. Rev.* xxxi (1928), p. 533

[3] p. 136 *supra*

[4] *ZS. f. P.* xxxvii (1926), p. 263

[5] *Proc. R.S.*(A), cxv (1927), p. 1

the mass by m, the position of the centre by (x, y, z) and the potential energy at distance r by $V(r)$, then the motion of revolution contributes terms

$$\tfrac{1}{2}m\left\{\left(\frac{dx}{dt}\right)^2+\left(\frac{dy}{dt}\right)^2+\left(\frac{dz}{dt}\right)^2\right\}+eV(r)$$

to the Lagrangean function. There is also a contribution

$$\tfrac{1}{2}I(\omega_x^2+\omega_y^2+\omega_z^2)$$

where I is the moment of inertia and $(\omega_x, \omega_y, \omega_z)$ the components of spin about (x, y, z). The orbital motion in the magnetic field H along z gives a contribution

$$-\frac{eH}{2c}\left(x\frac{dy}{dt}-y\frac{dx}{dt}\right).$$

The spin gives

$$-\frac{eHI\omega_z}{mc}$$

since e/mc is the ratio of magnetic moment to angular momentum. Lastly there is the interaction of the spin and the motion. The electric force has components $-(x/r)V'$ etc., so we obtain a term

$$-\frac{eIV'}{mc^2r}\left\{\omega_x\left(y\frac{dz}{dt}-z\frac{dy}{dt}\right)+\omega_y\left(z\frac{dx}{dt}-x\frac{dz}{dt}\right)+\omega_z\left(x\frac{dy}{dt}-y\frac{dx}{dt}\right)\right\}.$$

All these terms were taken together and converted into Hamiltonian form, and the Schrödinger wave-equation was then deduced. From this equation by use of spherical harmonic analysis the proper-values and wave-functions were calculated and Landé's g-formula was obtained.[1]

Schrödinger followed up his work on perturbations and the Stark effect by another paper[2] in which he extended the perturbation theory to perturbations that explicitly involve the time, and succeeded in obtaining by wave-mechanical methods the Kramers-Heisenberg formula for scattering.[3] Let q represent the co-ordinates which specify the state of the atom at the instant t, and let $H_0(q, p)$ represent its energy, so the wave-function for the unperturbed atom is given by the equation

$$i\hbar\frac{\partial\psi}{\partial t}=H_0\left(q,\frac{\hbar}{i}\frac{\partial}{\partial q}\right)\psi.$$

[1] cf. K. Darwin, *Proc. R.S.*(A), cxviii (1928), p. 264

[2] *Ann. d. Phys.*(4) lxxxi (1926), p. 109; cf. O. Klein, *ZS. f. P.* xli (1927), p. 407

[3] cf. p. 206 *supra*

Let the atom be irradiated by light whose wave-length is so great compared with atomic dimensions that we can regard the electric vector of the light as constant over the atom: let this electric vector be

$$\mathbf{E} = \mathbf{V}^* e^{2\pi i\nu t} + \mathbf{V} e^{-2\pi i\nu t}$$

where $\mathbf{V}$ is a complex vector. The additional energy of the atom due to this perturbation is $-(\mathbf{M}\,.\,\mathbf{E})$, where $\mathbf{M}$ denotes the electric moment $\sum e\mathbf{r}$ of the atom: so the wave-function of the perturbed atom is determined by the equation

$$i\hbar\frac{\partial\psi}{\partial t} = \left\{\mathrm{H}_0\left(q, \frac{\hbar}{i}\frac{\partial}{\partial q}\right) - (\mathbf{M}\,.\,\mathbf{E})\right\}\psi.$$

The wave-function, when the atom is in the state P, may be denoted by

$$\psi_{\mathrm{P}}{}^0 = e^{\frac{\mathrm{H_P}t}{i\hbar}}\psi_{\mathrm{P}}(q)$$

and a general solution may be represented by a series of these wave-functions. Supposing that the atom is initially in the state P, let us solve the wave-equation of the perturbed atom by writing

$$\psi = \psi_{\mathrm{P}}{}^0 + \psi_{\mathrm{P}}{}^1, \text{ where } \psi_{\mathrm{P}}{}^1 \text{ is small.}$$

Substituting in the differential equation, and neglecting small quantities of the second order, we have

$$\mathrm{H}_0\left(q, \frac{\hbar}{i}\frac{\partial}{\partial q}\right)\psi_{\mathrm{P}}{}^1 - i\hbar\frac{\partial\psi_{\mathrm{P}}{}^1}{\partial t} = (\mathbf{M}\,.\,\mathbf{E})\psi_{\mathrm{P}}{}^0.$$

We solve this by substituting in it

$$\psi_{\mathrm{P}}{}^1 = \psi_{\mathrm{P}^+} e^{\frac{\mathrm{H_P}+h\nu}{i\hbar}t} + \psi_{\mathrm{P}^-} e^{\frac{\mathrm{H_P}-h\nu}{i\hbar}t}$$

and equating terms which have the same time-factor: the resulting equation for ψ_{P^+} is

$$\mathrm{H}_0\left(q, \frac{\hbar}{i}\frac{\partial}{\partial q}\right)\psi_{\mathrm{P}^+} - (\mathrm{H_P} + h\nu)\psi_{\mathrm{P}^+} = (\mathbf{M}\,.\,\mathbf{V})\psi_{\mathrm{P}}.$$

Now expand ψ_{P^+} and $\mathbf{M}\psi_{\mathrm{P}}$ as series of the wave-functions ψ_{R}, say

$$\psi_{\mathrm{P}^+} = \sum_{\mathrm{R}} a_{\mathrm{PR}}\psi_{\mathrm{R}}$$

and

$$\mathbf{M}\psi_{\mathrm{P}} = \sum_{\mathrm{R}} \mathbf{M}_{\mathrm{PR}}\psi_{\mathrm{R}} \quad \text{where} \quad \mathbf{M}_{\mathrm{PR}} = \int \psi_{\mathrm{R}}{}^*\mathbf{M}\psi_{\mathrm{P}}\,d\tau,$$

so $\mathbf{M}_{PR}$ is the element in the (matrix) electric moment of the atom, that corresponds to a transition from the level R to the level P. Thus

$$\sum_R a_{PR}\left\{H_0\left(q, \frac{\hbar}{i}\frac{\partial}{\partial q}\right) - H_P - h\nu\right\}\psi_R = \sum_R (\mathbf{M}_{PR} \cdot \mathbf{V})\psi_R$$

or

$$\sum_R a_{PR}(H_R - H_P - h\nu)\psi_R = \sum_R (\mathbf{M}_{PR} \cdot \mathbf{V})\psi_R$$

and therefore

$$a_{PR} = \frac{(\mathbf{M}_{PR} \cdot \mathbf{V})}{h(\nu_{RP} - \nu)}.$$

Similarly the coefficient of ψ_R in the expansion of ψ_P^- is determined : and thus

$$\psi_P^1 = \frac{1}{h}\sum_R \left\{\frac{(\mathbf{M}_{PR} \cdot \mathbf{V})}{\nu_{RP} - \nu}e^{\frac{H_P + h\nu}{i\hbar}t} + \frac{(\mathbf{M}_{PR} \cdot \mathbf{V}^*)}{\nu_{RP} + \nu}e^{\frac{H_P - h\nu}{i\hbar}t}\right\}\psi_R.$$

This equation gives the perturbed value of the wave-function. The electric moment of the classical oscillator which would emit the radiation emitted in the scattering process of an atom in the state P, associated with the transition to the state Q, is the real part of

$$\int \psi_Q^* \mathbf{M}\psi_P d\tau, \quad \text{or in this case} \quad \int (\psi_Q^0 + \psi_Q^1)^* \mathbf{M}(\psi_P^0 + \psi_P^1) d\tau.$$

Neglecting terms of the second order, this is found to be

$$\mathbf{M}_{PQ}e^{-2\pi i\nu_{PQ}t} + \frac{1}{h}\sum_R \left\{\frac{\mathbf{M}_{RQ}(\mathbf{M}_{PR} \cdot \mathbf{V})}{\nu_{RP} - \nu} + \frac{\mathbf{M}_{PR}(\mathbf{M}_{RQ} \cdot \mathbf{V})}{\nu_{RQ} + \nu}\right\}e^{-2\pi i(\nu_{PQ} + \nu)t}$$

$$+ \frac{1}{h}\sum_R \left\{\frac{\mathbf{M}_{RQ}(\mathbf{M}_{PR} \cdot \mathbf{V}^*)}{\nu_{RP} + \nu} + \frac{\mathbf{M}_{PR}(\mathbf{M}_{RQ} \cdot \mathbf{V}^*)}{\nu_{RQ} - \nu}\right\}e^{-2\pi i(\nu_{PQ} - \nu)t}.$$

The term $\mathbf{M}_{PQ}e^{-2\pi i\nu_{PQ}t}$ represents the spontaneous emission associated with the transition P→Q, and *the other terms are precisely those found by Kramers and Heisenberg for the electric moment corresponding to the scattered radiation associated with the transition* P→Q.

The Compton effect was investigated quantum-mechanically in 1926 by Dirac[1] by means of his symbolic representation of matrix-mechanics, and by W. Gordon[2] by wave-mechanics. Gordon found that the quantum-mechanical frequency and intensity are the

[1] *Proc. R.S.*(A), cxi (1926), p. 405

[2] *ZS. f. P.* xl (1926), p. 117 ; cf. G. Breit, *Phys. Rev.* xxvii (1926), p. 362 ; O. Klein, *ZS. f. P.* xli (1927), p. 407, *at* p. 436 ; G. Wentzel, *ZS. f. P.* xliii (1927), pp. 1, 779 ; G. Beck, *ZS. f. P.* xliii (1927), p. 658

geometric means of the corresponding classical quantities at the beginning and end of the process. Breit[1] remarked that according to Gordon, the radiation in the case of the Compton effect may be regarded as due to a $\psi\psi^*$ wave moving with the velocity of light. Similarly Breit showed[2] that if the motion of the centre of gravity of an atom is taken into account, emission takes place only by means of unidirectional quanta due to $\psi\psi^*$ waves moving with the velocity of light.

In 1927 Schrödinger[3] showed that by means of the conception of de Broglie waves, the Compton effect may be linked up with some other investigations of a quite different character. He began by recalling an investigation of Léon Brillouin[4] on the way in which (according to classical physics) an elastic wave, in a transparent medium, affects the propagation of light. Let there be an elastic compressional or longitudinal wave (i.e. a sound-wave) of wave-length Λ, which is propagated in a transparent medium. It reflects at its wave-front light-rays which traverse the medium; but the reflected ray has negligible intensity except when there is between the wave-lengths and the angles the relation

$$\lambda = 2\Lambda \cos i$$

where i is the angle of incidence and λ is the wave-length of the light in the medium (so it is $K^{-\frac{1}{2}}$ times the wave-length of the light *in vacuo*, where K is the dielectric constant). This is precisely the equation which had been found in 1913 by W. L. Bragg as the condition that X-rays of wave-length λ should be reflected by the parallel planes rich in atoms in a crystal, when i was the angle of incidence and Λ was the distance between the parallel planes in the crystal.[5] In Brillouin's theorem it is supposed that the velocity of propagation of the elastic wave is small compared with the velocity of light: more accurately, the formula must be modified as in the case of reflection at a *moving* mirror.[6]

Schrödinger now showed that the Compton effect can be assimilated to the Brillouin effect, if the electron of the Compton effect in its initial state of rest is replaced by a 'wave of electrical density' and also in its final state is replaced by another wave. The two waves form by their interference a system of intensity-maxime located in parallel planes, which correspond to Bragg's planes in a crystal. The analytical formulae are, as he showed, identical with those found by Compton in the particle-interpretation. Thus *a wave-explanation is obtained for the Compton effect.* Other cases are known in which a transition of a particle from one state of motion

[1] *Proc. N.A.S.* xiv (1928), p. 553

[2] *J. Opt. Soc. Amer.* xiv (1927), p. 374

[3] *Ann. d. Phys.* lxxxii (1927), p. 257

[4] *Ann. d. Phys.* xvii (1921), p. 88

[5] cf. p. 20 *supra*

[6] Brillouin's problem was treated quantum-mechanically in *Phys. ZS.* xxv (1924), p. 89 by Schrödinger, who found that quantum theory led to the same formula as classical physics.

to another may be translated into the superposition of two interfering matter-waves.

The year 1926 saw the publication of some important papers on the problem of atoms with more than one electron. It had been known for some years that the energy-levels or discrete stationary states, that account for the spectrum of neutral helium (which has two electrons), constitute two systems,[1] such that the levels belonging to one system (the *para*-system) do not in general combine with the levels of the other system (the *ortho*-system) in order to yield spectral lines. This fact led spectroscopists at one time to conjecture the existence of two different chemical constituents in helium, to which the names *parhelium* and *orthohelium* were assigned. The para-level system consists of singlets and the ortho-system consists of triplets.

Heisenberg[2] and Dirac[3] now investigated the general quantum-mechanical theory of a system containing several identical particles, e.g. electrons. If the positions of two of the electrons are interchanged, the new state of the atom is physically indistinguishable from the original one. In this case we should expect the wave-functions to be either symmetrical or skew in the co-ordinates of the electrons (including the co-ordinate which represents spin). Now a skew wave-function vanishes identically when two of the electrons are in states defined by the same quantum numbers : this means that in a solution of the problem specified by skew wave-functions there can be no stationary states for which two or more electrons have the same set of quantum numbers, which is precisely Pauli's exclusion principle. A solution with symmetrical wave-functions, on the other hand, allows any number of electrons to have the same set of quantum numbers, so this solution cannot be the correct one for the problem of several electrons in one atom.

In a second paper,[4] Heisenberg applied the general theory to the case of the helium atom. There are energy-levels corresponding to wave-functions which are symmetric in the space-co-ordinates and skew in the spins of the electrons (i.e. the spins of the two electrons are antiparallel) : these may be identified with the para-system. There are also wave-functions which are skew in the space-co-ordinates and symmetric in the spins (i.e. the spins of the two electrons are parallel) : these are associated with the ortho-system. In both systems the wave-functions involving both co-ordinates and spin change sign when the two electrons are interchanged. Now if a wave-function is symmetrical (or skew) at one instant, it must remain symmetrical (or skew) at all subsequent instants, and therefore the changes in the system cannot affect the symmetry (or

[1] For studies of the helium atom and the related ions Li^+ and Be^{++} from the standpoint of the earlier quantum theory, cf. A. Landé, *Phys. ZS.* xx (1919), p. 228 ; H. A. Kramers, *ZS. f. P.* xiii (1923), p. 312 ; J. H. van Vleck, *Phys. Rev.* xxi (1923), p. 372 ; M. Born and W. Heisenberg, *ZS. f. P.* xxvi (1924), p. 216.

[2] *ZS. f. P.* xxxviii (1926), p. 411

[3] *Proc. R.S.*(A), cxii (1926), p. 661

[4] *ZS. f. P.* xxxix (1926), p. 499

skewness), so the para-levels and ortho-levels cannot have transitions into each other ; which explains their observed property.

Heisenberg's first paper was notable for the discovery of the property of like particles known as *exchange interaction*, which had an important place in the physical researches of the years immediately succeeding : an account of these must be reserved for the next volume.

Index of Authors Cited

Subject Index